ASF-5470
UCF

Approximation Theory IX

Theoretical Aspects

Innovations in Applied Mathematics

An international series devoted to the latest research in modern areas of mathematics, with significant applications in engineering, medicine, and the sciences.

Series Editor:
Larry L. Schumaker
Stevenson Professor of Mathematics
Vanderbilt University

Previously published titles include

Mathematical Methods for Curves and Surfaces (1995)

Curves and Surfaces with Applications in CAGD (1997)

Surface Fitting and Multiresolution Methods (1997)

Mathematical Methods for Curves and Surfaces II (1998)

Mathematical Models in Medical and Health Science (1998)

Approximation Theory IX

Volume 1. Theoretical Aspects

EDITED BY

Charles K. Chui
Department of Mathematics
Texas A&M University

Larry L. Schumaker
Department of Mathematics
Vanderbilt University

VANDERBILT UNIVERSITY PRESS
Nashville & London

Copyright © 1998 by Vanderbilt University Press
All Rights Reserved
No part of this publication may be reproduced or transmitted in any form or by any means, electronic or mechanical, including photocopy, recording, or any information storage and retrieval system, without permission in writing from the publisher.

First Edition 1998
98 99 00 01 02   5 4 3 2 1

This publication is made from paper that meets
the minimum requirements of ANSI/NISO Z39.48 (R 1997)
Permanence of Paper for Printed Library Materials. ∞

Manufactured in the United States of America

CONTENTS

Preface viii

Contributors ix

The Degree of Convergence of Sequences of Multivariate
Wavelet Type Operators
   George A. Anastassiou 1

Best m-Term Approximation of Functions from Classes $MW^r_{q,\alpha}$
   Alexander V. Andrianov and Vladimir N. Temlyakov 7

Necessary Conditions for Abadie's Constraint Qualification
and Best Approximation of Vector-Valued Functions
   Martin Bartelt 15

n-Widths of Multiplier Operators on Two-Point Homogeneous Spaces
   B. Bordin, A. K. Kushpel, J. Levesley, and S. A. Tozoni 23

Some Old Problems on Polynomials with Integer Coefficients
   Peter Borwein 31

On the Irrationality of $\sum_{i=0}^{\infty} q^{-i} \prod_{j=0}^{i}(1 + q^{-j}r + q^{-2j}s)$
   Peter B. Borwein and Ping Zhou 51

On a Weighted Recursive Rule
   Michele Campiti 59

Approximation of the Inverse Frame Operator
   Peter Casazza and Ole Christensen 67

The Bernstein Operator is the Closest Positive Operator to a Projection
   Bruce L. Chalmers, Dany Leviatan, and Michael P. Prophet 75

Finite-Dimensional Action Constants
   B. L. Chalmers and K. Pan 83

Approximation by Sums of Exponentials to Decay Functions
Using Piecewise Linear Models
   Maurice G. Cox, Helen E. Joyce, and John C. Mason 89

A Differential Geometric Approach to Equidistributed Knots on
Riemannian Manifolds
   Uwe Depczynski and Joachim Stöckler 97

The Role of the Strong Conical Hull Intersection Property in Convex
Optimization and Approximation
   Frank Deutsch 105

Fractional Error Constants for Quadrature Formulas
   Kai Diethelm 113

Generalized Bernstein-Erdős Conjecture
  P. Dragnev, D. Legg, and D. Townsend 119

Non-convexity of the Set of Pólya Limits
  A. Egger, T. Lay, and G. D. Taylor 127

Estimates for Some Positive Operators via Step-Weight
Functions with Concave Squares
  Michael Felten 133

Polynomial Approximation on the m-Dimensional Ball
  Michael I. Ganzburg 141

Approximation Using Scale Transformations
  Christian Gout 149

Linear Discrete Operators and Recovery of Functions
  Ivan Ivanov and Boris Shekhtman 157

Chebyshev Quadrature Recognizes Algebraic Curves and Surfaces
  J. Korevaar 165

Ray Sequences of Best Rational Approximants to Entire Functions
  A. V. Krot, V. A. Prokhorov, and E. B. Saff 175

Extremal Problems for Logarithmic Potentials
  A. B. J. Kuijlaars 201

Monotone Approximation Estimates Involving the Third
Modulus of Smoothness
  Dany Leviatan and Igor A. Shevchuk 223

Infinite Dimensional Width of Function Class With
Non-symmetric Smoothness Conditions
  Yongping Liu 231

Rational Approximation in Logarithmic Capacity of Meromorphic
Maps from $\mathbb{C}^n$ to $\mathbb{C}^m$
  Clement Lutterodt and Stanley Einstein-Matthews 239

Best Approximation by Quadratic Algebraic Functions
  Allan W. McInnes 247

L_p-Convergence of the Lagrange-Type Interpolation
in the Rational Space
  G. Min 255

Approximation of Functions by Linear Matrix Operators
  M. L. Mittal and Neeraj Bhardwaj 263

Weighted Approximation on Compact Sets
  Igor E. Pritsker 271

Characterization of Best L_1-Coapproximation
  Geetha S. Rao and R. Saravanan 279

Contents

Algebraic Aspects of Polynomial Interpolation in Several Variables
  Thomas Sauer 287

Adaptive Approximation and Compression
  Bl. Sendov 295

On the Discrete Norms of Polynomials
  Boris Shekhtman 303

Generation of Weight Functions for Orthogonal Polynomials
  M.-R. Skrzipek 309

On Products of Positive Definite Functions
  Hans Strauss 317

On the Jackson Theorem for Approximation by
Algebraic Polynomials
  Gancho Tachev 325

On Approximation of Cauchy-Type Integrals by Sequences
of Rational Functions with Preassigned Poles
  Genrikh Ts. Tumarkin 331

Sharp Error Estimates for Multivariate Positive Linear Operators
  Shayne Waldron 339

Realizable Approximation Bounds for a Solvable Neural Network
  Sumio Watanabe 347

Preface

The *Ninth International Conference on Approximation Theory* was hosted by Vanderbilt University in Nashville, January 3 – 6, 1998. Previous conferences in this series were held in 1973, 1976, 1980, and 1992 in Austin, and 1983, 1986, 1989, and 1995 in College Station. The conference was attended by 190 mathematicians from 21 different countries.

The program included seven invited one-hour survey talks and the one-hour presentation of this year's Popov Prize winner, Arno Kuijlaars. In addition to a large number of contributed talks, there were eight special sessions on interdisciplinary topics of current research interest, arranged by Yang Wang (wavelets), Richard Varga (approximation in the complex plane), Joe Ward (radial basis functions), Boris Shekhtman (abstract approximation), Hrushikesh Mhaskar (neural networks), Günther Nürnberger (splines), Hans Hagen (computer aided geometric design), and Vladymir Temlyakov (nonlinear m-term approximation).

Because of the large number of submitted papers, this proceedings appears in two separate volumes. Volume I covers several areas of classical and modern approximation theory, while Volume II is devoted to recent applied and computational developments.

We are indebted to the National Science Foundation and to Vanderbilt University for their financial support of the conference. We would like to thank the members of the local organizing committee, Tom Hogan, Kirill Kopotun, and Mike Neamtu for their assistance with various aspects of the meeting, and in particular for creating and managing the web site. Thanks are also due to the invited speakers, organizers of the special sessions, presenters, and everyone who attended for making the conference a success. We would like to thank our reviewers who helped select articles for the proceedings, and Robin Campbell for her TEXnical assistance with various papers. Finally, our special thanks to Margaret Chui and Gerda Schumaker who also assisted in the preparation of these volumes.

Charles K. Chui
Larry L. Schumaker

Sept. 1, 1998

CONTRIBUTORS

Numbers in parentheses indicate pages on which authors' contributions begin. Articles marked with * and ** are in Volumes 1 and 2, respectively.

GEORGE A. ANASTASSIOU (*1), *Department of Mathematical Sciences, The University of Memphis, Memphis, TN 38152, USA* [anastasg@hermes.msci.memphis.edu]

ALEXANDER V. ANDRIANOV (*7), *Department of Mathematics, University of South Carolina, Columbia, SC 29208, USA* [andriano@math.sc.edu]

MARTIN BARTELT (*15), *Department of Mathematics, Christopher Newport University, Newport News, VA 23606, USA* [mbartelt@pcs.cnu.edu]

NEERAJ BHARDWAJ (*263), *Department of Mathematics, D.N. (P.G.) College, C.C.S. University, Meerut Uttar Pradesh, India*

KAI BITTNER (**1), *Institute of Biomathematics and Biometry, GSF - National Research Center for Environment and Health, D-85764 Neuherberg, Germany* [bittner@gsf.de]

B. BORDIN (*23), *IMECC-UNICAMP, Caixa Postal 6065, 13081-970, Campinas, SP, Brazil* [bordin@ime.unicamp.br]

PETER BORWEIN (*31,*51), *Department of Mathematics and Statistics, Simon Fraser University, Burnaby, B. C., Canada, V5A 1S6* [pborwein@cecm.sfu.ca]

MICHELE CAMPITI (*59), *Department of Mathematics, Polytechnic of Bari, Via E. Orabona, 4, 70125 Bari, Italy* [campiti@dm.uniba.it]

PETER G. CASAZZA (*67), *Department of Mathematics, University of Missouri, Columbia, MO 65211, USA* [pete@casazza.math.missouri.edu]

JOSÉ E. CASTILHO (**9), *Universidade Federal de Uberlândia, Campus Santa Monica, 38400-902, Uberlândia MG, Brazil* [jecastilho@ufu.br]

BRUCE L. CHALMERS (*75,*83), *Department of Mathematics, University of California, Riverside, CA 92507, USA* [blc@math.ucr.edu]

OLE CHRISTENSEN (*67), *Mathematics Institute, Building 303 Technical University of Denmark, 2800 Lyngby, Denmark* [olechr@mat.dtu.dk]

MAURICE G. COX (*89), *National Physical Laboratory, Teddington, Middlesex, TW11 0LW, UK* [mgc@npl.co.uk]

OLEG DAVYDOV (**17,**25), Universität Dortmund, Fachbereich Mathematik, Lehrstuhl VIII, 44221 Dortmund, Germany
[oleg.davydov@math.uni-dortmund.de]

UWE DEPCZYNSKI (*97), Institut für Angewandte Mathematik und Statistik, Universität Hohenheim, D-70593 Stuttgart, Germany
[depczyns@uni-hohenheim.de]

FRANK DEUTSCH (*105), Department of Mathematics, The Pennsylvania State University, University Park, PA 16802, USA
[deutsch@math.psu.edu]

KAI DIETHELM (*113), Institut für Mathematik, Universität Hildesheim, Marienburger Platz 22, 31141 Hildesheim, Germany
[diethelm@informatik.uni-hildesheim.de]

P. DRAGNEV (*119), Department of Mathematics, Indiana-Purdue University, Fort Wayne, IN 46805, USA [dragnevp@ipfw.edu]

NIRA DYN (**33), Tel Aviv University, Sackler Faculty of Exact Sciences, School of Mathematical Sciences, Tel Aviv 69978, Israel
[niradyn@math.tau.ac.il]

ALAN EGGER (*127), Department of Mathematics, Idaho State University, Pocatello, Idaho 83209, USA [eggealan@isu.edu]

STANLEY EINSTEIN-MATTHEWS (*239), Department of Mathematics, Howard University, Washington, DC 20059, USA
[smem@scs.howard.edu]

BENEDIKTE ELBEL (**39), TU Darmstadt, Fachbereich Mathematik, Schlossgartenstrasse 7, D-64289 Darmstadt, Germany
[elbel@mathematik.tu-darmstadt.de]

KATHLEEN W. FARMER (**47), Department of Mathematics, Northwestern State University, 3329 University Parkway, Leesville, Louisiana 71446, USA [kfarmer@cp-tel.net]

GREGORY E. FASSHAUER (**55), Department of Computer Science and Applied Mathematics, Illinois Institute of Technology, Chicago, IL 60616, USA [fass@amadeus.csam.iit.edu]

MICHAEL FELTEN (*133), University of Dortmund, Department of Mathematics, D-44221 Dortmund, Germany
[felten@zx2.hrz.uni-dortmund.de]

MICHAEL S. FLOATER (**63), SINTEF Applied Mathematics, Post Box 124, Blindern, 0314 Oslo, Norway
[Michael.Floater@math.sintef.no]

MICHAEL I. GANZBURG (*141), Department of Mathematics, Hampton University, Hampton, VA 23668, USA
[ganzbrgm@fusion.hamptonu.edu]

Contributors

SÔNIA M. GOMES (**9), *Universidade Estadual de Campinas, IMECC, Caixa Postal 6065, 13081-970 Campinas SP, Brazil* [soniag@ime.unicamp.br]

TIM N. T. GOODMAN (**71), *Department of Mathematics, Dundee University, Dundee, DD1 4HN, Scotland* [tgoodman@mcs.dundee.ac.uk]

CHRISTIAN GOUT (*149), *UPRES A 5033, Laboratoire de Mathematiques Appliquees, Universite de Pau - IPRA, Avenue de l'Universite, 64000 Pau, France* [christian.gout@univ-pau.fr]

HANS HAGEN (**243), *Department of Computer Science, University of Kaiserslautern, Postfach 3049, D-67653 Kaiserslautern, Germany* [hagen@informatik.uni-kl.de]

BIN HAN (**97), *Department of Mathematical Sciences, University of Alberta, Edmonton, Alberta, Canada T6G 2G1* [bhan@math.ualberta.ca]

MATTHEW HE (**105), *Department of Mathematics, Nova Southeastern University, Ft. Lauderdale, FL 33314, USA* [hem@polaris.nova.edu]

TIAN XIAO HE (**113), *Department of Mathematics, Illinois Wesleyan University, Bloomington, IL 61702-2900, USA* [the@sun.iwu.edu]

WENJIE HE (**121), *Department of Mathematics, Texas A&M University, College Station, TX 77843-3368, USA* [wjhe@math.tamu.edu]

THOMAS A. HOGAN (**97), *Department of Mathematics, Vanderbilt University, Nashville, TN 37240, USA* [hogan@math.vanderbilt.edu]

DON HONG (**129), *Department of Mathematics, East Tennessee State University, Johnson City, TN 37614-0663, USA* [hong@etsu.edu]

IVAN V. IVANOV (*157), *Department of Mathematics, University of South Florida, Tampa, FL 33620, USA* [ivanov@chuma.cas.usf.edu]

KURT JETTER (**137), *Institut für Angewandte Mathematik und Statistik, Universität Hohenheim, D–70593 Stuttgart, Germany* [kjetter@uni-hohenheim.de]

HELEN E. JOYCE (*89), *School of Computing and Mathematics, University of Huddersfield, Huddersfield, HD1 3DH, UK* [j.c.mason@hud.ac.uk]

CHANDRA KAMBHAMETTU (**105), *Dept. of Computer and Information Science, University of Delaware, Newark, DE 19716-2712* [chandra@eecis.udel.edu]

J. KOREVAAR (*165), Department of Mathematics, University of Amsterdam, Plantage Muidergracht 24, 1018 TV Amsterdam, Netherlands
[korevaar@wins.uva.nl]

ANDREI V. KROT (*175), Department of Mathematics and Mechanics, Belorussian State University, 4 Scorina Ave., Minsk, 220080, Belarus

ARNO KUIJLAARS (*201), Department of Mathematics, Katholieke Universiteit Leuven, Celestijnenlaan 200 B, 3001 Leuven, Belgium
[arno@wis.kuleuven.ac.be]

A. K. KUSHPEL (*23,**145), IMECC-UNICAMP, Caixa Postal 6065, 13081-970, Campinas, SP, Brazil
[ak99@ime.unicamp.br, (ak99@mcs.le.ac.uk)]

MING-JUN LAI (**47,**121,**153), Department of Mathematics, University of Georgia, Athens, GA 30602, USA
[mjlai@math.uga.edu]

JOSEPH D. LAKEY (**161), Department of Mathematical Sciences, New Mexico State University, Las Cruces, NM 88003-8001, USA
[jlakey@nmsu.edu]

TERRY L. LAY (*127), Department of Mathematics, Idaho State University, Pocatello, ID 83209, USA [layterr@isu.edu]

D. LEGG (*119), Department of Mathematics, Indiana-Purdue University, Fort Wayne, IN 46805, USA [legg@ipfw.edu]

J. LEVESLEY (*23,**145), Department of Mathematics and Computer Science, University of Leicester, Leicester LE1 7RH, UK
[jl1@mcs.le.ac.uk]

DANY LEVIATAN (*75,*223), School of Mathematical Sciences, Sackler Faculty of Exact Sciences, Tel Aviv University, Tel Aviv 69978, Israel
[leviatan@math.tau.ac.il]

JIAN-AO LIAN (**169,**179), Department of Mathematics, Prairie View A&M University, Prairie View, Texas 77446, USA
[jian-ao_lian@pvamu.edu]

XUE-ZHANG LIANG (**189), Institute of Mathematics, Jilin University, Changchun, 130023, P. R. China [xzliang@mail.jlu.edu.cn]

HUAN-WEN LIU (**129), Department of Mathematics, University of Wollongong, Wollongong, NSW 2522, Australia
[Huanwen_Liu@uow.edu.au]

YONGPING LIU (*231), Department of Mathematics, Beijing Normal University, Beijing 100875, P. R. China [ypliu@bnu.edu.cn]

R. A. LORENTZ (**197), GMD, Schloss Birlinghoven, 53757 St. Augustin, Germany [lorentz@gmd.de]

Contributors xiii

CHUN-MEI LU (**189), Department of Mathematics, University of South Carolina, Columbia, SC 29208, USA [clu@math.sc.edu]

CLEMENT LUTTERODT (*239), Department of Mathematics, Howard University, Washington, DC 20059, USA [clutterodt@fac.howard.edu]

TOM LYCHE (**33), University of Oslo, Institutt for Informatikk, P.O. Box 1080, Blindern 0316, Oslo [tom@ifi.uio.no]

W. R. MADYCH (**197), Department of Mathematics, U-9, University of Connecticut, Storrs, CT 06269, USA [madych@uconnvm.uconn.edu]

MOHSEN MAESUMI (**205), Mathematics Department, Lamar University, Beaumont, TX 77710, USA [maesumi@math.lamar.edu]

MILJENKO MARUŠIĆ (**213), Dept. of Mathematics, University of Zagreb, Bijenička 30, 10000 Zagreb, Croatia [miljenko.marusic@math.hr]

JOHN C. MASON (*89), School of Computing and Mathematics, University of Huddersfield, Huddersfield, HD1 3DH, UK [j.c.mason@hud.ac.uk]

PETER R. MASSOPUST (**161), Department of Mathematics, Sam Houston State University, Huntsville, TX 77341, USA [mth_prm@shsu.edu]

ALLAN W. MCINNES (*247), Department of Mathematics, University of Canterbury, Christchurch, New Zealand [A.McInnes@math.canterbury.ac.nz]

G. MIN (*255), Department of Mathematics and Statistics, Simon Fraser University, Burnaby, B. C., Canada V5A 1S6 [gmin@cecm.sfu.ca]

M. L. MITTAL (*263), Department of Mathematics, University of Roorkee, Roorkee 247 667, Uttar Pradesh, India

RAM MOHAPATRA (**129), Department of Mathematics, University of Central Florida, Orlando, FL 32816, USA [ramm@pegasus.cc.ucf.edu]

FRANCIS J. NARCOWICH (**221), Department of Mathematics, Texas A&M University, College Station, TX 77843-3368, USA [fnarc@math.tamu.edu]

ALEXA NAWOTKI (**243), Department of Computer Science, University of Kaiserslautern, Postfach 3049, D-67653 Kaiserslautern, Germany [nawotki@informatik.uni-kl.de]

ERICH NOVAK (**251), *University of Erlangen and Nürnberg, Mathematical Institute, Bismarckstr. 1 1/2, D-91054 Erlangen, Germany* [novak@mi.uni-erlangen.de]

GÜNTHER NÜRNBERGER (**17,**259), *Universität Mannheim, Fakultät für Mathematik und Informatik, 68131 Mannheim, Germany* [nuern@euklid.math.uni-mannheim.de]

TERESA H. O'DONNELL (**267), *ARCON Corporation, 260 Bear Hill Road, Waltham, MA 02154, USA* [terry@arcon.com]

PETER OSWALD (**275), *Bell Laboratories, Lucent Technologies, 600 Mountain Av., Rm. 2C403, Murray Hill, NJ 07974, USA* [poswald@research.bell-labs.com]

K. PAN (*83), *Department of Mathematics, Barry University, Miami Shores, FL 33161, USA* [pan@buvax.barry.edu]

MARIA C. PEREYRA (**161), *Department of Mathematics, University of New Mexico, Albuquerque, NM 87131, USA* [crisp@math.unm.edu]

IGOR E. PRITSKER (*271), *Department of Mathematics, Case Western Reserve University, 10900 Euclid Avenue, Cleveland, OH 44106-7058, USA* [iep@po.cwru.edu]

VASILIY A. PROKHOROV (*175), *Department of Mathematics, University of South Florida, Tampa, Florida 33620, USA* [prokhoro@math.usf.edu]

MICHAEL P. PROPHET (*75), *Department of Mathematics, Murray State University, Murray, KY 42071, USA* [mike@banach.mursuky.edu]

EWALD G. QUAK (**63), *SINTEF Applied Mathematics, Post Box 124, Blindern, 0314 Oslo, Norway* [Ewald.Quak@math.sintef.no]

GEETHA S. RAO (*279), *Ramanujan Institute for Advanced Study in Mathematics, University of Madras, Madras-600 005, India* [geetsr@unimad.ernet.in]

KLAUS RITTER (**251), *Fakultät für Mathematik und Informatik, Universität Passau, 94030 Passau, Germany* [Klaus.Ritter@fmi.uni-passau.de]

GEORGE A. ROBERTS (**179), *Department of Mathematics, Prairie View A&M University, Prairie View, Texas 77446, USA* [groberts@zeno.math.pvamu.edu]

AMOS RON (**283), *Department of Computer Sciences, University of Wisconsin - Madison, 1210 West Dayton, Madison, WI 57311, USA* [amos@cs.wisc.edu]

EDWARD B. SAFF (*175), Institute for Constructive Mathematics, Department of Mathematics, University of South Florida, Tampa, Florida 33620, USA [esaff@math.usf.edu]

R. SARAVANAN (*279), Ramanujan Institute for Advanced Study in Mathematics, University of Madras, Madras-600 005, India [geetsr@unimad.ernet.in]

THOMAS SAUER (*287), Mathematical Institute, University Erlangen–Nürnberg, Bismarckstr 1 1/2, D-91054 Erlangen, Germany [sauer@mi.uni-erlangen.de]

BL. SENDOV (*295), Bulgarian Academy of Sciences, 1113 Sofia, Bulgaria [sendov@amigo.acad.bg]

BORIS SHEKHTMAN (*157,*303), Department of Mathematics, University of South Florida, Tampa, FL 33620, USA [boris@chuma.cas.usf.edu]

IGOR A. SHEVCHUK (*223), Institute of Mathematics, NAS of Ukraine, 3, Tereshchenkivska str., Kyiv 252601, Ukraine [shevchuk@imath.kiev.ua]

M.-R. SKRZIPEK (*309), FernUniversität-GHS Hagen, Fachbereich Mathematik, Postfach 940, D-58084 Hagen, Germany [Michael.Skrzipek@FernUni-Hagen.de]

MANFRED SOMMER (**25), Katholische Universität Eichstätt, Mathematisch-Geographische Fakultät, 85071 Eichstätt, Germany [manfred.sommer@ku-eichstaett.de]

HUGH SOUTHALL (**267), Electromagnetics Technology Division, US Air Force Research Laboratory, 31 Grenier Street, Hanscom AFB, MA 01731-3010, USA [southall@maxwell.rl.plh.af.mil]

FRAUKE SPRENGEL (**319), CWI, Kruislaan 413, P. O. Box 94079, NL-1090 GB Amsterdam, The Netherlands [frauke.sprengel@cwi.nl]

GABRIELE STEIDL (**39), Universität Mannheim, Fakultät für Mathematik und Informatik, A5, D-68131 Mannheim, Germany [steidl@math.uni-mannheim.de]

ACHIM STEINBAUER (**251), University of Erlangen and Nürnberg, Mathematical Institute, Bismarckstr. 1 1/2, D-91054 Erlangen, Germany [steinbau@mi.uni-erlangen.de]

JOACHIM STÖCKLER (*97,**137), Department of Mathematics and Computer Science, University of Missouri - St. Louis, 8001 Natural Bridge Road, St. Louis, MO 63121-4499 [stockler@uni-hohenheim.de]

HANS STRAUSS (*317), Institut für Angewandte Mathematik, Universität Erlangen–Nürnberg, Martensstr. 3, 91058 Erlangen, Germany [strauss@am.uni-erlangen.de]

GANCHO TACHEV (*325), Department of Mathematics, University of Architecture, 1 Hristo Smirnenski blvd., 1421 Sofia, Bulgaria
[gtt_fte@bgace5.uacg.acad.bg]

G. D. TAYLOR (*127), Department of Mathematics, Colorado State University, Fort Collins, CO 80523, USA
[taylor@math.colostate.edu]

VLADIMIR N. TEMLYAKOV (*7), Department of Mathematics, University of South Carolina, Columbia, SC 29208, USA
[temlyak@math.sc.edu]

JUN TIAN (**327), Computation Mathematics Laboratory, Rice University, Houston, TX 77005-1892, USA [juntian@rice.edu]

D. TOWNSEND (*119), Department of Mathematics, Indiana-Purdue University, Fort Wayne, IN 46805, USA [townsend@ipfw.edu]

S. A. TOZONI (*23), IMECC-UNICAMP, Caixa Postal 6065, 13081-970, Campinas, SP, Brazil [tozoni@ime.unicamp.br]

GENRIKH TS. TUMARKIN (*331), 630 Rossmore Road, Goleta, CA 93117, USA [tumarkin@gte.net]

SHAYNE WALDRON (*339), Department of Mathematics, University of Auckland, Private Bag 92019, Auckland, New Zealand
[waldron@math.auckland.ac.nz]

GUIDO WALZ (**337), Department of Mathematics and Computer Science, University of Mannheim, D-68131 Mannheim, Germany
[walz@math.uni-mannheim.de]

JOSEPH D. WARD (**137), Department of Mathematics, Texas A&M University, College Station, TX 77843, USA
[jward@math.tamu.edu]

JOE WARREN (**345), Department of Computer Science, Rice University, P.O. Box 1892, Houston, TX 77005-1892, USA
[jwarren@rice.edu]

SUMIO WATANABE (*347), Advanced Information Processing Division, Precision and Intelligence Laboratory, Tokyo Institute of Technology, Tokyo, Japan [swatanab@pi.titech.ac.jp]

G. ALISTAIR WATSON (**353), Department of Mathematics, University of Dundee, Dundee DD1 4HN, Scotland
[gawatson@mcs.dundee.ac.uk]

HENRIK WEIMER (**345), Department of Computer Science, Rice University, P.O. Box 1892, Houston, TX 77005-1892, USA
[henrik@rice.edu]

Contributors xvii

RAYMOND O. WELLS, JR. (**327), *Computation Mathematics Laboratory, Rice University, Houston, TX 77005-1892, USA* [wells@rice.edu]

HOLGER WENDLAND (**361), *Institut für Numerische und Angewandte Mathematik, Universität Göttingen, Lotzestr. 16-18, D-37083 Göttingen, Germany* [wendland@math.uni-goettingen.de]

PAUL WENSTON (**153), *Department of Mathematics, University of Georgia, Athens, GA 30602, USA* [paul@math.uga.edu]

NORMAN WEYRICH (**369), *Synopsys, Inc., Kaiserstr. 100, 52134 Herzogenrath, Germany* [weyrich@synopsys.com]

MARY F. WHEELER (**377), *Texas Institute for Computational and Applied Mathematics, The University of Texas at Austin, Austin, Texas 78712, USA* [mfw@ticam.utexas.edu]

IVAN YOTOV (**377), *Texas Institute for Computational and Applied Mathematics, The University of Texas at Austin, Austin, Texas 78712, USA* [yotov@ticam.utexas.edu]

FRANK ZEILFELDER (**17,**259), *Universität Mannheim, Fakultät für Mathematik und Informatik, 68131 Mannheim, Germany* [zeilfeld@fourier.math.uni-mannheim.de]

PING ZHOU (*51), *Department of Mathematics and Statistics, Simon Fraser University, Burnaby, B. C., Canada, V5A 1S6* [pzhou@cecm.sfu.ca]

The Degree of Convergence of Sequences of Multivariate Wavelet Type Operators

George A. Anastassiou

Abstract. Higher order differentiable multivariate functions are approximated by wavelet-type operators, old and new. The higher order of this approximation is estimated by establishing some multivariate Jackson type inequalities. Sharpness of some of these inequalities over continuous functions is achieved nontrivially.

§1. Introduction

Lately there has been great interest in the wavelet type multivariate approximations. In [1, 2] the author and X. M. Yu introduced and studied the wavelet type multivariate operators A_k, B_k over continuous functions on $\mathbb{R}^r$, $r \geq 1$. There we especially studied their approximation properties. In this article two new wavelet-type multivariate operators C_k, D_k are introduced in a natural way.

For all of these multivariate operators A_k, B_k, C_k, D_k, we study their approximation to the unity over $f \in C^N(\mathbb{R}^r)$, $N \geq 0$. We produce related multivariate Jackson-type inequalities, which give very close upper bounds to the error of this higher order approximation. When $N = 0$, sharpness over continuity is established for A_k, C_k, D_k operators. The same was established for B_k operators earlier in [1, 2].

All the inequalities involve the multivariate first order modulus of continuity ω_1 of $f^{(N)}$, and the multivariate scale function of compact support φ is left without any assumption about its orthogonality.

We use the following multivariate first order modulus of continuity ω_1: Let $f \in C(\mathbb{R}^r)$ be bounded or uniformly continuous. We define, for $h > 0$,

$$\omega_1(f,h) = \sup_{\substack{\text{all } x_i, x_i' \in \mathbb{R} \\ |x_i - x_i'| \leq h, \text{ for } i=1,\ldots,r}} |f(x_1,\ldots,x_r) - f(x_1',\ldots,x_r')|.$$

§2. Results

Next we present our first main result.

Theorem 1. *Let $f \in C^N(\mathbb{R}^r)$, $N \in \mathbb{N}$ and $r \geq 1$; $\vec{x} \in \mathbb{R}^r$ and $k \in \mathbb{Z}$. Let $\varphi \geq 0$ be a bounded function on $\mathbb{R}^r$ compact support $\subseteq \prod_{i=1}^{r}[-a_i, a_i]$, $0 < a_i < +\infty$, $a := \max(a_1, \ldots, a_r)$. Assume that*

$$\sum_{j_1=-\infty}^{\infty} \cdots \sum_{j_r=-\infty}^{\infty} \varphi(x_1 - j_1, \ldots, x_r - j_r) = 1, \quad \text{all}$$
$$\vec{x} := (x_1, \ldots, x_r) \in \mathbb{R}^r, \tag{1}$$

or briefly,

$$\sum_{\vec{j}=-\infty}^{\infty} \varphi(\vec{x} - \vec{j}) = 1, \quad \text{all } \vec{x} \in \mathbb{R}^r, \tag{1'}$$

where $\vec{j} := (j_1, \ldots, j_r)$. Call

$$B_k(f)(x_1, \ldots, x_r) := \sum_{j_1=-\infty}^{\infty} \cdots \sum_{j_r=-\infty}^{\infty} f\left(\frac{j_1}{2^k}, \ldots, \frac{j_r}{2^k}\right);$$
$$\varphi(2^k x_1 - j_1, \ldots, 2^k x_r - j_r), \quad \text{any } k \in \mathbb{Z}, \tag{2}$$

all $(x_1, \ldots, x_r) \in \mathbb{R}^r$, or briefly

$$B_k(f)(\vec{x}) = \sum_{\vec{j}=-\infty}^{\infty} f\left(\frac{\vec{j}}{2^k}\right) \varphi(2^k \vec{x} - \vec{j}), \quad \text{any } k \in \mathbb{Z} \text{ all } \vec{x} \in \mathbb{R}^r. \tag{2'}$$

Furthermore, assume that all of the partial derivatives of f of order N, denoted by

$$f_{\tilde{\alpha}} := \frac{\partial^{\tilde{\alpha}} f}{\partial x^{\tilde{\alpha}}} \left(\tilde{\alpha} := (\alpha_1, \ldots, \alpha_r), \alpha_i \in \mathbb{Z}^+, i = 1, \ldots, r : |\tilde{\alpha}|\right.$$
$$\left. = \sum_{i=1}^{r} \alpha_i = N\right),$$

are continuous and bounded or uniformly continuous on $\mathbb{R}^r$. Then

$$|(B_k(f))(\vec{x}) - f(\vec{x})| \leq \sum_{j=1}^{N} \frac{a^j}{j! 2^{kj}} \left(\left(\sum_{i=1}^{r} \left|\frac{\partial}{\partial x_i}\right|\right)^j f(\vec{x})\right)$$
$$+ \frac{a^N r^N}{N! 2^{kN}} \max_{\tilde{\alpha}:|\tilde{\alpha}|=N} \omega_1\left(f_{\tilde{\alpha}}, \frac{a}{2^k}\right), \quad \text{any } k \in \mathbb{Z}, \tag{3}$$

and equality is attained by constant functions.

Remark 1.

i) Obviously, $B_k f \to f$ pointwise over $\mathbb{R}^r$, as $k \to \infty$.

ii) Given that $f \in C_b^N(\mathbb{R}^r)$ (*i.e.*, all f and its partial derivatives up to order N are continuous and bounded) we obtain

$$\|B_k f - f\|_\infty \le \sum_{j=1}^N \frac{a^j}{j! 2^{kj}} \left(\left(\sum_{i=1}^r \left\| \frac{\partial}{\partial x_i} \right\|_\infty \right)^j f \right)$$
$$+ \frac{a^N r^N}{N! 2^{kN}} \max_{\tilde{\alpha}: |\tilde{\alpha}| = N} \omega_1 \left(f_{\tilde{\alpha}}, \frac{a}{2^k} \right), \quad \text{any } k \in \mathbb{Z}. \qquad (4)$$

That is, $B_k f \to f$, uniformly over $\mathbb{R}^r$, as $k \to \infty$.

iii) For $N = 1$ in (3), we have

$$|(B_k f)(\vec{x}) - f(\vec{x})| \le \frac{a}{2^k} \left\{ \left(\sum_{i=1}^r \left| \frac{\partial f(\vec{x})}{\partial x_i} \right| \right) \right.$$
$$\left. + r \max_{i \in \{1, \ldots, r\}} \omega_1 \left(\frac{\partial f}{\partial x_i}, \frac{a}{2^k} \right) \right\}, \quad \text{any } k \in \mathbb{Z}. \qquad (5)$$

The second main result follows.

Theorem 2. *Consider f and φ as in Theorem 1 and use the same notations there. Additionally assume that φ is Lebesgue measurable. Define*

$$A_k(f)(\vec{x}) := \sum_{\vec{j} = -\infty}^\infty \alpha_{k\vec{j}}(f) \varphi(2^k \vec{x} - \vec{j}), \qquad (6)$$

where

$$\alpha_{k\vec{j}}(f) := \int_{\mathbb{R}^r} f\left(\frac{\vec{u}}{2^k} \right) \varphi(\vec{u} - \vec{j}) \, d\vec{u}, \quad k \in \mathbb{Z}. \qquad (7)$$

Then for any $k \in \mathbb{Z}$,

$$|A_k(f)(\vec{x}) - f(\vec{x})| \le \sum_{j=1}^N \frac{a^j}{j! 2^{(k-1)j}} \left(\left(\sum_{i=1}^r \left| \frac{\partial}{\partial x_i} \right| \right)^j f(\vec{x}) \right)$$
$$+ \frac{a^N r^N}{N! 2^{(k-1)N}} \max_{\tilde{\alpha}: |\tilde{\alpha}| = N} \omega_1 \left(f_{\tilde{\alpha}}, \frac{a}{2^{k-1}} \right), \qquad (8)$$

and equality is attained by constant functions.

Remark 2.

i) Obviously $A_k f \to f$ pointwise on $\mathbb{R}^r$, as $k \to \infty$.

ii) Given $f \in C_b^N(\mathbb{R}^r)$, we obtain

$$\|A_k f - f\|_\infty \le \sum_{j=1}^N \frac{a^j}{j! 2^{(k-1)j}} \left(\left(\sum_{i=1}^r \left\| \frac{\partial}{\partial x_i} \right\|_\infty \right)^j f \right)$$
$$+ \frac{a^N r^N}{N! 2^{(k-1)N}} \max_{\tilde\alpha: |\tilde\alpha|=N} \omega_1 \left(f_{\tilde\alpha}, \frac{a}{2^{k-1}} \right), \text{ any } k \in \mathbb{Z}. \qquad (9)$$

That is $A_k f \to f$ uniformly on $\mathbb{R}^r$, as $k \to \infty$.

iii) For $N = 1$ (8), we have

$$|(A_k f)(\vec{x}) - f(\vec{x})| \le \frac{a}{2^{k-1}} \left\{ \left(\sum_{i=1}^r \left| \frac{\partial f(\vec{x})}{\partial x_i} \right| \right) \right.$$
$$\left. + r \max_{i \in \{1,\dots,r\}} \omega_1 \left(\frac{\partial f}{\partial x_i}, \frac{a}{2^{k-1}} \right) \right\}, \text{ any } k \in \mathbb{Z}. \qquad (10)$$

A new operator is studied next.

Theorem 3. *Under the same assumptions and notations as in Theorem 1, and setting*

$$\gamma_{k\vec{j}}(f) := 2^{rk} \int_{2^{-k}\vec{j}}^{2^{-k}(\vec{j}+\vec{1})} f(\vec{t}) \, d\vec{t} = 2^{rk} \int_{\vec{0}}^{2^{-\vec{k}}} f\left(\vec{t} + \frac{\vec{j}}{2^k}\right) d\vec{t}, \qquad (11)$$

$$C_k(f)(\vec{x}) := \sum_{\vec{j}=-\infty}^{\infty} \gamma_{k\vec{j}}(f) \varphi(2^k \vec{x} - \vec{j}), \qquad (12)$$

all $\vec{x} \in \mathbb{R}^r$ and $k \in \mathbb{Z}$, then

$$|C_k(f)(\vec{x}) - f(\vec{x})| \le \sum_{j=1}^N \frac{(a+1)^j}{j! 2^{kj}} \left(\left(\sum_{i=1}^r \left| \frac{\partial}{\partial x_i} \right| \right)^j f(\vec{x}) \right)$$
$$+ \frac{(a+1)^N r^N}{N! 2^{kN}} \max_{\tilde\alpha: |\tilde\alpha|=N} \omega_1 \left(f_{\tilde\alpha}, \frac{a+1}{2^k} \right), \qquad (13)$$

and equality which is attained by constant functions.

Remark 3.

i) Obviously $C_k f \to f$ pointwise on $\mathbb{R}^r$, as $k \to \infty$.

ii) Given $f \in C_b^N(\mathbb{R}^r)$, we obtain

$$\|C_k f - f\|_\infty \le \sum_{j=1}^N \frac{(a+1)^j}{j! 2^{kj}} \left(\left(\sum_{i=1}^r \left\| \frac{\partial}{\partial x_i} \right\|_\infty \right)^j f \right)$$
$$+ \frac{(a+1)^N r^N}{N! 2^{kN}} \max_{\tilde\alpha: |\tilde\alpha|=N} \omega_1 \left(f_{\tilde\alpha}, \frac{a+1}{2^k} \right), \qquad (14)$$

any $k \in \mathbb{Z}$. That is, $C_k f \to f$, uniformly on $\mathbb{R}^r$, as $k \to \infty$.

iii) For $N = 1$ in (13), we have

$$|(C_k f)(\vec{x}) - f(\vec{x})| \le \left(\frac{a+1}{2^k}\right) \left\{ \left(\sum_{i=1}^{r} \left|\frac{\partial f(\vec{x})}{\partial x_i}\right|\right) + r \max_{i \in \{1,\ldots,r\}} \omega_1 \left(\frac{\partial f}{\partial x_i}, \frac{a+1}{2^k}\right) \right\}, \quad (15)$$

any $k \in \mathbb{Z}$.

The next quadrature wavelet-type multivariate operator is studied also for the first time.

Theorem 4. *Under the same assumptions and notations as in Theorem 1, and defining ($k \in \mathbb{Z}$, $\vec{j} \in \mathbb{Z}^r$, $\vec{x} \in \mathbb{R}^r$)*

$$(D_k f)(\vec{x}) := \sum_{\vec{j}=-\infty}^{\infty} \delta_{k\vec{j}}(f) \varphi(2^k \vec{x} - \vec{j}), \quad (16)$$

where

$$\delta_{k\vec{j}}(f) := \sum_{\vec{\ell}=\vec{0}}^{\vec{n}} \omega_{\vec{\ell}} f\left(\frac{\vec{j}}{2^k} + \frac{\vec{\ell}}{2^k \vec{n}}\right),$$

$$\vec{\ell} \in \mathbb{Z}_+^r, \quad \vec{n} \in \mathbb{N}^r, \quad \omega_{\vec{\ell}} \ge 0, \quad \sum_{\vec{\ell}=0}^{\vec{n}} \omega_{\vec{\ell}} = 1,$$

or equivalently,

$$\delta_{k,j_1,\ldots,j_r}(f) = \sum_{\ell_1=0}^{n_1} \sum_{\ell_2=0}^{n_2} \cdots \sum_{\ell_r=0}^{n_r} \omega_{\ell_1,\ldots,\ell_r}$$
$$\cdot f\left(\frac{j_1}{2^k} + \frac{\ell_1}{2^k n_1}, \frac{j_2}{2^k} + \frac{\ell_2}{2^k n_2}, \ldots, \frac{j_r}{2^k} + \frac{\ell_r}{2^k n_r}\right),$$

$$\omega_{\ell_1,\ell_2,\ldots,\ell_r} \ge 0, \quad \sum_{\ell_1=0}^{n_1} \cdots \sum_{\ell_r=0}^{n_r} \omega_{\ell_1,\ldots,\ell_r} = 1,$$

then

$$|(D_k f)(\vec{x}) - f(\vec{x})| \le \sum_{j=1}^{N} \frac{(a+1)^j}{j! 2^{kj}} \left(\left(\sum_{i=1}^{r} \left|\frac{\partial}{\partial x_i}\right|\right)^j f(\vec{x}) \right)$$
$$+ \frac{(a+1)^N r^N}{N! 2^{kN}} \max_{\tilde{\alpha}: |\tilde{\alpha}|=N} \omega_1\left(f_{\tilde{\alpha}}, \frac{a+1}{2^k}\right), \quad (17)$$

and equality is attained by constant functions.

Remark 4.
i) Obviously $D_k f \to f$ pointwise on $\mathbb{R}^r$, as $k \to \infty$.
ii) Given $f \in C_b^N(\mathbb{R}^r)$ we obtain

$$\|D_k f - f\|_\infty \leq \sum_{j=1}^N \frac{(a+1)^j}{j! 2^{kj}} \left(\left(\sum_{i=1}^r \left\| \frac{\partial}{\partial x_i} \right\|_\infty \right)^j f \right)$$
$$+ \frac{(a+1)^N r^N}{N! 2^{kN}} \max_{\tilde{\alpha}: |\tilde{\alpha}|=N} \omega_1 \left(f_{\tilde{\alpha}}, \frac{a+1}{2^k} \right), \text{ any } k \in \mathbb{Z}. \quad (18)$$

That is, $D_k f \to f$ uniformly on $\mathbb{R}^r$, as $k \to \infty$.

iii) For $N = 1$ in (17), we have

$$|(D_k f)(\vec{x}) - f(\vec{x})| \leq \left(\frac{a+1}{2^k} \right) \left\{ \left(\sum_{i=1}^r \left| \frac{\partial f(\vec{x})}{\partial x_i} \right| \right) \right.$$
$$\left. + r \max_{i \in \{1,\ldots,r\}} \omega_1 \left(\frac{\partial f}{\partial x_i}, \frac{a+1}{2^k} \right) \right\}, \text{ any } k \in \mathbb{Z}. \quad (19)$$

Note. In [1] and [2] we proved that

$$|(B_k f)(\vec{x}) - f(\vec{x})| \leq \omega_1 \left(f, \frac{a}{2^k} \right), \quad (20)$$

where $f \in C(\mathbb{R}^r)$, $a := \max(a_1, \ldots, a_r)$, any $k \in \mathbb{Z}$, all $\vec{x} \in \mathbb{R}^r$, and that (20) is sharp.

Next we treat the other operators similarly.

Proposition 1. *Let $f \in C(\mathbb{R}^r)$, $r \geq 1$, $a := \max(a_1, \ldots, a_r)$, $k \in \mathbb{Z}$, $\vec{x} \in \mathbb{R}^r$.*
i) *Under the notations and assumptions of Theorem 2,*

$$|A_k(f)(\vec{x}) - f(\vec{x})| \leq \omega_1 \left(f, \frac{a}{2^{k-1}} \right). \quad (21)$$

ii) *Under the notations and assumptions of Theorems 1, 3,*

$$|C_k(f)(\vec{x}) - f(\vec{x})| \leq \omega_1 \left(f, \frac{a+1}{2^k} \right). \quad (22)$$

iii) *Under the notations and assumptions of Theorems 1, 4,*

$$|D_k(f)(\vec{x}) - f(\vec{x})| \leq \omega_1 \left(f, \frac{a+1}{2^k} \right). \quad (23)$$

Proposition 2. *Inequalities (21), (22), (23) are sharp.*

References

[1] Anastassiou, G. A. and X. M. Yu, Bivariate probabilistic wavelet approximation, in *Proceedings of Sixth S.E. Approximation Theory International Conference*, Memphis, Marcel Dekker, Inc., New York, 1992, pp. 79–92.

[2] Anastassiou, G. A. and X. M. Yu, Multivariable probabilistic scale approximation, in *Fund. Sci. Appl.*, Plovdiv, Bulgaria, 1998, to appear.

Best m-Term Approximation of Functions from Classes $MW_{q,\alpha}^r$

Alexander V. Andrianov and Vladimir N. Temlyakov

Abstract. The paper is devoted to the study of best m-term approximation of functions from classes $MW_{q,\alpha}^r$, $r > 0$, by general bases in the L_p-metrics. The correct order of approximation is found for $1 < p, q < \infty$. These results demonstrate the advantage of wavelet type systems over the classical trigonometric system for this kind of approximation.

This paper is a development of nonlinear approximation of functions with bounded mixed derivatives. The classes $MW_{q,\alpha}^r$ of multivariate functions with bounded mixed derivative have a long and impressive history in approximation theory (widths, polynomial approximation), numerical integration, theory of complexity (see [7, 12]). The nonlinear m-term approximation of these classes has been developed for the case of approximation by the trigonometric system [8, 9]. Recent study of the nonlinear m-term approximation by systems of wavelets and their analogues revealed the advantage of these systems over uniformly bounded systems (see [1, 2, 4, 10]). In this paper, we study the nonlinear m-term approximation of classes $MW_{q,\alpha}^r$ in the spirit of paper [3] where linear approximation of these classes by general wavelet type systems has been developed. We consider the periodic case here. We recall the definition of classes $MW_{q,\alpha}^r$ (see [7]).

Let $r > 0$ and α be real numbers. Consider the following functions

$$F_r(x, \alpha) := 1 + 2 \sum_{k=1}^{\infty} k^{-r} \cos\left(2\pi k x - \frac{\alpha \pi}{2}\right), \quad F_r(\boldsymbol{x}, \boldsymbol{\alpha}) := \prod_{j=1}^{d} F_r(x_j, \alpha_j),$$

where $\boldsymbol{x} := (x_1, \ldots x_d)$, and $\boldsymbol{\alpha} := (\alpha_1, \ldots \alpha_d)$.

We say that a function f belongs to the class $MW_{q,\boldsymbol{\alpha}}^r$, $r > 0$, $1 \leq q \leq \infty$, if there exists a function $\phi \in L_q([0,1]^d)$, $\|\phi\|_q \leq 1$, such that

$$f(\boldsymbol{x}) = I_{\boldsymbol{\alpha}}^r \phi(\boldsymbol{x}) := \int_{[0,1]^d} F_r(\boldsymbol{x} - \boldsymbol{y}, \boldsymbol{\alpha}) \phi(\boldsymbol{y}) d\boldsymbol{y}.$$

The basic tool in our study is the Littlewood-Paley inequalities. Let us recall these inequalities for the Haar system (see [5], Chapter III, § 3). It is convenient for us to enumerate function systems by dyadic intervals. Let $\mathcal{D}_0 := \mathcal{D}_0([0,1]^d)$ be the set of all dyadic intervals contained in the unit cube $[0,1]^d$ so that $I \in \mathcal{D}_0([0,1]^d)$ if $I = I_1 \times \ldots \times I_d$, $I_i \in \mathcal{D}_0([0,1])$, $i = 1, \ldots, d$. Then for any function $f \in L_q([0,1]^d)$, $1 < q < \infty$,

$$\|f\|_q \asymp \left\|\left(\sum_{I \in \mathcal{D}_0}[(f,H_I)\chi_I]^2\right)^{1/2}\right\|_q, \tag{1}$$

where χ_I stands for the characteristic function of the interval I normalized in the $L_2([0,1]^d)$-norm, and $\{H_I\}$, $I \in \mathcal{D}_0([0,1]^d)$, is the Haar basis for $L_2([0,1]^d)$.

Given a system of functions $\{\psi_I\}$, $I \in \mathcal{A} \subset \mathcal{D}_0$, we say that $\{\psi_I\}$, $I \in \mathcal{A}$, satisfies the Littlewood-Paley property for q, $1 < q < \infty$, if, for any sequence $\{c_I\}_{I \in \mathcal{A}}$ of real numbers,

$$\left\|\sum_{I \in \mathcal{A}} c_I \psi_I\right\|_q \asymp \left\|\left(\sum_{I \in \mathcal{A}} [c_I \chi_I]^2\right)^{1/2}\right\|_q, \tag{2}$$

where χ has the same meaning as in (1). Throughout this paper, the inequality $A \ll B$ ($A \gg B$) means that there exists a positive constant C which may depend on p, d, and q such that $A \leq CB$ ($B \leq CA$), and $A \asymp B$ means that $A \gg B$ and $A \ll B$. If (2) holds with $\asymp$ replaced by $\ll$, then we write $\{\psi_I\}_{I \in \mathcal{A}} \prec \{H_I\}_{I \in \mathcal{A}}$ in $L_q([0,1]^d)$.

Let there be given a system of functions $\{a_I\}_{I \in \mathcal{A}} \subset L_q([0,1]^d)$ and let $\{b_I\}_{I \in \mathcal{A}} \subset L_{q'}([0,1]^d)$ be its biorthogonal system. We recall that the best m-term approximation of a function f from some Banach space X with respect to the system $\{a_I\}_{I \in \mathcal{A}} \subset X$ is defined as follows

$$\sigma_m(f, \{a_I\}_{I \in \mathcal{A}})_X := \inf \left\|f - \sum_{j=1}^m c_j a_{I^j}\right\|_X, \quad m = 1, 2, \ldots,$$

where the inf is taken over all elements $a_{I^j} \in \{a_I\}_{I \in \mathcal{A}}$ and coefficients c_j, $j = 1, \ldots, m$.

Let $X := L_p([0,1]^d)$, and $\sigma_m(MW^r_{q,\alpha}, \{a_I\}_{I \in \mathcal{A}})_p := \sup_{f \in MW^r_{q,\alpha}} \sigma_m(f, \{a_I\}_{I \in \mathcal{A}})_p$. Then we have the following:

Theorem. *Let the systems $\{a_I\} \subset L_q([0,1]^d)$ and $\{b_I\} \subset L_{q'}([0,1]^d)$, $I \in \mathcal{D}_0$ be such that*

(A1) $\{a_I\}_{I \in \mathcal{D}_0}$ span $L_q([0,1]^d)$ for any $1 < q < \infty$, and satisfy the Littlewood-Paley property (2),

(A2) $\{|I|^{-r} I^r_{-\alpha} b_I\}_{I \in \mathcal{D}_0} \prec \{H_I\}_{I \in \mathcal{D}_0}$ in $L_{q'}([0,1]^d)$ for some $r > 0$ and for any $1 < q' < \infty$ where $\{H_I\}_{I \in \mathcal{D}_0}$ is the Haar basis for $L_{q'}([0,1]^d)$.

Then for any $1 < p < \infty$,

$$\sigma_m(MW^r_{q,\alpha}, \{a_I\}_{I \in \mathcal{D}_0})_p \asymp m^{-r}(\log m)^{(d-1)r}, \quad m = 2, 3, \ldots,$$

where $r > \max\left(\frac{1}{2}, \frac{1}{q}\right) - \min\left(\frac{1}{2}, \frac{1}{p}\right)$.

Proof: We prove the upper estimate first. The proof uses ideas from [8] (see Ch. 4, Th. 2.1 and Th. 4.2). We will use the following notation. Let $\mathcal{D}_s$, $s \geq 0$, denote the set of dyadic intervals from $\mathcal{D}_0([0,1])$ which have length 2^{-s}, and for a vector $s = (s_1, \ldots, s_d)$ from $\mathbb{N}^d$, $\mathbb{N} := \{0, 1, \ldots\}$, let

$$\mathcal{D}_s = \{I = I_1 \times \ldots \times I_d, \ I_j \in \mathcal{D}_{s_j}, \ j = 1, \ldots, d\}, \quad Q_n = \cup_{\|s\|_1 \leq n} \mathcal{D}_s,$$

$$\|s\|_1 := s_1 + \ldots + s_d, \quad \mathcal{H}(Q_n) := \left\{f = \sum_{I \in Q_n} c_I a_I\right\}, \quad n \in \mathbb{N}.$$

We need the following

Lemma 1. *Under the assumptions of Theorem, for any function $f \in MW^r_{q,\alpha}$, $r > 0$, we have*

$$E_{Q_n}(f)_q \ll 2^{-rn}, \quad n \in \mathbb{N}, \tag{3}$$

where $E_{Q_n}(f)_q := \inf_{g \in \mathcal{H}(Q_n)} \|f - g\|_q$.

Proof: It is known (see [3]) that assumption (**A1**) implies that the system $\{a_I\}_{I \in \mathcal{D}_0}$ is an unconditional basis for $L_q([0,1]^d)$. Hence, for any $f \in L_q([0,1]^d)$,

$$f = \sum_{I \in \mathcal{D}_0} (f, b_I) a_I$$

in the sense of $L_q([0,1]^d)$-convergence and so

$$E_{Q_n}(f)_q \leq \|f - S_{Q_n}(f)\|_q, \quad n \in \mathbb{N},$$

where $S_{Q_n}(f) := \sum_{I \in Q_n} (f, b_I) a_I$, $n \in \mathbb{N}$. Moreover, assuming $f \in MW^r_{q,\alpha}$ to be of the form $f = I^r_\alpha \phi$, we have

$$(f, b_I) = (I^r_\alpha \phi, b_I) = (\phi, I^r_{-\alpha} b_I)$$

so that

$$\|f - S_{Q_n}(f)\|_q = \left\|\sum_{|I| < 2^{-n}} (f, b_I) a_I\right\|_q = \left\|\sum_{|I| < 2^{-n}} (\phi, I^r_{-\alpha} b_I) a_I\right\|_q \ll$$

$$\left\|\sum_{|I| < 2^{-n}} (\phi, I^r_{-\alpha} b_I) H_I\right\|_q = \left\|\sum_{|I| < 2^{-n}} |I|^r (\phi, |I|^{-r} I^r_{-\alpha} b_I) H_I\right\|_q \ll$$

$$2^{-rn} \left\|\sum_{|I| < 2^{-n}} (\phi, |I|^{-r} I^r_{-\alpha} b_I) H_I\right\|_q,$$

where property (**A1**) was used.

Assumption (**A2**) means that the operator T and its adjoint T^* defined by

$$T\phi := \sum_{I\in\mathcal{D}_0}(\phi,H_I)|I|^{-r}I_{-\alpha}^r b_I, \quad T^*\phi := \sum_{I\in\mathcal{D}_0}(\phi,|I|^{-r}I_{-\alpha}^r b_I)H_I$$

are bounded from $L_q([0,1]^d)$ to $L_q([0,1]^d)$ and from $L_{q'}([0,1]^d)$ to $L_{q'}([0,1]^d)$, respectively. Boundedness of T^* means that

$$\left\|\sum_{I\in\mathcal{D}_0}(\phi,|I|^{-r}I_{-\alpha}^r b_I)H_I\right\|_{q'} \leq C\|\phi\|_{q'}$$

for any function $\phi \in L_{q'}([0,1]^d)$, $1 < q' < \infty$. Hence,

$$\|f - S_{Q_n}(f)\|_q \ll 2^{-rn}\|\phi\|_q \leq 2^{-rn}$$

by definition of f. This proves Lemma 1.

We continue the proof of Theorem. It is easy to derive from the proof of Lemma 1 and assumption (**A1**) that the following relation holds:

$$\left\|\sum_{I\in Q_n\setminus Q_{n-1}}(f,b_I)a_I\right\|_q \asymp \left\|\left(\sum_{|I|=2^{-n}}[(f,b_I)\chi_I]^2\right)^{1/2}\right\|_q \ll 2^{-rn}. \qquad (4)$$

Also monotonicity of the L_q-norm implies that in the proof of the upper estimates, we can assume that $1 < q \leq 2$ and $2 \leq p < \infty$.

We recall the following simple corollary of the Littlewood-Paley theorem (1) for the system $\{H_I\}_{I\in\mathcal{D}_0}$, and the Littlewood-Paley property (2) for the system $\{a_I\}_{I\in\mathcal{D}_0}$ (see [8]), for $1 < q \leq 2 \leq p < \infty$,

$$\left(\sum_{s\geq 0}\|\delta_s(f)\|_q^2\right)^{1/2} \ll \|f\|_q \leq \|f\|_p \ll \left(\sum_{s\geq 0}\|\delta_s(f)\|_p^2\right)^{1/2}, \qquad (5)$$

where $\delta_s(f) := \sum_{I\in\mathcal{D}_s}(f,b_I)H_I$. Using (5) and assumption (**A1**), we see that (4) implies

$$\left(\sum_{\|s\|_1=n}\|\delta_s(f)\|_q^2\right)^{1/2} \ll 2^{-rn}. \qquad (6)$$

We take some fixed number $a \in \mathbb{Z}_+$ and put

$$m_n := [\#\{Q_a \setminus Q_{a-1}\}2^{-\kappa(n-a)}], \quad \kappa > 0; \quad N_n := \#\{s \in \mathbb{N}^d, \|s\|_1 = n\},$$
$$n = a+1,\ldots; \quad l_n := \max\left(\left[\frac{m_n}{N_n}\right], 1\right),$$

where κ will be chosen later. Then it is easy to see that

$$m := \#\{Q_a\} + \sum_{n=a+1}^{\infty} m_n \le C(\kappa,d)2^a a^{d-1}.$$

Let us consider $\delta_s(f), \|s\|_1 = n$. We have

$$\|\delta_s(f)\|_q^q = \int_{[0,1]^d} \left|\sum_{I \in \mathcal{D}_s}(f,b_I)H_I\right|^q d\boldsymbol{x} = 2^{n(\frac{q}{2}-1)}\sum_{I \in \mathcal{D}_s}|(f,b_I)|^q, \tag{7}$$

since intervals from $\mathcal{D}_s$ are disjoint.

The following simple lemma is well known (see, for instance, [8], p. 97.)

Lemma 2. *Let $a_1 \ge a_2 \ge \ldots \ge a_N \ge 0, 1 \le q \le p \le \infty$, and $a_1^q + \ldots + a_N^q \le A^q$. Then for any $M < N$, $(a_M^p + \ldots + a_N^p)^{\frac{1}{p}} \le AM^{\frac{1}{p}-\frac{1}{q}}$.*

We apply Lemma 2 to the sequence $\{|(f,b_I)|\}_{I \in \mathcal{D}_s}$ rearranged in decreasing order, with $N := 2^n = \#\{\mathcal{D}_s\}$, $A := 2^{n(\frac{1}{q}-\frac{1}{2})}\|\delta_s(f)\|_q$, and $M := l_n$, to deduce from (7) that

$$\left(\sum_{I \in \mathcal{D}_s^*}|(f,b_I)|^p\right)^{1/p} \le 2^{n(\frac{1}{q}-\frac{1}{2})}\|\delta_s(f)\|_q l_n^{\frac{1}{p}-\frac{1}{q}}$$

where $\mathcal{D}_s^*$ does not contain $\left\lceil\frac{m_n}{N_n}\right\rceil$ intervals corresponding to $\left\lceil\frac{m_n}{N_n}\right\rceil$ biggest coefficients $|(f,b_I)|$, $I \in \mathcal{D}_s$. Using the argument of (7), we see that

$$\|\delta_s^*(f)\|_p = 2^{n(\frac{1}{2}-\frac{1}{p})}\left(\sum_{I \in \mathcal{D}_s^*}|(f,b_I)|^p\right)^{1/p} \le 2^{n(\frac{1}{q}-\frac{1}{p})}\|\delta_s(f)\|_q l_n^{\frac{1}{p}-\frac{1}{q}}$$

where $\delta_s^*(f) := \sum_{I \in \mathcal{D}_s^*}(f,b_I)H_I$, and $\|s\|_1 = n$.

It follows then from (5) and (6) that

$$\left\|\sum_{\|s\|_1=n}\delta_s^*(f)\right\|_p \ll \left(\sum_{\|s\|_1=n}\|\delta_s^*(f)\|_p^2\right)^{1/2} \le$$

$$\left(\sum_{\|s\|_1=n} 2^{2n(\frac{1}{q}-\frac{1}{p})}\|\delta_s(f)\|_q^2 l_n^{2(\frac{1}{p}-\frac{1}{q})}\right)^{1/2} \ll 2^{n(\frac{1}{q}-\frac{1}{p})}l_n^{\frac{1}{p}-\frac{1}{q}}2^{-rn}. \tag{8}$$

Let t denote the function

$$t := \sum_{I \in Q_a}(f,b_I)a_I + \sum_{n=a+1}^{\infty}\sum_{\|s\|_1=n}\sum_{I \in \mathcal{D}_s \setminus \mathcal{D}_s^*}(f,b_I)a_I.$$

Then t contains $\asymp 2^a a^{d-1}$ terms, and for $\|f-t\|_p$, we have, using assumption (**A1**) and (8), that

$$\|f-t\|_p \ll \sum_{n=a+1}^{\infty} \left\|\sum_{\|s\|_1=n} \delta_s^*(f)\right\|_p \ll \sum_{n=a+1}^{\infty} 2^{n(\frac{1}{q}-\frac{1}{p})} 2^{-rn} l_n^{\frac{1}{p}-\frac{1}{q}}.$$

Taking into account the definitions of m_n and N_n, we get from here, for any $\kappa > 0$ satisfying the inequality $\frac{1}{q} - \frac{1}{p} - r + \kappa(\frac{1}{q}-\frac{1}{p}) < 0$,

$$\|f-t\|_p \ll 2^{a(\kappa+1)(\frac{1}{p}-\frac{1}{q})} a^{(d-1)(\frac{1}{p}-\frac{1}{q})} (2^{a(\frac{1}{q}-\frac{1}{p}-r+\kappa(\frac{1}{q}-\frac{1}{p}))} a^{(d-1)(\frac{1}{q}-\frac{1}{p})}) = 2^{-ra}$$

which proves the upper estimate in Theorem for $m \asymp 2^a a^{d-1}$ and hence, for all m.

Let us prove the lower estimate. The proof will be carried out the same way as developed in [9]. It will involve the entropy numbers ϵ_n of classes $MW^r_{q,\alpha}$ in the spaces L_p (see [6] for their definition). We use the following result from [9]. Recall that a system $\Psi = \{\psi_j\}_{j=1}^{\infty}$ of elements in some Banach space X is said to satisfy the (VP) condition if there exist three positive constants A_i, $i=1,2,3$, and a sequence of numbers $\{n_k\}_{k=1}^{\infty}$, $n_{k+1} \leq A_1 n_k$, $k = 1,2,\ldots$, such that there is a sequence of operators $\{V_k\}_{k=1}^{\infty}$ with the properties

$$V_k(\psi_j) = \lambda_{k,j}\psi_j, \quad \lambda_{k,j} = 1, \quad j=1,\ldots,n_k,$$
$$\lambda_{k,j} = 0, \quad j > A_2 n_k; \quad \|V_k\|_{X\to X} \leq A_3, \quad k=1,2,\ldots. \tag{9}$$

Then one can prove the following:

Theorem A. *(see [9]) Assume that for some $a > 0$, $b \in \mathbb{R}$. Then for some centrally symmetric compact subset F of the space X,*

$$\epsilon_m(F,X) \geq C_1 m^{-a}(\log m)^b, \quad m \in \mathbb{Z}_+.$$

Moreover, suppose that the system Ψ satisfies the condition (VP) and the following inequality holds:

$$E_n(F,\Psi)_X := \sup_{f\in F} \inf_{c_1,\ldots,c_n} \left\|f - \sum_{j=1}^n c_j \psi_j\right\|_X \leq C_2 n^{-a}(\log n)^b, \quad n \in \mathbb{Z}_+. \tag{10}$$

Then

$$\sigma_m(F,\Psi)_X \gg m^{-a}(\log m)^b, \quad m = 2,3,\ldots.$$

We consider $X := L_p$, $1 < p < \infty$, and $\Psi := \{a_I\}_{I\in\mathcal{D}_0}$. It is known (see [11]) that for any $1 < p,q < \infty$, and any $r > 0$,

$$\epsilon_m(MW^r_{q,\alpha}, L_p) \gg m^{-r}(\log m)^{r(d-1)}.$$

Consider
$$S_{Q_n}(f) := \sum_{I \in Q_n} (f, b_I) a_I, \quad n \in \mathbb{Z}_+.$$

Then it follows from property (**A1**) (see [3]) that $\|S_{Q_n}\|_{p \to p} \ll 1$. Let $V_k := S_{Q_k}$, $k = 1, 2, \ldots$, and $F := MW_{q,\alpha}^r$. Then the conditions in (9) are satisfied with $n_k \asymp 2^k k^{d-1}$, $k = 1, 2, \ldots$, $A_1 = C(d)$, and $A_2 = 1$. Property (10), with $a := r$ and $b := (d-1)r$, follows from (3) for any $p \le q$ due to monotonicity of the L_p-norm. Hence, the lower estimate follows from Theorem A for the same a, b, and $p \le q$. Using the obvious inequality

$$\sigma_m(MW_{q,\alpha}^r, \{a_I\}_{I \in \mathcal{D}_0})_p \ge \sigma_m(MW_{q,\alpha}^r, \{a_I\}_{I \in \mathcal{D}_0})_q, \quad p \ge q,$$

we deduce the lower estimate for other values of p. This proves the theorem.

We now give, without proof, some sufficient conditions for a system $\{|I|^{-r} I_{-\alpha}^r b_I\}_{I \in \mathcal{D}_0}$ to satisfy property (**A2**). They parallel those treated in [3]. We define the space $L_p^\circ([0,1]^d)$, $1 < p < \infty$, as a set of functions f from $L_p([0,1]^d)$ such that $\int_{[0,1]} f(x) dx_j = 0$, $j = 1, \ldots d$.

Lemma 3. *Suppose a system $\{\eta_I\}_{I \in \mathcal{D}_0} \subset L_p^\circ([0,1])$ satisfies the following assumptions:*

*(**A3**) There exists a constant $C_1 > 0$ such that for any interval $J \subset \mathcal{D}_0$,*

$$|\eta_J(\xi_J + t|J|)| \le C_1 |J|^{-1/2} (1 + |t|)^{-1-\epsilon}, \quad -1 \le t|J| \le 1$$

where ξ_J is the center of the interval $J \subset \mathcal{D}_0$.

*(**A4**) There exists a constant $C_2 > 0$ and a partition of $[0,1]$ into intervals $J_1, \ldots J_m$ that are dyadic with respect to $[0,1]$, such that for all $J \in \mathcal{D}_0$, $j \in \mathbb{Z}_+$, $0 \le j|J| \le 1$ and for any t_1, t_2 in the interior of the same interval J_k, $k = 1, \ldots m$,*

$$|\eta_J(\xi_J + j|J| + t_1|J|) - \eta_J(\xi_J + j|J| + t_2|J|)| \le C_2 |J|^{-1/2} (1+j)^{-1-\epsilon} |t_2 - t_1|^\epsilon.$$

Then the multidimensional system $\{\eta_I'\}_{I \in \mathcal{D}_0}$ defined for $I := I_1 \times \ldots \times I_d$ by the formula

$$\eta_I' := \prod_{j=1}^d \eta_{I_j}', \quad \{\eta_{I_j}'\}_{I_j \in \mathcal{D}_0} = \{\eta_{I_j}\}_{I_j \in \mathcal{D}_0} \cup \{\chi_{[0,1]}\}, \quad j = 1, \ldots d,$$

*satisfies condition (**A2**), namely*

$$\{\eta_I'\}_{I \in \mathcal{D}_0} \prec \{H_I\}_{I \in \mathcal{D}_0} \text{ in } L_p([0,1]^d).$$

References

1. Andrianov, A. V., Approximation of functions from classes MH_q^r by Haar polynomials, 1997, preprint.
2. DeVore, R. A., B. Jawerth, and V. Popov, Compression of wavelet decompositions, Amer. J. of Math. **114** (1992), 737–785.
3. DeVore, R. A, S. V. Konyagin, and V. N. Temlyakov, Hyperbolic wavelet approximation, Constr. Approx. **14** (1998), 1–26.
4. DeVore, R. A. and V. N. Temlyakov, Nonlinear approximation by trigonometric sums, J. Fourier Analysis and Applications **2** (1995), 29–48.
5. Kashin, B. S. and A. A. Saakyan, *Orthogonal Series*, American Math. Soc., Providence, R.I., 1989.
6. Lorentz, G. G., M. V. Golitschek, and Yu. Makovoz, *Constructive Approximation: Advanced Problems*, Springer Verlag, N.Y., 1996.
7. Temlyakov, V. N., *Approximation of Periodic Functions*, Nova Science Publishers Inc., N.Y., 1993.
8. Temlyakov, V. N., *Approximation of Periodic Functions with Bounded Mixed Derivative*, Proc. Steklov Institute (1989, Issue 1).
9. Temlyakov, V. N., Nonlinear Kolmogorov's widths, to appear.
10. Temlyakov, V. N., Greedy algorithms with regard to the multivariate systems with a special structure, 1998, preprint.
11. Temlyakov, V. N., Estimates of the asymptotic characteristics of classes of functions with bounded mixed derivative or difference, Trudy Mat. Inst. Steklov (Proc. Steklov Inst. Math.) **189** (1989), 138–168.
12. Wozniakowski, H., Average case complexity of linear multivariate problems, J. Complexity **8** (1992), 337–372.

Alexander Andrianov
Department of Mathematics
University of South Carolina
Columbia, SC, 29208
andriano@math.sc.edu

Vladimir Temlyakov
Department of Mathematics
University of South Carolina
Columbia, SC, 29208
temlyak@math.sc.edu

Necessary Conditions for Abadie's Constraint Qualification and Best Approximation of Vector-Valued Functions

Martin Bartelt

Abstract. This note gives a necessary condition and studies sufficient conditions for the Abadie Constraint Qualification which is related to Hausdorff strong uniqueness in the best approximation of vector-valued functions on finite sets.

§1. Introduction

Let $I = \{1, ..., r\}$ be a finite set with the discrete topology and $C(I, \mathbf{R}^m)$ be the space of vector-valued functions from the index set I to the m-dimensional Euclidean space $\mathbf{R}^m$. Since I has the discrete topology, any function from I to $\mathbf{R}^m$ is continuous. For f in $C(I, \mathbf{R}^m)$, we write $f = (f_1, ..., f_m)$, where f_j is the j-th component function of f. The value of f at i in I is a vector in $\mathbf{R}^m : f(i) = (f_1(i), ..., f_m(i))$. A natural norm for functions in $C(I, \mathbf{R}^m)$ is the following mixture of the l_2-norm and the l_∞-norm:

$$\|f\| := \max_{i \in I} \|f(i)\|_2 = \max_{i \in I} \left(\sum_{j=1}^m (f_j(i))^2 \right)^{\frac{1}{2}}, \tag{1}$$

where $\|\cdot\|_2$ denotes the Euclidean norm on $\mathbf{R}^m$.

Let $g^1, ..., g^n$ be n linearly independent functions in $C(I, \mathbf{R}^m)$ and let $G := span\{g^1, ..., g^n\}$ be the n-dimensional subspace of $C(I, \mathbf{R}^m)$ generated by $g^1, ..., g^n$. For a function $f \in C(I, \mathbf{R}^m)$, consider the best approximation problem of finding a function g^* in G that solves the following minimization problem:

$$\min_{g \in G} \|f - g\|. \tag{2}$$

One special case of the above best approximation problem is the best approximation of complex-valued functions on a finite set I, since $C(I, \mathbf{C})$ can be identified with $C(I, \mathbf{R}^2)$ by using the isometric mapping: $f_1(x) + i f_2(x) \to (f_1(x), f_2(x))$, where $i = \sqrt{-1}$. Brosowski [4, p. 215] also studied the metric projection in vector-valued function spaces with emphasis on its connection to parametric semi-infinite optimization. In [7], Pinkus gave a comprehensive analysis of uniqueness of best approximation in more general vector-valued function spaces.

The set of all best approximations of f in G is denoted as

$$P_G(f) := \{g^* \in G : \|f - g^*\| = \text{dist}(f, G)\}, \tag{3}$$

where $\text{dist}(f, G) := \min_{g \in G} \|f - g\|$ denotes the distance from f to G. In general, $P_G(f)$ contains infinitely many elements. In fact, $P_G(f)$ is a singleton for every f in $C(I, \mathbf{R}^m)$ if and only if G satisfies the generalized Haar condition introduced by Zukhovitskii and Stechkin [10] (cf. also [1]). In such a special case, we also have the so-called strong unicity of order 2 for $P_G(f)$ [1]:

$$\|f - g\|^2 \geq \text{dist}(f, G)^2 + \gamma \cdot \text{dist}(g, P_G(f))^2 \quad \text{for } g \in G,$$

where γ is a positive constant.

In the general case when G is a subspace of a normed linear space, we say that $P_G(f)$ is Hausdorff strongly unique of order α, if there exists a positive constant γ (depending on f, α, and G) such that

$$\|f - g\|^\alpha \geq \text{dist}(f, G)^\alpha + \gamma \cdot \text{dist}(g, P_G(f))^\alpha \quad \text{for } g \in G. \tag{4}$$

If (4) holds and $P_G(f)$ is a singleton, then we say that $P_G(f)$ is strongly unique of order α.

Note that each function in the n-dimensional subspace G of $C(I, \mathbf{R}^m)$ can be identified with a vector in $\mathbf{R}^n$. For convenience, for $x = (x_1, ..., x_n) \in \mathbf{R}^n$, define

$$g_x := \sum_{k=1}^{n} x_k g^k. \tag{5}$$

For each fixed f in $C(I, \mathbf{R}^m)$ and each fixed index i in I, we introduce the function

$$h_i(x) = \|f(i) - g_x(i)\|_2^2 - \text{dist}(f, G)^2 \quad \text{for } x \in \mathbf{R}^n. \tag{6}$$

It was shown in [2] that each $h_i, 1 \leq i \leq r$, is a convex quadratic function and that the convex quadratic feasibility problem

$$h_i(x) \leq 0 \quad \text{for } i = 1, ..., r, \tag{7}$$

is equivalent to the best approximation problem (2) in that, letting

$$S(f) = \{x^* \in \mathbf{R}^n : h_i(x^*) \leq 0 \quad \text{for } i = 1, ..., r\},$$

we have $P_G(f) = \{g_{x^*} : x^* \in S(f)\}$, i.e. $x^* \in S(f)$ if and only if $g_{x^*} \in P_G(f)$. Furthermore, it follows that (cf. [2]) for f in $C(I, \mathbf{R}^m) \setminus G$ and $x^* \in S(f)$, the convex quadratic inequality system (7) satisfies the Abadie Constraint Qualification if and only if $P_G(f)$ is Hausdorff strongly unique of order $\alpha = 1$. Thus we will study the Abadie Constraint Qualification for the general feasibility problem

$$h_i(x) \leq 0 \quad \text{for } i = 1, ..., r \tag{8}$$

for convex differentiable functions $h_i \in C(\mathbf{R}^n, \mathbf{R})$.

Consider both the Abadie Constraint Qualification and the Slater condition (cf [5], p. 307, 311 and [3], p. 193).

Definition 1. *The system (8) satisfies Slater's constraint qualification if there exists an x in $S(f)$ such that $h_i(x) < 0, i = 1, ..., r$.*

Definition 2. *The system (8) satisfies the Abadie Constraint Qualification (ACQ) at a point x in $S(f)$ if the normal cone of S at x is equal to*

$$\left\{ \sum_{i \in J} \lambda_i \nabla h_i(x) : \lambda_i \geq 0 \text{ for } i \in J \right\}. \tag{9}$$

Here, $\nabla h_i(x)$ denotes the gradient of h_i at x, and J denotes the active constraint set, i.e.,

$$J(x) := \{i \in I : h_i(x) = 0\}.$$

The normal cone of S at x is given by

$$N_S(x) = \{z \in \mathbf{R}^n :\, <z, y - x> \,\leq 0 \text{ for } y \in S\}. \tag{10}$$

This definition of the ACQ in this situation is equivalent to the usual definition of the ACQ ([6], lemma 2). It is well known (cf [3], p. 194) that the Slater condition implies the ACQ.

Example 3. It is easy to see that even in this situation the ACQ does not imply the Slater condition; this happens, for example, at $(x, y) = (0, 0)$ if $h_1(x, y) = y^2 - x, h_2(x, y) = y^2 + x, h_3(x, y) = x^2 - y$, and $h_4(x, y) = x^2 + y$.

§2. Results

First, let $H = \{h_i : 1 \leq i \leq r\}$, $S = S(f) = \{x : h_i(x) \leq 0 : 1 \leq i \leq r\}$, and $S_i = \{x : h_i(x) \leq 0\}$ for $1 \leq i \leq r$. Then $S = \bigcap_{i=1}^{r} S_i$. To ensure that no function h_i is superfluous, we consider sets H that are efficient. The definition of efficiency for H resembles the definition of efficient generators of a cone (cf [9], p. 38).

Definition 4. *The set $H = \{h_i : 1 \leq i \leq r\}$ is said to be efficient if $\bigcap_{i=1, i \neq j}^{r} S_i \neq S$ for every $j = 1, ..., r$; for $r = 1$, this means $\mathbf{R}^n \neq S$.*

Let $r = 1$ so that H consists of just one function h_1. It is easy to see that if h_1 is not constant and $S \neq \emptyset$, then there exists a point x such that $h_1(x) > 0$ and hence $H = \{h_1\}$ is efficient. Suppose for some $\bar{x}$ in S that $\nabla h_1(\bar{x}) = 0$ and $h_1(\bar{x}) = 0$. Then it is easy to see that $\bar{x}$ is a boundary point of S if $S = \{\bar{x}\}$, and if card$(S) > 1$ and $\bar{x}$ is not a boundary point, there would be a neighborhood N of $\bar{x}$ in S. If $h_1(x) = 0$ for all x in N, then h_1 being analytic would be identically zero, so there exists a point $x_1 \in N$ where $h_1(x_1) < 0$. Then by the convexity of h_1, we have $h_1(\lambda \bar{x} + (1 - \lambda) x_1) > 0$ for all $\lambda > 1$. Since $\bar{x}$ is a boundary point, there exists a supporting hyperplane to S at $\bar{x}$ so that $N_S(\bar{x}) \neq \{0\}$, and by (9) the ACQ is not satisfied. Thus a necessary condition for the ACQ for $r = 1$ and that h_1 is not constant is that $\nabla h_1(\bar{x}) \neq 0$ when $h_1(\bar{x}) = 0$. The next theorem shows that this holds in general and not just when the functions h_i are quadratic, but also when they are analytic. In case the only function h_1 is a non-positive constant, there will always be the ACQ, but then also H is not efficient; if h_1 is a positive constant, there is no ACQ at a point x in S, since $S = \emptyset$.

Theorem 5. *Suppose that $H = \{h_i\}_{i=1}^r$ is an efficient set of convex analytic functions in $C(\mathbf{R}^n, \mathbf{R})$. If the Abadie Constraint Qualification holds at $\bar{x} \in S := \{x : h_i(x) \leq 0, 1 \leq i \leq r\}$, then $\{\nabla h_i(\bar{x}) \neq 0 : \text{ for all } i \in J\}$, where $J = \{i \in I : h_i(x) = 0\}$*

Proof: The case $r = 1$ is done above. So we let $r > 1$. Assume, to the contrary, that, without loss of generality, $h_1(\bar{x}) = 0$ and $\nabla h_1(\bar{x}) = 0$, and let

$$S = \{x : h_i(x) \leq 0 \text{ for } 1 \leq i \leq r\},$$

$$S^1 := \{x : h_i(x) \leq 0 \text{ for } 2 \leq i \leq r\},$$

and $J_1 = J \setminus \{1\}$.

Then since the Abadie Constraint Qualification is satisfied at $\bar{x}$, we have

$$N_S(\bar{x}) = \left\{ \sum_{i \in J} \lambda_i \nabla h_i(\bar{x}) : \lambda_i \geq 0 \right\} = \left\{ \sum_{i \in J_1} \lambda_i \nabla h_i(\bar{x}) : \lambda_i \geq 0 \right\}.$$

Since $S \subset S^1$, we have $N_{S^1}(\bar{x}) \subset N_S(\bar{x})$, i.e.,

$$N_{S^1}(\bar{x}) \subset \left\{ \sum_{i \in J_1} \lambda_i \nabla h_i(\bar{x}) : \lambda_i \geq 0 \right\}. \tag{11}$$

Also, since the reverse inclusion in (8) always holds we have

$$N_{S^1}(\bar{x}) = \left\{ \sum_{i \in J_1} \lambda_i \nabla h_i(\bar{x}) : \lambda_i \geq 0 \right\} = N_S(\bar{x}).$$

Now it is easy to see that $S^1 - \{\bar{x}\} \subset$ the polar cone of $N_{S^1}(\bar{x})$, and from ([5], Cor 5.2.5 and Prop. 5.2.1), we have

$$S - \{\bar{x}\} \subset \text{ polar cone of } N_{S^1}(\bar{x}) = \text{ polar cone of } N_S(\bar{x})$$

Abadie's Constraint Qualification and Best Approximation

$$= \text{the tangent cone of } S \text{ at } \bar{x} = cl(\text{cone}(S - \{\bar{x}\})). \tag{12}$$

Let $x \in S^1$. Then by (12) there exist positive constants $\{\lambda_k\}_{k=1}^\infty$, and $\{x_k\}_{k=1}^\infty$ in S, such that

$$x - \bar{x} = \lim_{k \to \infty} \lambda_k(x_k - \bar{x}). \tag{13}$$

Thus from (13), we have

$$h_1(x) = \lim_{k \to \infty} h_1(\bar{x} + \lambda_k(x_k - \bar{x})).$$

Now if $x = \bar{x}$, then $x \in S$. Otherwise $x_k \neq \bar{x}$. Since h_1 is convex and $\nabla h_1(\bar{x}) = 0$, $h_1(\bar{x})$ is the minimum of h_1. But since $h_1(x_k) \leq 0 = h_1(\bar{x})$, we know that $h_1(x_k)$ is also the minimum of h_1. Hence,

$$h_1(\bar{x} + \lambda(x_k - \bar{x})) = 0, 0 < \lambda < 1,$$

and thus

$$h_1(\bar{x} + \lambda(x_k - \bar{x})) = 0 \text{ for all } \lambda \in \mathbf{R},$$

since $h_1(\bar{x} + \lambda(x_k - \bar{x}))$ is an analytic function of λ; thus $h_1(\bar{x} + \lambda_k(x_k - \bar{x})) = 0$ and so, $h_1(x) = 0$ and $x \in S$. Thus $S^1 = S$ and $h_1(x) \leq 0$ is redundant which contradicts the assumption that H was efficient and the proof is complete.

Remark 6. The condition of efficiency is needed in the previous theorem. For example, we get the ACQ at $(0,0)$ and $\nabla h_5(0,0) = (0,0)$ if we adjoin $h_5(x,y) = y^2$ to the functions given in Example 3.

Now we consider sufficiency for the ACQ. If there exists a point $(\bar{x}, \bar{y})$ in S such that $h_i(\bar{x}, \bar{y}) < 0$ for all $1 \leq i \leq r$, then this is Slater's condition and the ACQ follows. So we consider when, for at least one i, $h_i(\bar{x}, \bar{y}) = 0$. The following theorem give a sufficient condition when the functions h_i are functions of two variables, i.e., $n = 2$. Let $\text{ri}(S)$ and $\text{r}\delta(S)$ denote, respectively, the relative interior and relative boundary of the set S, and $\text{card}(S)$ denote the cardinality of S.

Lemma 7. *([8] p. 223) Let $S_1, ..., S_r$ be convex sets in $\mathbf{R}^n$. If either $\bigcap_{i=1}^r \text{ri}(S_i) \neq \emptyset$ or $S_1, ..., S_k$ are polyhedral, $1 \leq k \leq r$, and*

$$\left(\bigcap_{i=1}^k S_i\right) \cap \left(\bigcap_{i=k+1}^r \text{ri}(S_i)\right) \neq \emptyset,$$

then at any point p in $S = \bigcap_{i=1}^r S_i$,

$$N_S(p) = N_{S_1}(p) + \cdots + N_{S_r}(p),$$

where $N_S(p)$ is the normal cone to S at p and $N_{S_i}(p)$ is the normal cone to S_i at p, $1 \leq i \leq r$.

Theorem 8. Suppose $h_i, 1 \leq i \leq r$, are linear or convex quadratic functions of two variables and $(\bar{x}, \bar{y})$ is in $S = \{(x,y) : h_i(x,y) \leq 0, i = 1, ..., r\}$ with $\text{card}(S) > 1$ and

(i) for $i = 1, ..., k, h_i(\bar{x}, \bar{y}) = 0$ and $\nabla h_i(\bar{x}, \bar{y}) \neq 0$ and
(ii) for $i = k+1, ..., r$, $h_i(\bar{x}, \bar{y}) < 0$.

Then the ACQ holds at $(\bar{x}, \bar{y})$.

Proof: If $h_i(\bar{x}, \bar{y}) < 0$, then at $(\bar{x}, \bar{y}), N(S_i) = \{(0,0)\}$, so the conclusion of the above lemma is that at $(\bar{x}, \bar{y})$,

$$N_S = N_{S_1} + \cdots + N_{S_k}.$$

Now for $i = 1, ..., k$, N_{S_i} at $(\bar{x}, \bar{y})$ is $\{\lambda_i \nabla h_i(\bar{x}, \bar{y}) : \lambda_i \geq 0\}$ because the linear independence constraint qualification is satisfied (cf. [3] p. 190). Thus if the hypotheses of lemma 7 are satisfied, we have ACQ at $(\bar{x}, \bar{y})$.

Suppose that $\bigcap_{i=1}^{r} \text{ri}(S_i) = \emptyset$. Since $\text{card}(S) > 1$, S contains infinitely many points since it contains the line joining two points in S. Thus $\bigcap_{i=1}^{r} S_i$ contains infinitely many points. Since each

$$S_i = \text{ri}(S_i) \cup \text{rd}(S_i),$$

$i = 1, ..., r, \bigcap_{i=1}^{r} S_i$, consists of a finite union of terms of the form $\bigcap_{i=1}^{r} A_i$, where each A_i is either $\text{ri}(S_i)$ or $\text{rd}(S_i)$. So at least one of the $\bigcap_{i=1}^{r} A_i$, call it $\bigcap_{i=1}^{r} A_i$, contains at least two points. In $\bigcap_{i=1}^{r} A_i$, any given term $\text{rd}(S_j)$ for some $j = 1, ..., s$ contains at least two points and hence the line joining them lies in S_j. Then either some point on the line is in $\text{ri}(S_j)$ or $\text{rd}(S_j)$ contains a line. Since h_j is linear or convex quadratic it follows that S_j is polyhedral. Thus since $\text{card}(A_i) > 1$, the hypotheses of Lemma 7 are satisfied and there is ACQ at $(\bar{x}, \bar{y})$.

Remark 9. The previous theorem does not hold in general if card $(S) = 1$. For example, if $g_1(x,y) = x^2 - y$ and $g_2(x,y) = x^2 + y$, then $S = \{(0,0)\}$, $\nabla g_1(0,0) = (0,-1)$ and $\nabla g_2(0,0) = (0,1)$; but since $N(0,0) = \mathbf{R}^2$, there is no ACQ at $(0,0)$. Assuming card $(S) > 1$, Theorems 5 and 8 give necessary and sufficient conditions for the ACQ when $n = 2$.

The condition $\text{card}(S) > 1$ is equivalent to the condition that S contains a simplex of dimension 1. This leads to the following

Conjecture 10. Suppose h_i, $1 \leq i \leq r$ are linear or convex quadratic functions of n variables, $\bar{x}$ is in S, S contains a simplex of dimension $n-1$ and

(i) for $i = 1, ..., s$, $h_i(\bar{x}) = 0$ and $\nabla h_i(\bar{x}) \neq 0$

and

(ii) for $i = s+1, ..., r$, $h_i(\bar{x}) < 0$.

Then the ACQ holds at $\bar{x}$.

Acknowledgments. The author gratefully acknowledges the many contributions to the paper by Professor Wu Li of Old Dominion University in Norfolk, VA.

References

1. Bartelt, M. and W. Li, Haar theory in vector-valued continuous function spaces, in *Approximation Theory VIII-Vol. 1: Approximation and Interpolation*, C. K. Chui and L. L. Schumaker (eds.), World Scientific Publishing Co., Inc., New York, 1995, pp. 39–46.
2. Bartelt, M. and W. Li, Abadie's constraint qualification, Hoffman's error bounds, and Hausdorff strong unicity, J. Approx. Theory, to appear.
3. Bazaraa, M. S., H. D. Sherali, and C. M. Shetty, *Nonlinear Programming, Theory and Algorithms*, 2nd Edition, John Wiley and Sons, New York, 1993.
4. Brosowski, B., *Parametric Semi-Infinite Optimization*, Verlag Peter Lang, Frankfurt am Main, 1987.
5. Hiriart-Urruty, J.-B. and C. Lemaréchal, *Convex Analysis and Minimization Algorithms I*, Springer-Verlag, New York, 1993.
6. Li, W., Abadie's constraint qualification, metric regularity and error bounds for differentiable convex inequalities, SIAM Journal on Optimization, to appear.
7. Pinkus, A., Uniqueness in vector-valued approximation, J. Approx. Theory **73** (1993), 17–92.
8. Rockafellar, R. T., *Convex Analysis*, Princeton University Press, Princeton, NJ, 1970.
9. Steuer, R., *Multiple Criteria Optimization Theory, Computation and Practice*, John Wiley and Sons, New York, 1986.
10. Zukhovitskii, S. I. and S. B. Stechkin, On approximation of abstract functions with values in Banach space, Dokl. Adad. Nauk. SSSR **106** (1956), 773–776.

Martin Bartelt
Department of Mathematics
Christopher Newport University
Newport News, VA 23606
mbartelt@pcs.cnu.edu

n–Widths of Multiplier Operators on Two-Point Homogeneous Spaces

B. Bordin, A. K. Kushpel, J. Levesley, and S. A. Tozoni

Abstract. This work is concerned with approximation on manifolds M, more specifically on the two-point homogeneous spaces (i.e. on S^d, $P^d(\mathbb{R})$, $P^d(\mathbb{C})$, $P^d(\mathbb{H})$, $P^{16}(Cay)$). New estimates for the (p,q) norms of multiplier operators $\Lambda = \{\lambda_k\}_{k \in \mathbb{N}}$ on these spaces will be given. These results are applied to give sharp orders of n–widths and best polynomial approximations on different sets of smooth functions on manifolds.

§1. Introduction

Let X be a Banach space, $B_X := \{x \mid x \in X, \|x\| \leq 1\}$ the unit ball of X, and A a (convex, compact, centrally symmetric) subset of X. The linear n–width of A in X is defined by

$$\delta_n(A, B_X) := \delta_n(A, X) := \inf_{P_n} \sup_{f \in A} \|f - P_n f\|,$$

where P_n varies over all linear operators of rank at most n that map X into itself.

The Kolmogorov n–width of A in X is defined by

$$d_n(A, B_X) := d_n(A, X) := \inf_{X_n} \sup_{f \in A} \inf_{g \in X_n} \|f - g\|,$$

where X_n runs over all subspaces of X of dimension n or less.

The Bernstein n–width of A in X is defined by

$$b_n(A, B_X) := b_n(A, X) := \sup_{X_{n+1}} \sup\{\varepsilon > 0 : (\varepsilon B_X \cap X_{n+1}) \subset A\},$$

where X_{n+1} is any $(n+1)$-dimensional subspace of X. More information on n–widths can be found in [9].

§2. Elements of Harmonic Analysis

Two-point homogeneous spaces admit essentially only one invariant differential operator, the Laplace-Beltrami operator. A complete classification of the two-point homogeneous spaces was given by Wang [11]. We present here the complete list of such spaces.

i) The spheres S^d, $d = 1, 2, 3, \ldots$

ii) The real projective spaces $P^d(\mathbb{R})$, $d = 2, 3, 4, \ldots$

iii) The complex projective spaces $P^d(\mathbb{C})$, $d = 4, 6, 8, \ldots$

iv) The quaternionic projective spaces $P^d(\mathbb{H})$, $d = 8, 12, \ldots$

v) The Cayley elliptic plane $P^{16}(Cay)$.

The superscripts here denote the dimension over the reals, of the underlying manifolds M.

For information on this point, see e.g. the works of Cartan [1], Gangolli [3], and Helgason [4,5]. The geometry of these spaces is in many respects similar.

Let $\theta = d(x, e)$ be the distance of a point $x \in M$ from eK and L is the diameter of M. We may choose a geodesic polar coordinate system $(\theta, \mathbf{u})$ where $\mathbf{u}$ is an angular parameter. The radial part of the Laplace Beltrami operator Δ in this coordinate system has the expression

$$\Delta_\theta = (A(\theta))^{-1} \frac{d}{d\theta}\left(A(\theta)\frac{d}{d\theta}\right),$$

where $A(\theta)$ is the area of the sphere of radius θ in M. An explicit form of the function $A(\theta)$ has been obtained by Helgason [4, 5] (using some methods of Lie algebras):

$$A(\theta) = \omega_{p+q+1} \lambda^{-p}(2\lambda)^{-q}(\sin \lambda\theta)^p(\sin 2\lambda\theta)^q,$$

where ω_d is the area of the unit sphere in $\mathbb{R}^d$ and

i) S^d: $p = 0$, $q = d - 1$, $\lambda = \pi/2L$, $d = 1, 2, 3, \ldots$

ii) $P^d(\mathbb{R})$: $p = 0$, $q = d - 1$, $\lambda = \pi/4L$, $d = 2, 3, 4, \ldots$

iii) $P^d(\mathbb{C})$: $p = d - 2$, $q = 1$, $\lambda = \pi/2L$, $d = 4, 6, 8, \ldots$

iv) $P^d(\mathbb{H})$: $p = d - 4$, $q = 3$, $\lambda = \pi/2L$, $d = 8, 12, \ldots$

v) $P^{16}(Cay)$: $p = 8$, $q = 7$, $\lambda = \pi/2L$.

Using this expression we can represent the operator Δ_θ as

$$\Delta_\theta = ((\sin \lambda\theta)^p(\sin 2\lambda\theta)^q)^{-1} \frac{d}{d\theta}(\sin \lambda\theta)^p(\sin 2\lambda\theta)^q \frac{d}{d\theta},$$

and setting $x = \cos(2\lambda\theta)$, this operator takes the form, up to a positive multiple,

$$\Delta_x = (1-x)^{-\alpha}(1+x)^{-\beta}\frac{d}{dx}(1-x)^{1+\alpha}(1+x)^{1+\beta}\frac{d}{dx},$$

where $\alpha = (p+q-1)/2$ and $\beta = (q-1)/2$.

Though not explicit, important for our analysis will be the fact that Jacobi polynomials $\{P_n^{\alpha,\beta}\}_{n\geq 0}$ are eigenfunctions of the operator Δ_x with eigenvalues $-n(n+\alpha+\beta+1)$.

Jacobi polynomials are orthogonal with respect to $\omega^{\alpha,\beta}(x) = c^{-1}(1-x)^{\alpha}(1+x)^{\beta}$ on $(-1,1)$ where (due to the normalization condition $\int_M d\mu = 1$ of the invariant measure $d\mu$ on M)

$$c := \int_{-1}^{1}(1-x)^{\alpha}(1+x)^{\beta}dx = 2^{\alpha+\beta+1}\frac{\Gamma(\alpha+1)\Gamma(\beta+1)}{\Gamma(\alpha+\beta+2)},$$

$\alpha > -1$, $\beta > -1$. We normalize the Jacobi polynomials as follows:

$$P_n^{\alpha,\beta}(1) = \frac{\Gamma(n+\alpha+1)}{\Gamma(\alpha+1)\Gamma(n+1)}$$

(since the zeros of $P_n^{\alpha,\beta}(x)$ are in $-1 < x < 1$, so that $P_n^{\alpha,\beta}(1) \neq 0$). Then,

$$\int_{-1}^{1}(1-x)^{\alpha}(1+x)^{\beta}(P_n^{\alpha,\beta}(x))^2 dx =$$

$$= \frac{2^{\alpha+\beta+1}\Gamma(n+\alpha+1)\Gamma(n+\beta+1)}{(2n+\alpha+\beta+1)\Gamma(n+\alpha+\beta+1)\Gamma(n+1)} := h_n^{\alpha,\beta}.$$

In this way, spherical (or zonal) functions $Z^{(n)}$, $n \in \mathbb{N}$, $Z^{(0)} \equiv 1$ on M, can be identified in each of the five cases above (since the elementary spherical functions are the eigenfunctions of the Laplace Beltrami operator). In fact the univariate counterpart $\tilde{Z}^{(n)} : [0, L] \to \mathbb{R}$ of the zonal polynomial $Z^{(n)}$ is given by $\tilde{Z}^{(n)}(\theta) = C_n(M)P_n^{\alpha,\beta}(\cos 2\lambda\theta)$.

Suppose that $\varphi(x)$ is a function on M with finite $L_p(M) = L_p$ norm given by

$$\|\varphi\|_p = \|\varphi\|_{L_p(M)} = \begin{cases} (\int_M |\varphi(x)|^p d\mu(x))^{1/p}, & 1 \leq p < \infty, \\ \text{ess sup}\{|\varphi(x)| \mid x \in M\}, & p = \infty, \end{cases}$$

where $U_p := \{\varphi(x) \mid \|\varphi(x)\|_p \leq 1\}$.

The space $L_2(M)$ has the decomposition

$$L_2(M) = \bigoplus_{n=0}^{\infty} H_n,$$

where H_n contains the spherical polynomial of degree n. Let $\{Y_k^n\}_{k=1}^{d_n}$ be an orthonormal basis of H_n. The following addition formula is known (see, e.g., [8])

$$\sum_{k=1}^{d_n} Y_k^n(\xi)\overline{Y_k^n(\eta)} = C_n(M)P_n^{\alpha,\beta}(\cos\theta),$$

where $\theta = d(\xi, \eta)$.

Using the addition formula and integration over M with respect to the invariant normalized measure $d\mu$ we get the following.

Lemma 2.1. *As* $n \to \infty$,

$$C_n(M) = \frac{\Gamma(\beta+1)(2n+\alpha+\beta+1)\Gamma(n+\alpha+\beta+1)}{\Gamma(\alpha+\beta+2)\Gamma(n+\beta+1)} \asymp n^{(d+1)/2},$$

$$d_n(M) = \frac{\Gamma(\beta+1)(2n+\alpha+\beta+1)\Gamma(n+\alpha+1)\Gamma(n+\alpha+\beta+1)}{\Gamma(\alpha+1)\Gamma(\alpha+\beta+2)\Gamma(n+1)\Gamma(n+\beta+1)} \asymp n^{d-1}.$$

§3. Sets of Smooth Functions on M

Using multiplier operators we will introduce a wide range of smooth functions on M. Let φ be an arbitrary function $\varphi \in L_p(M)$, $1 \leq p \leq \infty$ with the formal Fourier expansion

$$\varphi(x) \sim \sum_{k=0}^{\infty} \sum_{m=1}^{d_k} c_{k,m}(\varphi)Y_m^{(k)}(x),$$

where

$$c_{k,m}(\varphi) = \int_M f\varphi(x)\overline{Y_k^n(x)}d\mu(x).$$

Let $\Lambda = \{\lambda_k\}_{k \in \mathbb{N}}$. If in $L_q(M)$ there is a function $f(x) := \Lambda\varphi(x)$ such that

$$f(x) \sim \sum_{k=0}^{\infty} \lambda_k \sum_{m=1}^{d_k} c_{k,m}(\varphi)Y_m^{(k)}(x)$$

and $\|\Lambda\|_{p,q} := \sup\{\|\Lambda\varphi\|_q \mid \|\varphi\|_p \leq 1\} < \infty$, then we shall say that the multiplier operator Λ is of (p,q)-type with norm $\|\Lambda\|_{p,q}$.

We shall say that the function f is in $\Lambda U_p \oplus \mathbb{R}$ if

$$\Lambda\varphi(x) = f(x) \sim c + \sum_{k=1}^{\infty} \lambda_k \sum_{m=1}^{d_k} c_{k,m}(\varphi)Y_m^{(k)}(x),$$

where $\varphi \in U_p = \{\varphi \mid \varphi \in L_p(M),\ \|\varphi\|_p \leq 1\}$, $c \in \mathbb{R}$.

n–Widths of Multiplier Operators

In particular, the r-th fractional integral ($r > 0$) of a function $\varphi \in L_1(M)$ is defined by the sequence $\lambda_k = (k(k+\alpha+\beta+1))^{-r/2}$. Sobolev's classes $W_p^r(M)$ on M are defined as sets of functions with formal Fourier expansions

$$c + \sum_{k=1}^{\infty}(k(k+\alpha+\beta+1))^{-r/2}\sum_{m=1}^{d_k} c_{k,m}(\varphi) Y_m^{(k)}(x),$$

where

$$\int_M \varphi d\mu = 0, \quad \|\varphi\|_{L_p(M)} \le 1, \quad c \in \mathbb{R}.$$

§4. Estimates of n–Widths

To produce our estimates we will need some information concerning Cesàro means. The Cesàro kernel can be defined by

$$S_n^\delta(x) = \frac{1}{C_n^\delta} \sum_{m=0}^n C_{n-m}^\delta (h_m^{\alpha,\beta})^{-1} P_m^{\alpha,\beta}(1) P_m^{\alpha,\beta}(x),$$

where C_n^δ are Cesàro numbers of order n and index δ, i.e.,

$$C_n^\delta = \frac{\Gamma(n+\delta+1)}{\Gamma(\delta+1)\Gamma(n+1)} \asymp n^\delta,$$

(see, e.g., [10]). The following technical result gives us useful estimates of L_p–norms of Cesàro kernels.

Lemma 4.1. Let $\alpha \ge \beta \ge -1/2$. (i) If $0 \le \delta \le \alpha + 3/2$, then

$$\|S_n^\delta\|_p := \left(c^{-1} \int_{-1}^1 (1-x)^\alpha (1+x)^\beta |S_n^\delta(x)|^p dx \right)^{1/p} \ll$$

$$\ll \begin{cases} n^{\alpha+1/2-\delta}, & 1 \le p < (2\alpha+2)/(\alpha+\delta+3/2), \\ n^{\alpha+1/2-\delta}(\ln n)^{(\alpha+\delta+3/2)/(2\alpha+2)}, & p = (2\alpha+2)/(\alpha+\delta+3/2), \\ n^{(2\alpha+2)(1-1/p)}, & p > (2\alpha+2)/(\alpha+\delta+3/2). \end{cases}$$

(ii) If $\alpha + 3/2 \le \delta \le \alpha + \beta + 2$, then

$$\|S_n^\delta\|_p \ll n^{(2\alpha+2)(1-1/p)}, \quad 1 \le p \le \infty.$$

The proof of Lemma 4.1 is based on some bounds for the Cesàro kernels S_n^δ which can be found, e.g., in [2], p.230.

Definition 4.1. Let

$$N := \begin{cases} (d+1)/2, & d=3, 5, \ldots \\ (d+2)/2, & d=2, 4, \ldots \end{cases}$$

Then the multiplier sequence $\{\lambda_k\}_{k\in\mathbb{N}}$ is said to be in A_p, if for all $0 \leq s \leq N$, the following conditions are satisfied:

a) When N is odd and $0 \leq s \leq N$, or N is even and $0 \leq s < N$,

$$\begin{array}{ll} \lim_{n\to\infty} |\Delta^s \lambda_n| n^{(d-1)/2} = 0, & 1 \leq p < 2d/(d+1+2s), \\ \lim_{n\to\infty} |\Delta^s \lambda_n| n^{(d-1)/2} (\ln n)^{(d+1+2s)/(2d)} = 0, & p = 2d/(d+1+2s), \\ \lim_{n\to\infty} |\Delta^s \lambda_n| n^{d(1-1/p)+s} = 0, & p > 2d/(d+1+2s) \end{array}$$

b) When N is even and $s = N$,

$$\lim_{n\to\infty} |\Delta^s \lambda_n| n^{d(1-1/p)+s} = 0, \ 1 \leq p \leq \infty.$$

The following statement gives us estimates of (p,q)–norms of multiplier operators. These are expressed in terms of sequences $\{\lambda_k\}_{k\in\mathbb{N}}$ which will prove to be important for our applications.

Theorem 4.1. Let $\Lambda = \{\lambda_k\}_{k\in\mathbb{N}} \in A_r$, $1/r = 1 - (1/p - 1/q)_+$ $1 \leq p, q \leq \infty$. Then

$$\|\Lambda\|_{p,q} \leq C \sum_{s=1}^{\infty} |\Delta^{N+1} \lambda_s| s^{N+d(1-1/r)},$$

where C is some absolute constant.

Remark 4.1. For Sobolev's classes $W_p^r(M)$, it follows that $\lambda_s = (s(s+\alpha+\beta+1))^{-r/2}$. If $r > d(1/p - 1/q)_+$, $s \in \mathbb{N}$, then from Theorem 4.1,

$$\sup_{f \in W_p^r(M)} \inf_{t_n \in \mathcal{T}_n} \|f - t_n\|_{L_q(M)} \ll n^{-r+d(1/p-1/q)_+}, \ 1 \leq p, q \leq \infty, \ n \to \infty,$$

where $\mathcal{T}_n$ is the space of polynomials of degree $\leq n$. In the case $M = S^d$ this result has been proved in [7].

Remark 4.2. Let $p = q$ and $1 < p < \infty$, then, for a wide range of commonly occurring multiplier sequences $\{\lambda_k\}_{k\in\mathbb{N}}$, Theorem 4.1 gives the same estimates as the result in [2] for the norm of the respective multiplier operators. Moreover, Theorem 4.1 remains valid for the limit values of p and q, i.e., for $p = 1, \infty$, and $q = 1, \infty$.

n–Widths of Multiplier Operators

Theorem 4.2. Let $1 \leq p \leq \infty$, $\{\lambda_k\}_{k \in \mathbb{N}} \in A_1$, and

$$\lambda_s^{(-1)} := \begin{cases} (\lambda_s)^{-1} & \text{if } 1 \leq s \leq n, \\ 0 & \text{if } s \geq n+1. \end{cases}$$

Then

$$C \left(\sum_{s=1}^{\infty} |\Delta^{N+1} \lambda_s^{(-1)}|s^N \right)^{-1} \leq b_{m-1}(\Lambda U_p \bigoplus \mathbb{R}, \ L_p(M)),$$

$$\delta_m(\Lambda U_p \bigoplus \mathbb{R}, \ L_p(M)) \leq C \left(\sum_{s=n+1}^{\infty} |\Delta^{N+1} \lambda_s| s^N \right),$$

where $m = \sum_{k=0}^{n} d_k(M) \asymp n^d$. In particular, for the Sobolev's classes $W_p^r(M)$, $r > 0$,

$$b_n(W_p^r(M), \ L_p(M)) \asymp d_n(W_p^r(M), \ L_p(M)) \asymp$$

$$\asymp \delta_n(W_p^r(M), \ L_p(M)) \asymp n^{-r/d}, \ 1 \leq p \leq \infty, \tag{1}$$

$$d_n(W_p^r(M), \ L_q(M)) \asymp n^{-r/d + 1/p - 1/q}, \ 1 \leq p \leq q \leq 2, \tag{2}$$

$$r > d(1/p - 1/q),$$

$$d_n(W_p^r(M), \ L_q(M)) \asymp n^{-r/d}, \ 2 \leq q \leq p < \infty, \ r > 0, \tag{3}$$

$$\delta_n(W_p^r(M), \ L_q(M)) \asymp n^{-r/d}, \tag{4}$$

$$\{1 < q \leq p \leq 2, \ r > 0\} \bigcup \{2 \leq q \leq p < \infty, \ r > 0\}.$$

Upper bounds for n–widths in the cases (1) - (4) follow from Theorem 4.1. We prove lower bounds for (1) using Theorem 4.1 and a well-known result concerning n–widths of the unit ball (see, e.g., [9]). The proof of the lower bounds in (2) and (3) is based on Theorem 4.1 and a theorem of Ismagilov [6]. The result (4) is the dire ct consequence of duality for linear n–widths (see, e.g., [9]).

Acknowledgments. The second author was supported in part by FAEP, Brazil, Grant 1266/97, while the fourth author was supported in part by FAPESP, Brazil, Grant 97/11270-4.

References

1. Cartan, E., Sur la determination d'un systeme orthogonal complet dans un espace de Riemann symetrique clos, Circolo matematico di Palermo, Rendiconti **53** (1929), 217–252.
2. Bonami, A. and J. L. Clerk, Sommes de Cesàro et multiplicateurs des developments en harmoniques spheriques, Trans. Amer. Math. Soc. **183** (1973), 223–263.
3. Gangolli, R., Positive definite kernels on homogeneous spaces and certain stochastic processes related to Lévy's Browian motion of several parameters, Ann. Inst. Henri Poincaré, vol. III **2** (1967), 121–225.
4. Helgason, S., Differential operators on homogeneous spaces, Acta Matematica **102** (1959), 239–299.
5. Helgason, S., The radon transform on Euclidean spaces, compact two-point homogeneous spaces and Grassmann manifolds, Acta Matematica **113** (1965), 153–180.
6. Ismagilov, R. S., n–dimensional widths of compacts in Hilbert Space, Funk. Anal. i Ego Prilog. **2** (2) (1968), 32–39 (in Russian).
7. Kamzolov, A. I., On the best approximation of sets of function $W_p^\alpha(S^d)$ by polynomials, Mat. Zametki **32** (3) (1982), 285–293 (in Russian).
8. Koornwinder, T., The addition formula for Jacobi polynomials and spherical harmonics, SIAM J. Appl. Math. **25** (2) (1973), 236–246.
9. Pinkus, A., *n–Widths in Approximation Theory*, Springer-Verlag, Berlin, 1985.
10. Szegö, G., *Orthogonal Polynomials*, AMS, New York, 1939.
11. Wang, H. C., Two-point homogeneous spaces, Annals of Mathematics **55** (1952), 177–191.

B. Bordin, A. K. Kushpel, S. A. Tozoni
IMECC-UNICAMP
Caixa Postal 6065
13081-970, Campinas, SP
Brazil
bordin@ime.unicamp.br, ak99@ime.unicamp.br
(ak99@mcs.le.ac.uk), tozoni@ime.unicamp.br

J. Levesley
Department of Mathematics and Computer Science
University of Leicester
Leicester LE1 7RH
UK
jl1@mcs.le.ac.uk

Some Old Problems on Polynomials with Integer Coefficients

Peter Borwein

Abstract. We survey a number of old and difficult problems all of which involve finding polynomials with integer coefficients with small norm. These problems include: the Integer Chebyshev Problem of Hilbert and Fekete; the Prouhet-Tarry-Escott problem; various conjectures of Littlewood and various conjectures of Erdös. These problems are unsolved and most are at least 35 years old. They do however lend themselves to partial solution and one suspects that they are not, in fact, totally intractable. They are also all amenable to being computed on and offer some interesting computational challenges.

§0. Introduction

We break the paper into three main sections as follows:
- Section 1: Integer Chebyshev Problems
- Section 2: Prouhet-Tarry-Escott Problems.
- Section 3: Littlewood Type Problems.

Each section is largely self-contained and there is a substantial bibliography that more than covers the material in the paper.

§1. Integer Chebyshev Problems

The basic problem is very fundamental. It is to find a polynomial with integer coefficients of minimum supnorm on an interval.

Problem 1.1. *For any interval $[\alpha, \beta]$ find*

$$\Omega[\alpha, \beta] := \lim_{N \to \infty} \Omega_N[\alpha, \beta]$$

where

$$\Omega_N[\alpha, \beta] := \left(\min_{a_i \in \mathcal{Z}, a_N \neq 0} \|a_0 + a_1 x + \cdots + a_N x^N\|_{[\alpha, \beta]} \right)^{\frac{1}{N}}.$$

From
$$(\Omega_{n+m}[a,b])^{n+m} \leq (\Omega_n[a,b])^n (\Omega_m[a,b])^m \qquad (*)$$
one can show that
$$\Omega[\alpha,\beta] := \lim_{N \to \infty} \Omega_N[\alpha,\beta]$$
exists. This quantity is called the integer Chebyshev constant for the interval or the integer transfinite diameter.

On $[-2,2]$ (or any interval with integer endpoints of length 4) this problem is solvable because the usual Chebyshev polynomials normalized to have lead coefficient 1 have integer coefficients and supnorm 2. So $\Omega[-2,2] = 1$. There are no other intervals were the explicit value is known.

For $b - a < 4$, Hilbert [49] showed that there exists an absolute constant c so that
$$\inf_{0 \neq p \in \mathcal{Z}_n} \|p\|_{L_2[a,b]} \leq cn^{1/2} \left(\frac{b-a}{4}\right)^{1/2},$$
and Fekete [44] showed that
$$(\Omega_n[a,b])^n \leq 2^{1-2^{-n-1}}(n-1)\left(\frac{b-a}{4}\right)^{n/2}.$$

Here $\mathcal{Z}_n$ denotes the polynomials of degree n with integer coefficients. See also Kashin [60].

One sees from $(*)$ above that
$$\Omega[a,b] \leq \Omega_n[a,b]$$
for any particular n. So upper bounds can be derived computationally from the computation of any specific $\Omega_n[a,b]$. For example, if we let

$$p_0(x) := x$$
$$p_1(x) := 1 - x,$$
$$p_2(x) := 2x - 1,$$
$$p_3(x) := 5x^2 - 5x + 1,$$
$$p_4(x) := 13x^3 - 19x^2 + 8x - 1,$$
$$p_5(x) := 13x^3 - 20x^2 + 9x - 1,$$
$$p_6(x) := 29x^4 - 58x^3 + 40x^2 - 11x + 1,$$
$$p_7(x) := 31x^4 - 61x^3 + 41x^2 - 11x + 1,$$
$$p_8(x) := 31x^4 - 63x^3 + 44x^2 - 12x + 1,$$
$$p_9(x) := 941x^8 - 3764x^7 + 6349x^6 - 5873x^5$$
$$+ 3243x^4 - 1089x^3 + 216x^2 - 23x + 1,$$

then we have

Proposition 1.2. *Let*

$$P_{210} := p_0^{67} \cdot p_1^{67} \cdot p_2^{24} \cdot p_3^9 \cdot p_4 \cdot p_5 \cdot p_6^3 \cdot p_7 \cdot p_8 \cdot p_9.$$

Then

$$\left(\|P_{210}\|_{[0,1]}\right)^{1/210} = \frac{1}{2.3543\ldots},$$

and hence

$$\Omega[0,1] \leq \frac{1}{2.3543\ldots}.$$

Here $\|\cdot\|_{[a,b]}$ denotes the supremum norm on $[a,b]$.

Refinements on the method in [21] give

$$\Omega[0,1] \leq \frac{1}{2.3605\ldots}.$$

This has been further improved in [53] to

$$\Omega[0,1] \leq \frac{1}{2.3612\ldots}.$$

Of course when the coefficients of the polynomials above are not required to be integers, this reduces to the usual problem of constructing Chebyshev polynomials, and the limit (provided $a_N = 1$) gives the usual transfinite diameter. From the unrestricted case we have the obvious inequality

$$\Omega_n[a,b] \geq 2^{1/n}\left(\frac{b-a}{4}\right).$$

However, inspection of the above example shows that the integer Chebyshev polynomial doesn't look anything like a usual Chebyshev polynomial. In particular, it has many multiple roots and, indeed, this must be the case since we have the following lemma.

Lemma 1.3. *Suppose $p_n \in \mathcal{Z}_n$ (the polynomials of degree n with integer coefficients) and suppose $q_k(z) := a_k z^k + \cdots + a_0 \in \mathcal{Z}_k$ has all its roots in $[a,b]$. If p_n and q_k do not have common factors, then*

$$\left(\|p_n\|_{[a,b]}\right)^{1/n} \geq |a_k|^{-1/k}.$$

From this lemma and the above mentioned bound we see that all of p_1 through p_9 must occur as high order factors of integer Chebyshev polynomials on $[0,1]$ for sufficiently large n.

There is a sequence of polynomials that Montgomery [72] calls the Gorshkov–Wirsing polynomials that arise from iterating the rational function

$$u(x) := \frac{x(1-x)}{1-3x(1-x)}.$$

These are defined inductively by

$$q_0(x) := 2x - 1, \quad q_1(x) := 5x^2 - 5x + 1$$

and

$$q_{n+1} := q_n^2 + q_n q_{n-1}^2 - q_{n-1}^4.$$

It transpires that

$$u^{(n)} = \frac{q_{n-1}^2 - q_n}{2q_{n-1}^2 - q_n}.$$

Each q_k is a polynomial of degree 2^k with all simple zeros in $(0,1)$, and if b_k is the lead coefficient of q_k, then

$$\lim b_k^{1/2^k} = 2.3768417062\ldots.$$

Wirsing has proved that these polynomials are all irreducible [72]. It follows now from Lemma 1.3 that

$$\Omega[0,1] \geq \frac{1}{2.3768417062\ldots}.$$

It is conjectured by Montogomery [72, p. 201] that if s is the least limit point of $|a_k|^{-1/k}$ (as in in Lemma 1.3) over polynomials with all their roots in $[0,1]$, then $\Omega[0,1] = s$. This was also conjectured by Chudnovsky in [34], and Chudnovsky further conjectured that the minimal s arises from the Gorshkov-Wirsing polynomials in which case s would equal $(2.3768417062\ldots)^{-1}$. In [21] we show that

$$\Omega[0,1] \geq \frac{1}{2.3768417062\ldots} + \epsilon.$$

This shows that either Montgomery's conjecture is false or the Gorshkov-Wirsing polynomials do not give rise to the minimal s. This leads us to

Conjecture 1.4. *The minimal s arising in Lemma 1.3 does not give the right value for $\Omega[0,1]$.*

In [21] we asked whether integer Chebyshev polynomials on $[0,1]$ have all their roots in $[0,1]$. In [53] Habsieger and Salvy show that this can fail, with the first non-totally-real-factor occurring for $n = 70$. This same paper computes extrema up to degree 75. This is a nontrivial computation and is quite likely NP hard. None-the-less, one suspects that there is a close relationship between $\Omega[0,1]$ and polynomials with integer coefficients and all roots in $[0,1]$. Sorting out this relationship would be of interest.

There is a somewhat related problem that we have called the Schur-Siegel-Smyth trace problem.

Problem 1.5. *Fix $\epsilon > 0$. Suppose*

$$p_n(z) = a_n z^n + \cdots + a_0, a_i \in \mathcal{Z}$$

has all real, positive roots and is irreducible. Then except for finitely many explicit exceptions,

$$|a_{n-1}| \geq (2-\epsilon)n.$$

There are some partial results. In the notation of Problem 1.5 except for finitely many (explicit) exceptions, $a_{n-1} \geq (1.771..)n$. This is due to Smyth [89]. Previously, in 1918, Schur had shown $a_{n-1} \geq e^{1/2}n$, and in 1943 Siegel had shown that $a_{n-1} \geq (1.733..)n$.

The relationship this has to integer Chebyshev problems is the following.

Lemma 1.6. *If*

$$C[0, 1/m] \leq \frac{1}{m+\delta},$$

then for totally positive polynomials (polynomials as in Problem 1.5),

$$a_{n-1} \geq \delta n,$$

with finitely many explicit exceptions.

This reduces finding better bounds in the Schur-Siegel-Smyth trace problem to computations on short intervals. From an example on $[0, 1/100]$ we derive

Corollary 1.7. $\delta > 1.744$.

Smyth has shown that this method can never give the full result of Problem 1.5, but it would be interesting to see how far it can be taken.

The papers by Aparicio and Montgomery's monograph provide a good additional entry point to this subject matter.

§2. Ideal Solutions of the Prouhet-Tarry-Escott Problem

This old conjecture states concisely as

Conjecture 2.1. *For any N there exists $p \in \mathcal{Z}[x]$ (the polynomials with integer coefficients) so that*

$$p(x) = (x-1)^N q(x) = \Sigma_k a_k x^k$$

and

$$l_1(p) := \Sigma_k |a_k| = 2N.$$

Note that the degree of the solution is not the issue. The problem is in terms of the size of the zero at 1. It is a reasonably simple exercise to see that $2N$ is a lower bound so this would be the best possible result for any N. It is probably equivalent (though not provably so) to restrict to polynomials with coefficients $\{0, -1, +1\}$, and in this case we are looking for a $p \in \mathcal{Z}[x]$ with a zero of order n at one and with

$$\|p\|_{L_2\{|z|=1\}} = \sqrt{2N}.$$

What is actually provable is that any solution of Problem 2.1 must have all coefficients in the set $\{0, -1, +1, -2, +2\}$.

An entirely equivalent form of Problem 2.1 asks to find two distint sets of integers $[\alpha_1, \cdots, \alpha_N]$ and $[\beta_1, \cdots, \beta_N]$ so that

$$\alpha_1 + \cdots + \alpha_N = \beta_1 + \cdots + \beta_N$$
$$\alpha_1^2 + \cdots + \alpha_N^2 = \beta_1^2 + \cdots + \beta_N^2$$
$$\vdots \qquad \vdots \qquad \vdots$$
$$\alpha_1^{N-1} + \cdots + \alpha_N^{N-1} = \beta_1^{N-1} + \cdots + \beta_N^{N-1}.$$

This equivalence is an easy exercise in Newton's equations. The later form is the usual form in which the problem arises and is stated.

Sets of integers (as above) are called ideal solutions of the Prouhet-Tarry-Escott problem. Non-ideal solutions are ones where the size of the sets is allowed to be greater than the number of equations plus one.

This conjecture explicitly goes back at least to Wright in 1935 [99]. It is not clear why the conjecture is made. There is not a convincing heuristic for it. Solutions exist for N up to and including 10, and no solutions are known for any $N > 10$. For the cases up to 10, except for 9, there are known to be infinitely many solutions. For $N = 9$ two solutions are known. (We do not count as distinct solutions that arise by linear transformation.)

The following gives solutions up to 10. Suppose

$$x^{\alpha_1} + \cdots + x^{\alpha_N} - x^{\beta_1} - \cdots - x^{\beta_N} = 0((x-1)^N).$$

We write the solutions, as is traditional, in the form

$$[\alpha_1, \ldots, \alpha_N] = [\beta_1, \ldots, \beta_N].$$

Solutions for $N = 2, 3, 4 \ldots, 10$ are given by
$[0, 3] = [1, 2]$
$[1, 2, 6] = [0, 4, 5]$
$[0, 4, 7, 11] = [1, 2, 9, 10]$
$[1, 2, 10, 14, 18] = [0, 4, 8, 16, 17]$
$[0, 4, 9, 17, 22, 26] = [1, 2, 12, 14, 24, 25]$
$[0, 18, 27, 58, 64, 89, 101] = [1, 13, 38, 44, 75, 84, 102]$
$[0, 4, 9, 23, 27, 41, 46, 50] = [1, 2, 11, 20, 30, 39, 48, 49]$
$[0, 24, 30, 83, 86, 133, 157, 181, 197] = [1, 17, 41, 65, 112, 115, 168, 174, 198]$
$[0, 3083, 3301, 11893, 23314, 24186, 35607, 44199, 44417, 47500]$
$\quad = [12, 2865, 3519, 11869, 23738, 23762, 35631, 43981, 44635, 47488].$

Polynomials with Integer Coefficients

The size 10 example above illustrates the problems inherent with searching for a solution. While it is not known whether this is the smallest size 10 solution, it is the smallest one known, and is far beyond a size findable by exhaustive searching.

The smaller solutions were found by Escott and Tarry in the early part of this century. The size 9 and 10 solutions are due to Letac and were found in the early forties (without the aid of computers). Indeed very little new on this problem has been found computationally. (See [25] for further survey material.) The following seems a reasonable but as yet unattainable goal.

Problem 2.2. *Design an algorithm to establish whether or not solutions exist of modest size (say $N \leq 15$) and modest height (say 1000).*

The following is Smyth's [91] elegant decoding of Letac's size 10 solution. Let

$$F_{10} := \left(t^2 - R_1^2\right)\left(t^2 - R_2^2\right)\left(t^2 - R_3^2\right)\left(t^2 - R_4^2\right)\left(t^2 - R_5^2\right)$$
$$- \left(t^2 - R_6^2\right)\left(t^2 - R_7^2\right)\left(t^2 - R_8^2\right)\left(t^2 - R_9^2\right)\left(t^2 - R_{10}^2\right).$$

A solution of size 10 will be given as

$$[\pm R_1, \pm R_2, \pm R_3, \pm R_4, \pm R_5] = [\pm R_6, \pm R_7, \pm R_8, \pm R_9, \pm R_{10}]$$

provided F_{10} expands to equal a constant (i.e. all the powers of t expand out). This is another equivalent form of the problem also deduced via Newton's equations. Now we choose

$$R_1 := (4n + 4m)^2 \qquad R_2 := (mn + n + m - 11)^2$$
$$R_3 := (mn - n - m - 11)^2 \qquad R_4 := (mn + 3n - 3m + 11)^2$$
$$R_5 := (mn - 3n + 3m + 11)^2 \qquad R_6 := (4n - 4m)^2$$
$$R_7 := (-mn + n - m - 11)^2 \qquad R_8 := (-mn - n + m - 11)^2$$
$$R_9 := (-mn + 3n + 3m + 11)^2 \qquad R_{10} := (-mn - 3n - 3m + 11)^2.$$

On expansion of F_{10}, the constant coefficient is

$$- 64\,mn \left(m^4n^4 - 10\,n^4m^2 + 9\,n^4 - 1210\,n^2 + 14641 \right.$$
$$\left. -524\,m^2n^2 + 726\,m^2 + 6\,m^4n^2 + 185\,m^4\right)$$
$$\times \left(m^4n^4 + 6\,n^4m^2 + 185\,n^4 + 726\,n^2 + 14641 \right.$$
$$\left. -524\,m^2n^2 - 1210\,m^2 - 10\,m^4n^2 + 9\,m^4\right).$$

The rest of the expansion is given as

$$+ 64\,mn \left(5\,m^4n^4 + 62\,n^4m^2 + 125\,n^4 + 62\,m^4n^2 + 1268\,m^2n^2\right.$$
$$\left. +7502\,n^2 + 125\,m^4 + 7502\,m^2 + 73205\right)$$
$$\times \left(m^2n^2 - 13\,n^2 + 121 - 13\,m^2\right) t^2$$
$$- 64\,mn \left(7\,m^2n^2 + 53\,n^2 + 847 + 53\,m^2\right)\left(m^2n^2 - 13\,n^2 + 121 - 13\,m^2\right) t^4$$
$$+ 192\,mn \left(m^2n^2 - 13\,n^2 + 121 - 13\,m^2\right) t^6,$$

and each coefficient of the above polynomial of t has a factor

$$m^2 n^2 - 13 n^2 + 121 - 13 m^2.$$

So any solution of the above biquadratic gives a size 10 solution. One such solution is given by $n := 153/61$ and $m = 191/79$. A second solution is given by $n := -296313/249661$ and $m = -1264969/424999$. It is an exercise in elliptic curves to see that the above biquadratic has infinitely many solutions, and hence so does the problem of size $N = 10$.

For sizes 1 through 8, parametric families of solutions exist. The following is a (homogenous) size 8 solution due to Chernick [33]

$$[\pm R_1, \pm R_2, \pm R_3, \pm R_4] = [\pm R_5, \pm R_6, \pm R_7, \pm R_8],$$

where

$$R_1 := 5 m^2 + 9 mn + 10 n^2 \qquad R_2 := m^2 - 13 mn - 6 n^2$$
$$R_3 := 7 m^2 - 5 mn - 8 n^2 \qquad R_4 := 9 m^2 + 7 mn - 4 n^2$$
$$R_5 := 9 m^2 + 5 mn + 4 n^2 \qquad R_6 := m^2 + 15 mn + 8 n^2$$
$$R_7 := 5 m^2 - 7 mn - 10 n^2 \qquad R_8 := 7 m^2 + 5 mn - 6 n^2.$$

One sees this by noting that if

$$F_8 := \left(t^2 - R_1{}^2\right)\left(t^2 - R_2{}^2\right)\left(t^2 - R_3{}^2\right)\left(t^2 - R_4{}^2\right)$$
$$- \left(t^2 - R_5{}^2\right)\left(t^2 - R_6{}^2\right)\left(t^2 - R_7{}^2\right)\left(t^2 - R_8{}^2\right)$$

then on expansion

$$F_8 = -10752 \, mn \, (2n + m)(n + m)(2n + 3m)(n + 2m)(4n - m)(5n + 4m)$$
$$\times (n - 2m)(3n + m)(n - m)(n + 5m)(3n^2 + 2mn - 2m^2)(n^2 + mn + m^2).$$

Now any integers n and m (provided the expression doesn't collapse) give rise to a solution of size 8.

There are only two non-equivalent solutions of size $N = 9$ known. They are given in symmetric form as

$$[98, 82, 58, 34, -13, -16, -69, -75, -99]$$
$$= [-98, -82, -58, -34, 13, 16, 69, 75, 99]$$

and

$$[174, 148, 132, 50, 8, -63, -119, -161, -169]$$
$$= [-174, -148, -132, -50, -8, 63, 119, 161, 169].$$

It would be of value to know whether there are infinitely many solutions of size 9, and it might be of interest to search for a parametric solution.

2.1. Variations on the Theme

The obvious question arises: If we can't make the l_1 norm of a polynomial with a zero of order N at 1 be $2N$, how small can we make it?

Problem 2.3. Find $0 \neq p_N \in \mathcal{Z}[x]$ where $p_N(x) = (x-1)^N q(x) = \Sigma a_k x^k$ so that
$$l_1(p_N) = \Sigma |a_i| = o(N^2)$$
or
$$l^2(p_N) = (\Sigma |a_i|^2)^{1/2} = o(N^2).$$

A fairly easy combinatorial argument shows that

$$l_1(p_N) \leq N^2/2$$

is possible for all N. (See [25]) However, this is where the problem is stuck (at least in terms of the principal term of the asymptotic), and even getting a bound like $N^2/(2+\epsilon)$ would be major progress.

This problem arises in the context of a problem Wright called the "easier Waring problem." The Waring problem asks how many positive Nth powers are required to write every sufficiently large integer as a sum of Nth powers. The "easier Waring problem" allows for differences as well as sums. The "easier" has proved to be a misnomer since currently the best approaches to the "easier Waring problem" all go through the Waring problem.

Fuchs and Wright [48] observed that if (as in Problem 2.3)

$$l_1(p_N) = O(A_N),$$

then the Easier Waring problem is also $O(A_N)$. (Here N is the power under investigation in Waring's problem.) At the moment Waring's problem is known to be $O(N \log N)$ (though it is suspected to be $O(N)$). So showing that

$$l_1(p_N) = O(N \log N)$$

would be a very major result.

If we demand that p has a zero of order N but not of order $N+1$ at 1, then
$$l_1(p) = O((\log N) N^2)$$

is possible, but this is all that is known [55]. And this argument is considerably harder than the one that gives $O(N^2)$ without the additional requirement that the multiplicity of the zero be *exactly* N. Any improvement on this would also be interesting.

2.2. Problem of Erdös and Szekeres (1958)

One approach to the Prouhet-Tarry-Escott problem is to construct products of the form

$$p(x) := \left(\prod_{k=1}^{N}(1-x^{\alpha_i})\right).$$

Obviously, such a product has a zero of order N at 1, and the trick is to minimize the l_1 norm.

Problem 2.4. *Minimize over* $\{\alpha_1, \ldots, \alpha_N\}$

$$l_1\left(\prod_{k=1}^{N}(1-x^{\alpha_i})\right).$$

Call this minimum E_N^*.

The following table shows what is known for N up to 13.

N	$\|p\|_{l_1}$	$\{\alpha_1, \ldots, \alpha_N\}$
1	2	$\{1\}$
2	4	$\{1,2\}$
3	6	$\{1,2,3\}$
4	8	$\{1,2,'3,4\}$
5	10	$\{1,2,3,5,7\}$
6	12	$\{1,1,2,3,4,5\}$
7	16	$\{1,2,3,4,5,7,11\}$
8	16	$\{1,2,3,5,7,8,11,13\}$
9	20	$\{1,2,3,4,5,7,9,11,13\}$
10	24	$\{1,2,3,4,5,7,9,11,13,17\}$
11	28	$\{1,2,3,5,7,8,9,11,13,17,19\}$
12	36	$\{1,\ldots,9,11,13,17\}$
13	48	$\{1,\ldots,9,11,13,17,19\}$

Note that for $N := 1,2,3,4,5,6,8$, this provides an ideal solution of the Prouhet-Tarry-Escott problem. And indeed the first known solutions were mostly of this form. Maltby [70] shows that for $N = 7, 9, 10, 11$, these kind of products cannot solve the Prouhet-Tarry-Escott problem. For $N = 7, 9, 10$, the above examples are provably optimal. This leads to the following conjecture.

Conjecture 2.5. *Except for* $N = 1, 2, 3, 4, 5, 6$ *and* 8,

$$E_N^* \geq 2N + 2.$$

Actually much more is likely to be true. Erdös and Szekeres [42] conjecture that E_N^* grows fairly rapidly. Specifically,

Polynomials with Integer Coefficients

Conjecture 2.6. *For any K,*
$$E_N^* \geq N^K,$$
for N sufficiently large.

Currently the only lower bounds known (except for Maltby's results for $N = 7, 9, 10$ and 11) are the trivial lower bounds $E_N^* \geq 2N$ of the Prouhet-Tarry-Escott problem.

Sub-exponential upper bounds in this problem of the form
$$E_N^* \ll \exp\left(O(\log^4 N)\right)$$
due to Belov and Konyagin are known [13]. See also [76], [62].

§3. Littlewood Type Problems

Here we are primarily concerned with polynomials with coefficients in the set $\{+1, -1\}$. Since many of these problems were raised by Littlewood, we denote the set of such polynomials by $\mathcal{L}_n$, and refer to them as Littlewood polynomials. Specifically
$$\mathcal{L}_n := \left\{ p : p(x) = \sum_{j=0}^{n} a_j x^j, \ a_j \in \{-1, 1\} \right\}.$$

The following conjecture is due to Littlewood, probably from some time in the fifties. It has been much studied, and has associated with it a considerable signal processing literature (see for example [31].)

Conjecture 3.1. *It is possible to find $p_n \in \mathcal{L}_n$ so that*
$$C_1 \sqrt{n+1} \leq |p_n(z)| \leq C_2 \sqrt{n+1}$$
for all complex z of modulus 1. Here the constants C_1 and C_2 are independent of n.

Such polynomials are often called "locally flat." Because the L_2 norm of a polynomial from $\mathcal{L}_n$ is exactly $\sqrt{n+1}$, the constants must satisfy $C_1 \leq 1$ and $C_2 \geq 1$. This is discussed in some detail in problem 19 of Littlewood's charming monograph [67]. Littlewood, in part, based his conjecture on computations of all such polynomials up to degree twenty. Odlyzko has now done extensive computations that tend to confirm the conjecture. However, it is still the case that no sequence is known that satisfies the lower bound.

A sequence of Littlewood polynomials that satisfies just the upper bound is given by the Rudin-Shapiro polynomials. These are defined by
$$p_0(z) := 1, \quad q_0(z) := 1$$
and
$$p_{n+1}(z) := p_n(z) + z^{2^n} q_n(z),$$

$$q_{n+1}(z) := p_n(z) - z^{2^n} q_n(z).$$

These have all coefficients ± 1 and are of degree $2^n - 1$. If $|z| = 1$, then

$$|p_{n+1}|^2 + |q_{n+1}|^2 = 2(|p_n|^2 + |q_n|^2)$$

and it is easy to deduce that

$$|p_n(z)| \leq 2\sqrt{2}^n = \sqrt{2}\sqrt{\deg(p_n)}$$

and

$$|q_n(z)| \leq 2\sqrt{2}^n = \sqrt{2}\sqrt{\deg(q_n)}$$

for all z of modulus 1.

This conjecture is complemented by a conjecture of Erdös [41].

Conjecture 3.2. *The constant C_2 in conjecture 3.1 is bounded away from 1 (independently of n).*

This is also still open. Though a remarkable result of Kahane's [59] shows that if the polynomials are allowed to have complex coefficients of modulus 1, then "locally flat" polynomials exist, and indeed it is possible to make C_1 and C_2 asymptotically arbitrarily close to 1. (Polynomials of this form are sometimes called "ultra-flat.") Another striking result due to Beck [10] proves that "locally flat" polynomials exist from the class of polynomials of degree n whose coefficients are 1200th roots of unity.

Of course, because of the monotonicity of the L_p norms, it is relevant to rephrase Erdös' conjecture in other norms. Newman and Byrnes speculate, too optimistically, in [74] that

$$\|p\|_4^4 \geq (6 - \delta)n^2/5$$

for $p \in \mathcal{L}_n$ and n sufficiently large. This, of course, would imply Erdös' conjecture above. Here and throughout this section

$$\|q\|_p = \left(\int_0^{2\pi} |q(\theta)|^p \, d\theta/(2\pi) \right)^{1/p}$$

is the normalized p norm on the boundary of the unit disc.

It is possible to find a sequence of $p_n \in \mathcal{L}_n$ so that

$$\|p_n\|_4^4 \asymp (7/6)n^2.$$

This sequence is constructed out of the Fekete polynomials

$$f_p(z) := \sum_{k=0}^{p-1} \left(\frac{k}{p}\right) z^k,$$

where $\left(\frac{\cdot}{p}\right)$ is the Legendre symbol. One now takes the Fekete polynomials and cyclically permutes the coefficients by about $p/4$ to get the above example due to Turyn [54]. Actually, computations suggest that even the 7/6 constant above may also be overly optimistic. Nonetheless, a variety of people conjecture the following.

Polynomials with Integer Coefficients

Problem 3.3. Show for some absolute constant $\delta > 0$ and for all $p_n \in \mathcal{L}_n$

$$||p||_4 \geq (1+\delta)\sqrt{n}$$

or even the much weaker

$$||p||_4 \geq \sqrt{n} + \delta.$$

This problem of finding Littlewood polynomials of minimal L_4 norm has a considerable literature. See [52, 51, 54, 56]. The engineering literature calls this the "merit factor" problem.

A Barker polynomial

$$p(z) := \sum_{k=0}^{n} a_k z^k$$

with each $a_k \in \{-1, +1\}$ is a polynomial where

$$p(z)\overline{p(z)} := \sum_{k=-n}^{n} c_k z^k$$

satisfies

$$|c_j| \leq 1, \qquad j = 1, 2, 3 \ldots.$$

Here

$$c_j = \sum_{k=0}^{n-j} a_k a_{k+j} \qquad \text{and} \qquad c_{-j} = c_j.$$

Note that if $p(z)$ is a Barker polynomial of degree n then

$$||p||_4 \leq ((n+1)^2 + 2n))^{1/4} < (n+1)^{1/2} + (n+1)^{-1/2}/2.$$

The nonexistence of Barker polynomials of degree n is now shown by showing

$$||p||_4 \geq (n+1)^{1/2} + (n+1)^{-1/2}/2.$$

This is even weaker than the weak form of Problem 3.3.

It is conjectured that no Barker polynomials exist for $n > 12$. See [81] for more on Barker polynomials and a proof of the nonexistence of self-inversive Barker polynomials. In [96] it is shown that no even degree Barker polynomials exist for $n > 12$ (and indeed none exist for any degree between 12 and 10^{12}.

The expected L_p norms of Littlewood polynomials and their derivatives are computed in [27]. For random $q_n \in \mathcal{L}_n$

$$\frac{\mathrm{E}(||q_n||_p)}{n^{1/2}} \to (\Gamma(1+p/2))^{1/p}$$

and

$$\frac{\mathrm{E}(||q_n^{(r)}||_p)}{n^{(2r+1)/2}} \to (2r+1)^{-1/2}(\Gamma(1+p/2))^{1/p}.$$

3.1. Lehmer's Conjecture

Mahler's Measure is defined as follows: if

$$p(z) := \prod_{i=1}^{n}(z - \alpha_i),$$

then

$$M(p) := \prod_{i=1}^{n} \max\{1, |\alpha_i|\},$$

or equivalently

$$M(p) := \exp\left\{\int_0^1 \log |p(e^{2\pi i t})|\, dt\right\}.$$

The problem commonly known as Lehmer's Conjecture is

Conjecture 3.4. *Suppose p is a monic polynomial with integer coefficients. Then either $M(p) = 1$ or $M(p) \geq 1.1762808\ldots$.*

See [30] for an exposition of this problem. This can be thought of as a generalization of Kronecker's theorem which can be stated as: $M(p) = 1$ implies that p is cyclotomic (that is, it has all its roots of modulus 1). Note that $M(p)$ is really the L_0 norm, so this too is a growth problem, and in fact for this conjecture it is sufficient to consider only polynomials with coefficients in the set $\{0, -1, +1\}$.

The minimal Mahler measure for a non-cylotomic p is speculated to be given by $p := x^{10} + x^9 - x^7 - x^6 - x^5 - x^4 - x^3 + x + 1$ for which $M(p) = 1.17628081825991750\ldots$. This is also speculated to be the smallest Salem number.

Problem 3.5. *Do there exist polynomials with coefficients $\{0, -1, +1\}$ with roots of arbitrarily high multiplicity inside the unit disk?*

A negative answer to the above would solve Lehmer's conjecture. It seems likely, however, that the answer to the above question is positive. See [7].

Mahler [69] raised the problem of finding the maximum Mahler measure over the polynomials of degree n with coefficients $\{0, +1, -1\}$.

Problem 3.6. *Does there exist a sequence of Littlewood polynomials $p_n \in \mathcal{L}_n$ so that*

$$\lim_n \frac{M(p_n)}{\sqrt{n}} = 1?$$

This is a weak form of the Erdös conjecture. The non-existence of a sequence, as in Problem 3.6, implies Conjecture 3.2.

3.2. Zeros of Littlewood and Related Polynomials

The following result concerning polynomials of height one is proved in [20].

Theorem 3.7. *Every polynomial p_n of the form*

$$p_n(x) = \sum_{j=0}^{n} a_j x^j, \quad |a_0| = 1, \quad |a_j| \leq 1, \quad a_j \in \mathbf{C} \qquad (**)$$

has at most $\lfloor \frac{16}{7} \sqrt{n} \rfloor + 4$ zeros at 1.

It is easy to prove the following:

Theorem 3.8. *There is an absolute constant $c > 0$ such that for every n, there is a polynomial p of degree n with coefficients in the set $\{0, -1, +1\}$ having at least $c\sqrt{n/\log(n+1)}$ zeros at 1.*

Theorems 3.7 and 3.8 show that the right upper bound for the number of zeros a polynomial p_n with coefficients in the set $\{0, -1, +1\}$ can have at 1 is somewhere between $c_1 \sqrt{n/\log(n+1)}$ and $c_2 \sqrt{n}$ with absolute constants $c_1 > 0$ and $c_2 > 0$.

Problem 3.9. *What is the maximum multiplicity of the zero at 1 for a polynomial of degree n with coefficients in $\{0, -1, +1\}$. In particular, is it $O(n^{1/2})$?*

This problem has substantial application to effective bounds in Roth's Theorem, particularly if the answer to the above conjecture is affirmative.

Boyd [29] shows that there is an absolute constant c such that every $p \in \mathcal{L}_n$ can have at most $c \log^2 n / \log \log n$ zeros at 1. Since it is easy to give polynomials $p \in \mathcal{L}_n$ with $c \log n$ zeros at 1, the following question is suggested.

Problem 3.10. *Prove or disprove that there is an absolute constant c such that every polynomial $p \in \mathcal{B}_n$ can have at most $c \log n$ zeros at 1.*

References

1. Amoroso F., Sur le diamètre transfini entier d'un intervalle réel, Ann. Inst. Fourier, Grenoble **40** (1990), 885–911.

2. Aparicio, E., Methods for the approximate calculation of minimum uniform Diophantine deviation from zero on a segment, Rev. Mat. Hisp.-Amer. **38** (1978), 259–270 (Spanish).

3. Aparicio, E., New bounds on the minimal Diophantine deviation from zero on $[0, 1]$ and $[0, 1/4]$, Actus Sextas J. Mat. Hisp.-Lusitanas (1979), 289–291.

4. Aparicio, E., On some systems of algebraic integers of D. S. Gorshkov and their application in calculus, Rev. Mat. Hisp.-Amer. **41** (1981), 3–17 (Spanish).

5. Aparicio, E., On some results in the problem of Diophantine approximation of functions by polynomials, Proc. Steklov Inst. Math. **163** (1985), 7–10.

6. Atkinson, F. A., On a problem of Erdös and Szekeres, Canad. Math. Bull. **1** (1961), 7–12.

7. Beaucoup, F., P. Borwein, D. Boyd, and C. Pinner, Multiple roots of $[-1,1]$ power series, J. London Math. Soc., to appear.

8. Beaucoup, F., P. Borwein, D. Boyd, and C. Pinner, Power series with restricted coefficients and a root on a given ray, Math. Computat, to appear.

9. Beck, J., Flat polynomials on the unit circle – Note on a problem of Littlewood, Bull. London Math. Soc. **23** (1991), 269–277.

10. Beck, J., The modulus of polynomials with zeros on the unit circle: A problem of Erdös, Annals of Math. **134** (1991), 609–651.

11. Bell, J., P. Borwein, and B. Richmond, Growth of the product $\prod_{j=1}^{n}(1-x^{a_j})$, Acta Arith., to appear.

12. Beenker, G., T. Claasen, and P. Hermes, Binary sequences with a maximally flat amplitude sequence, Philips J. Res. **40** (1985), 289–304.

13. Belov, A. S. and S. V. Konyagin, On estimates for the constant term of a nonnegative trigonometric polynomial with integral coefficients, Mat Zametki **59** (1996), 627–629.

14. Bharucha-Reid, A. T. and M. Sambandham, *Random polynomials*, Academic Press, Orlando, 1986.

15. Bloch, A. and G. Pólya, On the roots of certain algebraic equations, Proc. London Math. Soc **33** (1932), 102–114.

16. Boehmer, A. M., Binary pulse compression codes, IEEE Trans. Information Theory **13** (1967), 156–167.

17. Bombieri, E. and J. Vaaler, Polynomials with low height and prescribed vanishing, in *Analytic Number Theory and Diophantine Problems*, Birkhauser, Boston, 1987, pp. 53–73.

18. Borwein, P. and T. Erdélyi, Markov-Bernstein type inequalities under Littlewood-type coefficient constraints, submitted.

19. Borwein, P. and T. Erdélyi, Littlewood-type problems on subarcs of the unit circle, Indiana J. Math. **46** (1997), 1323–1346.

20. Borwein, P. and T. Erdélyi, On the zeros of polynomials with restricted coefficients, Illinois J. Math. **41** (1997), 667–675.

21. Borwein, P. and T. Erdélyi, The integer Chebyshev problem, Math. Computat. **65** (1996), 661–681.

22. Borwein, P. and T. Erdélyi, Markov and Bernstein type inequalities for polynomials with restricted coefficients, Ramanujan J. **1** (1997), 309–323.

23. Borwein, P. and T. Erdélyi, Questions about polynomials with $\{0, -1, +1\}$ coefficients, Constr. Approx. **12** (1996), 439–442.
24. Borwein, P. and T. Erdélyi, *Polynomials and Polynomial Inequalities*, Springer-Verlag, New York, 1995.
25. Borwein, P. and C. Ingalls, The Prouhet, Tarry, Escott problem, Ens. Math. **40** (1994), 3–27.
26. Borwein, P. and C. Pinner, Polynomials with $\{0, +1, -1\}$ coefficients and a root close to a given point, Canadian J. Math. **49** (1997), 887–915.
27. Borwein, P. and R. Lockhart, The expected L_p norm of random polynomials.
28. Bourgain, J., Sul le minimum d'une somme de cosinus, Acta Arith. **45** (1986), 381–389.
29. Boyd, D., On a problem of Byrnes concerning polynomials with restricted coefficients, Math. Comput. **66** (1977), 1697–1703.
30. Boyd, D., Variations on a theme of Kronecker, Canad. Math. Bull. **21** (1978), 1244–1260.
31. Byrnes, J. S. and D. J. Newman, Null steering employing polynomials with restricted coefficients, IEEE Trans. Antennas and Propagation **36** (1988), 301–303.
32. Carrol, F. W., D. Eustice, and T. Figiel, The minimum modulus of polynomials with coefficients of modulus one, J. London Math. Soc. **16** (1977), 76–82.
33. Chernick, J., Ideal solutions of the Tarry-Escott problem, Amer. Math. Monthly **44** (1937), 627-633.
34. Chudnovsky, G., Number theoretic applications of polynomials with rational coefficients defined by extremality conditions, in *Arithmetic and Geometry*, Vol. I, M. Artin and J. Tate, (eds.), Progress in Math., Vol. 35, Birkhäuser, Boston, 1983, pp. 61–105.
35. Clunie, J., On the minimum modulus of a polynomial on the unit circle, Quart. J. Math. **10** (1959), 95–98.
36. Cohen, P. J., On a conjecture of Littlewood and idempotent measures, Amer. J. Math. **82** (1960), 191–212.
37. Dickson, L. E., *History of the Theory of Numbers*, Chelsea Publishing Co., New York, 1952.
38. Dobrowolski, E., On a question of Lehmer and the number of irreducible factors of a polynomial, Acta Arithmetica **34** (1979), 341–401.
39. Dorwart, H. L. and O. E. Brown, The Tarry-Escot problem, Amer. Math. Monthly **44** (37), 613–626.
40. Erdös, P., Some old and new problems in approximation theory: research problems 95-1, Constr. Approx. **11** (1995), 419–421.
41. Erdös, P., An inequality for the maximum of trigonometric polynomials, Annales Polonica Math. **12** (1962), 151–154.

42. Erdös, P. and G. Szekeres, On the product $\prod_{k=1}^{n}(1 - z^{a_k})$, Publications de L'Institut Math. **12** (1952), 29–34.

43. Erdös, P. and P. Turán, On the distribution of roots of polynomials, Annals of Math. **57** (1950), 105–119.

44. Fekete, M., Über die Verteilung der Wurzeln bei gewissen algebraischen Gleichungen mit ganzzahligen Koeffizienten, Math. Zeit. **17** (1923), 228–249.

45. Ferguson, Le Baron O., *Approximation by Polynomials with Integral Coefficients*, Amer. Math. Soc., Rhode Island, 1980.

46. Fielding, G. T., The expected value of the integral around the unit circle of a certain class of polynomials, Bull. London Math. Soc. **2** (1970), 301–306.

47. Flammang, V., G. Rhin, and C. J. Smyth, The integer transfinite diameter of intervals and totally real algebraic integers, manuscript.

48. Fuchs, W. H. J. and E. M. Wright, The easier Waring problem, Quart. J. Math, **10** (1939), 190–209.

49. Hilbert, D., Ein Beitrag zur Theorie des Legendreschen Polynoms, Acta Math. **18** (1894), 155–159.

50. Gloden, A., *Mehrgradige Gleichungen*, Noordhoff, Groningen, 1944.

51. Golay, M. J., The merit factor of Legendre sequences, IEEE Trans. Information Theory **29** (1983), 934–936.

52. Golay, M. J., Sieves for low autocorrelation binary sequences, IEEE Trans. Information Theory **23** (1977), 43–51.

53. Habsieger, L. and B. Salvy, On integer Chebyshev polynomials, Math. Comp. **218** (1997), 763–770.

54. Hoholdt, T. and H. Jensen, Determination of the merit factor of Legendre sequences, IEEE Trans. Information Theory **34** (1988), 161–164.

55. Hua, L. K., *Introduction to Number Theory*, Springer-Verlag, Berlin, Heidelberg, New York, 1982.

56. Jensen, J., H. Jensen, and T. Hoholdt, The merit factor of binary sequences related to difference sets, IEEE Trans. Information Theory **37** (1991), 617–626.

57. Kac, M., On the average number of real roots of a random algebraic equation, II, Proc. London Math. Soc. **50** (1948), 390–408.

58. Kahane, J-P., *Some Random Series of Functions*, Vol. 5, Cambridge Studies in Advanced Mathematics, Cambridge, 1985; Second Edition.

59. Kahane, J-P., Sur les polynômes á coefficients unimodulaires, Bull. London Math. Soc **12** (1980), 321–342.

60. Kashin, B., Algebraic polynomials with integer coefficients with least deviation from zero on an interval, Mat. Zametki **50** (1991), 58–67.

61. Kleiman, H., A note on the Tarry-Escott problem, J. Reine. Angew. Math **278/279** (1975), 48–51.

62. Kolountzakis, M., *Probabilistic and Constructive Methods in Harmonic Anaysis and Additive Number Theory*, Ph.D. Thesis, Stanford University, 1994.
63. Konjagin, S., On a problem of Littlewood, Izv. A. N. SSSR, ser. mat. **45** (2) (1981), 243–265.
64. Körner, T. W., On a polynomial of J.S. Byrnes, Bull. London Math. Soc. **12** (1980), 219–224.
65. Littlewood, J. E., On the mean value of certain trigonometric polynomials, J. London Math. Soc. **36** (1961), 307–334.
66. Littlewood, J. E., On polynomials $\sum^n \pm z^m$ and $\sum^n e^{\alpha_m i} z^m$, $z = e^{\theta i}$, J. London Math. Soc. **41** (1966), 367–376.
67. Littlewood, J. E., *Some Problems in Real and Complex Analysis*, Heath Mathematical Monographs, Lexington, Massachusetts, 1968.
68. Littlewood, J. E. and A. C. Offord, On the number of real roots of a random algebraic equation, II, Proc. Cam. Phil. Soc. **35** (1939), 133–148.
69. Mahler, K., On two extremal properties of polynomials, Illinois J. Math. **7** (1963), 681–701.
70. Maltby, R., *Pure Product Polynomials of Small Norm*, Ph.D. Thesis, Simon Fraser University, 1996.
71. Melzak, Z. A., A note on the Tarry-Escott problem, Canad. Math. Bull **4** (1961), 233–237.
72. Montgomery, H. L., *Ten Lectures on the Interface Between Analytic Number Theory and Harmonic Analysis*, CBMS, Vol. 84, Amer. Math. Soc., R. I., 1994.
73. Montgomery, H. L., An exponential sum formed with the Legendre symbol, Acta Arithmetica **37** (1980), 375–380.
74. Newman, D. J. and J. S. Byrnes, The L^4 norm of a polynomial with coefficients ±1, MAA Monthly **97** (1990), 42–45.
75. Newman, D. J. and A. Giroux, Properties on the unit circle of polynomials with unimodular coefficients, in *Recent Advances in Fourier Analysis and its Applications*, J. S. Byrnes and J. F. Byrnes, (eds.), Kluwer, 1990, 79–81.
76. Odlyzko, A., Minima of cosine sums and maxima of polynomials on the unit circle, J. London Math. Soc. **26** (1982), 412–420.
77. Odlyzko, A. and B. Poonen, Zeros of polynomials with 0,1 coefficients, Ens. Math. **39** (1993), 317–348.
78. Pólya, G. and G. Szegö, *Problems and Theorems in Analysis*, Volume I, Springer-Verlag, New York, 1972.
79. Rees, E. and C. J. Smyth, On the Constant in the Tarry-Escott Problem, in *Cinquante Ans de Polynômes*, Fifty Years of Polynomials, Springer-Verlag, New York, 1988.
80. Robinson, L., M.Sc. Thesis, Simon Fraser University, 1997.

81. Saffari, B., *Barker Sequences and Littlewood's "Two-sided Conjectures" on Polynomials with ±1 Coefficients*, Séminaire d'Analyse Harmonique. Anneé 1989/90, Univ. Paris XI, Orsay, 1990, 139–151.

82. Saffari, B., Polynmes réciproques: conjecture d'Erdös en norme L^4, taille des autocorrélations et inexistence des codes de Barker, C. R. Acad. Sci., Paris Sér. I Math **308** (1989), 461–464.

83. Salem, R. and A. Zygmund, Some properties of trigonometric series whose terms have random signs, Acta Math **91** (1954), 254–301.

84. Schmidt, E., Über algebraische Gleichungen vom Pólya-Bloch-Typos, Sitz. Preuss. Akad. Wiss., Phys.-Math. Kl. (1932), 321.

85. Schur, I., Untersuchungen über algebraische Gleichungen, Sitz. Preuss. Akad. Wiss., Phys.-Math. Kl. (1933), 403–428.

86. Siegel, C. L., The trace of totally positive and real algebraic integers, Annals of Math. **46** (1945), 302–314.

87. Sinha, T. N., On the Tarry-Escott problem, Amer. Math. Monthly **73** (1966), 280–285.

88. Smyth, C. J., Cyclotomic factors of reciprocal polynomials and totally positive algebraic integers of small trace, manuscript.

89. Smyth, C. J., The mean values of totally real algebraic integers, Math. Comp **42** (1984), 663–681.

90. Smyth, C. J., Totally positive algebraic integers of small trace, Ann. Inst. Fourier **33** (1984), 1–28.

91. Smyth, C. J., Ideal 9th-order Multigrades and Letac's Elliptic Curve, Math. Comp. **57** (1991), 817–823.

92. Solomyak, B., On the random series $\sum \pm \lambda^n$ (an Erdös problem), Annals of Math. **142** (1995), 611–625.

93. Sudler, C., An estimate for a restricted partition function, Quart. J. Math. **2** (1964), 1–10.

94. Szegö, G., Bemerkungen zu einem Satz von E. Schmidt uber algebraische Gleichungen, Sitz. Preuss. Akad. Wiss., Phys.-Math. Kl. (1934), 86–98.

95. Turán, P., *On a New Method of Analysis and its Applications*, Wiley, New York, 1984.

96. Turyn, R. and J. Storer, On binary sequences, Proc. Amer. Math. Soc. **12** (1961), 394–399.

97. Wright, E. M., The Tarry-Escott and the "easier" Waring Problem, J. Reine Angew Math **311/312** (1972), 170–173.

98. Wright, E. M., Prouhet's 1851 solution of the Tarry-Escott Problem of 1910, Amer. Math. Monthly **66** (1959), 199–201.

99. Wright, E. M., On Tarry's Problem (I), Quart. J. Math. **6** (1935), 261–267.

100. Wright, E. M., An easier Waring's Problem, J. London Math. Soc. **9** (1934), 267–272.

On the Irrationality of
$\sum_{i=0}^{\infty} q^{-i} \prod_{j=0}^{i}(1+q^{-j}r+q^{-2j}s)$

Peter B. Borwein and Ping Zhou

Abstract. We prove that for any integer q greater than one, and positive rationals r and s,
$$\sum_{i=0}^{\infty} q^{-i} \prod_{j=0}^{i}(1+q^{-j}r+q^{-2j}s)$$
is irrational and is not a Liouville number.

§1. Introduction and Result

By using the the residue theorem and functional equation methods, we have proved the irrationality of various multivariate functions whose univeriate versions have been extensively investigated (see [2, 3, 5, 6, 7, 8, 9, 10]), namely, $\prod_{j=0}^{\infty}(1+q^{-j}r+q^{-2j}s)$; $\sum_{j=0}^{\infty} \frac{1}{1+q^j r - q^{2j}s}$; $\sum_{i,j=0}^{\infty} \frac{r^i s^j}{q^{i+j+1}-1}$, and $\sum_{j_1,\cdots,j_m=0}^{\infty} \frac{r_1^{j_1}\cdots r_m^{j_m}}{q^{j_1+\cdots+j_m+1}-1}$, where q is an integer greater than one, r, s and $r_1, \cdots, r_m$ are any positive rationals, see Borwein-Zhou [4], Zhou-Lubinsky [11] and Zhou [12] for details. The general approach has been to examine the Padé approximants to the appropriate functions and to show, with some modifications, that they provide rational approximations that are too rapid to be consistent with rationality. In this paper, we use similar techniques to prove a new irrationality result:

Theorem 1.1. *If q is an integer greater than one, and r, s are any positive rationals, then*
$$\sum_{i=0}^{\infty} q^{-i} \prod_{j=0}^{i}(1+q^{-j}r+q^{-2j}s)$$
is irrational, and is not Liouville number.

We refer to Borwein [2] for the concepts of standard q analogues of factorials $[n]!$ and binomial coefficients $\begin{bmatrix} n \\ k \end{bmatrix}$. We note that (see Zhou [12])

$$\prod_{h=0, h\neq k}^{n} (q^{-k} - q^{-h}) = (-1)^k q^{-k(k-1)/2 - n(n+1)/2}[n]![k]!(1-q)^n, \tag{1.1}$$

and for $|t| < |q|^{-n}$,

$$\frac{1}{\prod_{k=0}^{n}(t - q^{-k})} = (-1)^{n+1} q^{n(n+1)/2} \sum_{l=0}^{\infty} \begin{bmatrix} n+l \\ l \end{bmatrix} t^l. \tag{1.2}$$

We prove some properties of the approximants to the function

$$F(x,y) = \sum_{i=0}^{\infty} q^{-i} \prod_{j=0}^{i} (1 + q^{-j}x + q^{-2j}xy) \tag{1.3}$$

in Section 2, and use those properties to prove Theorem 1.1 in Section 3.

§2. Some Results On the Relevant Function

Theorem 2.1. Let $|q| > 1$, and $F(x,y)$ be defined by (1.3). Then $F(x,y)$ is entire in $\mathbf{C} \times \mathbf{C}$. Furthermore, writing

$$F(x,y) = \sum_{i,j=0}^{\infty} c_{ij} x^i y^j, \tag{2.1}$$

and

$$\prod_{j=0}^{\infty}(1 + q^{-j}x + q^{-2j}xy) = \sum_{i,j=0}^{\infty} a_{ij} x^i y^j, \tag{2.2}$$

then

$$\left(\prod_{j=1}^{r+s}(q^j - 1)\right) a_{rs} \in \mathbb{Z}[q], \quad r, s = 0, 1, 2 \cdots, \tag{2.3}$$

and

$$\left(\prod_{j=1}^{r+s}(q^j - 1)\right) c_{rs} \in \mathbb{Z}[q], \quad r, s = 0, 1, 2 \cdots, \tag{2.4}$$

where $\mathbb{Z}[q]$ is the set of polynomials in q with integer coefficients.

Proof: See Section 2 of Zhou-Lubinsky [11] for a proof of (2.3). (2.4) follows immediately from (2.3). □

Irrationality of a Series   53

Theorem 2.2. Let $q > 1$ be an integer, and $F(x,y)$ be defined by (1.3). Also, let $n \geq 0$ be an integer,

$$I(x,y) := \frac{1}{2\pi i} \int_\Gamma \frac{F(tx,ty)dt}{\left(\prod_{k=0}^n (t-q^{-k})\right) t^{n+1}}, \qquad (2.5)$$

where Γ is a circular contour containing $0, q^{-n}, \cdots, q^{-1}, q^0$, and

$$R_k(x,y) := \prod_{j=0}^{k-1} (q^{2j} + q^j x + xy), \qquad (2.6)$$

$$S_k(x,y) := q^{-2k} \sum_{i=1}^{k} q^{-i(2k-i-2)} \prod_{j=k-i}^{k} (q^{2j} + q^j x + xy), \qquad (2.7)$$

$$Q(x,y) := \frac{q^{n(n+1)/2}}{(1-q)^n [n]!} \sum_{k=0}^{n} (-1)^k \begin{bmatrix} n \\ k \end{bmatrix} q^{k(n+k)+k(k+1)/2} R_k^{-1}(x,y), \qquad (2.8)$$

and

$$P(x,y) := \frac{q^{n(n+1)/2}}{(1-q)^n [n]!} \sum_{k=0}^{n} (-1)^{k+1} \begin{bmatrix} n \\ k \end{bmatrix} q^{nk+k(k+1)/2} S_k(x,y)$$
$$+ \frac{1}{n!} \frac{d^n}{dt^n} \left\{ \frac{F(tx,ty)}{\prod_{k=0}^n (t-q^{-k})} \right\}_{t=0}. \qquad (2.9)$$

Then

i)
$$I(x,y) = Q(x,y)F(x,y) + P(x,y); \qquad (2.10)$$

ii)
$$I(x,y) = \sum_{i+j \geq 2n+1} d_{ij} x^i y^j, \quad d_{ij} \in \mathbb{C}; \qquad (2.11)$$

iii)
$$q^{-n(n+1)/2} \left(\prod_{j=1}^{n} (q^j - 1)\right) R_n(x,y) Q(x,y) \in \mathbb{Z}[q,x,y]; \qquad (2.12)$$

iv)
$$q^{-n(n+1)/2} \left(\prod_{j=1}^{n} (q^j - 1)\right) P(x,y) \in \mathbb{Z}[q,x,y]; \qquad (2.13)$$

v) For any integer l, with $0 \leq l < n$,

$$q^{-n(n+1)/2} \left(\prod_{j=1}^{n} (q^j - 1)\right) \cdot q^{l(l+1)} R_n\left(\frac{x}{q^l}, \frac{y}{q^l}\right) Q\left(\frac{x}{q^l}, \frac{y}{q^l}\right) \in \mathbb{Z}[q,x,y];$$
$$(2.14)$$

$$q^{-n(n+1)/2}\left(\prod_{j=1}^{n}(q^j-1)\right)\cdot q^{nl}P\left(\frac{x}{q^l},\frac{y}{q^l}\right)\in\mathbb{Z}[q,x,y]; \qquad (2.15)$$

vi) For fixed $n\in\mathbb{N}$ and

$$(x,y)\in\mathbb{P}:=(x,y)\in\mathbb{C}\times\mathbb{C}:0<|x|,|y|\leq 1/2\}, \qquad (2.16)$$

$$|I(x,y)|\leq c_q\cdot\frac{(n+1)v_{x,y}^{2n+1}}{q^{n^2}}, \qquad (2.17)$$

where c_q is a constant depending only on q, and

$$v_{x,y}:=\max\{|x|,|y|\}. \qquad (2.18)$$

Proof: Proof of (i). From (1.3), we have the functional relations for $F(x,y)$ for integers $k\geq 1$:

$$\begin{aligned}F(q^{-k}x,q^{-k}y)&=\sum_{i=0}^{\infty}q^{-i+k}\prod_{j=0}^{i-k}(1+q^{-j-k}x+q^{-2j-2k}xy)\\&\quad-\sum_{i=1}^{k}q^i\prod_{j=k-i}^{k}(1+q^{-j}x+q^{-2j}xy)\\&=q^{k^2}R_k^{-1}(x,y)F(x,y)-S_k(x,y).\end{aligned} \qquad (2.19)$$

By the residue theorem and the functional equation (2.19) and (1.1), we have

$$\begin{aligned}I(x,y)&=\\&\sum_{k=0}^{n}\frac{F(q^{-k}x,q^{-k}y)}{\left(\prod_{h=0,h\neq k}^{n}(q^{-k}-q^{-h})\right)q^{-k(n+1)}}+\frac{1}{n!}\frac{d^n}{dt^n}\left\{\frac{F(tx,ty)}{\prod_{k=0}^{n}(t-q^{-k})}\right\}_{t=0}\\&=\frac{q^{n(n+1)/2}}{(1-q)^n[n]!}\sum_{k=0}^{n}(-1)^k\begin{bmatrix}n\\k\end{bmatrix}q^{k(n+k)+k(k+1)/2}R_k^{-1}(x,y)F(x,y)\\&\quad-\frac{q^{n(n+1)/2}}{(1-q)^n[n]!}\sum_{k=0}^{n}(-1)^k\begin{bmatrix}n\\k\end{bmatrix}q^{nk+k(k+1)/2}S_k(x,y)\\&\quad+\frac{1}{n!}\frac{d^n}{dt^n}\left\{\frac{F(tx,ty)}{\prod_{k=0}^{n}(t-q^{-k})}\right\}_{t=0}=Q(x,y)F(x,y)+P(x,y).\end{aligned}$$

Proof of (ii). From (2.5) we observe that the denominator in the integral defining $I(x,y)$ is a polynomial of degree $2n+2$ in t, and any terms in

$$F(tx,ty)=\sum_{i,j=0}^{\infty}c_{ij}t^{i+j}x^iy^j$$

Irrationality of a Series

of order less than $2n+1$ in t vanish on integration. So (2.11) holds.

Proof of (iii). From (2.6) and (2.8), we have

$$R_n(x,y)Q(x,y) = \frac{q^{n(n+1)/2}}{(1-q)^n[n]!} \sum_{k=0}^{n}(-1)^k \begin{bmatrix}n\\k\end{bmatrix} q^{k(n+k)+k(k+1)/2}$$

$$\times \prod_{j=k}^{n-1}(q^{2j}+q^j x+xy). \quad (2.20)$$

As $\begin{bmatrix}n\\k\end{bmatrix}$ is a polynomial in q with integer coefficients, (2.12) holds.

Proof of (iv). From (1.2),

$$\frac{F(tx,ty)}{\prod_{k=0}^{n}(t-q^{-k})} = (-1)^{n+1} q^{n(n+1)/2} \sum_{r,s,l=0}^{\infty} c_{rs} x^r y^s t^{r+s+l} \begin{bmatrix}n+l\\l\end{bmatrix}, \quad (2.21)$$

and

$$P(x,y) = \frac{q^{n(n+1)/2}}{(1-q)^n[n]!} \sum_{k=0}^{n}(-1)^{k+1}\begin{bmatrix}n\\k\end{bmatrix}$$

$$\times \left(\sum_{i=1}^{k} q^{k(n-i)-i(k-i)+(k^2+4i-3k)/2} \prod_{j=k-i}^{k}(q^{2j}+q^j x+xy) \right)$$

$$+ (-1)^{n+1} q^{n(n+1)/2} \sum_{\mu=0}^{n} \left(\sum_{r+s=\mu} c_{rs} x^r y^s \right) \begin{bmatrix}2n-\mu\\n-\mu\end{bmatrix}. \quad (2.22)$$

As

$$k(n-i) - i(k-i) + (k^2+4i-3k)/2 \geq 0, \quad (2.23)$$

for $0 \leq k \leq n$ and $1 \leq i \leq k$, with (2.4), we have (2.13).

Proof of (v). For $0 \leq l \leq n-1$,

$$R_n(\frac{x}{q^l}, \frac{y}{q^l}) = q^{-l(l+1)}\left(\prod_{j=0}^{l}(q^{2l+2j}+q^{j+l}x+xy)\right). \quad (2.24)$$

So

$$q^{l(l+1)} R_n(\frac{x}{q^l}, \frac{y}{q^l}) \in \mathbb{Z}[q,x,y]. \quad (2.25)$$

From (2.12), (2.20), and (2.25), we have (2.14), and (2.15) follows from (2.22).

Proof of (vi). See the proof of (vi) of Theorem 2.2 in Zhou [12]. □

§3. Proof of Theorem 1.1

The proof is similar to the proof of (3.1) in Zhou [12]. We note that for $x, y > 0$,
$$I(x,y) > 0, \tag{3.1}$$
where $I(x,y)$ is defined by (2.5). Now let r, s be any fixed positive rational numbers, then $u := s/r$ is also a positive rational number such that

$$F(r,u) = \sum_{i=0}^{\infty} q^{-i} \prod_{j=0}^{\infty} (1 + q^{-j}r + q^{-2j}ru) > 0.$$

We now use the estimate (2.17) in (vi) of Theorem 2.2 to prove the main result. Let $n \geq 0$ be chosen so that n is a multiple of 4, and

$$X := \frac{r}{q^{n/4}} \leq 1/2 \quad \text{and} \quad Y := \frac{u}{q^{n/4}} \leq 1/2, \tag{3.2}$$

(now (X,Y) is in the set $\mathbb{P}$ defined by (2.16)) and let

$$v_{r,u} := \max\{r, u\}; \tag{3.3}$$

$$v_{X,Y} := \max\{X, Y\}. \tag{3.4}$$

Then
$$v_{X,Y} = \frac{v_{r,u}}{q^{n/4}}. \tag{3.5}$$

Now let
$$H_n(q) := q^{-n(n+1)/2} \prod_{j=1}^{n} (q^j - 1). \tag{3.6}$$

Then
$$0 < |H_n(q)| \leq 1. \tag{3.7}$$

From the functional equation (2.19), (3.2), and (2.6), we have

$$F(r,u) = q^{-(n/4)^2} R_{n/4}(r,u) \left[F(X,Y) + S_{n/4}(r,u)\right], \tag{3.8}$$

$$|R_n(X,Y)| \leq \prod_{j=0}^{n-1}(1 + q^j + q^{2j}) \leq a q^{n(n-1)}, \tag{3.9}$$

where $a := \prod_{j=0}^{\infty}(1 + q^{-j} + q^{-2j})$ is a constant depending only on q. Now in order to get integer coefficient approximants to $F(r,u)$, let

$$Q^*(r,u) := q^{n(3n+2)/8} H_n(q) R_n(X,Y) Q(X,Y); \tag{3.10}$$

$$\begin{aligned}P^*(r,u) := {}& q^{n(5n+4)/16} H_n(q) R_{n/4}(r,u) R_n(X,Y) \\ & \times \left(P(X,Y) - S_{n/4}(r,u) Q(X,Y)\right).\end{aligned} \tag{3.11}$$

Irrationality of a Series

Then from (v) of Theorem 2.2, (2.14), (2.15), and (3.6), we have
$$Q^*(r,u), P^*(r,u) \in \mathbb{Z}[q,r,u], \tag{3.12}$$
and from (3.8), (3.10), and (3.11), we have
$$\begin{aligned}\Delta &:= |Q^*(r,u)F(r,u) + P^*(r,u)| \\ &= q^{n(3n+2)/8} |H_n(q)| |R_n(X,Y)| \left|q^{-(n/4)^2} R_{n/4}(r,u)\right| |I(X,Y)|.\end{aligned} \tag{3.13}$$
For $\Delta > 0$, and from (2.17), (3.5), (3.7), (3.9) and the fact that
$$\left|q^{-(n/4)^2} R_{n/4}(r,u)\right| \leq q^{-(n/4)^2 + (n/4)(n/4-1)} \left|\prod_{j=0}^{\infty}(1 + q^{-j}r + q^{-2j}ru)\right|$$
$$=: bq^{-n/4},$$
where $b := \left|\prod_{j=0}^{\infty}(1 + q^{-j}r + q^{-2j}ru)\right|$ is a constant depending only on r, u and q, we have
$$\Delta \leq q^{3n^2/8 + n(n-1)} abc_q \frac{(n+1)v_{X,Y}^{2n+1}}{q^{n^2}} =: f_{q,r,u} \frac{(n+1)v_{r,u}^{2n+1}}{q^{n(n+10)/8}}, \tag{3.14}$$
where $f_{q,r,u} := abc_q$ is a constant depending only on q, r, and u. Finally, if
$$r := \frac{i}{l} \quad \text{and} \quad u := \frac{j}{m} \tag{3.15}$$
with i, j, l, m being positive integers, then
$$Q^{**}(r,u) := (lm)^{4n} Q^*(r,u), \tag{3.16}$$
and
$$P^{**}(r,u) := (lm)^{4n} P^*(r,u), \tag{3.17}$$
are integers, and
$$v_{r,u} = \frac{1}{lm} v_{mi,lj} := \frac{1}{lm} \max\{mi, lj\}. \tag{3.18}$$
Then by (3.14) to (3.18), we obtain
$$0 < |Q^{**}(r,u)F(r,u) + P^{**}(r,u)| \leq (lm)^{4n} f_{q,r,u}(n+1)\frac{v_{r,u}^{2n+1}}{q^{n(n+10)/8}}$$
$$= f_{q,r,u}(n+1)(lm)^{2n-1}\frac{v_{mi,lj}^{2n+1}}{q^{n(n+10)/8}},$$
which tends to zero as $n \to \infty$, This shows that $F(r,u)$ is irrational, and we complete the proof of Theorem 1.1. □

Now by the standard methods (as in chapter 11 of Borwein-Borwein [1]), the estimates in the proof of Theorem 1.1 gives under the assumption of the theorem, that
$$\left|F(r,u) - \frac{s}{t}\right| > \frac{1}{t^\alpha},$$
for some constant α and all integers s and t, and hence $F(r,u)$ is not a Liouville number.

References

1. Borwein, J. M. and P. B. Borwein, *Pi and the AGM—A Study in Analytic Number Theory and Computational Complexity*, New York, 1987.
2. Borwein, P. B., Padé approximants for the q-elementary functions, Constr. Approx. **4** (1988), 391–402.
3. Borwein, P. B., On the irrationality of $\sum(1/(q^n + r))$, J. of Number Theory, **37** (1991), 253–259.
4. Borwein, P. B. and P. Zhou, On the irrationality of a certain q series, Proc. Amer. Math. Soc., to appear.
5. Bundschuh, P., Arithmetische untersuchungen unendlicher produkte, Inventiones Math. **6** (1969), 275–295.
6. Chudnovsky, D. V. and G. V. Chudnovsky, Padé and rational approximation to systems of functions and their arithmetic applications, in *Lecture Notes in Mathematics*, Vol. 1052, Springer-Verlag, Berlin, 1984.
7. Gasper, G. and M. Rahman, Basic hypergeometric series, in *Encyclopedia of Maths and its Applications*, Vol. 35, Cambridge University Press, Cambridge, 1990.
8. Lototsky, A. V., Sur l'irrationalité d'un produit infini, Mat. Sb. **12** (54) (1943), 262–272.
9. Mahler, K., Zur approximation der exponentialfunktion und des logarithmus, J. Reine Angew. Math. **166** (1932), 118–150.
10. Wallisser, R., Rationale approximation des q-Analogons der exponentialfunktion und irrationalitätsaussagen für diese funktion, Arch. Math. **44** (1985), 59–64.
11. Zhou, P. and D. S. Lubinsky, On the irrationality of $\prod_{j=0}^{\infty}(1 \pm q^{-j}r + q^{-2j}s)$, Analysis **17** (1997), 129–153.
12. Zhou, P., On irrationality of $\prod_{j=0}^{\infty}(1 + q^{-j}r + q^{-2j}s)$, Math. Proc. Camb. Phil. Soc., to appear.

Peter Borwein and Ping Zhou
Department of Mathematics and Statistics
Simon Fraser University
Burnaby, BC.
Canada, V5A 1S6
pborwein@cecm.sfu.ca
pzhou@cecm.sfu.ca

On a Weighted Recursive Rule

Michele Campiti

Abstract. We continue the analysis of more general binomial-type coefficients in the sequence of Bernstein operators by introducing a recursive rule where the weight function p already considered in [2] may depend on $n \geq 1$. We deal with some approximation properties of the corresponding operators.

§1. Introduction and Notation

In [2, 3, 4] we introduced some generalizations of Bernstein operators where the binomial coefficients were replaced by more general ones satisfying a similar recursive rule. The motivation of these generalizations was the approximation of the solutions of suitable differential problems connected to these operators by a Voronovskaja-type formula via semigroup theory (see [2, 4] and [1, Chapter 6] for a more complete treatment).

Here, we are interested to a further generalization of the recursive rule from an approximation point of view.

We define our binomial coefficients by considering weight functions p and q as in [2], but in this case we allow these functions to depend on $n \geq 1$.

More precisely, consider two positive sequences $(\lambda_n)_{n \geq 1}$ and $(\rho_n)_{n \geq 1}$ of real numbers and two positive sequences $(p_n)_{n \geq 1}$ and $(q_n)_{n \geq 1}$ of continuous real functions on $[0, 1]$.

For every $n \geq 1$, we define

$$\alpha_{n,0}(x) = \lambda_n \cdot \prod_{j=1}^{n} q_j(x), \quad \alpha_{n,n}(x) = \rho_n \cdot \prod_{i=1}^{n} p_i(x), \quad x \in [0, 1] \quad (1)$$

and, for every $k = 1, \ldots, n$,

$$\alpha_{n+1,k}(x) = p_k(x)\alpha_{n,k-1}(x) + q_{n+1-k}(x)\alpha_{n,k}(x), \quad x \in [0, 1]. \quad (2)$$

The corresponding recursively defined operators are obtained by setting, for every $n \geq 1$,

$$C_n f(x) = \sum_{k=0}^{n} \alpha_{n,k}(x) \, x^k (1-x)^{n-k} f\left(\frac{k}{n}\right), \qquad f \in C[0,1], \quad x \in [0,1]. \quad (3)$$

Obviously, if we take $\lambda_n = \rho_n = 1$ and $p_n(x) = q_n(x) = 1$ for every $n \geq 1$ and $x \in [0,1]$, we obtain the classical Bernstein operators

$$B_n f(x) = \sum_{k=0}^{n} \binom{n}{k} x^k (1-x)^{n-k} f\left(\frac{k}{n}\right), \qquad f \in C[0,1], \quad x \in [0,1]. \quad (4)$$

In the sequel, we consider some simple cases of particular interest where we can ensure the convergence of the operators $(C_n)_{n \geq 1}$.

The following lemma generalizes Lemma 1.1 of [2], and establishes a recursive formula for binomial-type coefficients which is useful for our discussion.

Lemma 1. *For every $n \geq 2$ and $k = 1, \ldots, n-1$, we have*

$$\alpha_{n,k} = \left(\sum_{i=1}^{n-k} \binom{n-1-i}{k-1} \lambda_i + \sum_{j=1}^{k} \binom{n-1-j}{n-1-k} \rho_j \right) \prod_{i=1}^{k} p_i \prod_{j=1}^{n-k} q_j. \quad (5)$$

Proof: If $n = 2$ and $k = 1$, by (2) and (1)

$$\alpha_{2,1} = \alpha_{1,0} p_1 + \alpha_{1,1} q_1 = (\rho_1 + \lambda_1) p_1 q_1,$$

and therefore (5) holds.

If (5) holds for $n \geq 2$, then by (1) and (2) we obtain immediately (5) for $\alpha_{n+1,k}$ when $k = 1$ and $k = n$; further, if $k = 2, \ldots, n-1$,

$$\alpha_{n+1,k} = \left(\sum_{i=1}^{n-k} \left(\binom{n-1-i}{k-2} + \binom{n-1-i}{k-1} \right) \lambda_i + \lambda_{n+1-k} \right.$$

$$\left. + \sum_{j=1}^{k-1} \left(\binom{n-1-j}{n-k} + \binom{n-1-j}{n-1-k} \right) \rho_j + \rho_k \right) \prod_{i=1}^{k} p_i \prod_{j=1}^{n+1-k} q_j$$

$$= \left(\sum_{i=1}^{n+1-k} \binom{n-i}{k-1} \lambda_i + \sum_{j=1}^{k} \binom{n-j}{n-k} \rho_j \right) \prod_{i=1}^{k} p_i \prod_{j=1}^{n+1-k} q_j$$

and the proof is complete. □

As a justification of the assumption (1), we observe that if we use the convention $\prod_{i=1}^{0} a_i = 1$, Lemma 1 holds also for $k = 0$ and $k = n$. Obviously, $\alpha_{n,k}(x) = \binom{n}{k} \prod_{i=1}^{k} p_i(x) \times \prod_{j=1}^{n-k} q_j(x)$ if $\lambda_n = \rho_n = 1$ for every $n \geq 1$.

§2. Convergence Results

In [3], we have seen that in the case $p_n = q_n = 1$ a necessary and sufficient condition for the convergence of the sequence $(C_n)_{n\geq 1}$ is that the sequences $(\lambda_n)_{n\geq 1}$ and $(\rho_n)_{n\geq 1}$ converge; hence, for the sequel this last assumption will be quite natural; we also require these limits strictly positive to avoid that $(C_n 1)_{n\geq 1}$ tends to 0 at the endpoints. Then we put

$$\lim_{n\to+\infty} \lambda_n = \lambda_\infty > 0, \qquad \lim_{n\to+\infty} \rho_n = \rho_\infty > 0. \tag{6}$$

We also define, for every $n \geq 1$,

$$s_n = \max_{k=1,\ldots,n} \{\lambda_k, \rho_k\}. \tag{7}$$

In order to study the convergence of the sequence $(C_n)_{n\geq 1}$, it turns to be useful to recall the following consequence of Toeplitz's theorem (see, e.g., [6, Theorem 6, p. 75]), which can be stated as follows.

Proposition 2. *Let $(a_{n,k})_{n\geq 0, k=0,\ldots,n}$ be a system of positive real numbers satisfying the following conditions:*

a) *For every fixed $k \geq 0$, we have $\lim_{n\to+\infty} a_{n,k} = 0$ and $\lim_{n\to+\infty} a_{n,n-k} = 0$;*

b) $\lim_{n\to+\infty} \sum_{k=0}^{n} a_{n,k} = 1.$

Then, if $(u_n)_{n\geq 0}$ and $(v_n)_{n\geq 0}$ are sequences of real numbers converging to u and v, respectively, we have

$$\lim_{n\to+\infty} \sum_{k=0}^{n} a_{n,k} u_k v_{n-k} = u \cdot v.$$

At this point, we assume that the infinite products $\prod_{n=1}^{+\infty} p_n$ and $\prod_{n=1}^{+\infty} q_n$ converge, and consider the functions

$$p = \prod_{n=1}^{+\infty} p_n, \qquad q = \prod_{n=1}^{+\infty} q_n. \tag{8}$$

In order to obtain the convergence at the points 0 and 1, we shall make the assumption

$$p(0) = 1, \qquad q(1) = 1. \tag{9}$$

It is also useful to consider the operators A_n defined as in (3) with $p_n = q_n = 1$ for every $n \geq 1$ and denote by $\beta_{n,k}$ the corresponding binomial-type coefficients defined according to (1) and (2). Hence, for every $f \in C[0,1]$ and $x \in [0,1]$,

$$A_n f(x) = \sum_{k=0}^{n} \beta_{n,k} x^k (1-x)^{n-k} f\left(\frac{k}{n}\right),$$

and by Lemma 1,

$$C_n f(x) = \sum_{k=0}^{n} \beta_{n,k} \prod_{i=1}^{k}(x\, p_i(x)) \prod_{j=1}^{n-k}((1-x)\, q_j(x))\, f\left(\frac{k}{n}\right). \qquad (10)$$

We recall that the sequence $(A_n \mathbf{1})_{n\geq 1}$ converges to the function w defined by

$$w(x) = \begin{cases} \lambda_\infty, & \text{if } x = 0, \\ \sum_{m=1}^{+\infty} (\lambda_m x(1-x)^m + \rho_m x^m(1-x)), & \text{if } 0 < x < 1, \\ \rho_\infty, & \text{if } x = 1, \end{cases} \qquad (11)$$

(see [3, Theorem 2.1]).

At this point we can study the convergence of the sequence $(C_n)_{n\geq 1}$. For simplicity, a quantitative estimate will be obtained in terms of the quantities

$$\ell_n(x) = |C_n \mathbf{1}(x) - w(x)\, p(x)\, q(x)|, \qquad x \in [0,1] \qquad (12)$$

and the partial products

$$r_n(x) = \max_{k=1,\ldots,n} \left\{ \prod_{i=1}^{k} p_i(x),\ \prod_{j=1}^{k} q_j(x) \right\}, \qquad x \in [0,1] \qquad (13)$$

of the infinite products (8).

We have the following main result.

Theorem 3. *Assume that the sequences* $(\lambda_n)_{n\geq 1}$ *and* $(\rho_n)_{n\geq 1}$ *converge to* λ_∞ *and* ρ_∞, *respectively, and that the infinite products* $\prod_{n=1}^{+\infty} p_n$ *and* $\prod_{n=1}^{+\infty} q_n$ *converge uniformly to the functions* p *and* q, *respectively.*

Then, for every $f \in C[0,1]$ *and* $x \in [0,1]$,

$$|C_n f(x) - w(x)\, p(x)\, q(x)\, f(x)| \leq \frac{5}{4} r_n(x)\, s_n\, \omega\left(f, \frac{1}{\sqrt{n}}\right) + \ell_n(x)\, |f(x)| \qquad (14)$$

and further

$$\lim_{n \to +\infty} C_n f = w\, p\, q\, f \qquad (15)$$

uniformly on $[0,1]$.

Proof: Let $f \in C[0,1]$ and $x \in [0,1]$. We have

$$|C_n f(x) - w(x)\, p(x)\, q(x)\, f(x)| \leq |C_n f(x) - f(x)\, C_n \mathbf{1}(x)| \\ + |f(x)|\, |C_n \mathbf{1}(x) - w(x)\, p(x)\, q(x)|. \qquad (16)$$

Since $\beta_{n,k} \leq s_n \binom{n}{k}$ (see [3]), for every $\delta > 0$, by (10) and (13) the first term can be estimated as

$$|C_n f(x) - f(x) C_n \mathbf{1}(x)|$$
$$= \sum_{k=0}^{n} \beta_{n,k} \prod_{i=1}^{k}(x\, p_i(x)) \prod_{j=1}^{n-k}((1-x)\, q_j(x)) \left|f\left(\frac{k}{n}\right) - f(x)\right|$$
$$\leq r_n(x)\, s_n \sum_{k=0}^{n} \binom{n}{k} x^k (1-x)^{n-k} \left|f\left(\frac{k}{n}\right) - f(x)\right|$$
$$\leq r_n(x)\, s_n \sum_{k=0}^{n} \binom{n}{k} x^k (1-x)^{n-k} \left(1 + \frac{1}{\delta^2}\left(\frac{k}{n} - x\right)^2\right) \omega(f,\delta)$$
$$= r_n(x)\, s_n \left(1 + \frac{x(1-x)}{n\delta^2}\right) \omega(f,\delta),$$

and taking $\delta = 1/\sqrt{n}$, we get

$$|C_n f(x) - f(x) C_n \mathbf{1}(x)| \leq \frac{5}{4} r_n(x)\, s_n\, \omega(f, \frac{1}{\sqrt{n}}).$$

Taking (12) into account, this completes the proof of (14).

Finally, by (16) we have only to show that the sequence $(C_n \mathbf{1})_{n \geq 1}$ converges uniformly to $w\,p\,q$. Indeed, recalling that (see [3, (2.30), (2.34), and Proposition 2.5]), we can write $\beta_{n,k} = \binom{n}{k} b_{n,k}$. Also recalling that the continuous piecewise affine functions b_n satisfying $b_n(k/n) = b_{n,k}$ for $k = 0, \ldots, n$ converge uniformly to w, we have by (10)

$$|C_n \mathbf{1}(x) - w(x)\, p(x)\, q(x)|$$
$$\leq \left|\sum_{k=0}^{n} \binom{n}{k} x^k (1-x)^{n-k} \prod_{i=1}^{k} p_i(x) \prod_{j=1}^{n-k} q_j(x)\, b_{n,k} - w(x)\, p(x)\, q(x)\right|$$
$$\leq r_n(x) \sum_{k=0}^{n} \binom{n}{k} x^k (1-x)^{n-k} |b_{n,k} - b_n(x)|$$
$$+ \sum_{k=0}^{n} \binom{n}{k} x^k (1-x)^{n-k} \prod_{i=1}^{k} p_i(x) \prod_{j=1}^{n-k} q_j(x)\, |b_n(x) - w(x)|$$
$$+ |w(x)| \left(\sum_{k=0}^{n} \binom{n}{k} x^k (1-x)^{n-k} \prod_{i=1}^{k} p_i(x) \prod_{j=1}^{n-k} q_j(x) - p(x)\, q(x)\right).$$

The first addend in the last sum tends uniformly to 0 since it coincides with $r_n(x)\, |B_n b_n(x) - b_n(x)|$ and $(b_n)_{n \geq 1}$ is equicontinuous. Since $(b_n)_{n \geq 1}$ tends uniformly to w, also the second addend tends uniformly to 0. Finally, by Proposition 2 the same holds for the third addend. Thus, (15) is also proved. □

The different recursive rule considered in [2] corresponds to the case where the sequences $(p_n)_{n\geq 1}$ and $(q_n)_{n\geq 1}$ are constant. In this case the infinite products (8) converge if and only if $p_n = q_n = 1$. However, in [2] the convergence of the corresponding operators C_n under the more general condition $xp(x) + (1-x)q(x) = 1$ (or $xp(x) + (1-x)q(x) \leq 1$) is considered.

The following result generalizes the convergence theorem obtained in [2].

Corollary 4. *Assume that the sequences* $(\lambda_n)_{n\geq 1}$ *and* $(\rho_n)_{n\geq 1}$ *converge and that there exist continuous functions* $p, q, u,$ *and* v *such that*

a) $\lim\limits_{n\to+\infty} \dfrac{\prod_{i=1}^n p_i(x)}{u(x)^n} = p(x)$, $\lim\limits_{n\to+\infty} \dfrac{\prod_{j=1}^n q_j(x)}{v(x)^n} = q(x);$

b) *The function* $x \mapsto x\, u(x)$ *is strictly increasing and, for every* $x \in [0,1]$, $x\, u(x) + (1-x)\, v(x) = 1$.

Then, for every $f \in C[0,1]$,

$$\lim_{n\to+\infty} C_n f(x) = w(x\, u(x))\, p(x)\, q(x)\, f(x\, u(x)) \qquad (17)$$

uniformly in $x \in C[0,1]$.

Proof: For every $f \in C[0,1]$ and $x \in [0,1]$, we can write

$$C_n f(x) = \sum_{k=0}^n \beta_{n,k}(x\, u(x))^k ((1-x)\, v(x))^{n-k} \frac{\prod_{i=1}^k p_i(x)}{u(x)^k} \frac{\prod_{j=1}^{n-k} q_j(x)}{v(x)^{n-k}} f\left(\frac{k}{n}\right).$$

By assumption b), we have $(1-x)\, v(x) = 1 - x\, u(x)$ and, therefore, the operator C_n can be considered as the operator arising by (3) corresponding to the sequences

$$\left(\frac{p_n(\varphi(x))}{u(\varphi(x))}\right)_{n\geq 1}, \qquad \left(\frac{q_n(\varphi(x))}{v(\varphi(x))}\right)_{n\geq 1},$$

and evaluated at the point $x\, u(x)$, where φ denotes the inverse of $x \mapsto x\, u(x)$. At this point, by the assumption a) we may apply Theorem 3 and conclude the proof. □

A quantitative estimate of the convergence can be directly obtained from (14).

The assumption b) in Theorem 4 ensures that $C_n 1(x)$ does not tend to 0. Without this restriction, we may assume the weaker condition $x\, u(x) + (1-x)\, v(x) \leq 1$ $(0 \leq x \leq 1)$.

Examples 5. Besides the trivial examples where C_n coincides with well-known operators, we consider some other cases which may be of interest.

1) Let $m \geq 1$ be fixed and put $\lambda_m = 1$ and $\lambda_n = 0$ for $n \neq m$. Moreover, let $\rho_n = 0$ for every $n \geq 1$. By (11) we have $w(x) = x(1-x)^m$ (see also [3, (2.25)]) and consequently, if $(p_n)_{n\geq 1}$ and $(q_n)_{n\geq 1}$ satisfy the assumptions

of Theorem 3, the corresponding operators defined by (3) (see also [3, (1.8)])

$$L_{m,n}f(x) = \begin{cases} \sum_{k=0}^{n-m} \binom{n-m-1}{k-1} \prod_{i=1}^{k}(x\,p_i(x)) \\ \quad \times \prod_{j=1}^{n-k}((1-x)q_j(x))\,f\left(\frac{k}{n}\right), & \text{if } n > m, \\ \prod_{j=1}^{n}((1-x)q_j(x))\,f(0), & \text{if } n = m, \end{cases}$$

satisfy $\lim_{n \to +\infty} L_{m,n} f(x) = x\,(1-x)^m\,p(x)\,q(x)\,f(x)$ for every $f \in C[0,1]$ uniformly in $x \in [0,1]$.
Analogously, if $\rho_m = 1$, $\rho_n = 0$ for $n \neq m$ and $\lambda_n = 0$ for every $n \geq 1$, then $w(x) = x^m(1-x)$ and the corresponding operators are defined by

$$R_{m,n}f(x) = \begin{cases} \sum_{k=m}^{n-1} \binom{n-m-1}{k-m} \prod_{i=1}^{k}(x\,p_i(x)) \\ \quad \times \prod_{j=1}^{n-k}((1-x)q_j(x))\,f\left(\frac{k}{n}\right), & \text{if } n > m, \\ \prod_{i=1}^{n}(x\,p_i(x))\,f(1), & \text{if } n = m, \end{cases}$$

and satisfy $\lim_{n \to +\infty} R_{m,n} f(x) = x^m\,(1-x)\,p(x)\,q(x)\,f(x)$ for every $f \in C[0,1]$ uniformly in $x \in [0,1]$.

2) Assume that for every $n \geq 1$, $\lambda_n = 1$ and $\rho_n = 0$. In this case, by (11) we have $w(x) = 1-x$ and consequently, under the assumptions of Theorem 3, the corresponding operators (see also [3, Example 2])

$$C_n^l f(x) = \sum_{k=0}^{n-1} \binom{n-1}{k} \prod_{i=1}^{k}(x\,p_i(x)) \prod_{j=1}^{n-k}((1-x)q_j(x))\,f\left(\frac{k}{n}\right)$$

satisfy $\lim_{n \to +\infty} C_n^l f(x) = (1-x)\,p(x)\,q(x)\,f(x)$ for every $f \in C[0,1]$ uniformly in $x \in [0,1]$.
Analogously, if $\lambda_n = 0$ and $\rho_n = 1$ for every $n \geq 1$, we have $w(x) = x$ and

$$C_n^r f(x) = \sum_{k=1}^{n} \binom{n-1}{k-1} \prod_{i=1}^{k}(x\,p_i(x)) \prod_{j=1}^{n-k}((1-x)q_j(x))\,f\left(\frac{k}{n}\right).$$

In this case $\lim_{n \to +\infty} C_n^r f(x) = x\,p(x)\,q(x)\,f(x)$ for every $f \in C[0,1]$ uniformly in $x \in [0,1]$.

3) If $\lambda_n = \rho_n = 1$ for every $n \geq 1$ and $q_n = 1$, we have $w(x) = 1$ $(x \in [0,1])$ and the operators defined in (3) become (see also [3, Example 1])

$$P_n f(x) = \sum_{k=0}^{n} \binom{n}{k} \prod_{i=1}^{k}(x\,p_i(x))\,(1-x)^{n-k}\,f\left(\frac{k}{n}\right).$$

By Theorem 3, we have $\lim_{n\to+\infty} P_n f = p f$ for every $f \in C[0,1]$ uniformly in $[0,1]$.
Analogously, if $p_n = 1$, we obtain the operators

$$Q_n f(x) = \sum_{k=0}^{n} \binom{n}{k} x^k \prod_{j=1}^{n-k} ((1-x)x\, q_j(x)) f\left(\frac{k}{n}\right);$$

and again by Theorem 3, $\lim_{n\to+\infty} Q_n f = q f$ ($f \in C[0,1]$).

Remark 6. Finally, we observe that a simple condition which may ensure that the sequence $(C_n)_{n\geq 1}$ is equi-bounded is the existence of a constant $M > 0$ such that

$$\prod_{i=1}^{n} p_i(x) \leq M \prod_{i=1}^{n} p_{\sigma(i)}(x), \qquad \prod_{j=1}^{n} q_j(x) \leq M \prod_{j=1}^{n} q_{\sigma(j)}(x) \qquad (19)$$

for every $n \geq 1$, $x \in [0,1]$ and every permutation σ of $\{1, \ldots, n\}$.
Indeed, in this case it is easy to show that

$$|C_n \mathbf{1}(x)| \leq M \prod_{i=1}^{n} (x\, p_i(x) + (1-x)\, q_i(x)).$$

If $(p_n)_{n\geq 1}$ and $(q_n)_{n\geq 1}$ are both increasing condition (19) is satisfied. Further, if the product $\prod_{i=1}^{n} (x\, p_i(x) + (1-x)\, q_i(x))$ converges or tends to 0 (this happens if $(x\, p_n(x) + (1-x)\, q_n(x)) \leq 1$ for every $x \in [0,1]$ and $n \geq 1$), we have the convergence of the sequence $(C_n)_{n\geq 1}$. We omit the details for brevity.

References

1. Altomare, F. and M. Campiti, *Korovkin-Type Approximation Theory and its Applications*, De Gruyter Studies in Mathematics Vol. 17, W. De Gruyter, 1994.

2. Campiti, M., General binomial coefficients in the sequence of Bernstein operators, preprint.

3. Campiti, M. and G. Metafune, Approximation properties of recursively defined Bernstein-type operators, J. Approx. Theory **87** (1996), 243–269.

4. Campiti, M. and G. Metafune, Evolution equations associated with recursively defined Bernstein-type operators, J. Approx. Theory **87** (1996), 270–290.

5. Campiti, M. and G. Metafune, Approximation of solutions of some degenerate parabolic problems, Numer. Funct. Anal. and Optimiz. **17** (1996), 23–35.

6. Knoop, K., *Theory and Application of Infinite Series*, Hafner Publishing Company, New York, 1971.

Approximation of the Inverse Frame Operator

Peter Casazza and Ole Christensen

Abstract. A frame allows every element in a Hilbert space $\mathcal{H}$ to be written as a linear combination of the frame elements, with coefficients called frame coefficients. Calculation of the frame coefficients requires inversion of an operator S on $\mathcal{H}$. We show how the inverse of S can be approximated as close as we want using linear algebra. In contrast with previous methods, our approximation can be used for any frame. Various consequences for approximation of the frame coefficients are discussed.

§1. Introduction

In all what follows, $\mathcal{H}$ denotes a separable Hilbert space with the inner product $\langle \cdot, \cdot \rangle$ linear in the first entry.

A family of elements $\{f_i\}_{i=1}^{\infty} \subseteq \mathcal{H}$ is a frame if

$$\exists A, B > 0: \quad A\|f\|^2 \leq \sum_{i=1}^{\infty} |\langle f, f_i \rangle|^2 \leq B\|f\|^2, \quad \forall f \in \mathcal{H}.$$

The numbers A, B are called *frame bounds*.

If $\{f_i\}_{i=1}^{\infty}$ is a frame, the frame operator is defined by

$$S: \mathcal{H} \to \mathcal{H}, \quad Sf = \sum_{i=1}^{\infty} \langle f, f_i \rangle f_i.$$

The series defining Sf converges unconditionally for all $f \in \mathcal{H}$ and S is a bounded, invertible, and self-adjoint operator. This leads to the frame decomposition:

$$f = SS^{-1}f = \sum_{i=1}^{\infty} \langle f, S^{-1}f_i \rangle f_i, \quad \forall f \in \mathcal{H}.$$

The possibility of representing every $f \in \mathcal{H}$ in this way is the main feature of a frame. The coefficients $\{\langle f, S^{-1}f_i \rangle\}_{i=1}^{\infty}$ are called *frame coefficients*.

For general information about frames, we refer to the original paper [6] and [4, 5, 7, 8].

Frames can equally well be considered in finite-dimensional spaces. It is easy to see that every finite collection of elements in $\mathcal{H}$ is a frame for its span. Given a frame $\{f_i\}_{i=1}^\infty$, we let $n \in \mathbb{N}$ and consider the family $\{f_i\}_{i=1}^n$, which is a frame for $\mathcal{H}_n := \text{span}\{f_i\}_{i=1}^n$ with frame operator

$$S_n : \mathcal{H}_n \to \mathcal{H}_n, \quad S_n f = \sum_{i=1}^n \langle f, f_i \rangle f_i$$

and frame decomposition $f = \sum_{i=1}^n \langle f, S_n^{-1} f_i \rangle f_i$, $f \in \mathcal{H}_n$. It can be shown that the orthogonal projection P_n of $\mathcal{H}$ onto $\mathcal{H}_n$ is given by

$$P_n f = \sum_{i=1}^n \langle f, S_n^{-1} f_i \rangle f_i, \quad f \in \mathcal{H}.$$

It is very natural to ask whether

$$\langle f, S_n^{-1} f_i \rangle \to \langle f, S^{-1} f_i \rangle, \quad \forall i \in \mathbb{N}, \ \forall f \in \mathcal{H}. \tag{1}$$

This question is important for practical implementations of frames: whereas calculation of $\langle f, S^{-1} f_i \rangle$ requires inversion of S (which can be difficult when $\mathcal{H}$ is infinite-dimensional) calculation of $\langle f, S_n^{-1} f_i \rangle$ can be done using linear algebra. In [3] it is shown that the condition (1) is equivalent to

$$\forall j \in \mathbb{N}, \ \exists c_j > 0 : \ \|S_n^{-1} f_j\| \leq c_j \ \text{for} \ n \geq j.$$

This condition is satisfied for any Riesz basis $\{f_i\}_{i=1}^\infty$, but may fail if just one element is added to a Riesz basis:

Example 1.1. Let $\{e_i\}_{i=1}^\infty$ be an orthonormal basis for $\mathcal{H}$, and define

$$f_1 := e_1, \quad f_i = e_{i-1} + \frac{1}{i} e_i, \ i \geq 2.$$

Then $\{f_i\}_{i=1}^\infty$ is a frame for $\mathcal{H}$ consisting of the single element f_1 added to the Riesz basis $\{f_i\}_{i=2}^\infty$. But it can be shown that

$$\|S_n^{-1} f_1\| = \sqrt{\sum_{i=1}^n (i!)^2} \to \infty \ \text{for} \ n \to \infty. \quad \square$$

Thus, we need a more general principle for approximation of the frame coefficients. In the next section we present a new method that works for all frames.

Approximation of the Inverse Frame Operator

§2. Approximation of S^{-1}

In the whole section we let $\{f_i\}_{i=1}^\infty$ be a frame with bounds A, B.

Lemma 2.1. *Given $n \in \mathbb{N}$, there exists a number $m(n)$ such that*

$$\frac{A}{2}\|f\|^2 \leq \sum_{i=1}^{n+m(n)} |\langle f, f_i\rangle|^2, \quad \forall f \in \mathcal{H}_n.$$

Proof: Let $n \in \mathbb{N}$. Given $\epsilon) 0$, choose a finite set of elements $g_k \in \mathcal{H}_n$ such that $\|g_k\| = 1$ for all k, and such that the balls

$$B(g_k, \epsilon) := \{f \in \mathcal{H}_n \mid \|f - g_k\| \leq \epsilon\}$$

cover the compact set $\{f \in \mathcal{H}_n \mid \|f\| = 1\}$. Since $A \leq \sum_{i=1}^\infty |\langle g_k, f_i\rangle|^2$ for all k, we can choose $m(n)$ such that

$$A\frac{2}{3} \leq \sum_{i=1}^{n+m(n)} |\langle g_k, f_i\rangle|^2, \quad \forall k.$$

Now let $f \in \mathcal{H}_n, \|f\| = 1$. Choose k such that $f \in B(g_k, \epsilon)$. By the opposite triangle inequality applied to

$$\{\langle f, f_i\rangle\}_{i=1}^{n+m(n)} = \{\langle g_k, f_i\rangle\}_{i=1}^{n+m(n)} - \{\langle g_k - f, f_i\rangle\}_{i=1}^{n+m(n)},$$

we have

$$[\sum_{i=1}^{n+m(n)} |\langle f, f_i\rangle|^2]^{1/2} \geq [\sum_{i=1}^{n+m(n)} |\langle g_k, f_i\rangle|^2]^{1/2} - [\sum_{i=1}^{n+m(n)} |\langle g_k - f, f_i\rangle|^2]^{1/2}$$

$$\geq \sqrt{A\frac{2}{3}} - \sqrt{B}\|g_k - f\| \geq \sqrt{A\frac{2}{3}} - \sqrt{B}\epsilon.$$

By choosing ϵ small enough, $\sqrt{A\frac{2}{3}} - \sqrt{B}\epsilon \geq \sqrt{\frac{A}{2}}$, from which the result follows. □

In most of what follows, we choose $m(n)$ as in Lemma 2.1. The proof of Lemma 2.1 shows that it is very easy to find $m(n)$ in practice. The next lemma shows that for every frame $\{f_i\}_{i=1}^\infty$ we can construct a family of frames "approaching $\{f_i\}_{i=1}^\infty$," which have common frame bounds.

Lemma 2.2. For any $n \in \mathbb{N}$, choose $m(n)$ as in Lemma 2.1. Then $\{P_n f_i\}_{i=1}^{n+m(n)}$ is a frame for $\mathcal{H}_n$ with bounds $\frac{A}{2}, B$. The frame operator is $P_n S_{n+m(n)} : \mathcal{H}_n \to \mathcal{H}_n$, and

$$\|P_n S_{n+m(n)}\| \leq B, \quad \|(P_n S_{n+m(n)})^{-1}\| \leq \frac{2}{A}.$$

Proof: Let $f \in \mathcal{H}_n$. Then

$$\sum_{i=1}^{n+m(n)} |\langle f, P_n f_i \rangle|^2 = \sum_{i=1}^{n+m(n)} |\langle f, f_i \rangle|^2 \geq \frac{A}{2}\|f\|^2.$$

Also,

$$\sum_{i=1}^{n+m(n)} |\langle f, P_n f_i \rangle|^2 = \sum_{i=1}^{n+m(n)} |\langle f, f_i \rangle|^2$$

$$\leq \sum_{i=1}^{\infty} |\langle f, f_i \rangle|^2 \leq B\|f\|^2.$$

So $\{P_n f_i\}_{i=1}^{n+m(n)}$ is a frame for $\mathcal{H}_n$ with the claimed bounds. The frame operator is given by

$$f \longmapsto \sum_{i=1}^{n+m(n)} \langle f, P_n f_i \rangle P_n f_i = P_n S_{n+m(n)} f, \quad f \in \mathcal{H}_n.$$

The norm estimates now follows from [2], where it is shown that the norm of the frame operator is equal to the smallest upper frame bound, and that the norm of the inverse is equal to the reciprocal of the biggest lower frame bound. □

We are now ready to prove that S^{-1} can be approximated arbitrarily close in the strong operator topology using the operators

$$(P_n S_{n+m(n)})^{-1} P_n : \mathcal{H}_n \to \mathcal{H}_n, \quad n \in \mathbb{N}.$$

Observe that $(P_n S_{n+m(n)})^{-1} P_n$ can be found using finite-dimensional methods!

Theorem 2.3. For $n \in \mathbb{N}$, choose $m(n)$ as in Lemma 2.1. Then

$$(P_n S_{n+m(n)})^{-1} P_n f \to S^{-1} f \text{ for } n \to \infty, \quad \forall f \in \mathcal{H}.$$

Proof: Let $f \in \mathcal{H}$ and define

$$\phi_n := S^{-1} f - (P_n S_{n+m(n)})^{-1} P_n f$$

$$= P_n S^{-1} f - (P_n S_{n+m(n)})^{-1} P_n f + (I - P_n) S^{-1} f.$$

Approximation of the Inverse Frame Operator 71

Since $(I - P_n)S^{-1}f \to 0$ as $n \to \infty$, it is enough to show that

$$\psi_n := P_n S^{-1} f - (P_n S_{n+m(n)})^{-1} P_n f \to 0.$$

Since $\psi_n \in \mathcal{H}_n$, we can apply the operator $P_n S_{n+m(n)}$ to get

$$\begin{aligned}\psi_n &= (P_n S_{n+m(n)})^{-1} P_n S_{n+m(n)} \psi_n \\ &= (P_n S_{n+m(n)})^{-1} P_n S_{n+m(n)} (P_n S^{-1} f - (P_n S_{n+m(n)})^{-1} P_n f) \\ &= (P_n S_{n+m(n)})^{-1} (P_n S_{n+m(n)} P_n S^{-1} f - P_n f).\end{aligned}$$

Consequently,

$$\begin{aligned}||\psi_n|| &\leq ||(P_n S_{n+m(n)})^{-1}|| \cdot ||P_n S_{n+m(n)} P_n S^{-1} f - P_n f|| \\ &\leq \frac{2}{A} ||S_{n+m(n)} P_n S^{-1} f - f|| \to 0 \text{ for } n \to \infty. \quad \square\end{aligned}$$

Thus, the choice of $m(n)$ in lemma 2.1 is enough to guarantee that the sequence of operators $(P_n S_{n+m(n)})^{-1} P_n$ converge to S^{-1} in the strong operator topology. Sometimes a slightly different formulation of the result is useful:

Corollary 2.4. Given $n \in \mathbb{N}$, choose $m(n)$ such that

$$\sum_{i=n+m(n)+1}^{\infty} |\langle f, f_i \rangle|^2 \leq \frac{A}{2} ||f||^2, \quad \forall f \in \mathcal{H}_n.$$

Then

$$(P_n S_{n+m(n)})^{-1} P_n f \to S^{-1} f \text{ for } n \to \infty, \quad \forall f \in \mathcal{H}.$$

Proof: First we prove that $m(n)$ always exists such that the condition is satisfied. Fix $n \in \mathbb{N}$ and consider the sequence of operators $Q_k : \mathcal{H}_n \to \mathcal{H}_n$, $k \in \mathbb{N}$, defined by $Q_k = P_n S - P_n S_{n+k}$. For $f \in \mathcal{H}_n$,

$$\begin{aligned}\langle Q_k f, f \rangle &= \langle (P_n S - P_n S_k) f, f \rangle \\ &= \sum_{i=1}^{\infty} |\langle f, f_i \rangle|^2 - \sum_{i=1}^{k} |\langle f, P_n f_i \rangle|^2 \\ &= \sum_{i=k+1}^{\infty} |\langle f, f_i \rangle|^2 \geq 0.\end{aligned}$$

That is, Q_k is a positive operator on $\mathcal{H}_n$. So

$$\begin{aligned}||Q_k|| &= \sup_{f \in \mathcal{H}_n, ||f||=1} \langle (P_n S - P_n S_{n+m(n)}) f, f \rangle \\ &= \sup_{f \in \mathcal{H}_n, ||f||=1} \sum_{i=k+1}^{\infty} |\langle f, f_i \rangle|^2.\end{aligned}$$

Since $\mathcal{H}_n$ is finite dimensional and $Q_k f \to 0$ for $k \to \infty$, $\forall f \in \mathcal{H}_n$, it follows that $\|Q_k\| \to 0$ for $k \to \infty$. So by choosing $m(n)$ big enough,

$$\sum_{i=n+m(n)+1}^{\infty} |\langle f, f_i \rangle|^2 \leq \frac{A}{2}\|f\|^2, \qquad \forall f \in \mathcal{H}_n.$$

The rest follows from the definition: with the above choice of $m(n)$, for all $f \in \mathcal{H}_n$ we have

$$\sum_{i=1}^{n+m(n)} |\langle f, f_i \rangle|^2 = \sum_{i=1}^{\infty} |\langle f, f_i \rangle|^2 - \sum_{i=n+m(n)+1}^{\infty} |\langle f, f_i \rangle|^2$$

$$\geq (A - \frac{A}{2})\|f\|^2 = \frac{A}{2}\|f\|^2,$$

i.e., the condition in Theorem 2.3 is satisfied. □

A "decay condition" on the inner products $\langle f_i, f_j \rangle$ makes it easy to apply Corollary 2.4:

Theorem 2.5. Let $\{f_i\}_{i=1}^{\infty}$ be a frame with lower bound A and suppose that there exists a constant $C \rangle 0$ such that

$$|\langle f_i, f_j \rangle|^2 \leq \frac{C}{1+|i-j|^2}, \qquad \forall i, j.$$

Let A_n denote a lower frame bound for $\{f_i\}_{i=1}^n$. Given $n \in \mathbb{N}$, choose $m(n)$ such that

$$\frac{\pi^2}{6} - \sum_{i=1}^{m(n)} \frac{1}{i^2} \leq \frac{A}{2C} \frac{A_n}{n}.$$

Then

$$(P_n S_{n+m(n)})^{-1} P_n f \to S^{-1} f \quad \text{for} \quad n \to \infty, \qquad \forall f \in \mathcal{H}.$$

Proof: Every $f \in \mathcal{H}_n$ can be written $f = \sum_{j=1}^{n} \langle f, S_n^{-1} f_j \rangle f_j$; thus

$$|\langle f, f_i \rangle|^2 = |\sum_{j=1}^{n} \langle f, S_n^{-1} f_j \rangle \langle f_j, f_i \rangle|^2$$

$$\leq \sum_{j=1}^{n} |\langle f, S_n^{-1} f_j \rangle|^2 \cdot \sum_{j=1}^{n} |\langle f_j, f_i \rangle|^2$$

$$\leq \frac{1}{A_n}\|f\|^2 \sum_{j=1}^{n} |\langle f_j, f_i \rangle|^2.$$

It follows that for $f \in \mathcal{H}_n$,

$$\sum_{i=n+m(n)+1}^{\infty} |\langle f, f_i \rangle|^2 \leq \frac{C}{A_n}\|f\|^2 \sum_{j=1}^{n} \sum_{i=n+m(n)+1}^{\infty} \frac{1}{1+|i-j|^2}$$

$$\leq \frac{C}{A_n}\|f\|^2 n \sum_{i=m(n)+1}^{\infty} \frac{1}{i^2} = \frac{nC}{A_n}(\frac{\pi^2}{6} - \sum_{i=1}^{m(n)} \frac{1}{i^2})\|f\|^2.$$

Approximation of the Inverse Frame Operator 73

Now apply Corollary 2.4. □

As discussed in Section 1, approximation of the frame coefficients $\langle f, S^{-1}f_i\rangle$, $f \in \mathcal{H}$ is a very important problem in frame theory. Theorem 2.3 shows that we can approximate $\langle f, S^{-1}f_i\rangle$ as close as we want using finite-dimensional methods, since

$$\langle f, (P_n S_{n+m(n)})^{-1} P_n f_i\rangle \to \langle f, S^{-1}f_i\rangle \ \ for \ \ n \to \infty, \quad \forall f \in \mathcal{H}.$$

Actually, much more is true: as $n \to \infty$, the *sequence* of coefficients $\{\langle f, (P_n S_{n+m(n)})^{-1} P_n f_i\rangle\}_{i=1}^{n+m(n)}$ converges to $\{\langle f, S^{-1}f_i\rangle\}_{i=1}^{\infty}$ in the ℓ^2-sense i.e.,

$$\sum_{i=1}^{n+m(n)} |\langle f, (P_n S_{n+m(n)})^{-1} P_n f_i\rangle - \langle f, S^{-1}f_i\rangle|^2$$
$$+ \sum_{i=n+m(n)+1}^{\infty} |\langle f, S^{-1}f_i\rangle|^2 \to 0 \ \ for \ \ n \to \infty.$$

This is the content of the following Theorem. Observe that the second term above trivially converges to 0 as $n \to \infty$, so we can concentrate on the first term.

Theorem 2.6. *For $n \in \mathbb{N}$, choose $m(n)$ as in Lemma 2.1 or Corollary 2.4. Then, for all $f \in \mathcal{H}$,*

$$\sum_{i=1}^{n+m(n)} |\langle f, (P_n S_{n+m(n)})^{-1} P_n f_i\rangle - \langle f, S^{-1}f_i\rangle|^2 \to 0 \ \ for \ \ n \to \infty.$$

Proof:

$$\sum_{i=1}^{n+m(n)} |\langle f, (P_n S_{n+m(n)})^{-1} P_n f_i\rangle - \langle f, S^{-1}f_i\rangle|^2$$
$$= \sum_{i=1}^{n+m(n)} |\langle (P_n S_{n+m(n)})^{-1} P_n f - S^{-1}f, f_i\rangle|^2$$
$$\leq B\|S^{-1}f - (P_n S_{n+m(n)})^{-1} P_n f\|^2 \to 0 \ \ for \ \ n \to \infty. \quad \square$$

Acknowledgments. The first author was supported by NSF grant 970618 and the second author by the Danish Research Council.

References

1. Casazza, P. G. and O. Christensen, Riesz frames and approximation of the frame coefficients, Approximation Theory and Its Applications, to appear.
2. Christensen, O., Frames and pseudo-inverses, J. Math. Anal. Appl. **195** (1995), 401–414.
3. Christensen, O., Frames and the projection method, Appl. Comp. Harm. Anal. **1** (1993), 50–53.
4. Daubechies, I., *Ten Lectures on Wavelets*, SIAM Conf. Series in Applied Math., Boston, 1992.
5. Daubechies, I., The wavelet transformation, time-frequency localization and signal analysis, IEEE Trans. Inform. Theory **36** (1990), 961–1005.
6. Duffin R. and A. Schaeffer, A class of nonharmonic Fourier series, Trans. Amer. Math. Soc. **72** (1952), 341–366.
7. Heil, C. and D. Walnut, Continuous and discrete wavelet transforms, SIAM Review **31** (1989), 628–666.
8. Young, R., *An Introduction to Nonharmonic Fourier Series*, Academic Press, New York, 1980.

Peter G. Casazza
Department of Mathematics
University of Missouri
Columbia
MO 65211
USA
pete@casazza.math.missouri.edu

Ole Christensen
Mathematics Institute
Building 303
Technical University of Denmark
2800 Lyngby, Denmark
lechr@mat.dtu.dk

The Bernstein Operator is the Closest Positive Operator to a Projection

Bruce L. Chalmers, Dany Leviatan, and Michael P. Prophet

Abstract. We give a geometric description of positive linear operators from $C[0,1]$ onto π_n which are supported on $\{\frac{k}{n}\}_{k=0}^n$. We then show that in the sense of this geometry, the Bernstein operator is the closest positive operator to a projection.

§1. Introduction and Preliminaries

Let X denote a Banach space and V an n-dimensional subspace of X. We will use the following notation: an n-tuple from X is to be considered a column vector while an n-tuple from X^* will be a row vector. Elements of $\mathbb{R}^n$ will be column vectors. Let $S \subset X$ denote the set of all elements that possess a specified "shape." For example, S might denote the set of convex functions or the set of monotone functions in $C[0,1]$. The problems involved with preserving the "shape" (i.e., leaving S invariant) while approximating elements of X by elements of V have been the object of much study, especially in case of best approximation (see, for example, [3, 6, 7, 8, 10, 11]). Best approximation operators that are invariant on S are, in general nonlinear, and their existence is usually not an issue. It is in the attempt to preserve a "shape" using *linear* operators that existence becomes problematic. Small variations in the "action" of a linear operator on V may greatly influence the ability of that operator to leave S invariant.

Denote by $\mathcal{B} = \mathcal{B}(X,V)$ the space of bounded linear operators from X to V. Given $P \in \mathcal{B}$, there exists $\boldsymbol{u} = (u_1, \ldots, u_n) \in (X^*)^n$ and basis $\boldsymbol{v} = (v_1, \ldots, v_n)^T \in (V)^n$ such that the representation $P = \boldsymbol{u} \otimes \boldsymbol{v} = \sum_{i=1}^n u_i \otimes v_i$ is valid, where $Pf = \sum_{i=1}^n \langle f, u_i \rangle v_i$.

Definition 1.1. *For a given $n \times n$ nonsingular matrix A, $P \in \mathcal{B}$ is said to be an A-action operator if P can be written as $P = \sum_{i=1}^n u_i \otimes v_i$ such that $(\langle v_i, u_j \rangle) = A$; i.e., $P\boldsymbol{v} = A\boldsymbol{v}$.*

Note 1.1. There is an entire equivalence class of matrices associated with a particular A-action operator. That is to say, if $P = \boldsymbol{u} \otimes \boldsymbol{v}$ is an A-action operator, then P is also an MAM^{-1}-action operator, for any nonsingular matrix M, since $P = \boldsymbol{u}M \otimes M^{-1}\boldsymbol{v}$ and $(\langle (M^{-1}\boldsymbol{v})_i, (\boldsymbol{u}M)_j \rangle) = MAM^{-1}$. In the following, it will frequently be advantageous for us to rewrite an operator's representation, as above. To this end, we will resist fixing a particular nonsingular matrix A, and instead simply refer to a given 'action' and use A to denote a representative from the equivalence class.

We will now consider the existence of A-action operators that preserve the "shape" of elements of X in the following sense (see [1, 2, and 4] for related considerations). We will take the term cone to mean a convex set, closed under nonnegative scalar multiplication. A pointed cone is a cone that contains no lines. (In the following $B(Z)$ denotes the unit ball of the Banach space Z.)

Definition 1.2. *Let S^* be a pointed weak*-closed cone in X^*. Then $f \in X$ is said to have shape (in the sense of S^*) if $\langle f, u \rangle \geq 0$ for all $u \in S^*$. Let S be the set of all elements of X with shape. (Note that S is also a cone.) Let $S_1^* = S^* \cap B(X^*)$ and let S_0^* denote the set of extreme points of S_1^* less 0. (Note that S_1^* is the closed convex hull of $S_0^* \cup \{0\}$ by the Krein-Milman theorem.) In order to emphasize the geometric flavor of our discussion, we will sometimes refer to S_0^* as "corners" of S_1^* and to $E(S^*) := \pi^{-1}(S_0^*)$ as the "edges" of the cone S^*, where $\pi(z) := z/\|z\|$. We will also say that S^* is generated by S_0^* or by $E(S^*)$ and write $S^* = \overline{\mathrm{cone}}(S_0^*)$ or $S^* = \overline{\mathrm{cone}}(E(S^*))$. Finally, we will sometimes refer to the edge of a cone as the ray generated by all positive scalar multiples of a particular nonzero element of the edge and sometimes identify such an element with the edge itself.*

Note 1.2. $f \in X$ has shape (in the sense of S^*) if and only if $\langle f, u \rangle \geq 0$, $\forall u \in S_0^*$ if and only if $\langle f, u \rangle \geq 0$, $\forall u \in S_1^*$. (S^* is the closed convex cone generated by S_0^*.)

Assumptions. We assume that S^* is total over V; that is, we assume that $S^*_{|V}$ contains n independent elements. Furthermore, we assume that $S \cap \sim (S^*)^\perp \neq \emptyset$ and that S contains at least n independent elements.

Lemma 1.1. (See [5]). *S and S^* are "dual" cones in the sense that if $\langle f, u \rangle \geq 0, \forall f \in S$, then $u \in S^*$.*

Example 1. Let $X = C[0,1]$ and let $S_0^* = \{\delta_t : t \in [0,1]\}$. Then S is the cone consisting of all nonnegative functions in $C[0,1]$.

Definition 1.3. *$P \in \mathcal{B}$ is said to be shape-preserving (in the sense of S^*) if whenever f has shape, Pf has shape (i.e., $f \in S$ implies $Pf \in S$). Denote the set of all shape-preserving A-action operators (of $\mathcal{B}$) by $\mathcal{A}_{S^*}$.*

From the above assumptions the following lemma is immediate. We will say that a basis $v_1, \ldots, v_n$ for V has shape if every basis element has shape.

Lemma 1.2. *If there does not exist a basis for V with shape, then $\mathcal{A}_{S^*} = \emptyset, \forall A$.*

Proof: We prove the contrapositive. Let $P = \sum_{i=1}^n u_i \otimes v_i \in \mathcal{A}_{S^*}$ for some A and let $\boldsymbol{f} = (f_1, \ldots, f_n)^T \in S^n$ be an n-tuple of independent elements. Then $P\boldsymbol{f} = \langle \boldsymbol{f}, \boldsymbol{u} \rangle \boldsymbol{v}$ is a basis that has shape. □

In the following we will therefore assume that V contains a basis with shape. As seen in [4], $\mathcal{A}_{S^*}$ may be empty for certain (standard) S^*, where $A = I_n$. In the following section, we attempt to characterize when $\mathcal{A}_{S^*} \neq \emptyset$.

§2. Characterization

Lemma 2.1. *Let $P \in \mathcal{B}$. Then $PS \subset S \iff P^*S^* \subset S^*$.*

Proof: The proof is an immediate consequence of the duality equation $\langle Pf, u \rangle = \langle f, P^*u \rangle$ and Lemma 1.1. □

Note 2.1. If $P = \boldsymbol{u} \otimes \boldsymbol{v} = \sum_{i=1}^n u_i \otimes v_i$ preserves shape and has range V, then from Lemma 2.1 we see that without loss (after a possible change of basis), we may assume that $u_i \in S^*$, $i = 1, \ldots, n$. This fact already gives us much insight into the make-up of shape-preserving operators; i.e., the functionals of an operator preserving shape S^* must be, without loss, in S^* themselves.

Theorem 2.1. (Characterization). (See [5]). *$\mathcal{A}_{S^*} \neq \emptyset$ if and only if there exists $\boldsymbol{u} = (u_1, \ldots, u_n) \in (S^*)^n$ such that $\boldsymbol{u} A \boldsymbol{\lambda}_u \in S^*$ for all $u \in S^*$, where A denotes an action matrix and $u_{|V} = \boldsymbol{u}_{|V} \boldsymbol{\lambda}_u$ (where $\boldsymbol{\lambda}_u$ is a (column) vector of scalars).*

Proof: ($\Rightarrow$) Suppose $P = \boldsymbol{u}' \otimes \boldsymbol{v}' = \sum_{i=1}^n u_i' \otimes v_i' \in \mathcal{A}_{S^*}$ and note that, for each $u \in S^*$, $\langle f, P^*u \rangle \geq 0$ for all $f \in S$. Then by Lemma 1.1 we have that $P^*u \in S^*$ for all $u \in S^*$. Now since $P^*u = \boldsymbol{u}'\langle \boldsymbol{v}', u \rangle \in S^*$ for $u \in S^*$, it follows that, via a change of basis, we may rewrite P as $P = \sum_{i=1}^n u_i \otimes v_i$, where $u_i \in S^*$, $i = 1, \ldots, n$. Thus, for $u \in S^*$, $\boldsymbol{u} A \boldsymbol{\lambda}_u = \boldsymbol{u} \langle \boldsymbol{v}, \boldsymbol{u} \rangle \boldsymbol{\lambda}_u = \boldsymbol{u} \langle \boldsymbol{v}, u \rangle = P^*u \in S^*$. ($\Leftarrow$) Let $\boldsymbol{v} \in V^n$ such that $\langle \boldsymbol{v}, \boldsymbol{u} \rangle = A$. Then $P = \sum_{i=1}^n u_i \otimes v_i$ is shape-preserving, since for $f \in S$ and $u \in S^*$, we have $\langle Pf, u \rangle = \langle f, P^*u \rangle = \langle f, \boldsymbol{u} A \boldsymbol{\lambda}_u \rangle \geq 0$. □

Note 2.2. The preceding characterization theorem has an interesting geometric interpretation. For a fixed cone S^*, the question of whether or not a particular action A preserves this shape is actually a question concerning the existence of subcones of S^* that have a particular set of n generators. Specifically, $\mathcal{A}_{S^*} \neq \emptyset$ if and only if there exists a subcone S_A^* of S^*, possessing n "A-cone" edges; i.e., n elements $(u_1, \ldots, u_n) = \boldsymbol{u} \in (S^*)^n$ such that $S_A^* = \{\boldsymbol{u} A \boldsymbol{\lambda}_u \mid u \in S^*\}$. It is often the case that the set of extreme points, S_0^*, of S_1^* forms an independent set (i.e., every element of S_1^* has a unique reprsentation as a (possibly infinite) nonnegative linear combination of elements from S_0^*); in this setting, it is of interest to consider $P = \boldsymbol{u} \otimes \boldsymbol{v} \in \mathcal{A}_{S^*}$ such that $\boldsymbol{u} \in (S_0^*)^n$ (see [2, 3, and 5]). It is in this case that the notion of "A-cone" edges simplifies to actual edges (in the sense that *nonnegative* linear combinations of $\{u_1, \ldots, u_n\}$ will recover S_A^*). This is expressed in the following characterization.

Corollary 2.1. *Let S_0^* be an independent set and let $\boldsymbol{u} \in (S_0^*)^n$. Then $P = \boldsymbol{u} \otimes \boldsymbol{v} \in \mathcal{A}_{S^*}$ if and only if $A\boldsymbol{\lambda}_u$ has nonnegative entries for all $u \in S^*$.*

Proof: ($\Rightarrow$) Let $u \in S^*$. Then by Theorem 2.1, $\boldsymbol{u}A\boldsymbol{\lambda}_u \in S^*$. Since $\boldsymbol{u} \in (S_0^*)^n$, the components of $\boldsymbol{u}$ are extreme points of S_1^* and thus $A\boldsymbol{\lambda}_u$ must have nonnegative entries.
($\Leftarrow$) This direction follows immediately from the proof of Theorem 2.1. □

Example 2.1. Consider the "quadratics" $V = [1, t, t^2]$ in $C[0,1]$ and let $S_0^* = \{\delta_t, t \in [0,1]\}$. Setting $u = \delta_t$, we have $u_{|V} = \boldsymbol{u}_{|V}\boldsymbol{\lambda}_u$, where

$$\boldsymbol{u} = (\delta_0, \delta_{\frac{1}{2}}, \delta_1)$$

and
$$\boldsymbol{\lambda}_u = (1 - 3t + 2t^2, 4(t - t^2), 2t^2 - t)^T. \tag{1}$$

Thus, if $A = I$, the (interpolating at t_i, $i = 1, 2, 3$) projection operator $P = \sum_{i=1}^{3} \delta_{t_i} \otimes v_i$ ($v_i(t) = \Pi_{j \neq i}(t - t_j)/\Pi_{j \neq i}(t_i - t_j)$) : $C[0,1] \to V$ does not preserve positivity if $(t_1, t_2, t_3) = (0, \frac{1}{2}, 1)$, since both the first and third elements of $\boldsymbol{\lambda}_u$ are sometimes negative on $[0,1]$. (In fact the argument works for any choice of t_i to show that there is no interpolating projection onto the quadratics which preserves positivity. Actually more is known; there is no projection onto the quadratics which preserves positivity (see, e.g., [4]).)
On the other hand if
$$A = \begin{pmatrix} 1 & 1/4 & 0 \\ 0 & 1/2 & 0 \\ 0 & 1/4 & 1 \end{pmatrix},$$
then
$$A\boldsymbol{\lambda}_u = (1 - 2t + t^2, 2(t - t^2), t^2)^T,$$

and all three $(A\boldsymbol{\lambda}_u)_i$ are always nonnegative on $[0,1]$. Thus, by Theorem 2.1, the operator $P = \sum_{i=1}^{3} u_i \otimes v_i$, where $(\langle v_i, u_j \rangle) = A$ (where $\boldsymbol{v} = ((1-t)^2, 2t(1-t), t^2)^T$), preserves positivity. (Note that P is the classical Bernstein operator onto the quadratics.)

Note 2.3. The example above illustrates the observation that in order to determine a set of action operators preserving a certain given shape, one may proceed as follows: for each $\boldsymbol{u} \in (S_0^*)^n$ consider $\Lambda_{\boldsymbol{u}} := \{\boldsymbol{\lambda}_u : u \in S_0^*\}$ and suppose $R_{\boldsymbol{u}} := \{\boldsymbol{a} : \boldsymbol{a} \cdot \boldsymbol{\lambda}_u \geq 0, \forall \boldsymbol{\lambda}_u \in \Lambda_{\boldsymbol{u}}\}$ is non-empty. Then $\mathcal{A}_{S^*}$ is not empty for any "action" matrix A whose rows are members of $R_{\boldsymbol{u}}$.

§3. Application

We now use the theory of the preceeding section to investigate the set $\mathcal{C}_n$ of those positive linear operators from $C[0,1]$ onto π_n which are supported on $\{\frac{k}{n}\}_{k=0}^{n}$. Clearly $\mathcal{C}_n$ is a cone.

For illustrative purposes we treat first the case $\mathcal{C}_2$. The following theorem describes the set $R_{\boldsymbol{u}}$ determined in Note 2.3 above.

The Bernstein Operator

Theorem 3.1. *In the case of C_2, $R_{\boldsymbol{u}}$ is a cone in $\boldsymbol{R}^3_+$ (the first octant of $\boldsymbol{R}^3$) consisting of*

$$\{(a, b, c) : a(1 - 3t + 2t^2) + 4b(t - t^2) + c(2t^2 - t) \geq 0, \quad \forall t \in [0,1]\} \quad (2)$$

and this cone is generated by the point $(0, 1, 0)$ and the curved edges (a, b, c), where

$$b = \frac{\frac{1}{2}(a+c) - \sqrt{ac}}{2}. \quad (3)$$

Proof: The proof follows from (1) and Note 2.3. Here $S_0^* = \{\delta_t,\ t \in [0,1]\}$ and $\boldsymbol{u} = (\delta_0, \delta_{\frac{1}{2}}, \delta_1)$. Defining $p(t) := a(1 - 3t + 2t^2) + 4b(t - t^2) + c(2t^2 - t)$, we see that $(p(0), p(\frac{1}{2}), p(1)) = (a, b, c)$ and thus the cone $R_{\boldsymbol{u}}$ lies in the first octant. Furthermore, setting $p(t) = r(t - t_0)^2$, for an arbitrary $r > 0$ and $t_0 \in [0,1]$, we obtain

$$(a, b, c) = r\left(t_0^2, \left(t_0 - \frac{1}{2}\right)^2, (1 - t_0)^2\right), \quad t_0 \in [0,1],$$

whence (3) follows. Finally, setting $p(t) = t(1-t)$, we obtain that $(a, b, c) = (0, 1, 0)$ defines another edge of the cone. Since $(t - t_0)^2$ and $t(1-t)$ clearly exhaust the extreme cases, we are done. $\square$

Note 3.1. The cone $R_{\boldsymbol{u}}$ above has three sharp edges containing, respectively, points making up the rows of the action matrix of the Bernstein operator B_2, with respect to the basis $\boldsymbol{v}(t) = (v_k(t))_{k=0}^2 = (\binom{2}{k}t^k(1-t)^{2-k})_{k=0}^2$ (see Example 2.1). Now to each of the points (a, b, c) in $\{e_k\}_{k=0}^2 \cup R_{\boldsymbol{u}} \subset \boldsymbol{R}^3_+$, where e_k is the $(k+1)^{th}$ standard basis vector, it is natural to associate the nonnegative quadratic polynomial spline $p_{(a,b,c)}$ of maximal smoothness which has values a, b, c at $0, 1/2, 1$, respectively. Thus we have that $p_{e_1}(t) = 4t(1-t)$ on $[0,1]$, $p_{e_0}(t) = 4(1/2-t)^2$ on $[0, 1/2]$ and $= 0$ elsewhere, $p_{e_2}(t) = 4(t-1/2)^2$ on $[1/2, 1]$ and $= 0$ elsewhere, while, for (a, b, c) in $R_{\boldsymbol{u}}$, $p_{(a,b,c)} = p(t)$, where $p(t) = rt(1-t)$ or $p(t) = r(t - s_1)^2$, some $r > 0$ and $s_1 \in [0,1]$ for all $(a, b, c) \in R_{\boldsymbol{u}}$. Thus in measuring "closeness" of e_k to $R_{\boldsymbol{u}}$, we measure closeness of p_{e_k} to $\pi R_{\boldsymbol{u}} = \{p_{(a,b,c)} : (a, b, c) \in R_{\boldsymbol{u}}\}$. Furthermore, since a primary characteristic of p_{e_k} is that it increases until it peaks at t_k and then decreases thereafter, we seek a "closest" element in the convex cone $\pi R_{\boldsymbol{u}}$ which is also co-monotone with p_{e_k}. But since such an element must reside on the boundary of $\pi R_{\boldsymbol{u}}$ with respect to any measure of closeness and since the only elements of the boundary which are co-monotone with p_{e_k} are positive scalar multiples of v_k (which peaks at $t_k = k/2$), we conclude that a positive scalar multiple of $v_k(t)$ is the unique closest such element with respect to *any measure of closeness* that respects co-monotonicity. I.e., the "closest" point on $R_{\boldsymbol{u}}$ to e_k is a positive scalar multiple of the point $(v_k(0), v_k(1/2), v_k(1))$ on the $(k+1)^{th}$ edge of the cone $R_{\boldsymbol{u}}$. We conclude that the Bernstein action is closest to a projection (with action matrix I (whose $(k+1)^{th}$ row is e_k)) among all positive operators T such that $T\mathbf{1} = \mathbf{1}$. It is interesting to observe further that when one measures

$$\text{dist}(p_{e_k}, \pi_{R_{\mathbf{u}}}) = \sup_{p \in \pi_{R_{\mathbf{u}}}} \|p_{e_k} - p\|_{w,\infty},$$

where, for example $w(t) = \frac{1}{\sqrt{t(1-t)}}$, one finds a unique b_k such that

$$\text{dist}(p_{e_k}, \pi_{R_{\mathbf{u}}}) = \|p_{e_k} - b_k\|_{w,\infty},$$

where b_k is a positive scalar multiple of v_k. This provides an example of a measurement of closeness that does not require co-monotonicity, yet still identifies the "sharp" edges of R_u as closest to the e_k's.

We now consider the case of general n.

Lemma 3.1. [9]. *Let π_n^+ denote the cone of nonnegative nth degree polynomials on $[0,1]$. $x(t) \in \pi_n^+$ has n zeros counting multiplicities in $[0,1]$ if and only if $x(t)$ is contained in an extreme ray of π_n^+.*

Theorem 3.2. *In the case of $\mathcal{C}_n$, $R_{\mathbf{u}}$ is a cone in $\mathbf{R}_+^{n+1}$ (the first orthant of $\mathbf{R}^{n+1}$) consisting of*

$$\{(a_0, a_2, ..., a_n) : a_0\lambda_0(t) + a_1\lambda_1(t) + \cdots + a_n\lambda_n(t) \geq 0 \ \forall t \in [0,1]\}, \quad (4)$$

where

$$\lambda_k(t) = (W^{-1}(1, t, ..., t^n)^T)_k = L_k(t),$$

W being the $(n+1) \times (n+1)$ Vandermonde matrix at the points $\{t_i\} = \{\frac{i}{n}\}_{i=0}^n$ and $L_k(t) = \Pi_{i \neq k}(t - t_i)/\Pi_{i \neq k}(t_k - t_i)$, the Lagrange functions with respect to $\{\frac{i}{n}\}_{i=0}^n$. Furthermore, setting $\mathbf{v} = (v_0, ..., v_n)$ with $v_k(t) = \binom{n}{k}t^k(1-t)^{n-k}$, this cone is generated by the points $(v_k(t_0), ..., v_k(t_n))$, $k = 0, ..., n$ and the curved edges

$$(a_0, a_1, ..., a_n) = (p(t_0), p(t_1), ..., p(t_n))$$

where

$$p(t) = rt^{n_1}(1-t)^{n_2}(t-s_1)^2(t-s_2)^2 \cdots (t-s_j)^2, \quad (5)$$

for an arbitrary $r > 0$ and $s_1, ..., s_j \in [0,1]$ and n_1, n_2 nonnegative integers such that $n_1 + n_2 + 2j = n$.

Proof: Setting $u = \delta_t$, we have $u_{|V} = \mathbf{u}_{|V}\boldsymbol{\lambda}_u$, where

$$\mathbf{u} = (\delta_0, \delta_{\frac{1}{n}}, ..., \delta_{\frac{n-1}{n}}, \delta_1)$$

and

$$\boldsymbol{\lambda}_u = W^{-1}(1, t, ..., t^n)^T = (\lambda_0(t), \lambda_1(t), ..., \lambda_n(t))^T.$$

Note that, since $(\boldsymbol{\lambda}_u)_i(t_j) = \delta_{ij}$, we have $(\boldsymbol{\lambda}_u)_i = L_i$. Representing $p \in \mathcal{C}_n$ by $p(t) := a_0\lambda_0(t) + a_1\lambda_1(t) + \cdots + a_n\lambda_n(t)$, we see that

$$\left(p(0), p\left(\frac{1}{n}\right), ..., p\left(\frac{n-1}{n}\right), p(1)\right) = (a_0, a_1, ..., a_n)$$

The Bernstein Operator 81

and thus the cone R_u lies in R_+^{n+1}. Furthermore, according to the above lemma, we set $p(t) = rt^{n_1}(1-t)^{n_2}(t-s_1)^2(t-s_2)^2 \cdots (t-s_j)^2$, for an arbitrary $r > 0$ and $s_1, ..., s_j \in [0,1]$, to obtain

$$(a_0, a_1, ..., a_n) = (p(t_0), p(t_1), ..., p(t_n))$$

whence (5) follows.

Finally, note that if $s_i \neq 0$ or 1, $\forall i = 1, ..., j$, then $(a_0, a_1, ..., a_n)$ defines a curved portion of the surface of the cone with continuous curvature in a direction corresponding to each s_i. A point on the surface of the cone with no directions of continuous curvature through it (except for the direction given by the point itself) will be called a point on a sharp edge of the cone. Thus, setting $p(t) = \binom{n}{k}t^k(1-t)^{n-k}$, we obtain that $(a_0, a_1, ..., a_n) = (v_k(t_0), ..., v_k(t_n))$, where $v_k(t) = \binom{n}{k}t^k(1-t)^{n-k}$, defines a sharp edge of the cone for each $k = 0, ..., n$. Since by the lemma these exhaust all the extreme cases, we are done. □

Note 3.2. The cone R_u above has $n+1$ sharp edges containing, respectively, points making up the rows of the action matrix of the Bernstein operator B_n (with respect to the basis $\{v_k(t)\}_{k=0}^n = \{\binom{n}{k}t^k(1-t)^{n-k}\}_{k=0}^n$). Now to each of the points $\boldsymbol{a} = (a_0, ..., a_n)$ in $\{e_k\}_{k=0}^n \cup R_u \subset R_+^{n+1}$, where $e_k = (0, ..., 0, 1, 0, ..., 0)$ is the $(k+1)^{th}$ standard basis vector, it is natural to associate the nonnegative n^{th}-degree polynomial spline $p_{\boldsymbol{a}}$ of maximal symmetric smoothness which has values a_i at the knots $t_i = i/n$, $i = 0, ..., n$. Thus we have that, for some $r > 0$, $p_{e_k}(t) = r(t-t_{k-1})^{[n/2]}(t_{k+1}-t)^{[n/2]}$ on $[t_{k-1}, t_{k+1}]$ and $= 0$ elsewhere, for $k = 1, ..., n-1$, $p_{e_0}(t) = r(t_1-t)^n$ on $[t_0, t_1]$ and $= 0$ elsewhere, $p_{e_n}(t) = r(t-t_{n-1})^n$ on $[t_{n-1}, t_n]$ and $= 0$ elsewhere, while, for $\boldsymbol{a}$ in R_u, $p_{\boldsymbol{a}} = p(t)$, where $p(t) = rt^{n_1}(1-t)^{n_2}(t-s_1)^2(t-s_2)^2 \cdots (t-s_j)^2$, some $r > 0$ and $s_1, ..., s_j \in [0,1]$ for all $\boldsymbol{a} \in R_u$. Thus in measuring "closeness" of e_k to R_u, we measure closeness of p_{e_k} to $\pi_{R_u} = \{p_{\boldsymbol{a}} : \boldsymbol{a} \in R_u\}$. Furthermore, since a primary characteristic of p_{e_k} is that it increases until it peaks at t_k and then decreases thereafter, we seek an element in the cone π_{R_u} which is closest to p_{e_k} with respect to any measure that respects co-monotonicity. But since such an element must reside on the boundary of π_{R_u} with respect to any measure of closeness and since the only elements of the boundary which are co-monotone with p_{e_k} are positive scalar multiples of $v_k(t)$ (which peaks at $t_k = k/n$), we conclude that a positive scalar multiple of $v_k(t)$ is the unique closest such element. I.e., the "closest" point on R_u to $e_k = (0, ..., 0, 1, 0, ..., 0)$ is a positive scalar multiple of the point $(v_k(t_0), ..., v_k(t_n))$ on the $(k+1)^{th}$ edge of the cone R_u. We conclude that the Bernstein action is closest to a projection (with action matrix I (whose $(k+1)^{th}$ row is e_k)) among all positive operators T such that $T\mathbf{1} = \mathbf{1}$. It is still under investigation whether, when one measures

$$\text{dist}(p_{e_k}, \pi_{R_u}) = \sup_{p \in \pi_{R_u}} \|p_{e_k} - p\|_{w,\infty},$$

for some natural weight w, one finds, as in the $n = 2$ case of Note 3.1, that $\text{dist}(p_{e_k}, \pi_{R_u}) = \|p_{e_k} - b_k\|_{w,\infty}$, where b_k is a positive scalar multiple of v_k.

References

1. Berens, H. and R. DeVore, A characterization of Bernstein polynomials, in *Approximation Theory III*, E. W. Cheney (ed.), Academic Press, New York 1980, pp. 213–219.
2. Chalmers, B. L., D. Leviatan, and M. P. Prophet, Optimal interpolating spaces preserving shape, J. Approx. Theory, to appear.
3. Chalmers, B. L. and G. D. Taylor, Uniform approximation with constraints, Jber. d. Dt. Math.-Verein. **81** (1979), 49–86.
4. Chalmers, B. L. and M. P. Prophet, Minimal shape-preserving projections onto Π_n, Numer. Funct. Anal. and Optimiz. **18** (1997), 507–520.
5. Chalmers, B. L. and M. P. Prophet, The existence of shape-preserving operators with a given action, Rocky Mountain J. Math., to appear.
6. DeVore, R. A., Monotone approximation by polynomials, SIAM J. Math. Anal. **8** (1977), 906–921.
7. Ditzian, Z., D. Jiang, and D. Leviatan, Shape preserving polynomial approximation in $C[0,1]$, Proc. Cambridge Phil. Soc. **112** (1992), 309–316.
8. Hu, Y. K., X. M. Yu, and D. Leviatan, Convex polynomial and spline approximation in $C[-1,1]$, Constructive Approx. **10** (1994), 31–64.
9. Karlin, S. and W. Studden, *Tchebyshev Systems: with Applications in Analysis and Statistics*, Interscience Publishers, New York, 1966.
10. Leviatan, D., Recent developments in shape preserving approximation, in *Approximation Theory, Proc. IDoMAT 1995*, M. W. Müller, M. Felten, and D. H. Mache (eds.), Akademie Verlag, Math. Research **86** (1995), 189–200.
11. Leviatan, D. and V. Operstein, Shape preserving approximation in L_p, Constr. Approx. **11** (1995), 299–320.

B. L. Chalmers
Department of Mathematics
University of California
Riverside CA 92507
blc@math.ucr.edu

D. Leviatan
School of Mathematics
Tel Aviv University
Tel Aviv, Israel 69978
leviatan@math.tau.ac.il

M. P. Prophet
Department of Mathematics
Murray State University
Murray KY 42071
mike@banach.mursuky.edu

Finite-Dimensional Action Constants

B. L. Chalmers and K. Pan

Abstract. We extend the results of [3] from projection (identity-action) constants to $\tilde{A}$-action constants, where $\tilde{A}$ is a similarity equivalence class of matrices. These action constants are (linearly) isometric invariants.

§1. Introduction

Let V_k be a k-dimensional subspace of an n-dimensional space X_n. Fix a basis $\{v_i\}_{i=1}^k$ in V_k, and let $A := (a_{ij})$ be a $k \times k$ matrix. Then denote a fixed A-action operator P_A by

$$P_A : X_n \to V_k, \quad P_A = \sum_{i=1}^k u_i \otimes v_i,$$

where $u_i \in X_n^*$ and $\langle v_i, u_j \rangle = a_{ij}$, and denote the relative A-action constant of $\{v_i\}$ in X_n by

$$\lambda_A(\{v_i\}, X_n) := \inf\{\|P_A\| : P_A \text{ is an A-action operator }\}.$$

(Note that if $A = I$ (the identity matrix), then P_I is a projection and λ_I denotes the usual (relative) projection constant.)

For a fixed basis $\{v_i\}$ in V_k, define

$$\mathcal{D} := \left\{ \begin{array}{l} D(T) \text{ is a } k \times k \text{ matrix} : T \in L(X_n),\ T(V_k) \subseteq V_k, \\ \nu(T) = 1,\ T|_{V_k} = \sum_{i=1}^k w_i \otimes v_i,\ \langle v_i, w_j \rangle =: d_{ij} \\ \text{and } D(T) = (d_{ij}) \end{array} \right\},$$

where $L(X)$ denotes all linear operators of X and ν denotes the nuclear norm.

We follow an argument in [3] which utilizes the following.

Lemma 1.1. For a fixed basis $\{v_i\}$ in V_k,
$$\lambda_A(\{v_i\}, X_n) = \sup_{D(T) \in \mathcal{D}} |\text{tr}(D(T)A)|.$$

Proof: We only have to prove the inequality "$\leq$," since the other one is immediate. With both V_k and X_n being finite dimensional, there exists an A-action $P_A : X_n \to V_k$, $P_A = \sum_{i=1}^k u_i \otimes v_i$, $\langle v_i, u_j \rangle = a_{ij}$, such that $\|P_A\| = \lambda_A(\{v_i\}, X_n)$. Consider
$$\mathcal{A} := \{S \in L(X_n) \, : \, \|S\| < \|P_A\|\}$$
and
$$\mathcal{B} := \{P \in L(X_n) \, : \, P = P_A + \sum_{i=1}^m x_i^* \otimes x_i, \, m \in \mathbf{N}, x_i^* \in V_k^\perp \subset X_n^*, \, x_i \in V_k\}.$$

Since $\mathcal{B}$ consists of all A-actions, then $\mathcal{A} \cap \mathcal{B} = O$. Moreover, $\mathcal{A}$ and $\mathcal{B}$ are convex sets in $L(X_n)$ which can be separated. Thus, by the trace duality there is $T \in L(X_n)$ such that
$$\Re(\text{tr}(TS)) \leq \|P_A\| \leq \Re(\text{tr}(TP)), \; S \in \mathcal{A}, \; P \in \mathcal{B}.$$

This implies $\|P_A\| = \text{tr}(TP_A)$ and $\nu(T) = \sup |\text{tr}(TS)|/\|S\| = 1$.

To prove that $T(V_k) \subseteq V_k$, i.e., $\langle x^*, Tx \rangle = 0$ for all $x^* \in V_k^\perp$ and $x \in V_k$, we take $P = P_A + x^* \otimes x$; then
$$\|P_A\| \leq \Re(\text{tr}(TP_A)) + \text{tr}(T(x^* \otimes x)) = \|P_A\| + \Re\langle x^*, Tx\rangle.$$

Hence $\Re\langle x^*, Tx \rangle \geq 0$ for all $x^* \in V_k^\perp$, $x \in V_k$. Since V_k, $V_k^\perp$ are linear spaces, this yields $\langle x^*, Tx \rangle = 0$.

Now we can write
$$T = \sum_{i=1}^k w_i \otimes v_i + \sum_{i=k+1}^n w_i \otimes v_i,$$
where $w_{k+1}, ..., w_n \in V_k^\perp$ and $v_{k+1}, ..., v_n$ is a basis of $X_n \backslash V_k$. So
$$TP_A = \sum_{i=1}^k w_i \otimes P_A v_i + \sum_{i=k+1}^n w_i \otimes P_A v_i.$$

Since $w_{k+1}, ..., w_n \in V_k^\perp$ and $P_A v_i \in V_k$, $i = 1, ..., n$, we have
$$\text{tr}(TP_A) = \sum_{i=1}^k \langle P_A v_i, w_i \rangle$$
$$= \sum_{i=1}^k \left\langle \sum_{j=1}^k \langle v_i, u_j \rangle v_j, w_i \right\rangle$$
$$= \sum_{i=1}^k \sum_{j=1}^k \langle v_i, u_j \rangle \langle v_j, w_i \rangle$$
$$= \text{tr}(D(T)A). \; \square$$

Finite-Dimensional Action Constants

When the basis $\{v_i\}_{i=1}^k$ changes to $\{v_i'\}_{i=1}^k$ in the sense $v_i' = \sum_{j=1}^k q_{ij} v_i$, where $Q := (q_{ij}) \in \mathcal{M}_k$ and $\mathcal{M}_k$ denotes all nonsingular $k \times k$ matrices, then $P_A = \sum_{i=1}^k u_i' \otimes v_i'$ where $\langle v_i', u_j' \rangle =: a_{ij}'$ and $A' := (a_{ij}') = QAQ^{-1}$. So, if we define

$$\tilde{A} := \{B : \text{ there is a } Q \in \mathcal{M}_k \text{ such that } B = QAQ^{-1}\},$$

then we have the following property.

Lemma 1.2. *Define*

$$\lambda_{\tilde{A}}(\{v_i\}, X_n) := \sup_{D(T) \in \mathcal{D}} \inf_{B \in \tilde{A}} |\text{tr}(D(T)B|.$$

Then

$$\lambda_{\tilde{A}}(\{v_i\}, X_n) = \lambda_{\tilde{A}}(\{v_i'\}, X_n).$$

From Lemma 1.2, we can define the relative $\tilde{A}$-action constant of V_k in X_n by

$$\lambda_{\tilde{A}}(V_k, X_n) := \lambda_{\tilde{A}}(\{v_i\}, X_n),$$

and the absolute $\tilde{A}$-action constant of V_k by

$$\lambda_{\tilde{A}}(V_k) := \sup\{\lambda_{\tilde{A}}(V_k, X) : V_k \subseteq X\},$$

which are independent of the basis. We can also define

$$\lambda_{\tilde{A}}(k, n) := \sup\{\lambda_{\tilde{A}}(V_k, X_n) : V_k \subseteq X_n\}$$

and

$$\lambda_{\tilde{A}}(k) := \sup_n \{\lambda_{\tilde{A}}(k, n)\}.$$

Note 1.1. From Lemma 1.1 it follows $\lambda_{\tilde{A}}(V_k, X_n) \leq \inf_{A \in \tilde{A}} \|P_A\|$.

Lemma 1.3. ([3]) $1 < r < \infty$, $\mathbf{K} \in \{\mathbf{R}, \mathbf{C}\}$ *and* $Z_r = (\mathbf{K}^n, \|\cdot\|_{Z_r})$, *where* $\|(\xi_i)_{i=1}^n\|_{Z_r} := \max(|\sum_{i=1}^n \xi_i|, \|(\xi_i)_{i=1}^n\|_r)$. *Then the dual norm is given by*

$$\|(\mu_i)_{i=1}^n\|_{Z_r^*} = \inf_{t \in \mathbf{K}} \{|t| + \|(\mu_i - t)_{i=1}^n\|_{r'}\}, \qquad 1/r + 1/r' = 1.$$

§2. Main Theorem

Theorem 2.1.
$$\lambda_{\tilde{A}}(k,n) \le f_{\tilde{A}}(k,n),$$

where $f_{\tilde{A}}(k,n) := \rho/n + \sqrt{(n-1)(n\beta - \rho^2)}/n$, $\rho := |\text{tr}(A)|$, $\beta := \sum_{i=1}^{k} |d_i|^2$, and d_i are the eigenvalues of A, for all $A \in \tilde{A}$.

Proof: For any $V_k \subset X_n$, by Lemma 1.2,

$$\lambda_{\tilde{A}}(V_k, X_n)\lambda_{\tilde{A}}(\{v_i\}, X_n)$$
$$= \sup_{D(T) \in \mathcal{D}} \inf_{B \in \tilde{A}} |\text{tr}(D(T)B)|$$
$$= \sup_{D(T) \in \mathcal{D}} \inf_{Q \in \mathcal{M}_k} |\text{tr}(Q^{-1}D(T)QA)|.$$

So it suffices to show $|\text{tr}(D(T)QAQ^{-1})| \le f_{\tilde{A}}(k,n)$ for all $D(T) \in \mathcal{D}$ and for some $Q \in \mathcal{M}_k$. Given any $D(T) \in \mathcal{D}$, let $\eta_1, ..., \eta_n$ be the eigenvalues of T (counted according to their multiplicity). k of them (WLoG $\eta_1, ..., \eta_k$) are the eigenvalues of $T|_{V_k} : V_k \to V_k$. Let $Q_2 \in \mathcal{M}_k$ change A to a Jordan form $Q_2^{-1}AQ_2$ with d_i in the diagonal, and also let $Q_1 \in \mathcal{M}_k$ be such that $Q_1^{-1}D(T)Q_1$ is a Jordan form for $D(T)$ with $Q = Q_1 Q_2^{-1}$. Then

$$|\text{tr}(D(T)Q_1 Q_2^{-1} A Q_2 Q_1^{-1})| = |\text{tr}(Q_1^{-1}D(T)Q_1 Q_2^{-1} A Q_2)| = \left| \sum_{i=1}^{k} \eta_i d_i \right|.$$

From [3], we know $\|(\eta_i)_{i=1}^n\|_{Z_2} \le 1$; so

$$|\text{tr}(D(T)Q_1 Q_2^{-1} A Q_2 Q_1^{-1})| \le \left| \sum_{i=1}^{k} \eta_i d_i \right| / \|(\eta_i)_{i=1}^n\|_{Z_2}$$
$$\le \|(d_1, ..., d_k, 0, ..., 0)\|_{Z_2^*}$$
$$= \inf_{\mathbf{K}} \left\{ |t| + \left(\sum_{i=1}^{k} |d_i - t|^2 + (n-k)|t|^2 \right)^{1/2} \right\}.$$

In order to get the inf, we let $g(t) = |t| + (\sum_{i=1}^{k} |d_i - t|^2 + (n-k)|t|^2)^{1/2}$ and $t = re^{i\theta}$, $d_j = \rho_j e^{i\phi_j}$. So

$$g(t) = g(r, \theta) = |r| + \left\{ \sum_{j=1}^{k} (r^2 - 2\Re[d_j \bar{t}] + \rho_j^2) + (n-k)r^2 \right\}^{1/2}$$
$$= |r| + \left\{ nr^2 - 2\Re\left[\sum_{j=1}^{k} d_j \right] \bar{t} + \sum_{j=1}^{k} \rho_j^2 \right\}^{1/2}.$$

Finite-Dimensional Action Constants 87

Also let $\text{tr}(A) =: \rho e^{i\phi}$; then $g(r,\theta) = r + \{nr^2 - 2\rho r \cos(\theta - \phi) + \beta\}^{1/2}$. Letting $g_\theta(r,\theta) = 0$, we have $\sin(\theta - \phi) = 0$ and so pick $\theta = \phi$. Letting $g_r(r,\theta) = 0$ and $\theta = \phi$, we get $(nr - \rho)^2 = (nr^2 - 2\rho r + \beta)$; solve for r: $r_0 = \rho/n \pm \sqrt{(n\beta - \rho^2)/(n-1)}/n$. Thus we have

$$\lambda_{\tilde{A}}(V_k, X_n) \leq f_{\tilde{A}}(k,n) = g(r_0, \phi) = \rho/n + \sqrt{(n-1)(n\beta - \rho^2)}/n. \quad \square$$

Theorem 2.2. $\lambda_{\tilde{A}}(V_k) \leq \sqrt{\sum_{i=1}^k |d_i|^2}$, where d_i are the eigenvalues of A for any $A \in \tilde{A}$.

Proof: $\lim_{n\to\infty} f_{\tilde{A}}(k,n) = \sqrt{\beta} = \sqrt{\sum_{i=1}^k |d_i|^2}$. $\quad \square$

§3. Observations and Conjectures

Following [2], we make the following conjecture.

Conjecture 1. *For any k-dimensional space V_k,*

$$\lambda_{\tilde{A}}(V_k) \leq f_{\tilde{A}}(k, n(k)) \text{ with } n(k) := \begin{cases} k(k+1)/2 & \mathbf{R} \\ k^2 & \mathbf{C} \end{cases}.$$

The answer is yes for $A = I$, as is shown in [4].

Now, analogously as with $\lambda_{\tilde{A}}$, we can define the relative A-action constant of V_k in X_k by

$$\lambda_A(V_k, X_n) := \inf_{B \in \tilde{A}} \sup_{D(T) \in \mathcal{D}} |\text{tr}(D(T)B|.$$

Note that $\lambda_A(V_k, X_n) = \inf_{A \in \tilde{A}} \|P_A\|$. Further, we can also analogously define the absolute A-action constant of V_k by

$$\lambda_A(V_k) := \sup\{\lambda_A(V_k, X) : V_k \subseteq X\},$$

$$\lambda_A(k,n) := \sup\{\lambda_A(V_k, X_n) : V_k \subseteq X_n\},$$

and

$$\lambda_A(k) := \sup_n \{\lambda_A(k,n)\}.$$

Recall that from Note 1.1 in all cases $\lambda_{\tilde{A}} \leq \lambda_A$ with equality if $A = I$.

Conjecture 2. *If A is positive definite, then*
i) $\lambda_A(V_k, X_n) \leq f_{\tilde{A}}(k,n)$.
ii) $\lambda_A(V_k) \leq f_{\tilde{A}}(k, n(k))$.

Note that Conjecture 2 (ii) is a strengthening of Conjecture 1. The truth of Conjecture 2 (ii) and the fact that it is best possible, for $k = 2$, $\mathbf{R}$, A diagonalizable, and V_2 an unconditional space, has been shown in [1].

It is immediate from the definition of $\lambda_{\tilde{A}}$ that if V_k is (linearly) isometric to W_k ($V_k \cong W_k$), then $\lambda_{\tilde{A}}(V_k) = \lambda_{\tilde{A}}(W_k)$; i.e., $\lambda_{\tilde{A}}$ is an isometric invariant among all k-dimensional spaces.

Conjecture 3. $\{\lambda_{\tilde{A}}\}$ provides a complete set of isometric invariants for all k-dimensional symmetric spaces; i.e., if V_k and W_k are k-dimensional symmetric spaces, then

$$\lambda_{\tilde{A}}(V_k) = \lambda_{\tilde{A}}(W_k), \text{ for all } \tilde{A} \iff V_k \cong W_k.$$

Conjecture 4. $\{\lambda_A\}$ provides a complete set of isometric invariants for all k-dimensional symmetric spaces.

References

1. Chalmers, B. and B. Shekhtman, Extension constants of unconditional two-dimensional operators, Lin. Alg. and Appl. **240** (1996), 173–182.

2. König, H., Spaces with large projection constants, Israel J. Math. **50** (1985), 181–188.

3. König, H., D. R. Lewis, and P.-K. Lin, Finite dimensional projection constants, Stud. Math. **75** (1983), 341–358.

4. König, H. and N. Tomczak-Jaegermann, Bounds for projection constants and 1-summing norms, Trans. Amer. Math. Soc. **320** (1990), 799–823.

B. L. Chalmers
Department of Mathematics
University of California
Riverside, CA 92521
blc@math.ucr.edu

K. Pan
Department of Mathematics
Barry University
Miami Shores, FL 33161
pan@buvax.barry.edu

Approximation by Sums of Exponentials to Decay Functions Using Piecewise Linear Models

Maurice G. Cox, Helen E. Joyce, and John C. Mason

Abstract. The approximation of discrete data representing a rapidly decaying function by a linear sum of n undetermined negative exponentials is a nonlinear and inherently ill-conditioned problem. A sensible approach is to predict the exponential constants by an initial algorithm, and then solve the resulting linear problem for the multipliers. An algorithm for the prediction stage based on a best approximation to the data by a piecewise linear function with free breakpoints is presented. This piecewise linear function can immediately be re-expressed as a sum of so-called R-functions. Each R-function is itself a piecewise linear function, but with only one breakpoint, where to the right of the knot the function is identically zero. A best approximation by a negative exponential on $[0, \infty)$ is found for each R-function. The algorithm has three parts: i) a procedure for determining the best piecewise linear approximant (based on dynamic programming), ii) a best approximation problem for an R-function, and iii) a linear approximation problem for a sum of exponentials. Each of these parts is described briefly, and part ii) is solved explicitly. The total algorithm is tested on a variety of problems based on sums of three exponentials, and good approximations are consistently obtained, even though the values of the negative exponential powers are not always close to the true values defined by the model data.

§1. Introduction

Consider a data set $D = \{(x_i, y_i)\}_1^N$, with $0 \leq x_1 < \cdots < x_N$, that represents a rapidly decaying function. The best approximation to the data of form

$$g(x) = \sum_{j=1}^n a_j e^{-\lambda_j x}, \tag{1}$$

with free parameters λ_j and coefficients a_j, is nonlinear and generally ill-conditioned [6]. Indeed, a wide variety of parameter sets can typically yield good approximations [4] - this is also confirmed by the numerical results at the end of this paper. An approach is therefore considered in which $\{\lambda_j\}$ are fixed in advance, and a simple but effective procedure for achieving this is described.

The latter procedure has, as one key step, the relatively simple problem of approximating in L_2 on $[0, \infty)$ the function $e^{-\lambda x}$ by what is called here an "R-function," namely a piecewise linear function with only one breakpoint, here to the right of the breakpoint the function is identically zero. This is described very briefly here, but a fuller discussion for a variety of standard decay functions $f(\lambda; x)$ and various norms is given in Cox and Mason [5], where the general case of approximating $e^{-\lambda x}$ by any number of R-functions is also discussed.

The second key step in the algorithm is the use of dynamic programming to find the best (ℓ_2) approximation to D by a PLZ-function, i.e., a piecewise linear function followed by the zero function. A PLZ-function can be expressed as a sum of R-functions, each of which may be related to a unique exponential $e^{-\lambda x}$ (as its best approximation). The resulting set of λ's provides the starting parameters for (1).

§2. Best L_2 Approximation to $e^{-\lambda x}$ by an R-function

Consider an R-function defined by

$$R(m, t; x) = m(t - x)_+, \qquad (2)$$

where $u_+ = \max(u, 0)$. R is a single-knot first-degree spline expressed in truncated power-function form [4]. Introduce the y-axis intercept,

$$c = R(m, t, 0) = mt, \qquad (3)$$

so that the linear portion of the R-function for $x \leq t$ is $y = -mx + c$.

The best L_2 approximation to $e^{-\lambda x}$ by $R(m, t; x)$ can be determined from

$$\min_{m,t} E = \int_0^t [e^{-\lambda x} - m(t - x)]^2 dx + \int_t^\infty (e^{-\lambda x})^2 dx. \qquad (4)$$

It is necessary to solve

$$\frac{\partial E}{\partial m} = \frac{\partial E}{\partial t} = 0.$$

Thus,

$$\frac{\partial E}{\partial m} = -2 \int_0^t (t - x)[e^{-\lambda x} - m(t - x)] dx = 0, \qquad (5)$$

$$\frac{\partial E}{\partial t} = -2m \int_0^t [e^{-\lambda x} - m(t - x)] dx + e^{-2\lambda t} - e^{-2\lambda t} = 0. \qquad (6)$$

Approximation by Sums of Exponentials

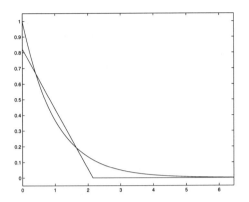

Fig. 1. The R-function and the function $e^{-\lambda x}$ it approximates for the case $\lambda = 1$.

From (5) and (6), it is found that

$$\int_0^t x e^{-\lambda x} dx = \int_0^t m x(t-x) dx,$$

and

$$\int_0^t e^{-\lambda x} dx = \int_0^t m(t-x) dx.$$

Hence,

$$\lambda^{-2}[1 - e^{-\lambda t}(1 + \lambda t)] = mt^3/6, \tag{7}$$

and

$$\lambda^{-1}(1 - e^{-\lambda t}) = mt^2/2. \tag{8}$$

By eliminating m between (7) and (8), and writing

$$T = \lambda t, X = \lambda x, M = m/\lambda, C = c, \tag{9}$$

it is found that

$$(3 + 2T)e^{-T} = 3 - T. \tag{10}$$

It is easy to see that (10) has precisely one positive zero, which may readily be obtained by Newton-Raphson (for example) as

$$T = 2.1491258... . \tag{11}$$

It is also readily found, from (3), (7), (8) and (9), that

$$C = 6/(3 + 2T) = 0.8221147..., \tag{12}$$

$$M = C/T = 0.3825345... . \tag{13}$$

Thus M, C, and T are constants independent of λ.

Note that the best approximation problem has been solved for all values of λ. The value of the intercept $c = C$ on the y-axis is independent of λ, while the values of m and t may readily be deduced from (9) in terms of λ.

The R-function and the exponential function $e^{-\lambda x}$ it approximates are shown for the case $\lambda = 1$ in Figure 1. Note that, consistent with the above comments, the figure for any other positive value of λ is identical with Figure 1 except that the x-axis scale is different.

The connection between R-functions and exponentials may be reversed. Specifically, given $R(m,t;x)$, the best L_2 approximation by the function $ae^{-\lambda x}$ is achieved with $m = aM\lambda$ and $c = aC$.

Hence,

$$a = c/C, \qquad (14)$$
$$\lambda = mC/(Mc), \qquad (15)$$

where C, M are the constants (12) and (13).

§3. Best ℓ_2 Approximation to Data by a Sum of R-functions

Our approach to determining $\{D_k\}$, and hence the approximant, follows closely that given in Section 13 of Cox [2], based on the dynamic programming method of Bellman [1], but modified to include the zero function.

The first step in the algorithm is to seek a best $(n+1)$-segment PLZ-function approximant to D of the form

$$y^* = \begin{cases} m_k^*(t_k^* - x), & \text{on } D_k = [t_{k-1}^*, t_k^*] \quad (k = 1, \ldots, n), \\ 0, & \text{on } D_{n+1} = [t_n^*, \infty), \end{cases} \qquad (16)$$

where $t_0^* = 0$ and $\{D_k\}$ is an undetermined partition of $[0, \infty)$ into a given number $n+1$ of sets corresponding to a proper partition of the abscissae $\{x_i\}$ of D.

Note that such a piecewise linear approximant is not in general continuous. However, in this case the data of concern represent a sum of decaying exponentials, which is a decreasing convex function. Hence, the approximant will consist of a sequence of straight-line segments of negative gradient, each being less negative than the previous one. Such a function will automatically be continuous, the breakpoints being defined by the intersections of adjacent segments. A continuous analogue, that of the approximation of convex functions by piecewise linear functions, has been considered by Cox [3].

Define $S_{j\ell}$ to be the sum of squares of pointwise errors of the best j-segment PLZ-function over the last ℓ data points, namely $x = z_1, \ldots, z_\ell$ where $x_i = z_{m-i+1}(i = 1, \ldots, m)$. Also let $\sigma_{h\ell}$ be the sum of squares of pointwise errors of the best straight line approximation over $z_{h+1}, \ldots, z_\ell$. ($z_1, \ldots, z_m$ have been introduced to reverse the order of the points $\{x_i\}$ to be consistent with Cox [2]. Whereas [2] recursively adds a line at the right end of the interval, a line is to be added here at the left end.) Then,

$$S_{j\ell} = \min_{2j-4 \leq h \leq \ell-2}(S_{j-1,h} + \sigma_{h\ell}), \quad \ell = 2j-2, \ldots, m-2(n-j+1); \; j = 3, \ldots, n,$$

with, finally,

$$S_{n+1,m} = \min_{2n-2 \leq h \leq m-2}(S_{nh} + \sigma_{hm}).$$

Starting values are given by

$$S_{2\ell} = \min_{0 \leq h \leq \ell-2}(\tau_h + \sigma_{h\ell}), \qquad \ell = 2, \ldots, m-2n+2,$$

Approximation by Sums of Exponentials 93

where τ_h is the sum of squares of the last h y-values. They correspond to the best lines over $z_{h+1}, \ldots, z_\ell$ and zero over $z_1, \ldots, z_h$, minimized over h. The latter formula provides an essential difference from the algorithm in Cox [2].

The complexity of the algorithm is clearly based on a triple loop and proves to be $\mathcal{O}(m^2 n)$ operations assuming $m \gg n$ and that care is taken to use updating formulae for calculating the residual sums of squares associated with least-squares lines. To determine the final explicit best approximation, it is necessary to keep track of the recursions of the developing data partitions (defined by h) and corresponding line segments $\sigma_{h\ell}$ that result from the minimizations.

Once the optimal $(n+1)$-segment PLZ-approximant with form (16) has been determined, it can be decomposed as a sum of R-functions of form (2) and (3):

$$y^*(x) = \sum_{j=1}^{n} R(m_j, t_j; x). \tag{17}$$

Defining $c_j = R(m_j, t_j; 0) = m_j t_j$, we have, from (2), (3), (16), (17),

$$m_n = m_n^*, \qquad c_n = c_n^*, \tag{18}$$
$$m_j = m_j^* - (m_{j+1} + \ldots + m_n), \tag{19}$$
$$c_j = c_j^* - (c_{j+1} + \ldots + c_n), \tag{20}$$
$$t_j = c_j/m_j, \qquad t_j^* = c_i^*/m_j^*. \tag{21}$$

Each of the R-functions may now be uniquely associated with a scaled exponential $b_j e^{-\lambda_j x}$, where, by (14) and (15),

$$b_j = c_j/C, \qquad \lambda_j = m_j C/(M c_j). \tag{22}$$

§4. Linear Exponential Fit and Numerical Results

The values of λ_j are now used as fixed exponents in the approximation of the original data set $D = \{(x_i, y_i)\}_1^N$ by

$$y^{**}(x) = \sum_{j=1}^{n} a_j e^{-\lambda_j x}. \tag{23}$$

The form (23) is linear and the ℓ_2 problem is elementary.

It remains to test the full algorithm, not only to check that $y^{**}(x)$ comes close to the data set D, but also to see whether the estimated λ_j are appropriate exponents.

To achieve this, the data are generated as discrete values of exact exponential sums,

$$y(x) = \sum_{j=1}^{n} u_j e^{-v_j x}, \tag{24}$$

and we compare not only y^{**} with y, but also a_j, λ_j with u_j, v_j. The result of various tests for a sum of $n = 3$ exponentials is shown in Table 1. The entry "rms" is the root mean square error over the data, while rms/max is rms scaled by max$|y|$ and expressed as a percentage. Good approximations, with rms/max below 1% in every case, are consistently obtained. Note that a_j, λ_j are sometimes quite close to u_j, v_j (e.g., Problem 3) but that they are not in general especially close, and this confirms the ill-conditioning of the problem. However, the spread of the λ_j is generally rather appropriate for practical purposes, and it is therefore believed that the procedure is effective.

	Problem 1	Problem 2	Problem 3	Problem 4	Problem 5	Problem 6
D	0(0.1)4	0(0.5)10	0(0.5)20	0(0.5)20	0(0.05)5	0(0.25)5
u_1	1	1	1	1	0.15	0.15
u_2	1	1	1	1	0.42	0.42
u_3	1	1	1	1	0.37	0.37
v_1	1	1	0.1	0.1	1	1
v_2	2	2	0.4	1	3	3
v_3	4	4	1.6	7	7	7
a_1	0.508	0.087	1.034	1.023	0.116	0.045
a_2	2.052	1.503	0.888	0.420	0.547	0.4054
a_3	0.444	1.410	1.098	1.563	0.286	0.491
λ_1	0.733	0.572	0.108	0.108	0.915	0.697
λ_2	2.024	1.246	0.347	0.501	2.94	1.97
λ_3	6.247	3.558	1.737	3.498	9.616	6.822
rms	0.0069	0.00053	0.0146	0.0215	0.0022	0.0021
$\frac{rms}{max}\%$	0.23	0.018	0.49	0.72	0.23	0.22

Tab. 1. Numerical results for six three-exponential problems.

§5. Summary and Properties of the Approach

The approach presented is now summarized and some of its properties given.

An approach has been presented to the problem of approximating rapidly decaying data by a sum of negative exponentials, viz.,

$$g(x) = \sum_{j=1}^{n} a_j e^{-\lambda_j x}, \quad (25)$$

where the multipliers a_j and the exponential constants λ_j are all to be determined.

Since the problem is generally inherently ill-conditioned, in that a small change in the data points tends to introduce large perturbations in the approximation parameters, the approach seeks to estimate *sensible* values for the exponential constants. The problem is then linear in the multipliers and thus readily solved for them.

The manner in which the exponential constants are determined is to approximate the data by a PLZ-function. A PLZ-function is a sum of so-called R-functions, an R-function being a function that is piecewise linear with only one breakpoint, where to the right of the breakpoint the function is identically zero. This piecewise approximation is constructed to be optimal in the sense of least squares with respect to the breakpoints and the parameters of the straight-line segments. There can exist many local minima to this nonlinear problem [2]. The global solution is ensured, however, using an algorithm based on dynamic programming which delivers the best solution that would have been obtained using full combinatorial search.

The PLZ-function so constructed is then decomposed into its constituent R-functions. This decomposition is unique and is trivially obtained.

Then, each R-function is approximated by a single decaying exponential term; this approximation is optimal in the sense of least squares.

Finally, the exponential constants so obtained are taken as defining sensible values for the data set. Moreover, the *spread* of their values can be expected to be reasonable because it relates to the spread of the breakpoints of the PLZ-function. For these constants the values of the multipliers are then determined that are best in the sense of least squares in terms of approximating the data.

It follows that the approach can be seen as "stepwise optimal" in the following sense:

a) in the approximation of the data by a PLZ-function,

b) in the approximation by single exponentials of the R-functions of which the PLZ-function is composed, and

c) in the approximation of the data by a sum of exponentials whose constants have been so estimated.

It is of course not claimed that the process is optimal overall. Indeed, the ill-conditioned nature of exponential fitting problems would indicate that an optimal solution is not necessarily desirable. The process here selects a solution from the space of solutions that are "nearby" in terms of the ℓ_2 norm (but probably not so in parameter space).

In posing the problem, it is implicitly assumed that the number n of exponential terms is specified. However, the dynamic programming approach used to determine the exponential constants essentially solves a generalization of the problem. In fact, it can be used to generate all optimal PLZ-function approximations to the data, having two or more segments. Once these approximations have been determined, the solution of the corresponding linear problems for the multipliers permits approximations by $n = 1, 2, \ldots, n_{\max}$ exponentials to be provided, where $n_{\max}$ denotes some prescribed maximum value.

Finally, a comment is made on the nature of the errors in the data being approximated. The data are used twice, once to determine the exponential constants, and then to compute the multipliers. If the data have errors which can all be regarded as drawn from a Gaussian distribution or from some other distribution that has zero mean and does not depart too far from normality, it can be expected that both of these stages will provide reasonable estimates of the corresponding parameters, relative to the conditioning of the problem. If, however, the data contain wild points, the construction of the PLZ-function could be adversely affected. For instance, a single poor data point, sufficiently removed from the other points, could result in an unduly short linear segment in the PLZ-function. Indeed, if the point is very "wild," the resulting PLZ-function might not be convex or continuous, thus invalidating the approach. It is intended that this and other aspects be investigated in further research. For data that are more reasonable, the approach tends to work very satisfactorily.

References

1. Bellman, R., *Dynamic Programming*, University Press, Princeton, 1957.
2. Cox, M. G., Curve fitting with piecewise polynomials, J.I.M.A. **8** (1971), 36–52.
3. Cox, M. G., An algorithm for approximating convex functions by means of first degree splines. Comput. J. **14** (1971), 272–275.
4. Cox, M. G., Practical spline approximation, in *Notes in Mathematics 965: Topics in Numerical Analysis*, P. R. Turner (ed.), Springer-Verlag, Berlin, 1982, pp. 79–112.
5. Cox, M. G. and J. C. Mason, Best approximation to decay functions by free-knot piecewise linear functions on $[0, \infty)$, 1998, submitted.
6. Varah, J. M., On fitting exponentials by nonlinear least squares, S.I.A.M. J. Stat. Comput. **6** (1985), 30–44.

Helen E. Joyce and John C. Mason
School of Computing and Mathematics
University of Huddersfield
Huddersfield
HD1 3DH, UK
j.c.mason@hud.ac.uk

Maurice G. Cox
National Physical Laboratory
Teddington
Middlesex
TW11 0LW, UK
mgc@npl.co.uk

A Differential Geometric Approach to Equidistributed Knots on Riemannian Manifolds

Uwe Depczynski and Joachim Stöckler

Abstract. We introduce a differential geometric approach to the numerical generation of equidistributed knots on closed compact Riemannian manifolds. The algorithms are based on minimization of a potential function by intrinsic versions of gradient or BFGS methods. Our approach finds also applications to equidistribution on spheres and is able to reproduce already known results. Numerical examples are given at the end of the paper.

§1. Introduction

The problem of finding equidistributed knot sets on a sphere has a long history. Applications of such knots can be found, e.g., in interpolation or numerical quadrature on spheres, in geophysics, meteorology, astrophysics, and elsewhere.

A common way to generate equidistributed points on spheres is the minimization of a given potential function E (see, e.g., [3, 7, 8, 11]) which is motivated by physics. In case of the 2-sphere $S^2 \subset \mathbb{R}^3$, this potential is the electromagnetic potential

$$E(x_1, \ldots, x_N) = \sum_{1 \leq i < j \leq N} \|x_i - x_j\|_2^{-1}, \qquad (1)$$

where the points can be interpreted as electrons influenced by repulsing forces. For optimal points an equilibrium position is reached (in physics, this problem is known as Thomson's problem [10]). Here, $x_i \in \mathbb{R}^3$ are the coordinates of the electrons and $\|\cdot\|_2$ is the usual Euclidean norm in $\mathbb{R}^3$. Because all electrons should stay on the sphere, we have the additional constraints $\|x_i\|_2 = 1$ with $i = 1, \ldots, N$, and minimization of E is a constrained optimization problem.

Note that in this approach the distance between two points $\|x_i - x_j\|_2$ is measured in Euclidean space $\mathbb{R}^3$.

The numerical minimization of E is usually done by mapping the sphere first on some flat 2-dimensional parameter space (e.g., on $\mathbb{R}^2$ when using the stereographic projection, or on a rectangle $[0, \pi] \times [0, 2\pi]$ when using spherical coordinates), and applying a numerical minimization scheme then. Therefore, minimization does not take place on the sphere S^2 itself, but on some flat subset of $\mathbb{R}^2$. Note that the topological structure of these parameter spaces is different from the topological structure of the sphere (e.g., the plane $\mathbb{R}^2$ is not compact and the rectangle $[0, \pi] \times [0, 2\pi]$ has a boundary, while the sphere does not). Further, some methods (like the stereographic projection) are restricted to the 2-sphere and cannot be generalized to higher dimensional spheres S^n with $n > 2$.

In our new method we avoid these problems by formulating everything intrinsically, i.e., we work in terms of Riemannian geometry. We consider the sphere no longer as a subset of a Euclidean space, but as a closed compact Riemannian manifold. Using this approach, all points stay on the manifold by definition (yielding an unconstrained optimization problem) and distances are measured *inside* the manifold. Denoting the intrinsic distance between two points p, q by $d(p, q)$, the intrinsic version of (1) reads

$$E(x_1, \ldots, x_N) = \sum_{1 \leq i < j \leq N} d(x_i, x_j)^{-1}.$$

In case of the n-sphere $S^n \subset \mathbb{R}^{n+1}$, the intrinsic distance between two points $p, q \in \mathbb{R}^{n+1}$, $\|p\|_2 = \|q\|_2 = 1$, is the *spherical distance*, which can be written $d(p, q) = \arccos(p^T \cdot q)$, with $p^T \cdot q$ the usual inner product in $\mathbb{R}^{n+1}$.

Minimization of E now requires modified intrinsic minimization schemes. In this paper, we show how to adapt the *gradient* and the *BFGS method* to the Riemannian case. It is also possible (although very technical) to construct an intrinsic *Newton method*, but we will describe this in a forthcoming paper.

Because the starting point of our method is a closed compact Riemannian manifold, we are no longer restricted to spheres. Moreover, it is possible to apply these algorithms directly to other manifolds, e.g., tori or ellipsoids, in any finite dimension.

§2. Notations

Many of our notations come from differential geometry. The reader who is not familiar with the concepts addressed below may consult [2, 4, 6].

We denote by (M, g) the Riemannian manifold M with metric tensor (first fundamental form) g and finite dimension n. For $p \in M$, the tangent space at p is written T_pM. T_pM is a linear space and has the same dimension as M. Because we restrict ourselves to real manifolds, T_pM is isomorphic to $\mathbb{R}^n$. The inner product of two vectors $v, w \in T_pM$ is written $\langle v, w \rangle := g_p(v, w)$, where g_p is the metric tensor at the point p (therefore, T_pM is in general not isometric to $\mathbb{R}^n$). The length or norm of a vector $v \in T_pM$ is $\|v\| := \sqrt{\langle v, v \rangle}$.

A Differential Geometric Approach to Equidistributed Knots

Let $\gamma : [a, b] \to M$ be a smooth path in M. Differentiating γ results in the vector field $\dot\gamma$. For every $t \in [a, b]$, we have $\dot\gamma(t) \in T_{\gamma(t)}M$. If γ is parameterized by arclength, then $\|\dot\gamma\| \equiv 1$. The length of γ is

$$L(\gamma) = \int_a^b \|\dot\gamma\|\, dt.$$

The distance $d(p, q)$ between two points $p, q \in M$ is defined as the length of the shortest path in M joining p and q. $d : M \times M \to \mathbb{R}$ is a metric in the usual sense.

Given two vector fields X and Y on M, the covariant derivative of Y in the direction X is $\nabla_X Y$. Here ∇ is the Riemannian connection. For three vector fields X, Y, Z, the Riemannian curvature tensor is $R(X, Y, Z) = \nabla_X \nabla_Y Z - \nabla_Y \nabla_X Z - \nabla_{[X,Y]} Z$, with $[X, Y] := XY - YX$ the Lie bracket. A vector field Y along the path γ is said to be parallel along γ if $\nabla_{\dot\gamma} Y \equiv 0$. The path γ is called a geodesic if $\dot\gamma$ is parallel along γ, i.e., $\nabla_{\dot\gamma} \dot\gamma \equiv 0$. Geodesics will play an important role in our approach. In Euclidean space, geodesics are straight lines; on spheres, geodesics are the great circles. Unfortunately, on most manifolds there are no explicit formulas for geodesics. But in principle they can always be calculated (at least numerically) by solving the nonlinear differential equation $\nabla_{\dot\gamma} \dot\gamma \equiv 0$. Public domain software for doing this is available in the internet [1].

If $p \in M$ and $X \in T_p M$, $\|X\| = 1$, there exists a unique geodesic γ (parameterized by arclength) with $\gamma(0) = p$ and $\dot\gamma(0) = X$. In general, γ is defined only in a neighborhood of 0, i.e., $\gamma : (-\varepsilon, \varepsilon) \to M$, with some $\varepsilon > 0$. If γ is defined in all of $\mathbb{R}$, M is said to be complete. Because we only consider closed compact Riemannian manifolds, our manifolds are always complete. Thus, we can define the exponential map at $p \in M$ as follows:

$$\exp|_p : T_p M \to M, \qquad \exp|_p(X) := \gamma_{p, \frac{X}{\|X\|}}(\|X\|),$$

where $\gamma_{p, \frac{X}{\|X\|}}$ is the unique geodesic with $\gamma_{p, \frac{X}{\|X\|}}(0) = p$ and $\dot\gamma_{p, \frac{X}{\|X\|}}(0) = \frac{X}{\|X\|}$.

In case of the n-sphere S^n, there is an explicit expression for the exponential map (see, e.g., [4], p. 117):

$$\exp|_p(X) = p\cos(\|X\|) + \frac{X}{\|X\|}\sin(\|X\|), \qquad X \neq 0.$$

§3. Potential Function

For $N \in \mathbb{N}$ points (knots) $x_1, \ldots, x_N \in M$ define the potential function

$$E(x_1, \ldots, x_N) := \sum_{1 \leq i < j \leq N} U(d(x_i, x_j)), \tag{2}$$

with $U : (0, \infty) \to \mathbb{R}$ a nonnegative, strictly decreasing, smooth function. In our experiments we used $U(r) = r^{-1}$. Further, we require that there is one and only one minimal length geodesic joining p and q (i.e., we avoid conjugated points). In case of the sphere, e.g., the north and south pole are conjugated points. In order to include conjugated points on the sphere, one has to employ a potential function which vanishes at π together with its first derivative.

Our aim is to solve the minimization problem

$$\min_{x_1,\ldots,x_N \in M} E(x_1,\ldots,x_N). \tag{3}$$

To develop numerical methods (like gradient or BFGS methods) we need the first derivative of E (the gradient). Due to linearity, it is enough to consider the first derivative of $U(d(p,q)) : M \times M \to \mathbb{R}$ (*two* variables only).

Proposition 1. *Let $p, q \in M$, $p \neq q$, two points in M which are not conjugated, $\ell := d(p,q)$, and $\gamma : [0,\ell] \to M$ the unique minimal length geodesic joining p and q, parameterized by arclength, $\gamma(0) = p$, $\gamma(\ell) = q$. With $f(p,q) := U(d(p,q))$ we have*

$$\frac{\partial f}{\partial p}(p,q) = -U'(d(p,q)) \cdot \dot{\gamma}(0) \quad \text{and} \quad \frac{\partial f}{\partial q}(p,q) = U'(d(p,q)) \cdot \dot{\gamma}(\ell).$$

Proof: For $v \in T_pM$ and $w \in T_qM$ $(v, w \neq 0)$ we denote by Z the Jacobi field along γ (*i.e.*, $\nabla_{\dot{\gamma}}\nabla_{\dot{\gamma}}Z + R(Z,\dot{\gamma},\dot{\gamma}) \equiv 0$) with $Z(p) = v$ and $Z(q) = w$. If $\varepsilon > 0$ is sufficiently small, we can define the *variation* of γ by

$$V(s,t) : [0,\ell] \times [-\varepsilon,\varepsilon] \to M, \qquad V(s,t) := \exp|_{\gamma(s)} (t \cdot Z(\gamma(s))).$$

V satisfies $V(s,0) = \gamma(s)$ for all $s \in [0,\ell]$, and

$$V(0,t) = \exp|_p(t\,v), \qquad V(\ell,t) = \exp|_q(t\,w), \qquad t \in [-\varepsilon,\varepsilon].$$

Further, we have $d(\exp|_p(t\,v),\exp|_q(t\,w)) = L(V(\cdot,t))$ with $L(V(\cdot,t))$ the length of $V(\cdot,t)$. Therefore

$$\frac{d}{dt} U\left(d(\exp|_p(t\,v),\exp|_q(t\,w))\right)|_{t=0} = U'(d(p,q)) \cdot \frac{d}{dt} L(V(\cdot,t))|_{t=0}.$$

But from [4, p. 122], we get

$$\frac{d}{dt} L(V(\cdot,t))|_{t=0} = \langle w, \dot{\gamma}(\ell) \rangle - \langle v, \dot{\gamma}(0) \rangle. \quad \square$$

We only remark, that the second derivative of $U(d(p,q))$ can be calculated in a similar way, giving the Hessian of U (which we need to develop Newton methods). Details of the computations will be given in a forthcoming paper by the authors.

Using the gradient as given in Proposition 1, we can build intrinsic gradient and BFGS methods.

A Differential Geometric Approach to Equidistributed Knots

3.1. Gradient Method

The simplest minimization procedure we can construct is the gradient method. The intrinsic version of the algorithm works as follows:

(1) Set $k = 0$ and start with N different points $x_1^{(0)}, \ldots, x_N^{(0)} \in M$ which are not conjugated.

(2) For $i = 1, \ldots, N$, compute $v_i := \dfrac{\partial E}{\partial x_i}(x_1^{(k)}, \ldots, x_N^{(k)})$.

(3) For $i = 1, \ldots, N$, compute $x_i^{(k+1)} := \exp|_{x_i^{(k)}}(h_k v_i)$, h_k stepsize.

(4) If $E(x_1^{(k+1)}, \ldots, x_N^{(k+1)})$ is not small enough, set $k := k+1$ and go to step (2).

3.2. BFGS Method

To improve the convergence rate, we modify the gradient method to obtain a BFGS method.

Before starting, we need the notation of *parallel displacement* or *pullbacks*. To define this, fix $b \in M$. If $p \in M$, $p \neq b$, $v \in T_pM$, and $\gamma : [\alpha, \beta] \to M$ the minimal length geodesic with $\gamma(\alpha) = p$ and $\gamma(\beta) = b$, then there exists one and only one parallel vector field X along γ (i.e., $\nabla_{\dot\gamma} X \equiv 0$) with $X(p) = v$. We call $v^* := X(b)$ the parallel displacement or pullback of v (from T_pM to T_bM). For $v \in T_bM$ (i.e., $p = b$) set $v^* := v$. Using this notation, we obtain the following BFGS algorithm, which is an intrinsic version of the Euclidean BFGS method (described e.g., in Stoer [9, p. 281]):

(1) Start with N different points $x_1^{(0)}, \ldots, x_N^{(0)} \in M$ which are not conjugated. Let $H^{(0)} := I$, the $(N \cdot \dim M) \times (N \cdot \dim M)$ identity matrix, b a fixed point in M, and
$$v_i^{(0)} := \frac{\partial E}{\partial x_i}(x_1^{(0)}, \ldots, x_N^{(0)}), \quad i = 1, \ldots, N.$$

(2) For $k = 0, 1, 2, \ldots$ iterate as follows:

(3)
$$\begin{pmatrix} w_1^{(k)} \\ \vdots \\ w_N^{(k)} \end{pmatrix} := H^{(k)} \cdot \begin{pmatrix} (v_1^{(k)})^* \\ \vdots \\ (v_N^{(k)})^* \end{pmatrix}$$

(4) Let $u_i^{(k)}$ be the parallel displacement of $w_i^{(k)}$ from T_bM to $T_{x_i^{(k)}}M$, $i = 1, \ldots, N$ (i.e., $(u_i^{(k)})^* = w_i^{(k)}$).

(5) Minimize (as a function of $\lambda \geq 0$)
$$E\left(\exp|_{x_1^{(k)}}(\lambda u_1^{(k)}), \ldots, \exp|_{x_N^{(k)}}(\lambda u_N^{(k)})\right),$$
and name the minimal solution $\lambda_{\min}^{(k)}$.

(6) Calculate the new points $x_i^{(k+1)} := \exp|_{x_i^{(k)}}(\lambda_{\min}^{(k)} u_i^{(k)})$, $i = 1, \ldots, N$.
If $E(x_1^{(k+1)}, \ldots, x_N^{(k+1)})$ is small enough, then stop.

(7) Let $p_i^{(k)} := \left(\lambda_{\min}^{(k)} u_i^{(k)}\right)^*$, $i = 1, \ldots, N$, and set $p^{(k)} := \begin{pmatrix} p_1^{(k)} \\ \vdots \\ p_N^{(k)} \end{pmatrix}$.

(8) Calculate $v_i^{(k+1)} := \dfrac{\partial E}{\partial x_i}(x_1^{(k+1)}, \ldots, x_N^{(k+1)})$, $i = 1, \ldots, N$.

(9) Let $q_i^{(k)} := \left(v_i^{(k+1)}\right)^* - \left(v_i^{(k)}\right)^*$, $i = 1, \ldots, N$, and $q^{(k)} := \begin{pmatrix} q_1^{(k)} \\ \vdots \\ q_N^{(k)} \end{pmatrix}$.

(10) Compute the matrix $H^{(k+1)} := \Psi(H^{(k)}, p^{(k)}, q^{(k)})$, where

$$\Psi(H,p,q) = H + \left(1 + \frac{q^T H q}{p^T q}\right) \cdot \frac{pp^T}{p^T q} - \frac{1}{p^T q}(pq^T H + H q p^T).$$

§4. Application to Spheres

Because most numerical results involve minimizing the (Euclidean) $1/r$ potential (1) of point sets on the 2-sphere S^2, we apply our method to this special case. The algorithm also works for higher dimensional spheres S^n with $n > 2$. To compare our results with already known methods, we use a modified version of our intrinsic potential function U in (2), such that $U(d(p,q)) = 1/\|p - q\|_2$, with $\|\cdot\|_2$ the Euclidean norm in $\mathbb{R}^3$.

The numerical computations were performed by a MatLab implementation of the intrinsic BFGS method. We always start with random points and perform 20 iteration steps. 192 points generated by this method are plotted in Figure 1.

We compare our results for $N = 50, 100, 192, 212, 272, 282$ points with results from a list of Hardin, Sloane, and Smith, which can be found in the internet [5]. The (conjecturally) minimal $1/r$ potentials found there are given in column "HSS" in Table 1. Our best results are listed in column "min E."

N Points	min E	HSS	Difference
50	1055.51281	1055.18231	0.0313 %
100	4449.80132	4448.35063	0.0326 %
192	16968.41162	16963.33838	0.0299 %
212	20773.49271	20768.05308	0.0262 %
272	34527.45917	34515.19329	0.0355 %
282	37160.01633	37147.29441	0.0342 %

Tab. 1. Numerical results after 20 iterations.

In Figure 2, we plotted the minimum potential E which was reached after $k = 1, \ldots, 20$ iterations in case of 192 points. As can be seen in the figure, after only a few iteration steps we are quite near the optimal minimum.

A Differential Geometric Approach to Equidistributed Knots 103

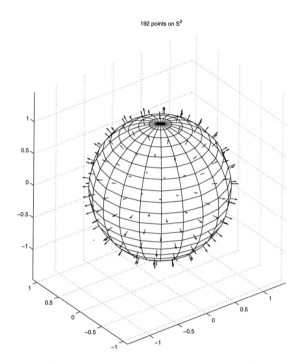

Fig. 1. 192 "equidistributed" points on S^2.

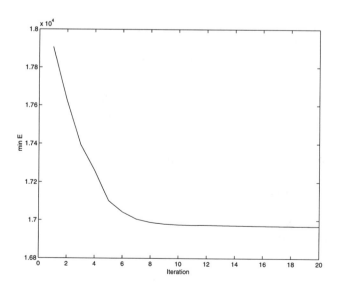

Fig. 2. Minimal potential E after $k = 1, \ldots, 20$ iterations (192 points).

Acknowledgments. Support by the Deutsche Forschungsgemeinschaft DFG is gratefully acknowledged. We also thank Bill Anderson from Elements Research for supporting us with C software for computing geodesics and exponential maps.

References

1. Anderson, W. L., geodes software, Elements Research, Charlotte, NC, Web address: http://www.netcom.com/~elements/, 1997.
2. do Carmo, M. P., *Differential Geometry of Curves and Surfaces*, New Jersey, 1976.
3. Fliege, J. and U. Maier, A two-stage approach for computing cubature formulae for the sphere, Ergebnisberichte Angewandte Mathematik Nr. 139T, Universität Dortmund, 1996.
4. Gromoll, D., W. Klingenberg, and W. Meyer, *Riemannsche Geometrie im Großen*, Lecture Notes in Mathematics 55, Springer, Berlin, 1967.
5. Hardin, R. H., N. J. A. Sloan, and W. D. Smith, Minimal energy arrangements of points on a sphere, WWW page available in the internet, http://www.research.att.com/~njas/electrons/index.html, 1994.
6. Kobayashi, S. and K. Nomizu, *Foundations of Differential Geometry*, vol. I and II, Wiley, New York, 1996.
7. Saff, E. B. and A. B. J. Kuijlaars, Distributing many points on a sphere, Math. Intelligencer **19**(1) (1997), 5–11.
8. Steinacker, J., E. Thamm, and U. Maier, Efficient integration of intensity functions on the sphere, J. Quantitative Spectroscopy and Radiative Transfer **56**(1) (1996), 97–107.
9. Stoer, J. S., *Einführung in die Numerische Mathematik I*, Springer, Berlin, 1983.
10. Thomson, J. J., Philos. Mag. **41** (1921), p. 510.
11. Zhou, Y., *Arrangements of Points on the Sphere*, Ph.D. Thesis, University of South Florida, Tampa, 1995.

Uwe Depczynski and Joachim Stöckler
Institut für Angewandte Mathematik und Statistik
Universität Hohenheim
D-70593 Stuttgart
depczyns@uni-hohenheim.de
stockler@uni-hohenheim.de

The Role of the Strong Conical Hull Intersection Property in Convex Optimization and Approximation

Frank Deutsch

Abstract. Let $\{C_0, C_1, \ldots, C_m\}$ be a collection of closed convex subsets with nonempty intersection C in a Hilbert space X. A geometric characterization is given in terms of those collections for which the Karush-Kuhn-Tucker (or Lagrange multiplier) conditions are valid for characterizing optimal minimal solutions for any convex continuous function $f : C \to \mathbb{R}$ over C. They are precisely those sets $\{C_0, C_1, \ldots, C_m\}$ which have the "strong conical hull intersection property" that was introduced in [5].

§1. Introduction

Let X be a (real) Hilbert space, C a nonempty closed convex subset of X, and $f : X \to \mathbb{R}$ a convex continuous function. One of the major problems in convex optimization theory is to *characterize* those points $x_0 \in C$ which minimize f over C:
$$f(x_0) = \min\{f(x) \mid x \in C\}.$$

The general characterization of such optimal solutions is well-known and easy to prove (see [9; Theorem 47.C, p. 391] or [1; p. 153]), and can be stated as follows.

Theorem 1.1. *(Characterization of Optimal Solutions).* A point $x_0 \in C$ minimizes f over C if and only if
$$0 \in \partial f(x_0) + (C - x_0)^0. \tag{1.1.1}$$

Here ∂f (resp., D^0) denotes the *subdifferential of f* (resp., the *dual cone of D*). That is,
$$\partial f(x_0) := \{x^* \in X \mid \langle x^*, x - x_0 \rangle + f(x_0) \leq f(x) \text{ for all } x \in X\}$$

and
$$D^0 := \{x^* \in X \mid \langle x^*, x \rangle \leq 0 \text{ for all } x \in D\}.$$

The set $(C - x_0)^0$ is also called the normal cone to C at x_0.

We are interested in what improvements or strengthening of Theorem 1.1 can be made in the case when C is the intersection of a finite number of closed convex sets C_i:

$$C = \bigcap_0^m C_i. \tag{1.1.2}$$

This is a case that often arises in practice and, at least for certain special convex subsets C_i, has been the main subject of the Karush-Kuhn-Tucker and Lagrange multiplier theories.

The main goal of this paper is to show that a certain geometric condition on $\{C_0, C_1, \ldots, C_m\}$ (called the strong conical hull intersection property, or strong CHIP for short) is the weakest property satisfied by $\{C_0, C_1, \ldots, C_m\}$ such that the Karush-Kuhn-Tucker and Lagrange multiplier conditions hold.

We note that strong CHIP was first introduced in [5] as a strengthening of a property in [3] and has already been used in approximation theory to characterize precisely when the problem of determining the best approximation to an element x from the set C can be replaced by an equivalent (but simpler) problem of determining the best approximation of a certain perturbation of x from C_0 (see [5, 6]). Also, the importance of strong CHIP for certain convex optimization problems was shown in [4]; there strong CHIP was seen to characterize when strong duality held. Finally, the special case when the C_i are given by

$$C_i = \{x \in X \mid f_i(x) \leq 0\},$$

where the f_i are convex functions on X, was considered by Bauschke, Borwein, and Li [2]. They showed that if these functions f_i satisfied the *basic constraint qualification*, or equivalently *Abadie's constraint qualification*, then $\{C_0, C_1, \ldots, C_m\}$ has strong CHIP. It follows that if the Slater point (or weak Slater point) condition holds, then $\{C_0, C_1, \ldots, C_m\}$ has strong CHIP (see [2; Corollary 5.7]).

§2. Main Result

Throughout the rest of the paper, X will denote a Hilbert space, C_i a closed convex subset for $i = 0, 1, \ldots, m$, and

$$C := \bigcap_0^m C_i \neq \emptyset. \tag{2.0.1}$$

Given any $x_0 \in C$, what is the dual cone of $C - x_0$ (i.e., the normal cone to C at x_0)? It is easy to see that

$$(C - x_0)^0 \supset \sum_{i=0}^m (C_i - x_0)^0 \tag{2.0.2}$$

Convex Optimization 107

always holds. In general, however, the reverse inclusion fails. The importance of the dual cone $(C - x_0)^0$ (see Theorem 1.1 above) provides the main motivation for the following definition.

Definition 2.1. ([5]) *The collection* $\{C_0, C_1, \ldots, C_m\}$ *is said to have the strong conical hull intersection property (or strong CHIP) at a point* $x_0 \in C = \bigcap_0^m C_i$ *iff*

$$(C - x_0)^0 = \sum_0^n (C_i - x_0)^0. \tag{2.1.1}$$

Further, $\{C_0, C_1, \ldots, C_m\}$ *is said to have* **strong CHIP** *if it has strong CHIP at each point in* C.

In [5], there were given several equivalent formulations for strong CHIP. One of them, a dual version of (2.1.1), provided motivation for the strong CHIP terminology.

Examples 2.2.
(1) ([4; Lemma 4.1]) The collection $\{C_0, C_1, \ldots, C_m\}$ has strong CHIP if the interior of $\bigcap_0^m C_i$ is nonempty.
(2) ([5; Theorem 3.6]) Every collection of polyhedral sets (e.g., closed half-spaces or closed hyperplanes) has strong CHIP.

Our main results are stated in the next two theorems. The first characterizes strong CHIP at a point, the second characterizes strong CHIP. These characterizations describe precisely when the Karush-Kuhn-Tucker or Lagrange multiplier conditions are applicable.

Theorem 2.3. *For any* $x_0 \in C = \bigcap_0^m C_i$, *the following statements are equivalent.*

(1) $\{C_0, C_1, \ldots, C_m\}$ *has strong CHIP at* x_0;
(2) *If* $f : X \to \mathbb{R}$ *is a continuous convex function, then* f *attains its minimum over* C *at* x_0 *if and only if*

$$0 \in \partial f(x_0) + \sum_{i=0}^m (C_i - x_0)^0; \tag{2.3.1}$$

(3) *If* $f : X \to \mathbb{R}$ *is a continuous affine function, then* f *attains its minimum over* C *at* x_0 *if and only if* (2.3.1) *holds.*

Proof: (1) $\Longrightarrow$ (2). Suppose $\{C_0, C_1, \ldots, C_m\}$ has strong CHIP at x_0 and $f : X \to \mathbb{R}$ is a continuous convex function. If f attains its minimum at x_0, then Theorem 1.1 implies that

$$0 \in \partial f(x_0) + (C - x_0)^0. \tag{2.3.2}$$

By strong CHIP at x_0, (2.3.2) becomes

$$0 \in \partial f(x_0) + \sum_0^m (C_i - x_0)^0 \tag{2.3.3}$$

and hence (2.3.1) holds. Conversely, if (2.3.1) holds, then

$$0 \in \partial f(x_0) + \sum_{0}^{m}(C_i - x_0)^0 \subset \partial f(x_0) + (C - x_0)^0,$$

where (2.0.2) was used for the last inclusion. By Theorem 1.1, x_0 minimizes f.

(2) $\Longrightarrow$ (3). This is obvious since each affine function is convex.

(3) $\Longrightarrow$ (1). Assume (3) holds. To prove (1), it suffices by (2.0.2) to verify that

$$(C - x_0)^0 \subset \sum_{0}^{m}(C_i - x_0)^0. \qquad (2.3.4)$$

Let $y \in (C - x_0)^0$ and define $f : X \to \mathbb{R}$ by

$$f(x) := \langle x, -y \rangle + \langle x_0, y \rangle, \qquad x \in X. \qquad (2.3.5)$$

Then f is a continuous affine function and $\partial f(x_0) = f'(x_0) = -y$. Thus

$$0 = \partial f(x_0) + y \subset \partial f(x_0) + (C - x_0)^0$$

which, using Theorem 1.1, implies that f is minimized at x_0. By the hypothesis, we have

$$0 \in \partial f(x_0) + \sum_{0}^{m}(C_i - x_0)^0.$$

Thus

$$y = -\partial f(x_0) \in \sum_{0}^{m}(C_i - x_0)^0$$

and (2.3.4) is verified. $\square$

The *global* version of Theorem 2.3 is now an immediate consequence.

Theorem 2.4. *The following statements are equivalent.*
(1) $\{C_0, C_1, \ldots, C_m\}$ *has strong CHIP;*
(2) *If* $f : X \to \mathbb{R}$ *is a continuous convex function, the minimum of f over C is attained at some $x_0 \in C$ if and only if*

$$0 \in \partial f(x_0) + \sum_{0}^{m}(C_i - x_0)^0; \qquad (2.4.1)$$

(3) *If* $f : X \to \mathbb{R}$ *is a continuous affine function, the minimum of f over C is attained at some point $x_0 \in C$ if and only if (2.4.1) holds.*

Convex Optimization   109

§3. Applications

The practical usefulness of the preceding results arises in those cases where the dual cones $(C_i - x_0)^0$ have a relatively simple representation. This will be the case, for example, when C_i is a half-space, a hyperplane, or $C_i = \{x \in X \mid f_i(x) \leq 0\}$ for a continuous convex function f_i for which $\inf_{x \in X} f_i(x) < 0$. Indeed, let $h \in X \setminus \{0\}$, $\alpha \in \mathbb{R}$,

$$H := \{x \in X \mid \langle x, h \rangle \leq \alpha\}, \quad \text{and} \quad P := \{x \in X \mid \langle x, h \rangle = \alpha\}.$$

The following proposition is well known and easy to verify.

Proposition 3.1.
(1) If $x_0 \in H$, then

$$(H - x_0)^0 = \begin{cases} \{0\} & \text{if } \langle x_0, h \rangle < \alpha \\ \{\lambda h \mid \lambda \geq 0\} & \text{if } \langle x_0, h \rangle = \alpha. \end{cases}$$

(2) If $x_0 \in P$, then

$$(P - x_0)^0 = \{\gamma h \mid \gamma \in \mathbb{R}\} = \text{span}\{h\}.$$

If $f : X \to \mathbb{R}$ is a continuous convex function, let

$$L := \{x \in X \mid f(x) \leq 0\}. \tag{3.1.1}$$

Then it is easy to verify that

$$\overline{\text{con}}\,(\partial f(x_0)) \subset (L - x_0)^0 \tag{3.1.2}$$

for each $x_0 \in L$, where $\overline{\text{con}}\,(\partial f(x_0))$ denotes the intersection of all closed convex cones which contain $\partial f(x_0)$. The inclusion (3.1.2) is strict in general. (For example, if $f(x) = \frac{1}{2}\|x\|^2$, then $L = \{0\}$ and $\partial f(x) = \{x\}$ so $\overline{\text{con}}\,(\partial f(x_0)) = \{0\}$ and $(L - x_0)^0 = (0)^0 = X$ for all $x_0 \in L$.) However, by imposing one additional condition on f which guarantees that L is *proper* (i.e., L is not the zero set of f), we can represent the dual cone $(L - x_0)^0$ rather simply.

Proposition 3.2. *Let $f : X \to \mathbb{R}$ be a continuous convex function with*

$$L := \{x \in X \mid f(x) \leq 0\} \tag{3.2.1}$$

and $\inf_{x \in X} f(x) < 0$. Then for each $x_0 \in L$,

$$(L - x_0)^0 = \begin{cases} \{0\} & \text{if } f(x_0) < 0 \\ \overline{\text{con}}\,(\partial f(x_0)) & \text{if } f(x_0) = 0 \end{cases}$$
$$= \{\lambda x^* \mid \lambda \geq 0, x^* \in \partial f(x_0),\ \lambda f(x_0) = 0\}.$$

Variants of Proposition 3.2, at least when $\dim X < \infty$, are well-known (see, e.g., [8; Theorem 23.7 and Corollary 23.7.1, p. 222] and [7; Theorem 1.3.5, p. 245]).

While there are numerous applications that one can make of Theorem 2.4 and Propositions 3.1 and 3.2, page limitations require us to limit ourselves to just two.

The first is a theorem of Karush-Kuhn-Tucker type.

Theorem 3.3. Let $f_i : X \to \mathbb{R}$ be a continuous convex function for $i = 0, 1, \ldots, m$,

$$C_i := \{x \in X \mid f_i(x) \leq 0\}, \tag{3.3.1}$$

$C := \bigcap_0^m C_i \neq \emptyset$, and suppose the Slater condition holds; i.e., there exists $\overline{x} \in X$ such that

$$f_i(\overline{x}) < 0 \qquad (i = 0, 1, \ldots, m). \tag{3.3.2}$$

Then a continuous convex function $f : X \to \mathbb{R}$ attains its minimum over C at a point $x_0 \in C$ if and only if

$$0 \in \partial f(x_0) + \sum_0^m \lambda_i \partial f_i(x_0) \tag{3.3.3}$$

for some scalars $\lambda_i \geq 0$ with $\lambda_i f_i(x_0) = 0$ for all i.

Proof: The Slater condition implies the interior of C is not empty. Then 2.2 (1) implies that $\{C_0, C_1, \ldots, C_m\}$ has strong CHIP. By Theorem 2.4 and Proposition 3.2, the result follows. □

Finally we give an application of Theorem 2.4 and Proposition 2.1 to best approximation theory. We need the following well-known characterization of best approximations.

Theorem 3.4. Let K be a closed convex subset of the Hilbert space X, $x \in X$, and $x_0 \in K$. Then $x_0 = P_K(x)$ if and only if $x - x_0 \in (K - x_0)^0$. (Here $P_K(x)$ denotes the unique nearest point in K to x.)

For the final result let C_0 be a closed convex subset of X, $a_i \in X \backslash \{0\}$, $b = (b_1, b_2, \ldots, b_m) \in \mathbb{R}^m$,

$$C_i := \{x \in X \mid \langle x, a_i \rangle = b_i\} \qquad (i = 1, 2, \ldots, m),$$

and $C := \bigcap_0^m C_i$. In other words, $C = C_0 \cap A^{-1}(b)$, where $A : X \to \mathbb{R}^m$ is defined by

$$Ax := (\langle x, a_1 \rangle, \langle x, a_2 \rangle, \ldots, \langle x, a_m \rangle).$$

Since $\{C_1, \ldots, C_m\}$ has strong CHIP by Example 2.2 (2), it follows that $\{C_0, C_1, \ldots, C_m\}$ has strong CHIP if and only if $\{C_0, A^{-1}(b)\} = \{C_0, \bigcap_1^m C_i\}$ has strong CHIP. The next result shows that strong CHIP is the precise condition which allows one to compute $P_C(x)$ by computing the best approximation to a *perturbation* of x, $x - \sum_1^m \alpha_i a_i$, from the set C_0.

Convex Optimization

Theorem 3.5. ([5; Theorem 3.2]) *The following statements are equivalent:*
(1) $\{C_0, A^{-1}(b)\}$ *has strong CHIP;*
(2) *For each* $x \in X$, $P_C(x) = P_{C_0}(x - \sum_1^m \alpha_i a_i)$ *for some scalars* α_i.

Proof: (1) $\Longrightarrow$ (2). If $\{C_0, A^{-1}(b)\}$ has strong CHIP, then for each $x \in X$, Theorem 2.4 implies that the convex function $f(y) := \frac{1}{2}\|y - x\|^2$, $y \in X$, attains its minimum over C at x_0 (i.e., $x_0 = P_C(x)$) iff

$$0 \in \partial f(x_0) + \sum_0^m (C_i - x_0)^0$$

iff

$$0 \in x_0 - x + (C_0 - x_0)^0 + \sum_1^m \text{span}(a_i)$$

iff

$$x - \sum_1^m \alpha_i a_i - x_0 \in (C_0 - x_0)^0 \quad \text{for some scalars} \quad \alpha_i$$

iff (using Theorem 3.4) $x_0 = P_{C_0}(x - \sum_1^m \alpha_i a_i)$ for some scalars α_i.
(2) $\Longrightarrow$ (1). The proof of this implication is exactly the same as that given in [5; Theorem 3.2]. Suppose (2) holds, $x_0 \in C$, and let $y \in (C - x_0)^0$. Setting $x := x_0 + y$, we see that $x - x_0 \in (C - x_0)^0$ so by Theorem 3.4, $x_0 = P_C(x)$. The hypothesis implies that $x_0 = P_{C_0}(x - \sum_1^m \alpha_i a_i)$ for some scalars α_i. By Theorem 3.4 again, $x - \sum_1^m \alpha_i a_i - x_0 \in (C_0 - x_0)^0$ implies that

$$y = x - x_0 \in (C_0 - x_0)^0 + \sum_1^m \alpha_i a_i \subset (C_0 - x_0)^0 + \sum_1^m \text{span}\{a_i\}$$
$$= (C_0 - X_0)^0 + \sum_1^m (C_i - x_0)^0,$$

where Proposition 3.1(2) was used for the last equality. Thus

$$(C - x_0)^0 \subset (C_0 - x_0)^0 + \sum_1^m (C_i - x_0)^0$$

and hence $\{C_0, C_1, \ldots, C_m\}$ has strong CHIP which, by the remarks preceding this theorem, implies $\{C_0, A^{-1}(b)\}$ has strong CHIP. $\square$

Remarks. An analogous result was also established in [6] for the (more general) situation when the constraint set is of the form

$$C_0 \cap \{x \in X \mid Ax \leq b\}.$$

It is worth mentioning, however, that in both [5] and [6], a *perturbation-type* characterization of $P_C(x)$ was *always* attained (i.e., even without strong CHIP) provided one replaces C_0 by a certain convex extremal subset of C_0.

References

1. Barbu, V. and Th. Precupanu, *Convexity and Optimization in Banach Spaces*, Sijthoff & Noordhoff, The Netherlands, 1978.
2. Bauschke, H. H., J. M. Borwein, and W. Li, Strong conical hull intersection property, bounded linear regularity, Jameson's property (G), and error bounds in convex optimization, Sept. 1997, preprint.
3. Chui, C. K., F. Deutsch, and J. D. Ward, Constrained best approximation in Hilbert space, Constr. Approx. **6** (1990), 35–64.
4. Deutsch, F., W. Li, and J. Swetits, Fenchel duality and the strong conical hull intersection property, Nov. 1997, preprint.
5. Deutsch, F., W. Li, and J. D. Ward, A dual approach to constrained interpolation from a convex subset of Hilbert space, J. Approx. Theory, **90** (1997), 385–414.
6. Deutsch, F., W. Li, and J. D. Ward, Best approximation from the intersection of finitely many closed convex subsets in Hilbert space, and the strong conical hull intersection property, March 1997, preprint.
7. Hiriart-Urruty, J.-B. and C. Lemaréchal, *Convex Analysis and Minimization Algorithms I*, Springer-Verlag, New York, 1993.
8. Rockafellar, R. T., *Convex Analysis*, Princeton Univ. Press, Princeton, N. J., 1970.
9. Zeidler, E., *Nonlinear Functional Analysis and its Applications III: Variational Methods and Optimization*, Springer-Verlag, New York, 1985.

Frank Deutsch
Department of Mathematics
The Pennsylvania State University
University Park, PA 16802
deutsch@math.psu.edu

Fractional Error Constants for Quadrature Formulas

Kai Diethelm

Abstract. For a large class of quadrature formulas, we derive error bounds in terms of fractional derivatives of the integrand function. The asymptotic order of the error constants is determined.

§1. Introduction and Main Results

A quadrature formula Q_n for the integral $\int_a^b f(x)dx$ is a linear functional of the form

$$Q_n[f] = \sum_{j=1}^n a_{jn} f(x_{jn}),$$

where the weights a_{jn} are arbitrary real numbers and the nodes x_{jn} are assumed to satisfy $a \leq x_{1n} < x_{2n} < \cdots < x_{nn} \leq b$. It is well known [1, 3] that it is possible to give error bounds of the form

$$|R_n[f]| := \left| \int_a^b f(x)dx - Q_n[f] \right| \leq c_{ps}(R_n) \left\| f^{(s)} \right\|_p \tag{1}$$

if and only if $R_n[\pi] = 0$ whenever π is a polynomial of degree $s - 1$. Here and in the following we assume that $c_{ps}(R_n)$ is the smallest possible number such that (1) holds. The constant $c_{ps}(R_n)$ is called the sth error constant of the quadrature formula Q_n with respect to the p-norm. When we consider sequences of such quadrature formulas with increasing number of nodes, it is also known that for every choice of $s \in \mathbb{N}$ and $p \in [1, \infty]$, the optimal formula $Q_n^{ps\,\text{opt}}$ satisfies the relation

$$c_{ps}(R_n^{ps\,\text{opt}}) = O(n^{-s}). \tag{2}$$

Furthermore, many sequences of quadrature formulas built according to simple construction criteria share this optimal behaviour. In particular, this is true for positive interpolatory quadrature formulas, i.e. formulas with the two properties that (i) $R_n[\pi] = 0$ whenever π is a polynomial of degree $n-1$, and (ii) $a_{jn} \geq 0$ for all $j = 1, 2, \ldots, n$ and all n (the most prominent formula in this class is the Gaussian formula), and it is also true for compound quadrature methods satisfying the above-mentioned polynomial exactness assumption (like the midpoint, trapezoidal, or Simpson's method).

In this paper, we want to investigate this problem for the case that s is not an integer. For this purpose, we have to replace the sth derivative in (1) by the fractional derivative of order s in the sense of Riemann and Liouville [6, 10], denoted and defined by

$$(D_{a+}^s f)(x) := \frac{1}{\Gamma(\sigma - s)} \frac{d^\sigma}{dx^\sigma} \int_a^x f(t)(x-t)^{\sigma-s-1} dt \qquad (3)$$

where σ is an integer greater than s.

In [4, §5] we have shown that it is possible to give error bounds similar to those of the form (1), but we have to introduce a weighted norm of the fractional derivative of f. To be precise, using the notation

$$\widehat{D}_{a+}^r f := D_{a+}^r \left((\cdot - a)^{r - \lfloor r+1 \rfloor} f \right),$$

the following result has been shown [4, Theorem 5.1]:

Theorem 1. *Let $s \in \mathbb{N}$, and $r \notin \mathbb{N}$ such that $0 < r < s$. Let R_n be the remainder of a quadrature formula with n nodes, and assume that $R_n[\pi] = 0$ whenever π is a polynomial of degree $s - 1$. Then, assuming $\widehat{D}_{a+}^{r-1} f$ is absolutely continuous on $[a, b]$ and, if $r < 1$, additionally that $\widehat{D}_{a+}^r f \in L_p(a, b)$ for some $p > 1/r$, there exists a constant $c_{pr}(R_n) < \infty$ such that*

$$|R_n[f]| \leq c_{pr}(R_n) \left\| \widehat{D}_{a+}^r f \right\|_p.$$

The question that we want to address now is: What can we say about the behaviour of the $c_{pr}(R_n)$ as $n \to \infty$? Recalling (2), the following answer is not unexpected.

Theorem 2. *Let (Q_n) be either a sequence of positive interpolatory quadrature formulas or a sequence of compound quadrature formulas exact for all polynomials of degree at least $s - 1$. Let $r \notin \mathbb{N}$ and $1 < r < s$. Then, the constants $c_{pr}(R_n)$ of Theorem 1 behave as*

$$c_{pr}(R_n) = O(n^{-r}) \qquad (4)$$

for every $p \in [1, \infty]$.

This result settles a conjecture stated in [4] in the positive sense at least for $r > 1$. The proof will be given in Section 2.

For the case $0 < r < 1$, we had the additional condition $p > 1/r$ in Theorem 1. It is known from [4] that $c_{pr}(R_n) = \infty$ if $p \leq 1/r$. Therefore, we cannot expect the full statement of Theorem 2 to hold for $0 < r < 1$. We can however show that under the restriction on p mentioned above, the bound (4) remains valid.

Theorem 3. Let Q_n be as in Theorem 2, and let $0 < r < 1$. Furthermore, assume $p > 1/r$. Then, the constants $c_{pr}(R_n)$ of Theorem 1 behave as

$$c_{pr}(R_n) = O(n^{-r}).$$

Before we come to the proofs, we note that the reasoning behind Theorem 1 does not only work for quadrature formulas. It may be extended to a very large class of approximation methods, cf. [5].

§2. Proof of Theorem 2

In the classical case of error bounds involving integer-order error constants, it is well known that the corresponding error constants can be determined with the help of Peano kernels. The same is true in our situation. Indeed, from [4, §3] we derive that

$$c_{pr}(R_n) = \left\| \widehat{K}_r(R_n; \cdot) \right\|_q, \tag{5}$$

where $1/p + 1/q = 1$ and $\widehat{K}_r(R_n; \cdot)$ is the generalized Peano kernel of R_n, given by

$$\widehat{K}_r(R_n; x) = \frac{1}{\Gamma(r)} R_n[\phi_{r,x}], \tag{6}$$

$$\phi_{r,x}(t) = (t-a)^{\lfloor r+1 \rfloor - r}(t-x)_+^{r-1}.$$

Here, $u_+ := \max\{u, 0\}$ denotes the truncated power function. From these relations, it is evident why we now concentrate on the estimation of $R_n[\phi_{r,x}]$. We recall from [4, 5] the general property $\widehat{K}_r(R_n; x) = 0$ for $x = a$ or $x = b$. We thus have to investigate $R_n[\phi_{r,x}]$ only for $a < x < b$. In this case, we split up the function $\phi_{r,x}$ according to

$$\phi_{r,x}(t) = \phi_{r,x,1}(t) + \phi_{r,x,2}(t),$$

where for $k \in \{1, 2\}$,

$$\phi_{r,x,k}(t) := \frac{1}{2}(t-a)^{\lfloor r+1 \rfloor - r} |t-x|^{r-1} (\operatorname{sgn}(t-x))^k.$$

The behaviour of the sequences $(R_n[\phi_{r,x,k}])_n$ for fixed k has been investigated by Petras [8, 9] from whose results we may deduce that uniformly for the x under consideration,

$$|R_n[\phi_{r,x,k}]| \leq C \|K_s\|_\infty^{r-s+1} \|K_{s-1}\|_\infty^{s-r}. \tag{7}$$

Here C is an absolute constant, and K_s and K_{s-1} are the classical Peano kernels of order s and $s-1$ of the functional R_n, respectively. These are known to exist because of our assumption on the polynomial degree of exactness of

R_n. Indeed, relation (7) may be shown in almost the same way as the part (a) of the Corollary in Section 2 of [9].

Now, for compound quadrature, it is well known [1, Chapter V] that $\|K_\mu\|_\infty = O(n^{-\mu})$ for $\mu \in \{s-1, s\}$. For positive interpolatory quadrature methods, this follows from [2, Theorem 2]. Combining this estimate with (6) and (7), we derive

$$c_{pr}(R_n) = \left\|\widehat{K}_r(R_n;\cdot)\right\|_q \leq \frac{(b-a)^{1/q}}{\Gamma(r)} \sup_{a \leq x \leq b} |R_n[\phi_{r,x}]| = O(n^{-r}).$$

The proof of Theorem 2 is thus complete. □

§3. Proof of Theorem 3

The proof of Theorem 3 is also based on relations (5) and (6). However, in order to estimate the relevant expressions, we now use the bound

$$|R_n[\phi_{r,x}]| \leq \|K_1\|_\infty \left(\phi_{r,x}(x_{j+1,n}) + \int_{x_{j+1,n}}^b |\phi'_{r,x}(t)|\,dt\right) + \int_{x_{jn}}^{x_{j+1,n}} |\phi_{r,x}(t)|\,dt$$

that follows from part (a) of the Lemma in [9]. Here, x_{jn} and $x_{j+1,n}$ are the nodes of the quadrature formula Q_n satisfying $x_{jn} < x < x_{j+1,n}$, where without loss of generality, we assume $x_{0n} = a$ and $x_{n+1,n} = b$. As already mentioned in the proof of Theorem 2, for the quadrature formulas under consideration, we have $\|K_1\|_\infty = O(n^{-1})$. Furthermore, an explicit calculation immediately reveals that $\phi'_{r,x}$ is negative on the interval $(x_{j+1,n}, b)$. Thus,

$$|R_n[\phi_{r,x}]| \leq \frac{C_1}{n}\phi_{r,x}(x_{j+1,n}) + \int_x^{x_{j+1,n}} \phi_{r,x}(t)dt. \qquad (8)$$

Here and in the following, by $C_1, C_2, \ldots$ we denote constants depending neither on n nor on x. Recalling the definition of $\phi_{r,x}$, we estimate the last summand on the right-hand side by

$$\int_x^{x_{j+1,n}} \phi_{r,x}(t)dt \leq (b-a)^{1-r} \int_x^{x_{j+1,n}} (t-x)^{r-1} dt \leq C_2 (x_{j+1,n} - x_{jn})^r. \qquad (9)$$

For our quadrature formulas, we have the relation

$$\max_j (x_{j+1,n} - x_{jn}) = O(n^{-1}). \qquad (10)$$

In the case of compound methods, this is an immediate consequence of their construction; for the positive interpolatory rules, it may be deduced from the above-mentioned estimate $\|K_1\|_\infty = O(n^{-1})$. Combining (8), (9), and (10), we derive

$$|R_n[\phi_{r,x}]| \leq \frac{C_1(x_{j+1,n}-a)^{1-r}}{n}(x_{j+1,n}-x)^{r-1} + \frac{C_3}{n^r} \leq \frac{C_4}{n}(x_{j+1,n}-x)^{r-1} + \frac{C_3}{n^r}.$$

Fractional Error Constants

Thus, by equations (5) and (6),

$$c_{pr}(R_n) \le \frac{C_5}{n}\left(\sum_{j=0}^{n}\int_{x_{jn}}^{x_{j+1,n}}(x_{j+1,n}-x)^{q(r-1)}dx\right)^{1/q} + O(n^{-r}).$$

Under our assumptions on the relation between r and p (and thus r and q), the integrals exist, and they can easily be determined explicitly, giving

$$c_{pr}(R_n) \le \frac{C_6}{n}\left(\sum_{j=0}^{n}(x_{j+1,n}-x_{jn})^{1+q(r-1)}\right)^{1/q} + O(n^{-r}).$$

We combine this bound with (10) and obtain

$$c_{pr}(R_n) \le \frac{C_7}{n}\left(\sum_{j=0}^{n} n^{-1-q(r-1)}\right)^{1/q} + O(n^{-r}) = \frac{C_7}{n}n^{1-r} + O(n^{-r})$$

which is our desired result. $\square$

§4. Remarks

4.1. The Implied Constants

We have derived the order of convergence of the fractional error constants for a large class of quadrature methods. However, the precise values of the constants implicitly contained in the O-terms are still unknown. For the classical (integer order) error constants, these values are known, cf. [1, 3] for the compound methods and [7] for the positive interpolatory methods. It seems that a straightforward generalization of these expressions does not lead to the correct results. For example, consider the midpoint formula where we have, for $r \in \{1,2\}$,

$$c_{pr}(R_n) = \frac{(b-a)^r}{2^{2r-1}}\left(\frac{b-a}{rq+1}\right)^{1/q} n^{-r}$$

where $1 \le p \le \infty$ and $1/p + 1/q = 1$. A comparison with the numerical values of [4, §5] reveals that this relation is not true if r is not an integer.

4.2. Centering of the Riemann-Liouville Derivatives

In eq. (3), we have introduced the Riemann-Liouville derivatives with respect to the point a, i.e. the left end point of the interval of integration. A completely symmetric theory can of course be constructed when this point is replaced by the opposite end point b.

4.3. The Singularity of the Argument of $\widehat{D}_{a+}^r$

Recalling the definition of the differential operator $\widehat{D}_{a+}^r$, we see that the function f is multiplied by a function with an algebraic singularity at a. Then we used this differential operator in connection with quadrature formulas being exact for polynomials of degree $\lfloor r \rfloor$. Thus, we may remove possible problems arising from this singularity as follows. Let T be the $\lfloor r \rfloor$th degree Taylor polynomial of f centered at a. By construction, $R_n[f] = R_n[f - T]$, so we may replace f in all our considerations by $f - T$. Since $f(x) - T(x) = o((x-a)^{-\lfloor r \rfloor})$ as $x \to a$, the strength of the singularity is reduced. Depending on the particular value of r, it may even be removed completely.

References

1. Braß, H., *Quadraturverfahren*, Vandenhoeck & Ruprecht, Göttingen, 1977.

2. Braß, H., Bounds for Peano kernels, in *Numerical Integration IV*, H. Braß and G. Hämmerlin (eds.), Birkhäuser, Basel, 1993, pp. 39–55.

3. Davis, P. J. and P. Rabinowitz, *Methods of Numerical Integration*, 2nd Edition, Academic Press, Orlando, 1984.

4. Diethelm, K., Peano kernels of non-integer order, Z. Anal. Anwendungen **16** (1997), 727–738.

5. Diethelm, K., A fractional version of the Peano-Sard theorem, Numer. Funct. Anal. Optim. **18** (1997), 745–757.

6. Oldham, K. B. and J. Spanier, *The Fractional Calculus*, Academic Press, New York, 1974.

7. Petras, K., Asymptotic behaviour of Peanokernels of fixed order, in *Numerical Integration III*, H. Braß and G. Hämmerlin (eds.), Birkhäuser, Basel, 1988, pp. 186–198.

8. Petras, K., Asymptotics for the remainder of a class of positive quadratures for integrands with an interior singularity, Numer. Math. **65** (1993), 121–133.

9. Petras, K., On the integration of functions having singularities, Z. Angew. Math. Mech. **75** (1995), S655–S656.

10. Samko, S. G., A. A. Kilbas, and O. I. Marichev, *Fractional Integrals and Derivatives: Theory and Applications*, Gordon and Breach, Yverdon, 1993.

Kai Diethelm
Institut für Mathematik
Universität Hildesheim
Marienburger Platz 22
31141 Hildesheim, Germany
diethelm@informatik.uni-hildesheim.de
http://www.informatik.uni-hildesheim.de/~diethelm

Generalized Bernstein-Erdős Conjecture

P. Dragnev, D. Legg, and D. Townsend

Abstract. We introduce a generalization of the Lebesgue function and Lebesgue constant and investigate some properties of these quantities. In this relation, a generalization of the Bernstein-Erdős conjecture is considered, and quantitative results for $n = 2$ and $n = 3$ are derived.

§1. Generalized Lebesgue Function - Definition and Properties

In this section we introduce the problem and the notations used to present our results. Let $n \leq m$ be two positive integers and let $\tau_m = \{a = t_0 < t_1 < \cdots < t_m = b\}$ be a partition of the interval $[a, b]$. Define the class

$$\mathcal{P}_n(\tau_m) := \{p \in \mathcal{P}_n : |p(t_i)| \leq 1, \ i = 0, \ldots, m\}, \tag{1}$$

where $\mathcal{P}_n$ denotes the class of polynomials of degree at most n. When $n \leq m$ the quantity $\|p(x)\|_{\tau_m} := \max_i |p(t_i)|$ defines a norm over the space $\mathcal{P}_n$ of polynomials of degree n. So then $\mathcal{P}_n(\tau_m)$ is the unit ball in $\mathcal{P}_n$ with the above norm.

Definition 1. *The* generalized Lebesgue function *(GLF) of order n with nodes τ_m is defined by*

$$B_n(x, \tau_m) := \max\{p(x) \ : \ p \in \mathcal{P}_n(\tau_m)\}, \quad x \in [a, b] \tag{2}$$

and subsequently, the generalized Lebesgue constant *is given by*

$$B_n(\tau_m) := \|B_n(x, \tau_m)\|_\infty, \tag{3}$$

where $\| \cdot \|_\infty$ denotes the usual sup norm on $[a, b]$.

We note that in the case of $m = n$, the function above coincides with the classical Lebesgue function. Indeed, if $\tau_n = \{a = x_0 < x_1 < \cdots < x_n = b\}$, then

$$\Lambda_n(x, \tau_n) := \sum_{i=0}^{n} |l_{i,n}(x)|,$$

where $l_{i,n}(x) = w(x)/(x-x_i)w'(x_i)$, $i = 0, 1, \ldots, n$, and $w(x) = (x-x_0)(x-x_1)\ldots(x-x_n)$. If $p \in \mathcal{P}_n(\tau_n)$, then by the Lagrange interpolation formula

$$L_n(p,x) = p(x) = \sum_{i=0}^{n} p(x_i) l_{i,n}(x),$$

we derive that $|p(x)| \leq \Lambda_n(x, \tau_n)$. Since on every subinterval $[t_{i-1}, t_i]$, the Lebesgue function is a polynomial, we obtain

$$\Lambda_n(x, \tau_n) = \max\{p(x) : p \in \mathcal{P}_n(\tau_n)\}.$$

Next we list some properties of $B_n(x, \tau_m)$ that resemble the properties of the classical Lebesgue function.

Properties of the GLF:

(i) $B_n(x, \tau_m)$ is a piecewise-polynomial function, namely $B_n(x, \tau_m) = p_i(x)$ for $x \in [t_{i-1}, t_i]$, $i = 1, 2, \ldots, m$.
(ii) $B_n(x, \tau_m) \geq 1$ with equality only at the nodes t_i, $i = 0, 1, \ldots, m$.
(iii) $B_n(x, \tau_m)$ has precisely one maximum point, say s_i, on every subinterval (t_{i-1}, t_i), $i = 1, 2, \ldots, m$. We denote the maximum values by $b_i(\tau_m) = p_i(s_i)$.

Definition 2. *Let $p \in \mathcal{P}(\tau_m)$. We mark a node $t_i \in \tau_m$ with "+" if $p(t_i) = 1$, and with "−" if $p(t_i) = -1$. A maximal string of consecutive nodes with only positive or unmarked nodes, starting and ending with positive nodes, is called a positive group. Similarly we define negative groups.*

We now prove an important property of the extremal polynomials.

Proposition 3. *Let $x \in (t_{i-1}, t_i)$ be fixed and p_x be an extremal polynomial, i.e., $p_x \in \mathcal{P}_n(\tau_m)$ and $p_x(x) = B_n(x, \tau_m)$ for this fixed x. Then p_x has at least n distinct groups with $\{t_{i-1}, t_i\}$ being a positive group.*

Proof: It is clear from a compactness argument that if x is a fixed point in (t_{i-1}, t_i), there will exist an extremal polynomial p_x for this point. Suppose such an extremal polynomial has k distinct groups (which must alternate in sign). Since p_x has degree at most n, $k \leq n+1$. Suppose $k < n$. Then we can choose numbers $u_1 < u_2 < \cdots < u_{k-1}$ which separate the groups and $\epsilon = \pm 1$ such that $q(t) := \epsilon(t - u_1)(t - u_2) \cdots (t - u_{k-1})$ has sign opposite to that of the groups. We now distinguish three cases.

Case 1: If x lies between two nodes in a positive group, then the polynomial

$$r(t) := p_x(t) + \delta q(t)(t - t_{i-1})(t - t_i) \tag{4}$$

is in $\mathcal{P}_n(\tau_m)$ for sufficiently small $\delta > 0$, and $r(x) > p_x(x)$, which is a contradiction to the choice of p_x.

Case 2: If x lies between two nodes in a negative group, then the same contradiction could be derived by considering

$$r(t) = p_x(t) + \delta q(t). \tag{5}$$

Case 3: If x is between two distinct groups, then we can choose the u_i node which separates these two groups, so that $q(x) > 0$ and consider the polynomial in Case 2.

Thus, we proved our assumption wrong and therefore $k \geq n$. But observe that cases 2 and 3 work also if $k = n$ or $k = n+1$, so the only possibility is $k \geq n$ and x belongs to a positive group. Since in this group we could have only one single hump (because of the degree of p_x), and $p_x(t_s) \leq 1$, $s = i - 1, i$, we derive that the group starts at t_{i-1} and ends at t_i, which proves the proposition. □

As a corollary of this proposition we could derive the following property (iv) of the GLF, first proved by Schonhage [5] for equidistant nodes τ_m, and later proved by Micchelli and Rivlin [4] for general nodes. Both proofs follow different lines than ours.

(iv) *The GLF is the lower envelope of all Lebesgue functions of $n + 1$ points out of the set of nodes τ_m, i.e.*

$$B_n(x, \tau_m) = \min\{\Lambda_n(x, \tau'_n) : \tau'_n \subset \tau_m \text{ and } |\tau'_n| = n+1\}. \tag{6}$$

Proof: Indeed, for fixed x, by Proposition 3 we can choose $n - 1$ nodes in addition to t_{i-1} and t_i, such that $p_x(t_{i-1}) = p_x(t_i) = 1$ and p_x alternately changes sign at the other nodes. Denote these $n - 1$ nodes together with t_{i-1} and t_i by τ'_n. Then $p_x(t) = \Lambda_n(t, \tau'_n)$ for $t \in [t_{i-1}, t_i]$. On the other hand if $y \in (t_{i-1}, t_i)$ is another point, then p_y will be bounded by one at τ'_n, and therefore $p_y(t) \leq \Lambda_n(t, \tau'_n)$ for any $t \in [t_{i-1}, t_i]$. Thus $p_y(y) \leq p_x(y)$ and by the definition of p_y, we get $p_y(y) = p_x(y) = \Lambda_n(y, \tau'_n)$. But this is possible only if $p_y(t) = \Lambda_n(t, \tau'_n)$. Since the Lebesgue function is unique, we get $p_x \equiv p_y$.

Now let $\tau'_n \subset \tau_m$, and $|\tau'_n| = n+1$. Let $x \in [a,b]$ be fixed. Without loss of generality we can assume that $x \notin \tau_m$. Then there is an extremal polynomial p_x. Since p_x is bounded at the nodes of τ'_n, we have $p_x(x) \leq \Lambda_n(x, \tau'_n)$, thus proving that $B_n(x, \tau_m) = p_x(x) \leq \min\{\Lambda_n(x, \tau'_n)\}$. We already showed that equality holds for some choice of τ'_n. This proves the property. □

§2. Bernstein-Erdős Conjecture and its Generalization

The characterization of the optimal choice of nodes τ_n that minimize the Lebesgue constant was a long standing open question known as the Bernstein-Erdős conjecture. It was settled by Kilgore [3], and de Boor and Pinkus [1] in 1978, namely that the optimal solution is the unique configuration τ_n^* for which $\lambda_1 = \lambda_2 = \cdots = \lambda_n = \lambda(\tau_n^*)$, where

$$\lambda_i = \lambda_i(\tau_n) := \max_{x \in [x_{i-1}, x_i]} \Lambda_n(\tau_n, x),$$

and for any other configuration $\tau_n \neq \tau_n^*$ we have

$$\min \lambda_i(\tau_n) < \lambda(\tau_n^*).$$

The following formula is of major importance in the proof:

$$\frac{\partial \lambda_i}{\partial x_j} = -p_i'(x_j) \cdot \prod_{j \neq k} \frac{(s_i - x_k)}{(x_j - x_k)}, \qquad (7)$$

where p_i is the polynomial which coincides with $\Lambda_n(\tau_n, x)$ on $[x_{i-1}, x_i]$, and s_i is the point in (x_{i-1}, x_i) where p_i achieves its maximum. Then the authors show that

$$J_k = \det(\frac{\partial \lambda_i}{\partial x_j})_{i \neq k} \neq 0 \qquad \text{for every } k \qquad (8)$$

by showing that any $(n-1)$ polynomials of $\{q_i = p_i'(x)/(x - s_i)\}_{i=1}^n$ are linearly independent. Then from the implicit function theorem they derive $\lambda_1^* = \lambda_2^* = \cdots = \lambda_n^*$.

With regards to the similarity between the GLF and $\Lambda_n(x, \tau_n)$, one of the authors stated in [2] the generalization of the Bernstein-Erdős conjecture.

Conjecture. Let

$$b_i = b_i(\tau_m) := \max_{[t_{i-1}, t_i]} B_n(x, \tau_m), \qquad i = 1, 2, \ldots, m. \qquad (9)$$

Then there exists unique partition τ_m^*, such that

$$B_{n,m} := B_n(\tau_m^*) = \min_{\tau_m} B_n(\tau_m),$$

and the partition is uniquely determined by the condition

$$b_1(\tau_m^*) = b_2(\tau_m^*) = \cdots = b_m(\tau_m^*) = B_{n,m}. \qquad (10)$$

Moreover, for any other partition $\tau_m \neq \tau_m^*$, we have

$$\min_i b(\tau_m) < B_{n,m} < \max_i b(\tau_m). \qquad (11)$$

In Section 3 we present numerical evidence that supports this conjecture. However, the conjecture seems to be very difficult to prove. One of the reasons is that, unlike the classical case $m = n$, the local maximums $b_i = b_i(t_1, \ldots, t_{m-1})$ may not be differentiable functions of t_j. The continuity and differentiability is discussed in the theorem below. In what follows let $\tau_m^{j,\epsilon} := \{t_0, t_1, \ldots, t_j + \epsilon, \ldots, t_m\}$ (in particular we have $\tau_m^{j,0} = \tau_m$).

Generalized Bernstein-Erdős Conjecture

Theorem 4. Let $\mathcal{B}_i(\tau_m) := \{t_k : |p_i(t_k)| = 1, \ k = 0, 1, \ldots, m\}$ and $\beta_{i,j}^{\pm} = \lim_{\epsilon \to 0^{\pm}} \mathcal{B}_i(\tau_m) \cap \mathcal{B}_i(\tau_m^{j,\epsilon})$. Then $b_i(\tau_m) = b_i(t_1, \ldots, t_{m-1}), i = 1, \ldots, m$ are continuous functions in t_j, $j = 1, \ldots, m - 1$. Moreover, the one-sided partials exist, in particular, if $|\beta_{i,j}^{\pm}| = n$, then

$$\left(\frac{\partial b_i}{\partial t_j}\right)_{\pm} = \lim_{\epsilon \to 0^{\pm}} \frac{b_i(\tau_m^{j,\epsilon}) - b_i(\tau_m)}{\epsilon} = -p_i'(t_j) \prod_{t_k \in \beta_{i,j}^{\pm}} \frac{s_i - t_k}{t_j - t_k}, \qquad (12)$$

and if $|\beta_{i,j}^{\pm}| > n$, then $\left(\dfrac{\partial b_i}{\partial t_j}\right)_{\pm} = 0$ (for example if $t_j \notin \mathcal{B}_i$).

Proof: The continuity can be easily derived from property (iv) and the continuity with respect to the nodes of the Lebesgue function. Therefore, we shall focus on the partial derivatives. Let $1 \leq j \leq m - 1$ be fixed and let us consider the case $\epsilon \to 0^+$ (the other case is similar). By the alternation property, $\mathcal{B}_i$ has at least $n+1$ elements for any partition of nodes, and in particular for $\tau_m^{j,\epsilon}$. Then $|\beta_{i,j}^+| \geq n$. For the extremal polynomials $p_i^{j,\epsilon} := p_i(\tau_m^{j,\epsilon}, x)$ associated with $\tau_m^{j,\epsilon}$, we have $p_i^{j,\epsilon} \to p_i$ uniformly on $[a, b]$. Therefore, there is $\epsilon_0 > 0$, such that for any $0 < \epsilon < \epsilon_0$ we have $\mathcal{B}_i(\tau_m^{j,\epsilon}) \setminus \{t_j + \epsilon\} \subset \mathcal{B}_i$. Thus, $\beta_{i,j}^+ = \mathcal{B}_i(\tau_m^{j,\epsilon}) \setminus \{t_j + \epsilon\}$. Assume $|\beta_{i,j}^+| \geq n + 1$. Then $p_i^{j,\epsilon}(t_l) = p_i(t_l)$ for all $t_l \in \beta_{i,j}^+$, which means that $p_i^{j,\epsilon} = p_i$ for all $\epsilon < \epsilon_0$. This implies $\left(\dfrac{\partial b_i}{\partial t_j}\right)_{+} = 0$. If $|\beta_{i,j}^+| = n$, then $p_i^{j,\epsilon}(x) = \Lambda_n(\tau_n^{j,\epsilon}, x)$ for $x \in [t_{i-1}, t_i]$, where $\tau_n^{j,\epsilon} = \beta_{i,j}^+ \cup \{t_j + \epsilon\}$. Since $\tau_n^{j,\epsilon} \to \tau_n := \tau_n^{j,0}$, we obtain (12) from the corresponding result (7) for the Lebesgue function. □

We now derive the generalized Bernstein-Erdős conjecture for $n = 2$.

Theorem 5. Let $n = 2$. Then

$$B_{2,m} = \frac{1}{2}(l + \frac{1}{l}), \qquad (13)$$

where, if $m = 2k$ then $l = 2^{1/k}$, and if $m = 2k + 1$ then l is the solution of $l^{1/k} = 1 + l$. The optimal nodes τ_m^* are

$$t_s = -t_{m-s} = 1 - \frac{2}{l^s}, \quad s = 0, 1, \ldots, \left[\frac{m}{2}\right]. \qquad (14)$$

Proof: First, we prove that if τ_m^* is an extremal partition, i.e., one that minimizes $B_2(\tau_m)$, then $b_1(\tau_m^*) = \cdots = b_m(\tau_m^*)$. Among all extremal partitions (there is at least one), choose $\widehat{\tau_m}$ to be with smallest possible number k of local maxima $b_i(\widehat{\tau_m}) = B_{2,m}$. Assume that $k < m$. The GLF is a parabola p_i on every subinterval $[t_{i-1}, t_i]$. Therefore, by the alternation property $p_i(t_{i-1}) = p_i(t_i) = 1$, and $p_i(-1) \geq -1$, $p_i(1) \geq -1$, with equality in at least one of the inequalities. The maximum b_i is achieved at the vertex $s_i = (t_{i-1} + t_i)/2$. If $s_i < 0$, then $p_i(-1) > -1$ and $p_i(1) = -1$, and if $s_i > 0$, then $p_i(-1) = -1$ and $p_i(1) > -1$. Finally, if $s_i = 0$, then $p_i(-1) = p_i(1) = -1$. By the assumption that $k < m$, there are two adjacent single humps, with one being maximal $(= B_{2,m})$, and the other $< B_{2,m}$, say $b_{i-1} < b_i = B_{2,m}$. But then for $\epsilon > 0$ small enough, the partition $\tau_m^\epsilon = \{t_0 < \cdots < t_{i-1} < t_i + \epsilon < t_{i+1} < \cdots < t_m\}$ will have the same humps, except for $b_{i-1} < b_{i-1}^\epsilon < b_i^\epsilon < b_i$. This is a contradiction with the minimality of k. Thus, $k = m$ which proves the equioscillation property of the GLF of order two for the extremal partition τ_m^*.

Now we derive easily the uniqueness of τ_m^*. Indeed, there is unique t_1 such that $p_1(-1) = p_1(t_1) = 1$, $p_1(1) = -1$, and $p_1((-1+t_1)/2) = B_{2,m}$. Then we uniquely determine t_2 and so on. Note that the uniqueness of τ_m^* implies also the symmetry of this partition.

We finally obtain the quantitative result. Let $B_{2,m} = K$ and $l_i := t_i - t_{i-1}$, $i = 1, \ldots, m$. We can write

$$p_i(x) = A_i(x - s_i)^2 + K. \qquad (15)$$

Without loss of generality, assume $s_i \leq 0$. Then

$$A_i(l_i/2)^2 + K = 1$$

and

$$A_i(l_i/2 + l_{i+1} + \cdots + l_m)^2 + K = -1,$$

from which we derive

$$A_i = -\frac{2}{L_i L_{i+1}}, \quad K = \frac{1}{2}\left(\frac{L_i}{L_{i+1}} + \frac{L_{i+1}}{L_i}\right), \quad i = 1, \ldots \left[\frac{m}{2}\right], \qquad (16)$$

where $L_i := l_i + \cdots + l_m$.

Since $f(x) := (1/2)(x + x^{-1})$ is strictly increasing for $x > 1$, then

$$L_i/L_{i+1} = l, \quad i = 1, \ldots, [m/2], \qquad (17)$$

where $l > 1$ is the solution of $f(l) = K$.

Case 1. Let $m = 2k$. Then $L_{k+1} = 1$ and $L_1 = 2$. By multiplying the equations (17) we obtain $l^k = 2$, or $l = 2^{1/k}$. This proves (13) in this case. We can easily derive (14) from the fact that $1 - t_s = L_{s+1}$, $s = 0, \ldots, k$ and the symmetry of τ_m^*.

Case 2. Let $m = 2k + 1$. Then $1 - L_{k+1} = L_k - 1$ and $L_1 = 2$. We again multiply (17), but this time we get $l^k = 2/L_{k+1} = 1 + l$, which proves (13). We derive (14) as in Case 1. □

Remark. The authors have a rigorous proof of the generalized Bernstein-Erdős conjecture for $n = 3$ and $m = 4$, which will appear elsewhere, together with further investigation of this problem.

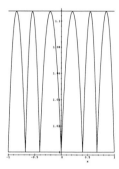
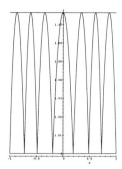

Fig. 1. The optimal GLF of order $n = 3$ for $m = 6$ and $m = 7$.

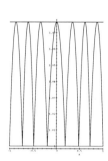
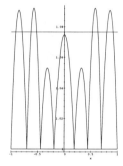

Fig. 2. The GLF of order $n = 3$ for τ_7^* and τ_7.

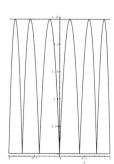
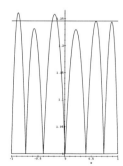

Fig. 3. The GLF of order $n = 4$ for τ_6^* and τ_6.

§3. Numerical Results

Here we present some numerical results that support the generalized Bernstein-Erdős conjecture. The computations are done with MAPLE. In Figure 1 we sketch the optimal GLF for $n = 3$ and $m = 6, 7$. Figures 2 and 3 support the conjecture that $\min_i b_i(\tau_m) < B_{n,m} < \max_i b_i(\tau_m)$ for $n = 3$ and $n = 4$.

References

1. de Boor, C. and A. Pinkus, Proof of the Conjectures of Bernstein and Erdős concerning the optimal nodes of interpolation, J. Approx. Theory **24** (1978), 289–303.
2. Dragnev, P. and E. B. Saff, Open problems in approximation theory, East J. Approx. **2** (1996), 499–517.
3. Kilgore, T., A Characterization of the lagrange interpolating projection with minimal tchebycheff norm, Journal Approx. Theory **24** (1978), 273–288.
4. Micchelli, C. and T. Rivlin, Optimal recovery of best approximation, Resultate der Mathematik **3** (1980), 25–32.
5. Schonhage, A., Fehlerfortpflanzung bei interpolation, Numerische Mathematik **3** (1961), 62–71.

P. Dragnev, D. Legg, and D. Townsend
Department of Mathematics
Indiana-Purdue University
Fort Wayne, IN 46805
dragnevp@ipfw.edu
legg@ipfw.edu
townsend@ipfw.edu

Non-convexity of the Set of Pólya Limits

A. Egger, T. Lay, and G. D. Taylor

Abstract. In 1963, J. Descloux described an example of a continuous function and a 1-dimensional approximating space which shows that the Pólya algorithm does not converge in general. A modification of this example is presented which demonstrates that the set of Pólya limits need not be convex. The effect of imposing smoothness conditions is discussed. We end with a conjecture.

§1. Introduction

Given an interval $[a, b]$, a finite-dimensional subspace V of $C[a, b]$ and a function f in $C[a, b]$ but not in V, let v_p, $1 < p < \infty$, denote the unique best L^p approximation to f from V. We say that the Pólya Algorithm converges if $\lim_{p \to \infty} v_p$ exists. An example due to J. Descloux [1] illustrates that, in this context, the Pólya Algorithm need not converge. For V, the subspace of lines through the origin, Descloux constructed a continuous function for which the set of L^p-best approximations from V on $[-2, 2]$ contains infinitely many limit points (Pólya limits).

While existence of a unique uniform-best approximation would be sufficient to ensure convergence of the Pólya Algorithm, it is certainly not necessary. It is natural to seek weaker conditions. It is known [2, 3] that the existence of a continuous selection ensures convergence. In fact, this condition is sufficient to establish *stable* convergence of the Pólya Algorithm. The question remains as to whether we can impose explicit conditions which ensure convergence of the Pólya Algorithm. For example, if f is a polynomial and V is a polynomial subspace, must the Pólya Algorithm converge?

§2. Insufficiency of Smoothness Conditions

The function approximated in the Descloux example was built as a limit of piecewise linear functions and was piecewise linear away from 0. Rice [3] observed that the construction could easily be modified to produce a function which is smooth everywhere except 0. There, a point of non-differentiability is an essential feature. Failure of smoothness turns out not to be a necessary condition for failure of the Pólya Algorithm. G. D. Taylor constructed an example where the approximating subspace and the function to be approximated are in $C^1[-1, 1]$ and the algorithm fails. In this example, the subspace consists of multiples of $x|x|$ and the function to be approximated lies between $1 - x^2$ and $1 - x^2/4$. In fact, this example can be modified to be C^∞.

§3. Convexity of the Pólya Limits

In search for conditions sufficient to ensure the convergence of the Pólya Algorithm it would be useful to characterize the set of all Pólya limits. This set is always compact. In the Descloux example it is convex as well, but convexity is not a necessary condition. To demonstrate this, we revisit the Descloux construction, make some modifications, and produce a continuous function and a 2-dimensional space of approximating functions such that the set of Pólya limits is not convex.

§4. Definitions and Terminology

The following functions will guide our constructions:

$$u(x) = \begin{cases} 6 - 2|x|, & -3 \leq x \leq 3, \\ 0, & \text{elsewhere}; \end{cases} \quad l(x) = \begin{cases} u(x), & |x| > 2, \\ 2, & |x| \leq 2; \end{cases}$$

$$t(x) = \begin{cases} -2x - 6, & -3 \leq x \leq -2, \\ x, & -2 < x < 2, \\ -2x + 6, & 2 \leq x \leq 3, \\ 0, & \text{elsewhere}. \end{cases}$$

We will be working with continuous functions f which satisfy

$$l(x) \leq f(x) \leq u(x); \quad f(x) < 6. \tag{1}$$

We examine best L^p-approximations to these functions from the set $V = \{at, \ a \in R\}$. For each f and for each $p > 1$, let $a_p = a_p(f)$ denote the value of a which minimizes

$$\int_{-3}^{3} |f(x) - at(x)|^p \, dx. \tag{2}$$

We say that f is skewed left at p provided $a_p < -1/2$ and skewed right at p provided $a_p > 1/2$. These skew properties can be detected by observing the value of the derivative of (2) with respect to a. A positive value when $a = -1/2$ indicates left-skew and a negative value when $a = 1/2$ indicates right-skew. For f satisfying (1), direct computations show that f is skewed left if and only if

$$\int_{-2}^{0} (f(x) + x/2)^{p-1} |x| \, dx > \int_{0}^{2} (f(x) + x/2)^{p-1} x \, dx + \Phi(p) \tag{3}$$

and skewed right if and only if

$$\int_0^2 (f(x) - x/2)^{p-1} x \, dx > \int_{-2}^0 (f(x) - x/2)^{p-1} |x| \, dx + \Phi(p), \quad (4)$$

where $\Phi(p) = \dfrac{2}{p+1} \left(3^{p-1} - 1\right)$.

§5. The Basic Step

The Descloux example [1] provides a function whose skew behavior oscillates as p becomes large. The construction which follows will do this, but will also allow us to establish and maintain skew behavior for specified intervals of p. We then weave two such functions together to obtain our main result. The key ingredient is embodied in the following lemma:

Lemma 1. *Let f be a continuous function satisfying (1), I a small closed interval containing 0 in its interior, and $1 < \hat{p} < p_*$. If f is skewed left (right) for all $p \geq \hat{p}$, then there is a continuous function $\tilde{f}$ satisfying (1), a value $\tilde{p} > p_*$ and a closed interval $\tilde{I} \subset I$ containing 0 in its interior such that*

(A) $\tilde{f}$ agrees with f outside of I.
(B) any function h which lies between $\tilde{f}$ and u and agrees with $\tilde{f}$ outside of $\tilde{I}$ is skewed left (right) for $p \in [\hat{p}, p_*]$.
(C) $\tilde{f}$ is skewed right (left) for all $p \geq \tilde{p}$.

In the proof, we consider the skewed-left setting. The case where f is skewed right is handled similarly. The intervals $\tilde{I} \subset I$ are important in that they control the regions where f (and iterates of f) may be modified. In the subsequent construction, we apply Lemma 1 repeatedly and (B) allows us to preserve in the limit function the skew properties constructed at each stage.

Proof: The strategy in the proof is to first increase the values of f over a small interval to the right of 0. This produces the eventual skewed-right behavior. The interval is chosen small enough so that the skewed-left behavior in $[\hat{p}, p_*]$ is retained. f is then modified to the left of 0 to regain continuity, but with enough control so that the eventual right skew will still happen (see Figure 1).

Since f is skewed left for $p \in [\hat{p}, p_*]$, the continuity of the integrals in (3) yields a positive number λ so that for all $p \in [\hat{p}, p_*]$

$$\int_{-2}^0 (f(x) + x/2)^{p-1} |x| \, dx > \lambda + \int_0^2 (f(x) + x/2)^{p-1} x \, dx + \Phi(p). \quad (5)$$

Choose positive ϵ with $[-\epsilon, \epsilon] \subset I$ such that $\int_0^\epsilon 7^{p_* - 1} x \, dx < \lambda$. Let R denote the line connecting $(0,6)$ to $(\epsilon, 0)$, let m denote the maximum of f and note that $m < 6$ and let (α, β) denote the intersection of R and the graph of f with α positive and minimal. Choose w such that $(0, w)$ is more than half

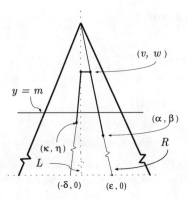

Fig. 1. The basic step.

way from $(0, m)$ to $(0, 6)$ and let (v, w) be the intersection of the horizontal line $y = w$ with R. For $0 \leq x$, $\tilde{f}(x)$ is defined as

$$\tilde{f}(x) = \begin{cases} w, & 0 \leq x \leq v, \\ R(x), & v < x \leq \alpha, \\ f(x), & \alpha < x. \end{cases}$$

If h is any function which satisfies (1), agrees with $\tilde{f}$ to the right of α and is bounded below by $\tilde{f}$ in $[0, \alpha]$, then for all $p \in [\hat{p}, p_*]$,

$$\int_0^2 (h(x) + x/2)^{p-1} x \, dx < \int_0^\alpha (6 + x/2)^{p-1} x \, dx + \int_\alpha^2 (\tilde{f}(x) + x/2)^{p-1} x \, dx$$
$$< \int_0^\epsilon 7^{p_*-1} x \, dx + \int_\alpha^2 (f(x) + x/2)^{p-1} x \, dx \qquad (6)$$
$$< \lambda + \int_0^2 (f(x) + x/2)^{p-1} x \, dx.$$

Next, let (s, m) denote the intersection of $y = m$ with the left half of the graph of $u(x)$. It follows that for x in $[-2, 0]$, the difference $f(x) - x/2$ is no more than $m_* = m + |s|/2$, which is smaller than w, chosen above. Choose $0 < \delta < v$ and let L denote the line connecting the point $(0, w)$ with the point $(-\delta, 0)$. Let (κ, η) denote the intersection of L and the graph of f with κ maximal. Note that $|\kappa| < v$. For $x \leq 0$, $\tilde{f}(x)$ is defined as

$$\tilde{f}(x) = \begin{cases} f(x), & x \leq \kappa, \\ L(x), & \kappa < x \leq 0. \end{cases}$$

The line L is sufficiently steep so that

$$\int_\kappa^0 (L(x) - x/2)^{p-1} |x| \, dx < \int_0^{|\kappa|} (w - x/2)^{p-1} x \, dx \qquad (7)$$

and it is easily seen that

$$\int_{|\kappa|}^{v} (w - x/2)^{p-1} x \, dx > (w - v/2)^{p-1} (v^2 - \kappa^2)/2. \tag{8}$$

The definition of m_* is such that

$$\int_{-2}^{\kappa} (\tilde{f} - x/2)^{p-1} |x| \, dx < 4 m_*^{p-1} \tag{9}$$

and w was chosen in such a way that $w - v/2$ is larger than either m_* or 3. This makes it possible to select $\tilde{p} > p_*$ so large that if $p > \tilde{p}$, then

$$4 m_*^{p-1} < (w - v/2)^{p-1} (v^2 - \kappa^2)/2 - \Phi(p). \tag{10}$$

Combining (7), (8), (9), and (10), it follows that for $p > \tilde{p}$,

$$\int_{-2}^{0} (\tilde{f}(x) - x/2)^{p-1} |x| \, dx < 4 m_*^{p-1} + \int_{\kappa}^{0} (L(x) - x/2)^{p-1} |x| \, dx$$
$$< (w - v/2)^{p-1} (v^2 - \kappa^2)/2 - \Phi(p)$$
$$+ \int_{0}^{|\kappa|} (w - x/2)^{p-1} x \, dx \tag{11}$$
$$\leq \int_{0}^{v} (\tilde{f} - x/2)^{p-1} x \, dx - \Phi(p)$$
$$< \int_{0}^{2} (\tilde{f} - x/2)^{p-1} x \, dx - \Phi(p).$$

Finally, set $\tilde{I} = [\kappa, v]$. The proof is completed by noting that (A) is a consequence of the values of κ and v, (B) follows from (6), the fact that $\tilde{f}$ dominates f, and (C) is immediate from (4) and (11). □

§6. The Example

To simplify matters, we want to view Lemma 1 as a procedure P with input $f, I, \hat{p}$, and p_*, and output $\tilde{f}, \tilde{I}$, and $\tilde{p}$. It is understood that P maintains the skew of the input function from $\hat{p}$ to p_* and "reverses" the skew of the input function beyond $\tilde{p}$. We will use the notation $(\tilde{f}, \tilde{I}, \tilde{p}) = P(f, I, \hat{p}, p_*)$.

To construct our main example, let f_1 and g_1 be continuous functions satisfying (1) and skewed left for all $p \geq p_0 > 1$. Set $I_1 = J_1 = [-1, 1]$ and $p_1 = q_1 = p_0$. We iteratively define f_n and g_n as

$$(f_n, I_n, p_n) = P(f_{n-1}, I_{n-1}, p_{n-1}, q_{n-1})$$
$$(g_n, J_n, q_n) = P(g_{n-1}, J_{n-1}, q_{n-1}, p_n).$$

It is easy to verify that the sequences $\{f_n\}$ and $\{g_n\}$ converge to continuous functions, say F and G. (B) follows from Lemma 1 and the way the intervals I_n and J_n are nested, which insure that F has exactly the same skew behavior as f_n for $p \in [p_n, q_n]$ and that G has exactly the same skew behavior as g_n for $p \in [q_n, p_{n+1}]$.

If $H(x) = F(x-3) + G(x+3)$ and

$$W = \{w_{ab} \mid w_{ab}(x) = at(x-3) + bt(x+3)\,;\, a, b \in R\},$$

then the L^p-best approximations to H from W wind essentially around an annulus in the complement of $[-1/2, 1/2] \times [-1/2, 1/2]$. From this it follows that there must be Pólya limits w_{ab} and w_{cd} with $a = c = 0$, $b > 0$ and $d < 0$. The set of Pólya limits cannot be convex since w_{00} cannot be a Pólya limit.

§7. A Conjecture

We end this discussion with the following conjecture: If V is a finite dimensional subspace of entire functions and f is an entire function not in V, then the Pólya Algorithm will converge. That is, if v_p is the L^p-best approximation to f from V, then $\lim_{p \to \infty} v_p$ exists.

References

1. Descloux, J., Approximations in L^p and Chebyshev approximations, J. Soc. Indust. Appl. Math. **11** (1963), 1017–1026.
2. Li, Wu, The convergence of the Pólya algorithm and continuous metric selections in the space of continuous Functions, J. Approx. Theory **80** (1995), 164–179.
3. Rice, John R., *The Approximation of Functions*, Vol. 2, Addison-Wesley, 1969.
4. Sommer, M., Continuous selections and convergence of the best L^p-approximations in subspaces of spline functions, Numer. Funct. Anal. Optim. **6** (1983), 213–234.

Alan Egger and Terry L. Lay
Department of Mathematics
Idaho State University
Pocatello, ID 83209
eggealan@isu.edu
layterr@isu.edu

G. D. Taylor
Department of Mathematics
Colorado State University
Fort Collins, CO 80523
taylor@math.colostate.edu

Estimates for Some Positive Operators via Step-Weight Functions with Concave Squares

Michael Felten

Abstract. Local and global estimates for exponential-type operators and Lototsky-Schnabl operators are bridged via Ditzian-Totik moduli of smoothness $\omega_\phi^2(f,\delta)$, whereby the step-weights ϕ are functions such that ϕ^2 are concave. In particular, the cases $\phi = \varphi^\lambda$, $\lambda \in [0,1]$, for the Bernstein operator ($\varphi(x) = \sqrt{x(1-x)}$) and the Szász-Mirakjan operator ($\varphi(x) = \sqrt{x}$) are included.

§1. Direct Results

In a series of papers [6–8] the gap between the Timan-type pointwise estimate and the norm estimate for polynomial approximation in $C[-1,1]$ has been bridged. In [10] it was shown that for the classical Bernstein operator on $C[0,1]$ given by

$$(B_n f)(x) = \sum_{k=0}^{n} \binom{n}{k} x^k (1-x)^{n-k} f\left(\frac{k}{n}\right), \quad x \in [0,1],$$

the pointwise estimate

$$|(B_n f)(x) - f(x)| \le C \omega_\phi^2 \left(f, n^{-\frac{1}{2}} \frac{\varphi(x)}{\phi(x)}\right), \quad x \in [0,1], \tag{1}$$

with $\varphi(x) := \sqrt{x(1-x)}$ holds true if $\phi\colon [0,1] \to \mathbb{R}$ is an admissible step-weight function of the Ditzian-Totik modulus and if ϕ^2 is a concave function. The Ditzian-Totik modulus of second order in (1) is defined by

$$\omega_\phi^2(f,\delta) := \sup_{0 < h \le \delta} \sup_{x \pm h\phi(x) \in [0,1]} |f(x - \phi(x)h) - 2f(x) + f(x + \phi(x)h)|.$$

Inequality (1) gives for $\phi = 1$ the classical local estimate (see Strukov and Timan [15]) and for $\phi = \varphi$ the global norm estimate developed by Ditzian and Totik [9, p. 117]. Therefore, inequality (1) yields a relation between the local estimate ($\phi = 1$) and the global estimate ($\phi = \varphi$) via step-weight functions ϕ whose squares are concave. Since $\varphi^{2\lambda}$, $\lambda \in [0,1]$, are concave functions, the step-weight functions $\phi = \varphi^\lambda$, $\lambda \in [0,1]$, are included in (1), that is

$$|(B_n f)(x) - f(x)| \leq C \omega^2_{\varphi^\lambda}(f, n^{-\frac{1}{2}}\varphi(x)^{1-\lambda}), \quad x \in [0,1], \qquad (2)$$

which was proved by Ditzian [5].

The purpose of this section is to investigate inequality (1) for other bounded positive linear operators which preserve constants and linear functions. We shall formulate the result in terms of the K-functional

$$K^2_\phi(f, \delta^2) := \inf_{g' \in AC_{loc}(I)} (\|f - g\| + \delta^2 \|\phi^2 g''\|), \quad \delta \geq 0,$$

in which $I \subseteq \mathbb{R}$ is an interval, $f: I \to \mathbb{R}$ and $\|f\| := \sup\{|f(x)| : x \in I\}$ denotes the uniform norm of f on I. The main result is

Theorem 1. *Let $I \subseteq \mathbb{R}$ be an interval and $A: C(I) \to C(I)$ be a bounded positive linear operator which preserves constants and linear functions. If $\phi: I \to \mathbb{R}$ is a function with ϕ^2 being concave then the pointwise approximation*

$$|f(x) - (Af)(x)| \leq 2K^2_\phi\left(f, \frac{A((\bullet - x)^2)(x)}{\phi^2(x)}\right)$$

holds true for $x \in I$ and $f \in C(I)$.

Proof: Let $x \in I$, $n \in \mathbb{N}$ be fixed and $g: I \to \mathbb{R}$ with $g' \in AC_{loc}(I)$ be arbitrary. We have

$$g(t) - g(x) = g'(x)(t-x) + \int_x^t g''(s)(t-s)\,ds, \quad t \in I.$$

Application of A on both sides with respect to t gives (using positivity of A and $A(t-x)(x) = 0$)

$$|(Ag)(x) - g(x)| \leq A\left(\left|\int_x^t g''(s)(t-s)\,ds\right|\right)(x)$$

$$\leq \|\phi^2 g''\| A\left(\left|\int_x^t \frac{|t-s|}{\phi^2(s)}\,ds\right|\right)(x).$$

For each $s = t + \tau(x - t)$, $\tau \in [0,1]$, we can estimate, by using the concavity of ϕ^2, that

$$\frac{|t-s|}{\phi^2(s)} = \frac{\tau|x-t|}{\phi^2(t + \tau(x-t))} \leq \frac{\tau|x-t|}{\phi^2(t) + \tau(\phi^2(x) - \phi^2(t))} \leq \frac{|x-t|}{\phi^2(x)}.$$

and therefore

$$|(Ag)(x) - g(x)| \leq \|\phi^2 g''\| \, A\left(\left|\int_x^t \frac{|t-x|}{\phi^2(x)} ds\right|\right)(x)$$
$$= \|\phi^2 g''\| \frac{A((t-x)^2)(x)}{\phi^2(x)}.$$

This gives

$$|(Af)(x) - f(x)| = |((A - \mathrm{id})(f - g))(x) + ((A - \mathrm{id})g)(x)|$$
$$\leq 2\left(\|f - g\| + \frac{A((t-x)^2)(x)}{\phi^2(x)} \|\phi^2 g''\|\right).$$

Taking the infimum on the right side over all $g\colon I \to \mathbb{R}$ with $g' \in AC_{loc}(I)$ we obtain our theorem. □

Theorem 1 extends a result for the Bernstein operator in [10] to positive operators which preserve constants and linear functions. For operators A which do not preserve linear functions, a consideration will be found in the forthcoming paper [11].

By reason of the fact that $K_\phi^2(f, \delta^2)$ and $\omega_\phi^2(f, \delta)$ are equivalent for admissible step-weight functions ϕ (see [9]), we obtain the following corollary.

Corollary 2. *Let A be given as in Theorem 1, and let $\phi\colon I \to R$ be an admissible step-weight function of the Ditzian-Totik modulus such that ϕ^2 is concave. Then*

$$|(Af)(x) - f(x)| \leq C\omega_\phi^2\left(f, \frac{\sqrt{A((\bullet - x)^2)(x)}}{\phi(x)}\right) \quad (3)$$

holds true for $x \in I$ and $f \in C(I)$ where C in (3) is independent of A, f and x.

If we let A in Corollary 2 be the Bernstein operators B_n, $n \in \mathbb{N}$, we obtain inequality (1) because $B_n((\bullet - x)^2)(x) = x(1-x)/n$.

We now apply the results to some positive operators which preserve constants and linear functions. We begin with exponential-type operators. We say

$$(L_n f)(x) = \int_I W(n, x, u) f(u) \, du, \quad x \in I, \ n \in \mathbb{N}, \quad (4)$$

is an exponential-type operator if $W(n, x, u) \geq 0$ and

$$\int_I W(n, x, u) \, du = 1, \quad x \in I, \ n \in \mathbb{N},$$

$$\frac{d}{dx} W(n, x, u) = \frac{n}{\varphi^2(x)} W(n, x, u)(u - x), \quad x \in I, \ n \in \mathbb{N},$$

where φ^2 is an analytic positive function in the interior of the interval $I \subseteq \mathbb{R}$ (see [12, 13]). L_n is a positive operator and simple calculations show that L_n preserves linear functions and

$$L_n((\bullet - x)^2)(x) = \frac{\varphi^2(x)}{n}. \tag{5}$$

If ϕ^2 is concave on I, then Corollary 2 gives the estimate

$$|f(x) - (L_n f)(x)| \leq C \omega_\phi^2\left(f, n^{-\frac{1}{2}} \frac{\varphi(x)}{\phi(x)}\right). \tag{6}$$

If, in addition, φ^2 in (5) is concave, then we can choose $\phi = \varphi$ as a step-weight function in inequality (6). Thus, inequality (6) bridges the local estimate ($\phi = 1$) and the global estimate ($\phi = \varphi$) via step-weight functions ϕ whose squares are concave.

If φ^2 is a polynomial of degree at most two, there are essentially six types of operators (see [13]), namely the Bernstein operator ($I = [0,1]$, $\varphi^2(x) = x(1-x)$), the operators of Szász-Mirakjan ($I = [0,\infty)$, $\varphi^2(x) = x$), Post-Widder ($I = [0,\infty)$, $\varphi^2(x) = x^2$), Gauss-Weierstrass ($I = (-\infty,\infty)$, $\varphi^2(x) = 1$) Baskakov ($I = [0,\infty)$, $\varphi^2(x) = x(1+x)$) and Ismail/May ($I = [-\infty,\infty)$, $\varphi^2(x) = 1 + x^2$). Therefore, for the Bernstein operator and the operator of Szász-Mirakjan, we obtain with (6) an inequality which combines local and global estimates for these operators.

Let $\mu: [0,1] \to [0,1]$ be a continuous function. For every $n \in \mathbb{N}$ the nth Lototsky-Schnabl operator on $C[0,1]$ associated with μ is the positive linear operator $L_{n,\mu}: C[0,1] \to C[0,1]$ defined by

$$(L_{n,\mu}f)(x) = \sum_{k=0}^{n} \sum_{j=0}^{k} \binom{n}{k}\binom{k}{j} \mu(x)^k (1-\mu(x))^{n-k} f\left(\frac{j}{n} + (1-\frac{k}{n})x\right) x^j (1-x)^{k-j}.$$

When $\mu = 1$, the operators $L_{n,\mu}$ are the classical Bernstein operators B_n. $L_{n,\mu}$ is a positive operator which preserves constant and linear functions. Moreover,

$$L_{n,\mu}((\bullet - x)^2)(x) = \frac{\mu(x)\, x(1-x)}{n}.$$

From Corollary 2, we obtain

$$|f(x) - (L_{\mu,n}f)(x)| \leq C \omega_\phi^2\left(f, n^{-\frac{1}{2}} \frac{\sqrt{\mu(x)\, x(1-x)}}{\phi(x)}\right) \tag{7}$$

for every admissible step-weight function ϕ whose square is concave. Hence, if $\mu(x)\, x(1-x)$ is concave then (7) combines local and global estimates.

§2. Converse Results

In this section we investigate converse results to (6). For a more detailed consideration the reader is referred to [11]. From the direct result (1) we obtain

$$\omega_\phi^2(f,\delta) = O(\delta^\alpha) \implies |(B_nf)(x) - f(x)| = O\left(\left(n^{-\frac{1}{2}}\frac{\varphi(x)}{\phi(x)}\right)^\alpha\right)$$

for $x \in [0,1]$ and $\alpha \in (0,2)$. In [10], it was shown that

$$|(B_nf)(x) - f(x)| = O\left(\left(n^{-\frac{1}{2}}\frac{\varphi(x)}{\phi(x)}\right)^\alpha\right) \implies \omega_\phi^2(f,\delta) = O(\delta^\alpha)$$

for those concave functions ϕ^2 for which, in addition, $\frac{\varphi^2}{\phi^2}$ is concave. In particular, this is fulfilled for all step-weight functions

$$\phi = \varphi^\lambda, \quad \lambda \in [0,1], \quad \varphi(x) = \sqrt{x(1-x)}.$$

For a converse result for exponential-type operators (4) we need that

$$\begin{aligned}&\varphi^2 \text{ is a polynomial of degree at most 2}\\&\text{without a double zero, } \varphi^2(x) \neq 0 \text{ inside}\\&\text{of } I \text{ and } \varphi(x) = 0 \text{ for finite endpoints of } I\end{aligned} \qquad (8)$$

and

$$J := \frac{n^2}{\varphi^4(x)} \int_\alpha^\beta \left[\frac{(\varphi^2(x))'}{n} - (u-x)\right](u-x)^3 W(n,x,u)\,du \leq M, \qquad (9)$$

where $\alpha = \min\{x, x + \frac{(\varphi^2(x))'}{n}\}$, $\beta = \max\{x, x + \frac{(\varphi^2(x))'}{n}\}$ and M must be an absolute constant independent of n and x. Condition (9) has been imposed by Sato [13].

For the proof of the converse result two lemmas are needed (see [11]).

Lemma 1. *Let L_n, $n \in \mathbb{N}$, be the exponential-type operator (4) satisfying (8) and (9). Then for every concave function $\phi^2: I \to \mathbb{R}$ the estimate*

$$\|\phi^2(L_ng)''\| \leq C \, \|\phi^2 g''\|, \quad g' \in AC_{loc}(I), \; n \in \mathbb{N}, \qquad (10)$$

holds for some constant C being independent of n, g, and ϕ.

Proof: By Taylor's formula we have

$$(L_ng)''(x) = \frac{d^2}{dx^2} \int_I W(n,x,u)\,g(u)\,du$$

$$= \int_I \left[\frac{d^2}{dx^2} W(n,x,u)\right] \left\{g(x) + g'(x)(u-x) + \int_x^u (u-s)g''(s)\,ds\right\} du$$

and by reason of

$$\int_I \left[\frac{d^2}{dx^2}W(n,x,u)\right](u-x)^i\, du = 0 \quad \text{for } i = 0, 1,$$

we obtain

$$(L_n g)''(x) = \int_I \left[\frac{d^2}{dx^2}W(n,x,u)\right]\int_x^u (u-s)g''(s)\, ds\, du.$$

By concavity of ϕ^2,

$$\frac{|t-s|}{\phi^2(s)} = \frac{\tau|x-t|}{\phi^2(t+\tau(x-t))} \leq \frac{\tau|x-t|}{\phi^2(t)+\tau(\phi^2(x)-\phi^2(t))} \leq \frac{|x-t|}{\phi^2(x)}$$

and by positivity of L_n

$$|\phi(x)^2 (L_n g)''(x)| \leq \|\phi^2 g''\| \int_I \left|\frac{d^2}{dx^2}W(n,x,u)\right| \left|\int_x^u \frac{|u-s|}{\phi^2(s)}\, ds\right|\, du\, \phi^2(x)$$

$$\leq \|\phi^2 g''\| \int_I \left|\frac{d^2}{dx^2}W(n,x,u)\right| \left|\int_x^u |u-x|\, ds\right|\, du$$

$$\leq \|\phi^2 g''\| \int_I \left|\frac{d^2}{dx^2}W(n,x,u)\right|(u-x)^2\, du$$

$$\leq \|\phi^2 g''\| \left(4 + \frac{(\varphi^2(x))''}{n} + 2J\right). \tag{11}$$

In view of conditions (8) and (9) the left-hand side of (11) is bounded by a constant times $\|\phi^2 g''\|$. □

In the case of Bernstein operators it has been shown (see [10]) that the constant C in inequality (10) can be replaced by 8.

Lemma 2. Let $I \subseteq \mathbb{R}$ be an interval and let $\phi: I \to \mathbb{R}$, $\phi \not\equiv 0$, be a function such that ϕ^2 is concave. Then for all $x \in I$, $h > 0$ with $x \pm h \in I$ the inequality

$$\int_{-\frac{h}{2}}^{\frac{h}{2}} \int_{-\frac{h}{2}}^{\frac{h}{2}} \frac{ds\, dt}{\phi^2(x+s+t)} \leq C \frac{h^2}{\phi^2(x)}$$

holds true, whereby $8\log 2$ can be chosen as the constant C.

Proof: See [11]. □

Lemma 2 generalizes a lemma of Becker [2]. As a converse to (6) we have

Theorem 3. Let L_n, $n \in \mathbb{N}$, be the exponential-type operator (4) satisfying (8) and (9) and let $\phi: I \to \mathbb{R}$ be an admissible step-weight function of the Ditzian-Totik modulus. Moreover, let ϕ^2, φ^2, and $\frac{\varphi^2}{\phi^2}$ be concave functions.

Then for $f \in C(I)$ and $\alpha \in (0,2)$ the pointwise approximation

$$|(L_n f)(x) - f(x)| = O\left(\left(n^{-\frac{1}{2}}\frac{\varphi(x)}{\phi(x)}\right)^\alpha\right) \tag{12}$$

implies

$$\omega_\phi^2(f,\delta) = O(\delta^\alpha). \tag{13}$$

Proof: Standard methods, using Lemmas 1 and 2 and a well-known lemma of Berens-Lorentz in [3], give the assertion. For details see [11]. □

In conjunction with inequality (6) we see that both conditions (12) and (13) are equivalent. In particular, this is fulfilled for the operators of

- Gauss-Weierstrass ($\varphi(x) = 1$)
- Szász-Mirakjan ($\varphi(x) = \sqrt{x}$)
- Bernstein ($\varphi(x) = \sqrt{x(1-x)}$).

We give two corollaries which combine direct and inverse estimates.

Corollary 4. Let $\varphi(x) := \sqrt{x(1-x)}$ and $\phi:[0,1] \to \mathbb{R}$ with ϕ^2 being concave. For $f \in C[0,1]$ and $\alpha \in (0,2)$, the following are equivalent for the Bernstein operator B_n:

(i) $|(B_n f)(x) - f(x)| = O\left(\left(n^{-\frac{1}{2}}\frac{\varphi(x)}{\phi(x)}\right)^\alpha\right)$,

(ii) $\omega_\phi^2(f,\delta) = O(\delta^\alpha)$.

If we set $\phi = \varphi^\lambda$ in Corollary 4 the equivalence was already known ([4, 16], see also [10]).

Corollary 5. Let $\varphi(x) := \sqrt{x}$ and $\lambda \in [0,1]$. For $f \in C[0,\infty)$ and $\alpha \in (0,2)$, the following are equivalent for the Szász-Mirakjan operator L_n:

(i) $|(L_n f)(x) - f(x)| = O\left((n^{-\frac{1}{2}}\varphi^{1-\lambda}(x))^\alpha\right)$,

(ii) $\omega_{\varphi^\lambda}^2(f,\delta) = O(\delta^\alpha)$.

It cannot yet be said whether the inverse also holds true for the operators of

- Baskakov ($\varphi(x) = \sqrt{x(1+x)}$)
- Post-Widder ($\varphi(x) = x$)
- Ismail/May ($\varphi(x) = \sqrt{1+x^2}$).

In these cases $\varphi^2(x)$ is not concave. The question is also still open as to whether a converse of (7) for the Lototsky-Schnabl operators in the sense of Theorem 3 is fulfilled.

References

1. Altomare, F., Lototsky-Schnabl operators on the unit interval, C. R. Acad. Sci., Paris, Sér. I **313** (1991), 371–375.

2. Becker, M., Global approximation theorems for Sász-Mirakjan and Baskakov operators in polynomial weighted spaces, Indiana Univ. Math. J. **27** (1978), 127–142.

3. Berens, H. and G. G. Lorentz, Inverse theorems for Bernstein polynomials, Indiana Univ. Math. J. **21** (1972), 693–708.

4. Ditzian, Z., Rate of convergence for Bernstein polynomials revisited, J. Approx. Theory **50** (1987), 40–48.

5. Ditzian, Z., Direct estimate for Bernstein polynomials, J. Approx. Theory **79** (1994), 165–166.

6. Ditzian, Z. and D. Jiang, Approximation of functions by polynomials in $C(-1,1)$, Can. J. Math. **44** (1992), 924–940.

7. Ditzian, Z., D. Jiang, and D. Leviatan, Shape-preserving polynomial approximation in $C[-1,1]$, Math. Proc. Camb. Phil. Soc. **112** (1992), 309–316.

8. Ditzian, Z., D. Jiang, and D. Leviatan, Simultaneous polynomial approximation, SIAM J. Math. Anal. **24** (1993), 1652–1661.

9. Ditzian, Z. and V. Totik, *Moduli of Smoothness*, Springer-Verlag, New York, 1987.

10. Felten, M., Direct and inverse estimates for Bernstein polynomials, Constructive Approximation **14** (1998), 459–468.

11. Felten, M., Local and global approximation theorems for positive linear operators, J. Approx. Theory, to appear.

12. Ismail, M. E. H. and C. P. May, On a family of approximation operators, J. Math. Anal. Appl. **63** (1978), 446–462.

13. Satô, K., Global approximation theorems for some exponential-type operators, J. Approx. Theory **32** (1981), 32–46.

14. Schnabl, R., Zum globalen saturationsaproblem der Folge der Bernsteinoperatoren, Acta Math. Szeged **31** (1970), 351–358.

15. Strukov, L. I. and A. F. Timan, Mathematical expectation of contiuous functions of random variables, smoothness and variance, Siberian Math. J. **18** (1977), 469–474.

16. Zhou, D.-X., On a conjecture of Z. Ditzian, J. Approx. Theory **69** (1992), 167–172.

Michael Felten
University of Dortmund
Department of Mathematics
D-44221 Dortmund, Germany
felten@zx2.hrz.uni-dortmund.de

Polynomial Approximation on the m-dimensional Ball

Michael I. Ganzburg

Abstract. Structural descriptions of two classes of functions defined on the m-dimensional ball B^m and satisfying the inequalities $E_n(f, B^m)_p \leq Cn^{-\alpha}$, $\alpha > 0$, $n \in \mathbb{N}$, $1 \leq p \leq \infty$, are obtained. Here $E_n(f, B^m)_p$ is the error of best polynomial approximation of a function f in a weighted L_p-metric.

§1. Introduction

This paper is devoted to constructive and structural descriptions of some classes of functions defined on the m-dimensional ball.

Let $\mathbb{R}^m$ be the m-dimensional euclidean space; $B^m = \{x \in \mathbb{R}^m : |x| \leq 1\}$ the unit ball in $\mathbb{R}^m$; S^m the unit sphere in $\mathbb{R}^{m+1}$; $L_{p,v}(\Omega)$, $0 < p < \infty$, the quasi-normed space of all measurable functions f on a compact set $\Omega \subseteq \mathbb{R}^m$ with finite quasi-norm $\|f\|_{L_{p,v}(\Omega)} = (\int_\Omega |f(x)|^p v(x)\, dx)^{1/p}$, where v is an integrable weight; $L_p(\Omega) = L_{p,1}(\Omega)$; $L_\infty(\Omega)$ the Banach space of all continuous functions on a compact set $\Omega \subset \mathbb{R}^m$ with the uniform norm; $C^k(B^m)$ the class of all k times differentiable functions on B^m with continuous partial derivatives of order k; and $\mathcal{P}_{n,m}$ the class of all algebraic polynomials in m variables of degree n or less. Let us denote by $E_n(f,\Omega)_{p,v} = \inf_{P \in \mathcal{P}_{n,m}} \|f - P\|_{L_{p,v}(\Omega)}$, the error of best polynomial approximation in $L_{p,v}(\Omega)$, $0 < p \leq \infty$; $E_n(f,\Omega)_p = E_n(f,\Omega)_{p,1}$. Throughout, C, C_1, and C_2 will denote possibly different positive constants independent on n, f, x, y, t.

The following two problems play an important role in classic Approximation Theory:

Problem 1. Find a constructive description of a given class of functions defined by Lipschitz conditions and differentiability properties.

Problem 2. Find a structural description of a class of functions satisfying the inequalities $E_n(f,\Omega)_{p,v} \leq Cn^{-\alpha}$, $\alpha > 0$, $0 < p \leq \infty$, $n \in \mathbb{N}$.

Jackson, 1911; Bernstein, 1912, 1946; Krein, 1938; Zygmund, 1945; Nikolskii, 1951; Stechkin, 1951; and others showed that Problems 1 and 2 coincide for classes of functions defined on the m-dimensional torus or $\mathbb{R}^m$ and for approximation by multivariate trigonometric polynomials or entire functions of exponential type. For all this see [5, 18].

The situation with polynomial approximation is different. In particular, the classic Jackson's inequality for $E_n(f, [-1,1])_\infty$ has no converse because of so-called "endpoints effect" as was explained by Nikolskii, 1946. Problem 1 for polynomial approximation in $L_\infty(-1,1)$ was solved by Timan, 1951; Dzyadyk, 1956, 1959; Brudnyi, 1963; and others [5, 18], while the solution of the corresponding L_p-problem, $1 \leq p < \infty$, was obtained by Operstein, 1987-1990.

There are three approaches to Problem 2. The first of them goes back to Bernstein [2] who remarked that, "... a Lipschitz condition of order α at the interior points of an interval has the same meaning for least deviation as a Lipschitz condition of order $\alpha/2$ at the endpoints." Developing this approach Fuksman [9], in particular, proved the following result.

Theorem A. *The inequalities $E_n(f, [-1,1])_\infty \leq Cn^{-\alpha}$, $0 < \alpha < 1$, $n \in \mathbb{N}$, hold if and only if for all $x, y \in [-1,1]$,*

$$|f(x) - f(y)| \leq C_1 (|x-y|/(\sqrt{1-x^2} + \sqrt{1-y^2} + \sqrt{|x-y|}))^\alpha.$$

The second approach was developed in the 1970's-1990's by Ibragimov and Mamedhanov, Ditzian, Totik, Ivanov, DeVore, Leviatan, and others [5, 7]. They have introduced some variable-step moduli of smoothness $w_r(f,t)_p$ and proved that for $0 < \alpha < r$, $0 < p \leq \infty$,

$$E_n(f, [-1,1])_p \leq Cn^{-\alpha} \iff w_r(f,t)_p \leq C_1 t^\alpha.$$

In the 1970's-1980's Potapov, Butzer, Stens and their students [4, 15] came up with a new approach to Problem 2. Using generalized shifts which are generated by certain differential operators, they defined moduli of smoothness $\omega_r(f,t)_{p,v}$ and obtained the results of the form

$$E_n(f, [-1,1])_{p,v} \leq Cn^{-\alpha} \iff \omega_r(f,t)_{p,v} \leq C_1 t^\alpha,$$

where $v(x) = (1-x)^r (1+x)^\lambda$ is a weight, $0 < \alpha < r$, $1 \leq p \leq \infty$. This approach is connected with some areas of Analysis and Algebra, such as Orthogonal Polynomials, Differential Equations, Functional Spaces, and Representation of Lie Groups, and seems very promising. In particular, it is very fruitful in polynomial approximation on S^m. The spherical shift and the spherical difference can be defined by

$$\tau_h^* f(x) = \frac{\Gamma(m/2)}{2\pi^{m/2}(1-h^2)^{(m-1)/2}} \int_{xy=h} f(y)\,dy, \tag{1}$$

$$\Delta_h^* f(x) = f(x) - \tau_h^* f(x), \tag{2}$$

where $h \in (-1,1), x, y \in S^m$, and dy is the "curve" element of the circle $xy = h$ on S^m.

The direct and converse theorems of polynomial approximation on the sphere were obtained by Kushnirenko, 1958; Ragozin, 1967; Butzer and Johnen, 1971; Pawelke, 1972; Wherens, 1981; Zhidkov, 1981, 1983; Lizorkin and Nikolskii, 1981–1988; Lizorkin, 1983; Petrova, 1984; Kalyabin, 1987; Rustamov, 1985–1993; Lizorkin and Rustumov, 1993, and others (see [12, 14, 16, 17, 19] and the references given there). In particular, Wherens [19] defined a modulus of smoothness

$$\omega_r^*(f,t)_p = \sup_{1-t<h_j<1, 1\leq j \leq r} \|\Delta_{h_1}^* \ldots \Delta_{h_r}^* f\|_{L_p(S^m)} \qquad (3)$$

and obtained the following result.

Theorem B. For $f \in L_p(S^m)$, $1 \leq p \leq \infty$ and $0 < \alpha < r$, $r \in \mathbb{N}$,

$$E_n(f, S^m)_p \leq Cn^{-2\alpha} \iff \omega_r^*(f,t)_p \leq C_1 t^\alpha.$$

Ragozin [16] found a constructive description of the spherical Lipschitz class (in the spirit of Problem 1.)

Theorem C. For $f \in L_\infty(S^m)$, $0 < \alpha < 1$,

$$E_n(f, S^m)_\infty \leq Cn^{-\alpha} \iff |f(x) - f(y)| \leq C_1 |x-y|^\alpha, \qquad x, y \in S^m.$$

There are a few results on the polynomial approximation of functions in several variables. The following authors obtained Jackson-type theorems on convex bodies in $\mathbb{R}^m$: Newman and Shapiro [13], Brudnyi [3], Ganzburg [10, 11], Ditzian and Totik [7], Dubiner [8], Ditzian [6], and others. In particular, using Approach 2, Ditzian [6] for a simple polytope $V \in \mathbb{R}^m$ and Dubiner [8] for a convex body V, established the direct and converse theorems of polynomial approximation of the form ($p > 0$, $0 < \alpha < r$)

$$E_n(f, V)_p \leq Cn^{-\alpha} \iff w_r(f,t)_p \leq C_1 t^\alpha,$$

where $w_r(f,t)_p$ is a measure of smoothness of f. Unfortunately, there is not an explicit connection between this measure of smoothness and differential properties of f.

In the paper we present two results on application of Approach 1 (Theorem 1) and Approach 3 (Theorem 2) to polynomial approximation on B^m. The proofs of these theorems outlined in Section 5 are based on the method of associated functions developed for $p = \infty$ in [13, 18].

§2. A Fuksman-type Theorem for Multivariate Approximation

Theorem 1. *If f is a continuous function on B^m, then*
$$E_n(f, B^m) \leq Cn^{-\alpha}, \quad 0 < \alpha < 1, \tag{4}$$
if and only if for all x and y from B^m
$$|f(x) - f(y)| \leq C_1(|x-y| + (||x| - |y||)/(\sqrt{1-|x|^2} + \sqrt{1-|y|^2}))^\alpha. \tag{5}$$

Remark 1. It is easy to show that for all $x, y \in [-1, 1]$,
$$\frac{(1/2)|x-y|}{\sqrt{1-x^2} + \sqrt{1-y^2} + \sqrt{|x-y|}} \leq |x-y| + \frac{||x|-|y||}{\sqrt{1-x^2} + \sqrt{1-y^2}}$$
$$\leq \frac{21|x-y|}{\sqrt{1-x^2} + \sqrt{1-y^2} + \sqrt{|x-y|}},$$

and consequently Theorem A follows from Theorem 1 for $m = 1$. Note that the analogues of these inequalities are not valid for $m > 1$.

§3. The Generalized Chebyshev Operator and Generalized Moduli of Smoothness on the Ball

Here and in the sequel, we consider the multi-dimensional Chebyshev weight $v(x) = (1 - |x|^2)^{-1/2}$. Let
$$D = \sum_{i=1}^m (1-x_i^2)\partial^2/\partial x_i^2 - 2 \sum_{1 \leq i < j \leq m} x_i x_j \partial^2/\partial x_i \partial x_j - m \sum_{i=1}^m x_i \partial/\partial x_i$$

be the generalized Chebyshev operator [1]. The one-dimensional operator $D = (1-x^2)d^2/dx^2 - xd/dx$ is the classic Chebyshev operator.

Lemma 1. *The operator D has the following properties:*

(a) *D is an elliptic differential operator with a degeneracy on the boundary of B^m;*

(b) *all eigenfunctions of D are polynomials of the form*
$$\Phi_{k,N}(x) = P_k(x) C_{2N}^{k+(m-1)/2}(\sqrt{1-|x|^2}),$$
where $P_k \in \mathcal{P}_{k,m}$ is a harmonic polynomial, and $C_{2N}^\lambda \in \mathcal{P}_{2N,1}$ is the Gegenbauer polynomial, $k, N = 0, 1, ...$;

(c) *for all eigenfunctions $\Phi_{k,N}$, $D\Phi_{k,N}(x) = n(n+m-1)\Phi_{k,N}$, $n = k + 2N$, i.e. all eigenvalues of D are $n(n+m-1)$, $n = 0, 1, ...$;*

(d) *the set $\{\Phi_{k,N}\}_{k+2N \leq n}$ forms an orthogonal basis for $\mathcal{P}_{n,m}$ on B^m with respect to the weight v, and $\{\Phi_{k,N}\}_{k \geq 0, N \geq 0}$ forms an orthogonal basis for $L_{2,v}(B^m)$;*

Approximation on the Ball 145

(e) there is an extension of D^r, $r \in \mathbb{N}$, to a weighted Sobolev-type space $W_{p,D,v}^r \subseteq L_{p,v}(B^m)$ with the "derivatives" D^r.

We define the generalized translation of $f \in L_{p,v}(B^m)$ by

$$\tau_h f(x) = (1/|\Omega(x)|) \int_{\Omega(x)} f(xh + z\sqrt{1-h^2}) dz, \qquad h \in [-1, 1],$$

where $\Omega(x)$ is the m-dimensional ellipsoid in B^m centered at zero with axes of lengths $2\sqrt{1-|x|^2}$, 2, ..., 2, and with $x/|x|$ the directional vector of the first axis. We also define the generalized difference $\Delta_h f(x) = f(x) - \tau_h f(x)$.

Lemma 2. *The operators τ_h and Δ_h have the following properties:*
(a) *if $f \in C^2(B^m)$, then $Df(x) = \lim_{h \to 1^-} (\Delta_h f(x))/(1-h)$;*
(b) *for any eigenfunction $\Phi_{k,N}(x)$, $k \geq 0$, $N \geq 0$,*

$$\tau_h \Phi_{k,N}(x) = (C_n^{(m-1)/2}(1))^{-1} C_n^{(m-1)/2}(h) \Phi_{k,N}(x), \qquad n = k + 2N.$$

The generalized modulus of smoothness of $f \in L_{p,v}$ is defined by

$$\omega_r(f, t)_p = \sup_{1-t < h_j < 1,\, 1 \leq j \leq m} \|\Delta_{h_1} \ldots \Delta_{h_r}\|_{L_{p,v}(B^m)}. \tag{6}$$

Lemma 3. *The modulus of smoothness has the following properties:*
(a) $\lim_{t \to 0+} \omega_r(f, t)_p = 0$;
(b) $\omega_r(f, t_1)_p \leq \omega_r(f, t_2)_p$, $\quad 0 < t_1 < t_2 \leq 2$;
(c) $\omega_r(f, t)_p \leq 2^{r-k} \omega_k(f, t)_p$, $\quad 1 \leq k < r$;
(d) *if $f \in W_{p,D,v}^k(B^m)$, then*

$$\omega_r(f, t)_p \leq C t^k \omega_{r-k}(D^k f, t)_p, \qquad k < r,$$
$$\omega_k(f, t)_p \leq C t^k \|D^k f\|_{L_{p,v}(B^m)};$$

(e) $C \omega_r(f, t)_p \leq K(t^r, f, L_{p,v}(B^m), W_{p,D,v}^r(B^m)) \leq C_1 \omega_r(f, t)_p$, *where*

$$K(t^r, f, L_{p,v}(B^m), W_{p,D,v}^r(B^m)) = \inf_{g \in W_{p,D,v}^r(B^m)} (\|f - g\|_{L_{p,v}(B^m)} + t^r \|D^r g\|_{L_{p,v}(B^m)})$$

is the K-functional;
(f) $\omega_r(f, t_1)_p \leq \max(1, (t_1/t_2)^r) \omega_r(f, t_2)$, $\quad 0 < t_1, t_2 < 2$.

§4. Best Approximation of Functions from $L_{p,v}(B^m)$

Theorem 2. *If $f \in L_{p,v}(B^m)$, then the following assertions are equivalent to each other:*
(a) $E_n(f, B^m)_{p,v} \leq C n^{-2(r+\alpha)}$, $\quad r \in \mathbb{N}$, $0 < \alpha \leq 1$;
(b) $\omega_k(f, t)_p \leq C_1 t^{-(r+\alpha)}$, $\quad r \in \mathbb{N}$, $0 < \alpha \leq 1$, $k > r + \alpha$;
(c) $\omega_l(D^r f, t)_p \leq C_2 t^{-\alpha}$, $\quad l, r \in \mathbb{N}$, $0 < \alpha \leq 1$, $l > \alpha$.

For $m = 1$ these results were proved by Butzer and Stens [4], while for $m > 1$, $r = 0$, and $k = l = 1$ the theorem was obtained by Ganzburg [11].

§5. Proofs

We define the operator $F(x_1, \ldots, x_m, x_{m+1}) = Lf(x) = f(x_1, \ldots, x_m)$, where $f \in L_{p,v}(B^m)$, $x = (x_1, \ldots, x_m) \in B^m$, $(x_1, \ldots, x_m, x_{m+1}) \in S^m$, i.e., $\sum_{i=1}^{m+1} x_i^2 = 1$).

Lemma 4. L has the following properties:

(a) L is a linear operator from $L_{p,v}(B^m)$ to $L_p(S^m)$, $1 \leq p \leq \infty$ with $\|L\| = 1$;

(b) if $f \in W_{p,D}^r(B^m)$, $1 \leq p \leq \infty$, and δ is the Beltrami-Laplace operator on S^m, then $LD^r f(x) = \delta^r Lf(x)$;

(c) the function $L\Phi_{k,N}(x)$ is a spherical harmonic of degree $k + 2N$, $k \geq 0$, $N \geq 0$;

(d) $L\tau_h f = \tau_h^* Lf$, where $f \in L_{p,v}$, and τ_h^* is spherical shift (1);

(e) if P is the restriction of a polynomial from $\mathcal{P}_{n,m}$ to B^m, then LP is the restriction of a polynomial from $\mathcal{P}_{n,m+1}$ to S^m.

Proof: The function F does not depend on x_{m+1}, consequently Lemma 4(a) follows from the identity

$$\left(\int_{S^m} |F(y)|^p dy\right)^{1/p} = \left(\int_{B^m} |f(x)|^p (1 - |x|^2)^{-1/2} dx\right)^{1/p}, \quad 1 \leq p \leq \infty.$$

Statements (b), (c), and (d) can be proved by the straightforward calculations, and Lemma 4(e) is a simple consequence of the inclusion $\mathcal{P}_{n,m} \subseteq \mathcal{P}_{n,m+1}$. □

It is easy to see that Lemmas 1, 2, and 3 follow from Lemma 4 and the corresponding spherical results [19].

Lemma 5. For $f \in L_{p,v}(B^m)$, $1 \leq p \leq \infty$, the following relations hold:

$$L\Delta_{h_1} \ldots \Delta_{h_r} f(x) = \Delta_{h_1}^* \ldots \Delta_{h_r}^* Lf(x), \quad x \in B^m; \tag{7}$$

$$\omega_r(f, t)_p = \omega_r^*(F, t)_p, \quad t \in [0, 1]; \tag{8}$$

$$E_n(f, B^m)_{p,v} = E_n(F, S^m)_p \tag{9}$$

where Δ_h^*, ω_r^*, and ω_r are defined by (2), (3), and (6), respectfully.

Proof: Note first that the set $\bigcup_{n=0}^\infty \text{span}\{\Phi_{k,N}\}_{k+2N \leq n}$ is dense in $L_{p,v}(B^m)$, by Lemma 1(d). Consequently it suffices to prove (7) for $f = \Phi_{k,N}$. Using Lemma 2(b) we obtain

$$L\Delta_{h_1} \ldots \Delta_{h_r} \Phi_{k,N}(x) = \prod_{i=1}^r (1 - (C_n^{(m-1)/2}(1))^{-1} C_n^{(m-1)/2}(h_i)) \Phi_{k,N}(x), \tag{10}$$

where $n = k + 2N$. On the other hand, $L\Phi_{k,N}$ is a spherical harmonic of degree n, by Lemma 4(b), consequently [19]

$$\Delta_{h_1}^* \ldots \Delta_{h_r}^* L\Phi_{k,N}(x) = \prod_{i=1}^r (1 - (C_n^{(m-1)/2}(1))^{-1} C_n^{(m-1)/2}(h_i)) \Phi_{k,N}(x). \tag{11}$$

Approximation on the Ball 147

Then (7) follows from (10) and (11). Since (8) is an immediate consequence of (7), it remains to prove (9). Let $P_n \in \mathcal{P}_{n,m}$ satisfy the equality $E_n(f, B^m)_{p,v} = \|f - P_n\|_{L_{p,v}(B^m)}$. Using statements (a) and (e) of Lemma 4, we have

$$E_n(F, S^m)_p \leq \|Lf - LP_n\|_{L_p(S^m)} \leq \|f - P_n\|_{L_{p,v}(B^m)} = E_n(f, B^m)_{p,v}. \quad (12)$$

Next, let $P_n^* \in \mathcal{P}_{n,m+1}$ satisfy the equality $E_n(F, S^m)_p = \|F - P_n^*\|_{L_p(S^m)}$. For $Q(x_1, ..., x_m, x_{m+1}) = (1/2)(P_n^*(x_1, ..., x_m, x_{m+1}) + P_n^*(x_1, ..., x_m, x_{m+1}))$ we obtain $Q_n \in \mathcal{P}_{n,m+1}$ and $H_n(x) = Q_n(x_1, ..., x_m, \sqrt{1 - |x|^2}) \in \mathcal{P}_{n,m}$. Taking account of the equality $F(x_1, ..., x_m, x_{m+1}) = F(x_1, ..., x_m, -x_{m+1})$, it is easy to show that

$$E_n(F, S^m)_p = \|F - Q_n\|_{L_p(S^m)} = \|f - H_n\|_{L_{p,v}(B^m)} \geq E_n(f, B^m)_{p,v}. \quad (13)$$

Now (9) follows from (12) and (13). □

Proof of Theorem 1: Using (8) and (9) for $p = \infty$ we obtain that (4) holds if and only if for all $x, y \in B^m$,

$$|f(x) - f(y)| \leq C_1 \left(\sum_{i=1}^m (x_i - y_i)^2 + ((1 - |x|^2)^{1/2} - (1 - |y|2)^{1/2})^2 \right)^{\alpha/2}.$$

This inequality is equivalent to (5). □

The Theorem 2 is an immediate consequence of (8), (9), and the corresponding spherical results like Theorem B [19].

Remark 2. Using this approach and the corresponding spherical results [12, 17] it is possible to obtain the analogues of Theorem 2 for other moduli of smoothness. In particular, Theorem 2 also holds for

$$\omega_r(f, t)_p = \sup_{1-t < h < 1} \|\Delta_h^r\|_{L_{p,v}(B^m)}.$$

Acknowledgments. Thanks to Roald Trigub, Igor Pritsker, and the referee for inspiring discussions and provision of references.

References

1. Appell, P. and J. Kampe de Feriet, *Fonctions hypergeometriques et hyperspheriques, Polynomes d'Hermite*, Gauthier-Villars, 1926.

2. Bernstein, S. N., On the best approximation of continuous functions by polynomials of given degree, Communications of Charkov Math. Soc. **13**(2) (1912), 49–194 (Russian).

3. Brudnyi, Yu. A., Approximation of functions defined on a convex polyhedron, Dokl. Akad. Nauk. SSSR **195** (1970), 1007–1009 (Russian).

4. Butzer, P. L. and R. L. Stens, Chebyshev transform methods in the theory of best algebraic approximation, Abhandlungen. Math. Sem. Univ. Hamburg **45** (1976), 165–190.

5. DeVore, R. A. and G. G. Lorentz, *Constructive Approximation*, Springer-Verlag, Berlin, Heidelberg, New York, 1993.

6. Ditzian, Z., Polynomial approximation in $L_p(S)$ for $p > 0$, Constr. Approx. **12** (1996), 241–269.

7. Ditzian, Z. and V. Totik, *Moduli of Smoothness*, Springer-Verlag, New York, 1987.

8. Dubiner, M., The theory of multi-dimensional polynomial approximation, J. Analyse Math. **67** (1995), 39–116.

9. Fuksman, A. L., Structural characteristics of functions for which $e_n(f, [-1, 1]) \leq Mn^{-(k+\alpha)}$, Uspehi Mat. Nauk **20** (1965), 187–190 (Russian).

10. Ganzburg, M. I., Multidimensional Jackson theorems, Siberian Math. J. **23** (1981), 316–331.

11. Ganzburg, M. I., A direct and inverse theorem on approximation by polynomials on the m-dimensional ball, Proc. Steklov Inst. Math. **3** (1989), 103–104.

12. Lizorkin, P. I. and Kh. P. Rustamov, Nikol'skii-Besov spaces on the sphere in connection with approximation theory, Proc. Steklov Inst. Math. **3** (1994), 149–172.

13. Newman, D. J. and H. S. Shapiro, Jackson's theorem in higher dimensions, in *On Approximation Theory V*, Birkhauser-Verlag, Basel and Stuttgart, 1964, pp. 208–219.

14. Pawelke, S., Uber die Approximationsordrung bei Kugelfunktionen und algebraischen Polynomen, Tohoku Math. J. **24** (1972), 473–486.

15. Potapov, M. K., Approximation by algebraic polynomials in an integral metric with Jacobi weights, Moscow Univ. Math. Bull. **38**, 48–57.

16. Ragozin, D. L., *Approximation Theory on Compact Manifolds, and Lie Groups, with Applications to Harmonic Analysis*, Ph.D. Thesis, Harvard Univ., Cambridge, MA, 1967.

17. Rustamov, Kh. P., On approximation of functions on the sphere, Dokl. Akad. Nauk SSSR **320** (1991), 1319–1325 (Russian).

18. Timan, A. F., *Theory of Approximation of Functions of a Real Variable*, McMillan, New York, 1963.

19. Wherens, M., Best approximation on the unit sphere in $\mathbb{R}^k$, in *Functional Analysis and Approximation (Oberwolfach, 1980)*, Birkhauser, Basel, 1981, pp. 233–245.

Approximation Using Scale Transformations

Christian Gout

Abstract. Curve and surface fitting using spline functions from rapidly varying data is a difficult problem. Without information about the location of the variations, the usual approximation methods lead to instability phenomenae or undesirable oscillations that can locally and even globally hinder the approximation. So, we propose a new method which uses scale transformations (see Gout [17]). The originality of the method consists in a pre-processing and a post-processing of the data. Instead of trying to find directly an approximant, we first apply a scale transformation φ_d to the z-values of the function. In the particular case of the approximation of surfaces, the originality of the method consists in removing the variations of the unknown function using a scale transformation in the pre-processing. And so, the "new" data do not have great variations. So, we now could use a usual approximant T^d which will not create oscillations. We then go back to the "real" value of the function and apply an inverse scale transformation ψ_d. The convergence of the method is proved when the number of the data increases to infinity.

§1. Introduction

The goal of this work is to give an approximation of surfaces from rapidly varying data knowing that usual methods lead to humps and oscillations. We have to take into account that

- We do not have any idea about the location of the large variations;
- The data file contains several thousands points of data (Geophysics...);
- CPU time equivalent with the one using usual methods.

Several methods exist to treat this type of problem. When having the knowledge of the location of the large variations, Salkauskas [21], Foley [14] have proposed methods that use spline under tension with a non-constant smoothing parameter, Hsieh and Chang [18] propose a concept of virtual nodes inserted at the level of the large variations in the case of CAGD approximant. Without any knowledge about the location of the large variations, Franke [15], Le Méhauté and Bouhamidi [7] have proposed spline under tension belonging to more general space; these last methods give good results in the case of curve fitting but the result could be improved for surface fitting. The proposed method in this work uses scale transformations, this method is applied without any knowledge about the data, in particular, about the location of the large variations.

From a Lagrange data set of a function f, indexed with the real d (when $d \to 0$, it means that the number of data points increase to infinity), we have: $f \colon \Omega \longrightarrow [a,b], \varphi_d \colon [a,b] \longrightarrow [\alpha, \beta]$, and $T^d \colon (\varphi_d \circ f) \in H^m(\Omega, [\alpha, \beta]) \to T^d(\varphi_d \circ f) \in H^m(\Omega, [\alpha, \beta])$, and finally $\psi_d \circ (T^d(\varphi_d \circ f)) \in H^m(\Omega, [a,b])$, where f is an explicit unknown function, φ_d(pre-processing) and ψ_d(post-processing) are scale transformations families and $T^d \in \mathcal{L}_c(H^m(\Omega, [\alpha, \beta]), H^m(\Omega, [\alpha, \beta]))$ is an approximation operator (spline for example).

The goals of this work consist in constructing and establishing the convergence of the approximation $\psi_d \circ T^d(\varphi_d \circ f)$ in a certain space when d comes to 0.

§2. Notations

We introduce :
- $m', m \in \mathbb{N}, m' > m + n/2, n \in \mathbb{N}^*$.
- Ω a bounded nonempty connected set with a Lipschitz-continuous boundary of $\mathbb{R}^n$.
- $f \in H^{m'}(\Omega, [a,b])$, $\left(H^{m'}(\Omega) \hookrightarrow C^m(\Omega) \right)$ an unknown function.
- a subset D of reals > 0 where 0 is an accumulation point.
- for any $d \in D$, a subset A^d of $N = N(d)$ distinct points $\overline{\Omega}$ such that

$$\sup \delta\left(x, A^d\right) = d,$$

where δ is the Euclidean distance of $\mathbb{R}^2$.
- for any $d \in D$, the set Z_1^d of $N = N(d)$ reals such that

$$\forall x \in A^d, \quad f(x) \in Z_1^d.$$

- for any $d \in D$, the sequence $\widetilde{Z}_2^d$ of $p(d)$ distinct z-values ordered from Z_1^d.

§3. Construction and Convergence of the Approximation

3.1. Scale Transformations Families

In this section, we propose a construction for the scale transformation families (φ_d) (pre-processing) and (ψ_d) (post-processing) (see also Apprato [1], Torrens [25], which define such families). We recall that $\widetilde{Z}_2^d$ is the set of $p(d)$ reals from Z_1^d, included in $[a,b]$ $(= Im\ f)$:

$$\forall \widetilde{z}_i \in Z_2^d, a = \widetilde{z}_1 < \widetilde{z}_2 < \widetilde{z}_3 < ... < \widetilde{z}_{p(d)-1} < \widetilde{z}_{p(d)} = b.$$

For any $d \in D$, we subdivide $[\alpha, \beta]$ using a regular repartition

$$\{u_1, u_2, ..., u_i, ..., u_{p(d)}\}_{i=1,...,p(d)}.$$

Approximation Using Scale Transformations 151

This subdivision satisfies: $u_{i+1} - u_i = (\beta - \alpha)/(p(d) - 1)$. For all $d \in D$, $i = 1, ..., p(d) - 1$, and $x \in [\tilde{z}_i, \tilde{z}_{i+1}]$, we define

$$\varphi_d(x) = u_i q_{0m}^0 [(x - \tilde{z}_i)/(\tilde{z}_{i+1} - \tilde{z}_i)] + u_{i+1} q_{0m}^1 [(x - \tilde{z}_i)/(\tilde{z}_{i+1} - \tilde{z}_i)]$$
$$+ \alpha_1(\tilde{z}_i)(\tilde{z}_{i+1} - \tilde{z}_i) q_{1m}^0 [(x - \tilde{z}_i)/(\tilde{z}_{i+1} - \tilde{z}_i)]$$
$$+ \alpha_1(\tilde{z}_{i+1})(\tilde{z}_{i+1} - \tilde{z}_i) q_{1m}^1 [(x - \tilde{z}_i)/(\tilde{z}_{i+1} - \tilde{z}_i)],$$

where the (q_{*m}^*) are the Hermite's finite element basis function in dimension 1 and where $\alpha_1(\tilde{z}_i) = (u_{i+1} - u_i)/(\tilde{z}_{i+1} - \tilde{z}_i)$ and $\alpha_1(\tilde{z}_{p(d)}) = (\tilde{z}_{p(d)} - \tilde{z}_{p(d)-1})/(u_{p(d)} - u_{p(d)-1})$. Then we have (see Gout [17]):

(i) $\forall d \in D, \forall i = 1, ..., p(d), \varphi_d(\tilde{z}_i) = u_i$,
(ii) $\forall d \in D, \varphi_d \in C^m[a, b]$,
(iii) $\exists C > 0, C$ no dependant of d, such that $\|\varphi_d\|_{C^m[a,b]} \leq C$,
(iv) $\lim \varphi_d = \varphi$ in $C^0([a, b])$ when $d \to 0$, where $\varphi : [a, b] \to [0, 1]$ is defined by

$$\varphi(z) = [(\beta - \alpha)(z - a)]/(b - a) + \alpha.$$

For all $d \in D$, $i = 1, ..., p(d) - 1$, and $x \in [u_i, u_{i+1}]$, we now define

$$\psi_d(x) = \tilde{z}_i q_{0m}^0 [(x - u_i)/(u_{i+1} - u_i)] + \tilde{z}_{i+1} q_{0m}^1 [(x - u_i)/(u_{i+1} - u_i)]$$
$$+ (u_{i+1} - u_i) \beta_1(u_i) q_{1m}^0 [(x - u_i)/(u_{i+1} - u_i)]$$
$$+ (u_{i+1} - u_i) \beta_1(u_{i+1}) q_{1m}^1 [(x - u_i)/(u_{i+1} - u_i)],$$

where the q_{lm}^i, for any $i = 0, 1$, $l = 0, 1$, represent the Hermite's finite element basis function in dimension 1 and where

$$\beta_1(u_i) = (\tilde{z}_{i+1} - \tilde{z}_i)/(u_{i+1} - u_i) \quad \text{and} \quad \beta_1(u_{p(d)}) = \beta_1(u_{p(d)-1}).$$

Then, we have (see Gout [17]) :

(i) $\forall d \in D, \forall i = 1, ..., p(d), \psi_d(u_i) = \tilde{z}_i$;
(ii) $\forall d \in D, \psi_d \in C^m[\alpha, \beta]$;
(iii) $\exists C > 0, C$ no dependant with d, such that $\|\psi_d\|_{C^m[\alpha,\beta]} \leq C$
(iv) $\lim \psi_d = \varphi^{-1}$ in $C^0([\alpha, \beta])$ when $d \to 0$.

3.2. Operator T^d: D^m Smoothing Spline

Spline functions first appeared in Schoenberg [15]. In one-dimension, I. J. Schoenberg defined a spline function to be a piecewise polynomial function whose derivatives are continuous to order m. In one-dimension, the same function could be obtained by minimizing a functional $\int_a^b (f^m(t))^2 dt$ under interpolation constraints of Lagrange or Hermite type. Two different approaches occur in dimension 2 or higher.

The first approach (called local interpolation) needs the triangulation of a domain of $\mathbb{R}^n$, and the calculation of some polynomial functions on each n-simplex, under some boundary conditions. The problem of this approach consists in dealing with this boundary conditions and in the choice of the triangulation. The obtained spline function does not satisfy minimization criterions. Many authors have worked on this approach, resulting in many CAGD algorithms and methods. For example, see Bézier [6], Chui [8], de Boor [11], de Casteljau [12], Schumaker [23], etc.

The second approach (called global interpolation) involvess the minimization of a functional under some constraints (Lagrange or Hermite). To insure the existence-uniqueness of a spline function and to insure the smoothness of this spline function, we consider that this function belongs to a (semi)-Hilbertian space, and we minimize its norm (called energy in the case of thin plates) in this space. This method uses Hilbertian spaces and reproducing kernels properties (see Aronsajn [4]) and also Hilbertian kernel (see Schwarz [24]). The problem of this approach is to obtain a characterization of the spline which could be used. It is necessary to determine the considered kernel in the case of a two-dimensional space (or higher), because this determination comes from a partial differential equation which is hard to solve in this case.

Spline function in Hilbert spaces appeared in 1958. Its generalization was done by Attéia [5], Laurent [19], Arcangéli [2], Duchon [13], and others. In this paper, we will use a smoothing D^m-spline (cf. Arcangéli [3]) which allows us to work with large sets of data. We consider the problem of minimizing

$$J_\varepsilon^d(\Phi) = \langle \rho^d (\Phi - \varphi_d \circ f) \rangle_d^2 + \varepsilon |\Phi|_{m,\Omega}^2, \qquad \forall\, \Phi \in H^m(\Omega).$$

We call σ_ε^d the D^m-*smoothing spline* on Ω relative to $\rho^d(\varphi_d \circ f)$, which is the unique solution of the minimization problem:

$$\forall\, \Phi \in H^m(\Omega), \quad \sigma_\varepsilon^d \in H^m(\Omega), \qquad J_\varepsilon^d(\sigma_\varepsilon^d) \leq J_\varepsilon^d(\Phi).$$

The solution σ_ε^d of the previous problem is also the solution of the variational problem: $\forall\, \Phi \in H^m(\Omega)$,

$$\sigma_\varepsilon^d \in H^m(\Omega), \langle \rho^d \sigma_\varepsilon^d, \rho^d \Phi \rangle_d + \varepsilon\left(\sigma_\varepsilon^d, \Phi\right)_{m,\Omega} = \langle \rho^d(\varphi_d \circ f), \rho^d \Phi \rangle_d.$$

This problem has a unique solution (using Lax-Milgram Lemma and Necas' result [20] to establish norm equivalence).

3.3 Convergence of the Approximation

We are now going to establish the convergence of the approximation when the number of data increases to infinity.

Approximation Using Scale Transformations

Theorem. *We keep the notations of the Sections 3.1 and 3.2. Then*
i) *The family* $\left(\sigma_\varepsilon^d \left(\varphi_d \circ f\right)\right)_{(d,\varepsilon) \in D \times]0, \varepsilon_0[}$ *is bounded in* $H^m(\Omega)$.
ii) $\lim \left(\sigma_\varepsilon^d \left(\varphi_d \circ f\right)\right)_{\varepsilon \in]0, \varepsilon_0[} = \varphi \circ f$ *in* $C^0(\overline{\Omega})$ *when d comes to 0.*

Proof: To prove these relations, we distinguish 4 steps.

Step 1) We show that the family $\left(\sigma_\varepsilon^d \left(\varphi_d \circ f\right)\right)$ is bounded in $H^m(\Omega)$. We take $\Phi = \sigma_\varepsilon^d \left(\varphi_d \circ f\right) - \varphi_d \circ f$ in the variational problem (cf. Section 2.2) in which we obtain the smoothing D^m spline. Thus, we get

$$\left\langle \rho^d \left(\sigma_\varepsilon^d \left(\varphi_d \circ f\right) - \varphi_d \circ f\right) \right\rangle_d^2 + \varepsilon \left(\sigma_\varepsilon^d \left(\varphi_d \circ f\right), \sigma_\varepsilon^d \left(\varphi_d \circ f\right) - \varphi_d \circ f\right)_{m,\Omega} = 0,$$

for all $(d, \varepsilon) \in D \times]0, \varepsilon_0[$. We deduce that

$$\varepsilon \left(\sigma_\varepsilon^d \left(\varphi_d \circ f\right), \sigma_\varepsilon^d \left(\varphi_d \circ f\right) - \varphi_d \circ f\right)_{m,\Omega} \leq 0,$$

and so $\forall (d, \varepsilon) \in D \times]0, \varepsilon_0[$, $\left|\sigma_\varepsilon^d \left(\varphi_d \circ f\right)\right|_{m,\Omega} \leq \left|\varphi_d \circ f\right|_{m,\Omega}$, because $\varepsilon > 0$. So, considering that the family $(\varphi_d \circ f)_{d \in D}$ is bounded in $H^m(\Omega)$, we show that $\left|\sigma_\varepsilon^d \left(\varphi_d \circ f\right)\right|_{m,\Omega}$ is bounded. We also deduce that

$$\exists\, C > 0, \forall (d, \varepsilon) \in D \times]0, \varepsilon_0[: \left\langle \rho^d \left(\sigma_\varepsilon^d \left(\varphi_d \circ f\right) - \varphi_d \circ f\right) \right\rangle_d^2 \leq C.$$

The norm equivalence between $|||.|||_0^d$ and $\|.\|_{m,\Omega}$ implies that

$$\exists\, C > 0, \exists\, \eta > 0, \forall (d, \varepsilon) \in D \times]0, \varepsilon_0[, d \leq \eta : \left\|\sigma_\varepsilon^d \left(\varphi_d \circ f\right)\right\|_{m,\Omega} \leq C.$$

Thus, the family $\left(\sigma_\varepsilon^d \left(\varphi_d \circ f\right)\right)_{(d,\varepsilon) \in D \times]0, \varepsilon_0[,\, d \leq \eta}$ is bounded in $H^m(\Omega)$, independently of d so i) is verified.

Step 2) There exists a subsequence $\left\{\sigma_{\varepsilon_n}^{d_n} \left(\varphi_{d_n} \circ f\right)\right\}_{n \in \mathbb{N}}$ extracted from the family $\left(\sigma_\varepsilon^d \left(\varphi_d \circ f\right)\right)_{d \in D}$, with (when $n \to +\infty$): $\lim d_n = 0$ and $\{\varepsilon_n\}_{n \in \mathbb{N}} \subset]0, \varepsilon_0[$, and there exists an element $f^* \in H^m(\Omega)$ such that

$$\lim weak\; \sigma_{\varepsilon_n}^{d_n} \left(\varphi_{d_n} \circ f\right) = f^* \text{ in } H^m(\Omega).$$

Thus, from the continuous embedding $H^m(\Omega) \subset C^0(\overline{\Omega})$, because $m \geq 2$, , when $n \to +\infty$:

$$\lim \left(\sigma_{\varepsilon_n}^{d_n} \left(\varphi_{d_n} \circ f\right)\right) = f^* in C^0(\overline{\Omega}).$$

Step 3) We show that $f^* = \varphi \circ f$. If we suppose that $f^* \neq \varphi \circ f$, there exists a non-empty open ω included in Ω and a real $\alpha > 0$ such that $\forall x \in \omega, |f^*(x) - (\varphi \circ f)(x)| > \alpha$, using the continuous embedding $H^m(\Omega) \subset C^0(\overline{\Omega})$. We take $\nu = \left(1 + E(C^2/\alpha^2)\right)$. Let B_0 be a subset constituted with ν distinct points of ω. Then (cf Arcangéli [2] for an equivalent proof),

$$\forall\, b \in B_0, \exists\, (b_d)_{d \in D} : \left(\forall\, d \in D, b_d \in A^d\right) \text{ and } (b = \lim b_d) \text{ when } d \to 0.$$

For any $d \in D$, we denote by B_d^0 the resulting set of points of A^d associated with B_0. It is clear that for any d small enough, B_d^0 is constituted with distinct points. We have

$$\forall n \in \mathbb{N}: \sum_{b^{d_n^*} \in B_0^{d_n^*}} \left[\sigma_{\varepsilon_n^*}^{d_n^*}(\varphi_{d*n} \circ f)\left(b^{d_n^*}\right) - (\varphi_{d_n^*} \circ f)\left(b^{d_n^*}\right) \right]^2 \leq C^2.$$

Using the Hölderian embedding theorem $\left(H^m(\Omega) \hookrightarrow C^{0,\alpha}(\overline{\Omega})\right)$, the Sobolev continuous embedding theorem $\left(H^m(\Omega) \subset C^0(\overline{\Omega})\right)$, and knowing that the families $\left(\sigma_\varepsilon^d(\varphi_d \circ f)\right)_d$ and $(\varphi_d \circ f)_d$ are bounded in $H^m(\Omega)$ following the point i), we obtain: (when $n \to +\infty$)

$$\forall b \in B_0, \quad \lim \sigma_{\varepsilon_n^*}^{d_n^*}(\varphi_{d*n} \circ f)\left(b^{d_n^*}\right) = f^*(b)$$

$$\forall b \in B_0, \quad \lim \varphi_{d_n^*}\left[f\left(b^{d_n^*}\right)\right] = \varphi(f(b)).$$

When $n \to +\infty$, $\sum_{b \in B_0} (f^*(b) - \varphi(f(b)))^2 \leq C^2$, and so

$$\alpha^2 \, card(B_0) \leq C^2,$$

which makes a contradiction with the definition of ν. Thus: $f^* = \varphi \circ f$.

Step 4) We show that the family $(\sigma_d^\varepsilon(\varphi_d \circ f))_{(d,\varepsilon) \in D \times]0,\varepsilon_0[, \, d \leq \eta}$ converges itself to $\varphi \circ f$ in $C^0(\overline{\Omega})$ when $d \to 0$. Suppose that

$$\exists \, \mu > 0, \forall \, \beta > 0, \exists \, d \in D, d \leq \beta \text{ and } \exists \varepsilon \in \,]0, \varepsilon_0[\text{ such that}$$

$$\left\| \sigma_\varepsilon^d(\varphi_d \circ f) - \varphi \circ f \right\|_{C^0(\overline{\Omega})} > \mu.$$

In particular, we have $\exists \, \mu > 0, \forall n \in \mathbb{N}, \exists d_n^{**} \in \mathbb{N}, \, d_n^{**} \leq 1/n$ and $\exists \varepsilon \in \,]0, \varepsilon_0[:$

$$\left\| \sigma_{\varepsilon_n^{**}}^{d_n^{**}}(\varphi_{d**n} \circ f) - \varphi \circ f \right\|_{C^0(\overline{\Omega})} > \mu.$$

But the sequence $\sigma_{\varepsilon_n^{**}}^{d_n^{**}}(\varphi_{d**n} \circ f)_{n \in \mathbb{N}}$ is bounded in $H^m(\Omega)$ because the family $\left(\sigma_\varepsilon^d(\varphi_d \circ f)\right)_{(d,\varepsilon) \in D \times]0,\varepsilon_0[, \, d \leq \eta}$ is bounded in $H^m(\Omega)$ and $d_n^{**} \to 0$ when $n \to +\infty$. Thus, following the the reasoning in 1) and 2), we can extract a subsequence which converges to $(\varphi \circ f)$ in $C^0(\overline{\Omega})$, which gives a contradiction, and the result follows. □

From this last result, it is possible to establish the convergence of the approximation.

Theorem. *(Convergence of the approximation)* $\lim \left\{ \psi_d \circ \sigma^\varepsilon_{\varphi_d \circ f} \right\} = \varphi^{-1} \circ \varphi \circ f = f$ in $H^{m-\theta}(\Omega)$ when $d \to 0$, for any $\theta > 0$ such that $\theta < m - 1$ $\left(\Rightarrow H^{m-\theta}(\Omega) \hookrightarrow C^0(\overline{\Omega}) \right)$.

Proof: Cf. Gout [16]. □

If we take $n = 2$ and $m = 3$, the convergence of the approximation takes place in $H^{2-\theta}$ for any $\theta \in]0, 1[$.
Numerical results are given in Gout [17] in the case of C^1 approximant (cases of both curves and surfaces), and we show that the rate of convergence is higher in this method than the one without pre and post-processing in Gout [16].

Acknowledgments. The author is grateful to the "Region Aquitaine" which supported this work.

References

1. Apprato, D., *Approximation de Surfaces Paramétrées Par Éléments Finis*, Thèse d'Etat, Université de Pau et des Pays de l'Adour, 1987.
2. Arcangéli, R., D^m-splines sur un domaine borné de $\mathbb{R}^n$, Publication UA 1204 CNRS numéro: 1986/2, 1986.
3. Arcangéli, R., Some applications of discrete D^m splines, in *Mathematical Methods in Computer Aided Geometric Design*, T. Lyche and L. L. Schumaker (eds.), Academic Press, 1989, pp. 35–44.
4. Aronsajn, N., Theory of reproducing kernels, Trans. Amer. Math. Soc. (1950).
5. Atteia, M., Fonctions "spline" définies sur un ensemble convexe, Numer. Math. **12** (1968), 192–210.
6. Bézier, P., Procédé de définition numérique des courbes et des surfaces non mathématiques, Système Unisurf. Automatisme **13** (1968).
7. Bouhamidi, A., *Interpolation et Approximation par des Fonctions Splines Radiales à Plusieurs Variables*, Thèse, Université de Nantes, 1992.
8. Chui, C. K., Lectures on multivariate spline: Theory and applications, Center of Approximation Theory, Dept. of Mathematics, Texas A&M, College Station, Texas, 1987.
9. Ciarlet, P. G., *The Finite Element Method for Elliptic Problems*, North Holland, Amsterdam, 1978.
10. Cohen, E. and T. Lyche, Discrete B-splines and subdivision techniques in computer-aided geometric design and computer graphics, Comput. Graphics Image Process. **14** (1980), 87–111.
11. de Boor, C., *A Practical Guide to Splines*, Springer Verlag, Berlin-Heidelberg, 1978.

12. De Casteljau, P., *Outillage Méthodes de Calcul*, André Citroën Automobiles, S. A., Paris, 1959.
13. Duchon, J., Splines minimizing rotation-invariant semi-norms in Sobolev spaces, Lecture Notes in Math. **571** Springer, 1977, pp. 85–100.
14. Foley, T. A., Weighted bicubic spline interpolation to rapidly varying data, ACM Trans. Graphics **6** (1987), 1–18.
15. Franke, R., Thin plate spline with tension, Computer Aided Geometric Design **2** (1985), 87–95.
16. Gout, C., *Etude de Changements D'échelle en Approximation - Ajustement Spline sur des Surfaces*, Thèse de Doctorat, Pau, 1997.
17. Gout, C., Approximation of curves and surfaces from rapidly varying data, Center for Pure and Applied Mathematics, PAM 1001, UC Berkeley, 1998.
18. Hsieh, H. C. and W. T. Chang, Virtual knot technique for curve fitting of rapidly varying data, Computer Aided Geometric Design **11** (1994), 71–95.
19. Laurent, P. J., *Approximation et Optimisation*, Hermann, Paris, 1972.
20. Necas, J., *Les Méthodes Directes en Théorie des Équations Elliptiques*, Masson, Paris, 1967.
21. Salkauskas, K., C^1 splines for interpolation of rapidly varying data, Rocky Mountain J. Math. **14** (1975), 239–250.
22. Schoenberg, I. J., Contribution to the problem of approximation of equidistant data by anlytic functions, Quart. of Appl. Math. **4** (1960), 45–99 and 112–141.
23. Schumaker, L. L., Fitting surfaces to scattered data, in *Approximation Theory II*, G. G. Lorentz, C. K. Chui, and L. L. Schumaker (eds.), Academic Press, 1976, pp. 203–269.
24. Schwarz, L., Sous espaces hilbertiens d'espaces vectoriels topologiques et noyaux associés (noyaux reproduisants), J. Anal. Math., Paris, 1964.
25. Torrens, J. J., *Interpolacion de Superficies Parametricas Con Discontinuidades Mediante*, Thesis, Universidad de Zaragoza, 1991.

Christian GOUT
UPRES A 5033
Laboratoire de Mathematiques Appliquees
Universite de Pau - IPRA
Avenue de l'Universite
64000 Pau - FRANCE
christian.gout@univ-pau.fr

Linear Discrete Operators and Recovery of Functions

Ivan Ivanov and Boris Shekhtman

Abstract. The problem of the recovery of a function from a given uniform algebra $\mathcal{A}$ by means of its values at a given collection of points is considered. A necessary condition is given in terms of the distribution of the points used for obtaining the functions values. It is shown that the recovery is not possible for all $f \in \mathcal{A}$ in the case when those points belong to a certain boundary of $\mathcal{A}$.

§1. Introduction

The main purpose of this paper is to discuss some necessary conditions for the recovery of a function f from a given uniform algebra $\mathcal{A}$, provided one knows the values of that function at a given collection of points on a boundary of $\mathcal{A}$. Here by a uniform algebra $\mathcal{A}$ we mean a closed subalgebra of $C(X)$ (the space of all complex-valued continuous functions on a compact Hausdorff space) that separates points and contains the constants. By a recovery we mean a sequence of linear operators $L_n : \mathcal{A} \to \mathcal{A}$ defined by

$$L_n f = \sum_{k=1}^{n} f(z_{k,n}) l_{k,n}, \quad f \in \mathcal{A}, \quad \{z_{k,n}\} \in X \tag{1}$$

and such that

$$L_n f \to f, \quad as \quad n \to \infty$$

for every $f \in \mathcal{A}$. One particular case of the above described situation was considered by Somorjai in 1980. In that paper [8] he considers the disk algebra $\mathcal{A}$ of the functions analytic inside the unit disk X and continuous on the boundary of X. In this special case all different boundaries for $\mathcal{A}$ coincide with the topological boundary of X, and the points $\{z_{k,n}\}$ belong to the boundary of the unit disk. The result that Somorjai proved was quite surprising at that time: *No matter how we choose the sequence* $\{L_n\}$ *satisfying (1), there exits* a function f in $\mathcal{A}$ such that $L_n f$ does not recover f. A different approach to Somorjai's theorem was given in [7], where it was pointed out that the reason for such a phenomenon is a certain Banach space property of the disk algebra, namely that it is not an $\mathcal{L}_\infty$ space (cf. [5]). In 1990 Kisliakov [1] proved that any proper uniform algebra is not an $\mathcal{L}_\infty$ space. In this paper we combine Kisliakov's result with a technique from [7] to extend Somorjai's theorem to arbitrary uniform algebras. We then apply it to the uniform algebras of holomorphic functions of several complex variables. These results are part of a Ph.D. thesis prepared by the first-named author.

§2. Preliminaries

Let l_∞^n be $\mathbf{C}^n$ equipped with the *sup* norm, i.e., $\|(x_i)\| = \max\{|x_i|, \ i = 1, ..., n\}$. The Banach-Mazur distance from an arbitrary n-dimensional Banach space E to l_∞^n is defined as

$$d(E, l_\infty^n) = \inf\{\| T \| \| T^{-1} \|: T \text{ is an isomorphism from } E \text{ onto } l_\infty^n\}.$$

Definition 1. [5] Let $1 \leq \lambda \leq \infty$. A Banach space $\boldsymbol{X}$ is called a $\mathcal{L}_{\infty,\lambda}$ space iff for every finite dimensional subspace $E \subset \boldsymbol{X}$ there exists a finite dimensional subspace $F \subset \boldsymbol{X}$ such that $E \subset F$ and $d(F, l_\infty^{\dim F}) \leq \lambda$. A Banach space $\boldsymbol{X}$ is called a $\mathcal{L}_\infty$ space iff $\boldsymbol{X}$ is a $\mathcal{L}_{\infty,\lambda}$ space for some $\lambda < \infty$.

Remark 1. [4] For every $\epsilon > 1$ the spaces l_∞^n are $\mathcal{L}_{\infty,\epsilon}$ spaces, $n \in \mathbb{N}$.

In the following definitions, e.g., [3], we assume that X is a locally compact topological space, and $\mathcal{A}$ is a separating function algebra on X.

Definition 2. A set $E \subset X$ is called a boundary for $\mathcal{A}$ iff for each $f \in \mathcal{A}$ there exists some $t \in E$ such that $|f(t)| = \| f \|_\infty = sup_{s \in X}|f(s)|$.

One can consider different kinds of boundaries for a separating function algebra $\mathcal{A}$ on X. First we have the so-called Bishop boundary:

Definition 3. A point $t \in X$ is called a peak point for $\mathcal{A}$ iff there exists some $f \in \mathcal{A}$ such that $|f(t)| = \| f \|_\infty = 1$ and $|f(s)| < 1$ for every $s \neq t$. The set of all peak points for $\mathcal{A}$ is called the Bishop boundary for $\mathcal{A}$ and is denoted by $\rho\mathcal{A}$.

Definition 4. The minimal closed boundary for $\mathcal{A}$ is called the Silov boundary and is denoted by $\partial\mathcal{A}$.

Definition 5. If, in addition to the above assumptions, the space X is compact and $\mathcal{A}$ contains the constants, then for every $t \in X$, set $\tau_t(f) = f(t)$, $f \in \mathcal{A}$, and $B_1^1 = \{x^* \in \mathcal{A}^*, \| x^* \|= x^*(1) = 1\}$.

The set $\chi\mathcal{A} = \{t \in X, \tau_t \in ext(B_1^1)\}$ is called the Choquet boundary for $\mathcal{A}$.

Recovery of Functions

In the above definition $\mathcal{A}^*$ denotes the dual space of $\mathcal{A}$, $\|\ \|$ is the usual norm in that dual space, and $ext(B_1^1)$ is the set of all extreme points of the unit ball in $\mathcal{A}^*$.

Here are some properties of the boundaries defined above, e.g., [3].

- $\rho\mathcal{A} \subset \partial\mathcal{A}$
- If X is a metrizable compact space and $\mathcal{A}$ is a uniform algebra on X, then $\rho\mathcal{A} = \chi\mathcal{A}$
- $\chi\mathcal{A} \neq \emptyset$
- $cl(\chi\mathcal{A}) = \partial\mathcal{A}$. Here $cl(\ )$ denotes the closure of the corresponding set.

Next we give a characterization of a $\mathcal{L}_\infty$ space, e.g., [7].

Theorem 1. *A Banach space X is an $\mathcal{L}_\infty$ space if and only if there exist $\lambda \geq 1$, $K \geq 1$, a sequence of $\mathcal{L}_{\infty,\lambda}$ spaces Y_n, and a sequence of linear operators $A_n : X \to Y_n$ and $U_n : Y_n \to X$ such that $U_n A_n x \to x$ for all $x \in X$ and $\|U_n\| \|A_n\| \leq K$.*

§3. Recovery - Necessary Conditions

In this section we state and prove our main theorem.

Theorem 2. *Let $\mathcal{A}$ be a uniform algebra on a compact Hausdorff space X, and let $\{L_n\}_{n\in\mathbb{N}}$ be a sequence of linear discrete operators $L_n : \mathcal{A} \to \mathcal{A}$, based on some points $\{z_{k,n}\}_{k=1,n\in\mathbb{N}}^n$ and functions $\{l_{k,n}\}_{k=1,n\in\mathbb{N}}^n$ as described in (1). Furthermore, assume that the points $\{z_{k,n}\}_{k=1,n\in\mathbb{N}}^n$ belong to the Choquet boundary $\chi\mathcal{A}$ of $\mathcal{A}$. Then there is a function $f \in \mathcal{A}$ that can not be recovered by $\{L_n\}$, i.e., $L_n f \not\to f$, as $n \to \infty$ in the topology of $\mathcal{A}$.*

Proof: Assume to the contrary that there is a sequence $\{L_n\}$ of linear discrete operators, based on the points $\{z_{k,n}\}$ that recovers every function f in $\mathcal{A}$, i.e. $L_n f \to f$, as $n \to \infty$ for every $f \in \mathcal{A}$. Now we construct a bounded factorization for L_n, as it is described in Theorem 1. Take an $\epsilon \in (0,1)$, and for every $n \in \mathbb{N}$ use the fact that $\{z_{i,n}\}_{i=1}^n \subset \chi\mathcal{A}$, e.g., [3], to find functions $\varphi_{i,n} \in \mathcal{A}$ and neighborhoods $U_{i,n}$ of $z_{i,n}$ such that

$$U_{i,n} \cap U_{j,n} = \emptyset$$
$$\|\varphi_{i,n}\| = |\varphi_{i,n}(z_{i,n})| = 1 \qquad (2)$$
$$|\varphi_{i,n}(z)| < \frac{\epsilon}{n}, \quad z \in X \setminus U_{i,n}, \quad i = 1, ..., n.$$

Let $\delta > 0$ be given. Then one can easily see from (2) that if ϵ is small enough the linear transformation $\Psi_n : l_\infty^{2n} \to l_\infty^{2n}$ given by its matrix $\Psi_n = (\varphi_{i,n}(z_{k,n}))_{i=1,k=1}^n$, satisfies $\|I - \Psi_n\| \leq \delta$, which in turn implies that the inverse transformation Ψ_n^{-1} exists, and moreover $\|\Psi_n^{-1}\|$ is bounded by a constant that depends on δ only. Thus, the matrix equation

$$I = \Psi_n C_n,$$

where I is the identity matrix and $C_n = (c_{k,i})_{k=n,i=n}^n$ has a unique solution for C_n. Moreover, $\|C_n\|$ is bounded by a constant that depends on δ only.

Next define

$$\psi_{i,n}(z) = c_{1,i}\varphi_{1,n}(z) + \cdots + c_{n,i}\varphi_{n,n}(z)$$

and U_n as $U_n = L_n V_n$, where

$$V_n : l_\infty^n \to \mathcal{A}$$

$$V_n(w_1, ..., w_n) := \sum_{i=1}^n w_i \psi_{i,n}(z).$$

Notice that

$$\psi_{i,n}(z_{k,n}) = \delta_{ik}, \quad i,k = 1,...,n.$$

Furthermore,

$$\sum_{i=1}^n |\psi_{i,n}(z)| \leq \|C_n\| \max\left\{\sum_{i=1}^n |\varphi_{i,n}(z)|, \ z \in X\right\}.$$

From (2) we can easily see that $\max\{\sum_{i=1}^n |\varphi_{i,n}(z)|, z \in X\}$ is achieved in one of the sets $U_{i,n}$, and moreover it is bounded by $1 + \epsilon$. Thus, we have

$$\|V_n\| \leq (1+\epsilon)M_\delta,$$

where the constant M_δ depends on δ only. Now combining this with the fact that $L_n f \to f$ for all $f \in \mathcal{A}$, we have a constant M for which $\|U_n\| \leq M$ for all $n \in \mathbb{N}$. Next define

$$A_n : \mathcal{A} \to l_\infty^n$$

$$A_n f := (f(z_{1,n}), ..., f(z_{n,n})).$$

Obviously $\|A_n\| = 1$ for every $n \in \mathbb{N}$. Thus, $L_n = U_n A_n$ and $\|U_n\|\|A_n\| \leq M$. Now Theorem 1 implies that $\mathcal{A}$ is an $\mathcal{L}_\infty$ space which already contradicts Kisliakov's theorem, and the proof of Theorem 2 is completed. $\square$

The notion of the Choquet boundary makes sense not only for uniform algebras, but in a more general situation when $\mathcal{A}$ is a commutative Banach algebra with an identity. Consider the Gelfand representation $\hat{\mathcal{A}}$ of $\mathcal{A}$, e.g., [3]. Thus the Choquet boundary for $\mathcal{A}$ is defined as $\chi\mathcal{A} = \chi\hat{\mathcal{A}}$. Therefore we have

Theorem 3. *Let $\mathcal{A}$ be a commutative Banach algebra with an identity, and such that $\hat{\mathcal{A}} \neq C(\Delta\mathcal{A})$, where $\Delta\mathcal{A}$ is the maximal ideal space for $\mathcal{A}$. Furthermore, let $\{\tau_{k,n}\}_{k=1,n\in\mathbb{N}}^{n} \subset \Delta\mathcal{A}$ be a collection of complex homomorphisms for $\mathcal{A}$. Then, for every sequence of linear operators $L_n : \mathcal{A} \to \mathcal{A}$ of the form $L_n f = \sum_{k=1}^{n} \tau_{k,n}(f) l_k, l_k \in \mathcal{A}$, there is a function $f_0 \in \mathcal{A}$ such that $L_n f_0 \not\to f_0$, as $n \to \infty$.*

One may apply Theorem 2 and Theorem 3 to many particular situations. Here we provide few examples. The only thing one needs in such applications is the description of the Choquet boundary for the corresponding uniform algebra.

Consider the polydisk $D_n := \{z : z \in \mathbb{C}^n, |z| \leq 1\}$, where $z = (z_1, ..., z_n)$ and $|z| = \max\{|z_i| : i = 1, .., n\}$ and let $\mathcal{P}(D_n)$ denote the closure of the space of all polynomials in z in $C(D_n)$. It is easy to see that $\mathcal{P}(D_n)$ is a uniform algebra, and it is well known, e.g., [3], that $\chi\mathcal{P}(D_n) = \{z : |z| = 1\}$, which is obviously a subset of the topological boundary of D_n. Thus, Theorem 2 gives the following result:

Corollary 1. *Suppose $\{z_{i,m}\}_{|i|\leq nm, m\in\mathbb{N}} \subset \mathbb{C}^n$ are such that $|z_{i_j,m}| = 1$ for all $i_1, ..., i_n = 1, .., m$, $m \in \mathbb{N}$, (here $i = (i_1, ..., i_n)$ is a multiindex). Then for every sequence of linear discrete operators L_m of the form*

$$L_m f(z) = \sum_{|i|\leq nm} f(z_i) l_{i,m}(z), \quad l_{i,m} \in \mathcal{P}(D_n),$$

there is a function $f \in \mathcal{P}(D_n)$ such that $L_m f \not\to f$, uniformly as $m \to \infty$.

The case of $\mathcal{A}(B_n)$ - the uniform algebra of functions holomorphic inside the closed unit ball $B_n \subset \mathbb{C}^n$ and continuous on the unit sphere $S_n \subset \mathbb{C}^n$ is even simpler, since the Bishop boundary for $\mathcal{A}(B_n)$ is S_n, e.g., [6]. Since B_n is metrizable, S_n is also the Choquet boundary for $\mathcal{A}(B_n)$, which gives us a natural extension of Somorjai's theorem in the case of $\mathcal{A}(B_n)$.

Corollary 2. *Suppose $\{z_{i,m}\}_{|i|\leq nm, m\in\mathbb{N}} \subset S_n$. Then for every sequence of linear discrete operators L_m of the form*

$$L_m f(z) = \sum_{|i|\leq nm} f(z_i) l_{i,m}(z), \quad l_{i,m} \in \mathcal{A}(B_n),$$

there is a function $f \in \mathcal{A}(B_n)$ such that $L_m f \not\to f$, uniformly as $m \to \infty$.

In the next example, we consider a situation somewhat different from the one we just described. The difference is that the polynomials we consider must have their exponents satisfying a certain cone condition. For simplicity we consider the case $\mathbb{C}^2$, which has an obvious extension to $\mathbb{C}^n$.

Corollary 3. *Consider the closure $\mathcal{B}_\alpha$ in $C(D_2)$ of the set*

$$\left\{ \sum_{i,j=0}^{m} a_{i,j} z^i w^j, \quad (z,w) \in \mathbb{C}^2, \quad |z| \leq 1, \quad |w| \leq 1, \quad i \leq \alpha j \right\},$$

where $0 < \alpha \leq 1$, and let $\{(z_{i,m}, w_{j,m})\}_{i,j=1,m\in\mathbb{N}}^m$ be a sequence of points in D_2 such that $|z_{i,m}| = |w_{j,m}| = 1$. Then for every sequence of linear discrete operators

$$L_m : \mathcal{B}_\alpha \to \mathcal{B}_\alpha$$

$$L_m f(z,w) = \sum_{i,j=1}^m f(z_{i,m}, w_{j,m}) l_{i,j,m}(z,w), \quad l_{i,j,m} \in \mathcal{B}_\alpha,$$

there is a function $f \in \mathcal{B}_\alpha$ such that $L_m f \not\to f$, uniformly as $m \to \infty$.

Proof: In order to prove this corollary one has to verify that $\chi \mathcal{B}_\alpha = \{(z,w) \in \mathbb{C} : |z| = |w| = 1\}$. It is trivial to show that the set $\{(z,w) \in \mathbb{C} : |z| = |w| = 1\}$ is a boundary for $\mathcal{B}_\alpha$, which immediately gives $\chi \mathcal{B}_\alpha \subset \{(z,w) \in \mathbb{C} : |z| = |w| = 1\}$. To prove the inverse inclusion, we need to show that every point in $\{(z,w) \in \mathbb{C} : |z| = |w| = 1\}$ is a peak point for $\mathcal{B}_\alpha$, which together with the fact that D_2 is metrizable gives us the desired result. Thus, we consider a point $(\xi, \eta) \in \mathbb{C}, |\xi| = |\eta| = 1$, and define

$$P(z,w) = \frac{(\xi + z)w^m}{2} + \frac{\eta + w}{2}, \quad 1 \leq \alpha m.$$

It is easy to check that $|P(z,w)| < 1, (z,w) \neq (\xi, \eta)$, and $|P(\xi, \eta)| = 1$, and since $P(z,w) \in \mathcal{B}_\alpha$ this shows that $(\xi, \eta) \in \rho \mathcal{B}_\alpha$, the Bishop boundary for $\mathcal{B}_\alpha$. □

The next example is given in an abstract setting. In order to state it we need some definitions, e.g., [2].

Definition 6. *Let X be a compact Hausdorff space and τ a homeomorphism on X such that τ^2 is the identity map on X. Then τ is called an involution on X.*

Definition 7. *Let τ be an involution on X, and consider the real commutative Banach algebra $C(X,\tau) = \{f \in C(X) : f(\tau(x)) = \overline{f}(x), \quad x \in X\}$.*

A real function algebra on (X, τ) is a real subalgebra $\mathcal{A}$ of $C(X, \tau)$ such that

- $\mathcal{A}$ is uniformly closed in $C(X, \tau)$;
- $\mathcal{A}$ contains the real constants;
- $\mathcal{A}$ separates the points in X.

In [2] one can find a detailed discussion about the concept of boundary for a real function algebra $\mathcal{A}$, and in particular a definition and proofs of some of the properties of the Choquet boundary and computations of that boundary for some particular examples of real function algebras $\mathcal{A}$. In the same paper there is introduced the so-called complexification $\mathcal{B}$ of $\mathcal{A}$:

$$\mathcal{B} = \{f + ig : f, g \in \mathcal{A}\}.$$

It is shown that $\mathcal{B}$ is an uniform algebra on X which has its Choquet boundary exactly the same set as the Choquet boundary for $\mathcal{A}$ is. Thus we arrive at the following corollary:

Corollary 4. *Let $\mathcal{A}$ be a real function algebra on a compact Hausdorff space X, and let $\{z_{k,n}\}_{k=1, n \in \mathbb{N}}^{n}$ be a sequence of points that belong to the Choquet boundary for $\mathcal{A}$. Then for every sequence of linear discrete operators $\{L_n\}$*

$$L_n : \mathcal{A} \to \mathcal{A}$$

$$L_n f(z) = \sum_{k=1}^{n} f(z_{k,n}) l_{k,n}, \quad l_{k,n} \in \mathcal{A},$$

there exists a function $f \in \mathcal{A}$ such that $L_n f \not\to f$, as $n \to \infty$ in the topology of $\mathcal{A}$.

The last example shows that the conclusion of Theorem 2 holds even when the points $\{z_{k,n}\}$ belong to the Shilov boundary for $\mathcal{A}$. Consider the disk algebra $\mathcal{A}$ as a uniform algebra on the unit circle T, i.e., $\mathcal{A} \subset C(T)$. Set $\mathcal{B} = \{f \in \mathcal{A} : f(t) = f(0) = 0\}$, where $t \in T$ is a fixed point on the unit circle. It is easy to check that $\mathcal{B}$ is an uniform algebra on T, and that $\chi \mathcal{B} = T \setminus \{t\}$. Now we can easily add the point t to any collection $\{z_{k,n}\}$ of points from the Choquet boundary for $\mathcal{B}$ that is used to construct the corresponding sequence of linear discrete operators $\{L_n\}$ in (1). For example, one can set

$$L_n f(z) = f(t) l(z) + \sum_{k=1}^{n} f(z_{k,n}) l_{k,n}(z),$$

where $l, l_{k,n} \in \mathcal{A}$. Thus, in this situation we do not have a solution for our recovery problem when the collection of points $\{z_{k,n}\}$ is taken from the Shilov boundary T for $\mathcal{B}$.

This example leads to the following conjecture:

Conjecture. *Theorem 2 holds when the points $\{z_{k,n}\}$ used to construct the operators L_n in (1) are taken from the Shilov boundary for the considered uniform algebra $\mathcal{A}$.*

References

1. Kisliakov, S. V., Proper uniform algebras are uncomplemented, Soviet Math. Dokl. **40(3)** (1990), 584–586.
2. Kulkarni, S. H. and S. Arundhathi, Choquet boundary for real function algebras, Canad. J. Math. **40(5)** (1988), 1084–1104.
3. Larsen, R., *Banach Algebras*, Marcel Dekker, Inc, New York, 1973.
4. Lindenstrauss, T. and H. Rosenthal, The $\mathcal{L}_p$ spaces, Israel J. Math. **7** (1969), 325–349.
5. Lindenstrauss, T. and L. Tsafriri, *Classical Banach Spaces*, Lecture Notes in Mathematics, no. **338**, Berlin, Springer-Verlag, 1973.

6. Rudin, W., *Function Theory in the Unit Ball of* $\mathbb{C}^n$, A series of Comprehensive Studies in Mathematics, no. 241, New York, Springer-Verlag, 1980.
7. Shekhtman, B., Discrete approximating operators on function algebras, Constr. Approx. **8** (1992), 371–377.
8. Somorjai, G., On discrete operatorsw in the function space $\mathcal{A}$, Constr. Approx. Theory **77** (1980), 489–496.

Ivan V. Ivanov
Department of Mathematics
University of South Florida
Tampa, FL 33620
ivanov@chuma.cas.usf.edu

Boris Shekhtman
Department of Mathematics
University of South Florida
Tampa, FL 33620
boris@chuma.cas.usf.edu

Chebyshev Quadrature Recognizes Algebraic Curves and Surfaces

J. Korevaar

Abstract. Let Γ be any singularity-free algebraic Jordan curve in $\mathbb{R}^d$ supplied with normalized arc length or a similar probability measure μ. Evaluating the integral $\int_\Gamma f d\mu$ by the arithmetic mean of the values of f on any cycle of N equally spaced nodes on Γ (relative to μ), the quadrature error will be bounded by $Ae^{-bN} \sup_\Gamma |f|$ for all polynomials $f(x)$ of degree $\leq cN$. It is plausible that small shifts of the nodes would give quadrature error zero for such polynomials. There is a related result for algebraic Jordan arcs. For certain algebraic surfaces Γ and measures μ of product type, and for product sets of nodes, the quadrature error is bounded by $Ae^{-b\sqrt{N}} \sup_\Gamma |f|$ for all polynomials of degree $\leq c\sqrt{N}$. One would like to obtain a similar bound for the sphere. The situation is completely different for nonalgebraic curves and surfaces. There, by recent work of L. Bos, cf. [6], the corresponding quadrature errors are at least of order $1/N$. The strong estimates for algebraic sets are based on complex potential theory.

§1. Introduction and Basic Examples

This paper deals with Chebyshev-type quadrature on compact sets Γ in $\mathbb{R}^d$ with associated probability measures μ. The aim of such quadrature is to evaluate the integral $\int_\Gamma f d\mu$ by the arithmetic mean of the values of f at fixed nodes in Γ for a large class of functions f. Thus we look for good approximate formulas of the form

$$\int_\Gamma f(x) d\mu(x) \approx \frac{1}{N} \sum_{k=1}^{N} f(\zeta_k), \quad \zeta_1, \ldots, \zeta_N \in \Gamma. \tag{1.1}$$

As can be seen from the examples below, Chebyshev quadrature works well for unrestricted N only for special pairs (Γ, μ).

The problem is to find N-tuples Z_N of nodes ζ_k for which the quadrature remainder

$$R(f, Z_N) \stackrel{\text{def}}{=} \int_\Gamma f d\mu - \frac{1}{N} \sum_{k=1}^{N} f(\zeta_k) \tag{1.2}$$

is *very small* for a large class of functions f. Because of their good approximation properties, it is convenient to use as test functions the *polynomials*

$$f(x) = f(x_1, \ldots, x_d)$$

up to a high degree related to N.

Ideally, we would like to make $R = 0$ for such f to get an *exact* quadrature formula.

The following examples serve as background and are of use later on.

Example 1.1: *The Circle.* Let Γ be the unit circle $C(0,1)$ in $\mathbb{R}^2 \cong \mathbb{C}$ and let $d\mu$ be normalized arc measure $d\lambda/L$, so that

$$\int_\Gamma f(x)\frac{d\lambda}{L} = \int_0^{2\pi} f(\cos\theta, \sin\theta)\frac{d\theta}{2\pi}.$$

On Γ, a polynomial f of degree q will correspond to a trigonometric polynomial F of order q:

$$F(\theta) = f(\cos\theta, \sin\theta) = \sum_{-q}^{q} c_p e^{ip\theta}.$$

Hence it is natural to choose as nodes equally spaced points corresponding to $\theta_k = k\frac{2\pi}{N}$, $0 \leq k \leq N-1$. The resulting approximation

$$\frac{1}{2\pi}\int_0^{2\pi} e^{ip\theta} d\theta \approx \frac{1}{N}\sum_{k=0}^{N-1} e^{ipk\frac{2\pi}{N}} = \frac{1}{N}\frac{e^{ip2\pi}-1}{e^{ip\frac{2\pi}{N}}-1} = \begin{cases} 0 & \text{for } p \not\equiv 0(N) \\ 1 & \text{for } p \equiv 0(N) \end{cases}$$

is exact for $|p| < N$. One may also use shifted nodes given by $\theta_k = \beta + k\frac{2\pi}{N}$.

Conclusion: for the circle, the quadrature formulas

$$\int_\Gamma f(x_1, x_2)\frac{d\lambda}{L} \approx \frac{1}{N}\sum_{k=0}^{N-1} f\left\{\cos\left(\beta + k\frac{2\pi}{N}\right), \sin\left(\beta + k\frac{2\pi}{N}\right)\right\} \qquad (1.3)$$

(with real β) are exact for all polynomials f of degree $\leq N-1$.

More generally, the approximations (1.3) will be *exponentially good* whenever the restriction $f|_\Gamma$ is given by an exponentially convergent Fourier series $\sum_{-\infty}^{\infty} c_p e^{ip\theta}$. This will be the case if f is real analytic on Γ.

Example 1.2: *The Interval* $I = [-1, 1]$. For the measure

$$d\mu(x) = dx/\pi\sqrt{1-x^2},$$

Gauss quadrature provides N-point Chebyshev quadrature which is exact to degree $2N - 1$. Alternatively one may go back to the previous case:

$$\frac{1}{\pi}\int_{-1}^{1} f(x)\frac{dx}{\sqrt{1-x^2}} = \frac{1}{\pi}\int_0^\pi f(\cos\theta)d\theta = \frac{1}{2\pi}\int_0^{2\pi} f(\cos\theta)d\theta.$$

This approach shows that *any* N equally spaced nodes relative to μ on the closed curve $I \cup -I$ give polynomial exactness to degree $N - 1$.

The uniform measure $\frac{1}{2}dx$ on I is not well suited to Chebyshev quadrature. Here Korevaar and Meyers [8] have shown that for every N and every N-tuple of real nodes, there is a polynomial f of degree $< 4\sqrt{N}$ such that

$$|R(f, Z_N)| > \frac{1}{100N}\sup_I |f|.$$

By a sophisticated result of Bernstein [1], well-chosen N-tuples of nodes do actually give polynomial exactness to a degree comparable to $\sqrt{N}$ (cf. Kuijlaars [10] and see [6] for additional references).

§2. Results for Curves

Our results show that Chebyshev quadrature works much better on algebraic curves than on other curves. With regard to such quadrature, algebraic Jordan curves in $\mathbb{R}^d$ with appropriate measure resemble the circle, algebraic Jordan arcs resemble the interval I!

Theorem 2.1. *Let Γ be a singularity-free algebraic Jordan curve in $\mathbb{R}^d$. Furthermore, let the probability measure μ on Γ be proportional to arc length λ, or more generally, have the form $d\mu = \alpha d\lambda$ where α is a positive real-analytic function on Γ. Then there are positive constants A, b, c such that for every N, and every cycle Z_N of N equally spaced nodes on Γ relative to μ, the Chebyshev quadrature remainder (1.2) satisfies the inequality*

$$|R(f, Z_N)| \le Ae^{-bN}\sup_\Gamma |f| \qquad (2.1)$$

for all polynomials f of degree $\le cN$.

It seems plausible that the exponentially small remainders can be made equal to zero by small perturbations of the nodes. For a special class of algebraic curves Γ, including ellipses, this was recently proved by Kuijlaars [12].

The proof of Theorem 2.1 is based on complex potential theory (see Section 3).

Theorem 2.2. *There is a result similar to Theorem 2.1 for singularity-free algebraic Jordan arcs* Γ *from* a *to* $b \neq a$ *in* $\mathbb{R}^d$, *with*

$$d\mu(x) = \frac{\alpha(x)\, d\lambda(x)}{\sqrt{|x-a||x-b|}},$$

(cf. Example 1.2 and [7].)

Special singularities are permissible:

Example 2.3: The *Square Curve* with Vertices $[\pm 1, \pm 1]$ in $\mathbb{R}^2$. Using the measure $d\mu = d\lambda/4\pi\sqrt{2-x^2-y^2}$, one may apply the first result in Example 1.2 to each of the sides. Avoiding the vertices, one obtains exact N-point Chebyshev quadrature up to degree $\approx \frac{1}{4}N$.

For nonalgebraic curves the situation is completely different. Counting linearly independent monomials, one finds that on *nonalgebraic* curves, N-point Chebyshev formulas (with large N) can never be exact to degree cN (or even $C\sqrt{N}$ with large C). Just think of squares $f = p^2$, where p is a polynomial which vanishes on Z_N.

The stronger result below follows from the work of L. Bos (cf. [7] and Theorem 4.6).

Theorem 2.4. *Let* Γ *be any compact nonalgebraic curve in* $\mathbb{R}^d$ *and let* μ *be any probability measure with support equal to* Γ. *Then for every* N, *there exist a polynomial* f *of degree* $\leq 4\sqrt{N}$ *and a cycle* Z_N *of* N *equally spaced nodes on* Γ *(relative to* μ*) such that*

$$|R(f, Z_N)| > \frac{1}{2N} \sup_\Gamma |f|.$$

§3. Proof of Theorem 2.1

Step 1) We use

$$d\sigma = 2\pi d\mu = 2\pi \alpha d\lambda$$

as arc measure, σ as (signed) arc length from a fixed point on Γ.

Locally, Γ is the common zero set of $d-1$ polynomials $H_i(x)$ whose Jacobi-matrix $[\partial H_i/\partial x_j]$ has rank $d-1$. Thus around a given point c, the coordinates x_j of $x \in \Gamma$ are real analytic functions of one of the coordinates, of x_k, say. There σ is real analytic in x_k, and conversely, x_k is real analytic in σ and so are the other x_j. Combining patches one obtains a *global* parametrization for Γ of the form

$$x_j = X_j(\sigma), \ \ 0 \leq \sigma \leq 2\pi, \ \ X_j(\sigma) \text{ real analytic, of period } 2\pi.$$

It follows that Γ has a singularity-free *complexification* Γ^* in a neighborhood U of Γ in $\mathbb{C}^d$, given by analytic functions of period 2π in a neighborhood of $\mathbb{R}$:

$$z_j = X_j(s), \ \ s = \sigma + i\tau \text{ with small } |\tau|. \tag{3.1}$$

The complexification is part of a 1-dimensional complex algebraic variety $V = V(\Gamma)$ of pure dimension 1, given by a finite number of polynomial equations.

Step 2) We now need a growth result for polynomials on this complexification:

Proposition 3.1. *There are positive constants δ and M such that for all polynomials $f(z) = f(z_1, \ldots, z_d)$ which are bounded by 1 on Γ, the inequality*

$$|f(z)| < e^{M \deg f} \tag{3.2}$$

holds throughout the δ-neighborhood of Γ in the algebraic variety $V(\Gamma)$.

The proof uses complex potential theory in $\mathbb{C}^d$. The basic tool is a *(pre-)Green function* $g_\Gamma(z)$ for Γ in $\mathbb{C}^d$ with logarithmic pole at infinity. This function is the logarithm of Siciak's extremal function which he introduced and studied in [15, 16, 17]. (See also Klimek [3] and Korevaar [4, 5].)

Proof: By a fundamental result of Zaharjuta [18] (cf. [16, 17]),

$$g_\Gamma(z) = \sup \frac{1}{\deg f} \log |f(z)|, \tag{3.3}$$

where the supremum is taken over all polynomials $f(z)$ which are bounded by 1 on Γ. One has $g_\Gamma(z) = 0$ on Γ, but $g_\Gamma(z) = +\infty$ outside $V(\Gamma)$!

Fortunately, the work of Sadullaev [13] and Bos, Levenberg, Milman and Taylor [2] implies a *minor miracle*: the pre-Green function g_Γ is locally bounded on the algebraic variety $V = V(\Gamma)$. Thus on any δ-neighborhood of Γ in V, the supremum of g_Γ is equal to a finite constant M. Proposition 3.1 now follows from (3.3).

(Because g_Γ is continuous on V at the points of Γ, one may actually take M small when δ is small (cf. [7]).)

Compare the way in which one uses the classical Green function g_I for $I = [-1, 1]$ in $\mathbb{C}$ to obtain a good bound for the polynomials $f(z)$ which are bounded by 1 on I: the subharmonic functions $(1/\deg f) \log |f(z)|$ are majorized by the harmonic function $g_I(z) = \log|z + \sqrt{z^2 - 1}|$.

Step 3) Let $f(x)$ be any polynomial in $x_1, \ldots, x_d$. We set $\deg f = q$ and assume without loss of generality that $\sup_\Gamma |f| = 1$. Using the complexification $\Gamma^* : z = X(s)$ of Γ (3.1), we now define $f \circ X(s) = F(s)$. This will also be an analytic function around $\mathbb{R} \subset \mathbb{C}$ of period 2π. By Proposition 3.1, there are a constant M and a closed ρ-neighborhood of $\mathbb{R}$ independent of f and q on which

$$|F(s)| = |f(z)| \leq e^{Mq}.$$

By Cauchy's theorem, the Fourier coefficients c_p of $F(\sigma)$ may be represented by integrals over segments of $s = \sigma \pm i\rho$. Thus one finds that

$$|c_p| = \left| \frac{1}{2\pi} \int_0^{2\pi} F(\sigma) e^{-ip\sigma} d\sigma \right| \leq e^{qM - |p|\rho}.$$

Hence,

$$|F(\sigma) - S_{N-1}(\sigma)| \leq \sum_{|p| \geq N} |c_p| \leq \frac{2 e^{qM - N\rho}}{1 - e^{-\rho}} \leq A e^{-bN} \overset{\text{def}}{=} \frac{2}{1 - e^{-\rho}} e^{-\frac{1}{2}\rho N},$$

provided we take $q = \deg f \leq (\rho/2M)N$. We will limit the degree of f in this way from here on.

Since the Chebyshev quadrature formula of order N with equally spaced nodes is exact for S_{N-1} (see Example 1.1), it has remainder $\leq Ae^{-bN}$ for F.

Returning to the function f on Γ one obtains (2.1). □

§4. Results for Surfaces

The theorems for curves in Section 2 can be extended to sets of dimension $m \geq 2$. The positive results below involve special algebraic surfaces of dimension $m = 2$.

Surfaces of Torus Type

Let Γ be a singularity-free bounded closed algebraic surface in $\mathbb{R}^d$ ($d \geq 3$) which can be parametrized in the form

$$x_j = X_j(\theta, \phi), \; j = 1, \cdots, d, \quad 0 \leq \theta, \phi \leq 2\pi$$

with real-analytic functions X_j of period 2π in θ and ϕ. Furthermore, let μ be a probability measure on Γ of the form $d\mu = \alpha(\theta)d\theta \cdot \beta(\phi)d\phi$, where α and β are positive real-analytic functions of period 2π. Using constant multiples, we may assume that $\alpha d\theta$ and $\beta d\phi$ are probability measures on $[0, 2\pi]$.

For good Chebyshev-type quadrature on Γ we will use as nodes ζ_k the vertices of a "square grid" over Γ in which every mesh has the same "area" relative to μ. In particular our N-tuples Z_N will contain $N = n^2$ nodes, where $n = 1, 2, \cdots$.

The method of Section 3 will now give

Theorem 4.1. *Under the above conditions on (Γ, μ), there are positive constants A, b, c such that the Chebyshev quadrature remainder satisfies the inequality*

$$|R(f, Z_N)| \leq Ae^{-b\sqrt{N}} \sup_\Gamma |f| \tag{4.1}$$

for all N-tuples Z_N as described and all polynomials f of degree $\leq c\sqrt{N}$.

Theorem 4.1 applies in particular to the torus $T \subset \mathbb{R}^3$ and ordinary area measure:

$$x_1 = (r + \rho\cos\theta)\cos\phi, \; x_2 = (r + \rho\cos\theta)\sin\phi, \; x_3 = \rho\sin\theta, \quad 0 \leq \theta, \phi \leq 2\pi,$$

$$d\lambda = (r + \rho\cos\theta)\rho \, d\theta d\phi, \quad r > \rho > 0.$$

For this torus, Kuijlaars [11] has proved the existence of N-tuples of nodes Z_N such that $R(f, Z_N)$ becomes zero for all polynomials f of degree $\leq c\sqrt{N}$.

Surfaces Like the Sphere

The Chebyshev quadrature problem is especially challenging for the *unit sphere* $\Gamma = S = S(0,1)$ in $\mathbb{R}^3$, supplied with normalized area measure $\sigma = \lambda/4\pi$ (cf. Saff and Kuijlaars [14]). Here an N-point formula (1.1) cannot be exact for all polynomials $f(x)$ up to degree $2\sqrt{N}$ (cf. Korevaar and Meyers [9]). However, it seems plausible that there are N-tuples Z_N which work up to degree $c\sqrt{N}$ for some $c > 0$ (cf. Korevaar [6]).

Using spherical coordinates and treating the sphere as the quasi-product of an interval and a circle, one readily obtains the following weaker result (cf. [9]).

Theorem 4.2. *On (S,σ) there are special N-tuples Z_N such that $R(f, Z_N) = 0$ for all polynomials f of degree $\leq c\sqrt[3]{N}$ with $c \approx 1/\sqrt[6]{2}$.*

To investigate if there is at least a result of the form (4.1) for the sphere, one may use the following equivalence between the Chebyshev quadrature problem for (S,σ) and a question involving electrostatic fields due to N point charges $1/N$ on S (see [6, 9]).

Proposition 4.3. *Let $\rho > 0$. Suppose that there exist constants $A, b, c > 0$ and N-tuples Z_N on $\Gamma = S$ with $N \to \infty$ such that*

$$|R(f, Z_N)| \leq Ae^{-bN^\rho} \sup_S |f| \qquad (4.2)$$

for all polynomials $f(x)$ of degree $\leq cN^\rho$. Then point charges $1/N$ at the points of Z_N produce a Coulomb field $\mathcal{E}(x, Z_N)$ such that for some constants $B(r), c(r) > 0$,

$$|\mathcal{E}(x, Z_N)| \leq B(r)e^{-c(r)N^\rho} \quad \text{on the ball } |x| \leq r < 1, \qquad (4.3)$$

and conversely.

By the preceding, (4.2) holds with $\rho = \frac{1}{3}$, but not with $\rho > \frac{1}{2}$. Can one use $\rho = \frac{1}{2}$? Here the author has conjectured the following (see [6]).

Conjecture 4.4. *There exist N-tuples $Z_N \subset S$ for which (4.3) and hence (4.2) hold with $\rho = \frac{1}{2}$.* Suitable N-tuples are provided by minimizing a modified energy

$$\frac{1}{N^2} \sum_{j,k=1, j\neq k}^{N} \frac{1}{\sqrt{(x_j - x_k)^2 + a^2}}.$$

The truth of Conjecture 4.4 would follow by (multidimensional) complex analysis if one could show that the points in the minimizing N-tuples for $N \to \infty$ are well separated:

$$|\zeta_{Nj} - \zeta_{Nk}| \geq \frac{\delta}{\sqrt{N}}, \quad j \neq k$$

for some constant $\delta > 0$ independent of N (see Korevaar [6]).

Another approach to the case $\rho = \frac{1}{2}$ may be based on the Coulomb potential $U(x, Z_N)$ due to charges $1/N$ at the points of $Z_N \subset S$. For each N of a sequence tending to infinity, let the N-tuple Z_N minimize the deviation of $U(x, Z_N)$ from 1 on the sphere $S(0, r)$ with $r < 1$. If there is a constant $\beta > 0$ such that $U(x, Z_N)$ has at least βN well-separated relative extreme points on every spherical cap of the sphere $S(0, r)$ of diameter r (say), then (4.3) and (4.2) hold with $\rho = \frac{1}{2}$ (cf. [6]).

Question 4.5. For which algebraic surfaces Γ of the type of the sphere and which measures μ is there a result of the form (4.2)? Mapping Γ onto the sphere, one finds that an ellipsoid of revolution with normalized area measure is all right. Which measures $\mu = \alpha\lambda$ with real analytic $\alpha > 0$ can one use on S itself?

Nonalgebraic Sets

On nonalgebraic surfaces one cannot expect quadrature remainders of lower order than $1/N$. The following general result of L. Bos (cf. [7]) implies Theorem 2.4 when $m = 1$.

Theorem 4.6. *Let Γ be a compact connected (curve or) surface in $\mathbb{R}^d$ of dimension m and let μ be a probability measure with support equal to Γ. Suppose furthermore that for some positive integer N and every point $x_0 \in \Gamma$, there is an N-tuple $Z_N \subset \Gamma$ of points including x_0 such that the Chebyshev quadrature based on Z_N has remainder*

$$|R(f, Z_N)| \leq \frac{1}{2N} \sup_\Gamma |f| \tag{4.4}$$

for all polynomials f of degree not exceeding $(2m+2)N^{1/(m+1)}$. Then Γ must be algebraic, i.e., Γ is a subset of an algebraic variety of the same dimension m.

References

1. Bernstein, S. N., On quadrature formulas with positive coefficients, (Russian), Izv. Akad. Nauk SSSR Ser. Mat. **1** (1937), 479–503. *Collected Works*, vol. 2, Izdat. Akad. Nauk SSSR, Moscow, 1954, pp. 205–227, MR **16**:433.
2. Bos, L., N. Levenberg, P. Milman, and B. A. Taylor, Tangential Markov inequalities characterize algebraic submanifolds of $\mathbb{R}^N$, Indiana Univ. Math. J. **44** (1995), 115–138, MR **96i**:41009.
3. Klimek, M., *Pluripotential Theory*, London Math. Soc. Monogr., N.S., vol. 6, Oxford Univ. Press, 1991, MR **93h**:32021.
4. Korevaar, J., Polynomial approximation numbers, capacities and extended Green functions for $\mathbb{C}$ and $\mathbb{C}^N$, in *Approximation Theory V*,

C. K. Chui, L. L. Schumaker, and J. D. Ward (eds.), Academic Press, New York, 1986, pp. 97–127, MR **89b**:32022.

5. Korevaar, J., Green functions, capacities, polynomial approximation numbers and applications in real and complex analysis, Nieuw Arch. Wisk. (4) **4** (1986), 133–153, MR **88a**:32022.

6. Korevaar, J., Chebyshev-type quadratures: use of complex analysis and potential theory, in *Complex Potential Theory*, P.M. Gauthier and G. Sabidussi (eds.), Kluwer, Dordrecht, 1994, pp. 325–364, MR **96g**:41029.

7. Korevaar, J. and L. Bos, Characterization of algebraic curves by Chebyshev quadrature, J. Anal. Math., to appear.

8. Korevaar, J. and J. L. H. Meyers, Massive coalescence of nodes in optimal Chebyshev-type quadrature on $[-1, 1]$, Indag. Math. N.S. **4** (1993), 327–338, MR **94h**:41064.

9. Korevaar, J. and J. L. H. Meyers, Spherical Faraday cage for the case of equal point charges and Chebyshev-type quadrature on the sphere, Integral Trans. Spec. Funct. **1** (1993), 105–117, MR **97g**:41046.

10. Kuijlaars, A. B. J., The minimal number of nodes in Chebyshev type quadrature formulas, Indag. Math., N.S. **4** (1993), 339–362, MR **94h**:41065.

11. Kuijlaars, A. B. J., Chebyshev-type quadrature and partial sums of the exponential series, Math. Comp. **64** (1995), 251–263, MR **95c**:65043.

12. Kuijlaars, A. B. J., Chebyshev-type quadrature for analytic weights on the circle and the interval, Indag. Math., N.S., **6** (1995), 419–432, MR **96j**:41027.

13. Sadullaev, A., Estimates of polynomials on analytic sets, (Russian), Izv. Akad. Nauk SSSR Ser. Mat. **46** (1982), 524–534, 671, MR **83k**:32018. [English translation USSR Izv. **20** (1983), 493–502].

14. Saff, E. B. and A. B. J. Kuijlaars, Distributing many points on a sphere, Math. Intelligencer **19** (1997), 5–11.

15. Siciak, J., On some extremal functions and their applications in the theory of analytic functions of several complex variables, Trans. Amer. Math. Soc. **105** (1962), 322–357, MR **26**–1495.

16. Siciak, J., Extremal plurisubharmonic functions in $\mathbb{C}^n$, Ann. Polon. Math. **39** (1981), 175–211, MR **83e**:32018.

17. Siciak, J., Extremal plurisubharmonic functions and capacities in $\mathbb{C}^n$, Sophia Kokyuroku in Math. no. 14, Dept of Math., Sophia Univ., Tokyo, 1982.

18. Zaharjuta, V. P., Extremal plurisubharmonic functions, orthogonal polynomials, and the Bernstein-Walsh theorem for analytic functions of several complex variables, (Russian), Ann. Polon. Math. **33** (1976), 137–148, MR **56**–3333.

J. Korevaar
Department of Mathematics
University of Amsterdam
Plantage Muidergracht 24
1018 TV Amsterdam, Netherlands
korevaar@wins.uva.nl

Ray Sequences of Best Rational Approximants to Entire Functions

A. V. Krot, V. A. Prokhorov, and E. B. Saff

Abstract. This article is devoted to results relating to the theory of rational approximation of entire functions. An analysis is made of the rate of decrease of the best approximation $\rho_{n,m}$ of an entire function by rational functions of type (n,m) in the uniform metric. It is assumed that the indices (n,m) progress along a ray sequence of the Walsh table, i.e. the sequence of indices (n,m) satisfies $m/n \to \theta \in [0,1]$ as $m+n \to \infty$.

§1. Introduction

Let E be an arbitrary compact set, $E \subset \mathbf{C}$, and let f be an entire function. For any nonnegative integers n and m denote by $\mathcal{R}_{n,m}$ the class of all rational functions with complex coefficients of order (n,m):

$$\mathcal{R}_{n,m} = \{r : r = p/q, \deg p \leq n, \deg q \leq m, q \not\equiv 0\}.$$

The deviation of f from $\mathcal{R}_{n,m}$ (in the uniform metric on E) is denoted by $\rho_{n,m}$:

$$\rho_{n,m} = \rho_{n,m}(f, E) = \inf_{r \in \mathcal{R}_{n,m}} \|f - r\|_E,$$

where $\|\cdot\|_E$ is the supremum norm on E.

We assume that $m = m(n)$ and the sequence of positive integers $\{m(n)\}$, $m(n) \leq n$, $n = 0, 1, 2, \ldots$, tending to infinity satisfies the following conditions:

$$m(n-1) \leq m(n) \leq m(n-1) + 1, \quad n = 1, 2, \ldots, \tag{1}$$

and

$$\lim_{n \to \infty} \frac{m(n)}{n} = \theta, \ 0 \leq \theta \leq 1. \tag{2}$$

One of the main results of this paper is Theorem 1 characterizing the rate of decrease of the ray sequence $\{\rho_{n,m(n)}\}_{n=0}^{\infty}$ of the Walsh table $\{\rho_{n,m}\}_{n,m=0}^{\infty}$ of the best rational approximations of an entire function of finite order.

The case of polynomial approximation ($m(n) = 0$, $n = 0, 1, 2, \ldots$) of entire functions has been thoroughly investigated. We mention works of Varga [18], Shah [15], and Winiarski [20] relating to this direction. The methods of the theory interpolation by polynomials and Walsh inequality (see [19]) give us, in terms related to the degree of decrease of the values $\rho_{n,0}$, necessary and sufficient conditions for a continuous function on E to admit a continuation to an entire function of finite order. More precisely the following result is known.

Suppose that f is continuous on E, where E is a compact set with the positive logarithmic capacity $\mathrm{cap}(E)$. Then f can be prolonged to an entire function of order $\sigma \geq 0$ if and only if

$$\limsup_{n\to\infty} \frac{\ln \rho_{n,0}}{n \ln n} = -\frac{1}{\sigma}.$$

It is to be noted that the condition

$$\limsup_{n\to\infty} \rho_{n,0}^{\sigma/n} n = \sigma e \tau (\mathrm{cap}(E))^{\sigma}$$

allows us to describe the class of entire functions of finite order $\sigma > 0$ and finite type $\tau > 0$.

We now point out the following estimates characterizing the behavior of the ray sequence $\{\rho_{n,m(n)}\}_{n=0}^{\infty}$ and following immediately from the results in polynomial approximation.

If f is an entire function of finite order $\sigma \geq 0$, then

$$\limsup_{n\to\infty} \frac{\ln \rho_{n,m(n)}}{n \ln n} \leq -\frac{1}{\sigma}; \tag{3}$$

and if f is an entire function of finite order $\sigma > 0$ and finite type $\tau > 0$, then

$$\limsup_{n\to\infty} \rho_{n,m(n)}^{\sigma/n} n \leq \sigma e \tau (\mathrm{cap}(E))^{\sigma}.$$

An important role in the theory of rational approximation of analytic functions is played by methods of rational interpolation of analytic functions, especially Padé approximants (interpolation sequences of rational functions with free poles) (see, for example, [5, 7, 16]). In addition to constructive methods, the theory of Hankel operators has been widely used in recent years in studying the degree of rational approximation of analytic functions (see [8, 9, 10, 13, 14]). These methods are based on the Adamyan-Arov-Kreĭn theorem [1] (see also [12]). This theorem makes it possible to reduce the investigation of the degree of rational approximation of analytic functions to an investigation of the rate of decrease of the sequence $\{s_n\}, n = 0, 1, \ldots,$ of singular numbers of the Hankel operator constructed from the function to be approximated.

Application of the methods of the theory of Hankel operators allowed one of the authors (see [13]) to estimate the rate of decrease of the diagonal sequence (in this case $m(n) = n$, $n = 0, 1, 2, \ldots$) of the best rational approximations of an entire function of finite order $\sigma \geq 0$. The following upper estimate was established:

$$\liminf_{n\to\infty} \frac{\ln \rho_{n,n}}{n \ln n} \leq -\frac{2}{\sigma}.$$

In connection with the last inequality we point out that the limit in this relation exists for certain entire functions, for example, $f(z) = e^z$ (see [2]):

$$\lim_{n\to\infty} \frac{\ln \rho_{n,n}}{n \ln n} = -\frac{2}{\sigma}.$$

In the present article we use methods employing the theory of Hankel operators and a generalization of the Adamyan–Arov–Kreĭn theorem to prove Theorem 1 concerning the ray sequence $\{\rho_{n,m(n)}\}_{n=0}^{\infty}$ of the deviations in best rational approximations. In this theorem the rate of convergence of the product $\prod_{i=0}^{m(n)} \rho_{n-i,m(n)-i}$ to zero is estimated.

Theorem 1. *If E is an arbitrary compact set in $\mathbb{C}$ and f is an entire function of finite order $\sigma \geq 0$, then*

$$\limsup_{n\to\infty} \frac{\ln(\rho_{n,m(n)}\rho_{n-1,m(n)-1}\cdots\rho_{n-m(n),0})}{nm(n)\ln n} \leq -\frac{1}{\sigma}. \tag{4}$$

The next assertion, which follows from Theorem 1, gives an estimate for $\liminf_{n\to\infty} \ln \rho_{n,m(n)}/n \ln n$.

Corollary 1.

$$\liminf_{n\to\infty} \frac{\ln \rho_{n,m(n)}}{n \ln n} \leq -\frac{2}{(2-\theta)\sigma}.$$

The next assertion enables us to characterize the behavior of the ray sequence $\{\rho_{n,m(n)}\}_{n=0}^{\infty}$ for functions for which equality is attained in (3).

Corollary 2. *If*

$$\limsup_{n\to\infty} \frac{\ln \rho_{n,m(n)}}{n \ln n} = -\frac{1}{\sigma},$$

then

$$\liminf_{n\to\infty} \frac{\ln \rho_{n,m(n)}}{n \ln n} \leq -\frac{1}{(1-\theta)\sigma}.$$

An investigation of the asymptotic behavior of the singular numbers of the Hankel operator constructed from the function being approximated enables us to also prove other results in the theory of rational approximation of entire functions.

The following theorem relates to the case when f is an entire function of finite order $\sigma > 0$ and finite type $\tau > 0$.

Theorem 2. *Suppose that E is an arbitrary compact set in $\mathbb{C}$, and f is an entire function of finite order $\sigma > 0$ and finite type $\tau > 0$. Then*

$$\limsup_{n\to\infty}(\rho_{n,m(n)}\rho_{n-1,m(n)-1}\cdots\rho_{n-m(n),0})^{\sigma/n(m(n)+1)}n \leq \sigma e\tau(\mathrm{cap}(E))^\sigma, \quad (5)$$

where $\mathrm{cap}(E)$ is the logarithmic capacity of E.

We now state corollaries of this theorem.

Corollary 3. *If*
$$\lim_{n\to\infty}(m(n)/n - \theta)\ln n = 0,$$
then
$$\liminf_{n\to\infty}\rho_{n,m(n)}^{\sigma(2-\theta)/2n}n \leq \sigma e^{1/2}\tau(\mathrm{cap}(E))^\sigma e^{-\frac{(1-\theta)^2}{\theta(2-\theta)}\ln(1-\theta)}. \quad (6)$$

Corollary 4. *If*
$$\limsup_{n\to\infty}\rho_{n,m(n)}^{\sigma/n}n = \sigma e\tau(\mathrm{cap}(E))^\sigma,$$
then for any λ with $1 - \theta < \lambda \leq 1$
$$\liminf_{n\to\infty}\rho_{n,m(n)}^{\lambda\sigma/n}n \leq \lambda\sigma e\tau(\mathrm{cap}(E))^\sigma.$$

The outline of this paper is as follows. Results needed below from the theory of Hankel operators are presented in Section 2. In Section 3 we investigate the degree of rational approximation of functions having $s \geq 1$ essential singularities of finite order. Theorem 1 is a consequence of the results obtained there. In Section 4 the proof of Theorem 2 is given.

§2. Some Results from the Theory of Hankel Operators

2.1. A Generalization of the Adamyan-Arov-Kreĭn Theorem

Let G be a bounded domain whose boundary Γ consists of N disjoint closed analytic Jordan curves. It will be assumed that Γ is positively oriented with respect to G, $0 \in G$.

Denote by $E_p(G)$, $1 \leq p \leq \infty$, the Smirnov class of analytic functions on G. We note that the condition

$$\int_\Gamma \frac{\varphi(\xi)\,d\xi}{\xi - z} = 0 \qquad \text{for all } z \in \overline{\mathbb{C}} \setminus \overline{G} \quad (7)$$

is necessary and sufficient for a function $\varphi(\xi)$, $\xi \in \Gamma$, belonging to $L_p(\Gamma)$, to be the boundary value of a function in the Smirnov class $E_p(G)$ (see [3, 6, 11, 17] for more details about the classes $E_p(G)$).

Fix a nonnegative integer l. Denote by $H_l = H_l(G)$ the class of functions q representable in the form $q = \varphi/\xi^l$, where $\varphi \in E_2(G)$.

Denote by $L_{p,l}(\Gamma)$, $1 \leq p < \infty$, the Lebesgue space of functions φ measurable on Γ such that

$$\|\varphi\|_{p,l} = \left(\int_\Gamma |\varphi(\xi)|^p |\xi|^l |d\xi|\right)^{1/p} < \infty.$$

The inner product in the Hilbert space $L_{2,l}(\Gamma)$ is denoted by

$$(\varphi, \psi)_{2,l} = \int_\Gamma (\varphi\overline{\psi})(\xi)|\xi|^l |d\xi|, \quad \varphi, \psi \in L_{2,l}(\Gamma).$$

For $l = 0$ we will write $L_{p,l}(\Gamma) = L_p(\Gamma)$ and $\|\cdot\|_{p,l} = \|\cdot\|_p$.

$L_\infty(\Gamma)$ is the space of essentially bounded functions, with the norm

$$\|\varphi\|_\infty = \operatorname*{ess\,sup}_\Gamma |\varphi(\xi)|.$$

Let $C(\Gamma)$ be the space of continuous functions on Γ, with the norm

$$\|\varphi\|_\Gamma = \max_{\xi \in \Gamma} |\varphi(\xi)|.$$

We represent $L_{2,l}(\Gamma)$ as direct sum $L_{2,l}(\Gamma) = H_l \oplus H_l^\perp$ of subspaces, where $H_l^\perp$ is the orthogonal complement of H_l in $L_{2,l}(\Gamma)$. Here and in what follows we will consider H_l and $E_2(G)$ as subspaces of $L_{2,l}(\Gamma)$.

Assume that f is continuous on Γ. The Hankel operator $A_f : E_2(G) \to H_l^\perp$ is defined as follows. For any function $q \in E_2(G)$ let $A_f q = \mathbb{P}_-(qf)$, where $\mathbb{P}_-$ is the orthogonal projection of $L_{2,l}(\Gamma)$ onto $H_l^\perp$. It is not hard to see that A_f is a compact operator.

Denote by $\{s_{n,l}\}$, $s_{n,l} = s_{n,l}(f;G)$, $n = 0, 1, 2, \ldots$, the sequence of singular numbers (counting multiplicity) of the operator A_f ($s_{n,l}$ is an eigenvalue of the operator $(A_f^* A_f)^{1/2}$, where $A_f^* : H_l^\perp \to E_2(G)$ is the adjoint operator of A_f). Assume that $s_{0,l} \geq s_{1,l} \geq \ldots \geq s_{n,l} \geq \ldots$ (for the properties of singular numbers see [4]).

For any nonnegative integer n denote by $\mathcal{M}_{n+l,n} = \mathcal{M}_{n+l,n}(G)$ the class functions representable in the form $h = p/q\xi^l$, where $p \in E_\infty(G)$ and q is a polynomial of degree at most n, $q \not\equiv 0$. We remark that $h \in \mathcal{M}_{n+l,n}$ has no more than $n+l$ poles and no more than n free poles. The deviation of f from the class $\mathcal{M}_{n+l,n}$ in the space $L_\infty(\Gamma)$ is denoted by

$$\Delta_{n+l,n} = \Delta_{n+l,n}(f;G) = \inf_{h \in \mathcal{M}_{n+l,n}} \|f - h\|_\infty.$$

Using the same arguments as in [12] it is not hard to prove a theorem establishing a connection between the singular numbers of the Hankel operator and the quantities $\Delta_{n+l,n}$.

Let G be a bounded domain whose boundary Γ consists of N disjoint closed analytic Jordan curves, and let f be a continuous function on Γ. Then

$$\Delta_{n+N-1+l,n+N-1} \leq s_{n,l} \leq \Delta_{n+l,n} \qquad (8)$$

for all integers $n \geq N - 1$.

This theorem is a generalization of the well-known Adamyan-Arov-Kreĭn theorem, which relates to the case when $G = \{z : |z| < 1\}$, $l = 0$. We then have $s_{n,0} = \Delta_{n,n}$, $n = 0, 1, 2, \ldots$ (see [1] and [12] for more details).

2.2. Auxiliary Results

In this subsection we point out some useful relations having to do with singular numbers of a Hankel operator.

We first prove a lemma giving necessary and sufficient conditions for a function u belonging to the space $L_{2,l}(\Gamma)$ to be an element of the subspace $E_2^{\perp}(G)$, where $E_2^{\perp}(G)$ is the orthogonal complement of $E_2(G)$ in $L_{2,l}(\Gamma)$.

Lemma 1. *Suppose that $u \in L_{2,l}(\Gamma)$. Then $u \in E_2^{\perp}(G)$ if and only if there exists a function $v \in E_2(G)$ such that*

$$\overline{u}(\xi)|\xi|^l \, |d\xi| = v(\xi) \, d\xi. \tag{9}$$

almost everywhere on Γ.

Proof: Assume the relation (9) for the function $u \in L_{2,l}(\Gamma)$, where v is some function in $E_2(G)$. We show that

$$(q, u)_{2,l} = 0 \quad \text{for any } q \in E_2(G), \tag{10}$$

where $(q, u)_{2,l}$ is the inner product in $L_{2,l}(\Gamma)$. We have that

$$(q, u)_{2,l} = \int_{\Gamma} q(\xi)\overline{u}(\xi)|\xi|^l \, |d\xi| = \int_{\Gamma} q(\xi)v(\xi) \, d\xi = 0.$$

The last equality in this relation follows from the fact that both functions q and v belong to $E_2(G)$. It remains to see that by (10), $u \in E_2^{\perp}(G)$.

Assume now that $u \in E_2^{\perp}(G)$. Then for any function $q \in E_2(G)$

$$(q, u)_{2,l} = \int_{\Gamma} q(\xi)\overline{u}(\xi)|\xi|^l \, |d\xi| = 0.$$

In particular, for $q(\xi) = 1/(\xi - z)$, (z is an arbitrary point in $\overline{\mathbb{C}} \setminus \overline{G}$,) we get

$$\int_{\Gamma} \frac{1}{\xi - z} \overline{u}(\xi) |\xi|^l \, |d\xi| = 0.$$

This implies (see (7)) that there exists a function $v \in E_2(G)$ such that (9) holds. □

The next lemma is established along the same lines presented above. This lemma gives us necessary and sufficient conditions for a function u belonging to the space $L_{2,l}(\Gamma)$ to be an element of subspace $H_l^{\perp}$.

Lemma 2. *Suppose that* $u \in L_{2,l}(\Gamma)$. *Then* $u \in H_l^\perp$ *if and only if there exists a function* $v \in E_2(G)$ *such that*

$$\bar{u}(\xi)|\xi|^l|d\xi| = v(\xi)\xi^l d\xi \tag{11}$$

almost everywhere on Γ.

Let $\{q_{n,l}\}$, $n = 0, 1, 2, \ldots$, be an orthonormal system of eigenfunctions of the operator $(A_f^* A_f)^{1/2}$, corresponding to the sequence of singular numbers $\{s_{n,l}\}$, $n = 0, 1, 2, \ldots$. We fix a nonnegative integer n. Since $s_{n,l}$ is an eigenvalue of $(A_f^* A_f)^{1/2}$,

$$A_f q_{n,l} = s_{n,l} u_{n,l}, \tag{12}$$

$$A_f^* u_{n,l} = s_{n,l} q_{n,l}, \tag{13}$$

where $u_{n,l} \in H_l^\perp(G)$.

Let us write these relations in another form. For this we note first that since $A_f q_{n,l} = \mathbb{P}_-(q_{n,l}f)$, there exists a unique function $p_{n,l} \in E_2(G)$ such that $A_f q_{n,l} = q_{n,l} f - p_{n,l}/\xi^l$; therefore, by (12),

$$q_{n,l} f - p_{n,l}/\xi^l = s_{n,l} u_{n,l}. \tag{14}$$

Second, it follows from the definition of A_f that the adjoint operator A_f^* of A_f is the composition of the operator of multiplication by the function $\bar{f}$ and the orthogonal projection $\mathbb{P}_+$ of $L_{2,l}(\Gamma)$ onto $E_2(G)$; namely, for any function $u \in H_l^\perp$ we have $A_f^* u = \mathbb{P}_+(u\bar{f})$. Therefore, $A_f^* u_{n,l}$ can be represented in the form $A_f^* u_{n,l} = u_{n,l}\bar{f} - v_{n,l}$, where $v_{n,l} \in E_2^\perp(G)$. The equality (13) thus implies that

$$u_{n,l}\bar{f} - v_{n,l} = s_{n,l} q_{n,l}. \tag{15}$$

We now use the fact that the functions $u_{n,l} \in H_l^\perp$ and $v_{n,l} \in E_2^\perp(G)$. Then we can assert (see (9) and (11)) that there exist functions $\alpha_{n,l}, \beta_{n,l} \in E_2(G)$ such that

$$\bar{u}_{n,l}(\xi)|\xi|^l|d\xi| = \alpha_{n,l}(\xi)\xi^l d\xi$$

and

$$\bar{v}_{n,l}(\xi)|\xi|^l|d\xi| = \beta_{n,l}(\xi)d\xi$$

almost everywhere on Γ. Therefore, by (14) and (15) we get

$$(q_{n,l}f - p_{n,l}/\xi^l)(\xi)\xi^l d\xi = s_{n,l}\bar{\alpha}_{n,l}(\xi)|\xi|^l|d\xi|, \tag{16}$$

$$(\alpha_{n,l}f - \beta_{n,l}/\xi^l)(\xi)\xi^l d\xi = s_{n,l}\bar{q}_{n,l}(\xi)|\xi|^l|d\xi| \tag{17}$$

almost everywhere on Γ.

The system of functions $\{q_{n,l}\}$, $n = 0, 1, 2, \ldots$, is an orthonormal system of functions; therefore, by (17),

$$\int_\Gamma (q_{i,l}\alpha_{j,l}f)(\xi)\xi^l d\xi = s_{j,l}\delta_{i,j}, \qquad i, j = 0, 1, 2, \ldots,$$

where $\delta_{i,j}$ is the Kronecker symbol.

Thus, the following formula holds for the product of singular numbers

$$s_{0,l}s_{1,l}\cdots s_{k,l} = \left|\int_\Gamma (q_{i,l}\alpha_{j,l}f)(\xi)\xi^l d\xi\right|_{i,j=0}^k, \quad k=0,1,2,\ldots, \tag{18}$$

(the right-hand side is a determinant of order $k+1$).

We mention also that the relations (16) and (17), together with $\|q_{n,l}\|_{2,l} = 1$, $n = 0,1,2,\ldots$, imply that $\|\alpha_{n,l}\|_{2,l} = 1$ for all $n = 0,1,\ldots$.

§3. Rational Approximation of Functions Having Finitely Many Essential Singularities

3.1. The Statement of Theorem 3

In this section we introduce and study a situation more general than that presented above. Here it will be assumed that the function to be approximated has $s \geq 1$ essential singularities of finite order in the extended complex plane.

Theorem 3. *Suppose that E is an arbitrary compact set in $\mathbf{C}$, f is a holomorphic function on $\bar{\mathbf{C}}\setminus\{a_1,\ldots,a_s\}$, $a_i \in \mathbf{C}\setminus E$, $i = 1,\ldots,s-1$, $a_s \in \overline{\mathbf{C}}\setminus E$, and the point a_i, $i = 1,\ldots,s$, is an essential singularity of f of finite order $\sigma_i \geq 0$. Then*

(i)
$$\limsup_{n\to\infty} \frac{\ln(\rho_{n,m(n)}\rho_{n-1,m(n)-1}\cdots\rho_{n-m(n),0})}{nm(n)\ln n}$$
$$\leq -\frac{(1+\theta)^2}{4\theta(\sigma_1+\cdots+\sigma_s)} + \frac{(1-\theta)^2}{4\theta\sigma_s},$$

for $a_s = \infty$ and $\sigma_s/(\sigma_1+\cdots+\sigma_s) > (1-\theta)/(1+\theta)$;

(ii)
$$\limsup_{n\to\infty} \frac{\ln(\rho_{n,m(n)}\rho_{n-1,m(n)-1}\cdots\rho_{n-m(n),0})}{nm(n)\ln n} \leq -\frac{\theta}{\sigma_1+\cdots+\sigma_{s-1}}$$

for $a_s = \infty$ and $\sigma_s/(\sigma_1+\cdots+\sigma_s) \leq (1-\theta)/(1+\theta)$;

(iii)
$$\limsup_{n\to\infty} \frac{\ln(\rho_{n,m(n)}\rho_{n-1,m(n)-1}\cdots\rho_{n-m(n),0})}{nm(n)\ln n} \leq -\frac{\theta}{\sigma_1+\cdots+\sigma_s},$$

for $a_s \neq \infty$.

Here and what follows we will use the following notation. Denote by $1/\sigma$ the expression

$$\frac{(1+\theta)^2}{4\theta(\sigma_1+\cdots+\sigma_s)} - \frac{(1-\theta)^2}{4\theta\sigma_s}$$

for $a_s = \infty$ and $\sigma_s/(\sigma_1 + \cdots + \sigma_s) > (1-\theta)/(1+\theta)$,

$$\frac{\theta}{\sigma_1 + \cdots + \sigma_{s-1}}$$

for $a_s = \infty$ and $\sigma_s/(\sigma_1 + \cdots + \sigma_s) \le (1-\theta)/(1+\theta)$, and

$$\frac{\theta}{\sigma_1 + \cdots + \sigma_s}$$

for $a_s \ne \infty$.

We mention that according to Theorem 3

$$\limsup_{n \to \infty} \frac{\ln(\rho_{n,m(n)} \rho_{n-1,m(n)-1} \cdots \rho_{n-m(n),0})}{nm(n) \ln n} \le -\frac{1}{\sigma}.$$

We have the following consequence to the theorem.

Corollary 5.

$$\liminf_{n \to \infty} \frac{\ln \rho_{n,m(n)}}{n \ln n} \le -\frac{2}{(2-\theta)\sigma}.$$

From Theorem 3 and the fact that the sequence $\{\rho_{n,m(n)}\}$, $n = 0, 1, \ldots$, is nonincreasing, we immediately have the following.

Corollary 6.

$$\limsup_{n \to \infty} \frac{\ln \rho_{n,m(n)}}{n \ln n} \le -\frac{1}{\sigma}.$$

The next corollary concerns functions for which equality holds in the last relation.

Corollary 7. If

$$\limsup_{n \to \infty} \frac{\ln \rho_{n,m(n)}}{n \ln n} = -\frac{1}{\sigma},$$

then

$$\liminf_{n \to \infty} \frac{\ln \rho_{n,m(n)}}{n \ln n} \le -\frac{1}{(1-\theta)\sigma}.$$

It is not hard to see from the proof of Theorem 3 that this theorem is valid under more general assumptions on f. Namely, it can be assumed that the point $a_i, i = 1, \ldots, s$, is an essential singularity of f of finite order no greater then σ_i.

Let us mention the scheme of proof of Theorem 3. It will be assumed that $\sigma_i > 0$ for $i = 1, \ldots, s$. The general case can be obtained from this case with help of the corresponding limit transition $\sigma_i \to 0$. In Subsections 3.2 and 3.3 we consider the situation when the function to be approximated has exactly one essential singularity of finite order at infinity, and by this we actually prove Theorem 1. In Subsection 3.4 we apply the results obtained in Subsections 3.2 and 3.3 to consider the general case.

Before proving Theorem 3 we note that the diagonal case when $m(n) = n$, $n = 0, 1, 2, \ldots$, was investigated in the paper [13]. In this case the following estimate is valid:

$$\limsup_{n \to \infty} \frac{\ln(\rho_{n,n}\rho_{n-1,n-1}\cdots\rho_{0,0})}{n^2 \ln n} \leq -\frac{1}{\sigma_1 + \cdots + \sigma_s}. \tag{19}$$

3.2. The Case When f is an Entire Function of Finite Order

In this subsection we shall start with the assumption that E is an arbitrary compact set in $\mathbf{C}$, and f is an entire function of finite order σ.

In order to continue with the proof of the corresponding assertion, let us begin with some remarks.

First, seeing that $\rho_{n,m} = \rho_{n,m}(f;E)$ is nondecreasing as the compact set E expands ($\rho_{n,m}(f;E) \leq \rho_{n,m}(f;E')$ for $E \subseteq E'$), we can assume that the complement G of E is connected, and E is bounded by finitely many disjoint closed analytic Jordan curves Γ.

Second, with help of an appropriate linear fractional transformation we can reduce the original theorem to the situation when E contains ∞ and f has exactly one essential singularity $a = 0$, $0 \in G = \bar{\mathbf{C}} \setminus E$, of finite order σ.

Precisely, for this situation we prove that

$$\limsup_{n \to \infty} \frac{\ln(\rho^*_{n,m(n)}\rho^*_{n-1,m(n)-1}\cdots\rho^*_{n-m(n),0})}{nm(n) \ln n} \leq -\frac{1}{\sigma}, \tag{20}$$

where

$$\rho^*_{n-j,m(n)-j} = \rho^*_{n-j,m(n)-j}(f;E) = \inf_{r \in \mathcal{R}^*_{n-j,m(n)-j}} \|f - r\|_E,$$

$$j = 0, 1, \ldots, m(n).$$

Here and what follows we will use the following notation:

$$\mathcal{R}^*_{n-j,m(n)-j} = \{r : r = p/qz^{n-m(n)}, \deg p \leq n-j, \deg q \leq m(n)-j, q \not\equiv 0\}.$$

In this subsection we prove the inequality

$$\limsup_{n \to \infty} \frac{\ln(\Delta_{n,m(n)}\Delta_{n-1,m(n)-1}\cdots\Delta_{n-m(n),0})}{nm(n) \ln n} \leq -\frac{1}{\sigma}, \tag{21}$$

where

$$\Delta_{n-j,m(n)-j} = \Delta_{n-j,m(n)-j}(f;G) = \inf_{h} \|f - h\|_\infty, \quad j = 0, 1, \ldots, m(n),$$

is the best approximation of f in $L_\infty(\Gamma)$ in the class $\mathcal{M}_{n-j,m(n)-j}$ of functions h such that $h = p/qz^{n-m(n)}$, $p \in E_\infty(G)$ and q is a polynomial of degree at most $m(n) - j$, $q \not\equiv 0$.

For this it suffices to show (see the estimate (8)) that

$$\limsup_{n\to\infty} \frac{\ln(s_{0,n-m(n)} s_{1,n-m(n)} \cdots s_{m(n),n-m(n)})}{nm(n) \ln n} \leq -\frac{1}{\sigma'}, \tag{22}$$

where $\{s_{k,n-m(n)}\}$, $s_{k,n-m(n)} = s_{k,n-m(n)}(f;G)$, $k = 0, 1, 2, \ldots$, is the sequence of singular numbers of the Hankel operator $A_f : E_2(G) \to H^\perp_{n-m(n)}$, constructed from the function f, and $H_{n-m(n)}$ is the class of functions q representable in the form $q = \varphi/\xi^{n-m(n)}$, where $\varphi \in E_2(G)$.

It is not difficult to pass from the estimate (21) to (20) (see Subsection 3.3); therefore, we now restrict ourselves to proving the inequality (22).

We fix an arbitrary domain G_1, $\overline{G}_1 \subset G$, $0 \in G_1$, bounded by a finite number of closed analytic Jordan curves, and $\sigma' > \sigma$. Let $t_n = n^{-1/\sigma'}$. Denote by l_n the circle of radius t_n with the center at 0. It will be assumed that l_n is positively oriented with respect to the open disk of radius t_n about 0 and n is a sufficiently large positive integer, $n \geq n_0$, such that the closed disk of radius t_n with center at 0 belongs to G_1.

Let us use (18), with $k = m(n)$, $l = n - m(n)$. Since the functions $q_{i,n-m(n)}, \alpha_{j,n-m(n)}$, $i, j = 0, 1, 2, \ldots$, belong to $E_2(G)$ and f is holomorphic on and outside the circle l_n, the relation

$$s_{0,n-m(n)} s_{1,n-m(n)} \cdots s_{m(n),n-m(n)}$$
$$= \left| \int_{l_n} (q_{i,n-m(n)} \alpha_{j,n-m(n)} f)(\xi) \xi^{n-m(n)} d\xi \right|_{i,j=0}^{m(n)}$$

can be written for the product of singular numbers. From the last relation

$$(m(n)+1)! s_{0,n-m(n)} s_{1,n-m(n)} \cdots s_{m(n),n-m(n)}$$
$$= \int_{l_n} \cdots \int_{l_n} f(\xi_0) \cdots f(\xi_{m(n)}) B_1(\xi_0, \ldots, \xi_{m(n)}) B_2(\xi_0, \ldots, \xi_{m(n)}) \tag{23}$$
$$\times \xi_0^{n-m(n)} \cdots \xi_{m(n)}^{n-m(n)} d\xi_0 \cdots d\xi_{m(n)},$$

where

$$B_1(\xi_0, \xi_1, \ldots, \xi_{m(n)}) = \left| \alpha_{j,n-m(n)}(\xi_i) \right|_{i,j=0}^{m(n)} \tag{24}$$

and

$$B_2(\xi_0, \xi_1, \ldots, \xi_{m(n)}) = \left| q_{j,n-m(n)}(\xi_i) \right|_{i,j=0}^{m(n)}. \tag{25}$$

We estimate the determinants B_1 and B_2. By the Cauchy formula,

$$\alpha^2_{j,n-m(n)}(\xi) \xi^{n-m(n)} = \frac{1}{2\pi i} \int_\Gamma \frac{\alpha^2_{j,n-m(n)}(t) t^{n-m(n)}}{t-\xi} dt, \; \xi \in \overline{G}_1,$$
$$j = 0, 1, 2, \ldots.$$

Since
$$\int_\Gamma |\alpha_{j,n-m(n)}(t)|^2 |t|^{n-m(n)} |dt| = 1, \quad j = 0, 1, 2, \ldots,$$
it follows from the last formula that
$$\max_{\xi \in \overline{G}_1} |\alpha^2_{j,n-m(n)}(\xi)\xi^{n-m(n)}| \leq C, \quad j = 0, 1, 2, \ldots \qquad (26)$$

(here and in what follows $C, C_1, C_2, \ldots$ will denote positive quantities not depending on n).

Similarly, since
$$\int_\Gamma |q_{j,n-m(n)}(t)|^2 |t|^{n-m(n)} |dt| = 1, \quad j = 0, 1, 2, \ldots,$$
it follows that
$$\max_{\xi_i \in \overline{G}_1} |q^2_{j,n-m(n)}(\xi)\xi^{n-m(n)}| \leq C, \quad j = 0, 1, 2, \ldots. \qquad (27)$$

Using the inequalities (26) and (27), we can write
$$\max_{\xi_i \in \overline{G}_1} |B_1(\xi_0, \ldots, \xi_{m(n)}) \times B_2(\xi_0, \ldots, \xi_{m(n)}) \xi_0^{n-m(n)} \cdots \xi_{m(n)}^{n-m(n)}|$$
$$\leq ((m(n)+1)!)^2 C^{m(n)+1}. \qquad (28)$$

For $\xi \in G_1$ let $g(z, \xi)$ be the Green's function of the domain G_1 with singularity at ξ. We estimate the product $B_1 B_2 \xi_0^{n-m(n)} \cdots \xi_n^{n-m(n)}$, in the case when the variables ξ_i, $i = 0, \ldots, m(n)$, belong to l_n. The next equality easily follows from (24) and (25):

$$D(\xi_0, \ldots, \xi_{m(n)})$$
$$:= B_1(\xi_0, \ldots, \xi_{m(n)}) B_2(\xi_0, \ldots, \xi_{m(n)}) \xi_0^{n-m(n)} \cdots \xi_{m(n)}^{n-m(n)}$$
$$= \prod_{0 \leq i < j \leq m(n)} (\xi_i - \xi_j)^2 \cdot \Psi(\xi_0, \xi_1, \ldots, \xi_{m(n)}) \xi_0^{n-m(n)} \cdots \xi_{m(n)}^{n-m(n)},$$

where the function $\Psi(\xi_0, \xi_1, \ldots, \xi_{m(n)})$ is a holomorphic function of $m(n)+1$ complex variables in the domain $G \times \cdots \times G$ (with $m(n)+1$ factors in the Cartesian product).

Consider now the function
$$\ln|D(\xi_0, \xi_1, \ldots, \xi_{m(n)})| + 2 \sum_{0 \leq i < j \leq m(n)} g(\xi_i, \xi_j) + (n - m(n)) \sum_{i=0}^{m(n)} g(\xi_i, 0).$$

This function is subharmonic in the domain G_1 with respect to the variable ξ_i, $i = 0, \ldots, m(n)$, when the remaining variables $\xi_j \in G_1$, $j \neq i$, $j \in \{0, 1, \ldots, m(n)\}$, are fixed.

We now employ the maximum principle for subharmonic functions successively with respect to each variable, together with the inequality (28), to get

$$\ln |D(\xi_0, \xi_1, \ldots, \xi_{m(n)})| + 2 \sum_{0 \leq i < j \leq m(n)} g(\xi_i, \xi_j)$$
$$+ (n - m(n)) \sum_{i=0}^{m(n)} g(\xi_i; 0)$$
$$\leq \ln(((m(n) + 1)!)^2 C^{m(n)+1}),$$

where $\xi_i \in l_n$, $i = 0, 1, \ldots, m(n)$.

In view of the formula for a product of singular numbers (see (23)), the last inequality implies

$$s_{0,n-m(n)} s_{1,n-m(n)} \cdots s_{m(n),n-m(n)}$$
$$\leq (m(n) + 1)! C_1^{m(n)} \left(\max_{\xi \in l_n} |f(\xi)| \right)^{m(n)+1} \exp(-w_n), \qquad (29)$$

where

$$w_n = \min_{\xi_i \in l_n} \left(2 \sum_{0 \leq i < j \leq m(n)} g(\xi_i, \xi_j) + (n - m(n)) \sum_{i=0}^{m(n)} g(\xi_i, 0) \right). \qquad (30)$$

We now use the following representation of the Green's function

$$g(z, \xi) = \ln \frac{1}{|z - \xi|} + d(\xi) + u(z, \xi), \qquad z, \xi \in G_1, \qquad (31)$$

where $d(\xi)$ is a quantity dependent on ξ, and $u(z, \xi)$ is a function harmonic in G_1 with respect to z, $u(\xi, \xi) = 0$, to establish a lower estimate of w_n. It can be shown that

$$d(\xi) \to d(\zeta) \quad \text{as} \quad \xi \to \zeta, \quad \zeta \in G_1,$$

and

$$u(z, \xi) \to 0,$$

if both the variables z and ξ tend to a point $\zeta \in G_1$.

We will consider the points $\xi_{0,n}, \xi_{1,n}, \ldots, \xi_{m(n),n}$, where the expression (30) takes its minimum on l_n. It is not hard to see that all the points $\xi_{i,n}$ are distinct. Taking into account (31), we get

$$w_n = 2 \sum_{0 \le i < j \le m(n)} g(\xi_{i,n}, \xi_{j,n}) + (n - m(n)) \sum_{i=0}^{m(n)} g(\xi_{i,n}, 0)$$

$$= 2 \sum_{0 \le i < j \le m(n)} \ln \frac{1}{|\xi_{i,n} - \xi_{j,n}|} + (n - m(n)) \sum_{i=0}^{m(n)} \ln \left| \frac{1}{\xi_{i,n}} \right| \quad (32)$$

$$+ 2 \sum_{i=1}^{m(n)} id(\xi_{i,n}) + (n - m(n))(m(n) + 1)d(0)$$

$$+ 2 \sum_{0 \le i < j \le m(n)} u(\xi_{i,n}, \xi_{j,n}) + (n - m(n)) \sum_{i=0}^{m(n)} u(\xi_{i,n}, 0).$$

Next we investigate how the last expression in (32) changes as $n \to \infty$. We know that the radius t_n tends to zero; therefore

$$\frac{1}{n(m(n)+1)} \left(2 \sum_{i=1}^{m(n)} id(\xi_{i,k}) + (n - m(n))(m(n)+1)d(0) \right) \to d(0)$$

and

$$\frac{1}{n(m(n)+1)} \left(2 \sum_{0 \le i < j \le m(n)} u(\xi_{i,n}, \xi_{j,n}) + (n - m(n)) \sum_{i=0}^{n} u(\xi_{i,n}, 0) \right) \to 0$$

as $n \to \infty$. This leads us to write

$$w_n = 2 \sum_{0 \le i < j \le m(n)} \ln \frac{1}{|\xi_{i,n} - \xi_{j,n}|} + (n - m(n)) \sum_{i=0}^{m(n)} \ln \frac{1}{|\xi_{i,n}|} + n(m(n)+1)d_n,$$

where $d_n \to d(0)$ as $n \to \infty$. From this it follows that

$$w_n \ge \min_{\xi_i \in l_n} 2 \sum_{0 \le i < j \le m(n)} \ln \frac{1}{|\xi_i - \xi_j|} + (n - m(n)) \sum_{i=0}^{m(n)} \ln \frac{1}{|\xi_{i,n}|} \quad (33)$$

$$+ n(m(n)+1)d_n.$$

The estimate of the first term on the right-hand side of (33) proceeds in several steps. First, from the representations $\xi_i = t_n x_i$ for any point $\xi_i \in l_n$, where $|x_i| = 1$, we compute that

$$\min_{\xi_i \in l_n} \sum_{0 \le i < j \le m(n)} \ln \frac{1}{|\xi_i - \xi_j|} = \frac{m(n)(m(n)+1)}{2} \ln \frac{1}{|t_n|}$$

$$+ \min_{|x_i|=1} \sum_{0 \le i < j \le m(n)} \ln \frac{1}{|x_i - x_j|}.$$

Second, taking into account of

$$\min_{|x_i|=1} \sum_{0 \leq i < j \leq m(n)} \ln \frac{1}{|x_i - x_j|} \geq \ln \frac{1}{(m(n)+1)!},$$

we obtain

$$w_n \geq n(m(n)+1) \ln \frac{1}{|t_n|} + n(m(n)+1)d_n + 2\ln \frac{1}{(m(n)+1)!}. \quad (34)$$

Thus, the estimate (see (29), (34))

$$s_{0,n-m(n)} s_{1,n-m(n)} \cdots s_{m(n),n-m(n)}$$
$$\leq ((m(n)+1)!)^3 C_1^{m(n)} \left(\max_{\xi \in l_n} |f(\xi)| \right)^{m(n)+1} \quad (35)$$
$$\times \exp(-n(m(n)+1)d_n)|t_n|^{n(m(n)+1)}$$

holds for $n \geq n_0$.

Recall that we have assumed that f has an essential singularity at a of order σ; this implies that

$$\max_{\xi \in l_n} |f(\xi)| \leq e^{1/t_n^{\sigma'}} = e^n \quad (36)$$

for sufficiently large n, $n \geq n_1$.

Finally, we obtain from (35), (36), and the equality $t_n = n^{-1/\sigma'}$, that

$$\limsup_{n \to \infty} \frac{\ln(s_{0,n-m(n)} s_{1,n-m(n)} \cdots s_{m(n),n-m(n)})}{nm(n) \ln n} \leq -\frac{1}{\sigma'}. \quad (37)$$

Letting $\sigma' \to \sigma$ on the right-hand side of the inequality (37) we obtain (22), which implies the inequality (21).

3.3. Proof of the Inequality (20)

We now apply the estimate (21) to get the inequality (20).

Fix an arbitrary domain G_1, $\bar{G}_1 \subset G$, $0 \in G_1$. We assume that the boundary Γ_1 of the domain G_1 consists of disjoint closed analytic Jordan curves; moreover, we assume that Γ_1 is positively oriented with respect to G_1.

Fix also nonnegative integers n and j, $0 \leq j \leq m(n)$. Using the Cauchy formula for an arbitrary function h representable in the form $h = p/(qz^{n-m(n)})$, where $p \in E_\infty(G_1)$, q is a polynomial of degree at most $m(n) - j$, with zeros outside Γ_1, $q \not\equiv 0$, we obtain

$$(r' - f)(z) + f(\infty) = \frac{1}{2\pi i} \int_{\Gamma_1} \frac{(f-h)(\xi)d\xi}{\xi - z}, \quad z \in E, \quad (38)$$

where r' is the sum of the principal parts of h corresponding to poles of h lying in G_1.

Consider the integral on the left-hand side. We have that

$$\|f - f(\infty) - r'\|_E \leq C\|f - h\|_\infty, \tag{39}$$

where the positive quantity C is independent of h, n, and j, and $\|\cdot\|_\infty$ is the norm in the space $L_\infty(\Gamma_1)$.

From the definition of the quantity $\rho^*_{n-j,m(n)-j}$ and the fact that the rational function $r' + f(\infty)$ belongs to the class $\mathcal{R}^*_{n-j,m(n)-j}$, the inequality (39) becomes

$$\rho^*_{n-j,m(n)-j} \leq C\|f - h\|_\infty.$$

Now, using the fact that h is an arbitrary function in $\mathcal{M}_{n-j,m(n)-j}(G_1)$ we obtain

$$\rho^*_{n-j,m(n)-j} \leq C \inf_{h \in \mathcal{M}_{n-j,m(n)-j}} \|f - h\|_\infty = C\Delta_{n-j,m(n)-j}(f; G_1).$$

Following the results in Subsection 3.2 (see the relation (21)), applied to the region G_1, we get

$$\limsup_{n \to \infty} \frac{\ln(\rho^*_{n,m(n)} \rho^*_{n-1,m(n)-1} \cdots \rho^*_{n-m(n),0})}{nm(n)\ln n} \leq -\frac{1}{\sigma}.$$

Thus, Theorem 3 has been proved for the case when the function being approximated is an entire function of finite order.

3.4. The Case When the Number of Singularities $s \geq 2$

Assume that the number s of singularities is ≥ 2. For $i = 1, \ldots, s-1$, let

$$f_i(z) = \varphi_i(z), \qquad z \in \bar{\mathbb{C}} \setminus \{a_i\},$$

where φ_i is the principal part of the Laurent expansion of f in a neighborhood of a_i; the function $f - (\varphi_1 + \cdots + \varphi_{s-1})$ can be extended to an analytic function f_s on $\bar{\mathbb{C}} \setminus \{a_s\}$. It is not hard to see that $f = f_1 + f_2 + \cdots + f_s$, where each of the functions f_i is holomorphic in $\bar{\mathbb{C}} \setminus \{a_i\}$, and a_i is an essential singularity of order σ_i for f_i, $i = 1, \ldots, s$.

It will be assumed that $\theta > 0$. For $\theta = 0$ the corresponding assertion is obvious.

First of all we consider the case when $a_s = \infty$.

The main goal of this subsection is to obtain an upper estimate of the product

$$\prod_{j=0}^{m(n)} \rho_{n-j,m(n)-j}(f; E)$$

by means of the product of the quantities $\rho_{j,j}(f_i; E)$, $i = 1, 2, \ldots, s-1$, and $\rho_{n-m(n)+j,j}(f_s; E)$.

Lemma 3. For any nonnegative integer n, there exist nonnegative integers $\chi_i(n)$, $i = 1, 2, \ldots, s$, such that $\chi_1(n) + \chi_2(n) + \cdots + \chi_s(n) = m(n) + 1$ and

$$\prod_{j=0}^{m(n)} \rho_{n-j,m(n)-j}(f;E) \leq s^{m(n)+1} \prod_{i=1}^{s-1} \prod_{j=0}^{\chi_i(n)-1} \rho_{j,j}(f_i;E) \tag{40}$$

$$\times \prod_{j=0}^{\chi_s(n)-1} \rho_{n-m(n)+j,j}(f_s;E),$$

(let $\prod_{j=0}^{\chi_i(n)-1} \rho_{j,j}(f_i;E) = 1$, in the product (40) if $\chi_i(n) = 0$ and respectively $\prod_{j=0}^{\chi_s(n)-1} \rho_{n-m(n)+j,j}(f_s;E) = 1$ if $\chi_s(n) = 0$).

Before proving the lemma, for any nonnegative integer n we define the quantities

$$\mu^*_{n-m(n)+j,j} = \min_{k_i}\left(\max\left(\max_{1\leq i\leq s-1} \rho_{k_i,k_i}(f_i;E), \rho_{n-m(n)+k_s,k_s}(f_s;E)\right)\right),$$

$$j = 0, 1, 2, \ldots,$$

where the minimum is over all tuples of nonnegative integers k_i, $i = 1, \ldots, s$, such that $k_1 + k_2 + \cdots + k_s \leq j$.

Proof of Lemma 3: Let us show that for any nonnegative integers n and j,

$$\rho_{n-m(n)+j,j}(f;E) \leq s\mu^*_{n-m(n)+j,j}. \tag{41}$$

For this purpose, choose arbitrary nonnegative integers k_i, $i = 1, \ldots, s$, such that $k_1 + k_2 + \cdots + k_s \leq j$. Also, for each $i \in \{1, \ldots, s-1\}$ choose an arbitrary rational function r_i in $\mathcal{R}_{k_i,k_i}$. Let r_s be an arbitrary rational function in $\mathcal{R}_{n-m(n)+k_s,k_s}$ and $r = r_1 + r_2 + \cdots + r_s$. Since $k_1 + k_2 + \cdots + k_s \leq j$, $r \in \mathcal{R}_{n-m(n)+j,j}$. We have

$$\rho_{n-m(n)+j,j}(f;E) \leq \|f-r\|_E \leq \sum_{i=1}^{s}\|f_i - r_i\|_E. \tag{42}$$

We now use the fact that r_i, $i = 1, 2, \ldots, s-1$, is an arbitrary rational function in $\mathcal{R}_{k_i,k_i}$ and r_s is an arbitrary rational function in $\mathcal{R}_{n-m(n)+k_s,k_s}$. From (42),

$$\rho_{n-m(n)+j,j}(f;E) \leq \sum_{i=1}^{s-1}\rho_{k_i,k_i}(f_i;E) + \rho_{n-m(n)+k_s,k_s}(f_s;E).$$

In turn, we obtain the inequality (41) from the last inequality and the fact that $k_1, \ldots, k_s$ are arbitrary nonnegative integers with $k_1 + k_2 + \cdots + k_s \leq j$.

We remark that, from (41),

$$\prod_{j=0}^{m(n)} \rho_{n-m(n)+j,j}(f;E) \leq s^{m(n)+1} \prod_{j=0}^{m(n)} \mu^*_{n-m(n)+j,j}. \tag{43}$$

The desired estimate (40) is an immediate consequence of the inequality (43) and an assertion about numerical sequences. We state the corresponding assertion (see [13]).

Suppose that we are given s nonincreasing sequences of nonnegative numbers $\{a_{j,i}\}_{j=0}^{\infty}$, $\lim_{j\to\infty} a_{j,i} = 0$, $i = 1, 2, \ldots, s$. Let

$$a_j^* = \min_{k_i} \left(\max_{1 \le i \le s} (a_{k_i,i}) \right), \quad j = 0, 1, 2, \ldots,$$

where min is taken over all possible tuples of nonnegative integers k_i, $i = 1, \ldots, s$, such that $k_1 + k_2 + \cdots + k_s \le j$.

Lemma 4. *For any nonnegative integer k there exist nonnegative integers $\chi_i(n)$, $i = 1, 2, \ldots, s$, such that $\chi_1(n) + \chi_2(n) + \cdots + \chi_s(n) = k + 1$ and*

$$a_0^* a_1^* \cdots a_k^* \le \prod_{i=1}^{s} \prod_{j=0}^{\chi_i(n)-1} a_{j,i}$$

(let $\prod_{j=0}^{\chi_i(n)-1} a_{j,i} = 1$ in the product if $\chi_i(n) = 0$).

Fix a nonnegative integer n. Using Lemma 4, with nonincreasing sequences $\{\rho_{j,j}(f_i; E)\}_{j=0}^{\infty}$, $i = 1, \ldots, s-1$, $\{\rho_{n-m(n)+j,j}(f_s; E)\}_{j=0}^{\infty}$, and $k = m(n)$, by (43), we get (40). $\square$

According to Lemma 3,

$$\prod_{j=0}^{m(n)} \rho_{n-j,m(n)-j}(f;E) \le s^{m(n)+1} \prod_{i=0}^{s-1} \prod_{j=0}^{\chi_i(n)-1} \rho_{j,j}(f_i;E) \\ \times \prod_{j=0}^{\chi_s(n)-1} \rho_{n-m(n)+j,j}(f_s;E), \quad (44)$$

where $\chi_1(n) + \cdots + \chi_s(n) = m(n) + 1$.

We take a sequence Λ of positive integers such that

$$\lim_{n\to\infty, n\in\Lambda} \frac{\ln \prod_{j=0}^{m(n)} \rho_{n-j,m(n)-j}}{nm(n)\ln n} = \limsup_{n\to\infty} \frac{\ln \prod_{j=0}^{m(n)} \rho_{n-j,m(n)-j}}{nm(n)\ln n} \quad (45)$$

and

$$\lim_{n\to\infty, n\in\Lambda} \frac{\chi_i(n)}{m(n)} = \omega_i, \quad i = 1, 2, \ldots, s. \quad (46)$$

We note that $\omega_i \ge 0$ and $\omega_1 + \omega_2 + \cdots + \omega_s = 1$.

Since for all $i \in \{1, \ldots, s-1\}$, a_i is an essential singularity of f_i of order σ_i, it follows from (19), with $s = 1$, that

$$\limsup_{n\to\infty, n\in\Lambda} \frac{\ln(\prod_{j=0}^{\chi_i(n)-1} \rho_{j,j}(f_i;E))}{nm(n)\ln n} \le -\frac{\omega_i^2 \theta}{\sigma_i}, \quad i = 1, 2, \ldots, s-1.$$

Ray Sequences of Best Rational Approximants to Entire Functions

By the relations (2), (4), and (46), we have

$$\lim_{n\to\infty, n\in\Lambda} \frac{\ln \prod_{j=0}^{\chi_s(n)-1} p_{n-m(n)+j,j}(f_s;E)}{nm(n)\ln n} \leq -\frac{\omega_s(1-\theta+\omega_s\theta)}{\sigma_s}.$$

From the last inequalities with the aid of (44) and (45) we obtain that

$$\limsup_{n\to\infty} \frac{\ln \prod_{j=0}^{m(n)} p_{n-j,m(n)-j}}{nm(n)\ln n} \leq -\left(\sum_{i=1}^{s-1} \frac{\omega_i^2 \theta}{\sigma_i} + \frac{\omega_s(1-\theta+\omega_s\theta)}{\sigma_s}\right)$$
$$= -\left(\sum_{i=1}^{s} \frac{\omega_i^2 \theta}{\sigma_i} + \frac{\omega_s(1-\theta)}{\sigma_s}\right). \tag{47}$$

To estimate the last expression we use the following simple assertion. This assertion can be obtained, for example, by the method of Lagrange multipliers.

Lemma 5. *If $\theta_i \geq 0$, $i = 1,2,\ldots,s$, and $\theta_s/(\theta_1+\cdots+\theta_s) > (1-\theta)/(1+\theta)$, then*

$$\frac{(1+\theta)^2}{4\theta^2 \sum_{i=1}^{s} \theta_i} - \frac{(1-\theta)^2}{4\theta^2 \theta_s} \leq \sum_{i=1}^{s} \frac{\omega_i^2}{\theta_i} + \frac{\omega_s(1-\theta)}{\theta\theta_s}$$

for all $\omega_i \geq 0$, $i = 1,2,\ldots,s$, $\sum_{i=1}^{s}\omega_i = 1$. If $\theta_i \geq 0$, $i = 1,2,\ldots,s$ and $\theta_s/(\theta_1+\cdots+\theta_s) \leq (1-\theta)/(1+\theta)$, then

$$\frac{1}{\sum_{i=1}^{s-1} \theta_i} \leq \sum_{i=1}^{s} \frac{\omega_i^2}{\theta_i} + \frac{\omega_s(1-\theta)}{\theta\theta_s}$$

for all $\omega_i \geq 0$, $i=1,2,\ldots,s$, $\sum_{i=1}^{s}\omega_i = 1$.

It remains to employ Lemma 5 with σ_i/θ ($i=1,\ldots,s$) instead of θ_i, use the inequality (47), and get the required relation

$$\limsup_{n\to\infty} \frac{\ln(p_{n,m(n)}p_{n-1,m(n)-1}\cdots p_{n-m(n),0})}{nm(n)\ln n} \leq -\frac{(1+\theta)^2}{4\theta(\sigma_1+\cdots+\sigma_s)} + \frac{(1-\theta)^2}{4\theta\sigma_s}$$

for the case when $\sigma_s/(\sigma_1+\cdots+\sigma_s) > (1-\theta)/(1+\theta)$.

In the situation when $\sigma_s/(\sigma_1+\cdots+\sigma_s) \leq (1-\theta)/(1+\theta)$ we have

$$\limsup_{n\to\infty} \frac{\ln(p_{n,m(n)}p_{n-1,m(n)-1}\cdots p_{n-m(n),0})}{nm(n)\ln n} \leq -\frac{\theta}{\sigma_1+\cdots+\sigma_{s-1}}.$$

We now consider the case $a_s \neq \infty$. Since

$$\prod_{j=0}^{m(n)} p_{n-j,m(n)-j} \leq \prod_{j=0}^{m(n)} p_{m(n)-j,m(n)-j} = \prod_{j=0}^{m(n)} p_{j,j},$$

we get with help of (2) and (19),

$$\limsup_{n\to\infty} \frac{\ln(\prod_{j=0}^{m(n)} p_{n-j,m(n)-j})}{nm(n)\ln n} \leq -\frac{\theta}{\sigma_1+\ldots+\sigma_s}.$$

Theorem 3 is proved.

3.5. Proof of the Corollaries to Theorem 3

We start with the proofs of the corollaries to Theorem 3. Here

$$\frac{1}{\sigma} = \frac{(1+\theta)^2}{4\theta(\sigma_1 + \ldots + \sigma_s)} - \frac{(1-\theta)^2}{4\theta\sigma_s}$$

for $a_s = \infty$ and $\sigma_s/(\sigma_1 + \cdots + \sigma_s) > (1-\theta)/(1+\theta)$,

$$\frac{1}{\sigma} = \frac{\theta}{\sigma_1 + \cdots + \sigma_{s-1}}$$

for $a_s = \infty$, $\sigma_s/(\sigma_1 + \cdots + \sigma_s) \leq (1-\theta)/(1+\theta)$, and

$$\frac{1}{\sigma} = \frac{\theta}{\sigma_1 + \ldots + \sigma_s}$$

for $a_s \neq \infty$. It will be assumed that $\theta > 0$ and $\sigma > 0$. For $\theta = 0$ and $\sigma = 0$ the corresponding assertions are obvious.

To prove Corollary 5, suppose to the contrary that

$$\liminf_{n \to \infty} \frac{\ln \rho_{n,m(n)}}{n \ln n} \geq -\frac{2}{2-\theta} \cdot \frac{\lambda}{\sigma} \qquad (48)$$

where $0 < \lambda < 1$.

It follows from the relation (1) that

$$m(n) - j \leq m(n-j), \quad j = 0, 1, \ldots, m(n).$$

Therefore, we have

$$\rho_{n-j,m(n-j)} \leq \rho_{n-j,m(n)-j}, \quad j = 0, 1, \ldots, m(n)$$

and

$$\prod_{k=n-m(n)}^{n} \rho_{k,m(k)} = \prod_{j=0}^{m(n)} \rho_{n-j,m(n-j)} \leq \prod_{j=0}^{m(n)} \rho_{n-j,m(n)-j}. \qquad (49)$$

It follows from Theorem 3 that

$$\limsup_{n \to \infty} \frac{\ln \prod_{j=0}^{m(n)} \rho_{n-j,m(n)-j}}{nm(n) \ln n} \leq -\frac{1}{\sigma}. \qquad (50)$$

By the relations (48), (49) and

$$\lim_{n \to \infty} \frac{\sum_{k=n-m(n)}^{n} k \ln k}{nm(n) \ln n} = \frac{2-\theta}{2},$$

we get

$$\liminf_{n\to\infty} \frac{\ln \prod_{j=0}^{m(n)} \rho_{n-j,m(n)-j}}{nm(n)\ln n} \geq -\frac{\lambda}{\sigma} > -\frac{1}{\sigma},$$

which contradicts the inequality (50). □

We now prove Corollary 7. Let Λ be a sequence of positive integers such that

$$\lim_{n\to\infty, n\in\Lambda} \frac{\ln \rho_{n,m(n)}}{n\ln n} = -\frac{1}{\sigma}. \tag{51}$$

Fix an arbitrary $1-\theta < \lambda \leq 1$. Denote by $\{k_n\}$, $n=1,2,\ldots$, the sequence of integers such that $n - m(n) \leq k_n \leq n$ and $k_n/n \to \lambda$ as $n \to \infty$. Since the sequence $\{\rho_{n,m(n)}\}$, $n=1,2,\ldots$, is nonincreasing,

$$\rho_{n,m(n)}^{m(n)+1} \leq \rho_{k_n,m(k_n)}^{k_n-n+m(n)+1} \rho_{n,m(n)}^{n-k_n} \leq \prod_{k=n-m(n)}^{n} \rho_{k,m(k)}.$$

From this and from the relations (49), (50), and (51), we get

$$\lim_{n\to\infty, n\in\Lambda} \frac{(k_n - n + m(n) + 1)\ln \rho_{k_n,m(k_n)}}{nm(n)\ln n} = -\frac{1}{\sigma} + \frac{1}{\sigma}\left(\frac{1-\lambda}{\theta}\right),$$

which implies that

$$\lim_{n\to\infty, n\in\Lambda} \frac{\ln \rho_{k_n,m(k_n)}}{k_n \ln k_n} = -\frac{1}{\sigma\lambda}.$$

It follows from the last relation that

$$\liminf_{n\to\infty} \frac{\ln \rho_{n,m(n)}}{n\ln n} \leq -\frac{1}{\sigma\lambda}.$$

Finally, we let λ tend to $(1-\theta)$ and obtain

$$\liminf_{n\to\infty} \frac{\ln \rho_{n,m(n)}}{n\ln n} \leq -\frac{1}{\sigma(1-\theta)}. \quad \square$$

§4. Proof of Theorem 2

4.1. Proof of Theorem 2

The specific setting of our problem is this. Using the fact that the quantity $\rho_{n,m(n)}(f;E)$ does not decrease with the widening of the compact set E, we assume that the complement of the compact set E is connected (we can replace the compact set E by the compact set $E_1 = \bar{C}\setminus U$, where U is the connected component of $\bar{C}\setminus E$ containing ∞). Moreover, it can be assumed that E is bounded by a finite number of closed analytic Jordan curves Γ. The transition from such compact sets to arbitrary compact sets is not difficult (cf. Subsection 3.3). With help of a corresponding fractional-linear transformation, we reduce the theorem to the situation when E is a compact set in $\overline{\mathbb{C}}$ containing ∞, with the function f having one essential singularity at $0 \in G = \overline{\mathbb{C}} \setminus E$ of finite order $\sigma > 0$ and finite type $\tau > 0$. We must show that

$$\limsup_{n \to \infty} (\rho^*_{n-m(n),0} \rho^*_{n-m(n)+1,1} \cdots \rho^*_{n,m(n)})^{\sigma/n(m(n)+1)} n \leq \sigma e \tau e^{-d(0)\sigma},$$

where

$$\rho^*_{n-j,m(n)-j} = \rho^*_{n-j,m(n)-j}(f;E) = \inf_{r \in \mathcal{R}^*_{n-j,m(n)-j}} \|f - r\|_E, \quad j = 0, 1, \ldots, m(n),$$

and $d(0)$ is determined from the representation of the Green's function (see (31) with 0 in place of ξ). To do this it suffices to show that

$$\limsup_{n \to \infty} ((s_{0,n-m(n)} s_{1,n-m(n)} \cdots s_{m(n),n-m(n)})^{\sigma/n(m(n)+1)} n) \qquad (52)$$
$$\leq \sigma e \tau e^{-d(0)\sigma},$$

where $\{s_{k,n-m(n)}\}$, $s_{k,n-m(n)} = s_{k,n-m(n)}(f;G)$, $k = 0,1,2,\ldots$, is the sequence of singular numbers of the Hankel operator $A_f : E_2(G) \to H^\perp_{n-m(n)}$ constructed from f, $H_{n-m(n)}$ is the space of the function q representable in the form $q = \varphi/\xi^{n-m(n)}$, where $\varphi \in E_2(G)$.

To prove Theorem 2 we employ the same arguments as in the proof of Theorem 1. We only sketch the proof.

Fix an arbitrary domain G_1, $\overline{G}_1 \subset G, 0 \in G_1$, bounded by a finite number of closed Jordan curves, and $\tau' > \tau$. Let $t_n = (\sigma\tau')^{1/\sigma} n^{-1/\sigma}$. Denote by l_n the circle of radius t_n with center at 0. It will be assumed that l_n is positively oriented with respect to the open disk of radius t_n about 0, and n is a sufficiently large positive integer, $n \geq n_0$, such that the closed disk of radius t_n with center at 0 belongs to G_1. The rest of the arguments are analogous to the corresponding arguments in Subsection 3.2. For sufficiently large n it immediately follows from the inequality (35) that

$$s_{0,n-m(n)} s_{1,n-m(n)} \cdots s_{m(n),n-m(n)}$$
$$\leq ((m(n)+1)!)^3 C_1^{m(n)} \|f\|_{l_n}^{m(n)+1} \qquad (53)$$
$$\times \exp(-n(m(n)+1)d_n)|t_n|^{n(m(n)+1)},$$

where $d_n \to d(0)$ as $n \to \infty$.

The fact that f has at 0 an essential singularity of order σ and type τ implies

$$\|f\|_{l_n} \leq e^{\tau'/t_n^\sigma} = e^{n/\sigma}$$

for $n \geq n_1$. This allows us to use the formula $t_n = (\sigma\tau')^{1/\sigma} n^{-1/\sigma}$ for the radius, to obtain from (53)

$$\limsup_{n \to \infty} (s_{0,n-m(n)} s_{1,n-m(n)} \cdots s_{m(n),n-m(n)})^{\sigma/n(m(n)+1)} n \leq \sigma e \tau' e^{-d(0)\sigma}.$$

It remains to pass to the limit as $\tau' \to \tau$ on the right-hand side of the last inequality and obtain (52). □

4.2. Proof of the Corollaries to Theorem 2

We now turn our attention to the corollaries to Theorem 2. We assume that $\operatorname{cap}(E) > 0$ and $\theta > 0$. For $\operatorname{cap}(E) = 0$ and $\theta = 0$ the corresponding assertions are obvious. We first show that the inequality (6) holds. For simplicity we denote

$$\sigma e^{1/2} \tau (\operatorname{cap}(E))^{\sigma} e^{-\frac{(1-\theta)^2}{\theta(2-\theta)} \ln(1-\theta)}$$

by w. Assume that

$$\liminf_{n \to \infty} \rho_{n,m(n)}^{\sigma(2-\theta)/2n} n > \lambda w, \quad \lambda > 1,$$

in order to reach a contradiction. Therefore, for sufficiently large positive integers n, $n \geq n_0$, we have

$$\ln \rho_{n,m(n)} \geq \frac{2n}{\sigma(2-\theta)} \ln(\lambda w) - \frac{2n}{\sigma(2-\theta)} \ln n,$$

which implies (see (49))

$$\ln \prod_{j=0}^{m(n)} \rho_{n-j,m(n)-j} \geq \ln \prod_{k=n-m(n)}^{n} \ln \rho_{k,m(k)}$$

$$\geq \frac{2}{\sigma(2-\theta)} \ln(\lambda w) \sum_{k=n-m(n)}^{n} k$$

$$- \frac{2}{\sigma(2-\theta)} \sum_{k=n-m(n)}^{n} k \ln k, \quad n - m(n) \geq n_0,$$

and

$$\ln \prod_{j=0}^{m(n)} \rho_{n-j,m(n)-j}$$

$$\geq \frac{(2n - m(n))(m(n) + 1)}{\sigma(2-\theta)} \ln(\lambda w)$$

$$- \frac{(2n - m(n))(m(n) + 1)}{\sigma(2-\theta)} \ln n$$

$$- \frac{n(m(n) + 1)}{\sigma(2-\theta)} \left(-\frac{(1-\theta)^2}{\theta} \ln(1-\theta) - 1 + \frac{\theta}{2} + \delta_n \right),$$

where $\delta_n \to 0$ as $n \to \infty$. From this we obtain

$$\ln \prod_{j=0}^{m(n)} \rho_{n-j,m(n)-j}^{\sigma/n(m(n)+1)} n$$

$$\geq \ln(\lambda w) \frac{2 - m(n)/n}{2 - \theta} + \ln n \frac{m(n)/n - \theta}{2 - \theta}$$

$$- \frac{1}{2 - \theta} \left(-\frac{(1-\theta)^2}{\theta} \ln(1-\theta) - 1 + \frac{\theta}{2} + \delta_n \right).$$

By the relation
$$\lim_{n\to\infty} (m(n)/n - \theta) \ln n = 0,$$
we now get the the inequality
$$\liminf_{n\to\infty} \prod_{j=0}^{m(n)} \rho_{n-j,m(n)-j}^{\sigma/n(m(n)+1)} n \geq \lambda \sigma e \tau (\text{cap}(E))^\sigma,$$
which contradicts the relation (5). □

We now prove Corollary 4. Denote by S the following expression
$$S = \sigma e \tau (\text{cap}(E))^\sigma.$$

Let Λ be a sequence of positive integers such that
$$\lim_{n\to\infty, n\in\Lambda} \rho_{n,m(n)}^{\sigma/n} n = S. \tag{54}$$

Fix $1 - \theta < \lambda \leq 1$. Denote by $\{k_n\}$, $n = 1, 2, \ldots$, the sequence of integers $k_n = [n\lambda]$.

Since the sequence $\{\rho_{n,m(n)}\}$, $n = 1, 2, \ldots$, is nonincreasing, for sufficiently large n we get
$$\rho_{n,m(n)}^{m(n)+1} \leq \rho_{k_n,m(k_n)}^{k_n-n+m(n)+1} \rho_{n,m(n)}^{n-k_n} \leq \prod_{k=n-m(n)}^{n} \rho_{k,m(k)}.$$

From this and from the relation (54), it follows that
$$\lim_{n\to\infty, n\in\Lambda} \left(\rho_{k_n,m(k_n)}^{k_n-n+m(n)+1} \rho_{n,m(n)}^{n-k_n} \right)^{\frac{\sigma}{n(m(n)+1)}} n = S$$
and
$$\lim_{n\to\infty, n\in\Lambda} \rho_{k_n,m(k_n)}^{\frac{\sigma(k_n-n+m(n)+1)}{n(m(n)+1)}} n^{\frac{k_n-n+m(n)+1}{m(n)+1}} = S^{\frac{\theta-1+\lambda}{\theta}}.$$

Thus,
$$\lim_{n\to\infty, n\in\Lambda} \left(\rho_{k_n,m(k_n)}^{\sigma/n} k_n \right)^{\frac{k_n-n+m(n)+1}{m(n)+1}} \left(\frac{n}{k_n} \right)^{\frac{k_n-n+m(n)+1}{m(n)+1}} = S^{\frac{\theta-1+\lambda}{\theta}}$$
and
$$\liminf_{n\to\infty, n\in\Lambda} \rho_{k_n,m(k_n)}^{\lambda\sigma/k_n} k_n \leq \lambda S.$$

Then, using the last inequality, we deduce that
$$\liminf_{n\to\infty} \rho_{n,m(n)}^{\lambda\sigma/n} n \leq \lambda \sigma e \tau (\text{cap}(E))^\sigma. \quad \square$$

Acknowledgments. The research of E. B. Saff was supported in part by the U.S. National Science Foundation grant DMS-9501130. The research of V. A. Prokhorov was conducted while visiting the Institute for Constructive Mathematics at the University of South Florida. V. A. Prokhorov was supported in part by INTAS Grant 93-0219.

References

1. Adamyan, V. M., D. Z. Arov, and M. G. Kreĭn, Analytic properties of Schmidt pairs, Hankel operators, and the generalized Schur-Takagi problem, Mat. Sb. **86 (128)** (1971), 34–75; English transl. in Math USSR Sb. **15** (1971).

2. Braess, D., On the conjecture of Meinardus on rational approximation of e^x, J. Approx. Theory **36** (1982), 317–320.

3. Garnett, J. B., *Bounded Analytic Functions*, Academic Press, New York, 1981.

4. Gokhberg, I. Ts. [Israel Gohberg] and M. G. Kreĭn, *Introduction to the Theory of Linear Nonselfadjoint Operators in Hilbert Space*, "Nauka", Moscow, 1965; English transl., Amer. Math. Soc., Providence, RI, 1969.

5. Gonchar, A. A. and E. A. Rakhmanov, Equilibrium distributions and the degree of rational approximation of analytic functions, Mat. Sb. **134 (176)** (1987), 306–352; English transl. in Math. USSR Sb. **62** (1989).

6. Koosis, P., *Introduction to H^p Spaces*, London Math. Soc. Lecture Note Ser., vol. **40**, Cambridge Univ. Press, Cambridge, 1980.

7. Levin, A. L. and E. B. Saff, Optimal ray sequences of rational functions connected with the Zolotorev problem, Constr. Approx. **10** (1994), 235–273.

8. Parfenov, O. G., Estimates of the singular numbers of a Carleson operator, Mat. Sb. **131 (173)** (1986), 501–518; English transl. in Math. USSR Sb. **59** (1988).

9. Peller, V. V. and S. V. Khrushchev, Hankel operators, best approximations, and stationary Gaussian processes, Uspekhi Mat. Nauk **37** (1982), no. 1 (223), 53–124; English transl. in Russian Math. Surveys **37** (1982).

10. Peller, V. V., A description of Hankel operators of class σ_p for $p > 0$, an investigation of the rate of rational approximation, and other applications, Mat. Sb. **122 (164)** (1983), 481–510; English transl. in Math. USSR Sb. **50** (1985).

11. Privalov, I. I., *Boundary Properties of Analytic Functions*, Moscow, 1950; German transl., VEB Deutscher Verlag Wiss., Berlin, 1956.

12. Prokhorov, V. A., On a theorem of Adamyan, Arov, Kreĭn , Mat. Sb. **184** (1993), 89–104; English transl. in Russian Acad. Sci. Sb. Math. **78** (1994).

13. Prokhorov, V. A., Rational approximation of analytic function, Mat. Sb. **184** (1993), 3–32; English transl. in Russian Acad. Sci. Sb. Math. **78** (1994).
14. Prokhorov, V. A., On the degree of rational approximation of meromorphic functions, Mat. Sb. **185** (1994), 3–26; English transl. in Russian Acad. Sci. Sb. Math. **81** (1995).
15. Shah, S. M., Polynomial approximation of an entire function and generalized orders, J. Approx. Theory **19** (1977), 315–324.
16. Stahl, H., Orthogonal polynomials with complex-valued weight functions, I, II, Constr. Approx. **2** (1986), 225–240, 240–251.
17. Tumarkin, G. Ts. and S. Yu. Khavinson, On the definition of analytic functions of class E_p in multiply connected domains, Uspekhi Mat. Nauk **13** (1958), no. 1 (79), 201–206, (Russian).
18. Varga, R. S., On an extension of a result of S. N. Bernstein, J. Approx. Theory **1** (1968), 176–179.
19. Walsh, J. L., *Interpolation and Approximation by Rational Functions in the Complex Domain*, 2nd ed., Amer. Math. Soc., Providence, RI, 1956.
20. Winiarski, T., Approximation and interpolation of entire functions, Ann. Polon. Math. **23** (1970/71), 259–273.

Andrei V. Krot
Department of Mathematics and Mechanics
Belorussian State University
4 Scorina ave.
Minsk, 220080, Belarus

Vasiliy A. Prokhorov
Department of Mathematics
University of South Florida
Tampa, Florida 33620, U.S.A.
prokhoro@math.usf.edu

Edward B. Saff
Institute for Constructive Mathematics
Department of Mathematics
University of South Florida
Tampa, Florida 33620, U.S.A.
esaff@math.usf.edu

Extremal Problems for Logarithmic Potentials

A. B. J. Kuijlaars

Abstract. In the theory of logarithmic potential theory with external fields, a central role is played by the equilibrium measure, which is characterized as the solution of an extremal problem. We present an introduction to this theory, with applications to weighted approximation with varying weights and fast decreasing polynomials. In addition, we discuss a new method to determine the support of the equilibrium measure.

§1. Introduction

Logarithmic potential theory has played a role in approximation theory for a long time. Potential theoretic notions such as Green function and capacity are used in classical works such as [1] and [49]. Since the early eighties there has been a renewed interest, starting with the works of Mhaskar and Saff [33, 34] in the U.S.A. and of Gonchar and Rakhmanov [15, 39] in the former Soviet Union. Several important open problems were solved and a new theory was developed, called logarithmic potential theory with external fields, which is extensively treated in the recent monograph of Saff and Totik [43]. Other useful books are [30] and [32], which contain some parts that deal with the use of potential theory in approximation theory.

Here we want to present an overview of some recent results in which the author has been actively involved. We start with an introduction to the classical extremal problem in potential theory, which leads to the notion of the equilibrium measure of a compact set in the complex plane. In Section 3 we introduce external fields, and discuss the equilibrium measure in the presence of an external field. This is used in Section 4 in the study of weighted approximation with varying weights and in Section 5 in the study of fast decreasing polynomials. The material in Sections 2–5 can be found in [43], and we hope that the presentation will be useful to the non-expert as an introduction to [43].

Then, in Sections 6-8 we focus on a new method to determine the support of the equilibrium measure in the presence of an external field. The existing methods were basically restricted to cases where the support consists of a single interval. The new method consists of a sequence of extremal problems, situated on a decreasing sequence of compact sets K_n. Each K_n contains the support of the equilibrium measure. In certain cases, the intersection of the sets K_n is equal to the support (or rather to the extended support, see below). The solution of the extremal problem on K_n is a signed measure μ_n on K_n, and these measures are related through a process called balayage. That is the reason why we have called this iterative method the "iterated balayage algorithm" The new method was used in [23] and [7] to determine the support for certain non-convex external fields where the support consists of at most two intervals.

Finally, in Section 9 we briefly discuss a number of new developments, which are of interest for future research.

General references on potential theory include [12, 17, 28, 43].

§2. The Classical Extremal Problem

A basic notion in logarithmic potential theory is the equilibrium measure of a compact set in the complex plane.

For a closed set $K \subset \mathbf{C}$, we denote by $\mathcal{P}(K)$ the collection of Borel probability measures on K. The classical extremal problem in logarithmic potential theory is to minimize

$$I(\mu) := \int \int \log \frac{1}{|x-y|} d\mu(x) d\mu(y)$$

among all measures $\mu \in \mathcal{P}(K)$ where K is a compact set. The double integral is called the (logarithmic) energy of the measure μ. It is well defined as an element of $(-\infty, \infty]$ for any finite nonnegative Borel measure with compact support.

There are two possibilities depending on the set K. Firstly, it is possible that all measures in $\mathcal{P}(K)$ have infinite energy. This happens for example if K is a finite or countable set. Then K is said to have zero (logarithmic) capacity

$$\mathrm{cap}\,(K) = 0.$$

In potential theoretic works, such a set K is usually called a polar set. More generally, a Borel set E is called polar if every compact subset of E is polar. The polar sets play a similar role in potential theory as sets of measure zero do in measure theory.

The second possibility is that the logarithmic energy is finite for certain measures in $\mathcal{P}(K)$. Then K is called a non-polar set and the capacity of K is defined by

$$\mathrm{cap}\,(K) = \exp\left(-\inf_{\mu \in \mathcal{P}(K)} I(\mu)\right),$$

Extremal Problems

so that K has positive capacity. A fundamental result, due to Frostman [16], is that for compact sets K with positive capacity, there exists a unique measure $\omega(K) \in \mathcal{P}(K)$ such that

$$I(\omega(K)) = \inf_{\mu \in \mathcal{P}(K)} I(\mu).$$

This extremal measure is called the **equilibrium measure** for the set K and it plays a very special role in potential theory.

The proof of existence and uniqueness follows easily from three general principles: compactness, strict convexity and lower semi-continuity. These principles also apply to other situations below and are therefore mentioned here.

1) For a compact set K, the collection $\mathcal{P}(K)$ is **compact in the weak* topology**. The weak* topology on $\mathcal{P}(K)$ arises from the duality of the space of Borel measures on K with the Banach space of continuous functions on K. Thus a sequence of Borel measures $\{\mu_j\}$ converges in the weak* topology to μ if and only if

$$\lim_j \int f d\mu_j = \int f d\mu$$

for every continuous function f on K.

2) The logarithmic energy functional $\mu \mapsto I(\mu)$ is **strictly convex** on $\mathcal{P}(K)$. That is, for $\mu, \nu \in \mathcal{P}(K)$, we have

$$I((\mu + \nu)/2) \leq (I(\mu) + I(\nu))/2$$

and equality holds if and only if $\mu = \nu$.

3) The logarithmic energy functional is **lower semi-continuous** on $\mathcal{P}(K)$. That is, if $\{\mu_j\}$ is a sequence in $\mathcal{P}(K)$ converging in weak* sense to μ, then

$$I(\mu) \leq \liminf_{j \to \infty} I(\mu_j).$$

In electrostatics the equilibrium measure describes the distribution of a unit charge placed on the conductor K. It satisfies the following condition

$$\int \log \frac{1}{|x-y|} d\omega(K)(y) = I(\omega(K)), \qquad x \in K \setminus E, \tag{2.1}$$

where E is a polar subset of K. This condition actually characterizes the equilibrium measure, in the sense that if for some $\mu \in \mathcal{P}(K)$, there exist a constant F and a polar set $E \subset K$, such that

$$\int \log \frac{1}{|x-y|} d\mu(y) = F, \qquad x \in K \setminus E,$$

then $\mu = \omega(K)$. The function on the left-hand side is known as the logarithmic potential of μ and is usually denoted by U^μ:

$$U^\mu(x) := \int \log \frac{1}{|x-y|} d\mu(y), \qquad x \in \mathbb{C}.$$

The condition (2.1) can be seen as an equilibrium condition. The equilibrium measure of K has a potential which is constant on K, except for a small set E. In many important cases, the set E is empty, so that equality holds throughout the entire set K. In such a case we will call the set K regular. For a regular set K, the equilibrium potential $U^{\omega(K)}$ is continuous on $\mathbb{C}$.

[Actually, regularity of K is equivalent to the property that the Dirichlet problem for harmonic functions in the complement $\overline{\mathbb{C}} \setminus K$ is uniquely solvable for continuous functions on K. That is, in the usual sense of potential theory, the complement of K in the extended plane is a regular domain. For simplicity we will say that K regular, instead of saying that the complement of K is regular for the Dirichlet problem.]

In the following we will consider sets $K \subset \mathbb{R}$. Then it is good to know that closed intervals (not a single point), or finite unions of closed non-degenerate intervals, are regular sets. Hence, in those cases one need not bother about possible exceptional polar sets.

§3. External Fields

From now on we will assume that K is a regular compact subset of the real line.

In this paper, an external field on K is simply a continuous function on K. In the literature, more general definitions of external field are used, see [43]. For example, the set K could be unbounded, in which case the external field is required to satisfy a certain growth condition at ∞. It is also possible to consider external fields which are not continuous (but lower semi-continuous) and to consider non-regular sets. The more restricted definition, however, is enough for the present purposes.

The extremal problem in the presence of the external field Q on K is the problem to minimize

$$\int\int \log \frac{1}{|x-y|} d\mu(x) d\mu(y) + 2\int Q(x) d\mu(x) \qquad (3.1)$$

among all measures $\mu \in \mathcal{P}(K)$. Again, there is a unique extremal measure $\mu = \mu_Q$. This follows from the same three principles, (compactness, strict convexity and lower semi-continuity) as before.

The extremal measure is characterized by the equilibrium conditions

$$U^\mu(x) + Q(x) = F, \qquad x \in \text{supp}(\mu), \qquad (3.2)$$
$$U^\mu(x) + Q(x) \geq F, \qquad x \in K, \qquad (3.3)$$

Extremal Problems

for some constant F. In a more general situation, where K is not regular or Q is not continuous, it is possible that there are exceptional polar sets in (3.2) and (3.3). Regularity of K together with continuity of Q implies that there are no such exceptional sets. It then also follows that the potential U^μ is a continuous function on $\mathbb{C}$. All this was first observed by Totik [46, 47].

An important distinction with the classical case is that the equality (3.2) need not hold on the full set K. The support of the extremal measure can be a proper subset of K. We use $S = S_Q$ to denote the support of μ_Q. We also define the *extended support* by

$$S^* = S_Q^* := \{x \in K : U^\mu(x) + Q(x) = F\}. \tag{3.4}$$

The determination of S_Q is in many cases an essential step towards the solution of the extremal problem. Once the support is known, the measure μ is the solution of an integral equation on S_Q with logarithmic kernel for which solution methods are known. For example, if the support is an interval $[a, b]$, then it follows from (3.2) that

$$\int_a^b \log|x - y| d\mu(y) = Q(x) - F, \qquad x \in [a, b].$$

Formally differentiating with respect to x (which can be justified under weak conditions on Q, see Section IV.3 of [43]) yields

$$PV \int_a^b \frac{1}{x - y} d\mu(y) = Q'(x), \qquad a < x < b,$$

where PV denotes the Cauchy principal value. This singular integral equation has the general solution

$$\frac{d\mu(y)}{dy} = \frac{1}{\pi\sqrt{(b-y)(y-a)}} \left[C + \frac{1}{\pi} PV \int_a^b \frac{Q'(s)}{s - y} \sqrt{(b-s)(s-a)} ds \right] \tag{3.5}$$

with an arbitrary constant $C \in \mathbb{R}$. Since

$$C = \int_a^b d\mu,$$

the solution of the extremal problem corresponds to $C = 1$. More complicated formulas exist if the support of μ consists of several intervals.

§4. Weighted Approximation with Varying Weights

As an example of the application of extremal problems with external fields to a problem in approximation theory, we shall consider here the so-called generalized Weierstrass problem. We will not go into too much detail here since it is treated extensively in [43].

The approximation problem is the following. Assuming we are given a continuous positive function w on the regular compact set $K \subset \mathbb{R}$, we are asked to determine the class of those continuous functions f on K for which there exists a sequence of polynomials $\{p_n\}$ with $\deg p_n \leq n$, such that

$$\lim_{n \to \infty} \|w^n p_n - f\|_K = 0.$$

The norm $\|\cdot\|_K$ denotes the uniform norm on K. This type of weighted approximations with varying weights has important applications in the theory of orthogonal polynomials and Padé approximation, see [43, 48].

Let $\mathcal{A}_w$ denote the class of functions f that can be approximated in the above sense. It is fairly easy to show that $\mathcal{A}_w$ has the following properties:

1) $\mathcal{A}_w$ is a vector space;
2) $\mathcal{A}_w$ is closed under multiplication;
3) $\mathcal{A}_w$ is closed in the uniform norm;
4) if $x_0 \in K$ is such that there exists $f_0 \in \mathcal{A}_w$ with $f_0(x_0) \neq 0$, then, for every $x_1 \in K \setminus \{x_0\}$, there is $f_1 \in \mathcal{A}_w$ with $f_1(x_0) \neq f_1(x_1)$.

From the properties 1)–4) we deduce the following result by virtue of a version of the Stone-Weierstrass theorem.

Theorem 1. ([20]) *There is a closed set $Z_w \subset K$ such that a continuous function f belongs to $\mathcal{A}_w$ if and only if $f(x) = 0$ for all $x \in Z_w$.*

Thus the generalized Weierstrass theorem reduces to the determination of the set Z_w. The points in Z_w are the bad points for approximation. A function has to vanish on Z_w in order to be approximable, while there are no restrictions for points in $K \setminus Z_w$.

The external field that plays a role here is given by $Q = -\log w$. Let μ_w be the extremal measure in the presence of the external field Q and let S_w denote its support. Mhaskar and Saff [34] showed that

$$\|w^n p_n\|_K = \|w^n p_n\|_{S_w} \tag{4.1}$$

for every n and every polynomial p_n of degree $\leq n$. The set S_w is the smallest closed set such that (4.1) holds. Using (4.1) and Theorem 1, one obtains the following result.

Theorem 2. ([47]) *If $f \in \mathcal{A}_w$, then $f \equiv 0$ outside S_w. In other words,*

$$(K \setminus S_w) \subset Z_w.$$

More refined results depend on the behavior of the measure μ_w on its support. These results are due to Totik and the author. For the proofs, the reader is referred to the original papers or to [43]. The topological notions (internal point, boundary point, neighborhood) in Theorems 3 and 4 are with respect to the space $\mathbb{R}$.

Extremal Problems

Theorem 3. Internal points ([47, 19, 21]). *Let x_0 be an internal point of K. Suppose that the extremal measure μ_w has a density $v(x)$ in a neighborhood U of x_0 satisfying*

a) *$v(x)$ is positive and continuous in $U \setminus \{x_0\}$, and*
b) *there exist $\beta > -1$ and $L > 0$ such that*

$$\lim_{x \to x_0} |x - x_0|^{-\beta} v(x) = L.$$

Then

$$x_0 \notin Z_w \iff \beta = 0.$$

Theorem 4. Boundary points ([19, 21]). *Let $x_0 \in K$ be such that for some $\delta > 0$, we have $K \cap (x_0 - \delta, x_0 + \delta) = (x_0 - \delta, x_0]$ or $K \cap (x_0 - \delta, x_0 + \delta) = [x_0, x_0 + \delta)$. Suppose there is a neighborhood U of x_0 such that the extremal measure μ_w has a density $v(x)$ in $K \cap U$ satisfying*

a) *$v(x)$ is positive and continuous in $(K \cap U) \setminus \{x_0\}$, and*
b) *there exist $\beta > -1$ and $L > 0$ such that*

$$\lim_{x \to x_0, x \in K} |x - x_0|^{-\beta} v(x) = L.$$

Then

$$x_0 \notin Z_w \iff \beta = -1/2.$$

To illustrate these results we mention two applications. For $w \equiv 1$ on $K = [a, b]$, the extremal measure μ_w is the classical equilibrium measure of K with density

$$\frac{d\mu_w}{dx} = \frac{d\omega([a,b])}{dx} = \frac{1}{\pi\sqrt{(b-x)(x-a)}}, \quad a < x < b.$$

By Theorems 3 and 4, the set Z_w is empty, which is nothing but the classical result of Weierstrass on polynomial approximation of continuous functions on a compact interval.

For the weight $w(x) = e^{x^2}$ on $[-1, 1]$, the extremal measure has density

$$\frac{d\mu_w}{dx} = \frac{2x^2}{\pi\sqrt{1-x^2}}, \quad -1 < x < 1.$$

Here the density vanishes at $x = 0$. Thus by Theorems 3 and 4, we have $Z_w = \{0\}$. This means that a sequence of polynomials $\{p_n\}$, $\deg p_n \leq n$, such that

$$\lim_{n \to \infty} e^{nx^2} p_n(x) = f(x) \quad \text{uniformly for } x \in [-1, 1]$$

exists, if and only if f is continuous and $f(0) = 0$.

Extensions of the results in Theorems 3 and 4 to cases where the extremal measure has a logarithmic-type singularity at an interior point are due to Simeonov [44] and Totik [48]. Such points turn out to be good points for the approximation problem, i.e., they do not belong to Z_w.

Rational approximation with varying weights on the real line is considered in the papers [3, 41]. These papers discuss approximation with weighted rational functions of the form $w^n r_n$ where $r_n = p_n/q_n$ with p_n, q_n polynomials of degree at most n. For the model case $w(x) = e^x$ on $[0, R]$, it was shown that for $R \in (0, 2\pi)$, there exists a sequence of rational functions $\{r_n\}$ such that
$$\lim_{n\to\infty} \|e^{nx} r_n(x) - 1\|_{[0,R]} = 0,$$
while for $R > 2\pi$ there is no such sequence. Furthermore, if $R \in (0, 2\pi)$, then every continuous function f on $[0, R]$ can be approximated uniformly by a sequence of weighted rationals of the form $e^{nx} r_n(x)$.

The situation for the critical value $R = 2\pi$ is open. The special role of 2π is also illustrated in [22]. For the approximation problem on $[0, 2\pi]$, I offer the following conjecture.

Conjecture. *Let $f \in C([0, 2\pi])$. There exists a sequence of rational functions $\{r_n\}$, $r_n = p_n/q_n$ with $\deg p_n \leq n$, $\deg q_n \leq n$, such that*
$$\lim_{n\to\infty} \|e^{nx} r_n(x) - f\|_{[0,2\pi]} = 0$$
if and only if
$$f(x_0) = 0 \text{ for some } x_0 \in [0, \pi]. \tag{4.2}$$

At first sight the condition (4.2) may seem a little surprising. The reason for it is the following. Extremal problems behind rational approximations involve signed measures. In this case the extremal problem is to minimize
$$\int\int \log \frac{1}{|x-y|} d\mu(x) d\mu(y) - 2 \int x\, d\mu(x)$$
among all signed Borel measures μ on $[0, 2\pi]$ of the form $\mu = \mu^+ - \mu^-$ with $\mu^\pm \in \mathcal{P}([0, 2\pi])$. The solution of this extremal problem is
$$\frac{d\mu(x)}{dx} = \frac{x - \pi}{\pi\sqrt{x(2\pi - x)}}, \qquad 0 < x < 2\pi.$$

Thus the positive part of μ is supported on $[\pi, 2\pi]$ and the negative part on $[0, \pi]$. The conjecture says that for the function f to be approximable it is necessary and sufficient that it vanishes somewhere on the support of the negative part of μ. I am able to show that the condition (4.2) is sufficient. Its necessity is open.

Weighted approximations with varying weights in the complex plane are considered by Borwein and Chen [2] and in a sequence of papers by Pritsker and Varga [35–38].

Extremal Problems

§5. Fast Decreasing Polynomials

Logarithmic potential theory with external fields plays a role in the construction of polynomials satisfying an estimate of the form

$$|p_n(x)| \leq M e^{nQ(x)}, \qquad x \in K,$$

where Q is a given continuous function on K that determines the possible exponential growth (or decay) of the polynomials p_n.

An important special case is when K is an interval containing 0 and Q is a continuous function on K which vanishes at 0 and is negative on $K \setminus \{0\}$. Then a sequence of polynomials $\{p_n\}$, $\deg p_n \leq n$, satisfying

$$p_n(0) = 1, \qquad |p_n(x)| \leq e^{n(Q(x)+o(1))}, \qquad x \in K \tag{5.1}$$

is fast decreasing on K except at the origin, with a rate given by Q. The $o(1)$ in (5.1) is a quantity that tends to 0 uniformly in x as $n \to \infty$.

The connection with potential theory was revealed by Totik, who proved the following fundamental result.

Theorem 5. ([46]) *Let K be a regular compact subset of the real line with $0 \in K$, and let Q be a continuous function on K with $Q(0) = 0$. Let μ be the extremal measure in the presence of the external field Q, and let S^* denote the extended support, cf. (3.4). Then the following are equivalent.*

a) *There exists a sequence of polynomials $\{p_n\}$, $\deg p_n \leq n$, such that (5.1) holds.*

b) $0 \in S^*$.

Thus the existence of fast decreasing polynomials comes down to the determination of the extended support.

Of special interest is the case where Q depends on a parameter as follows:

$$Q_c(x) = cQ(x), \qquad c > 0,$$

and one asks about the largest c such that polynomials satisfying

$$p_n(0) = 1, \qquad |p_n(x)| \leq e^{n(cQ(x)+o(1))}, \qquad x \in K \tag{5.2}$$

exist. If we denote the solution of the extremal problem by μ_c, its support by S_c and the extended support by S_c^*, then it follows from Theorem 5 that the best constant is given by

$$c_{opt} = \sup\{c > 0 : 0 \in S_c^*\}.$$

It is known that the sets S_c and S_c^* decrease as c increases, and that $S_d^* \subset S_c$ whenever $d < c$. Therefore, we have

$$c_{opt} = \sup\{c > 0 : 0 \in S_c\}.$$

See [6, 7, 43] for more results on S_c and S_c^*.

The family of extended supports $(S_c^*)_c$ has a nice interpretation in terms of a contact problem in elasticity, see [26, 43]. A rigid punch with profile $Q(x)$ if $x \in K$, and $Q(x) \equiv +\infty$ if $x \notin K$, is pushed into the elastic lower half-plane. We assume that there is no friction and that the laws of linear elasticity hold (which actually only hold if the deformations are infinitesimally small). Then there is a region of contact between the punch and the elastic material, which depends on the force. In proper units, if the force is $1/c$ then the region of contact is equal to S_c^*. On the region of contact there is a contact pressure or normal stress, given by the measure $c\mu_c$. From this point of view, the optimal constant c_{opt} is equal to the reciprocal of the smallest force which is necessary to establish contact between the punch and the elastic half plane at the point 0.

The cases $Q(x) = -x^\alpha$, $x \in [0, 1]$ with $\alpha > 0$ were considered in [18, 25, 46]. In [18] it was shown that for $\alpha \leq 1/2$, we have $c_{opt} = 0$, and in [46] the optimal constant was found for $1/2 < \alpha \leq 1$:

$$c_{opt} = \frac{2\sqrt{\pi}\Gamma(\alpha)}{\Gamma(\alpha - 1/2)}.$$

A more complicated (and less explicit) formula was given in [25] for the case $\alpha > 1$.

The reason that the case $\alpha \leq 1$ is simpler, is that in that case, the function $-cx^\alpha$ is convex on $[0, 1]$. For convex external fields, the support of the extremal measure (and also the extended support) is always an interval. For $\alpha > 1$, the external field is not convex, and the support may consist of more than one interval, which makes the analysis more complicated. For this case, the nature of the support was determined in [23], by a new method which will be described in the next sections. It was shown that the supports S_c consist of at most two intervals. A simple quadratic transformation then gives that for the case $Q(x) = -|x|^\alpha$ on $[-1, 1]$, the supports consist of at most three intervals.

The method was extended in a non-trivial way in [7] to the cases $Q(x) = -x^{2m+1}$ on $[-1, 1]$ with m a nonnegative integer. Here the supports S_c consist of at most two intervals.

§6. Determination of the Support

We assume that Q is a continuous function on $[0, 1]$. As before, we denote by μ_c the extremal measure with external field cQ and by S_c its support. If Q is convex, then S_c is an interval for every $c > 0$ as was first proved by Mhaskar and Saff [34]. If $xQ'(x)$ is increasing in $[0, 1]$, then it is known that S_c is an interval containing 0 for every $c > 0$. This condition applies for example to the external fields $Q(x) = x^\alpha$, $\alpha > 0$, which are not convex if $\alpha < 1$. The most general condition on Q such that all supports S_c are intervals was given by Buyarov [5].

Extremal Problems 211

The external fields
$$Q(x) = -x^\alpha, \qquad x \in [0,1] \tag{6.1}$$

are related to fast decreasing polynomials as described in Section 5. If $0 < \alpha \leq 1$, then Q is convex and it follows that the support S_c is an interval $[b(c), 1]$ for every c. Here we used the fact that $1 \in S_c$, since Q has its minimum at $x = 1$.

For $\alpha > 1$, the external field is not convex. The determination of the support was done in [23], with partial results in [9, 25].

Theorem 6. ([23]) *For $\alpha > 1$, there exist two critical values c_1, c_2 with $0 < c_1 < c_2 < \infty$, such that the following hold.*

a) *For $c > c_2$, we have*
$$S_c = S_c^* = [b(c), 1],$$
with $0 < b(c) < 1$.

b) *For $c = c_2$, we have*
$$S_c = [b(c), 1], \qquad S_c^* = \{0\} \cup [b(c), 1],$$
with $0 < b(c) < 1$.

c) *For $c_1 < c < c_2$, we have*
$$S_c = S_c^* = [0, a(c)] \cup [b(c), 1]$$
with $0 < a(c) < b(c) < 1$.

d) *For $c \leq c_1$, we have*
$$S_c = S_c^* = [0, 1].$$

Thus the support consists of at most two intervals for every c. We see that $1 \in S_c$ for every c, and that for large values, S_c is an interval containing 1. If we decrease c, the support grows and remains an interval, until we reach the critical value c_2. Then 0 enters into the support. Decreasing c further, the support consists of two intervals, one containing 0 and the other containing 1. These intervals grow as we decrease c until at the critical value c_1, the two intervals meet. For smaller values of c, the support is then the full interval $[0, 1]$.

The proof of Theorem 6 is based on a new technique called the iterated balayage algorithm. It consists of a sequence of extremal problems on a decreasing sequence of compact sets $\{K_n\}$. The first extremal problem is on $K_1 = [0, 1]$ and it is to minimize
$$\int\int \log \frac{1}{|x-y|} d\mu(x) d\mu(y) + 2 \int Q(x) d\mu(x) \tag{6.1}$$

among all finite signed Borel measures μ satisfying

$$\operatorname{supp}(\mu) \subset K_1, \qquad \int d\mu = 1. \tag{6.2}$$

The solution of this extremal problem is denoted by μ_1. Observe that we do not require that $\mu \geq 0$, so that the solution of this extremal problem is a priori different from the extremal measure with external field Q, which we denote by μ_Q. Of course if μ_1 is a nonnegative measure then it is also the extremal measure with external field. We have the equilibrium condition

$$U^{\mu_1}(x) + Q(x) = F_1, \qquad x \in K_1,$$

which characterizes the measure μ_1.

We denote the positive and negative parts of μ_1 by μ_1^+ and μ_1^-, respectively, so that $\mu_1 = \mu_1^+ - \mu_1^-$. It is basic for the method that

$$\mu_Q \leq \mu_1^+.$$

This property follows from a deep result in potential theory, see Theorem IV.4.5 of [43]. It follows that

$$S_Q = \operatorname{supp}(\mu_Q) \subset \operatorname{supp}(\mu_1^+).$$

Hence, in order to find μ_Q we can restrict our attention from now on to measures supported on $K_2 := \operatorname{supp}(\mu_1^+)$. This is the second extremal problem. We minimize (6.1) among all finite signed Borel measures μ satisfying

$$\operatorname{supp}(\mu) \subset K_2, \qquad \int d\mu = 1.$$

The solution of this extremal problem is denoted by μ_2 and we put $K_3 := \operatorname{supp}(\mu_2^+)$. Then we have $S_Q \subset K_3$ and the next step is to consider the minimization problem on K_3.

Continuing in this way we find a sequence of compact sets $\{K_n\}$ such that

$$K_1 \supset K_2 \supset \cdots \supset S_Q$$

and a sequence of signed measures $\{\mu_n\}$ with $\operatorname{supp}(\mu_n) \subset K_n$, $\int d\mu_n = 1$, $\operatorname{supp}(\mu_n^+) = K_{n+1}$ and

$$\mu_1^+ \geq \mu_2^+ \geq \cdots \geq \mu_Q.$$

As the sequence $\{\mu_n^+\}$ decreases, it converges to a measure and one expects

$$\lim_{n \to \infty} \mu_n^+ = \mu_Q, \tag{6.3}$$

where the convergence is in the weak* topology. One also expects

$$\bigcap_{n=1}^{\infty} K_n = S_Q^*. \tag{6.4}$$

Extremal Problems

The above procedure is called the iterated balayage algorithm and if (6.3) and (6.4) hold, then the iterated balayage algorithm is said to converge.

The convergence was shown in [23] for the cases $Q(x) = -cx^\alpha$ on $[0,1]$, and in [7] for the case where Q is a polynomial. These are the only cases for which convergence has been established so far. It would be very interesting to have a more general convergence result.

How can the convergence of the iterated balayage algorithm help to determine the nature of S_Q? The idea is that one has to control the sets K_n. If these sets are all intervals, then their intersection S_Q^* is also an interval and maybe, with a little extra work, one can also show that S_Q is an interval. Similarly, one might be able to prove that K_n consists of at most two intervals for every n. Then S_Q^* consists of at most two intervals and one might prove that also S_Q has at most two intervals.

In the iterated balayage algorithm as presented above, one would have to solve a minimization problem for every n. It is important for the method that these problems are not independent of each other and that one can use the solution of the nth problem to find the solution of the $(n+1)$st problem without explicitly solving the minimization problem. This is accomplished by a process known as balayage, which is the reason why we called the method the iterated balayage algorithm. Since balayage is a useful but not so well-known notion, I will discuss it in the next section.

§7. What is Balayage?

Balayage was considered by Poincaré in connection with the Dirichlet problem for the Laplace operator. He defined balayage as an operation on functions. Balayage of measures was developed by de la Vallée Poussin, Frostman, and Brelot in the thirties and forties. The reader is refered to [12, 17, 28] for potential theoretic discussions of the notion of balayage. Here we will focus on balayage of measures as discussed in [43].

Let us start with the definition. For simplicity we restrict ourselves to measures with a continuous potential.

Definition. *Let μ be a finite nonnegative Borel measure on $\mathbf{C}$ with a continuous potential U^μ and let K be a closed subset of the Riemann sphere $\mathbf{C} \cup \{\infty\}$. If ν is a nonnegative Borel measure such that*

1) $\mathrm{supp}\,(\nu) \subset K$,
2) $\int d\nu = \int d\mu$,
3) *for some constant c and for some polar set E,*

$$U^\nu(x) = U^\mu(x) + c, \qquad x \in K \setminus E, \tag{7.1}$$

then ν is called the balayage of μ onto K.

If the set K is clear from the context, we use $\hat{\mu}$ to denote the balayage of μ onto K. If we want to emphasize the set K as well, we use the notation $Bal(\mu; K)$.

The French word "balayer" means "to sweep". The process of balayage comes down to sweeping a measure μ onto a set K in such a way that the potential on K is changed by an additive constant only.

The problem of finding a balayage measure may be viewed in the light of the extremal problem in the presence of an external field. Indeed, let K be a regular compact set and let μ be a positive Borel measure with $\int d\mu = 1$ and a continuous potential. Then consider the external field

$$Q(x) = -U^\mu(x), \qquad x \in K. \qquad (7.2)$$

Note that the measure μ need not be supported on K. However, we consider the negative of its potential as an external field on K only. If ν is the extremal measure on K with external field Q then we have according to (3.2), (3.3),

$$U^\nu(x) - U^\mu(x) = c, \qquad x \in \mathrm{supp}\,(\nu), \qquad (7.3)$$
$$U^\nu(x) - U^\mu(x) \geq c, \qquad x \in K, \qquad (7.4)$$

for some constant c. Next, we note that $U^\nu - U^\mu$ is a subharmonic function on $\mathbb{C} \setminus \mathrm{supp}\,(\nu)$ which vanishes at infinity, since ν and μ have the same total mass. By the maximum principle for subharmonic functions, it follows that $U^\nu - U^\mu$ assumes its maximum on $\mathrm{supp}\,(\nu)$. In view of (7.3) the maximum is equal to the constant c and it follows that equality holds in (7.4) for all $x \in K$. Hence (7.1) is satisfied and ν is the balayage of μ onto K.

The above argument gives the existence of a balayage measure and also its uniqueness. It may be noted that this approach to balayage can be found already in Frostman's thesis [16] of 1935. It does not give any idea, however, what the measure ν looks like.

A useful way of thinking about balayage is in terms of a hitting distribution for Brownian motion in the plane. The connection between analytic and probabilistic potential theory is discussed extensively in [12]. Most potential theoretic notions have a probabilistic counterpart and balayage is one of them.

Brownian motion is a stochastic process, modeling a continuous random walk. Imagine that a particle starts a Brownian motion at a point $x_0 \notin K$ and that it continues until it meets the set K, where it is absorbed. If K is non-polar, then the Brownian motion particle meets K in finite time almost surely. Then there exists a probability measure ρ on K (the hitting distribution of K) such that for a Borel subset B of K, $\rho(B)$ is the probability that the particle first meets K in the set B. The measure ρ, which depends on the starting point x_0, is equal to the balayage of the Dirac point mass δ_{x_0} onto K. For a general measure μ, one has the formula

$$\hat{\mu} = \int \hat{\delta}_{x_0} d\mu(x_0) \qquad (7.5)$$

and it may be interpreted as the hitting distribution of the set K when the initial position of the particle is a random variable distributed according to μ.

Extremal Problems

The Brownian motion point of view gives some idea what the balayage measure looks like. Consider for example, the case $K = [a, b]$ and $x_0 < a$. Starting the Brownian motion at x_0, it is clear that, in principle, points of K closest to x_0 have the highest probability of being hit first. Thus one expects a high probability density at a and naively one might think that the density decreases on $[a, b]$. This is, however, not the case. The other end point b has a relatively high hitting probability as well, since the Brownian particle can approach the point b from a full angle, while it can hit a point in the interior of $[a, b]$ from above or from below only. Thus the density at b is higher than the density at neighboring points. This is confirmed by the explicit expression of the density of the balayage measure, see [43],

$$\frac{d\hat{\delta}_{x_0}(t)}{dt} = \frac{1}{\pi\sqrt{(b-t)(t-a)}} \frac{\sqrt{(b-x_0)(a-x_0)}}{|t-x_0|}, \qquad a < t < b.$$

The density is infinite at both end points a and b and therefore it is not decreasing on $[a, b]$. Only after multiplication with the factor $\pi\sqrt{(b-t)(t-a)}$, it is true that the density decreases on the interval $[a, b]$ if $x_0 < a$. More generally, if μ is a positive measure on $(-\infty, a]$, then one has by (7.5) that

$$\pi\sqrt{(b-t)(t-a)}\frac{d\hat{\mu}}{dt} \quad \text{decreases for } t \in (a, b). \tag{7.6}$$

§8. The Convergence of the Iterated Balayage Algorithm

We will now continue the discussion of Section 6, and we restrict our attention to the specific case

$$Q(x) = -cx^\alpha, \qquad x \in [0, 1],$$

with $\alpha > 1$. As explained in Section 6, the first step is to minimize (6.1) among all finite signed Borel measures on $[0, 1]$ with integral 1. The solution of this extremal problem has the density

$$\frac{d\mu_1}{dt} = \frac{1}{\pi\sqrt{t(1-t)}}\left[1 + \frac{1}{\pi}PV\int_0^1 \frac{Q'(s)}{s-t}\sqrt{s(1-s)}ds\right] \tag{8.1}$$

see also (3.5). The measure μ_1 satisfies the relation

$$U^{\mu_1}(x) + Q(x) = F_1, \qquad x \in K_1, \tag{8.2}$$

for some constant F_1.

In the second step of the iterated balayage algorithm we put $K_2 = \operatorname{supp}(\mu_1^+)$. Then μ_2 is the solution of the extremal problem on K_2. It satisfies the equilibrium condition for some constant F_2. Combining this with (8.2), we see that

$$U^{\mu_2}(x) = U^{\mu_1}(x) + F_2 - F_1, \quad x \in K_2.$$

Keeping in mind the definition of balayage, we conclude that

$$\mu_2 = \mu_1^+ - Bal(\mu_1^-; K_2). \tag{8.3}$$

Here we see the relation with balayage. The next measure μ_2 is equal to the positive part of μ_1 minus the balayage of the negative part onto K_2. This will hold in general and we have

$$\mu_{n+1} = \mu_n^+ - Bal(\mu_n^-; K_n)$$

for every n.

Now in order to proceed, the following property of the density (8.1) for the external field $Q(x) = -cx^\alpha$ was proved in [23].

Lemma. *There is a $t_0 \in (0,1)$ such that*

$$\pi\sqrt{t(1-t)}\frac{d\mu_1}{dt} \tag{8.4}$$

decreases on the interval $[0, t_0]$ and increases on $[t_0, 1]$. (Note that this property is independent of the parameter c.)

From the Lemma it follows that K_2 consists of at most two intervals. It is either equal to $[b_2, 1]$ for some $b_2 > 0$, or to the union of two intervals of the form $[0, a_2] \cup [b_2, 1]$, or to the full interval $[0, 1]$. This depends on the parameter c. Recall that $1 \in S_c$ and $S_c \subset K_2$, so that $1 \in K_2$ for every c.

If $K_2 = [0, 1]$, then μ_1 is a nonnegative measure and we are in case d) of Theorem 6.

If $K_2 = [b_2, 1]$, then we have that (8.4) increases on K_2. Since μ_1^- is supported on $[0, b_2]$, it follows from (7.6) that

$$\pi\sqrt{t(1-t)}\frac{dBal(\mu_1^-; K_2)}{dt}$$

decreases on K_2. Thus in view of (8.3), $\pi\sqrt{t(1-t)}\frac{d\mu_2}{dt}$ increases on K_2. It follows that K_3 is an interval of the form $K_3 = [b_3, 1]$ and we can continue. Inductively, we find that $K_n = [b_n, 1]$, with $\{b_n\}$ an increasing sequence. Then it can be shown that the iterated balayage algorithm converges and $S_c = S_c^* = [b(c), 1]$ with $b(c) = \lim_n b_n$. Thus we are in case a) of Theorem 6.

The case $K_2 = [0, a_2] \cup [b_2, 1]$ is somewhat more complicated, but the idea is the same. We are going to take the balayage of μ_1^- onto K_2, so that we need to study balayage onto two intervals. It can be shown that

$$\pi\sqrt{t(1-t)}\frac{dBal(\mu_1^-; K_2)}{dt}$$

increases on $[0, a_2]$ and decreases on $[b_2, 1]$. For this it is important that μ_1^- is supported on the gap $[a_2, b_2]$. Then it follows that $\pi\sqrt{t(1-t)}\frac{d\mu_2}{dt}$ decreases on $[0, a_2]$ and increases on $[b_2, 1]$. Hence K_3 is either a single interval of the form $[b_3, 1]$ or a union of two intervals $[0, a_3] \cup [b_3, 1]$. In either case we can continue as before. It follows that we are either in case a) of Theorem 6, namely if K_n is a single interval for some $n \geq 3$, or that we are in case b) or c) of Theorem 6, if K_n is a union of two intervals for every n. The case b) is a borderline case.

§9. Outlook

We conclude with an overview of some exciting recent work in which extremal problems for potentials play a role.

9.1 Further Properties of the Extremal Measures

Many new results on the extremal measure in the presence of an external field are obtained in [9]. For example, it is shown that if the set K is an interval and the external field Q is real analytic in a neighborhood of K, then the support of the extremal measure consists of a finite number of intervals. This result is sharp in the sense that real analytic cannot be replaced by C^∞. This property was used in the paper [10], in which very sharp asymptotics for orthogonal polynomials are derived.

Interesting new results on the sets S_c and the measures μ_c are derived in [6].

9.2 Random Matrices

The extremal problem with external field appears in the theory of random matrices [31]. In the space of $n \times n$ Hermitean matrices M one considers the probability distribution

$$Z_n^{-1} \exp(-n \operatorname{Tr}(V(M))) dM, \tag{9.1}$$

where dM is Lebesgue measure on the vector space of Hermitean matrices, Z_n is a normalization constant and V is a function on $\mathbf{R}$ that grows sufficiently fast at $\pm\infty$ (for example, an even polynomial). The probability measure (9.1) on the matrices induces a probability measure on the eigenvalues. This in turn gives a measure μ_n on the real line, such that $\mu_n([a, b])$ is $1/n$ times the expected number of eigenvalues in $[a, b]$. It can be shown [4] that as $n \to \infty$, the measures μ_n converge in weak* sense to a probability measure μ, called the density of states in random matrix literature. The measure μ is equal to the extremal measure in the external field $V(x)/2$.

9.3 Extremal Problems with a Constraint

To describe asymptotics of polynomials which are orthogonal on discrete sets, Rakhmanov [40] introduced and studied an extremal problem with a constraint. The problem is to minimize

$$\int\int \log\frac{1}{|x-y|} d\mu(x) d\mu(y)$$

among all measures $\mu \in \mathcal{P}(K)$ satisfying the condition $\mu \leq \sigma$. Here the constraint σ is a given nonnegative measure on K with total mass bigger than one. In [13] a constraint was combined with an external field and many interesting properties were given. Other contributions in this direction are [8, 14, 24, 27].

9.4 Singular Limits of Integrable Systems

The extremal problem with constraint and external field was found by Deift and McLaughlin [11] to govern the continuum limit of the Toda lattice. The problem is to minimize

$$\int\int \log \frac{1}{|x-y|} d\mu(x)d\mu(y) + 2\int (Q(x) - tx)d\mu(x)$$

among measures μ on $[-1, 1]$ satisfying

$$0 \leq \mu \leq \sigma, \qquad \int d\mu = \xi.$$

Here Q and σ are given, and t and ξ are parameters in the problem, representing time and position, respectively.

If the support of the extremal measure is an interval, say $[a(\xi, t), b(\xi, t)]$, then the end points satisfy the so-called continuous Toda equations

$$\frac{\partial a}{\partial t} = -\frac{b-a}{2}\frac{\partial a}{\partial \xi}, \qquad \frac{\partial b}{\partial t} = \frac{b-a}{2}\frac{\partial b}{\partial \xi}.$$

This is a hyperbolic system of PDEs which may develop shocks. A shock would correspond to a transition to an extremal measure which is supported on more than one interval. Then the end points of those intervals satisfy a more complicated system of PDEs, which may be seen as a continuation of the continuous Toda equations beyond shock time.

The transition from one to two or more intervals has a physical interpretation for the continuum limit – it implies a phase transition. Once a new interval has appeared, the particles of the Toda lattice exhibit sustained oscillations, and the continuum limit exists only in a weak sense.

The continuum limit of the Toda lattice is an example of a singular limit of an integrable system. The prototype of such a singular limit is the zero dispersion limit of the Korteweg-de Vries equation. Lax and Levermore [29] showed that the zero dispersion limit is governed by a minimization problem for a Green potential. Using a normalization somewhat different from [29], the problem is to minimize

$$\int\int \log \left|\frac{x+y}{x-y}\right| d\mu(x)d\mu(y) + 2\int (Q(x) + \xi x - tx^3)d\mu(x) \qquad (9.2)$$

among all measures μ on $[0, 1]$ such that $0 \leq \mu \leq \sigma$. The double integral in (9.2) is known as the Green potential of μ for the right half-plane. Here the external field depends both on time and position.

Acknowledgments. I want to thank Steve Damelin, Peter Dragnev, Ken McLaughlin, Vilmos Totik, and Walter Van Assche for reading the paper and for making valuable remarks.

References

1. Akhiezer, N. I., *Theory of Approximation*, Ungar, New York, 1956.
2. Borwein, P. and W. Chen, Incomplete rational approximation in the complex plane, Constr. Approx. **11** (1995), 85–106.
3. Borwein, P., E. A. Rakhmanov, and E. B. Saff, Rational approximation with varying weights I, Constr. Approx. **12** (1996), 223–240.
4. Boutet de Monvel, A., L. A. Pastur, and M. Shcherbina, On the statistical mechanics approach to random matrix theory: integrated density of states, J. Stat. Phys. **79** (1995), 585–611.
5. Buyarov, V. S., Logarithmic asymptotics of polynomials orthogonal on the real axis, Dokl. Akad. Nauk USSR **318** (1991), 781–784. English transl.: Soviet Math. Dokl. **43** (1991), 743–746.
6. Buyarov, V. S. and E. A. Rakhmanov, Families of equilibrium measures with external field on the real axis, manuscript.
7. Damelin, S. B. and A. B. J. Kuijlaars, The support of the extremal measure in the presence of a monomial external field, Trans. Amer. Math. Soc., to appear.
8. Damelin, S. B. and E. B. Saff, Asymptotics of weighted polynomials on varying sets, manuscript.
9. Deift, P., T. Kriecherbauer, and K. T-R McLaughlin, New results on the equilibrium measure for logarithmic potentials in the presence of an external field, J. Approx. Theory, to appear.
10. Deift, P., T. Kriecherbauer, K. T-R McLaughlin, S. Venakides, and X. Zhou, Asymptotics for polynomials orthogonal with respect to varying exponential weights, Int. Math. Res. Notices **16** (1997), 759–782.
11. Deift, P. and K. T-R McLaughlin, A continuum limit of the Toda lattice, Mem. Amer. Math. Soc. **624** (1998).
12. Doob, J. L., *Classical Potential Theory and its Probabilistic Counterpart*, Grundl. math. Wiss. 262, Springer-Verlag, New York, 1984.
13. Dragnev, P. D. and E. B. Saff, Constrained energy problems with applications to orthogonal polynomials of a discrete variable, J. Analyse Math. **72** (1997), 223–259.
14. Dragnev, P. D. and E. B. Saff, A problem in potential theory and zero asymptotics of Krawtchouk polynomials, manuscript.
15. Gonchar, A. A. and E. A. Rakhmanov, Equilibrium measure and the distribution of zeros of extremal polynomials, Mat. Sb. **125** (1984), 117–127. English transl.: Math. USSR-Sb. **53**, 119–130.

16. Frostman, O., *Potentiel d'equilibre et capacité des ensembles avec quelques applications a la théorie des fonctions*, Thesis, Lund, 1935.
17. Helms, L. L., *Introduction to Potential Theory*, Wiley-Interscience, New York, 1969.
18. Ivanov, K. G. and V. Totik, Fast decreasing polynomials, Constr. Approx. **6** (1990), 1–20.
19. Kuijlaars, A. B. J., The role of the endpoint in weighted polynomial approximation with varying weights, Constr. Approx. **12** (1996), 287–301.
20. Kuijlaars, A. B. J., A note on weighted polynomial approximation with varying weights, J. Approx. Theory **87** (1996), 112–115.
21. Kuijlaars, A. B. J., Weighted approximation with varying weights: the case of a power-type singularity, J. Math. Anal. Appl. **204** (1996), 409–418.
22. Kuijlaars, A. B. J., Best constants in one-sided weak-type inequalities, Meth. Appl. Anal. **5** (1998), 95–108.
23. Kuijlaars, A. B. J. and P. D. Dragnev, Equilibrium problems associated with fast decreasing polynomials, Proc. Amer. Math. Soc., to appear.
24. Kuijlaars, A. B. J. and E. A. Rakhmanov, Zero distributions for discrete orthogonal polynomials, manuscript.
25. Kuijlaars, A. B. J. and W. Van Assche, A problem of Totik on fast decreasing polynomials, Constr. Approx. **14** (1998), 97–112.
26. Kuijlaars, A. B. J. and W. Van Assche, A contact problem in elasticity related to weighted approximations on the real line, Rend. Circ. Mat. Palermo **52** (1998), 575–587.
27. Kuijlaars, A. B. J. and W. Van Assche, Extremal polynomials on discrete sets, Proc. London Math. Soc., to appear.
28. Landkof, N. S., *Foundations of Modern Potential Theory*, Springer-Verlag, Berlin, 1972.
29. Lax, P. D. and C. D. Levermore, The small dispersion limit of the Korteweg-de Vries equation I, II, III, Comm. Pure Appl. Math. **36** (1983), 253–290, 571–593, 809–829.
30. Lorentz, G. G., M. von Golitschek, and Y. Makovoz, *Constructive Approximation, Advanced Problems*, Grundl. math. Wiss. 304, Springer-Verlag, Berlin, 1996.
31. Mehta, M. L., *Random Matrices*, 2nd ed., Academic Press, San Diego, 1991.
32. Mhaskar, H. N., *Introduction to the Theory of Weighted Polynomial Approximation*, World Scientific, Singapore, 1996.
33. Mhaskar, H. N. and E. B. Saff, Extremal problems for polynomials with exponential weights, Trans. Amer. Math. Soc. **285** (1984), 204–234.

34. Mhaskar, H. N. and E. B. Saff, Where does the sup norm of a weighted polynomial live? Constr. Approx. **1** (1985), 71–91.
35. Pritsker, I. E., Polynomial approximation with varying weights on compact sets of the complex plane, Proc. Amer. Math. Soc., to appear.
36. Pritsker, I. E. and R. S. Varga, The Szegő curve, zero distribution and weighted approximation, Trans. Amer. Math. Soc. **349** (1997), 4085–4105.
37. Pritsker, I. E. and R. S. Varga, Weighted polynomial approximation in the complex plane, Constr. Approx., to appear.
38. Pritsker, I. E. and R. S. Varga, Weighted rational approximation in the complex plane, manuscript.
39. Rakhmanov, E. A., On asymptotic properties of polynomials orthogonal on the real axis, Mat. Sb. **119**, 163–203. English transl.: Math. USSR-Sb. **47** (1984), 155–193.
40. Rakhmanov, E. A., Equilibrium measure and the distribution of zeros of extremal polynomials of a discrete variable, Mat. Sb. **187** (1996), 109–124. English transl.: Sb. Math. **187** (1996), 1213–1228.
41. Rakhmanov, E. A., E. B. Saff, and P. C. Simeonov, Rational approximation with varying weights II, J. Approx. Theory **92** (1998), 331–338.
42. Ransford, T., *Potential Theory in the Complex Plane*, Cambridge University Press, Cambridge, 1995.
43. Saff, E. B. and V. Totik, *Logarithmic Potentials with External Fields*, Grundl. math. Wiss. 316, Springer-Verlag, Berlin, 1997.
44. Simeonov, P. C., Weighted polynomial approximation with weights with logarithmic singularity in the extremal measure, Acta Math. Hungarica, to appear.
45. Totik, V., Approximation by algebraic polynomials, in *Approximation Theory VII*, E. W. Cheney, C. Chui, and L. Schumaker (eds.), Academic Press, New York, 1992, pp. 227–249.
46. Totik, V., Fast decreasing polynomials via potentials, J. Analyse Math. **62** (1994), 131–154.
47. Totik, V., *Weighted Approximation with Varying Weights*, Lect. Notes in Math. 1569, Springer-Verlag, Berlin, 1994.
48. Totik, V., Weighted polynomial approximation for weights with slowly varying extremal density, J. Approx. Theory, to appear.
49. Walsh, J. L., *Interpolation and Approximation by Rational Functions in the Complex Domain*, 3rd ed., AMS Colloquim Publications XX, Amer. Math. Soc., Providence, 1960.

Arno Kuijlaars
Department of Mathematics, Katholieke Universiteit Leuven
Celestijnenlaan 200 B, 3001 Leuven, Belgium
arno@wis.kuleuven.ac.be

Monotone Approximation Estimates Involving the Third Modulus of Smoothness

Dany Leviatan and Igor A. Shevchuk

Abstract. For each nondecreasing function $f \in \mathbf{C}[-1,1]$ and $k = 1, 2$, Lorentz and Zeller ($k = 1$), and DeVore ($k = 2$) proved the estimates $E_n^{(1)}(f) \leq C\omega_k(1/n, f)$, $n \geq N$, where $N = 1$ and C is an absolute constant. Here $E_n^{(1)}(f)$ denotes the error of best uniform approximation of f by nondecreasing algebraic polynomials of degree $\leq n$ and $\omega_k(t, f)$ is the modulus of smoothness of f of order k. On the other hand, Shvedov showed that the above estimates fail for $k = 3$ with C and N absolute constants. We will nevertheless prove that it holds for $k = 3$ with $N = 2$ and $C = C(f)$, or with C an absolute constant, but $N = N(f)$. The same estimates are valid if we replace ω_3, with the Ditzian-Totik modulus of smoothness ω_3^φ. We construct a counterexample to show that the corresponding pointwise estimates fail to hold even with $C(f)$ and $N(f)$.

§1. Jackson-Zygmund-Stechkin Type Estimates

Let $\mathbf{C}$ be the space of continuous functions f on the closed interval $I := [-1, 1]$ endowed with the uniform norm, and let $\mathbb{P}_n$, $n \in \mathbb{N}$, be the space of algebraic polynomials of degree $\leq n$. Set

$$E_n(f) := \inf_{p_n \in \mathbb{P}_n} \|f - p_n\|,$$

the error of the best uniform approximation of f, and denoting $\Delta^{(1)}$, the set of nondecreasing functions $f \in \mathbf{C}$, put

$$E_n^{(1)}(f) := \inf_{p_n \in \mathbb{P}_n \cap \Delta^{(1)}} \|f - p_n\|,$$

the error of the uniform monotone approximation of f. Finally, let

$$\Delta_h^k(f, x) := \sum_{j=0}^{k} (-1)^{k-j} \binom{k}{j} f(x + jh)$$

be the kth difference of order $k \in \mathbb{N}$ of f, at the point x with step h, and

$$\omega_k(t;f) := \sup_{h \in [0,t]} \|\Delta_h^k(f,x)\|_{[-1,1-kh]}, \quad t \geq 0,$$

its modulus of smoothness of order k. Here and in the sequel we write

$$\|g\|_{[a,b]} := \max_{x \in [a,b]} |g(x)|,$$

and $\|g\| := \|g\|_I$. Also we will use c to denote absolute constants which may differ in different occurrences, even in the same line.

In 1968 Lorentz and Zeller [10] proved that for any $f \in \Delta^{(1)}$,

$$E_n^{(1)}(f) \leq c\omega_1(1/n, f), \quad n \in \mathbb{N}. \tag{1.1}$$

In 1976 DeVore [1] strengthened (1.1) by proving for such an f, that

$$E_n^{(1)}(f) \leq c\omega_2(1/n, f), \quad n \in \mathbb{N}. \tag{1.2}$$

Later Shvedov [13] showed that in (1.2), one cannot replace ω_2 with ω_k for $k \geq 3$. Namely, for each $n \in \mathbb{N}$ and $A > 0$, he constructed a function $f = f_{n,A} \in \Delta^{(1)}$, satisfying

$$E_n^{(1)}(f) > A\omega_3(1/3, f)$$

(specifically we may take $f_{n,A}(x) = (x+1-b)_+^2$ with sufficiently small b). Finally, for $k \geq 4$, Wu and Zhou [14] have constructed a function $f \in \Delta^{(1)}$, satisfying

$$\limsup_{n \to \infty} \frac{E_n^{(1)}(f)}{\omega_k(1/n, f)} = \infty, \tag{1.3}$$

and they have conjectured that (1.3) is valid for $k = 3$ as well. Our first result disproves this conjecture. Namely,

Theorem 1. For each function $f \in \Delta^{(1)}$ we have

$$E_n^{(1)}(f) \leq c\omega_3\left(\frac{1}{n}, f\right) + \frac{c}{n^4}\omega_2(1, f), \quad n \in \mathbb{N}.$$

Corollary. For each function $f \in \Delta^{(1)}$ we have

$$E_n^{(1)}(f) \leq c\omega_3\left(\frac{1}{n}, f\right), \quad n \geq N(f),$$

and

$$E_n^{(1)}(f) \leq C(f)\omega_3\left(\frac{1}{n}, f\right), \quad n \geq 2,$$

where $N(f)$ and $C(f)$ are constants depending only on f.

§2. Relations Between $E_n(f)$ and $E_n^{(1)}(f)$

Let $\alpha > 0$, $\alpha \neq 2$. Then it follows by Leviatan [7] ($0 < \alpha < 2$) and by Kopotun and Listopad [6] ($\alpha > 2$) that if $f \in \Delta^{(1)}$ and

$$E_n(f) \leq n^{-\alpha}, \quad \forall n > \alpha - 1, \tag{2.1}$$

then

$$E_n^{(1)}(f) \leq C(\alpha) n^{-\alpha}, \quad \forall n > \alpha - 1,$$

where $C(\alpha)$ is a constant, depending only on α. For $\alpha = 2$ this statement fails to hold, since Kopotun and Listopad [6] constructed, for each $m \in \mathbb{N}$ and $A > 0$, a function $f = f_{m,A} \in \Delta^{(1)}$ satisfying (2.1) with $\alpha = 2$, and $E_m^{(1)}(f) > Am^{-2}$. Nevertheless, we are now able to show

Theorem 2. *For each function $f \in \Delta^{(1)}$, satisfying*

$$E_n(f) \leq \frac{1}{n^2}, \quad \forall n \in \mathbb{N}, \tag{2.2}$$

we have

$$E_n^{(1)}(f) \leq \frac{c}{n^2}, \quad \forall n \in \mathbb{N}.$$

§3. Ditzian-Totik Type Estimates

For $\varphi(x) := \sqrt{1 - x^2}$, recall the definition of Ditzian-Totik [4] modulus of smoothness:

$$\omega_k^\varphi(t, f) := \sup_{h \in [0,t]} \max_x \left| \Delta_{h\varphi(x)}^k \left(x - \frac{k}{2} h\varphi(x), f \right) \right|,$$

where the maximum is taken over all x, satisfying $x \pm (k/2)h\varphi(x) \in I$.

We also need, for a fixed $n \geq 3$, the set of continuous quadratic piecewise polynomials with knots at $x_j := -\cos(j\pi/n)$, $j = 0, \ldots, n$, and only there, which we denote by $\mathbf{S}_n$. We recall the following result which is a special case of Proposition 3 of [9]. Namely,

Proposition 1. *For every n, if $s_n \in \mathbf{S}_n \cap \Delta^{(1)}$, then*

$$E_{cn}^{(1)}(s_n) \leq c\omega_3^\varphi\left(\frac{1}{n}, s_n\right).$$

Now, if $f \in \Delta^{(1)}$, then it is well known (see [7]) that

$$E_n^{(1)}(f) \leq c\omega_2^\varphi(1/n, f), \quad n \in \mathbb{N}.$$

Here we prove

Theorem 3. *For each function $f \in \Delta^{(1)}$ we have*
$$E_n^{(1)}(f) \leq c\omega_3\left(\frac{1}{n}, f\right) + \frac{c}{n^4}\omega_2^\varphi(1,f), \qquad n \in \mathbb{N}.$$

Evidently Theorem 1 is weaker than Theorem 3, so let us indicate how Theorem 3 yields the proof of Theorem 2.

Proof of Theorem 2. Applying the inverse estimates for the D-T moduli (see [4]), we see that (2.2) implies
$$\omega_3^\varphi(t,f) \leq ct^3 \sum_{0<n\leq 1/t} n^2 E_n(f) \leq ct^2,$$
and
$$\omega_2^\varphi(1,f) \leq cE_1(f) \leq c. \quad \Box$$

Proof of Theorem 3. Let $f \in \Delta^{(1)}$ and $n \geq 3$; and let l_1 and l_n be the linear functions which interpolate f at the endpoints of the intervals $I_1 := [-1, x_1]$ and $I_n := [x_{n-1}, 1]$, respectively. Then Lemma 1 of [8] guarantees the existence of a piecewise quadratic
$$s_n \in \mathbf{S}_n \cap \Delta^{(1)} \tag{3.1}$$
which satisfies $s_n(x) = l_1(x)$, $x \in I_1$, $s_n(x) = l_n(x)$, $x \in I_n$, and
$$\|f - s_n\|_{[x_1, x_{n-1}]} \leq c\omega_3^\varphi(1/n, f). \tag{3.2}$$

Since the length of I_1 and I_n is less than cn^{-2}, then the Burkill-Whitney inequality (see, e.g., (3.9) in [12]) implies
$$\|f - s_n\|_{I_1} = \|f - l_1\|_{I_1} \leq c\omega_2(1/n^2, f), \quad \|f - s_n\|_{I_n} \leq c\omega_2(1/n^2, f). \tag{3.3}$$
The proof of Marchaud inequality (see, e.g., (3.6) in [12]) yields
$$\omega_2(t,f) \leq ct^2 \int_t^1 \frac{\omega_3(u,f)}{u^3} du + ct^2 \omega_2(1,f), \qquad t \in (0,1].$$
Hence applying the inequalities ([4], see also (18.6) and (18.12) in [12])
$$\omega_3(u,f) \leq c\omega_3^\varphi(\sqrt{u}, f) \leq cn^3 u^{3/2} \omega_3^\varphi(1/n, f), \quad u \geq \frac{1}{n^2},$$
we get
$$\omega_2(1/n^2, f) \leq c\omega_3^\varphi(1/n, f) + \frac{c}{n^4}\omega_2(1,f) = c\omega_3^\varphi(1/n, f) + \frac{c}{n^4}\omega_2^\varphi(1,f). \tag{3.4}$$

Combining (3.2) through (3.4), we have
$$\|f - s_n\| \leq c\omega_3^\varphi(1/n, f) + cn^{-4}\omega_2^\varphi(1,f), \tag{3.5}$$
where
$$\omega_3^\varphi(1/n, s_n) \leq \omega_3^\varphi(1/n, f) + 8\|f - s_n\| \leq c\omega_3^\varphi(1/n, f) + cn^{-4}\omega_2^\varphi(1,f). \tag{3.6}$$
Hence for $n > c$, Theorem 3 follows by Proposition 1 combined with (3.5) and (3.6), while for $n \leq c$ it follows from the trivial inequality $E_1^{(1)}(f) = E_1(f) \leq c\omega_2^\varphi(1,f)$. $\Box$

§4. Nikolskii-Timan-Dzjadyk-Freud-Brudnyi Type Estimates

Pointwise estimates for monotone approximation are completely different for $k = 3$.

DeVore and Yu [3] have proved that for every function $f \in \Delta^{(1)}$ and all $n \in \mathbb{N}$, there is a polynomial $p_n \in \mathbb{P}_n \cap \Delta^{(1)}$, such that

$$|f(x) - p_n(x)| \leq c\omega_2\left(\frac{1}{n}\varphi(x), f\right), \quad x \in I. \tag{4.1}$$

Set $\rho_n(x) := \frac{1}{n^2} + \frac{1}{n}\varphi(x)$, and for $f \notin \mathbb{P}_{k-1}$, let

$$D_{n,k}(f) := \inf_{p_n \in \mathbb{P}_n} \max_{x \in I} \frac{|f(x) - p_n(x)|}{\omega_k(\rho_n(x), f)};$$

and

$$D_{n,k}^{(1)}(f) := \inf_{p_n \in \mathbb{P}_n \cap \Delta^{(1)}} \max_{x \in I} \frac{|f(x) - p_n(x)|}{\omega_k(\rho_n(x), f)}.$$

For convenience in notation, we let $D_{n,k}(f) = D_{n,k}^{(1)}(f) = 0$, if $f \in \mathbb{P}_{k-1}$. Then, in view of the classical Nikolskii-Timan-Dzjadyk-Freud-Brudnyi direct theorems, for each $f \in C$ and $k \in \mathbb{N}$, the estimate

$$D_{n,k}(f) \leq C(k), \quad n \geq k - 1,$$

holds, where $C(k) = const$, depends only on k; and (4.1) implies, for each $f \in \Delta^{(1)}$,

$$D_{n,2}^{(1)}(f) \leq c, \quad n \in \mathbb{N}.$$

For $k = 3$, applying ideas of Lorentz and Zeller [11], Shvedov [13], DeVore, Leviatan and Shevchuk [2], Dzyubenko, Gilewicz and Shevchuk [5], we prove

Theorem 4. *There is a function $f \in \Delta^{(1)}$ such that*

$$\limsup_{n \to \infty} D_{n,3}^{(1)}(f) = \infty. \tag{4.2}$$

Proof: Our proof is based on the following Dzjadyk type inequality for the polynomials $p_n \in \mathbb{P}_n$ (see, e.g., Lemma 14.1 and (14.9) in [12]). If for all $x \in I$,

$$|p_n(x)| \leq (1 + n^2(1+x))^2 = n^4(1/n^2 + (1+x))^2,$$

then

$$|p_n'(-1)| \leq cn^4(1/n^2)^{2-1} = cn^2.$$

Observe that $\varphi^2(x) \leq 2(1+x)$, $x \in I$. Hence if for $d \in (0, 2]$, we set

$$A := \|p_n(1 + n\varphi)^{-3}\|_{[-1, -1+d]},$$

then for $x \in [-1, -1+d]$,

$$|p_n(x)| \leq A(1 + n\varphi(x))^3 \leq A(1 + n\varphi(x))^4 \leq 4A(1 + n^2\varphi^2(x))^2$$
$$\leq 16A(1 + n^2(1+x))^2 \leq 16A(1 + n^2 2(1+x)/d)^2.$$

Finally, a change of variable, $y =: -1 + 2(x+1)/d$, gives $y \in I$, and for $q_n(y) := p_n(x)$, we have

$$|q_n(y)| \leq 16A(1 + n^2(1+y))^2.$$

Thus by Dzjadyk's inequality,

$$|p'_n(-1)| = \frac{2}{d}|q'_n(-1)| \leq \frac{2}{d}c_* A n^2 = \frac{2c_* n^2}{d}\|p_n(1+n\varphi)^{-3}\|_{[-1,-1+d]}, \quad (4.3)$$

where c_* is an absolute constant.

Now for $b \in (0,1)$, let $S_b(x)$ be a function so that

$$S_b \in \mathbf{C}^{(3)}; \quad S''_b(x) \geq 0, \ x \in I; \quad S_b(x) = 0, \ x < b-1; \quad S_b(x) = 1, \ x > 2b-1,$$

and let

$$Q_b(x) := (x+1-b)^2, \quad g_b(x) := Q_b(x)S_b(x).$$

Evidently,

$$g_b \in \Delta^{(1)} \cap \mathbf{C}^{(3)}, \quad (4.4)$$

$$\omega_3(t, g_b) = \omega_3(t, g_b - Q_b) \leq 8\|g_b - Q_b\| = 8b^2. \quad (4.5)$$

Let $\mathbb{P}_n^*$ denote the set of polynomials $p_n \in \mathbb{P}_n$, satisfying $p'_n(-1) \geq 0$. Then for $d \in (0, 2]$ and every $p_n \in \mathbb{P}_n^*$, it follows by (4.3) that

$$2b = -Q'_b(-1) \leq p'_n(-1) - Q'_b(-1) \leq \frac{2c_* n^2}{d}\|(p_n - Q_b)(1+n\varphi)^{-3}\|_{[-1,-1+d]}$$
$$\leq \frac{2c_* n^2}{d}(\|(p_n - g_b)(1+n\varphi)^{-3}\|_{[-1,-1,+d]} + \|(g_b - Q_b)(1+n\varphi)^{-3}\|),$$

where

$$\|(p_n - g_b)(1+n\varphi)^{-3}\|_{[-1,-1+d]} \geq \frac{bd}{c_* n^2} - b^2. \quad (4.6)$$

We are now in a position to define f which will satisfy (4.2). To this end, let $b_n := n^{-3}$ and set $n_1 := 1$, $d_0 := 1$, and

$$d_j := b_{n_j}^2 d_{j-1} = b_{n_1}^2 \cdot \ldots \cdot b_{n_j}^2, \quad f_j(x) := g_{b_{n_j}}(x), \quad j \geq 1,$$

where the sequence $\{n_\nu\}$ is defined by induction as follows. Suppose that $\{n_1, \ldots, n_{\sigma-1}\}$, $\sigma > 1$, has been defined. Then define

$$F_{\sigma-1} := \sum_{j=1}^{\sigma-1} d_{j-1} f_j,$$

Monotone Approximation Estimates

and take $n_\sigma > n_{\sigma-1}$ so big that

$$\frac{1}{c_*}d_{\sigma-1}b_{n_{\sigma-1}}n_\sigma - 9d_{\sigma-1} > (\|F^{(3)}_{\sigma-1}\| + 72d_{\sigma-1})\sqrt{n_\sigma} =: B_{\sigma-1}\sqrt{n_\sigma}. \quad (4.7)$$

Now let $\Phi_\sigma := \sum_{j=\sigma}^\infty d_{j-1}f_j$, where the convergence of the series is guaranteed by the definition of the d_j's and the fact that $\|g_b\| < 4$. Actually for $\sigma > 1$, we have

$$\|\Phi_\sigma\| \leq 4d_{\sigma-1}(1 + n_\sigma^{-6} + n_\sigma^{-6}n_{\sigma+1}^{-6} + \ldots) < 8d_{\sigma-1}. \quad (4.8)$$

Finally, we take $f := \Phi_1 = \sum_{j=1}^\infty d_{j-1}f_j$, and (4.4) implies that $f \in \Delta^{(1)}$. In order to prove (4.2) we estimate $\omega_3(t, f)$. It follows by (4.5) and (4.8) that

$$\omega_3(t, f) \leq \omega_3(t, F_{\sigma-1}) + \omega_3(t, d_{\sigma-1}f_\sigma) + \omega_3(t, \Phi_{\sigma+1})$$
$$\leq t^3\|F^{(3)}_{\sigma-1}\| + 8d_{\sigma-1}b_{n_\sigma}^2 + 8\|\Phi_{\sigma+1}\|$$
$$\leq t^3\|F^{(3)}_{\sigma-1}\| + 72d_{\sigma-1}b_{n_\sigma}^2,$$

where

$$\omega_3(\rho_{n_\sigma}(x), f) \leq \frac{B_{\sigma-1}}{n_\sigma^6}(1 + n_\sigma\varphi(x))^3. \quad (4.9)$$

Let $p_{n_\sigma} \in \mathbb{P}_{n_\sigma} \cap \Delta^{(1)} \subset \mathbb{P}^*_{n_\sigma}$, and set $P_{n_\sigma} := d_{\sigma-1}^{-1}p_{n_\sigma}$. Then by (4.6) through (4.8) we have,

$$\left\|\frac{\Phi_\sigma - p_{n_\sigma}}{(1+n_\sigma\varphi)^3}\right\|_{[-1,-1+b_{n_{\sigma-1}}]} \geq \left\|\frac{d_{\sigma-1}f_\sigma - p_{n_\sigma}}{(1+n_\sigma\varphi)^3}\right\|_{[-1,-1+b_{n_{\sigma-1}}]} - \left\|\frac{\Phi_{\sigma+1}}{(1+n_\sigma\varphi)^3}\right\|$$

$$\geq d_{\sigma-1}\left\|\frac{f_\sigma - P_{n_\sigma}}{(1+n_\sigma\varphi)^3}\right\|_{[-1,-1+b_{n_{\sigma-1}}]} - \|\Phi_{\sigma+1}\|$$

$$\geq \frac{d_{\sigma-1}b_{n_{\sigma-1}}b_{n_\sigma}}{c_* n_\sigma^2} - d_{\sigma-1}b_{n_\sigma}^2 - \|\Phi_{\sigma+1}\|$$

$$> \frac{d_{\sigma-1}b_{n_{\sigma-1}}}{c_*}\frac{1}{n_\sigma^5} - \frac{9d_{\sigma-1}}{n_\sigma^6} > \frac{B_{\sigma-1}}{n_\sigma^6}\sqrt{n_\sigma}.$$

We observe that for $x \in [-1, -1 + b_{n_{\sigma-1}}]$, it follows from the definition of f that $f(x) = \Phi_\sigma(x)$. Hence the last inequality together with (4.9), implies that for each polynomial $p_{n_\sigma} \in \mathbb{P}_{n_\sigma} \cap \Delta^{(1)}$,

$$\left\|\frac{f - p_{n_\sigma}}{\omega_3(\rho_{n_\sigma}, f)}\right\| \geq \frac{n_\sigma^6}{B_{\sigma-1}}\left\|\frac{f - p_{n_\sigma}}{(1+n_\sigma\varphi)^3}\right\|$$
$$\geq n_\sigma^6 B_{\sigma-1}^{-1}\|(\Phi_\sigma - p_{n_\sigma})(1+n_\sigma\varphi)^{-3}\|_{[-1,-1+b_{n_{\sigma-1}}]}$$
$$> \sqrt{n_\sigma}.$$

Therefore, $D^{(1)}_{n_\sigma,3}(f) \geq \sqrt{n_\sigma}$ for all σ and (4.2) follows. $\square$

Acknowledgments. The second author is supported by the Organizing Committee of the AT IX, Ukrainian Charitable Foundation for Furthering Development of Mathematical Sciences and NSF grant DMS-9705638.

References

1. DeVore, R. A., Degree of approximation, in *Approximation Theory, II*, G. G. Lorentz, C. K. Chui, and L. L. Schumaker (eds.), Academic Press, New York, 1976, pp. 117–162.
2. DeVore, R. A., D. Leviatan, and I. A. Shevchuk, Approximation of monotone functions: a counter example, in *Curves and Surfaces with Applications in CAGD*, A. Le Méhauté, C. Rabut, and L. L. Schumaker (eds.), Vanderbilt Univ. Press, Nashville, 1997, pp. 95–102.
3. DeVore, R. A. and X. M. Yu, Pointwise estimates for monotone polynomial approximation, Constr. Approx. **1** (1985), 323–331.
4. Ditzian, Z. and V. Totik, *Moduli of Smoothness*, Springer-Verlag, Berlin, 1987.
5. Dzyubenko, G. A., J. Gilewicz, and I. A. Shevchuk, Piecewise monotone pointwise approximation, Constr. Approx., to appear.
6. Kopotun, K. A. and V. V. Listopad, Remarks on monotone and convex approximation by algebraic polynomials, Ukrainian Math. J. **46** (1994), 1266–1270.
7. Leviatan, D., Monotone and comonotone approximation revisited, J. Approx. Theory **53** (1988), 1–16.
8. Leviatan, D. and I. A. Shevchuk, Nearly comonotone approximation, J. Approx. Theory, to appear.
9. Leviatan, D. and I. A. Shevchuk, Some positive results and counterexamples in comonotone approximation II, J. Approx. Theory, submitted.
10. Lorentz, G. G. and K. L. Zeller, Degree of approximation by monotone polynomials I, J. Approx. Theory **1** (1968), 501–504.
11. Lorentz, G. G. and K. L. Zeller, Degree of approximation by monotone polynomials II, J. Approx. Theory **2** (1969), 265–269.
12. Shevchuk, I. A., *Polynomial Approximation and Traces of Functions Continuous on a Segment*, Naukova Dumka, Kyiv, 1992 (in Russian).
13. Shvedov, A. S., Orders of coapproximation of functions by algebraic polynomials, Math. Notes **30** (1981), 36–70.
14. Wu, X. and S. P. Zhou, On a counterexample in monotone approximation, J. Approx. Theory **69** (1992), 205–211.

Dany Leviatan
School of Mathematical Sciences
Sackler Faculty of Exact Sciences
Tel Aviv University
Tel Aviv 69978, ISRAEL
leviatan@math.tau.ac.il

Igor A. Shevchuk
Institute of Mathematics
NAS of Ukraine
3, Tereshchenkivska str.
Kyiv 252601, UKRAINE
shevchuk@imath.kiev.ua

Infinite Dimensional Widths of Function Classes with Non-symmetric Smoothness Conditions

Yongping Liu

Abstract. The infinite dimentional width problems of some classes of functions defined on the real line and with some non-symmetric smoothness conditions are considered. Some exact results are obtained.

§1. Introduction

In [2], Chun Li introduced the concept of infinite (in fact, locally finite) dimensional widths of some classes of smooth functions defined on the whole real line and obtained some exact results of the widths of Sobolev classes $W_p^r(\mathbb{R})$ in the spaces $L_p(\mathbb{R})$ for $1 \leq p \leq \infty$. In [4], the author introduced some classes $W_{pq}^r(\mathbb{R})$, called Sobolev-Wiener classes, of smooth functions defined on the whole real line, and together with the other authors in [4, 5] in succession obtained some exact results of the widths of the classes $W_{pq}^r(\mathbb{R})$ in the metric $L_q(\mathbb{R})$ for the cases $1 \leq q \leq p = \infty$ and $1 = q \leq p \leq \infty$. But the same kinds of results of some function classes with some non-symmetric smoothness conditions have not been studied. In this paper, we consider this problem and obtain some exact results.

For $1 \leq p \leq \infty$ and a natural number r, as [4, 5], let $W_{p1}^{r,+}(\mathbb{R})$ denote the Sobolev-Wiener class of the smooth functions $f \in L(\mathbb{R}) =: L_1(\mathbb{R})$ with non-symmetric smoothness condition for which their $(r-1)$th derivatives $f^{(r-1)}(x)$ are locally absolutely continuous on $\mathbb{R}$ and their rth derivatives $f^{(r)}(x)$ satisfy the condition: $\|f_+^{(r)}\|_{p1} \leq 1$. Here the norm $\|\cdot\|_{p1}$ is defined by

$$\|g\|_{p1} = \sum_{j \in \mathbb{Z}} \|g(\cdot + j)\|_{L_p[0,1]},$$

if they are finite for some Lebesgue measurable functions g defined on $\mathbb{R}$, while $\mathbb{Z}$ denotes the set of all integers, and the function $g_+(x)$ is the positive part of g defined by $g_+(x) = \max\{g(x), 0\}$. When $p = 1$, we simply write $\|\cdot\|_1$, $W_1^{r,+}(\mathbb{R})$ and $L[a,b]$ instead of $\|\cdot\|_{11}$ and $W_{11}^{r,+}(\mathbb{R})$ and $L_1[a,b]$, respectively.

For any two subsets A and B of a normed space X with $\|\cdot\|_X$, we set

$$E(A, B, X) = \sup_{f \in A} \inf_{g \in B} \|f - g\|_X$$

if the quantity is finite.

For any subspace M of $L(\mathbb{R})$ and some $\sigma > 0$, we say that M is of an infinite (locally finite) dimension with the index σ, if the quantity

$$\widetilde{\dim}(M, L(\mathbb{R})) =: \lim_{a \to +\infty} \frac{\dim M|_{I_a}}{2a} = \sigma$$

is finite. Here $M|_{I_a}$ is the restriction of M on the interval $I_a =: [-a, a]$, $a > 0$. For a subset $\mathcal{M}$ of $L(\mathbb{R})$, we call the quantity

$$\widetilde{d}_\sigma(\mathcal{M}, L(\mathbb{R})) =: \inf\{E(\mathcal{M}, M, L(\mathbb{R})) : \widetilde{\dim}(M, L(\mathbb{R})) \leq \sigma, M \subset L(\mathbb{R})\}$$

the infinite (locally finite) dimensional $\sigma - K$ width of $\mathcal{M}$ in $L(\mathbb{R})$.

Let m and n be two natural numbers. Denote by $S_{nm}(\mathbb{R})$ the space of polynomial splines of degree m with the knots $\Delta_n =: \{\frac{j}{n}, j \in \mathbb{Z}\}$ and with the defect 1.

In this paper, our main results are

Theorem 1. Let $1 \leq p \leq \infty$, $\frac{1}{p} + \frac{1}{p'} = 1$ and r, n be two natural numbers. Then

$$E(W_{p1}^{r,+}(\mathbb{R}), S_{nr}(\mathbb{R}) \cap L(\mathbb{R}), L(\mathbb{R})) \leq n^{-r} \|\varphi_r(\cdot + \alpha_r) + \|\varphi_r\|_\infty\|_{L_{p'}[0,1]}. \quad (1)$$

Here the function $\varphi_r(x)$ is the 2-periodic Euler perfect spline of degree r, α_r is a number so chosen that

$$\varphi_r(\alpha_r) = -\|\varphi_r\|_\infty = \inf_{x \in R} \varphi_r(x).$$

Theorem 2. Let r, n be two natural numbers. Then

$$\widetilde{d}_n(W_{p1}^{r,+}(\mathbb{R}), L(\mathbb{R})) = E(W_{p1}^{r,+}(\mathbb{R}), S_{nr}(\mathbb{R}) \cap L(\mathbb{R}), L(\mathbb{R}))$$
$$= n^{-r} \|\varphi_r(\cdot + \alpha_r) + \|\varphi_r\|_\infty\|_{L_{p'}[0,1]}, \quad (2)$$

for the cases $p = 1, 2, \infty$.

§2. The Upper Estimate

Let $W^m_\infty(\mathbb{R})$ denote the Sobolev class of smooth functions $f(x)$ for which $f^{(m-1)}(x)$ are locally absolutely continuous on $\mathbb{R}$ and satisfy $\|f^{(m)}\|_\infty =: \sup_{x \in \mathbb{R}} |f^{(m)}(x)| \leq 1$. As in [1], let $r(f; a, b; t)$ denote the non-increasing rearrangement of f on $[a, b]$.

Lemma 1. *Let f be a nonnegative continuous function on $[a, b]$ and possess a continuous derivative f' on $[a, b]$. Set $F_0 = \{t \in (a, b) : f'(t) = 0\} \cup \{a, b\}$, $m(f) = \min_{a \leq x \leq b} f(x)$, $M(f) = \max_{a \leq x \leq b} f(x)$. Then, for any $\xi \in [a, b]$ satisfying $r(f; a, b; \xi) \in f[(a, b) - F_0]$, the set*

$$e_\xi = \{x \in (a, b) : f(x) = r(f; a, b; \xi)\}$$

is only a finite set, say $\{x_1, x_2, ..., x_k\}$, and $r(f; a, b; t)$ at the point $t = \xi$ has a finite derivative $r'(f; a, b; \xi)$, defined by

$$r'(f; a, b; \xi) = -\frac{1}{\sum_{j=1}^k |f'(x_j)|^{-1}}. \qquad (3)$$

Proof: First, it is easy to see that the set e_ξ is finite. Hence, we may assume that e_ξ has k distinct elements $x_1, x_2, ..., x_k$. Notice that the set F_0 is compact, and has Lebesgue measure zero. Then, there exists $h > 0$ such that the intervals $(x_j - h, x_j + h) \subset (0, 1) - F_0$ $(j = 1, 2, ..., k)$ are non-intersecting. At the same time, there exists a number $\delta > 0$, such that for any $\eta \in (\xi - \delta, \xi + \delta)$,

$$e_\eta = \{y \in (a, b) : f(y) = r(f; a, b; \eta)\} \subset \bigcup_{j=1}^k (x_j - h, x_j + h).$$

Of course, the set e_η also has only k distinct elements $y_1, y_2, ..., y_k$. This means that $f(y_j) = r(f; a, b; \eta), j = 1, 2, ..., k$. Without loss of generality, we may think that $y_j \in (x_j - h, x_j + h), j = 1, 2, ..., k$. It is easy to see that $|\xi - \eta| = \sum_{j=1}^k |y_j - x_j|$. Assume that $\eta > \xi$. Since $r(f; a, b; \eta) < r(f; a, b; \xi)$, we have

$$\frac{r(f, \eta) - r(f, \xi)}{\eta - \xi} = \frac{r(f; a, b; \eta) - r(f; a, b; \xi)}{\sum_{j=1}^k |y_j - x_j|} = -\frac{1}{\sum_{j=1}^k |\frac{f(y_j) - f(x_j)}{y_j - x_j}|^{-1}}. \qquad (4)$$

Notice that $y_j \to x_j, j = 1, 2, ..., k$, as $\eta \to \xi$. Then, by (4), we have

$$r'(f; a, b; \xi) = -\frac{1}{\sum_{j=1}^k |f'(x_j)|^{-1}},$$

which is (3). □

Set
$$W_\infty^{r+1}(\mathbb{R}, \Delta_n) = \{g \in W_\infty^{r+1}(\mathbb{R}) : g(\frac{j}{n}) = 0, \forall j \in \mathbb{Z}\}. \tag{5}$$

Lemma 2. *For* $g \in W_\infty^{r+1}(\mathbb{R}, \Delta_n)$, *set*

$$f_n(x) = \|\varphi_{nr}\|_\infty + g'(x), \quad \psi_n(x) = \|\varphi_{nr}\|_\infty + \varphi_{nr}(x + \alpha_{nr}), \tag{6}$$

where the constants α_{nr} *are chosen so that* $\psi_n(\frac{2i}{n}) = 0$ *for all* $i \in \mathbb{Z}$. *Let* $F_0 =: \{x \in (0,1) : f_n'(x) = 0\}$. *For each* $j = 1, 2, ..., n$, *let* $\eta, \xi \in (0, \frac{1}{n}) - F_0$ *satisfy* $r(f_n; \frac{j-1}{n}, \frac{j}{n}; , \xi) = r(\psi_n; \frac{j-1}{n}, \frac{j}{n}; \eta)$. *Then*

$$r'(f_n; \frac{j-1}{n}, \frac{j}{n}; \xi) \le r'(\psi_n; \frac{j-1}{n}, \frac{j}{n}; \eta). \tag{7}$$

Proof: Without loss of generality, we prove the lemma only for the case $j = 1$ and write $r_1(f_n, t)$ and $r_1(\psi_n, t)$ instead of $r(f_n; 0, \frac{1}{n}; t)$ and $r(\psi_n; 0, \frac{1}{n}; t)$ respectively. Notice the fact that $\psi_n(\frac{2i}{n}) = 0, \forall i \in \mathbb{Z}$. There exists only a $y \in (0, \frac{1}{n})$ such that $r_1(\psi_n, \eta) = \psi_n(y)$, and hence, $r_1'(\psi_n, \eta) = -|\psi_n'(y)|$. Suppose that the set $\{t \in (0, \frac{1}{n}) : r_1(f_n, \xi) = f_n(t)\}$ has only k distinct elements $x_1, x_2, ..., x_k$. By the Komogorov's Comparitive Theorem in [2], we have $|\psi_n'(y)| \ge |f_n'(x_j)|, j = 1, 2, ..., k$. Hence, we obtain

$$r_1'(\psi_n, \eta) = -|\psi_n'(y)| \le -\frac{1}{\sum_{j=1}^k |f'(x_j)|^{-1}} = r_1'(f_n, \xi),$$

which is (7) for the case $j = 1$. □

Lemma 3. *Let the functions* $f_n(t)$ *and* $\psi_n(t)$ *be definited as in Lemma 2. Then*

$$\int_0^x r(f_n; \frac{j-1}{n}, \frac{j}{n}; t) dt \le \int_0^x r(\psi_n; \frac{j-1}{n}, \frac{j}{n}; t) dt, \quad x \in [0, \frac{1}{n}], \tag{8}$$

$$\|f_n\|_{L_p[\frac{j-1}{n}, \frac{j}{n}]} \le \|\psi_n\|_{L_p[\frac{j-1}{n}, \frac{j}{n}]}, \tag{9}$$

for $j = 1, 2, ..., n$.

Proof: Without loss of the generality, we prove (8) only for the case $j = 1$. We continue to use the notations used in Lemma 2. Set $\delta(t) = r_1(\psi_n, t) - r_1(f_n, t)$. Notice that

$$\max_{x \in [0, \frac{1}{n}]} f_n(x) = \|\varphi_{nr}\|_\infty + \max_{x \in [0, \frac{1}{n}]} g'(x) \le \max_{x \in [0, \frac{1}{n}]} \psi_n(x) = 2\|\varphi_{nr}\|_\infty.$$

It is easy to see that $\delta(0) \ge 0$. In the following, we prove that if there exists a point $\xi_0 \in (0, \frac{1}{n})$ such that $\delta(\xi_0) \le 0$, then we have $\delta(\xi) \le 0$ for all $\xi \in (\xi_0, \frac{1}{n})$. Set $\xi_1 = inf\{t \in [0, \frac{1}{n}] : r_1(f_n, t) = m_1(f_n) =: inf_{0 \le x \le \frac{1}{n}} f_n(x)\}$. Then, it is easy to see that $r_1(f_n, t) = m_1(f_n)$, for all $t \in (\xi_1, \frac{1}{n})$, and hence $r_1'(f_n, t) = 0$

for $\xi \in (\xi_1, \frac{1}{n})$. Hence, by Lemma 2, we may see that if there exists a point $\xi_0 \in (0, \frac{1}{n})$ such that $\delta(\xi_0) = 0$; thus, $\delta(\xi) \le 0$ for all $\xi \in (\xi_0, \frac{1}{n})$. Notice that $\int_0^{\frac{1}{n}} \delta(\xi) d\xi = 0$. Thus, by the above-mentioned fact, we have $\int_0^x \delta(\xi) d\xi \ge 0$, for all $x \in [0, \frac{1}{n}]$, which shows (8). From Koneichuk [1], (9) follows (8). □

Proof of Theorem 1. Now we give the upper estimate of $E(W_{p1}^{r,0}(R)$, $S_{nr}(\mathbb{R}) \cap L(\mathbb{R}), L(\mathbb{R}))$. By the duality theorem of best approximation of functions in [5, 9], we have

$$E(W_{p1}^{r,+}(\mathbb{R}), S_{nr}(\mathbb{R}) \cap L(\mathbb{R}), L(\mathbb{R})) \le \sup\{\|c + g'\|_{p',\infty} : \quad (10)$$
$$g \in W_\infty^{r+1}(\mathbb{R}), g(\frac{j}{n}) = 0, \inf\{c + g'(t) : t \in R\} = 0\}.$$

For any $g \in W_\infty^{r+1}(\mathbb{R}, \Delta_n)$ and for each $i \in \mathbb{Z}$, let $f_n(t+i) = \|\varphi_{nr}\|_\infty + g'(t+i)$ and $\psi_n(x)$ be defined as in Lemma 2. Then, by the definition, it follows that

$$\|\|\varphi_{nr}\|_\infty + g'\|_{p',\infty} = \sup_{i \in Z} \|f_n(\cdot + i)\|_{L_{p'}[0,1]}, \quad (11)$$

and hence, by Lemma 3, for each $i \in \mathbb{Z}$, we have

$$\|f_n(\cdot + i)\|_{L_{p'}[0,1]}^{p'} = \sum_{j=1}^n \|f_n(\cdot + i)\|_{L_{p'}[\frac{i-1}{n}, \frac{i}{n}]}^{p'} \le \sum_{j=1}^n \|\psi_n\|_{L_{p'}[\frac{i-1}{n}, \frac{i}{n}]}^{p'} \quad (12)$$
$$= \|\psi_n\|_{L_{p'}[0,1]}^{p'} \le \|\|\varphi_{nr}\|_\infty + \varphi_{nr}(\cdot + \alpha_{nr})\|_{L_{p'}[0,1]}^{p'}.$$

From (10) to (12), we obtain the upper estimate of $E(W_{p1}^{r,+}(\mathbb{R}), S_{nr}(\mathbb{R}) \cap L(\mathbb{R}), L(\mathbb{R}))$. □

§3. The Lower Estimate of $\widetilde{d}_n(W_{p1}^{r,+}(\mathbb{R}), L(\mathbb{R}))$

For $\delta > 0$, denote by $\widetilde{W}_{p;1,\delta}^r[a,b]$ the set of functions defined on $[a,b]$ with the absolutely continuous (r-1)-th derivative $f^{(r-1)}(x)$, such that $f^{(i)}(a) = f^{(i)}(b), i = 0, 1, ..., r-1$, and $\|f^{(r)}\|_{L_p[a,b],1,\delta} =: \|f_+^{(r)} + \delta f_-^{(r)}\|_{L_p[a,b]} \le 1$. Here $f_-^{(r)}(x) =: (-f^{(r)})_+(x)$ is the negative part of $f^{(r)}(x)$. Let $W_{p;1,\delta}^{r,0}[a,b]$ denote the set of functions $f \in \widetilde{W}_{p;1,\delta}^r[a,b]$ with $f^{(j)}(a) = 0, j = 0, 1, ..., r-1$. Let N be any natural number. It is easy to see that $(2N)^{-1+\frac{1}{p}} W_{p;1,\delta}^{r,0}(I_N) \subset W_{p1}^{r,+}(\mathbb{R})$. Here we consider the functions $f(x)$ in $W_{p;1,\delta}^{r,0}(I_N)$ to be defined as 0 for all $x \notin I_N$. Denote by $\Pi_m^r[a,b]$ the set of the algebraic polynomial prefect splines (see [11]) of degree r on $[a,b]$ with the number of free knots $\le m$, and by $\mathcal{P}_r$ the set of all algebraic polynomials of degree $\le r$.

Let M be a subspace of $L(\mathbb{R})$ with $\widetilde{\dim}(M, L(\mathbb{R})) \le \sigma$, and set $K_N = \dim(M|_{I_N})$. For any N, by the duality theorem of best approximation of functions in [1,7], we have

$$E(W_{p;1,\delta}^{r,0}(I_N), M|_{I_N}, L(I_N))$$
$$= \sup\{\int_{I_N} f(t)h(t)dt : h \perp M|_{I_N}, \|h\|_{L_\infty(I_N)} \leq 1, f \in W_{p;1,\delta}^{r,0}(I_N)\}$$
$$= \sup\{\int_{I_N} f^{(r)}(t)h(t)dt : h^{(r)} \perp M|_{I_N}, h \in W_\infty^r(I_N), f \in W_{p;1,\delta}^{r,0}(I_N)\} \quad (13)$$
$$= \sup\{\inf_{g \in \mathcal{P}_r} \|h + g\|_{L_p(I_N),1,\delta^{-1}} : h^{(r)} \perp M|_{I_N}, h \in W_\infty^r(I_N)\}.$$

By (13) and letting $\delta \to 0^+$, we have

$$E(W_{p1}^{r;+}(\mathbb{R}), M, L(\mathbb{R})) \geq (2N)^{-1+\frac{1}{p}} \sup\{\inf\{\|h + g\|_{L_{p'}(I_N)} : \\ h(t) + g(t) \geq 0, g \in \mathcal{P}_r\} : h^{(r)} \perp M|_{I_N}, h \in \widetilde{W}_\infty^r(I_N)\}. \quad (14)$$

Notice that for any $h(t) \in \Pi_{K_N}^r(I_N)$ and for any $p(t) \in \mathcal{P}_r$ there exist $h_1(t) \in \Pi_{K_N}^r(I_N)$ and a constant c such that $h(t) + p(t) = c + h_1(t)$. Then, by Lemma 2 and (14), we have

$$E(W_{p1}^{r;+}(\mathbb{R}), M, L(\mathbb{R})) \geq (2N)^{-1+\frac{1}{p}} \inf\{\|c + h\|_{L_{p'}(I_N)} : h \in \Pi_{K_N}^r(I_N), \\ \inf\{c + h(t) : t \in I_N\} = 0, c \in R\}. \quad (15)$$

Now, we give the lower estimate of $\widetilde{d_\sigma}(W_{p1}^{r;+}(\mathbb{R}), L(\mathbb{R}))$ for $p = 1, 2, \infty$.

Let $p = 2$. From [10], it is easy to see that there exists at least a polynomial perfect spline $f_*(t)$ on I_N with the number of knots $\leq K_N + 2r$ and the degree $\leq r$ such that (i) $f_*^{(j)}(N) = f_*^{(j)}(-N), j = 0, 1, 2, ..., r-1$; (ii) $f_*^{(r)} \perp M|_{I_N}$; (iii) $\int_{I_N} D_j(2Nt - N)f_*(t)dt = 0, j = 1, 2, ..., r$. Here the function $D_j(t)$ is the Bernoulii polynomial of degree j defined on [0,1] for each $j = 1, 2, ..., r$. Notice that for any constant c the function $c + f_*(t)$ satisfies the conditions (i)–(iii) by taking the place of $f_*(t)$ and noticing that $\|f_*^{(r)}\|_{L_2(I_N)} = (2N)^{\frac{1}{2}}$. Then, by (15), we have

$$E(W_{2,1}^{r;+}(\mathbb{R}), M, L(\mathbb{R})) \geq \|c + f_*\|_{L_2(I_N)}. \quad (16)$$

By a comparison theorem of Ligun [3] and a proper change of variables, we have

$$E(W_{2,1}^{r;+}(R), M, L(R)) \geq \left(\frac{2N}{K_N + 2r}\right)^r \|\varphi_r(\cdot + \alpha_r) + \|\varphi_r\|_\infty\|_{L_2[0,1]}. \quad (17)$$

Next, consider $p = 1$, so that $p' = \infty$. By [8], we have

$$\inf\{\|h\|_{L_\infty(I_N)} : h \in \Pi_{K_N}^r(I_N)\} = d_{K_N}(W_\infty^r(I_N), L_\infty(I_N)) \\ \geq d_{K_N + r}(\widetilde{W}_\infty^r(I_N), L_\infty(I_N)) \geq \left(\frac{2N}{K_N + 2r}\right)^r \|\varphi_r\|_{L_\infty[0,1]}. \quad (18)$$

… Hence, by (18), it is easy to see that

$$\inf\{\|c+h\|_{L_\infty(I_N)} : h \in \Pi^r_{K_N}(I_N), \inf\{c+h(t) : t \in I_N\} = 0, c \in R\}$$
$$\geq \left(\frac{2N}{K_N+2r}\right)^r \|\varphi_r(\cdot+\alpha_r)+\|\varphi_r\|_\infty\|_{L_\infty[0,1]}. \tag{19}$$

Finally, let $p = \infty$, so that $p' = 1$. Denote by $\mathcal{M}^{0,2}_{r,K_N}(I_N)$ the set of all monosplines of the following forms

$$s(x) = \frac{t^r}{r!} - \sum_{k=0}^{r-1}\frac{c_k t^{r-k-1}}{(r-k-1)!} - \sum_{j=1}^{m}\left(a_j\frac{(t-x_j)_+^{r-2}}{(r-2)!} + b_j\frac{(t-x_j)_+^{r-3}}{(r-3)!}\right)$$

$$(-N < x_1 < x_2 < \ldots < x_m < N), \quad m \leq \left[\frac{K_N+2r}{2}\right],$$

and the condition: $s^{(j)}(-N) = s^{(j)}(N) = 0, j = 1,\ldots,r-1$. By [8], we have

$$\inf\{\|g\|_{L(I_N)} : \min\{g(t) : t \in I_N\} = 0, g \in \Pi^r_{K_N}(I_N)\}$$
$$\geq \inf\{\|S\|_{L(I_N)} : S \in \mathcal{M}^{0,2}_{r,K_N}(I_N)\}$$
$$= \inf\left\{\sup\left\{\int_{I_N} f(t)dt - \sum_{j=1}^{m}(a_j f(x_j) + b_j f'(x_j)) : f \in W^r_\infty(I_N)\right\} :\right.$$
$$\left. a_j, b_j \in R, j = 1,\ldots,m, -N < x_1 < x_2 < \ldots < x_m < N, m \leq \left[\frac{K_N+2r}{2}\right]\right\}$$
$$\geq \inf\left\{\sup\left\{\int_{I_N} f(t)dt - \sum_{j=1}^{m}(a_j f(x_j) + b_j f'(x_j)) : f \in \widetilde{W}^r_\infty(I_N)\right\} :\right.$$
$$\left. a_j, b_j \in R, j = 1,\ldots,m, -N < x_1 < x_2 < \ldots < x_m < N, m \leq \left[\frac{K_N+2r}{2}\right]\right\}$$
$$\geq \left(\frac{2N}{2[\frac{K_N+2r}{2}]}\right)^r \|\varphi_r(\cdot+\alpha_r)+\|\varphi_r\|_\infty\|_{L[0,1]}. \tag{20}$$

Hence, by (14) and (20), we have

$$E(W^{r,+}_{\infty 1}(R), M, L(R)) \geq \left(\frac{2N}{2[\frac{K_N+2r}{2}]}\right)^r \|\varphi_r(\cdot+\alpha_r)+\|\varphi_r\|_\infty\|_{L[0,1]}. \tag{21}$$

By (14),(17),(19),(21), and letting $N \to \infty$, we obtain

$$E(W^{r,+}_{\infty 1}(\mathbb{R}), M, L_p(\mathbb{R})) \geq \sigma^{-r}\|\varphi_r(\cdot+\alpha_r)+\|\varphi_r\|_\infty\|_{L_{p'}[0,1]}, \tag{22}$$

for $p = 1, 2, \infty$. Thus, by letting $\sigma = n$ be a natural number in (22), we give the lower estimate $\widetilde{d_n}(W^{r,+}_{p1}(\mathbb{R}), L(\mathbb{R}))$, for the cases $p = 1, 2, \infty$. Combining these results with what we discuss in the previous section, we complete the proof of the theorem.

Acknowledgments. This project was supported partially by the National Natural Science Foundation of China (No. 19671012) and by the Doctoral Programme Foundation of Institution of Higher Education of National Education Committee of China. This paper was completed by the author during his visit at Wright State University, U.S.A., from September 1997 to June 1998.

References

1. Kouneichuk, N. P., *Exact Constants in Approximation Theory*, translated from the Russian by Ivanov, K., Cambridge University Press, Cambridge, 1991.
2. Li, Chun, Infinite dimensional widths in the spaces of Functions II, J. Approx. Theory **69** (1992), 15–34.
3. Ligun, A. A., Exact inequalities for perfect splines and its applications, Izv. Vyssh. Uchben. Zaved. Mat. **5** (1985), 32–38.
4. Liu, Yongping, Infinite dimensional widths and optimal recovery of some smooth function classes of $L_p(R)$ in the metric $L(R)$, Analysis Mathematica **19** (1993), 169–182.
5. Liu, Y. and Y. S. Sun, Infinite-dimensional widths and optimal interpolation problems on Sobolev-Wiener spaces, Sci. Sinica Ser. A **6** (1992), 582–591.
6. Motornyi, V. P., On the best quadrature formula of the form $\sum p_k f(x_k)$ for some classes of differentiable periodic functions, (in Russian), Izv. Akad. Nauk SSSR Ser. Mat. **38(3)** (1974), 583–614.
7. Sun, Y. S., *Approximation Theory of Functions I*, Beijing Normal University Press, Beijing, 1989. (In Chinese).
8. Sun, Y. S. and G. Fang, *Approximation Theory of Function II*, Beijing Normal University Press, Beijing, 1990. (In Chinese).
9. Sun, Y. S. and C. Li, Best approximation of some classes of smooth functions on the whole real axis by cardinal splines of higher order (in Russian), Mat. Zametki **48:4** (1990), 100–109.
10. Sun, Y. S. and Y. P. Liu, N-widths for some classes of periodic functions with boundary conditions, Approx. Theory Appl. **8** (1992), 21–27.

Yongping Liu
Department of Mathematics
Beijing Normal University
Beijing 100875
People's Republic of China
ypliu@bnu.edu.cn

Rational Approximation in Logarithmic Capacity of Meromorphic Maps from $\mathbb{C}^n$ to $\mathbb{C}^m$

Clement Lutterodt and Stanley Einstein-Matthews

Abstract. We investigate the convergence behavior of diagonal rational approximation to a class of meromorphic maps in a balanced domain $\mathcal{G}$ in $\mathbb{C}^n$ using logarithmic capacity defined by the Robin's constant. In this paper all series expansions of holomorphic and polynomial maps and functions are given in terms of homogeneous polynomials.

§1. Introduction

The main result of this paper, Theorem 4.7, gives a criterion for convergence in logarithmic capacity of diagonal rational sequences approximating a certain class of meromorphic maps defined on a balanced domain in $\mathbb{C}^n$ with their ranges in $\mathbb{C}^m$. The description of the rest of the paper is self evident from the special subheadings. The approach taken in this paper contrasts with that used in [4].

§2. Meromorphic Maps and Index Interpolation Set

Let $\mathcal{G}$ be a domain in $\mathbb{C}^n$ and $\mathcal{D}$ a balanced relatively compact subdomain of $\mathcal{G}$. A map $F \colon \mathcal{G} \to \mathbb{C}^m$ is said to be meromorphic, if its restriction to a dense subset W of $\mathcal{G}$, $F|_W := F_W$, is holomorphic and in addition satisfies:
(1) The closure $\overline{Gra(F_W)}$ of the graph of F is an analytic subset of $\mathcal{G} \times \mathbb{C}^m$;
(2) the restriction of the natural projection map $Proj : \mathcal{G} \times \mathbb{C}^m \to \mathcal{G}$ to $\overline{Gra(F_W)}$, i.e., $Proj|_{\overline{Gra(F_W)}} : \overline{Gra(F_W)} \to \overline{\mathcal{D}}$ is proper, (see [5]).

We let $\mathcal{O}^m(W)$ denote the space of holomorphic maps on W. This means that $F_W := (f_{1W}, f_{2W}, ..., f_{mW}) \in \mathcal{O}^m(W)$, where the component functions, $f_{jW}, 1 \leq j \leq m$, of F_W, are holomorphic functions. We assume throughout this paper that the origin $0 \in \mathbb{C}^n$ is contained in W and that in some neighborhood U_0 of the origin, each component function f_{jW} has a power series represntation: $f_{jW}(z) = \sum_{\lambda=0}^{\infty} \Phi_{j\lambda}(z)$, with each $\Phi_{j\lambda}, 1 \leq j \leq m$, a homogeneous polynomial on $\mathbb{C}^n$.

Let $\mathcal{M}^m(\overline{\mathcal{D}})$ be the space of meromorphic maps in an open neighborhood of $\overline{\mathcal{D}}$ satisfying the conditions (1) and (2) given above. We let $\mathcal{M}_{[1]}^m(\overline{\mathcal{D}})$ denote a subspace of $\mathcal{M}^m(\overline{\mathcal{D}})$ satisfying the following properties:

(a) Each element of the subspace is holomorphic at $z = 0 \in \mathbb{C}^n$;

(b) Each component function f_j of F in the subspace has its polar set in $\mathcal{G} \setminus W$ which is completely determined by the zero set of some nonhomogeneous polynomial q_{jw_j}, of highest degree w_j, such that $q_{jw_j} f_j \in \mathcal{O}(\overline{\mathcal{D}}), 1 \leq j \leq m$, i.e., $q_{jw_j} f_j$ is holomorphic in an open neighborhood of $\overline{\mathcal{D}}$ for each j.

Let $\mathbf{I} = \mathbb{N} \cup \{0\}$, where $\mathbb{N}$ is the set of natural numbers. Set $\mathbf{I}^n = \mathbf{I} \times ... \times \mathbf{I}$, n copies of $\mathbf{I}$. We introduce the following partial ordering on $\mathbf{I}^n$. For any $\alpha = (\alpha_1, \alpha_2, ..., \alpha_n)$ and $\beta = (\beta_1, \beta_2, ..., \beta_n) \in \mathbf{I}^n$; $\alpha \preceq \beta \Leftrightarrow \alpha_j \leq \beta_j$, for $j = 1, ..., n$. Let $|\alpha| = \sum_{j=1}^{n} \alpha_j$ and denote by $\mathbf{E}_\mu^n := \{\alpha \in \mathbf{I}^n : 0 \leq |\alpha| \leq \mu, \mu \in \mathbf{I}\}$, a subset of $\mathbf{I}^n$ having the following properties:

(I) $\mathbf{E}_0^n = \{(0, ..., 0)\} \subset \mathbf{E}_\mu^n, \forall \mu \geq 1$;

(II) $\mathbf{E}_\lambda^n \subset \mathbf{E}_\mu^n$, for $\lambda, \mu \in \mathbf{I} \Leftrightarrow 0 \leq \lambda \leq \mu$;

(III) $|\mathbf{E}_\mu^n| := \binom{n+\mu}{n}$ represents the cardinality of $\mathbf{E}_\mu^n$.

The following index set, also a subset of $\mathbf{I}^n$, facilitates the definition of a rational approximant to some holomorphic map near $z = 0$.

Definition 2.1. A subset $\mathbf{E}_{\mu\nu}^n \subset \mathbf{I}^n$ is called an index interpolation set if

(1) $\mathbf{E}_\mu^n \subset \mathbf{E}_{\mu\nu}^n$ for each $\mu, \nu \in \mathbf{I}$;

(2) $\beta \in \mathbf{E}_{\mu\nu}^n \Longrightarrow \alpha \in \mathbf{E}_{\mu\nu}^n, \forall\ 0 \preceq \alpha \preceq \beta$;

(3) $\exists \lambda_{\mu\nu}$ with $\mu + 1 \leq \lambda_{\mu\nu} \leq \mu + \nu$, $\nu > 1$, such that
$$\binom{n+\lambda_{\mu\nu}-1}{n} \leq |\mathbf{E}_{\mu\nu}^n| \leq \binom{n+\lambda_{\mu\nu}}{n} \text{ and } |\mathbf{E}_{\mu\nu}^n| \leq \binom{n+\mu}{n} + \binom{n+\nu}{n} - 1.$$

We say that $\mathbf{E}_{\mu\nu}^n$ is maximal if $|\mathbf{E}_{\mu\nu}^n| \geq \binom{n+\mu}{n} + \binom{n+\nu}{n} - 1$.

Let $\mathcal{R}_{\mu\nu}$ denote the family of rational functions, $\{R_{\mu\nu} := \frac{P_\mu}{Q_\nu}\}$ with P_μ and Q_ν nonhomogeneous polynomials of degrees atmost μ and ν respectively. P_μ and Q_ν are relatively prime polynomials in $\mathbb{C}^n$. Note that the relative primeness of P_μ and Q_ν breaks down on the subvariety $\mathcal{A}(P,Q) := \{z \in \mathbb{C}^n : Q_\nu(z) = P_\mu(z) = 0\}$ in $\mathbb{C}^n$ of codim ≥ 2.

The concept of rational approximants to holomorphic map given in the following definitions is achieved componentwise.

Definition 2.2. Let U be an open neighborhood of $z = 0$ in a dense subset W of the domain $\mathcal{G}$ and let $F_W \in \mathcal{O}^m(W \cap U)$. We say that $\frac{P_{\mu_j}}{Q_{\nu_j}} \in \mathcal{R}_{\mu_j \nu_j}$ is a jth component rational approximant to F_W if it is a rational approximant at $z = 0$ to the jth component function f_{jW} of F_W and the following conditions hold:

(1) $\partial^{|\alpha_j|}(Q_{\nu_j}(\xi)f_{jW}(\xi) - P_{\mu_j}(\xi))|_{\xi=0} = 0$, $\forall \alpha_j \in \mathbf{E}_{\mu_j}^n$;
(2) $\partial^{|\alpha_j|}(Q_{\mu_j}(\xi)f_{jW}(\xi))|_{\xi=0} = 0)$, $\forall \alpha_j \in \mathbf{E}_{\mu_j \nu_j}^n \setminus \mathbf{E}_{\mu_j}^n$;
(3) $\mathbf{E}_{\mu_j \nu_j}^n$ is maximal.

Here $\partial^{|\alpha_j|} = \frac{\partial^{|\alpha_j|}}{\partial \xi_1^{\alpha_{1j}} \ldots \partial \xi_m^{\alpha_{mj}}}$ and $|\alpha_j| = \sum_{k=1}^m \alpha_{jk}$. The condition (3) in the above definition guarantees the unicity of each component rational approximant to the map F_W at $z = 0$ with respect to a choice of $\mathbf{E}_{\mu_j \nu_j}^n$ that is maximal. (See [1].) We shall call this type of rational approximants unisolvent component rational approximants and denote them by

$$\pi_{\mu_j \nu_j}(z) = \frac{P_{\mu_j \nu_j}(z)}{Q_{\mu_j \nu_j}(z)} \quad \forall z \in \mathcal{D} \setminus \{\zeta \in \mathbf{C}^n : P_{\mu_j \nu_j}(\zeta) = Q_{\mu_j \nu_j}(\zeta) = 0\}.$$

§3. Some Useful Estimates

For each $F \in \mathcal{M}_{[1]}^m(\overline{\mathcal{D}})$, we can find a nonhomogeneous polynomial $q_{jw_j}(z)$ of degree w_j which determines the polar set of the jth component f_j of F such that $f_j q_{jw_j} \in \mathcal{O}^m(\overline{\mathcal{D}})$ for $j = 1, 2, \ldots, m$. Consequently,

$$H_{\mu_j \nu_j}(z) = Q_{\mu_j \nu_j}(z) f_j(z) q_{jw_j}(z) - P_{\mu_j \nu_j}(z) q_{jw_j}(z)$$

is a holomorphic function in some open neighborhood of $\overline{\mathcal{D}}$, with $\nu_j \leq \mu_j$, $1 \leq j \leq m$. From now on, we shall confine ourselves to the diagonal sequences of the component rational approximants $\{\pi_{\mu_j \mu_j}\}_{j=1}^m$ which are also unisolvent. Next we provide an estimate which plays a crucial role in every type of convergence associated with the rational approximants we consider in $\mathbf{C}^n$.

Theorem 3.1. Fix ϵ, $0 < \epsilon < 1$, and define $\mathcal{D}_\epsilon := \{z \in \mathcal{D} : \text{dist}(z, \partial \mathcal{D}) > \epsilon\}$ such that $\mathcal{D}_\epsilon \subset \mathcal{D}$ and set $\rho_\epsilon := \text{dist}(0, \partial \mathcal{D}_\epsilon) < \delta := \text{dist}(0, \partial \mathcal{D})$. Let $\sigma_\epsilon := \frac{\rho_\epsilon}{\delta} < 1$. Then

$$\max_j \|H_{\mu_j \mu_j}(z)\|_{\overline{\mathcal{D}_\epsilon}} \leq \frac{M}{\epsilon} \sigma_\epsilon^{\mu+w+1},$$

where $\mu + \omega := \min_j(\mu_j + \omega_j)$ and $M := \sup_{\partial \mathcal{D}}\{\max_j |H_{\mu_j \mu_j}(z)|\}$.

The following three Lemmas are essential in the proof of Theorem 3.1. The first of the Lemmas is due to Rudin [7].

Lemma 3.2. *Let $\mathcal{D} \subset \mathbf{C}^n$ be a balanced domain. Let $\{F_l(z)\}$ be a sequence of homogeneous polynomials of degrees $l = 0, 1, \ldots$. Suppose further that for each $l = 0, 1, \ldots, \sup_l |F_l(z)| < \infty$, for every $z \in \mathcal{D}$. Then $\sum_{l \geq 0} F_l(z)$ is compactly convergent in $\mathcal{D}$.*

Next, we introduce the notion of a slice of holomorphic functions in $\mathbf{C}^n$, expanded formally in terms of homogeneous polynomials in the form: $F(z) = \sum_{s \geq 0} F_s(z)$, where F_s are homogeneous polynomials of degree s. The slice of $F(z)$ is defined to be the function $F^\lambda(z) := F(\lambda z)$, where $\lambda \in \mathbf{C}$. Formally, $F^\lambda(z) = \sum_{s \geq 0} F_s(z) \lambda^s$. Hence we get

$$\frac{d^s}{d\lambda^s} F^\lambda(z)|_{\lambda=0} = F_s(z).s! = \sum_{|\alpha|=s} (\partial^{|\alpha|} F(\xi)|_{\xi=0}) z^\alpha.$$

We expand each of the holomorphic functions, $H_{\mu_j \mu_j}(z), 1 \leq j \leq m$, in an open neighborhood of $\overline{\mathcal{D}}$ as a jth component power series:

$$H_{\mu_j \mu_j}(z) = \sum_{k \geq 0} F_{k\mu_j \mu_j}(z), \quad 1 \leq j \leq m,$$

where $F_{k\mu_j \mu_j}(z) = \frac{1}{k!} \sum_{|\alpha|=k} (\partial^{|\alpha|} H_{\mu_j \mu_j}(\xi)|_{\xi=0}) z^\alpha$ is a homogeneous polynomial of degree k.

Lemma 3.3. *For each $j, 1 \leq j \leq m$, we let $F_{k\mu_j \mu_j}(z) = F_{k\mu_j \mu_j}$. Then*

$$F_{k\mu_j \mu_j} = \begin{cases} 0, & 0 \leq k \leq \mu_j + w_j; \\ \frac{1}{k!} \sum_{|\alpha|=k} (\partial^{|\alpha|} (Q_{\mu_j \mu_j}(\xi) f_j(\xi) q_{jw_j}(\xi))_{\xi=0}) z^\alpha, & k \geq \mu_j + w_j + 1. \end{cases}$$

Proof: This is achieved via the use of the slice function of $H_{\mu_j \mu_j}(z)$ and a careful application of Leibnitz's rule together with Definition 2.2, (1) and (2). $\square$

Lemma 3.4. *Let $F_{k\mu_j w_j}(z)$ be as given in Lemma 3.3. Then*

$\sup_k |F_{k\mu_j w_j}(z)| < \infty$, *for $k \geq \mu_j + w_j + 1$ and $z \in \overline{\mathcal{D}}$.*

For a detailed proof of Lemma 3.4 and Theorem 3.1, see [1].

§4. Logarithmic Capacity and Convergence

In this section we develop the background from the theory of plurisubharmonic functions needed to define logarithmic capacity. A detailed exposition of this material can be found in Siciak's lecture notes [8] and Klimek [2].

Definition 4.1. *A function u with values in $[-\infty, +\infty)$ defined in an open set $X \subset \mathbf{C}^n$ is called plurisubharmonic if*

(1) *u is upper semi-continuous;*

(2) for arbitrary z and w in $\mathbb{C}^n$, the function $\tau \mapsto u(z + \tau w)$ is subharmonic in an open subset $\mathbf{D}$ of $\mathbb{C}$ where it is defined.

Definition 4.2. Let $Psh(\mathbb{C}^n)$ be the convex $\mathbb{R}_+$-cone of plurisubharmonic functions on $\mathbb{C}^n$. Let $\mathcal{L} := \mathcal{L}(\mathbb{C}^n)$ represent the class of all plurisubharmonic (psh) functions u on $\mathbb{C}^n$ satisfying the condition:

$$u(z) \leq \log(1 + ||z||), \quad \text{as } ||z|| \to \infty.$$

For any bounded set $E \subset \mathbb{C}^n$, define the $\mathcal{L}$-extremal psh function of E by setting

$$V_E(z) := \sup\{v(z) \in \mathcal{L} : v(\zeta) \leq 0, \forall \zeta \in E\}.$$

Definition 4.3. A set $E \subset \mathbb{C}^n$ is pluripolar if for every point $w \in E$ there is an open set U containing w, and an u in $Psh(U)$, not identically equal to $-\infty$, such that $E \cap U \subset \{u = -\infty\}$.

Now let $K \subset \mathbb{C}^n$ be any nonpluripolar compact subset. The $\mathcal{L}$-extremal function associated with K is

$$V_K(z) := \sup\{v(z) \in \mathcal{L} : v(z) \leq 0, z \in K\},$$

and let $V_K^*(z) = \limsup_{\zeta \to z} V_K(\zeta)$ be the upper semi-continuous regularization of V_K. The function V_K is in general not smooth on $\mathbb{C}^n \setminus K$ when $n > 1$. It is a theorem of Siciak, that either $V_K^* \equiv +\infty$, in which case the set K is pluripolar, or else V_K^* is plurisubharmonic and $V_K^* \in \mathcal{L}$. If V_K is already continuous then $V_K = V_K^* \in \mathcal{L}$.

Definition 4.4. Let $K \subset \mathbb{C}^n$ be a compact nonpluripolar set. The Robin constant $\gamma(K)$ of the compact set K is defined by:

$$\gamma(K) := \lim_{||z|| \to \infty}(V_K^*(z) - \log(1 + ||z||)).$$

The logarithmic capacity, also known as the $\mathcal{L}$-capacity in $\mathbb{C}^n$, is defined by

$$\text{Cap}_{\mathcal{L}}(K) := \exp(-\gamma(K)).$$

From the point of view of ease of application we introduce an equivalent description of logarithmic capacity in $\mathbb{C}^n$ due to Plésniak [6] (see also [2]).

Definition 4.5. Let $K \subset \mathbb{C}^n$ be a fixed compact set. For any set $E \subset \mathbb{C}^n$, define its capacity by

$$\alpha(E) := \alpha_K(E) = (\exp ||V_E^*||)^{-1}.$$

The capacity $\alpha_K(E)$ has the following properties:
(i) $\alpha_K(E) \geq 0$;
(ii) If $E \subset F$, then $\alpha_K(E) \leq \alpha_K(F)$.

It is related to the logarithmic capacity by the following result due to Taylor [9] (see also [3]):

Proposition 4.6. Let $E \subset \mathbb{C}^n$ be a compact nonpluripolar set. Let V_E^* be the upper semi-continuous regularization of V_E with $V_E^* \in \mathcal{L}$. Then for some constant $M > 0$ and $\delta \in]0,1[$, the Plésniak capacity $\alpha(E) = \exp(-\|V_E^*\|)$ is related to the logarithmic capacity $\text{Cap}_{\mathcal{L}}(E)$ by the inequality :

$$\alpha(E) \leq \text{Cap}_{\mathcal{L}}(E) \leq M(\alpha(E))^\delta,$$

where $\text{Cap}_{\mathcal{L}}(E) = \exp(-\gamma(E))$ and $\gamma(E)$ is the Robin constant.

Let $\eta > 0$ be given, and K be a compact nonpluripolar and holomorphically convex subset of $\mathbb{C}^n$. Set $X = K \cap \overline{\mathcal{D}}$, where $\mathcal{D}$ is a domain in $\mathbb{C}^n$. Define for each j, $1 \leq j \leq m$,

$$\Omega_{\eta,\mu_j} := \left\{ z \in X : |f_j(z) - \pi_{\mu_j\mu_j}(z)|^{1/\mu_j} > \frac{1}{\eta} \right\}.$$

By expressing $F = (f_1, ..., f_m)$ and $\Pi_{MM} = (\pi_{\mu_1\mu_1}, ..., \pi_{\mu_m\mu_m})$, we let the norm on $\mathbb{C}^m$ with respect to K be $\|\|\cdot\|\|_K = \max_{1 \leq j \leq m} \|\cdot\|_K$, i.e., the maximal component norm. Then the set

$$\hat{\Omega}_{\eta M} := \left\{ z \in X : \|\|F(z) - \Pi_{MM}(z)\|\|_{\overline{\mathcal{D}}}^{1/\mu} > \frac{1}{\eta} \right\},$$

is contained in $\bigcup_{j=1}^m \Omega_{\eta,\mu_j}$, where $\mu = \min_{1 \leq j \leq m} \mu_j$.

Now recall that the polar set of a meromorphic map on $\overline{\mathcal{D}}$ is contained in $\overline{\mathcal{D}} \backslash W$, for any $\mathcal{D} \subset\subset \mathcal{G}$ and dense subset W. The polar set of the map F may be characterized in terms of the graph of F by using the restriction of the natural projection map

$$\text{Proj}|_{\overline{Gra(F_W)}} : \overline{Gra(F_W)} \to \overline{\mathcal{D}} \quad \text{as} \quad (\text{Proj}|_{\overline{Gra(F_W)}}(\overline{Gra(F_W)})) \cap \overline{\mathcal{D}} \backslash W.$$

This can be expressed explicitly in terms of the nonhomogeneous q_{jw_j} as

$$\bigcup_{j=1}^m \{z \in \mathbb{C}^n : q_{jw_j}(z) = 0\} \cap \overline{\mathcal{D}}.$$

Theorem 4.7. Let $F \in \mathcal{M}_{[1]}^m(\overline{\mathcal{D}})$ be such that its polar set is given as above. Suppose componentwise $\{\pi_{\mu_j\mu_j}(z)\}_{\mu_j}$ is a sequence of diagonal rational unisolvent approximants to the jth-component f_j of F at $z = 0$. Then for $\eta \in (0,1)$ and $X = \overline{\mathcal{D}} \cap K$, the logarithmic capacity of $\hat{\Omega}_{\eta M}$ satisfies

$$\text{Cap}_{\mathcal{L}}(\hat{\Omega}_{\eta M}) \leq mC\eta,$$

for some constant $C > 1$.

Proof: For each j, $1 \leq j \leq m$, that the set Ω_{η,μ_j} satisfies $\text{Cap}_{\mathcal{L}}(\Omega_{\eta,\mu_j}) \leq C_j\eta$ with $C_j > 1$ is proved in [1]. The desired result follows from taking $C = \max_j\{C_j\}$ and noting that

$$\hat{\Omega}_{\eta M} \subset \bigcup_{j=1}^m \Omega_{\eta,\mu_j}. \quad \square$$

References

1. Einstein-Matthews, S. M. and C. H. Lutterodt, Approximation and interpolation on complex algebraic varieties in a pseudoconvex domain, to appear.
2. Klimek, M., *Pluripotential Theory*, LMS Monographs New Series, Oxford Science Publications, Oxford, 1991.
3. Levenberg, N. and B. A. Taylor, Comparison of capacities in $\mathbb{C}^n$, Lecture Notes in Mathematics **1094** (1984), 162–201.
4. Lutterodt, C. H., Rational approximants to meromorphic maps in $\mathbb{C}^n$, Approximation Theory **IV** (1983), 593–598.
5. Noguchi, J. and T. Ochiai, *Geometric Function Theory in Several Complex Variables*, Translations of Mathematical Monographs **80**, Amer. Math. Soc., Providence, 1984.
6. Plésniak, W., Characterization of quasi-analytic functions of several variables by rational approximation, Annales Polos. Mat **27** (1973), 149–157.
7. Rudin, W. *Function Theory in the Unit Ball of* $\mathbb{C}^n$, Springer Verlag, N.Y., 1980.
8. Siciak, J. *Extremal Plurisubharmonic Functions and Capacities on* $\mathbb{C}^n$, Sophia Kokyuroku Lectures in Mathematics **14**, Tokyo, 1982.
9. Taylor, B. A., An estimate for an extremal pluri-subharmonic functions on $\mathbb{C}^n$, Lecture notes in Mathmatics **1028** (1983), 318–328.

C. Lutterodt
Department of Mathematics
Howard University
Washington DC 20059
clutterodt@fac.howard.edu

S. Einstein-Matthews
Department of Mathematics
Howard University
Washington DC 20059
smem@scs.howard.edu

Best Approximation by Quadratic Algebraic Functions

Allan W. McInnes

Abstract. The quadratic algebraic function approximation is an approximation to a given function from the class of functions defined implicitly by a quadratic algebraic equation. Such approximations are commonly determined by collocation and are included in the general class of Hermite-Padé approximations. This paper considers the problem of finding a best uniform approximation to a given function from the class of quadratic algebraic functions.

§1. Introduction

The problem of approximating a real-valued function by an algebraic function has been considered by a number of authors [1, 2, 7] in recent years. However the approximation has invariably been determined by collocation, either at a single point [2, 3, 4, 7] or at a set of distinct points [6].

This paper considers the problem of determining the approximation using the criteria of best aproximation with the uniform norm. In an approach similar to earlier work [4, 6], this paper considers the approximation by quadratic algebraic functions. The extension of these results to the case of general algebraic functions will be the subject of subsequent work.

An outline of algebraic function approximation is considered in the next section. Then the classical result for best uniform approximation by polynomials is recalled, and the extension to approximation by rational functions is recast in terms of the notation used in algebraic function approximation.

With this background, the equioscillation criteria is extended to the algebraic functions, and a new condition on the algebraic form for the determination of the best uniform approximation is obtained. This condition is validated in the last two sections with the outline of a computational algorithm and a numerical example.

§2. Algebraic Function Approximation

The objective of this paper is to seek approximations to a given function from the class of algebraic functions.

Definition 1. *Let $\boldsymbol{n} = (n_0, n_1, \ldots, n_p)$ where for $i = 0(1)p$, the $n_i \geq -1$ are integers and $p \geq 1$. The class $\mathcal{A}_{\boldsymbol{n},p}$ of $\boldsymbol{n}$ algebraic functions of degree p is defined as those functions $Q(x)$ which satisfy*

$$P_{\boldsymbol{n},p}(Q, x) \equiv \sum_{i=0}^{p} a_i(x) Q(x)^i = 0, \tag{1}$$

where $a_i(x)$ is an algebraic polynomial with degree $(a_i(x)) \leq n_i$ for $i = 0(1)p$, with $a_i(x) \not\equiv 0$ for at least one value of i such that $i = 1(1)p$. (By convention a polynomial of degree -1 is identically zero.)

The case $p = 1$ gives rational functions and the case $p = 2$ gives quadratic algebraic functions. Note that this means that we can obtain approximations with a branch point structure.

Consider the approximation of $f(x)$ (which to simplify matters is assumed to belong to $C[-1, 1]$) by an algebraic function of this form. The approximation is determined by seeking an algebraic form which is the best approximation to (1) in some appropriate sense.

Definition 2. *The function*

$$P_{\boldsymbol{n},p}(f, x) \equiv \sum_{i=0}^{p} a_i(x) f(x)^i \tag{2}$$

is called an $\boldsymbol{n}$ algebraic form of degree p for the function $f(x)$, where the polynomial coefficients $a_i(x), \boldsymbol{n}, p$ are as defined previously.

The polynomial coefficients $a_i(x)$ are determined so that this expression has the same form or structure as an algebraic function. In general, the subscripts $\boldsymbol{n}, p$ will be omitted when the context is clear.

For example, if $N + 1 = \sum_{i=0}^{p} (n_i + 1)$, then the best approximation to zero in the space $C^N[\{0\}]$ is determined by the collocation conditions

$$D^k P(f, x) = 0, \quad k = 0(1)N - 1 \tag{3}$$

where $D = d/dx|_{x=0}$.

These conditions determine the coefficients $a_i(x)$, which then define the algebraic function $Q(x)$ using (1) and initial conditions [7]. This is an example of the Hermite-Padé approximation, analogous for $p = 1$ to the Taylor polynomial approximation if $n_1 = 0$, or the Padé rational approximation if $n_1 > 0$.

§3. Best Uniform Approximation

The classical theorem for best uniform approximation by polynomials is the Chebyshev Equioscillation Theorem [5, 8].

Theorem 3. *Let $f \in C[-1,1]$. Then p_n^* is the best approximation on $[-1,1]$ to f from the class $\mathcal{P}_n$ of polynomials of degree at most n if and only if there exists an alternating set Y_{n+2} for $f - p_n^*$ consisting of $n+2$ points. That is,*

$$f(x_k) - p_n^*(x_k) = (-1)^k \lambda$$

for $x_k \in Y_{n+2}$ and $|\lambda| = \|f - p_n^*\|$.

There is an equivalent theorem for rational functions $r(x) = p(x)/q(x)$ where $p \in \mathcal{P}_m$, $q \in \mathcal{P}_n$, and $r \in \mathcal{R}(m,n)$ [8]. In this case the alternating set consists of up to $m+n+2$ points and

$$f(x_k) - r^*(x_k) = (-1)^k \lambda$$

for x_k in the alternating set and $|\lambda| = \|f - r^*\|$.

However, there are some additional provisions in this form of the theorem:
(1) $r^*(x)$ is bounded in the interval.
(2) $p^*(x)$ and $q^*(x)$ do not have any common factors.

These provisions are the problem in extending this theorem to algebraic functions of higher degrees. But previous work [7] on algebraic function approximation provides the appropriate formulation of this problem to realize this extension. In particular, the concept of a normal form is required.

Definition 4. *The algebraic form $P^*(f,x)$ of best approximation is called normal if*

$$\partial P^*(f,x)/\partial f \neq 0 \quad \text{for } x \in [-1,1].$$

In the rational case this condition ensures that $r^*(x)$ remains bounded in the interval, and in the quadratic case it ensures that the two branches of the algebraic function do not intersect in the interval.

The problem of non-normality in the case of approximations determined by collocation has been considered in [7]. It results in a deficiency index, leading to a qualitative "under-approximation" property. In order to avoid these additional complications the property of normality will be assumed for the approximations in this paper.

The rational function form of the theorem may be rewritten in terms of the terminology established in this paper.

Theorem 5. *If $f \in C[-1,1]$, then $r^*(x) = a_0^*(x)/a_1^*(x)$ is the best approximation to f from the class $\mathcal{A}_{\mathbf{n},1}$ if and only if there is an alternating set Y_{N+1} for $f - r^*$ consisting of $n_0 + n_1 + 2 = N+1$ points, provided $P^*(f,x)$ is normal. That is,*

$$f(x_k) - r^*(x_k) = (-1)^k \lambda$$

for $x_k \in Y_{N+1}$ and $|\lambda| = \|f - r^*\|$.

§4. B.U.A. by Quadratic Algebraic Functions

If the equioscillation theorem is rewritten in terms of the quadratic algebraic functions the following theorem is obtained.

Theorem 6. Let $f \in C[-1,1]$. Then Q^* is the best approximation to f from the class $\mathcal{A}_{n,2}$ of quadratic algebraic functions if and only if there is an alternating set Y_{N+1} for $f - Q^*$ consisting of $N + 1 = \sum_{i=0}^{2}(n_i + 1)$ points, provided $P^*(f,x)$ is normal. That is,

$$f(x_k) - Q^*(x_k) = (-1)^k \lambda$$

for $x_k \in Y_{N+1}$ and $|\lambda| = \|f - Q^*\|$.

However, this form of the theorem does not directly give a set of conditions for determining the approximating algebraic form $P^*(f,x)$. The conditions for the algebraic form are obtained by noting that

$$0 = P^*(Q^*, x_k) = P^*f\big|_{x=x_k} + (f - Q^*)\frac{\partial P^*}{\partial f}\bigg|_{x=x_k} + \frac{1}{2}(f - Q^*)^2 \frac{\partial^2 P^*}{\partial f^2}\bigg|_{x=x_k}$$

which gives

$$0 = P^*f\big|_{x=x_k} + (-1)^k \lambda \frac{\partial P^*}{\partial f}\bigg|_{x=x_k} + \frac{1}{2}\lambda^2 \frac{\partial^2 P^*}{\partial f^2}\bigg|_{x=x_k} \qquad (4)$$

for $x_k \in Y_{N+1}$ and $|\lambda| = \|f - Q^*\|$.

Note that if $p = 1$ then $\partial^2 P^*/\partial f^2 \equiv 0$ and $\partial P^*/\partial f|_{x=x_k} = a_1^*(x_k)$ which is nonzero if $P^*(f,x)$ is normal.

Thus, since $r^*(x)$ is defined by $P^*(r^*, x) \equiv a_0^*(x) + a_1^*(x)r^*(x) = 0$, dividing through by $a_1^*(x_k)$ in (4) gives the form of the condition in Theorem 5. Furthermore, it is clear that Theorem 3 is also a special case.

The details of this formulation will be the subject of a forthcoming report. In this paper an outline of a computational algorithm to utilize the condition (4) will be given.

§5. Computational Algorithm

The primary unknowns are the $N + 1$ coefficients of the coefficient polynomials $a_i^*(x)$ and the error λ. The problem would then be straightforward if the alternating set Y_{N+1} was known. However, this set is also an unknown for the problem. In the case of rational best approximation ($p = 1$), Werner [9] formulated the problem as an eigenvalue problem and determined the alternating set Y_{N+1} by iteration. In this paper, this approach is extended to the case ($p = 2$) of approximation by quadratic algebraic functions.

Select an initial set Y_{N+1}. Then (4) may be rewritten as

$$\sum_{i=0}^{2} a_i(x_k)(f(x_k) + (-1)^k \lambda)^i = 0, \quad k = 0(1)N. \qquad (5)$$

Best Approximation by Quadratic Functions

Let

$$X_s = \begin{bmatrix} 1 & x_0 & \cdots & x_0^s \\ \vdots & \vdots & & \vdots \\ 1 & x_N & & x_N^s \end{bmatrix},$$

$$F = \mathrm{diag}(f_0, \ldots, f_N),$$
$$G = \mathrm{diag}(1, -1, \ldots, (-1)^N),$$
$$E = \mathrm{diag}(\eta_0, \ldots, \eta_N),$$

where $\eta_j = 1/w'(x_j)$ with $w(x) = \Pi_{i=0}^{N}(x - x_i)$.

Let $\boldsymbol{a}_i$ be the vector of coefficients of the polynomial coefficient $a_i(x)$ for $i = 0, 1, 2$.

Thus (5) may be written

$$X_{n_0}\boldsymbol{a}_0 + (F + \lambda G)X_{n_1}\boldsymbol{a}_1 + (F + \lambda G)^2 X_{n_2}\boldsymbol{a}_2 = \boldsymbol{0}. \tag{6}$$

Effectively the vector $\boldsymbol{a}_0$ is eliminated by applying the $(n_0 + 1)$st divided difference which eliminates the polynomial coefficient $a_0(x)$.

Write $m = n_1 + n_2 + 1$ and let O denote the zero matrix. Then it follows that $X_m^T E X_{n_0} = O$. Multiplying (6) on the left by $X_m^T E$ gives

$$(A + \lambda B)\boldsymbol{a}_1 + (C + \lambda D + O)\boldsymbol{a}_2 = \boldsymbol{0}$$

where

$$A = X_m^T E F\, X_{n_1},$$
$$B = X_m^T E G\, X_{n_1},$$
$$C = X_m^T E F^2 X_{n_2},$$
$$D = 2 X_m^T E F G\, X_{n_2},$$
$$O = X_m^T E G^2 X_{n_2} \quad \text{since } G^2 = I.$$

Thus

$$([A : C] + \lambda [B : D]) \begin{bmatrix} \boldsymbol{a}_1 \\ \boldsymbol{a}_2 \end{bmatrix} = \boldsymbol{0}$$

is an eigenvalue problem.

Only one of these eigenvalues leads to a real, continuous approximation to the real continuous function $f(x)$, and the corresponding eigenvector gives the coefficient vectors $\boldsymbol{a}_1, \boldsymbol{a}_2$. The remaining coefficient vector $\boldsymbol{a}_0$ is obtained by solving (6). The vectors $(\boldsymbol{a}_0, \boldsymbol{a}_1, \boldsymbol{a}_2)$ define the algebraic function $Q(x)$ by means of equation (1).

The abscissas of the extrema of the error function $e(x) = f(x) - Q(x)$ serve as the new values for the $x_k \in Y_{N+1}$, and the computation is iterated, similar to the scheme in the rational approximation case.

The results may be best illustrated by means of an example.

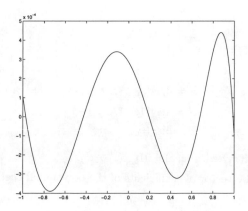

Fig. 1. Error function $e^x - Q_{1,2}(x)$.

§6. Numerical Example

Consider the best uniform approximation to $f(x) = \exp(x)$ on $[-1,1]$ by a $\mathbf{1} = (1,1,1)$ quadratic algebraic function. This result may be compared to the results obtained by Borwein [3] for the Hermite-Padé approximation determined by collocation at $x = 0$.

In this case $n_0 = n_1 = n_2 = 1$ and $N = 5$. The initial set Y_{N+1} was chosen as the abscissas of the extrema of the Chebyshev polynomial $T_5(x)$. That is

$$Y_6 = \{x_k : x_k = \cos(k-1)\pi/5, \ k = 1(1)6\}$$
$$= [-1, -0.8090, -0.3090, 0.3090, 0.8090, 1].$$

Solving the eigenvalue problem gives the eigenvalue $\lambda = 2.4429 \times 10^{-4}$, leading to the coefficients:

$$\mathbf{a}_0 = \begin{bmatrix} -0.6329 \\ -0.2083 \end{bmatrix} \ ; \ \mathbf{a}_1 = \begin{bmatrix} 0.0770 \\ -0.8092 \end{bmatrix} \ ; \ \mathbf{a}_2 = \begin{bmatrix} 0.5562 \\ -0.1730 \end{bmatrix}.$$

The extrema of the error function $e(x) = e^x - Q_{1,2}(x)$ are

$$[1, -3.9, 3.4, -3.3, 4.4, -1] \times 10^{-4}.$$

The abscissas of these extrema give the new set

$$Y_6 = [-1, -0.745, -0.113, 0.460, 0.875, 1].$$

The first iteration gives the coefficients

$$\mathbf{a}_0 = \begin{bmatrix} -0.5637 \\ -0.1865 \end{bmatrix} \ ; \ \mathbf{a}_1 = \begin{bmatrix} 0.0750 \\ -0.7153 \end{bmatrix} \ ; \ \mathbf{a}_2 = \begin{bmatrix} 0.4890 \\ -0.1519 \end{bmatrix}.$$

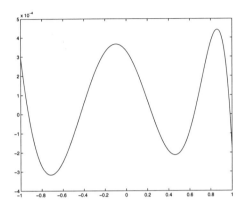

Fig. 2. Error function $e^x - Q_{1,2}(x)$ for first iteration.

The extrema of the error function are

$$[2.91, -3.164, 3.661, -2.118, 4.431, -1.83] \times 10^{-4}$$

whose abscissas form the new set

$$Y_6 = [-1, -0.716, -0.0964, 0.460, 0.862, 1].$$

The second iteration gives the coefficients

$$\boldsymbol{a}_0 = \begin{bmatrix} -0.5494 \\ -0.1819 \end{bmatrix} \; ; \; \boldsymbol{a}_1 = \begin{bmatrix} 0.0732 \\ -0.6971 \end{bmatrix} \; ; \; \boldsymbol{a}_2 = \begin{bmatrix} 0.4765 \\ -0.1418 \end{bmatrix}.$$

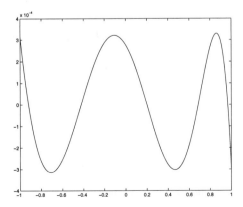

Fig. 3. Error function $e^x - Q_{1,2}(x)$ for second iteration.

In this case the plot of the error function (Figure 3) shows that there are $N + 1 = 6$ points at which the error function has approximately equal alternating extrema of a little over 3×10^{-4}.

These results may be compared to the Hermite-Padé approximation obtained by Borwein [3]. In this case

$$a_0 = (3,1) \; ; \; a_1 = (0,4) \; ; \; a_2 = (-3,1),$$

and the maximum error at $x = 1$ is 1.4×10^{-2}.

A more detailed account of this computation, together with further examples will be given in a subsequent report.

References

1. Baker, G. A. and P. R. Morris, *Padé Approximants Part II : Extensions and Applications*, Addison-Wesley, Reading MA, 1981.

2. Baker, G. A. and D. S. Lubinsky, Convergence theorems for rows of differential and algebraic Hermite-Padé approximants, J. Comput. Appl. Math. **18** (1987), 29–52.

3. Borwein, P. B., Quadratic Hermite-Padé approximations to the exponential function, Constr. Approx. **2** (1986), 291–302.

4. Brookes, R. G. and A. W. McInnes, The existence and local behaviour of the quadratic function approximation, J. Approx. Theory **62** (1990), 383–395.

5. Davis, P. J., *Interpolation and Approximation*, Blaisdell, Waltham, MA, 1963.

6. McInnes, A. W., Quadratic function approximation determined by collocation, *Approximation Theory VI*, C. Chui, L. Schumaker, and J. Ward (eds.), Academic Press, New York, 1989, 425–428.

7. McInnes, A. W., Existence and uniqueness of algebraic function approximations, Constr. Approx. **8** (1992), 1–21.

8. Rivlin, T. J., *An Introduction to the Approximation of Functions*, Blaisdell, Waltham, MA, 1969.

9. Werner, H., Rationale Tschebyscheff-Approximation, Eigenwerttheorie und Differenzenrechnung, Arch. Rational Mech. Anal. **13** (1963), 330–347.

Allan W. McInnes
Dept. of Mathematics
University of Canterbury
Christchurch, New Zealand
A.McInnes@math.canterbury.ac.nz

L_p Convergence of Lagrange-Type Interpolation in the Rational Space

G. Min

Abstract. This paper considers Lagrange-type interpolation in the rational space $\mathcal{P}_{n-1}(a_1,\ldots,a_n)$ with nonreal elements in $\{a_k\}_{k=1}^n \subset \mathbf{C} \setminus [-1,1]$ paired by complex conjugation. This Lagrange-type interpolation problem is based on the zeros of the Chebyshev polynomial of the first kind associated with the rational space $\mathcal{P}_n(a_1,\ldots,a_n)$. It is shown that mean convergence of the Lagrange-type interpolation holds when the poles stay outside an circle which contains the unit circle in its interior. It extends the Erdős-Feldheim theorem in classical polynomial interpolation.

§1. Introduction

Let

$$\mathcal{P}_m(a_1,\ldots,a_n) := \left\{ \frac{P(x)}{\prod_{k=1}^n (x-a_k)}, \quad P \in \mathcal{P}_m \right\}$$

with $\{a_k\}_{k=1}^n \subset \mathbf{C} \setminus [-1,1]$, where $\mathcal{P}_m$ is the set of all real algebraic polynomials of degree at most m. When all the poles $\{a_k\}_{k=1}^n$ are real and distinct, $\mathcal{P}_n(a_1, a_2,\ldots,a_n)$ is simply the algebraic span of

$$\left\{ 1, \frac{1}{x-a_1}, \frac{1}{x-a_2}, \ldots, \frac{1}{x-a_n} \right\}, \quad x \in [-1,1].$$

The explicit formulae for the Chebyshev polynomials for the rational space $\mathcal{P}_n(a_1, a_2,\ldots,a_n)$ with distinct real poles outside $[-1,1]$ are implicitly contained in Achiezer [1]. Recently, Borwein, Erdélyi, and Zhang [3] have derived analogous Chebyshev polynomials for $\mathcal{P}_n(a_1,\ldots,a_n)$. Moreover, they allow repeated poles and nonreal poles in this rational space with the nonreal poles in complex conjugate pairs. It was shown [3] that these Chebyshev polynomials preserve almost all the elementary properties of the classical Chebyshev polynomials. Recently, we constructed (see [6]) the generalized Lagrange interpolation using the zeros of Chebyshev polynomials of the first kind for $\mathcal{P}_n(a_1,\ldots,a_n)$. We showed that the corresponding Lebesgue constant is asymptotically of order $\ln n$ when the poles stay outside an interval which contains $[-1,1]$ in its interior for $\{a_k\}_{k=1}^n \subset \mathbb{R} \setminus [-1,1]$ being distinct. Therefore, the generalized Lagrange interpolation cannot converge uniformly to every continuous function on $[-1,1]$.

In this paper, we study mean convergence of this Lagrange-type interpolation. Here, we allow repeated poles and nonreal poles in the rational space $\mathcal{P}_n(a_1,\ldots,a_n)$ with the nonreal poles in complex conjugate pairs. We conclude that the corresponding L_p-convergence ($0 < p < \infty$) always holds for the continuous functions on $[-1,1]$ when the poles stay outside a circle which contains the unit circle in its interior. It extends the Erdős-Feldheim theorem for the classical polynomial interpolation (see [4, 12, 13]). For the studies of rational interpolation and its applications, the interested readers are referred to [5, 14, 15] and the references therein.

In this paper, we use $c(\alpha)$ and $c(\alpha,m)$ to denote positive constants depending only on α, and both α and m, respectively, but the values may be different from line to line. Also, denote

$$\|f\|_{v,p} := \left(\int_{-1}^1 \frac{1}{\sqrt{1-x^2}} |f(x)|^p \, dx \right)^{1/p}, \quad p > 0.$$

§2. Mean Convergence of Lagrange-type Interpolation

We use T_n and U_n to denote the corresponding Chebyshev polynomials of the first and second kinds for the rational space $\mathcal{P}_n(a_1,\ldots,a_n)$, respectively. Their construction can be found in [2, 3]. For the convenience, we include them here.

Given $\{a_k\}_{k=1}^n \subset \mathbb{C} \setminus [-1,1]$ such that the nonreal elements are paired by complex conjugation, so that $\mathcal{P}_n(a_1,a_2,\ldots,a_n)$ is a real rational space. We define the numbers $\{c_k\}_{k=1}^n$ by

$$c_k := a_k - \sqrt{a_k^2 - 1}, \quad |c_k| < 1, \tag{2.1}$$

throughout this paper, $\sqrt{a_k^2-1}$ will always be defined by (2.1).
Let

$$M_n(z) := \left(\prod_{k=1}^n (z-c_k)(z-\bar{c}_k) \right)^{1/2}, \tag{2.2}$$

where the square-root is so defined that $M_n^*(z) = z^n M_n(z^{-1})$ is analytic in a neighborhood of the closed unit disk, and let

$$f_n(z) := \frac{M_n(z)}{z^n M_n(z^{-1})}. \tag{2.3}$$

Then the Chebyshev polynomials of the first kind for the rational space

$\mathcal{P}_n(a_1, a_2, \ldots, a_n)$ is defined by

$$T_n(x) := \frac{f_n(z) + f_n^{-1}(z)}{2}, \quad x = \frac{z + z^{-1}}{2}, \quad |z| = 1, \tag{2.4}$$

and the Chebyshev polynomials of the second kind is defined by

$$U_n(x) := \frac{f_n(z) - f_n^{-1}(z)}{z - z^{-1}}, \quad x = \frac{z + z^{-1}}{2}, \quad |z| = 1. \tag{2.5}$$

It is shown in [2, 3] that these Chebyshev polynomials preserve almost all the elementary properties of the classical Chebyshev polynomials. Here we just state some of these properties which will be used later.

Theorem A. (cf. [3, Theorem 1.2, Corollary 4.9]). Let the nonreal elements in $\{a_k\}_{k=1}^\infty \subset \mathbb{R} \setminus [-1, 1]$ be paired by complex conjugation, and let $\{T_n\}_{n=1}^\infty$ and $\{U_n\}_{n=1}^\infty$ be Chebyshev polynomials of the first and second kinds with the rational space $\mathcal{P}_n(a_1, \ldots, a_n)$, respectively. Then

(a) $T_n \in \mathcal{P}_n(a_1, \ldots, a_n)$.

(b) $\|T_n\|_{[-1,1]} = \|\sqrt{1-x^2} U_n(x)\|_{[-1,1]} = 1$.

(c) $T_n(x)$ has exactly n zeros in $[-1, 1]$:

$$-1 < x_n < \cdots < x_1 < 1, \tag{2.6}$$

and

$$\sqrt{1 - x_k^2} U_n(x_k) = \epsilon(-1)^{k+1}, \quad k = 1, 2, \ldots, n, \tag{2.7}$$

where $\epsilon = 1$ or -1.

The conclusion that $T_n(x)$ has exactly n zeros can also be found in [11]. It should be mentioned that $\{T_n(x)\}_{n=1}^\infty$ are not orthogonal in general (cf. [3, Lemma 4.4]), this property is different from that of the classical Chebyshev polynomials. Let

$$B_n(x) := \sum_{k=1}^n \Re\left(\frac{\sqrt{a_k^2 - 1}}{a_k - x}\right) \quad (> 0). \tag{2.8}$$

$B_n(x)$ is called the Bernstein factor which plays an important role in [3]. Moreover, it is a convex function on $[-1, 1]$ and its maximum is attained at least at one of the endpoints ± 1 for $\{a_k\}_{k=1}^n \subset \mathbb{R} \setminus [-1, 1]$ (see [8]).

Let f be a function defined on $[-1, 1]$. We construct the Lagrange interpolation based on the zeros $\{x_k\}_{k=1}^n$ of $T_n(x)$ as follows:

$$L_n(f, x) := \sum_{k=1}^n f(x_k) l_k(x), \tag{2.9}$$

where $\{l_k(x)\}_{k=1}^n$ are the Lagrange fundamental functions:

$$l_k(x) := \frac{T_n(x)}{T_n'(x_k)(x-x_k)}, \quad k=1,\ldots,n.$$

From Theorem A (b) (c) and recalling $T_n'(x) = B_n(x)U_n(x)$, we have (see [6])

$$l_k(x) = \varepsilon \frac{(-1)^{k+1}\sqrt{1-x_k^2}T_n(x)}{B_n(x_k)(x-x_k)}, \quad k=1,\ldots,n. \tag{2.10}$$

It is easy to check that

$$L_n(f,x_k) = f(x_k), \quad k=1,\ldots,n,$$

and $L_n(f,x) \in \mathcal{P}_{n-1}(a_1,\ldots,a_n)$.

For the convenience of the statements of our results, we first introduce an assumption, which plays an important role in the proof of Lemma 3.1.

Definition 2.1. *Let the nonreal elements in $\{a_k\}_{k=1}^n \subset \mathbf{C} \setminus [-1,1]$ be paired by complex conjugation. If there exists some constant α such that*

$$|a_k| \geq \alpha > 1, \tag{2.11}$$

i.e, the poles must stay outside a circle which contains the unit circle in its interior, we say that $\{a_k\}_{k=1}^n \subset \mathbf{C} \setminus [-1,1]$ satisfy Assumption (A).

Remark. Recently, we have shown that the Assumption (A) is equivalent to the assertion: the orthonormal rational systems, which are based on $\mathcal{P}_n(a_1,\ldots,a_n)$ with the Chebyshev weight on the interval $[-1,1]$, are uniformly bounded on $[-1,1]$.

We use $E_n(f)$ to denote the best approximation of $f(x)$ from $\mathcal{P}_{n-1}(a_1,\ldots,a_n)$:

$$E_n(f) := \inf_{p_n \in \mathcal{P}_{n-1}(a_1,\ldots,a_n)} \|f - p_n\|_{[-1,1]}. \tag{2.12}$$

Now we state our main result.

Theorem 2.2. *Let $\{a_k\}_{k=1}^n \subset \mathbf{C} \setminus [-1,1]$ satisfy Assumption (A). Then for any $0 < p < \infty$,*

$$\|L_n(f) - f\|_{v,p} \leq c(p) E_n(f) \tag{2.13}$$

for $f \in C[-1,1]$.

When the nonreal elements of $\{a_k\}_{k=1}^\infty \subset \mathbf{C}\setminus[-1,1]$ are paired by complex conjugation, recall that $\{\mathcal{P}_{n-1}(a_1,\ldots,a_n)\}$ are dense in $C[-1,1]$ if and only if

$$\sum_{k=1}^\infty (1-|c_k|) = \infty,$$

where $\{c_k\}_{k=1}^\infty$ are defined by (2.1) (see [9], for the real and distinct $\{a_k\}_{k=1}^\infty$, and this can also be found in Achiezer [1, Problem 1, p. 254]). Then Theorem 2.2 yields

Corollary 2.3. Let the nonreal elements in $\{a_k\}_{k=1}^{\infty} \subset \mathbb{C} \setminus [-1,1]$ be paired by complex conjugation, and let $\{a_k\}_{k=1}^{\infty} \subset \mathbb{C} \setminus [-1,1]$ satisfy Assumption (A). Then

$$\|L_n(f) - f\|_{v,p} \to 0, \quad (n \to \infty)$$

for $f \in C[-1,1]$.

§3. Proofs

We need several lemmas.

Lemma 3.1. Let $\{a_k\}_{k=1}^{n} \subset \mathbb{C} \setminus [-1,1]$ satisfy Assumption (A). Then

$$\sum_{k=1}^{n} |l_k(x)|^m \le c(\alpha, m), \quad m = 2, 3, \ldots. \tag{3.1}$$

Proof: See [6, Lemma 3.4]. □

Lemma 3.2. Let the nonreal elements in $\{a_k\}_{k=1}^{n} \subset \mathbb{C} \setminus [-1,1]$ be paired by complex conjugation, and let $\rho_1, \ldots, \rho_k$ be distinct integers between 1 and n. Then, for even number k, we have

$$\int_{-1}^{1} \frac{1}{\sqrt{1-x^2}} l_{\rho_1}(x) \cdots l_{\rho_k}(x)\, dx = 0. \tag{3.2}$$

Proof: Note that

$$\prod_{i=1}^{s} \frac{1}{x-x_i} = \sum_{i=1}^{s} \frac{A_i}{x-x_i},$$

where

$$A_i = \prod_{j=1, j \ne i}^{s} \frac{1}{x_i - x_j}.$$

Hence, it follows from (2.10), that

$$l_{\rho_1}(x) \cdots l_{\rho_k}(x) = \varepsilon^k T_n^k(x) \prod_{j=1}^{k} \frac{(-1)^{\rho_j+1}\sqrt{1-x_{\rho_j}^2}}{B_n(x_{\rho_j})} \frac{1}{x - x_{\rho_j}}$$

$$= \varepsilon^k T_n^k(x) \prod_{j=1}^{k} \frac{(-1)^{\rho_j+1}\sqrt{1-x_{\rho_j}^2}}{B_n(x_{\rho_j})} \left(\sum_{i=1}^{k} \frac{A_{\rho_i}}{x - x_{\rho_i}} \right)$$

$$= \varepsilon^k \sum_{i=1}^{k} A_{\rho_i} \prod_{j=1, j \ne i}^{k} \frac{(-1)^{\rho_j+1}\sqrt{1-x_{\rho_j}^2}}{B_n(x_{\rho_j})} l_{\rho_i}(x) T_n^{k-1}(x).$$

Therefore, to prove (3.2), it is sufficient to prove that

$$\int_{-1}^{1} \frac{1}{\sqrt{1-x^2}} l_j(x) T_n^{k-1}(x)\, dx = 0, \quad j = 1, \ldots, n. \tag{3.3}$$

First we show that

$$\int_{-1}^{1} \frac{1}{\sqrt{1-x^2}} \frac{T_n^{k-1}(x)}{(x-a_m)^\rho} \, dx = 0, \quad \rho = 1, \ldots, r, \tag{3.4}$$

if $a_k \in \mathbb{R}$ with the multiplicity r in $\{a_k\}_{k=1}^n$. Using the transformation $x = (z + z^{-1})/2$, we have

$$\int_{-1}^{1} \frac{1}{\sqrt{1-x^2}} \frac{T_n^{k-1}(x)}{(x-a_m)^\rho} \, dx$$

$$= \frac{1}{2^{k-\rho}} \int_{C^+} \left(f_n(z) + f_n^{-1}(z)\right)^{k-1} \frac{1}{(-c_m - c_m^{-1} + z + z^{-1})^\rho} \frac{dz}{iz},$$

$$= \frac{1}{2^{k-1-\rho}i} \sum_{j=0}^{k-1} \binom{k-1}{j} \int_{C^+} f_n^{k-1-2j}(z) \frac{z^{s-1} dz}{((c_m - z)(c_m^{-1} - z))^\rho}$$

where C^+ is the upper half circle.

Note that since $k-1$ is an odd number, then applying the transformation $w = 1/z$ to convert the terms $f_n^{-1}(z), \ldots, f_n^{-(k-1)}(z)$ to the lower half circle C^-, we conclude that

$$\int_{-1}^{1} \frac{1}{\sqrt{1-x^2}} \frac{T_n^{k-1}(x)}{(x-a_m)^\rho} \, dx$$

$$= \frac{1}{2^{k-1-\rho}i} \sum_{j=0}^{\lfloor \frac{k}{2} \rfloor - 1} \binom{k-1}{j} \int_{C} f_n^{k-1-2j}(z) \frac{z^{\rho-1} dz}{((c_m - z)(c_m^{-1} - z))^\rho}.$$

Recall that the function

$$f_n^{k-1-2j}(z) \frac{z^{\rho-1}}{((c_m - z)(c_m^{-1} - z))^\rho}$$

is analytic in the unit disk for $\rho = 1, \ldots, r$ and $j = 0, \ldots, \lfloor \frac{k}{2} \rfloor - 1$, then (3.4) immediately follows.

For the case $\Im a_i \neq 0$ with the multiplicity q in $\{a_k\}_{k=1}^n$, we conclude, by the same fashion, for $\gamma = 1, \ldots, q$, that

$$\int_{-1}^{1} \frac{1}{\sqrt{1-x^2}} \frac{T_n^{k-1}(x)(ax+b)}{((x-a_i)(x-\overline{a}_i))^\gamma} \, dx = 0. \tag{3.5}$$

Note that since $l_j(x) \in \mathcal{P}_{n-1}(a_1, \ldots, a_n)$, by the partial fraction decomposition and combining (3.4) and (3.5), (3.2) follows. This finishes the proof of Lemma 3.2. □

Proof of Theorem 2.2. Obviously, for any given $0 < p < \infty$, there exists $r \in \mathbb{N}$ such that $2(r-1) \leq p \leq 2r$. Then, by the Hölder inequality we have

$$\left(\int_{-1}^1 \frac{1}{\sqrt{1-x^2}} |L_n(f) - f|^p \, dx\right)^{1/p}$$

$$\leq \left(\int_{-1}^1 \frac{1}{\sqrt{1-x^2}} |L_n(f) - f|^{p\frac{2r}{p}} \, dx\right)^{1/2r} \left(\int_{-1}^1 \frac{1}{\sqrt{1-x^2}} \, dx\right)^{\frac{2r-p}{2r}\frac{1}{p}}$$

$$\leq c(p) \left(\int_{-1}^1 \frac{1}{\sqrt{1-x^2}} |L_n(f) - f|^{2r} \, dx\right)^{1/2r}.$$

We suppose that $p_n \in \mathcal{P}_{n-1}(a_1, \ldots, a_n)$ is the best approximation to $f(x)$ on $[-1, 1]$:
$$\|p_n - f\|_{[-1,1]} = E_n(f).$$

Note that since $L_n(p_n, x) \equiv p_n$ (cf. [6, Lemma 3.7]), the Minkowski's' inequality yields

$$\|L_n(f) - f\|_{v,2r} \leq \|L_n(f - p_n)\|_{v,2r} + \|p_n - f\|_{v,2r}.$$

Hence, we only need to prove

$$\|L_n(f - p_n)\|_{v,2r} \leq c(r) E_n(f). \tag{3.6}$$

We denote $\Delta(x) := f(x) - p_n(x)$ and

$$I_n := \int_{-1}^1 \frac{1}{\sqrt{1-x^2}} |L_n(\Delta(x))|^{2r} \, dx.$$

Recall that

$$I_n = \sum \frac{(2r)!}{r_1! \cdots r_s!} \Delta^{r_1}(x_{i_1}) \cdots \Delta^{r_s}(x_{i_s}) \int_{-1}^1 \frac{1}{\sqrt{1-x^2}} l_{i_1}^{r_1}(x) \cdots l_{i_s}^{r_s}(x) \, dx, \tag{3.7}$$

where $\sum$ denotes multiple sum for $r_1 + \cdots + r_s = 2r$, $0 \leq r_i \leq 2r$, $i = 1, \ldots, s$. Then by Lemmas 3.1 and 3.2, and using the almost same fashion as the proofs in [4,10], we can conclude (3.6). □

References

1. Achiezer, N. I., *Theory of Approximation*, Ungar, New York, 1956.
2. Borwein, P. and T. Erdélyi, *Polynomials and Polynomial Inequalities*, Springer, New York, 1995.
3. Borwein, P., T. Erdélyi, and J. Zhang, Chebyshev polynomials and Markov-Bernstein type inequalities for the rational spaces, J. London Math. Soc. **50** (1994), 501–519.

4. Erdős, P. and E. Feldheim, Sur le mode de convergence pour l'interpolation de Lagrange, C. R. Acad. Sci. Paris **203** (1936), 913–915.
5. Gautschi, W., Gauss-type quadrature rules for rational functions, in *International Series of Numerical Mathematics*, Vol. 112, H. Brass and G. Hammerlin (eds.), Birkhäuser Verlag, Basel, 1993, pp. 111–130.
6. Min, G., Lagrange interpolation and quadrature formula in rational systems, J. Approx. Theory, to appear.
7. Min, G., Hermite-Fejér interpolation for rational systems, Constructive Approx., to appear.
8. Min, G., Inequalities for rational functions with prescribed poles, Canadian J. Math., to appear.
9. Min, G., On the denseness of rational systems, submitted.
10. Min, G., On weighted L^p-convergence of certain Lagrange interpolation, Proc. Amer. Math. Soc. **116** (1992), 1081–1087.
11. Pinkus, A. and Z. Ziegler, Interlacing properties of zeros of the error function in best L^p-approximation, J. Approx. Theory **27** (1979), 1–18.
12. Rivlin, T. J., *Chebyshev Polynomials*, 2nd Edition, John Wiley & Sons, 1990.
13. Szabados, J. and P. Vértesi, *Interpolation of Functions*, World Scientific, Singapore, 1990.
14. Van Assche, W. and I. Vanherwegen, Quadrature formulas based on rational interpolation, Math. Comp. **61** (1993), 765–783.
15. Walsh, J. L., *Interpolation and Approximation by Rational functions in the Complex Domain*, 4th Edition, Amer. Math. Soc. Coll. Publ., Vol. 20, Providence, R. I., 1965.

G. Min
Department of Mathematics and Statistics
Simon Fraser University
Burnaby, B. C., Canada V5A 1S6
gmin@cecm.sfu.ca

Approximation of Functions by Linear Matrix Operators

M. L. Mittal and Neeraj Bhardwaj

Abstract. In 1986, Chandra proved some theorems on approximation of functions by Nörlund operator N_p, using integral modulus of continuity. In this note, we extend these theorems of Chandra to certain linear matrix operators. This paper is dedicated to Professor K. V. Mittal.

§1. Introduction

Let $f(x)$ be 2π-periodic function in $L_1(-\pi, \pi)$. The Fourier series associated with f at the point x is defined by

$$f(x) \sim \frac{1}{2}a_o + \sum_{n=1}^{\infty}(a_n \cos nx + b_n \sin nx) = \sum_{n=0}^{\infty} A_n(x), \qquad (1.1)$$

with partial sums $s_n(f, x)$. Define

$$t_n(f; x) = \sum_{k=0}^{n} a_{n,k} s_k(f; x), \qquad n = 0, 1, 2, \ldots, \qquad (1.2)$$

where A is a linear matrix operator represented by the lower-triangular matrix $A = (a_{n,k})$.

The linear matrix operator A reduces to the Nörlund operator N_p, if

$$a_{n,k} = \frac{p_{n-k}}{P_n}, \qquad 0 \le k \le n,$$
$$= 0, \qquad k > n,$$

where $P_n = \sum_{k=0}^{n} p_k \ne 0$ and $P_{-1} = 0 = p_{-1}$. In this case, the transform (1.2) reduces to the Nörlund transform $N_n(f; x)$.

A linear operator A is said to be **regular** if it is limit-preserving over c, the space of convergent sequences. Each matrix A in this paper is nonnegative and has row sums one. Hence, if $\lim_n a_{n,k} = 0$ for each k, then A is regular.

§2. Notations

Throughout this note, we use $\downarrow$ (or $\uparrow$) for non-decreasing (or non-increasing). Also all norms considered here are L_p-norms ($p \geq 1$) which would be taken with respect to the variable x. The classes Lip α and Lip (α, p) will be as usual (see [4], p. 612, and pp. 42 and 45). We write

$$\phi_x(t) = \frac{1}{2}\{f(x+t) + f(x-t) - 2f(x)\}, \quad A_{n,k} = \sum_{r=k}^{n} a_{n,r},$$

$$\omega_p(t; f) = \sup_{|h| \leq t} \left(\int_0^{2\pi} |f(x+h) - f(x)|^p dx \right)^{1/p}, \quad p \geq 1$$

$$\Delta_k a_{n,k} = a_{n,k} - a_{n,k+1}, \quad \tau = [\pi/t]$$

where $[\pi/t]$ denotes the greatest integer contained in π/t.

§3. Previous Results

Chandra [2] proved the following theorems on N_p-operator.

Theorem A. Let $f \in L_p (p > 1)$ and let $0 \leq \{p_n\} \downarrow$. Then

$$\|N_n(f; x) - f(x)\|_p = O\left\{ (1/P_n) \sum_{k=1}^{n} k^{-1} P_k \omega_p(\pi/k; f) \right\}. \quad (3.1)$$

Theorem B. Let $f \in L_p (p > 1)$ and $0 \leq \{p_n\} \downarrow$. For $t \to 0+$, let $\omega_p(t; f)$ satisfy

$$\int_t^{\pi} u^{-2} \omega_p(u; f) du = O(H(t)), \quad (3.2)$$

where $H \geq 0$ and that

$$tH(t) = o(1), \quad (3.3)$$

$$\int_0^t H(u) du = O\{tH(t)\}. \quad (3.4)$$

Then

$$\|N_n(f; x) - f(x)\|_p = O\{(p_n/P_n) H(p_n/P_n)\}. \quad (3.5)$$

Theorem C. Let f and $\{p_n\}$ be defined as in Theorem B and let $\omega_p(t; f)$ satisfy (3.2) and (3.4). Then

$$\|N_n(f; x) - f(x)\|_p = O\{(p_n/P_n) H(\pi/n)\}. \quad (3.6)$$

Theorem D. If $f \in \text{Lip}(\alpha, p), 0 < \alpha \le 1, p > 1$, and if $\{p_n\} \ge 0$, then

$$\|N_n(f;x) - f(x)\|_p = O\left(\frac{1}{P_n}\sum_{k=1}^n k^{-1-\alpha}P_k\right), \{p_n\}\uparrow, \quad (3.7)$$

and, whenever $\{p_n\}\downarrow$,

$$\|N_n(f;x) - f(x)\|_p = \begin{cases} O\{(p_n/P_n)^\alpha\}, & 0 < \alpha < 1, \\ O\{(p_n/P_n)\log(p_n/P_n)\}, & \alpha = 1. \end{cases} \quad (3.8)$$

Theorem E. Let $f \in \text{Lip}(\alpha, p), 0 < \alpha \le 1, p > 1, \alpha p > 1$ and let $\{p_n\}$ be nonnegative. Then, uniformly in x almost everywhere,

$$\|N_n(f;x) - f(x)\|_p = \begin{cases} O\left\{P_n^{-1}\sum_{k=1}^n k^{\frac{1}{p}-1-\alpha}P_k\right\}, & \{p_n\}\uparrow, \\ O\{(p_n/P_n)^{\alpha-1/p}\}, & \{p_n\}\uparrow. \end{cases} \quad (3.9)$$

§4. Main Results

In this note, we extend Theorems A to E to linear matrix operators.

Theorem 1. Let $f \in L_p(p > 1)$ and $A = (a_{n,k})$ be an infinite triangular matrix satisfying

$$a_{n,k} \le a_{n,k+1}, \quad \forall\, 0 \le k \le n. \quad (4.1)$$

Then

$$\|t_n(f;x) - f(x)\|_p = O\left[\sum_{k=1}^n k^{-1}A_{n,n-k}\omega_p(\pi/k; f)\right]. \quad (4.2)$$

Theorem 2. Let $f \in L_p(p > 1)$ and let $A = (a_{n,k})$ be an infinite triangular matrix satisfying

$$a_{n,k} \ge a_{n,k+1}, \quad 0 \le k \le n. \quad (4.3)$$

If $\omega_p(t; f)$ satisfies condition (3.2), (3.3), and (3.4), then

$$\|t_n(f;x) - f(x)\|_p = O\{a_{n,0}H(a_{n,0})\}. \quad (4.4)$$

Theorem 3. Let f and $\{a_{n,k}\}$ be defined as in Theorem 2 and let $\omega_p(t; f)$ satisfy (3.2) and (3.4). Then

$$\|t_n(f;x) - f(x)\|_p = O\{a_{n,0}H(\pi/n)\}. \quad (4.5)$$

Theorem 4. If $f \in \text{Lip}(\alpha, p), 0 < \alpha \le 1, p > 1$, then

$$\|t_n(f;x) - f(x)\|_p = O\left(\sum_{k=1}^n k^{-1-\alpha}A_{n,n-k}\right), \quad \text{for } \{a_{n,k}\}\downarrow, \quad (4.6)$$

and, for $\{a_{n,k}\} \uparrow$

$$\|t_n(f;x) - f(x)\|_p = \begin{cases} O(a_{n,0})^\alpha, & 0 < \alpha < 1, \\ O\{(a_{n,0})\log_{(\pi/a_{n,0})}\}, & \alpha = 1. \end{cases} \quad (4.7)$$

Theorem 5. Let $f \in \text{Lip}(\alpha, p), 0 < \alpha \leq 1, p > 1, \alpha p > 1$. Then uniformly in x almost everywhere,

$$\|t_n(f;x) - f(x)\|_p = \begin{cases} O\left(\sum_{k=1}^{n} k^{(1/p)-\alpha-1} A_{n,n-k}\right), & \{a_{n,k}\} \downarrow \\ O(a_{n,0})^{\alpha-(1/p)}, & \{a_{n,k}\} \uparrow. \end{cases} \quad (4.8)$$

In the case of Nörlund operators, we observe that conditions (4.1) of Theorem 1 and (4.3) of Theorem 2 reduce to $\{p_n\} \uparrow$ of Theorem A and $\{p_n\} \downarrow$ of Theorem B, respectively. Therefore, Theorems 1 and 2 are extensions of Theorems A and B, respectively, while in Theorem 3, $\{a_{n,k}\}$ and f satisfy the condition of Theorem 2 which in turn satisfy all conditions of Theorem C in N_p-operator. Hence, Theorem 3 extends Theorem C. Moreover, Theorems 4 and 5 extend Theorems D and E, respectively, to linear matrix operators.

§5. Preliminary Lemmas

We shall require the following Lemmas.

Lemma 1. ([3], p. 148, 6.13.9). If $h(x,t)$ is a function of two variables defined for $0 \leq t \leq \pi, 0 \leq x \leq 2\pi$, then

$$\left\|\int H(x,t)dt\right\|_p \leq \int \|h(x,t)\|_p dt, \quad (p > 1).$$

Lemma 2. [5]. Let $A = (a_{n,k})$ be a lower triangular matrix that satisfies condition (4.1). Then

$$\left|\sum_{k=1}^{n} a_{n,k} \sin\left(k + \frac{1}{2}\right)t\right| = O\{A_{n,n-\tau+1}\}, \quad 0 \leq t < \pi.$$

Lemma 3. ([4] Theorem 5 (ii), p. 627). Suppose that $f \in \text{Lip}(\alpha, p)$, where $p \geq 1, 0 < \alpha \leq 1$, and $\alpha p > 1$. Then f is equal to a function $g \in \text{Lip}(\alpha - (1/p))$ almost everywhere.

§6. Proofs of Main Results

Proof of Theorem 1. From (1.1), we have ([1], p. 402)

$$s_n(f;x) = \frac{1}{\pi} \int_0^\pi \{f(x+t) + f(x-t)\} \frac{\sin(n+1/2)t}{2\sin(t/2)} dt. \quad (6.1)$$

Now using (6.1), (1.2) becomes

$$t_n(f;x) = \frac{1}{\pi}\int_0^\pi \{f(x+t)+f(x-t)\}\sum_{k=0}^n a_{n,k}\frac{\sin(k+1/2)t}{2\sin(t/2)}dt.$$

Since $(\sin(t/2))^{-1} \leq \pi t^{-1}$, for $0 \leq t \leq \pi$ and $t_n(1;f) = 1$, we have

$$t_n(f;x) - f(x) = \frac{1}{\pi}\int_0^\pi \phi_x(t)\sum_{k=0}^n a_{n,k}\frac{\sin(k+1/2)t}{\sin(t/2)}dt$$

$$\leq \int_0^\pi \frac{\phi_x(t)}{t}\sum_{k=0}^n a_{n,k}\sin(k+1/2)t\,dt$$

$$= \left(\int_0^{\pi/n} + \int_{\pi/n}^\pi\right)t^{-1}\phi_x(t)\left(\sum_{k=0}^n a_{n,k}\frac{\sin(k+1/2)t}{\sin(t/2)}\right)dt$$

$$= I_1 + I_2, \text{ say}.$$

Then by Minkowski's inequality, we have

$$\|t_n(f;x) - f(x)\|_p \leq \|I_1\|_p + \|I_2\|_p. \tag{6.2}$$

Now, by Lemma 1 and $\omega_p(\delta_1, f) \leq \omega_p(\delta_2, f)$, for $0 \leq \delta_1 \leq \delta_2$, we have

$$\|I_1\|_p \leq \int_0^{\pi/2} t^{-1}\|\phi_x(t)\|_p\left|\sum_{k=0}^n a_{n,k}\sin\left(k+\frac{1}{2}\right)t\right|dt$$

$$\leq (n+1)\int_0^{\pi/2}\omega_p(t;f)dt$$

$$\leq \omega_p(\pi/n;f)\sum_{k=0}^n a_{n,n-k}$$

$$= O\left(\omega_p(\pi/n;f)\sum_{k=1}^n \frac{A_{n,n-k}}{k+1}\right)$$

$$= O\left(\sum_{k=1}^n k^{-1}A_{n,n-k}\omega_p(\pi/k;f)\right), \tag{6.3}$$

in view of (4.1) which implies $a_{n,k} = O\left(\frac{A_{n,k}}{n-k+1}\right)$, for $0 \leq k \leq n$.

Again by Lemma 1 and Lemma 2, we have

$$\|I_2\|_p = O\left(\int_{\pi/n}^{\pi} t^{-1}\omega_p(t;f)A_{n,n-\tau}dt\right),$$

$$= O\left(\int_1^n k^{-1}\omega_p(\pi/k;f)A_{n,n-k}dk\right)$$

$$= O\left(\sum_{k=1}^n k^{-1}A_{n,n-k}\omega_p(\pi/k;f)\right). \tag{6.4}$$

Putting (6.3) and (6.4) in (6.2), we get (4.2).

Remark: The case $p = 1$ of Theorem 1 is due to Mazhar [6].

Proof of Theorem 2. Proceeding as in Theorem 1, we have

$$t_n(f;x) - f(x) = \frac{1}{\pi}\int_0^{\pi}\phi_x(t)\sum_{k=0}^n a_{n,k}\frac{\sin(k+1/2)t}{\sin(t/2)}dt,$$

$$\leq \left(\int_0^{a_{n,0}} + \int_{a_{n,0}}^{\pi}\right)t^{-1}\phi_x(t)\sum_{k=0}^n a_{n,k}\sin(k+1/2)t\, dt$$

$$= J_1 + J_2, \quad \text{say.}$$

Now by Minkowski's inequality, we get

$$\|t_n(f;x) - f(x)\|_p \leq \|J_1\|_p + \|J_2\|_p, \tag{6.5}$$

where, by Lemma 1,

$$\|J_1\|_p \leq \int_0^{a_{n,0}} t^{-1}\|\phi_x(t)\|_p \left|\sum_{k=0}^n a_{n,k}\sin(k+1/2)t\right| dt$$

$$\leq \int_0^{a_{n,0}} t^{-1}\omega_p(t;f)dt = O((a_{n,0})H(a_{n,0})), \tag{6.6}$$

integrating by parts and using conditions (3.2), (3.3), and (3.4). Again using Lemma 1, we have

$$\|J_2\|_p = O\left(\int_{a_{n,0}}^{\pi} t^{-1}\omega_p(t;f)\left(\frac{a_{n,0}}{t}\right)dt\right) = O(a_{n,0}H(a_{n,0})), \tag{6.7}$$

in view of (4.3) (which implies $\left|\sum_{k=0}^{n} a_{n,k} \sin\left(k+\frac{1}{2}\right)t\right| = O(a_{n,0}/t)$, by Abel's lemma) and (3.2). Now using (6.6) and (6.7) in (6.5), we get (4.4). This completes the proof of Theorem 2.

Proof of Theorem 3. Proceeding as in Theorem 1, we have

$$\|I_1\|_p \leq (n+1) \int_0^{\pi/n} \omega_p(t;f) dt.$$

Now integrating by parts and using (3.2) and (3.4), we have

$$\|I_1\|_p = O[n^{-1} H(\pi/n)] = O[(a_{n,0}) H(\pi/n)),$$

in view of (4.1) which implies $(n+1)a_{n,0} \geq 1$.

Once again, by Lemma 1 and Abel's Lemma, we have

$$\|I_2\|_p = O \int_{\pi/2}^{\pi} t^{-1} \omega_p(t;f) \left(\frac{a_{n,0}}{t}\right) dt = O(a_{n,0} H(\pi/n)),$$

in view of condition (3.2). This completes the proof of Theorem 3.

Proof of Theorem 4. Since $f \in \text{Lip}(\alpha, p), 0 < \alpha \leq 1, p > 1$, we have

$$\omega_p(t;f) = O(t^\alpha). \tag{6.8}$$

Using (6.8) in (4.2), we get (4.6) and again using (6.8) in (6.6) and (6.7), we get (4.7). This completes the proof of Theorem 4.

Proof of Theorem 5. By Lemma 3, we can suppose that there exists a function $\phi_x(t)$ such that

$$\phi_x(t) = O(t^{\alpha - (1/p)}) \tag{6.9}$$

almost everywhere.

We first consider the case $\{a_{n,k}\} \downarrow$. Now in I_1 of (6.2) using (6.9), we have

$$\|I_1\| = O(n^{(1/p)-\alpha}) = O\left(n^{(1-p)-\alpha} \sum_{k=1}^{n} \frac{A_{n,n-k}}{k+1}\right) = O\left(\sum_{k=1}^{n} k^{(1/p)-\alpha-1} A_{n,n-k}\right),$$

since $\{a_{n,k}\} \downarrow$. Moreover, in I_2 of (6.2) using (6.9) and Lemma 2, we get

$$\|I_2\|_p = O\left(\int_{\pi/n}^{\pi} t^{\alpha-(1/p)} \left(\frac{A_{n,n-\tau}}{t}\right) dt\right) = O\left(\sum_{k=1}^{n} k^{(1/p)-\alpha-1} A_{n,n-k}\right).$$

Combining I_1 and I_2, we get the first part of Theorem 5.

In the case when $\{a_{n,k}\} \uparrow$, using (6.9) in J_1 of (6.5), we have

$$\|J_1\|_p = O\left(\int_0^{a_{n,0}} t^{-1} t^{\alpha-(1/p)} dt\right) = O((a_{n,0})^{\alpha-(1/p)}).$$

Also, using (6.9) and Abel's Lemma in J_2 of (6.5), we have

$$\|J_2\| = O\left(\int_{a_{n,0}}^{\pi} t^{-1} t^{\alpha-(1/p)} \left(\frac{a_{n,0}}{t}\right) dt\right) = O((a_{n,0})^{\alpha-(1/p)}).$$

Combining J_1 and J_2, we establish the second part of Theorem 5.

References

1. Alexits, G., *Convergence Problems of Orthogonal Series*, Pergamon Press, New York, 1961.
2. Chandra, P., Approximation by Nörlund operators, Matemat. Bechn. **38** (1986), 263–269.
3. Hardy, G. H., J. E. Littlewood, and G. Polya, *Inequalities*, Cambridge, 1967.
4. Hardy, G. H. and J. E. Littlewood, A convergence criterion for Fourier series, Math. Zeit **28** (1928), 612–634.
5. Kishore, H. and G. C. Hotta, On absolute matrix summability of Fourier series, Indian J. Math. **13** (1971), 99–110.
6. Mazhar, S. M., On the degree of approximation of a class of functions by means of Fourier series, Proc. American Math. Soc. **88** (1983), 317–320.

M. L. Mittal
Department of Mathematics
University of Roorkee
Roorkee 247 667
Uttar Pradesh, India

Neeraj Bhardwaj
Department of Mathematics
D.N.(P.G.) College
C.C.S. University Meerut Uttar Pradesh
India

Weighted Approximation on Compact Sets

Igor E. Pritsker

Abstract. For a compact set E with connected complement in the complex plane, we consider a problem of the uniform approximation on E by the weighted polynomials $W^n(z)P_n(z)$, where $W(z)$ is a continuous nonvanishing weight function on E, analytic in the interior of E. Let $A(E, W)$ be the set of functions uniformly approximable on E by such weighted polynomials. If E has empty interior, then $A(E, W)$ is completely characterized by a zero set $Z_W \subset E$, where all functions from $A(E, W)$ must vanish. This generalizes recent results of Totik and Kuijlaars for the real line case. However, if E is a closure of Jordan domain, the description of $A(E, W)$ also involves an inner function. In both cases, we exhibit the role of the support of a certain extremal measure, which is the solution of a weighted logarithmic energy problem, played in the descriptions of $A(E, W)$.

§1. Introduction

Let E be a compact set in the complex plane $\mathbf{C}$ with the connected complement $\overline{\mathbf{C}} \setminus E$. We denote the uniform algebra of functions which are continuous on E and analytic in the interior of E by $A(E)$ (see, e.g., [4, p. 25]). Clearly, the corresponding uniform norm for any $f \in A(E)$ is defined by

$$\|f\|_E := \max_{z \in E} |f(z)|. \tag{1.1}$$

Consider a weight function $W \in A(E)$ such that $W(z) \neq 0$ for any $z \in E$, and define the weighted polynomials $W^n(z)P_n(z)$, where $P_n(z)$ is an algebraic polynomial in z with complex coefficients, $\deg P_n \leq n$. We are interested in a description of the function set $A(E, W)$, consisting of the uniform limits on E of sequences of the weighted polynomials $\{W^n(z)P_n(z)\}_{n=0}^{\infty}$, as $n \to \infty$. It is well known that if $W(z) \equiv 1$ on E then $A(E, 1) = A(E)$ by Mergelyan's theorem [4, p. 48]. In general, we have that $A(E, W) \subset A(E)$.

Our problem originated in the work of Lorentz [10] on incomplete polynomials on the real line. Surveys of results in this area, dealing with weighted approximation on the real line, can be found in [16] and [14, Ch. VI]. The most recent developments are in [6]–[8].

The questions of density of the weighted polynomials in the set of analytic functions in a domain have been considered in [3, 11, and 12]. In particular, [12] contains a necessary and sufficient condition such that any function analytic in a bounded open set is uniformly approximable by the weighted polynomials $W^n(z)P_n(z)$ on *compact subsets*. However, the description of $A(E,W)$ seems to be much more complicated, in that no general necessary and sufficient condition is known (in terms of the weight $W(z)$), even for the real interval case, i.e., for $E = [a,b] \subset \mathbf{R}$.

We shall approach the above mentioned problems on $A(E,W)$, using ideas of the theories of uniform algebras and of weighted potentials.

§2. $A(E,W)$ as a Closed Ideal and Weighted Potentials

Proposition 2.1. *$A(E,W)$, endowed with norm (1.1), is a closed function algebra (not necessarily containing constants and separating points).*

We have already remarked that $A(E,W) \subset A(E)$. To make this inclusion more precise, let us introduce the algebra $[W(z), zW(z)]$ generated by the two functions $W(z)$ and $zW(z)$, which is the uniform closure of all polynomials in $W(z)$ and $zW(z)$ (with constant terms included) on E. Clearly, $[W(z), zW(z)] \subset A(E)$. Since any weighted polynomial $W^n(z)P_n(z)$ is an element of $[W(z), zW(z)]$, then $A(E,W) \subset [W(z), zW(z)]$. Thus, we arrive at the following

Proposition 2.2. $A(E,W) \subset [W(z), zW(z)] \subset A(E)$.

Proposition 2.3. *$A(E,W)$ is a closed ideal of $[W(z), zW(z)]$.*

It turns out that in many cases $[W(z), zW(z)] = A(E)$, so that $A(E,W)$ becomes a closed ideal of $A(E)$ by Proposition 2.3.

Proposition 2.4. $[W(z), zW(z)] = A(E)$ *iff* $1/W(z) \in [W(z), zW(z)]$.

Unfortunately, we do not know any effectively verifiable necessary and sufficient condition on the weight $W(z)$, so that the equality $[W(z), zW(z)] = A(E)$ is valid. Nevertheless, a number of sufficient conditions can be given, guaranteeing that the two algebras $[W(z), zW(z)]$ and $A(E)$ coincide.

Proposition 2.5. *Each of the following conditions implies that* $[W(z), zW(z)] = A(E)$:
(a) *The point $\zeta = 0$ belongs to the unbounded component of* $\overline{\mathbf{C}} \setminus W(E)$;
(b) *E is the closure of a Jordan domain or a Jordan arc, and $W(z)$ is one-to-one on E;*
(c) *E is a Jordan arc and $W(z)$ is of bounded variation on E;*
(d) *E is a Jordan arc and $W(z)$ is locally one-to-one on E;*
(e) *$E = \overline{G}$, where G is a Jordan domain bounded by an analytic curve, and $W'(z) \in A(\overline{G})$.*

Assuming that E has positive logarithmic capacity, then

$$w(z) := \begin{cases} |W(z)|, & z \in E, \\ 0, & z \notin E, \end{cases} \quad (2.1)$$

is an admissible weight for the weighted logarithmic energy problem on E considered in Section I.1 of [14]. This enables us to use certain results of [14], which we summarize below for the convenience of the reader. Recall that the logarithmic potential of a compactly supported Borel measure μ is given by

$$U^\mu(z) := \int \log \frac{1}{|z-t|} d\mu(t). \quad (2.2)$$

Proposition 2.6. *There exists a positive unit Borel measure μ_w, with support $S_w := \operatorname{supp} \mu_w \subset \partial E$, such that for any polynomial $P_n(z), \deg P_n \leq n$, we have*

$$|W^n(z)P_n(z)| \leq \|W^n P_n\|_{S_w} \exp(n(F_w - U^{\mu_w}(z) + \log|W(z)|)), \quad (2.3)$$

where $z \in E$ and where F_w is a constant. Furthermore, the inequality

$$U^{\mu_w}(z) - \log|W(z)| \geq F_w \quad (2.4)$$

holds quasi-everywhere on E, and

$$U^{\mu_w}(z) - \log|W(z)| \leq F_w, \text{ for any } z \in S_w. \quad (2.5)$$

By saying quasi-everywhere (q.e.), we mean that a property holds everywhere, with the exception of a set of zero logarithmic capacity. The measure μ_w is the solution of a weighted energy problem, corresponding to the weight $w(z)$ of (2.1) (see Section I.1 of [14]). It follows from (2.3) and (2.4) that the norm of a weighted polynomial $W^n P_n$ essentially "lives" on S_w. In particular, the following is valid (see Corollary III.2.6 of [14]).

Proposition 2.7. *Suppose that for every point $z_0 \in E$, the set $\{z : |z - z_0| < \delta, z \in E\}$ has positive capacity for any $\delta > 0$. Then*

$$\|W^n P_n\|_E = \|W^n P_n\|_{S_w} \quad (2.6)$$

for any polynomial $P_n, \deg P_n \leq n$.

§3. Sets with Empty Interior

Let E be a compact set with connected complement and empty interior. Obviously, $A(E) = C(E)$ in this case. We characterize $A(E,W)$ in terms of a certain zero set.

Theorem 3.1. *Suppose that E has a connected complement and an empty interior, and that $W \in C(E)$ is a nonvanishing weight on E. Assume that $[W(z), zW(z)] = C(E)$. Then, there exists a closed set $Z_W \subset E$ such that $f \in A(E,W)$ if and only if $f \in C(E)$ and $f|_{Z_W} \equiv 0$.*

It is clear that $A(E,W) = C(E)$ if and only if the set Z_W is empty. This is true, for example, for $W(z) \equiv 1$ on E.

Theorem 3.1 generalizes a recent result of Kuijlaars (see Theorem 3 of [8]), related to polynomial approximation with varying weights on the real line. However, it has a new part even in the latter case, allowing us to consider the complex valued weights $W(z)$ on subsets of the real line.

A description of the set Z_W in terms of the weight $W(z)$ is unknown in general. We can only show that Z_W must contain the complement of S_w (see Proposition 2.6) in E.

Theorem 3.2. *Let E be an arbitrary compact set with the connected complement $\overline{\mathbf{C}} \backslash E$ and let $W \in A(E)$ be a nonvanishing weight on E. Suppose that for every point $z_0 \in E$, the set $\{z : |z - z_0| < \delta, z \in E\}$ has positive logarithmic capacity for any $\delta > 0$. Assume further that $\overline{\mathbf{C}} \backslash S_w$ is connected and $[W(z), zW(z)] = C(S_w)$ on S_w. If $f \in A(E,W)$, then $f(z) = 0$ for any $z \in \overline{E \backslash S_w}$. In particular, if E has empty interior, then $\overline{E \backslash S_w} \subset Z_W$.*

The proof of Theorem 3.2 is based on an idea of Kuijlaars (see Theorem 2 and its proof in [8]).

If E is a compact subset of the real line and the weight $W(z)$ is real valued, then condition (a) of Proposition 2.5 is clearly satisfied, so that $[W(z), zW(z)] = C(E)$. Therefore, the conclusion of Theorem 3.1 is valid, and coincides with that of Theorem 3 of [8]. Furthermore, if for any point in E, the intersection of its arbitrary neighborhood with E has positive logarithmic capacity, then $\overline{E \backslash S_w} \subset Z_W$. Since $[W(z), zW(z)] = C(S_w)$ on S_w by Proposition 2.5(a), Theorem 3.2 essentially reduces to Theorem 2 of [8] in this case, which in turn contains an earlier result of Theorem 4.1 of [16].

§4. Unit Disk and Jordan Domains

The first result of this section is a consequence of the well-known description of closed ideals of $A(\overline{D})$, where D is the unit disk, due to Beurling (unpublished) and Rudin [13] (see also [5, pp. 82–87] for a discussion). Recall that g is an inner function if it is analytic in D, with $\|g\|_{\overline{D}} \leq 1$, and $|g(e^{i\theta})| = 1$ almost everywhere on the unit circle (cf. [5, p. 62]). By the factorization theorem, every inner function can be uniquely expressed in the form

$$g(z) = B(z)S(z), \quad z \in D, \tag{4.1}$$

where $B(z)$ is a Blaschke product and $S(z)$ is a singular function, i.e.,

$$S(z) := \exp\left(-\int \frac{e^{i\theta}+z}{e^{i\theta}-z} d\nu_s(\theta)\right), \quad z \in D, \qquad (4.2)$$

with ν_s being a positive measure on the unit circle, singular with respect to $d\theta$ (see [5, pp. 63–67]).

Theorem 4.1. *Let a nonvanishing weight $W \in A(\overline{D})$ be such that $[W(z), zW(z)] = A(\overline{D})$. Assume that $A(\overline{D}, W)$ contains a function not identically zero. Then there exist a closed set $H_W \subset \partial D$ of Lebesgue measure zero and an inner function g_W satisfying*
  (i) *every accumulation point of the zeros of its Blaschke product is in H_W,*
  (ii) *the measure ν_s of its singular function is supported on H_W;*
  such that

  $f \in A(\overline{D}, W)$ *if and only if* $f = g_W h$, *where* $h \in A(\overline{D})$ *and* $h|_{H_W} \equiv 0$.

The case of a Jordan domain G can be reduced to that of the unit disk, using a canonical conformal mapping $\phi : G \to D$ and its inverse $\psi := \phi^{-1}$.

Our next goal is to exhibit the role of the set S_w (see Proposition 2.6) in the case of weighted approximation on Jordan domains. Since $W \in A(\overline{G})$ is analytic in G, then $S_w \subset \partial G$ by Theorem IV.1.10(a) of [14] and (2.1). The following result shows that $S_w = \partial G$ is necessary for nontrivial weighted approximation on $\overline{G}$.

Theorem 4.2. *Let G be a Jordan domain and let $W \in A(\overline{G})$ be a nonvanishing weight. Assume that S_w is a proper subset of ∂G and that $[W(z), zW(z)] = C(S_w)$ on S_w. Then $A(\overline{G}, W)$ contains the identically zero function only.*

§5. Proofs

Proof of Proposition 2.1: We have to show that $A(E, W)$ is closed under addition, multiplication by constants and by functions of $A(E, W)$, and under uniform limits. Suppose that $W^n P_n \to f \in A(E, W)$ and $W^n Q_n \to g \in A(E, W)$ uniformly on E, as $n \to \infty$. Then $W^n(P_n + Q_n) \to (f+g)$, as $n \to \infty$, so that $(f+g) \in A(E, W)$. If $\alpha \in \mathbf{C}$ then $W^n \alpha P_n \to \alpha f$, as $n \to \infty$, and $\alpha f \in A(E, W)$. Observe that

$$\|fg - W^{2n}P_n Q_n\|_E \leq \|fg - fW^n Q_n\|_E + \|fW^n Q_n - W^{2n} P_n Q_n\|_E \leq$$

$$\|f\|_E \, \|g - W^n Q_n\|_E + \|W^n Q_n\|_E \, \|f - W^n P_n\|_E \to 0,$$

as $n \to \infty$, i.e., $fg \in A(E, W)$. Applying the standard diagonalization argument, we see that $A(E, W)$ is closed in norm (1.1). □

Proof of Proposition 2.3: Assume that $f \in A(E,W)$ and $W^n P_n \to f$ uniformly on E, as $n \to \infty$. Then, for any pair of nonnegative integers k and ℓ such that $k \geq \ell$, we have

$$\|f(z)W^k(z)z^\ell - W^{n+k}(z)z^\ell P_n(z)\|_E \leq$$

$$\|W^k(z)z^\ell\|_E \, \|f - W^n P_n\|_E \to 0, \quad \text{as } n \to \infty,$$

which gives that $f(z)W^k(z)z^\ell \in A(E,W)$. Since $A(E,W)$ is closed under addition and multiplication by constants (by Proposition 2.1), the product of f and any polynomial in $W(z)$ and $zW(z)$ belongs to $A(E,W)$. Thus, if $g \in [W(z), zW(z)]$ then $fg \in A(E,W)$ follows immediately, because $A(E,W)$ is closed in the uniform norm on E (cf. Proposition 2.1). The proof is now complete in view of Propositions 2.1 and 2.2. □

Proof of Proposition 2.4: Obviously, if $[W(z), zW(z)] = A(E)$ then $1/W(z) \in A(E) = [W(z), zW(z)]$. Assume that $1/W(z) \in [W(z), zW(z)]$. It follows that $z \in [W(z), zW(z)]$ and, consequently, every polynomial in z is in $[W(z), zW(z)]$. Since $[W(z), zW(z)]$ is uniformly closed on E by definition, then $A(E) \subset [W(z), zW(z)]$ by Mergelyan's theorem [4, p. 48]. Thus, Proposition 2.2 implies that $A(E) = [W(z), zW(z)]$. □

Proof of Proposition 2.5: First, we remark that $W(z)$ and $zW(z)$ together separate points of any set E.

(a) Observe that $W(E)$, the image of E in ζ-plane under the mapping $\zeta = W(z)$, is compact. By assumption, function $1/\zeta$ is analytic on the polynomially convex hull of $W(E)$ and can be uniformly approximated there by polynomials in ζ (by Mergelyan's theorem). Returning to z-plane, we obtain that $1/W(z)$ is uniformly approximable on E by polynomials in $W(z)$. It follows that $[W(z), zW(z)] = A(E)$ by Proposition 2.4.

(b) The mapping $\zeta = W(z)$ can be extended to a homeomorphism between z-plane and ζ-plane (cf. [9, p. 535]). Since $W(z)$ does not vanish on E, $\zeta = 0$ belongs to the domain $\overline{\mathbf{C}} \setminus W(E) = W(\overline{\mathbf{C}} \setminus E)$, which contains $\zeta = \infty$. Hence, (b) follows from (a).

(c) If $E = [0,1]$ then (c) is a direct consequence of Theorem 2 of [2]. For E being a Jordan arc, we consider a homeomorphic parametrization of E by $\tau : [0,1] \to E$. Since $W \circ \tau(x)$ is of bounded variation on $[0,1]$, we have, as before, that $[W \circ \tau(x), \tau(x)(W \circ \tau)(x)] = C([0,1])$. Clearly, τ induces an isometric isomorphism between $C([0,1])$ and $C(E)$. Thus, the result follows after returning to E with the help of τ^{-1}.

(d) is implied by Theorem 1 of [1] for $E = [0,1]$. The case of a Jordan arc can be reduced to that of the interval as in the proof of (c).

(e) First, assume that $E = \overline{D}$. Then (e) follows at once from [17, p. 135]. It is well known that the conformal mapping $\phi : G \to D$ extends as a diffeomorphism between $\overline{G}$ and $\overline{D}$ (with nonvanishing derivatives of ϕ and $\psi := \phi^{-1}$), because G is bounded by an analytic Jordan curve. Using ϕ, the result for $E = \overline{G}$ is a consequence of [17, p. 135], too. □

Proof of Proposition 2.6: Since $W(z)$ is a continuous nonvanishing function on E and $w(z)$ of (2.1) is so too, the existence of μ_w and inequalities (2.4)–(2.5) follow from Theorem I.1.3 of [14]. Moreover, $W(z)$ is analytic in the interior of E, which implies that $S_w \subset \partial E$ by Theorem IV.1.10(a) of [14] and (2.1). The inequality (2.3) is a direct consequence of Theorem III.2.1 of [14]. □

Proof of Theorem 3.1: We have that $[W(z), zW(z)] = C(E)$ by the assumption of the theorem. Thus, $A(E, W)$ is a closed ideal of $C(E)$ (cf. Proposition 2.3), which is known to be described by its zero set (see [15, p. 32]). □

Proof of Theorem 3.2: We essentially follow the proof of Theorem 2 of [8]. Suppose that there exist $f_0 \in A(E, W)$ and $z_0 \in E \backslash S_w$ such that $f_0(z_0) \neq 0$ and $W^n P_n \to f_0$ uniformly on E, as $n \to \infty$. It is clear that $f_0|_{S_w} \in A(S_w, W)$. Recall that $S_w \subset \partial E$ by Proposition 2.6, i.e., S_w has empty interior. Applying Theorem 3.1, with E replaced by S_w, we obtain that $A(S_w, W)$ is described by the zero set $Z_W^* \subset S_w$. Observe that multiplying $A(S_w, W)$ by $(z - z_0)W(z)$, we obtain a closed ideal of $[W(z), zW(z)] = C(S_w)$ (cf. Proposition 2.3), which consists of all functions, uniformly approximable on S_w by the weighted polynomials $W^n(z)Q_n(z)$ such that $Q_n(z_0) = 0$, as $n \to \infty$. On the other hand, the zero set of the ideal $(z - z_0)W(z)A(S_w, W)$ coincides with that of $A(S_w, W)$. It follows that $(z - z_0)W(z)A(S_w, W) = A(S_w, W)$ (see [15, p. 32]) and that $f_0|_{S_w} \in (z - z_0)W(z)A(S_w, W)$.

Thus, there exists a sequence of the weighted polynomials $\{W^n Q_n\}_{n=0}^\infty$, with $Q_n(z_0) = 0$, uniformly convergent to f_0 on S_w, as $n \to \infty$. Since $W^n(z)(P_n(z) - Q_n(z))$ converges to zero uniformly on S_w and converges to $f_0(z_0) \neq 0$ for $z = z_0 \in E \backslash S_w$, as $n \to \infty$, we obtain a direct contradiction with (2.6) for some sufficiently large n. Consequently, if $f \in A(E, W)$ then $f(z) = 0$ for any $z \in E \backslash S_w$. Furthermore, the same is true for any $z \in \overline{E \backslash S_w}$ by the continuity of $f(z)$. □

Proof of Theorem 4.1: Since $[W(z), zW(z)] = A(\overline{D})$ by the assumption of the theorem, $A(\overline{D}, W)$ is a closed ideal of $A(\overline{D})$ by Proposition 2.3. The result now follows from the description of nontrivial closed ideals of the disk algebra (see [13] and [5, pp. 82–87]). □

Proof of Theorem 4.2: Since G is a Jordan domain, the set $\{z : |z - z_0| < \delta, z \in \overline{G}\}$ has positive logarithmic capacity for any $z_0 \in \overline{G}$ and $\delta > 0$. It is clear that S_w is contained in some Jordan arc, as a proper closed subset of ∂G, so that $\overline{\mathbb{C}} \backslash S_w$ is connected. Observe that all conditions of Theorem 3.2 are satisfied in this case, which yields that any function $f \in A(\overline{G}, W)$ must vanish on $\overline{(\overline{G} \backslash S_w)} = \overline{G}$. □

Acknowledgments. Supported in part by NSF Grant DMS-9707359.

References

1. Alexander, H., Polynomial approximation and analytic structure, Duke Math. J. **38** (1971), 123–135.

2. Alexander, H., Polynomial approximation and hulls in sets of finite linear measure in $\mathbf{C}^n$, Amer. J. Math. **93** (1971), 65–74.

3. Borwein, P. B. and W. Chen, Incomplete rational approximation in the complex plane, Constr. Approx. **11** (1995), 85–106.

4. Gamelin, T. W., *Uniform Algebras*, Chelsea Publ. Co., New York, 1984.

5. Hoffman, K., *Banach Spaces of Analytic Functions*, Prentice-Hall, Englewood Cliffs, 1962.

6. Kuijlaars, A. B. J., The role of the endpoint in weighted polynomial approximation with varying weights, Constr. Approx. **12** (1996), 287–301.

7. Kuijlaars, A. B. J., Weighted approximation with varying weights: the case of a power type singularity, J. Math. Anal. Appl. **204** (1996), 409–418.

8. Kuijlaars, A. B. J., A note on weighted polynomial approximation with varying weights, J. Approx. Theory **87** (1996), 112–115.

9. Kuratowski, K., *Topology*, vol. II, Academic Press, New York, 1968.

10. Lorentz, G. G., Approximation by incomplete polynomials (problems and results), in *Padé and Rational Approximations: Theory and Applications*, E. B. Saff and R. S. Varga (eds.), Academic Press, New York, 1997, pp. 289–302.

11. Pritsker, I. E. and R. S. Varga, The Szegő curve, zero distribution and weighted approximation, Trans. Amer. Math. Soc. **349** (1997), 4085–4105.

12. Pritsker, I. E. and R. S. Varga, Weighted polynomial approximation in the complex plane, Constr. Approx., to appear.

13. Rudin, W., The closed ideals in an algebra of analytic functions, Can. J. Math. **9** (1957), 426–434.

14. Saff, E. B. and V. Totik, *Logarithmic Potentials with External Fields*, Springer-Verlag, Heidelberg, 1997.

15. Stout, E. L., *The Theory of Uniform Algebras*, Bogden and Quigley, Belmont, 1971.

16. Totik, V., *Weighted Approximation with Varying Weights*, Lecture Notes in Math., vol. 1569, Springer-Verlag, Heidelberg, 1994.

17. Wermer, J., *Banach Algebras and Several Complex Variables*, Springer-Verlag, New York, 1976.

Igor E. Pritsker
Department of Mathematics
Case Western Reserve University
10900 Euclid Avenue
Cleveland, OH 44106-7058
iep@po.cwru.edu

Characterization of Best L_1 Coapproximation

Geetha S. Rao and R. Saravanan

Abstract. This paper deals with a characterization of best coapproximation with respect to L_1-norm. Some results concerning best coapproximation and best one-sided coapproximation with respect to L_1-norm are obtained.

§1. Introduction

A new kind of approximation was first introduced in 1979 by Franchetti and Furi [1] to characterize real Hilbert spaces among real reflexive Banach spaces. This was christened 'best coapproximation' by Papini and Singer [7]. This theory is largely concerned with the questions of existence, uniqueness, and characterizations of best coapproximation. The aim of this paper is to obtain a characterization of best L_1-coapproximation. Some fundamental concepts of best approximation and best coapproximation that are required in the sequel are provided in Section 2. A necessary and sufficient condition characterizing best L_1-coapproximation is obtained in Section 3. Some results concerning best one-sided L_1-coapproximation are established in Section 4.

§2. Preliminaries

Definition 2.1. Let G be a nonempty subset of a real normed linear space X. An element $g_f \in G$ is called a **best approximation** (respectively, **best coapproximation**) to $f \in X$ from G if for every $g \in G$, $\|f - g_f\| \leq \|f - g\|$ (respectively, $\|g - g_f\| \leq \|f - g\|$). The set of all best approximations (respectively, best coapproximations) to $f \in X$ from G is denoted by $P_G(f)$ (respectively, $R_G(f)$). The subset G is called an existence set if $R_G(f)$ contains at least one element, for every $f \in X$.

Proposition 2.2. [7]. *Let G be a linear subspace of a real normed linear space X. Then $R_G(\alpha f) = \alpha R_G(f)$, for all real α.*

Definition 2.3. *Let $C[a,b]$ be a space of continuous real valued functions on $[a,b]$. The sign of a function $g \in C[a,b]$ is defined by*

$$\operatorname{sgn} g(t) = \begin{cases} -1, & \text{if } g(t) < 0, \\ 0, & \text{if } g(t) = 0, \\ 1, & \text{if } g(t) > 0. \end{cases}$$

Best coapproximation problems can be considered with respect to various norms. In order to minimize the area between the functions $f \in C[a,b] \setminus G$ and $g_f \in G$, the L_1-norm is considered. Best approximation (respectively, best coapproximation) with respect to L_1-norm is called best L_1-approximation (respectively, best L_1-coapproximation).

Definition 2.4. *Let $f \in C[a,b]$. The zero set of f is defined as $Z(f) := \{t \in [a,b] : f(t) = 0\}$.*

Unless otherwise stated all normed linear spaces considered in this paper are real normed linear spaces, and all subsets and subspaces considered in this paper are existence subsets and existence subspaces with respect to best coapproximation. Since best coapproximation (respectively, best approximation) of an element in a subset from the same subset is the element itself i.e., if $G \subset X$, $f \in G \Rightarrow R_G(f) = f$ and $P_G(f) = f$, it is sufficient to deal with the element to which a best coapproximation (respectively, best approximation) to be found, which lies outside the subset i.e., $f \in X \setminus G$.

§3. Characterization of Best L_1-Coapproximation

The following theorem is a characterization of best L_1- approximation due to Kripke and Rivlin [6].

Theorem 3.1. *Let G be a subspace of $C[a,b]$, $f \in C[a,b] \setminus G$, and $g_f \in G$. Then the following statements are equivalent:*
(i). *The function g_f is a best L_1-approximation to f from G.*
(ii). *For every function $g \in G$, $\int_a^b g(t) \operatorname{sgn}(f(t) - g_f(t)) \, dt \leq \int_{Z(f-g_f)} |g(t)| \, dt$.*

A corresponding theorem is proved here for best L_1-coapproximation.

Theorem 3.2. *Let G be a subspace of $C[a,b]$, $f \in C[a,b] \setminus G$ and $g_f \in G$. Then the following statements are equivalent:*
(i). *The function g_f is a best L_1-coapproximation to f from G.*
(ii). *For every function $g \in G$,*

$$\int_a^b (f(t) - g_f(t)) \operatorname{sgn}(g(t)) \, dt \leq \int_{Z(g)} |f(t) - g_f(t)| \, dt.$$

Proof: (i) $\Rightarrow$ (ii). Assume that (i) holds.

Best L_1-Coapproximation

Claim. Let $h \in C[a,b] \setminus G$. If g_h is a best L_1-coapproximation to h from G such that $\|h - g_h\|_\infty = 1$, then for every $g \in G$,

$$\int_a^b (h(t) - g_h(t)) \operatorname{sgn} g(t)\, dt \le \int_{Z(g)} |h(t) - g_h(t)|\, dt.$$

Suppose for the moment that the claim is true. Let $g_f \in R_G(f)$. Then it follows from Proposition 2.2 that

$$\int_a^b (f(t) - g_f(t)) \operatorname{sgn} g(t)\, dt \le \int_{Z(g)} |f(t) - g_f(t)|\, dt.$$

Therefore, it remains to prove the claim. Suppose that the claim fails. Then there exists a function $g_1 \in G$ such that

$$\int_a^b (h(t) - g_h(t)) \operatorname{sgn} g_1(t)\, dt > \int_{Z(g_1)} |h(t) - g_h(t)|\, dt.$$

For each $\epsilon > 0$, define a subset $A(\epsilon)$ of $[a,b]$ by $A(\epsilon) = \{t \in [a,b] : |g_1(t)| \le \epsilon\}$. It is clear that for all $t \in [a,b] \setminus A(\epsilon)$,

$$\operatorname{sgn}(\epsilon(h(t) - g_h(t)) - g_1(t)) = \operatorname{sgn}(-g_1(t)) \tag{1}$$

and for all $t \in A(\epsilon) \setminus Z(g_1)$,

$$\{\epsilon(h(t) - g_h(t)) - g_1(t)\}\{\operatorname{sgn}(\epsilon(h(t) - g_h(t)) - g_1(t)) + \operatorname{sgn}(g_1(t))\} \le 2\epsilon.$$

Hence

$$\int_{A(\epsilon) \setminus Z(g_1)} \{\epsilon(h(t) - g_h(t)) - g_1(t)\}\{\operatorname{sgn}(\epsilon(h(t) - g_h(t)) - g_1(t))$$
$$+ \operatorname{sgn}(g_1(t))\}\, dt \le \epsilon \int_{A(\epsilon) \setminus Z(g_1)} 2\, dt. \tag{2}$$

Then it follows that

$$\|\epsilon h - (\epsilon g_h + g_1)\|_1 - \|\epsilon g_h + g_1 - \epsilon g_h\|_1$$

$$= \|\epsilon(h - g_h) - g_1\|_1 - \|g_1\|_1$$

$$= \int_a^b |\epsilon(h(t) - g_h(t)) - g_1(t)|\, dt - \int_a^b |g_1(t)|\, dt$$

$$= \int_{[a,b] \setminus A(\epsilon)} |\epsilon(h(t) - g_h(t)) - g_1(t)|\, dt$$

$$+ \int_{A(\epsilon)\setminus Z(g_1)} |\epsilon(h(t) - g_h(t)) - g_1(t)|\, dt + \int_{Z(g_1)} |\epsilon(h(t) - g_h(t))|\, dt$$

$$- \int_{[a,b]\setminus A(\epsilon)} |g_1(t)|\, dt - \int_{A(\epsilon)\setminus Z(g_1)} |g_1(t)|\, dt$$

$$= \int_{[a,b]\setminus A(\epsilon)} (\epsilon(h(t) - g_h(t)) - g_1(t))\operatorname{sgn}(\epsilon(h(t) - g_h(t)) - g_1(t))\, dt$$

$$+ \int_{A(\epsilon)\setminus Z(g_1)} |\epsilon(h(t) - g_h(t)) - g_1(t)|\, dt + \int_{Z(g_1)} |\epsilon(h(t) - g_h(t))|\, dt$$

$$- \int_{[a,b]\setminus A(\epsilon)} (g_1(t))\operatorname{sgn}(g_1(t))\, dt - \int_{A(\epsilon)\setminus Z(g_1)} |g_1(t)|\, dt$$

$$= \int_{[a,b]\setminus A(\epsilon)} (\epsilon(h(t) - g_h(t)) - g_1(t))\operatorname{sgn}(-g_1(t))\, dt$$

$$+ \int_{A(\epsilon)\setminus Z(g_1)} |\epsilon(h(t) - g_h(t)) - g_1(t)|\, dt + \int_{Z(g_1)} |\epsilon(h(t) - g_h(t))|\, dt$$

$$- \int_{[a,b]\setminus A(\epsilon)} (g_1(t))\operatorname{sgn}(g_1(t))\, dt - \int_{A(\epsilon)\setminus Z(g_1)} |g_1(t)|\, dt \qquad \text{by} \quad (1)$$

$$= - \int_{[a,b]\setminus A(\epsilon)} (\epsilon(h(t) - g_h(t)) - g_1(t) + g_1(t))\operatorname{sgn}(g_1(t))\, dt$$

$$+ \int_{A(\epsilon)\setminus Z(g_1)} |\epsilon(h(t) - g_h(t)) - g_1(t)|\, dt + \int_{Z(g_1)} |\epsilon(h(t) - g_h(t))|\, dt$$

$$- \int_{A(\epsilon)\setminus Z(g_1)} |g_1(t)|\, dt$$

$$= - \int_a^b \epsilon((h(t) - g_h(t))\operatorname{sgn}(g_1(t))\, dt$$

$$+ \int_{A(\epsilon)} \epsilon((h(t) - g_h(t))\operatorname{sgn}(g_1(t))\, dt$$

$$+ \int_{A(\epsilon)\setminus Z(g_1)} |\epsilon(h(t) - g_h(t)) - g_1(t)|\, dt$$

$$+ \int_{Z(g_1)} |\epsilon(h(t) - g_h(t))|\, dt - \int_{A(\epsilon)\setminus Z(g_1)} |g_1(t)|\, dt$$

$$= - \int_a^b \epsilon((h(t) - g_h(t))\operatorname{sgn}(g_1(t))\, dt$$

$$+ \int_{A(\epsilon)} \epsilon((h(t) - g_h(t))\operatorname{sgn}(g_1(t))\, dt$$

$$+ \int_{A(\epsilon)\setminus Z(g_1)} (\epsilon(h(t) - g_h(t)) - g_1(t))\operatorname{sgn}(\epsilon(h(t) - g_h(t)) - g_1(t))\, dt$$

$$+ \int_{Z(g_1)} |\epsilon(h(t) - g_h(t))|\, dt - \int_{A(\epsilon)\setminus Z(g_1)} (g_1(t))\operatorname{sgn}(g_1(t))\, dt$$

$$= \epsilon \left\{ \int_{Z(g_1)} |h(t) - g_h(t)|\, dt - \int_a^b ((h(t) - g_h(t))\operatorname{sgn}(g_1(t))\, dt \right\}$$

$$+\int_{A(\epsilon)\setminus Z(g_1)} (\epsilon(h(t) - g_h(t)) - g_1(t))$$
$$\{\operatorname{sgn}(\epsilon(h(t) - g_h(t)) - g_1(t)) + \operatorname{sgn} g_1(t)\}\, dt$$
$$\leq \epsilon \left\{ \int_{Z(g_1)} |h(t) - g_h(t)|\, dt - \int_a^b ((h(t) - g_h(t))\operatorname{sgn}(g_1(t))\, dt \right\}$$
$$+ \epsilon \int_{A(\epsilon)\setminus Z(g_1)} 2\, dt \quad \text{by (2).}$$

Since
$$\int_{Z(g_1)} |h(t) - g_h(t)|\, dt - \int_a^b ((h(t) - g_h(t))\operatorname{sgn}(g_1(t))\, dt < 0$$

and
$$\int_{A(\epsilon)\setminus Z(g_1)} 2\, dt \to 0 \quad \text{as } \epsilon \to 0,$$

for sufficiently small $\epsilon > 0$, it follows that
$$\|\epsilon h(t) - (\epsilon g_h(t) + g_1(t))\|_1 - \|\epsilon g_h(t) + g_1(t) - \epsilon g_h(t)\|_1 < 0.$$

This shows that ϵg_h is not a best L_1-coapproximation to ϵh, which implies that g_h is not a best L_1-coapproximation to h, a contradiction.

(ii) $\Rightarrow$ (i). Assume that (ii) holds. Let $g \in G$ be an arbitrary function. Since $g_f \in G$, $g - g_f \in G$. Hence by (ii) it follows that
$$\int_a^b (f(t) - g_f(t))\operatorname{sgn}(g(t) - g_f(t))\, dt \leq \int_{Z(g-g_f)} |f(t) - g_f(t)|\, dt.$$

Therefore, it follows that
$$\|f - g\|_1 - \|g - g_f\|_1$$
$$= \int_a^b |f(t) - g(t)|\, dt - \int_a^b |g(t) - g_f(t)|\, dt.$$
$$= \int_a^b (f(t) - g_f(t) + g_f(t) - g(t))\operatorname{sgn}(f(t) - g(t))\, dt$$
$$- \int_a^b (g(t) - f(t) + f(t) - g_f(t))\operatorname{sgn}(g(t) - g_f(t))\, dt.$$
$$= \int_a^b (f(t) - g_f(t))\operatorname{sgn}(f(t) - g_f(t) + g_f(t) - g(t))\, dt$$
$$+ \int_a^b (g_f(t) - g(t))\operatorname{sgn}(f(t) - g(t))\, dt$$
$$- \int_a^b (g(t) - f(t))\operatorname{sgn}(g(t) - g_f(t))\, dt$$

$$-\int_a^b (f(t) - g_f(t))\,\mathrm{sgn}\,(g(t) - g_f(t))\,dt$$

$$= \int_{[a,b]\setminus Z(g-g_f)} (f(t) - g_f(t))\,\mathrm{sgn}\,(f(t) - g(t))\,dt$$

$$+ \int_{Z(g-g_f)} (f(t) - g_f(t))\,\mathrm{sgn}\,(f(t) - g_f(t))\,dt$$

$$+ \int_{[a,b]\setminus Z(g-g_f)} (g_f(t) - g(t))\,\mathrm{sgn}\,(f(t) - g(t))\,dt$$

$$- \int_a^b (g(t) - f(t))\,\mathrm{sgn}\,(g(t) - g_f(t))\,dt$$

$$- \int_a^b (f(t) - g_f(t))\,\mathrm{sgn}\,(g(t) - g_f(t))\,dt$$

$$\geq \int_{[a,b]\setminus Z(g-g_f)} \{(f(t) - g_f(t)) + (g_f(t) - g(t))\}\,\mathrm{sgn}\,(f(t) - g(t))\,dt$$

$$- \int_a^b (g(t) - f(t))\,\mathrm{sgn}\,(g(t) - g_f(t))\,dt \geq 0.$$

Thus $\|g - g_f\|_1 \leq \|f - g\|_1$, for every $g \in G$. □

The following characterization can be considered as a criterion for best L_1-coapproximation.

Theorem 3.3. *Let G be a subset of $C[a,b]$, $f \in C[a,b] \setminus G$ and $g_f \in G$. Then the function g_f is a best L_1 coapproximation to f from G if and only if for all $g \in G$,*

$$\int_{[a,b]\setminus Z(f-g_f)} |g(t) - g_f(t)|\,dt \leq \int_{[a,b]\setminus Z(f-g_f)} |f(t) - g(t)|\,dt.$$

Proof: Since

$$\int_{Z(f-g_f)} |f(t) - g(t)|\,dt = \int_{Z(f-g_f)} |g(t) - g_f(t)|\,dt,$$

the proof follows immediately. □

Proposition 3.4. *Let G be a subset of $C[a,b]$, $f \in C[a,b] \setminus G$, and $g_f \in G$. Then g_f is a best L_1-coapproximation to f from G, if for every function $g \in G$ and all $t \in [a,b]$, either $(f(t) - g_f(t))(g_f(t) - g(t)) \geq 0$, or $g(t) - |f(t) - g(t)| \leq g_f(t) \leq g(t) + |f(t) - g(t)|$.*

Proof: The proof is obvious. □

Proposition 3.5. *Let G be a subset of $C[a,b]$, $f \in C[a,b] \setminus G$, and $g_f \in G$. If for every $g \in G$ and $t \in [a,b]$, $(g_f(t) - f(t)) \perp \mathrm{sgn}\,(g_f(t) - g(t))$, then g_f is a best L_1-coapproximation to f from G.*

Proof: Now

$$\|f-g\|_1 = \int_a^b |f(t)-g(t)|\,dt \geq \int_a^b (f(t)-g(t))\,\mathrm{sgn}\,(g_f(t)-g(t))\,dt$$

$$= \int_a^b (f(t)-g(t))\,\mathrm{sgn}\,(g_f(t)-g(t))\,dt$$

$$+ \int_a^b (g_f(t)-f(t))\,\mathrm{sgn}\,(g_f(t)-g(t))\,dt$$

$$= \int_a^b |g_f(t)-g(t)|\,dt = \|g_f-g\|_1. \quad \square$$

§4. Characterization of Best One-sided L_1-Coapproximation

Definition 4.1. Let G be a subset of $C[a,b]$ and $f \in C[a,b]$. A function $g_f \in G$ with $g_f \leq f$ (i.e., $g_f(t) \leq f(t)$, for all $t \in [a,b]$) is called a best one-sided L_1-approximation to f from G if for every $g \in G$ with $g \leq f$, $\|f - g_f\|_1 \leq \|f - g\|_1$. Similarly, a function $g_f \in G$ with $g_f \leq f$ is called a best one-sided L_1-coapproximation to f from G if for every $g \in G$ with $g \leq f$, $\|g - g_f\|_1 \leq \|f - g\|_1$.

Proposition 4.2. Let G be a subset of $C[a,b]$, $f \in C[a,b] \setminus G$, $g_f \in G$ and $g_f \leq f$. If for every $g \in G$ with $g \leq f$, $g - g_f \leq 0$, then g_f is a best one-sided L_1-approximation and best one-sided L_1-coapproximation to f from G.

Proof: The proof follows immediately by using the monotonicity of the integration. $\square$

Remark 4.3. In Proposition 4.2 the non-positivity of $g - g_f$ is not necessary, because in the case of approximation it is sufficient that $\int_a^b (g(t) - g_f(t))\,dt \leq 0$ and in the case of coapproximation it is sufficient that $g - g_f \leq f - g$. This can be verified easily.

Proposition 4.4. The set of best one-sided L_1-coapproximations is convex.

Proof: Let G be a subset of $C[a,b]$ and $f \in C[a,b] \setminus G$. If f has a unique best one-sided L_1-coapproximation, then the result is obvious. Let g_1 and g_2 be best one-sided L_1-coapproximations to f from G. It is sufficient to prove that $\alpha g_1 + (1-\alpha)g_2$, $0 \leq \alpha \leq 1$, is a best one-sided L_1-coapproximation to f from G. It can be easily checked that $\alpha g_1 + (1-\alpha)g_2 \leq f$ and for every $g \in G$ with $g \leq f$, $\|g - (\alpha g_1 + (1-\alpha)g_2)\| \leq \|f - g\|$. eop

Proposition 4.5. Let G be subset of $C[a,b]$ such that $0 \in G$ and $f \in C[a,b] \setminus G$. If 0 and $g_f \neq 0$ are best one-sided L_1-coapproximations to f from G, then $Z(f - \alpha g_f) \subset Z(g_f)$, $0 < \alpha < 1$. Furthermore, if f and g_f are continuously differentiable, then

$$Z(f' - \alpha g_f') \cap Z(f - \alpha g_f) \cap (a,b) \subset Z(g_f') \cap Z(g_f) \cap (a,b). \tag{1}$$

Proof: Since 0 and $g_f \neq 0$ are best one-sided L_1-coapproximations to f from G, by Proposition 4.4, αg_f, $0 \leq \alpha \leq 1$ is also a best one-sided L_1-coapproximation to f from G. It is clear that $f(t) - g_f(t) \geq 0$ and $f(t) \geq 0$, for all $t \in [a,b]$. Since $f - \alpha g_f = \alpha(f - g_f) + (1-\alpha)f$, $f(t) - \alpha g_f(t) = 0$ implies that $f(t) - g_f(t) = 0$ and $f(t) = 0$. Hence $g_f(t) = 0$. Thus $Z(f - \alpha g_f) \subset Z(g_f)$. Since $f(t) - g_f(t) \geq 0$, $f(t) \geq 0$ for all $t \in [a,b]$ and $f(t) = f(t) - g_f(t) = 0$ for all $t \in Z(f - \alpha g_f) \cap (a,b)$, it is clear that $f'(t) = f'(t) - g_f'(t) = 0$ for all $t \in Z(f' - \alpha g_f') \cap Z(f - \alpha g_f) \cap (a,b)$. Hence $g_f'(t) = 0$, for all $t \in Z(f' - \alpha g_f') \cap Z(f - \alpha g_f) \cap (a,b)$. Thus (1) is proved. □

Remark. Though the results in this paper have been proved for continuous functions, they can also be established for more general L_1 functions quite easily.

References

1. Franchetti, C. and M. Furi, Some characteristic properties of real Hilbert spaces, Rev. Roumaine Math. Pures Appl. **17** (1972), 1045–1048.

2. Rao, G. S. and K. R. Chandrasekaran, Best coapproximation in normed linear spaces with property (Λ), Math. Today **2** (1984), 33-40.

3. Rao, G. S., Best coapproximation in normed linear spaces, in *Approximation Theory V*, C. Chui, L. Schumaker, and J. Ward (eds.), Academic Press, New York, 1986, pp. 535–538.

4. Rao, G. S. and K. R. Chandrasekaran, Some properties of the maps R_G and R_G', Pure Appl. Math. Sci. **23** (1986), 21–27.

5. Rao, G. S. and K. R. Chandrasekaran, Characterizations of elements of best coapproximation in normed linear spaces, Pure Appl. Math. Sci. **26** (1987), 139–147.

6. Kripke, B. R. and T. J. Rivlin, Approximation in the metric of $L^1(x,\mu)$, Trans. Amer. Math. Soc.**119** (1965), 101–122.

7. Papini, P. L. and I. Singer, Best coapproximation in normed linear spaces, Mh. Math. **88** (1979), 27–44.

Geetha S. Rao and R. Saravanan
Ramanujan Institute for Advanced Study in Mathematics
University of Madras, Madras-600 005
India
geetsr@unimad.ernet.in

Algebraic Aspects of Polynomial Interpolation in Several Variables

Thomas Sauer

Abstract. This paper summarizes relations between the constructive theory of polynomial ideals and polynomial interpolation in several variables. The main ingredient is a generalization of the algorithm for division with remainder to "quotients" of polynomials and finite sets of polynomials.

§1. Introduction

When compared to the univariate case, polynomial interpolation in several variables turns out to be notoriously troublesome by providing various nontrivial new difficulties. To overcome some of these problems and to gain understanding of polynomial interpolation in several variables, it is necessary to adopt and apply techniques from algebraic geometry, in particular of the constructive theory of polynomial ideals. The main ingredient of this paper will be a concept closely related to the so-called Gröbner bases, which have been introduced by Buchberger [5, 6] in 1965 and which are an important tool in all Computer Algebra systems, especially, but not only, for algebraic techniques to solve polynomial systems of equations.

The issue to be considered here is a very simple observation in the univariate case: let $x_0, \ldots, x_n \in \mathbb{R}$ be distinct points and let

$$\omega(x) = (x - x_0) \cdots (x - x_n)$$

be the polynomial of degree $n + 1$ with normalized leading coefficient which vanishes in all the points. It is well known and probably taught in any class on Numerical Analysis that for any $f : \{x_0, \ldots, x_n\} \to \mathbb{R}$ there exists a unique interpolation polynomial $L_{x_0, \ldots, x_n} f$ of degree n. Moreover, if p is any polynomial, then there is an "algebraic" way to compute the interpolant with respect to p by doing division with remainder. Indeed, the polynomial p can be uniquely decomposed into

$$p(x) = q(x)\omega(x) + r(x), \tag{1}$$

where r is a polynomial of smaller degree than ω, *i.e.*, a polynomial of degree at most n. Since $q(x)\omega(x)$ vanishes at $x \in \{x_0, \ldots, x_n\}$, the remainder polynomial r coincides with p at these points and hence

$$r = L_{x_0, \ldots, x_n} p.$$

There are some simple remarks on this "algebraic approach" to interpolation: first, this approach immediately carries over to multiple interpolation points and yields a Hermite interpolation polynomial. Second, it is possible to interpolate at points which are given only *implicitly* as the zeros of a polynomial; the points themselves do not have to be computed.

In several variables this observation becomes much more involved. First, a finite set of distinct points does not define a "generic" associated interpolation space and the structure of this interpolation space depends not only on the number of points, but also on their geometric configuration (cf. [2, 4, 10]). Second, the multivariate decomposition corresponding to (1) needs additional assumptions on the "quotient." And, finally, not even the notion of "degree" is generic or unique in the multivariate setting.

We fix some notation for multivariate polynomials. Let $\mathbb{K}$ be an infinite field and let $\Pi = \mathbb{K}[x_1, \ldots, x_d]$ denote the ring of polynomials in d variables over $\mathbb{K}$. This ring is equipped with a **graded** structure by means of a totally ordered monoid Γ, *i.e.*, an (additive) semigroup with neutral element (zero) and an admissible well-ordering "$<$" on Γ (this means that zero is minimal element and that any strictly $<$-decreasing sequence in Γ stabilizes after finitely many steps). Γ is called a grading monoid if it induces a direct sum decomposition into **homogeneous subspaces** (abelian subgroups of Π)

$$\Pi = \bigoplus_{\gamma \in \Gamma} \Pi_\gamma$$

which is compatible with the semigroup operation on Γ in the sense that

$$\Pi_{\gamma+\gamma'} \subset \Pi_\gamma \, \Pi_{\gamma'}, \qquad \gamma, \gamma' \in \Pi.$$

Any polynomial $p \in \Pi$ can be represented by its homogeneous terms as

$$p = \sum_{\gamma \in \Gamma} p_\gamma, \qquad p_\gamma \in \Pi_\gamma, \, \gamma \in \Gamma,$$

where the sum only runs over *finitely* many terms. The **degree** of a polynomial p, written as $\delta(p) \in \Gamma$, is the ($<$-)maximal index of the nonzero homogeneous terms, *i.e.*,

$$\delta(p) = \max_{<} \{\gamma : p_\gamma \neq 0\}.$$

This notion naturally yields the leading term $\Lambda_\Gamma(p)$ of $p \in \Pi$ as the homogeneous polynomial $p_{\delta(p)}$. The "usual" gradings (which one should have in mind as "standard" examples) are $\Gamma = \mathbb{N}_0$ and grading by total degree or $\Gamma = \mathbb{N}_0^d$ together with a term order.

§2. Minimal Degree Interpolation

Let $\Theta \subset \Pi'$ be a finite set of linearly independent linear functionals defined on Π. Following [1] we say that Θ admits an ideal interpolation scheme if

$$\ker \Theta = \{p \in \Pi : \Theta(p) = (\theta(p) : \theta \in \Theta) = 0\} \subset \Pi,$$

is an ideal in Π. Ideal interpolation schemes can be characterized as Hermite interpolation schemes where the "local" differential operators to be interpolated are closed under taking derivatives; cf. [4, 11] for further aspects and references.

We say that a subspace $\mathcal{P} \subset \Pi$ is an **interpolation space** with respect to Θ, if for any $y \in \mathbb{K}^\Theta$ there exists a *unique* polynomial $p \in \mathcal{P}$ such that $\Theta(p) = y$. If $\mathcal{P}$ is an interpolation space with respect to Θ, then we write $L(\mathcal{P}; \cdot)$ for the (linear) interpolation operator $\mathbb{K}^\Theta \to \Pi$. If, in addition, the interpolation operator

$$L(\mathcal{P}; \Theta(\cdot)) : \Pi \to \mathcal{P}$$

is **degree reducing**, *i.e.*,

$$\delta(L(\mathcal{P}; \Theta(p))) \leq \delta(p), \qquad p \in \Pi,$$

then we call $\mathcal{P}$ a Γ-**minimal degree interpolation space** with respect to the grading monoid Γ. Clearly, these spaces depend on Θ and the grading structure, but still are not unique in general.

Let us look at the behavior of the minimal degree interpolation space $\mathcal{P}$ by considering the simple example of three points in $\mathbb{R}^2$.

1) If $\Gamma = \mathbb{N}_0$ and the grading is by total degree, then, as long as the points do not lie on a line, $\mathcal{P}$ is unique and consists of all linear polynomials. On the other hand, if the three points lie on a line, then $\mathcal{P}$ consists of the constants, one linear and one quadratic polynomial, where the latter two can be chosen in various ways.
2) If $\Gamma = \mathbb{N}_0^d$ and the term order is lexicographic with $x < y$, then the interpolation space depends on whether the x-components of the points coincide. If these are all distinct, then $\mathcal{P} = \text{span}_{\mathbb{K}}\{1, x, x^2\}$ is unique, in the other cases there is again a multitude of choices for $\mathcal{P}$ available.

For information on the practical construction of minimal degree interpolation spaces with respect to the "standard" gradings see [3, 10].

§3. Reduction and Syzygies – How to Construct "Good" Ideal Bases for Graded Polynomial Rings

For a (finite or infinite) set of polynomials $\mathcal{P} \subset \Pi$, the ideal generated by $\mathcal{P}$ is defined as

$$\langle \mathcal{P} \rangle = \left\{ \sum_{p \in \mathcal{P}} q_p\, p : q_p \in \Pi,\, p \in \mathcal{P} \right\}.$$

It is stated in Hilbert's well known *Basissatz* that any ideal in Π has a *finite* basis, *i.e.*, for any ideal $\mathcal{I} \subset \Pi$ there exists a *finite* set $\mathcal{P}$ of polynomials such that $\mathcal{I} = \langle \mathcal{P} \rangle$. However, one is usually interested in "good" or "standard" bases $\mathcal{G}$ which have the desirable (as we will, hopefully, see soon) additional property that any polynomial $p \in \langle \mathcal{G} \rangle$ can be represented as

$$p = \sum_{g \in \mathcal{G}} p_g \, g \qquad \delta(p) \geq \delta(p_g g), \qquad g \in \mathcal{G}. \tag{2}$$

The important requirement here is that all terms in the sum of the right-hand side of the representation have their degree bounded by $\delta(p)$. An ideal basis $\mathcal{G}$ which is able to represent any $p \in \langle \mathcal{G} \rangle$ as in (2) is called a Γ-basis.

Suppose that $\mathcal{G}$ is *not* a Γ-basis for $\langle \mathcal{G} \rangle$. Then there exists a $p \in \mathcal{G}$ such that, for any representation of the form $p = \sum_{g \in \mathcal{G}} p_g g$ at least one of the terms in the sum has degree strictly bigger than $\delta(p)$. Let γ be the maximal degree appearing in this representation and let $\mathcal{G}' \subset \mathcal{G}$ be given as

$$\mathcal{G}' = \{g \in \mathcal{G} : \, \delta(p_g g) = \gamma\}.$$

Since $\delta(p) < \gamma$ we have that

$$0 = \sum_{g \in \mathcal{G}'} \Lambda_\Gamma(p_g g) = \sum_{g \in \mathcal{G}'} \Lambda_\Gamma(p_g) \, \Lambda_\Gamma(g), \tag{3}$$

hence there exists a (nonlinear homogeneous) dependency relation between the homogeneous polynomials $\Lambda_\Gamma(g)$, $g \in \mathcal{G}$. This is what is called a syzygy, and the above exposition indicates that syzygies will play an important role for the construction of Γ-bases.

Definition 1. *Let $\mathcal{P} \subset \Pi$ be a finite set of polynomials. A vector $\boldsymbol{q} \in \Pi^\mathcal{P}$ is called a syzygy for $\mathcal{P}$ if*

$$\boldsymbol{q} \cdot \mathcal{P} := \sum_{p \in \mathcal{P}} q_p p = 0.$$

The module of all syzygies for $\mathcal{P}$ is denoted by $S(\mathcal{P})$.

The module of syzygies $S(\mathcal{P})$, which can be thought of as a "vector space" over the ring Π (one can add and subtract two syzygies and multiply them, component by component, with a polynomial), is known to be finitely generated, *i.e.*, there exists a finite set $\mathcal{S} \subset S(\mathcal{P})$ of syzygies such that any syzygy $\boldsymbol{q} \in S(\mathcal{P})$ can be written as

$$\boldsymbol{q} = \sum_{\boldsymbol{s} \in \mathcal{S}} q_{\boldsymbol{s}} \, \boldsymbol{s}, \qquad q_{\boldsymbol{s}} \in \Pi, \qquad \boldsymbol{s} \in \mathcal{S}.$$

Such a basis can even be computed in finitely many steps, making use of a reduced Gröbner basis as described in [7].

We can now reformulate (3) in terms of syzygies, saying that whenever $\mathcal{G}$ is *not* a Γ-basis, then there exist certain nontrivial syzygies of *leading terms* of $\mathcal{G}$, i.e., certain nontrivial elements of $S(\Lambda_\Gamma(\mathcal{G}))$.

Next, we turn to a generalization of the division-with-remainder algorithm to the multivariate setting, extending the "standard" reduction process which is the core part of the Gröbner basis construction (see, for example, [8], which is generally a nice introduction to constructive ideal theory). For that purpose we let $(\cdot,\cdot)$ be any scalar product defined on $\Pi \times \Pi$. Given a polynomial $f \in \Pi$ and a finite set $\mathcal{P} \subset \Pi$ of polynomials, we want to compute a decomposition

$$f = \sum_{p \in \mathcal{P}} q_p \, p + r, \qquad (4)$$

where r is of a special, reduced form. In the "standard" case of reduction ($\Gamma = \mathbb{N}_0^d$ and a term order), a polynomial is reduced if it contains no monomial term which is a multiple of any of the monomial leading terms $\Lambda_\Gamma(p)$, $p \in \mathcal{P}$. For general grading monoids, however, in particular for the grading by total degree, such a strong requirement cannot be satisfied. To extend the notion of reduced polynomials in suitable way, we define, for $\gamma \in \Gamma$, the homogeneous polynomial vector spaces

$$V_\gamma(\mathcal{P}) = \left\{ \sum_{p \in \mathcal{P}} q_p \, \Lambda_\Gamma(p) : q_p \in \Pi_{\gamma - \delta(p)} \right\} \subset \Pi_\gamma,$$

and call $p \in \mathcal{P}$ reduced with respect to $\mathcal{P}$ if each homogeneous term of p is $(\cdot,\cdot)$-orthogonal to the respective homogeneous subspace, i.e.,

$$(p_\gamma, V_\gamma(\mathcal{P})) = 0, \qquad \gamma \in \Gamma.$$

As shown in [12] a decomposition of the form (4) can be computed by successive orthogonal projections, yielding a remainder $r =: p_{\overrightarrow{\mathcal{P}}}$ which is reduced with respect to $\mathcal{P}$. However, this algorithm requires some ordering of $\mathcal{P}$, and in general different orderings lead to different remainders. Nevertheless, there is an important special case where any reduced remainder r in the representation (4) is *always* $p_{\overrightarrow{\mathcal{P}}}$, the one computed by the division-with-remainder algorithm.

Theorem 2. *Let $\mathcal{G}$ be a Γ-basis and suppose that $p \in \Pi$ can be written as*

$$p = \sum_{g \in \mathcal{G}} p_g \, g + r,$$

where r is reduced with respect to $\mathcal{G}$. Then $r = p_{\overrightarrow{\mathcal{G}}}$.

In particular, $p_{\overrightarrow{\mathcal{G}}}$ is independent of the ordering of $\mathcal{G}$.

Moreover, with the help of reduction we can now make the intuition that "Γ-bases have something to do with syzygies of leading terms" precise.

Theorem 3. *Let $\mathcal{G} \subset \Pi$ be a finite set of polynomials and let $\mathcal{S}$ be a (finite) basis of $S(\Lambda_\Gamma(\mathcal{G}))$. Then $\mathcal{G}$ is a Γ-basis if and only if*

$$(s \cdot \mathcal{G}) \underset{\mathcal{G}}{\to} = 0, \qquad s \in \mathcal{S}. \tag{5}$$

This theorem enables us to compute, from any given finite set of polynomials $\mathcal{P}$, a Γ-basis $\mathcal{G}$ for $\langle \mathcal{P} \rangle$ by a variant of Buchberger's Algorithm. The basic idea of that algorithm is very simple: take any element $s \in S(\Lambda_\Gamma(\mathcal{G}))$ and compute $h = (s \cdot \mathcal{G}) \underset{\mathcal{G}}{\to}$. If $h = 0$ for all basis syzygies s, then we have a Γ-basis by Theorem 3, if $h \neq 0$, then we proceed with $\mathcal{P} \cup \{h\}$, which yields a Γ-basis after a finite number of these extension steps.

Since the reduction $\underset{\mathcal{G}}{\to}$ modulo a Γ-basis $\mathcal{G}$ is a well-defined projection, we can use $p \underset{\mathcal{G}}{\to}$ as the "standard" representer of the equivalence class $[p] = p + \langle \mathcal{G} \rangle$. Therefore, we call $p \underset{\mathcal{G}}{\to}$ the normal form of p modulo $\mathcal{G}$.

Remark 4. It is important to realize that the notion of reduction defined in this section is subject to *two* intrinsic parameters: one is the grading monoid, the other one is the scalar product which defines the notion of orthogonality and therefore of being reduced. In general, changing one of these parameters leads to different Γ-bases and different reduced remainders. However, if $\Gamma = \mathbb{N}_0^d$ and the spaces Π_γ consist of single monomials, then the scalar product becomes irrelevant and everything boils down to the standard Gröbner bases theory.

§4. A Minimal Degree Interpolation Space

The notion of reduction introduced in the last section can now be connected to minimal degree interpolation in a way which yields a natural extension of equation (1).

Theorem 5. *Suppose that $\Theta \subset \Pi'$ admits an ideal interpolation scheme and let $\mathcal{G}$ be a Γ-basis for $\ker \Theta$. Then $\Pi \underset{\mathcal{G}}{\to}$ is a Γ-minimal degree interpolation space with respect to Θ and*

$$L\left(\Pi\underset{\mathcal{G}}{\to}, \Theta(p)\right) = p\underset{\mathcal{G}}{\to}, \qquad p \in \Pi.$$

Some of these "normal form" interpolation spaces are well known: if $\Gamma = \mathbb{N}_0^d$ together with a term order, then these are the minimal degree interpolation spaces with minimal monomials (cf. [3, 10, 11]); if $\Gamma = \mathbb{N}_0$ together with grading by total degree and scalar product $(p, q) = (p(D)q)(0)$, then one obtains the least interpolation space introduced by de Boor and Ron [2].

Indeed, it is no accident that least interpolation spaces show up here as normal forms, as long as the scalar product is chosen as $(p, q) = (p(D)q)(0)$.

Algebraic Aspects of Polynomial Interpolation

To see this, we call the grading induced by Γ a monomial grading if the homogeneous spaces Π_γ, $\gamma \in \Gamma$, have a vector space basis consisting of monomials, *i.e.*, there exist sets $I_\gamma \subset \mathbb{N}_0^d$, $\gamma \in \Gamma$, such that

$$\Pi_\gamma = \text{span}_{\mathbb{K}} \{x^\alpha : \alpha \in I_\gamma\}, \gamma \in \Gamma.$$

Let $\mathbb{K}[\![x_1, \ldots, x_d]\!]$ denote the formal power series with coefficients in $\mathbb{K}$; it has been pointed out by de Boor and Ron that Π' can be identified with $\mathbb{K}[\![x_1, \ldots, x_d]\!]$ via the assignment

$$\Pi' \ni \theta \leftrightarrow f_\theta \in \mathbb{K}[\![x_1, \ldots, x_d]\!] \quad \Leftrightarrow \quad \theta(p) = (p, f_\theta), \quad p \in \Pi.$$

The Γ-least term of a power series f, $\lambda_\Gamma(f)$ is then the Γ-minimal term nonzero term of the power series, *i.e.*,

$$\lambda_\Gamma(f) = \min_{\gamma \in \Gamma} \{f_\gamma : f_\gamma \neq 0\}, \qquad f = \sum_{\gamma \in \Gamma} f_\gamma.$$

Following [4], we associate to any $\Theta \subset \Pi'$ the least polynomial space

$$\lambda_\Gamma(f_\Theta) = \{\lambda_\Gamma(f) : f \in \text{span}_{\mathbb{K}} \{f_\theta : \theta \in \Theta\}\}.$$

We then have the following result.

Theorem 6. *Suppose that $\Theta \subset \Pi'$ admits an ideal interpolation scheme and let $\mathcal{G}$ be a Γ-basis for $\ker \Theta$, where Γ is a monomial grading. Then*

$$\Pi_{\vec{\mathcal{G}}} = \lambda_\Gamma(f_\Theta) = \bigcap_{g \in \mathcal{G}} \ker \Lambda_\Gamma(g)(D). \tag{6}$$

Corollary 7. *Under the assumptions of Theorem 6 any least polynomial space is a Γ-minimal degree interpolation space.*

Finally, (6) also yields a description of the joint kernel of Γ-homogeneous partial differential operators with constant coefficients.

Corollary 8. *Let $\mathcal{P}$ be a finite set of Γ-homogeneous polynomials and let $\mathcal{G}$ be a Γ-basis for $\langle \mathcal{P} \rangle$. Then*

$$\bigcap_{p \in \mathcal{P}} \ker p(D) = \Pi_{\vec{\mathcal{G}}}.$$

References

1. Birkhoff, G., The algebra of multivariate interpolation, in *Constructive Approaches to Mathematical Models*, C. V. Coffman and G. J. Fix (eds.), Academic Press 1979, pp. 345–363.

2. de Boor, C. and A. Ron, On multivariate polynomial interpolation, Constr. Approx. **6** (1990), 287–302.

3. de Boor, C. and A. Ron, Computational aspects of polynomial interpolation in several variables, Math. Comp. **58** (1992), 705–727.

4. de Boor, C. and A. Ron, The least solution for the polynomial interpolation problem, Math. Z. **210** (1992), 347–378.

5. Buchberger, B., *Ein Algorithmus zum Auffinden der Basiselemente des Restklassenrings nach einem Nulldimensionalen Polynomideal*, PhD Thesis, Innsbruck, 1965.

6. Buchberger, B., An Algorithmic criterion for the solvability of algebraic systems of equations (German), Aequationes Math. **4** (1970), 374–383.

7. Buchberger, B., Gröbner bases: An algorithmic method in polynomial ideal theory, in *Multidimensional Systems Theory*, N. K. Bose (ed), D. Reidel Publishing Company, 1985, pp. 184–232.

8. Cox, D., J. Little, and D. O'Shea, *Ideals, Varieties and Algorithms*, Springer, 1992.

9. Marinari, M. G., H. M. Möller, and T. Mora, Gröbner bases of ideals given by dual bases, in *Proceedings of ISAAC 1991*, Springer, 1991.

10. Sauer, T., Polynomial interpolation of minimal degree, Numer. Math. **78** (1997), 59–85.

11. Sauer, T., Polynomial interpolation of minimal degree and Gröbner bases, in *Groebner Bases and Applications (Proc. of the Conf. 33 Years of Groebner Bases)*, B. Buchberger and F. Winkler (eds), Cambridge University Press, 1998, pp. 483–494.

12. Sauer, T., Ideal bases for graded polynomial rings and applications to interpolation, 1998, preprint.

Thomas Sauer
Mathematical Institute
University Erlangen–Nuremberg
Bismarckstr $1\frac{1}{2}$
D-91054 Erlangen
Germany.
sauer@mi.uni-erlangen.de
http://www.mi.uni-erlangen.de/~sauer/

Adaptive Approximation and Compression

Bl. Sendov

Abstract. A new method is established for adaptive approximation of functions, defined on a compact dyadic topological group. This method is applied in signal and image compression, using the correspondence between the dyadic topological group and the unit interval or the unit square. Computer examples of adaptive approximation and compression are presented.

§1. Introduction

It is natural to choose the instrument for approximation according to the object to be approximated. In approximation theory, the instrument for approximation is usually tuned to the class of functions to be approximated. For a rough function, representing a signal or an image, an individual adaptation of the instrument for approximation is useful. For example, the wavelet packets and libraries of wavelets are used for individual adaptation in the methods for image compression [2].

In this paper, a method for individual adaptation of an instrument for approximation to a given function, from the set $\mathcal{C}$ of the continuous functions defined on a compact dyadic topological group, is established. The concept is the following:

1) A non-linear infinite dimensional subset $\mathcal{S} \subset \mathcal{C}$ of so-called starting functions is defined. From every starting function $\phi \in \mathcal{S}$ an orthonormal system $\phi_0 = \phi, \phi_1, \phi_2, \phi_3, \ldots$ may be constructed, which is closed in $\mathcal{C}$. This construction is simple and identical for every function from $\mathcal{S}$. A starting function is like a mother wavelet, from which an orthonormal system is constructed only by dilates and translates.

2) For every function $f \in \mathcal{C}$ and every natural number n, on the basis of a given criterion for "similarity" between functions in $\mathcal{C}$, n parameters $\lambda_1, \lambda_2, \lambda_3, \ldots, \lambda_n$ are calculated, which define a function $\phi = \phi(f; .) \in \mathcal{S}$.

3) An orthonormal system

$$\phi_0 = \phi(f;.), \phi_1, \phi_2, \phi_3, \ldots, \phi_{2^n-1} \qquad (1)$$

is constructed and the best L_2 approximation

$$\Phi_n(f;x) = \sum_{i=0}^{2^n-1} c_i \phi_i(x), \qquad (2)$$

where c_i, $i = 0, 1, \ldots, 2^n - 1$, are the Fourier coefficients of the function f for the orthonormal system (1), is calculated.

In this way $\Phi_n(f;.)$ is an approximation of the function f defined by $2^n + n$ numbers instead of 2^n numbers. But the adaptation of the orthonormal system (1) to the function f may provide Fourier coefficients with smaller entropy, which is significant for compression.

§2. Dyadic Topological Group

The dyadic topological group $\mathbf{G}$ is the set of all 0, 1 sequences $\mathbf{x} = (x_1, x_2, \ldots)$ in which the group operation is $\mathbf{x} + \mathbf{y} = (x_1 \dot{+} y_1, x_2 \dot{+} y_2, x_3 \dot{+} y_3, \ldots)$, where $\dot{+}$ is addition modulo 2 [3].

$\mathbf{G}$ is a compact metric space with distance

$$\beta(\mathbf{x}, \mathbf{y}) = \sum_{i=1}^{\infty} 2^{-i} |x_i - y_i| = \sum_{i=1}^{\infty} 2^{-i}(x_i \dot{+} y_i). \qquad (3)$$

We set the notations:

$$\mathbf{0} = (0,0,0,\ldots), \quad \mathbf{1} = (1,1,1,\ldots), \quad \mathbf{2}^{-k} = (\underbrace{0,0,0,\ldots,0}_{k-1},1,0,0,0,\ldots).$$

An element $\mathbf{x} \in \mathbf{G}$ is called left rational if it has finite number of coordinates equal to 0 and right rational if it has finite number of coordinates equal to 1.

The order in $\mathbf{G}$ is lexicographical. For every element $\mathbf{x} \in G$ we have $\mathbf{0} \leq \mathbf{x} \leq \mathbf{1}$.

Definition 1. Let

$$\mathbf{x} = (x_1, x_2, x_3, \ldots, x_k, 0, 0, 0, \ldots), \quad \mathbf{y} = (x_1, x_2, x_3, \ldots, x_k, 1, 1, 1, \ldots).$$

The segment $[\mathbf{x}, \mathbf{y}] = [k; p]$, where $p = 2^{k-1}x_1 + 2^{k-2}x_2 + \ldots + 2x_{k-1} + x_k$ is called pixel of rank k.

It is obvious that $[k;0] \bigcup [k;1] \bigcup [k;2] \bigcup \ldots \bigcup [k;2^k-1] = [0;0] = [\mathbf{0},\mathbf{1}] = \mathbf{G}$, where $[k;p] \bigcap [k;q] = \emptyset$ for $p \neq q$. Define the function $l : \mathbf{G} \to [0,1]$ as follows

$$l(\mathbf{x}) = \sum_{i=1}^{\infty} 2^{-i} x_i = x \in [0,1]. \qquad (4)$$

The function l does not have a single-valued inverse on the rational elements of $\mathbf{G}$. Consider the right rational in this case, with one exception.

Let $m : [0,1] \to \mathbf{G}$ be the inverse of the function l, with the agreement that if $x \in [0,1]$ is binary rational number, then $m(x)$ is the right rational element of $\mathbf{G}$ corresponding to x. The exception is for $1 \in [0,1]$, namely $m(1) = \mathbf{1} = (1,1,1,\ldots)$.

From (3) and (4), it follows that $|x - y| \leq \beta(\mathbf{x},\mathbf{y}) \leq 2|x - y|$, for $l(\mathbf{x}) = x \neq y = l(\mathbf{y})$.

§3. Real Dyadic Functions

The set of all dyadic functions $f : \mathbf{G} \to \mathbb{R}$ is denoted by $\mathcal{G}$. The set of points $(l(\mathbf{x}), f(\mathbf{x})) \in [0,1] \times \mathbb{R}$, where l is defined by (4), is used as a "graphical representation" of the dyadic function f. When the values of f, for the two different rational elements $\mathbf{x}, \mathbf{x}' \in \mathbf{G}$ corresponding to one real number $x = l(\mathbf{x}) = l(\mathbf{x}')$, are different, we join the points $(x, f(\mathbf{x}))$, $(x, f(\mathbf{x}'))$ with a vertical segment. The graphical representation of a dyadic function $f(\mathbf{x})$ is the *completed graph* [4] of the multi-valued function $\tilde{f}(x) = f(\mathbf{x})$, where $x = l(\mathbf{x})$.

The uniform norm of a bounded function $f \in \mathcal{G}$ is defined as : $\|f\| = \sup\{|f(\mathbf{x})|; \mathbf{x} \in \mathbf{G}\}$.

3.1. Continuous Functions

The uniform module of continuity of a function $f \in \mathcal{G}$ is

$$\omega(f; \delta) = \sup\{|f(\mathbf{x}) - f(\mathbf{y})| : \beta(\mathbf{x},\mathbf{y}) < \delta, \ \mathbf{x}, \mathbf{y} \in G\}.$$

The dyadic function f is continuous iff $\lim_{\delta \to +0} \omega(f;\delta) = 0$. The set of all continuous dyadic functions is denoted by $\mathcal{C}$. The integral in $\mathcal{C}$ may be defined as

$$\int_G f(\mathbf{x})\, d\mathbf{x} = \int_0^1 \tilde{f}(x)\, dx$$

and the L_2 norm as

$$\|f\|_2 = \left(\int_G f(\mathbf{x})^2\, d\mathbf{x}\right)^{1/2}.$$

Definition 2. Let $f, \phi \in \mathcal{C}$ and $2^{-k-1} < \delta \leq 2^{-k}$, then

$$\omega_\phi(f, \delta)_2 = \left\{\sum_{p=0}^{2^k-1} \inf_{c \in \mathbb{R}} \left[\int_{[k;p]} |f(\mathbf{x}) - c\phi(\mathbf{x})|^2\, d\mathbf{x}\right]\right\}^{1/2} \tag{5}$$

is called the L_2 relative module of continuity of the function f to the function ϕ.

For every pair $f, \phi \in \mathcal{C}$ we have $\lim_{\delta \to +0} \omega_\phi(f;\delta)_2 = 0$.

3.2. Pixel Functions

A function $f \in \mathcal{G}$ is called a pixel function of **rank** k if $f(\mathbf{x}) = c_p =$ constant for $\mathbf{x} \in [k;p]$; $p = 0, 1, 2, \ldots, 2^k - 1$. The set of all pixel functions of rank k is denoted by $\mathcal{P}_k$. It is easy to see that $\mathcal{P}_0 \subset \mathcal{P}_1 \subset \mathcal{P}_2 \subset \ldots \subset \mathcal{P}_s \subset \ldots$ and $\bigcup_{s=0}^{\infty} \mathcal{P}_s = \mathcal{C}$.

From the definition of the module of continuity it follows directly that the necessary and sufficient condition for a function $P \in \mathcal{G}$ to be a pixel function of rank k is $\omega(P;\delta) = 0$ for $\delta \leq 2^{-k}$. Every pixel function is a continuous dyadic function.

The L_2 relative module of continuity (5) of a function f to a function ϕ may be represented in the form

$$\omega_\phi(f,\delta)_2 = \inf\{\|f - P\phi\|_2 : P \in \mathcal{P}_k\}, \qquad (6)$$

where $2^{-k-1} < \delta \leq 2^{-k}$.

3.3. Dyadic Exponential Functions

Definition 3. Let $\Lambda = \{\lambda_k\}_{k=1}^{\infty}$ be a sequence, $\lambda_i \in \mathbb{R} \setminus 0$, $\overline{\lambda}_i = \max\{1, \lambda_i\}$, $\underline{\lambda}_i = \min\{1, \lambda_i\}$, such that

$$\lim_{k \to \infty} \prod_{i=k}^{\infty} \lambda_i = \lim_{k \to \infty} \prod_{i=k}^{\infty} \overline{\lambda}_i = \lim_{k \to \infty} \prod_{i=k}^{\infty} \underline{\lambda}_i = 1.$$

Let $\lambda_i(0) = 1$, $\lambda_i(1) = \lambda_i$, then

$$\Lambda(\mathbf{x}) = c \prod_{i=1}^{\infty} \lambda_i(x_i), \quad c = \Lambda(0), \quad \|\Lambda\|_2 = \prod_{i=1}^{\infty} \left(\frac{1+\lambda_i^2}{2}\right)^{1/2},$$

is called dyadic exponential function with sequence Λ.

The motivation for this name is the fact that a dyadic exponential function is $e^{\mathbf{x}} = \prod_{i=1}^{\infty} e^{2^{-i}x_i}$. A dyadic exponential function Λ with $\|\Lambda\|_2 = 1$ is called a starting function. The unit sphere $\mathcal{S}$ in the set of dyadic exponential functions is the set of starting functions. The set of dyadic exponential functions is not linear, but it is multiplicative.

If $\lambda_i \geq 1 + \xi^{-1}$, $\xi \in (1,2]$, then the dyadic exponential function Λ has fractal dimension $\geq 2 - \ln_2 \xi$. Every dyadic exponential function Λ is continuous since

$$\omega(\Lambda; 2^{-s}) = c\left(\prod_{i=s+1}^{\infty} \overline{\lambda}_i - \prod_{i=s+1}^{\infty} \underline{\lambda}_i\right) \to 0 \quad \text{for} \quad s \to \infty.$$

From Definition 3, it follows that,

$$\Lambda\left(\mathbf{x} + \mathbf{2}^{-k}\right) = \begin{cases} \lambda_k \Lambda(\mathbf{x}) & \text{for } x_k = 0, \\ \lambda_k^{-1} \Lambda(\mathbf{x}) & \text{for } x_k = 1. \end{cases} \qquad (7)$$

Adaptive Approximation

Let ϕ, ψ be dyadic exponential functions with sequences $\{\lambda_k\}_1^\infty$, $\{\mu_k\}_1^\infty$, then from Definition 3 and (7) there follows

$$\langle \phi, \psi \rangle = \int_G \phi(\mathbf{x})\psi(\mathbf{x})\, d\mathbf{x} = \phi(\mathbf{0})\psi(\mathbf{0}) \prod_{i=1}^\infty \frac{1+\lambda_i\mu_i}{2}. \tag{8}$$

Lemma 1. *Let ϕ be a starting function with sequence $\{\lambda_k\}_{k=1}^\infty$, and*

$$r_s(\phi; \mathbf{x}) = (-1)^{x_s}\phi(\mathbf{x} + 2^{-s}).$$

Then $\psi(\mathbf{x}) = r_s(\phi; \mathbf{x})$ is also a starting function and $\langle \psi, \phi \rangle = 0$.

Proof: Consider that ψ is a dyadic exponential function with sequence $\{\mu_k\}_{k=1}^\infty$, where $\mu_i = \lambda_i$ for $i \neq s$ and $\mu_s = -1/\lambda_s$. Then apply (8). □

3.4. Orthonormal Bases

Theorem 1. *Let $\phi = \phi_0$ be a starting function and*

$$\phi_{2^{s-1}+p}(\mathbf{x}) = r_s(\phi_p; \mathbf{x}); \quad s = 1, 2, 3, \ldots, \quad p = 0, 1, 2, \ldots, 2^{s-1} - 1.$$

Then $\{\phi_n\}_{n=0}^\infty$ is an orthonormal basis in $\mathcal{C}$ and for every function $f \in \mathcal{C}$, the equality

$$E_{2^k}(\phi; f)_2 = \|f - \Phi_k(f; \cdot)\|_2 = \omega_\phi(f; 2^{-k})_2 \tag{9}$$

holds, where $\Phi(f; .)$ is defined by (2).

Proof: The orthonormality of $\{\phi_n\}_{n=0}^\infty$ follows directly from Lemma 1. The proof of (9) is based on (6). □

If the starting function is $\phi(\mathbf{x}) = 1;\ \mathbf{x} \in \mathbf{G}$, then the orthonormal system in Theorem 1 is the Walsh system. It is possible to generate from a starting function also an orthonormal system of Haar type.

§4. Adaptation of a Starting Function

Consider the starting function ϕ_0 with sequence $\Lambda = \{\lambda_1, \lambda_2, \ldots, \lambda_n, 1, 1, \ldots\}$. Then $\phi_0 \in \mathcal{P}_n$ and the orthonormal system $\{\phi_n\}_{n=0}^{2^n-1}$, generated from ϕ_0 according to Theorem 1, is an orthonormal basis in $\mathcal{P}_n$ depending on n parameters. These parameters may be used to adapt this basis to a given function $f \in \mathcal{C}$.

4.1. Entropy Criterion

Let $\{\phi_i\}_{i=0}^{2^n-1}$ be an orthonormal basis in $\mathcal{P}_n$, and $c_i(\phi; f)$, $i = 0, 1, 2, \ldots, 2^n-1$ be the Fourier coefficients of the function $f \in \mathcal{P}_n$. The entropy criterion, for the adaptation of an orthonormal basis $\{\phi_i\}$ to a given function f with $\|f\|_2 = 1$, is to minimize the value of the entropy

$$\epsilon^2(\phi; f) = -\sum_{i=0}^{2^n-1} c_i(\phi; f)^2 \ln c_i(\phi; f)^2. \tag{10}$$

This criterion is used for adaptation of Malvar wavelets and wavelet packets [2].

Calculation of the parameters of the starting function ϕ_0, for a given function f ($\|f\|_2 = 1$), in such a way that the entropy (10) be minimal, is practically possible for small values of n.

4.2. First Coefficient Criterion

A criterion for adaptation is to maximize the first coefficient $c_0(\phi; f)^2$. That is to maximize the energy taken by the first Fourier coefficient from the expansion of f in this basis. The optimization problem is to find the maximum of the function

$$F(\xi_1, \xi_2, \ldots, \xi_n) = \left(\prod_{i=1}^{n} \frac{1+\xi_i^2}{2} \right)^{-1/2} \int_G f(\mathbf{x}) \prod_{i=1}^{n} \xi_i^{x_i} \, d\mathbf{x}. \qquad (11)$$

The necessary conditions for an extremum of (11)

$$\frac{\partial F}{\partial \xi_1} = \frac{\partial F}{\partial \xi_2} = \cdots = \frac{\partial F}{\partial \xi_n} = 0$$

are equivalent to the system

$$\xi_k = \int_{\mathbf{G}_k^{(1)}} f(\mathbf{x}) \xi_k^{-1} \prod_{i=1}^{n} \xi_i^{x_i} \, d\mathbf{x} \Big/ \int_{\mathbf{G}_k^{(0)}} f(\mathbf{x}) \prod_{i=1}^{n} \xi_i^{x_i} \, d\mathbf{x}, \quad k = 1, 2, \ldots, n. \qquad (12)$$

For $n = 3$, the equations (12) are

$$\xi_1 = \frac{f_4 + f_5\xi_3 + f_6\xi_2 + f_7\xi_2\xi_3}{f_0 + f_1\xi_3 + f_2\xi_2 + f_3\xi_2\xi_3}, \quad \xi_2 = \frac{f_2 + f_3\xi_3 + f_6\xi_1 + f_7\xi_1\xi_3}{f_0 + f_1\xi_3 + f_4\xi_1 + f_5\xi_2\xi_3},$$

$$\xi_3 = \frac{f_1 + f_3\xi_2 + f_5\xi_1 + f_7\xi_1\xi_2}{f_0 + f_2\xi_2 + f_4\xi_1 + f_3\xi_1\xi_2}; \quad f_k = \int_{[3;k]} f(\mathbf{x}) \, d\mathbf{x}, \quad k = 0, \ldots, 7.$$

The Newton method for solving (12) is working very well.

To illustrate the adaptation by first coefficient criterion, we consider the following:

Example 1. Let $g(\mathbf{x}) = \tilde{g}(x)$ be the fractal function - the fixed point of the Iterated Function System (IFS) [1]:

$$\tilde{g}(x) = (m+2)^{-1}\tilde{g}(4x-m) + (m+1)^2; \quad x \in [m/4, (m+1)/4), \quad m = 0, 1, 2, 3.$$

Figure 1 shows the completed graphs of $\tilde{g}$ and the starting function $\tilde{\phi}$, adapted to g on the base of the first coefficient criterion. The entropy of the Fourier coefficients of the fractal function g, ($\|g\|_2 = 1$, for 2^{12} pixels) are given on Table 1 for the pixel representation, for the orthonormal systems of Haar and Walsh and for the Adopted orthonormal system by 12 parameters and 28 parameters. The influence of the adaptation, according to first coefficient criterion, on the minimization of the entropy (10) is obvious.

Adaptive Approximation

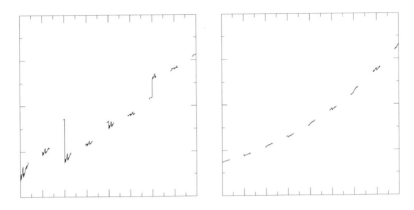

Fig. 1. Fractal function g and its adapted starting function.

Signal	Size	Pixels	Haar	Walsh	Adapt. 12 p.	Adapt. 28 p.
Function g	2^{12}	11.5448	0.9088	0.9166	0.1280	0.0437

Tab. 1. Entropy of the Fourier coefficients of g.

Image	Size	Pixels	Haar	Walsh	Ad.18 p.	Ad.27 p.	Ad.42 p.
Lenna	$2^9 \times 2^9$	17.6568	1.7563	1.7823	1.6395	1.5850	1.5417
Eye	$2^9 \times 2^9$	17.9431	0.3049	0.3109	0.2299	0.2163	0.1828

Tab. 2. Entropy of the Fourier coefficients of Lenna and the eye's retina.

 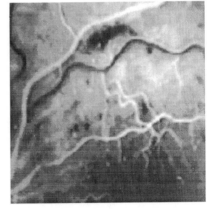

Fig. 2. Images of Lenna and ophtalmoscopia of an eye's retinae.

§5. Generalizations

Call the function

$$\Lambda(\mathbf{x}) = c \prod_{i=1}^{\infty} \lambda_i(x_{\nu(i-1)+1}, x_{\nu(i-1)+2}, \ldots, x_{\nu i}); \ \|\Lambda\|_2 = 1$$

a starting function of rank ν. The theory we presented for the starting functions of rank 1 may be generalized only for the starting functions of ranks 2 and 3. For every natural n, from a starting function ϕ_0 of rank ν we may construct very easily an orthonormal system $\{\phi_0, \phi_1, \ldots, \phi_{2^{\nu n}-1}\}$ with $(2^\nu - 1)n$ free parameters.

For functions of two variables, defined on the unit square $[0,1] \times [0,1]$, the correspondence $l : \mathbf{G} \to [0,1]^2$,

$$l(\mathbf{x}) = \left(\sum_{i=1}^{\infty} 2^{-i} x_{2i-1}, \sum_{i=1}^{\infty} 2^{-i} x_{2i} \right) = (x,y) \in [0,1]^2$$

may be used.

The images of standard Lenna and an image of an eye's ophtalmoscopia are depicted in Figure 2. The entropy of the Fourier coefficients of Lenna and the eye's retinae, for different orthonormal systems, are given in Table 2. The advantages of the adapted orthonormal systems, using 18, 27 and 42 parameters are bigger for the image of the eye's retinae than for Lenna.

Acknowledgments. Supported by Contract I-401/94 with Bulgarian National Research Foundation.

References

1. Barnsley, M. F. and L. P. Hurd, *Fractal Image Compression*, AK Peters, Ltd., Wellesley, Massachusetts, 1993.

2. Meyer, Y., *Wavelets*, SIAM, Philadelphia, 1993.

3. Pomtrjagin, L. S., *Topologishe Grouppen*, B. G. Teubner Ver., Leipzig, 1957.

4. Sendov, Bl., *Hausdorff Approximations*, Kluwer Academic Publ., Dordrecht, 1990.

Bl. Sendov
Bulgarian Academy of Sciences
1113 Sofia, Bulgaria
sendov@amigo.acad.bg

On the Discrete Norms of Polynomials

Boris Shekhtman

Abstract. Let Δ_m be a subset of the unit circle $\mathbb{T}$ consisting of m distinct points. Let p be a polynomial of degree n. We compare the uniform norm of the polynomial p to $\max\{|p(t)| : t \in \Delta_m\}$. The results are obtained in terms of the relationship between the integers m and n.

§1. Introduction

Let H_n be the space of trigonometric polynomials defined on the unit circle $\mathbb{T}$ endowed with the uniform norm. Let Δ_m be a set of m distinct points in $\mathbb{T}$.

In this note we compare the uniform norm of a polynomial $p \in H_n$ to its discrete norm on Δ_m. We distinguish two cases: $n \leq m$ and $n \geq m$. If $n \leq m$, we introduce the quantities

$$\nu(\Delta_m, n) = \sup\{\|p\| : p \in H_n; |p(t_j)| \leq 1 \text{ for all } t_j \in \Delta_m\}, \quad (1.1)$$

$$\nu(m, n) = \inf\{\nu(\Delta_m, n) : \Delta_m \in \mathbb{T}\}. \quad (1.2)$$

In the case $n \geq m$ let $H_n^0(\Delta_m) = \{p \in H_n : p|\Delta_m = 0\}$ and let $H_n(\Delta_m) = H_n/H_n^0(\Delta_m)$. Once again, we introduce

$$\rho(\Delta_m, n) = \sup\{\|\widetilde{p}\| : \widetilde{p} \in H_n(\Delta_m); |\widetilde{p}(t_j)| \leq 1; t_j \in \Delta_m\} \quad (1.3)$$

$$\rho(m, n) = \inf\{\rho(\Delta_m, n) : \Delta_m \subset \mathbb{T}\}. \quad (1.4)$$

The quantity $\rho(m,n)$ was studied extensively (cf. [5, 8, 9]). It was conjectured by Erdös that

$$\rho(m,n) \to \infty \text{ as } n \to \infty \text{ and } m = n - o(n). \quad (1.5)$$

This was further refined by Szabados who conjectured that

$$\rho(m,n) \sim \log\left(\frac{n}{|m-n|+1}\right). \quad (1.6)$$

Both conjectures are still open.

The quantity $\nu(m,n)$ was studied in [1, 2, 6]. Observe that $\rho(m,n) = \nu(m,n)$ if $m = n+1$. There seems to be an illusive "duality" between $\rho(m,n)$ and $\nu(m,n)$. In fact we propose

Conjecture 1. $\rho(m,n) \sim \nu(n,m)$.

Conjecture 2. $\nu(m,n) \sim \log\left(\frac{m}{|(m-n)|+1}\right)$.

The purpose of this paper is to prove the following

Theorem 1. Let $|m-n| = o\left(\log^2 n\right)$. Then

$$\rho(m,n) \to \infty \text{ as } n \to \infty \tag{1.7}$$

$$\nu(m,n) \to \infty \text{ as } n \to \infty. \tag{1.8}$$

This theorem verifies the Erdös Conjecture (1.5) in the case where $|m-n| = o\left(\log^2 n\right)$. It also gives further support for the existence of "duality" between ρ and ν.

§2. The Proofs

We will need the notions of projectional constant and of the Banach-Mazur distance as defined in the theory of Banach spaces (cf. [4]). Let E, F be two finite-dimensional Banach spaces of the same dimension. We define

$$d(E,F) = \inf\left\{\|T\|\,\|T^{-1}\| : T \text{ is an isomorphism } E \to F\right\}. \tag{2.1}$$

If X is a Banach space containing E, we define

$$\lambda(E,X) = \inf\{\|P\| : P \text{ projection from } X \text{ onto } E\}, \tag{2.2}$$

$$\lambda(E) = \sup\{\lambda(E,X) : X \supset E\}. \tag{2.3}$$

Some well-known properties of the mentioned constants are collected in

Proposition 1. Let E, F, and X be as above. Then

$$\lambda(E) \leq d(E,F) \cdot \lambda(F) \tag{2.4a}$$

$$\lambda(E,X) \leq \sqrt{\dim E} + 1 \tag{2.4b}$$

$$\lambda(E, \ell_\infty) = \lambda(E) \tag{2.4c}$$

$$\lambda(E_n) \leq d\left(E_n, \ell_\infty^{(n)}\right) \tag{2.4d}$$

$$\lambda(H_n) \sim d\left(H_n, \ell_\infty^{(n)}\right) \sim \log n. \tag{2.4e}$$

Proof: The proofs of the first four statements can be found in [4]. The proof of (2.4e) is in [7]. □

Discrete Norms

We are now ready to prove

Theorem 2. *There exists a constant $C > 0$ such that*
$$\nu(m,n) \geq C \log n / (\sqrt{m-n}+1). \tag{2.5}$$

In particular,
$$\nu(m,n) \to \infty \text{ if } m - n o\left(\log^2 n\right).$$

Proof: Let $\Delta_m = \{t_1, \ldots, t_m\} \subset \mathbb{T}$ Consider an operator
$$\mathcal{J} : H_n \to E_n \hookrightarrow \ell_\infty^m$$
defined by
$$\mathcal{J} p = (p(t_1), \ldots, p(t_m)) \in \ell_\infty^{(m)}.$$

Clearly $\|\mathcal{J}\| \leq 1$, and $\|\mathcal{J}^{-1} : E_n \to H_n\| = \nu(m,n)$. From Proposition 1 we have
$$\nu(m,n) \geq d(H_n, E_n) \geq \lambda(H_n)/\lambda(E_n)$$
$$= \lambda(H_n)/\lambda\left(E_n, \ell_\infty^{(m)}\right) \geq C \log n / \left(\sqrt{(m-n)}+1\right). \quad \square$$

We will now use a similar technique to prove

Theorem 3. *There exists a constant $C > 0$ such that*
$$\rho(m,n) \geq C \log n / \left(\sqrt{|m-n|}+1\right). \tag{2.6}$$

Proof: Let $\Delta_m = \{t_j\} \subset \mathbb{T}$ be given. Define functionals $\lambda_1, \ldots, \lambda_m \in H_n^*$ by $\lambda_j(p) = p(t_j)$. Since $m \leq n-1$, the functionals $\lambda_1, \ldots, \lambda_m$ are linearly independent and are of norm 1. Let $\alpha_1, \ldots, \alpha_m$ be an arbitrary sequence of complex numbers. Then
$$\left\|\sum \alpha_j \lambda_j\right\| \leq \sum |\alpha_j|. \tag{2.7}$$

Pick $\varepsilon_j = \operatorname{sgn} \alpha_j$. From the definition of $\rho(\Delta_m, n)$ there exists a polynomial $p \in H_n$ such that $p(t_j) = \varepsilon_j$ and $\|p\| \leq \rho(\Delta_m, n)$. Using Helly's theorem (cf. [3]) we obtain
$$\frac{1}{\rho(\Delta_m,n)} \sum |\alpha_j| = \frac{1}{\rho(\Delta_m,n)} \left|\sum \alpha_j \varepsilon_j\right| \leq \left\|\sum \alpha_j \lambda_j\right\|. \tag{2.8}$$

Let T be a map from $\Lambda := \operatorname{span}[\lambda_j]$ onto $\ell_1^{(m)}$ defined by
$$T\left(\sum \alpha_j \lambda_j\right) = (\alpha_j) \in \ell_1^{(m)}.$$

Using (2.7) and (2.8) we have
$$d\left(\Lambda, \ell_1^{(m)}\right) \leq \|T\| \|T^{-1}\| \leq \rho(\Delta_m, n). \tag{2.9}$$

Since Λ is an m-dimensional subspace of the n-dimensional space H_n^*, by (2.5) there exists a projection P from H_n^* into itself such that

$$\text{range } P = \Lambda; \quad \|P\| \leq \left(\sqrt{|n-m|}+1\right).$$

The adjoint P^* is once again a projection from H_n onto an m-dimensional subspace $V_m \subset H_n$. We now claim that

$$d\left(V_m, \ell_\infty^{(m)}\right) \leq \|P\|\rho(\Delta_m, n).$$

Indeed, let $P^*h = \sum h(t_j) v_j$. Consider an arbitrary sequence $(\alpha_j) \subset \mathbf{C}$ with $|\alpha_j| \leq 1$. Let $p \in H$ be such that

$$p(t_j) = \alpha_j; \quad \|p\| \leq \rho(\Delta_m, n). \tag{2.10}$$

Then

$$\left\|\sum \alpha_j v_j\right\| = \|Pp\| \leq \|P\|\|p\| \leq \rho(\Delta_m, n)\left(\sqrt{|n-m|}+1\right). \tag{2.11}$$

On the other hand, we have

$$\left\|\sum \alpha_j v_j\right\| \geq \left|\left(\sum \alpha_j v_j\right)(t_k)\right| \geq |\alpha_k|; \quad k = 1, \ldots, m. \tag{2.12}$$

Let S be a mapping from V onto $\ell_\infty^{(m)}$ defined by

$$S\left(\sum \alpha_j v_j\right) = (\alpha_j) \in \ell_\infty^{(m)}.$$

From (2.11) and (2.12) we have

$$d\left(V, \ell_\infty^{(m)}\right) \leq \|S\|\|S^{-1}\| \leq \rho(\Delta_m, n)\left(\sqrt{|n-m|}+1\right).$$

It now follows from (2.4) that there exists a projection

$$Q : C(\mathbb{T}) \to V \subset H_n,$$

such that

$$\|Q\| \leq \rho(\Delta_m, n)\left(\sqrt{|n-m|}+1\right).$$

On the other hand it was shown (cf. [5, 8]) that

$$\|Q\| \geq C_1 \log(n/(n-m+1)).$$

Hence

$$\rho(\Delta_m, n) \geq C_1(\log n - \log(n-m+1))/(\sqrt{n-m}+1)$$
$$\geq C \log n /(\sqrt{n-m}+1). \quad \square$$

Remark. The fact that the space $\Lambda \subset H_n^*$ is $\rho(\Delta_m, n)$ close to $\ell_1^{(m)}$ gives some hope that this space can be complemented in H_n^* by a projection with the norm depending on $\rho(\Delta_m, n)$ only. If that is the case, then the proof of Theorem 3 will provide the proof for the Erdös-Szabados conjecture without restriction on $(n - m)$.

References

1. Ehlich, M. and K. Zeller, Schwaukungen von polynomen zwischen gitterpunkten, Mathem. Zeitschaift **86** (1964), 41–44.
2. Golitschek, V. H., Polynomial approximation of discrete sets, in *Approximation Theory V*, 1986, pp. 351–354.
3. Helly, E., Über Systeme linearer gleichungen mit unendlich vleden unbekannten, Monatsh. für Math. u. Phys. **31** (1921), 60–91.
4. Pelczynski, A., Geometry of finite-dimensional Banach spaces and operator ideals, in *Notes in Banach Spaces*, E. Lacey (ed.), University of Texas Press, 1980, pp. 81–181.
5. Privaloff, A. A., On the growth of the degree of polynomial bases, Mat. Zametki **42** (1987), 207–214.
6. Rivlin, T. J. and E. W. Cheney, A comparison of uniform approximations on an interval and a finite subset theory, SIAM J. Numer. Anal. **3** (1966), 311–320.
7. Shekhtman, B., On geometry of real polynomials, in *Lecture Notes in Math.*, Vol. **1287**, Springer-Verlag, 1987, pp. 161–175.
8. Shekhtman, B., On the norms of interpolating operators, Israel J. Math. **64(1)** (1988), 39–48.
9. Szabados, J., On the norms of certain interpolating operators, Anal. Math., to appear.
10. Wojtaszyk, P., *Banach Spaces for Analysts*, Cambridge University Press, 1991.

Boris Shekhtman
Department of Mathematics
University of South Florida
Tampa, FL 33620-5700
boris@math.usf.edu

Generation of Weight Functions for Orthogonal Polynomials

M.-R. Skrzipek

Abstract. Let $\{P_\nu\}_{\nu \in \mathbb{N}_0}$ be a given sequence of polynomials, orthogonal on $I \subseteq \mathbb{R}$ with respect to a measure $d\omega$, and let $P_0^{(\Phi)}, P_1^{(\Phi)}, \ldots, P_m^{(\Phi)}$ be sequence of orthogonal polynomials, given, e.g., by their three term recurrence relation. In many cases their weight(s) $d\tilde{\omega}$ with respect to the fact that they are orthogonal, are unknown. We assume that both sequences of polynomials have the same interval I of orthogonality and that $d\tilde{\omega}$ can be written as $d\tilde{\omega} = \Phi \, d\omega$ where Φ has a.e. in I a uniformly convergent Fourier expansion in terms of the P_ν, $\nu \in \mathbb{N}_0$. By using the mth partial sum Φ_m of Φ we show how we get weight function(s) with respect to which the $\{P_\nu^{(\Phi)}\}_{\nu=0,\ldots,j}$, $2j \leq m$, are orthogonal. A connection between the Fourier coefficients of Φ_m resp. Φ and the recurrence coefficients of the $\{P_\nu\}_{\nu \in \mathbb{N}_0}$ is also given.

§1. Introduction

In many cases orthogonal polynomials are given by their three term recurrence

$$P_{-1}(x) = 0, \quad P_0(x) = \lambda_0 \neq 0,$$
$$P_{n+1}(x) = (a_n x + b_n) P_n(x) - c_n P_{n-1}(x), \quad n \in \mathbb{N}_0, \tag{1}$$

with real coefficients and $a_{n-1} a_n c_n > 0$, $n \geq 1$. On the other hand they satisfy an orthogonality relation

$$\langle P_l, P_k \rangle_\omega := \int_I P_l(x) P_k(x) \, d\omega(x) = d_l \delta_{k,l}, \quad d_l \neq 0, \; l,k \geq -1, \tag{2}$$

where $\langle \cdot, \cdot \rangle_\omega$ denotes the inner product with respect to which the $\{P_\nu\}_{\nu \geq -1}$ are orthogonal. By a theorem of Farvard [1, p. 7], [2, p. 21, 75], (1) and (2) are equivalent for characterizing orthogonal polynomials.

One way to connect the recurrence coefficients with the measure is the usage of Stieltjes transforms (see, e.g., [1, p. 11], [2, p. 90]). But in using them there are some disadvantages (see [3, 5]), which make it necessary to look for some 'suitable' approximations to the weight. In [5], [6, Theorem 4.9], [9, Theorem 12], various ways are given to find such approximations. However there were some assumptions which we want to avoid, e.g., that the weight satisfies the Szegő condition, or that there are some restrictions to the asymptotic behavior of the recurrence coefficients, etc.

Thus, we look for another method for determining the weights for orthogonal polynomials numerically. We assume that a sequence $\{P_\nu\}_{\nu \in \mathbb{N}_0}$ of polynomials, orthogonal on $I \subseteq \mathbb{R}$ with respect to a measure $d\omega$, is given. Furthermore, let a sequence of polynomials $P_0^{(\Phi)}, P_1^{(\Phi)}, \ldots, P_m^{(\Phi)}$ be given which are orthogonal on I with respect to an unknown weight $d\tilde{\omega} = \Phi \, d\omega$. The superscript at quantities denotes the absolutely continuous part of the measure to which the quantities belong. We assume that Φ can be written almost everywhere (a.e.) on I as

$$\Phi = \sum_{\nu=0}^{\infty} \gamma_\nu P_\nu, \quad \gamma_\nu = \tfrac{1}{d_\nu} \int_I P_\nu(x) \Phi(x) \, d\omega(x). \tag{3}$$

Note that $\tilde{\omega}$ has at least m points of increase. Even if the weight $\tilde{\omega}$ has infinite points of increase and if it is uniquely determined for the whole sequence $\{P_\nu^{(\Phi)}\}_{\nu=0,1,2,3,\ldots}$ (in the sense of the Lebesgue measure and modulo a constant factor), the polynomials $P_\nu^{(\Phi)}$, $\nu = 0, \ldots, m$, may be orthogonal with respect to different measures. This means that in general Φ is not uniquely determined for a finite sequence of orthogonal polynomials, if there are more points of increase in its spectrum than the number of polynomials we consider.

§2. Representations of Orthogonal Polynomials with Different Weights

We write the Fourier expansions of $P_n^{(\Phi)}$ and $\Phi P_n^{(\Phi)}$ in terms of $\{P_\nu\}_{\nu \in \mathbb{N}_0}$ as

$$P_n^{(\Phi)} = \sum_{\nu=0}^{n} \alpha_{\nu,n}^{(\Phi)} P_\nu, \tag{4}$$

$$\Phi P_n^{(\Phi)} = \sum_{\mu=n}^{\infty} \beta_{\mu,n}^{(\Phi)} P_\mu. \tag{5}$$

Since $\beta_{\mu,n}^{(\Phi)} \langle P_\mu, P_\mu \rangle_\omega = \langle \Phi P_n^{(\Phi)}, P_\mu \rangle_\omega$, we have $\beta_{\mu,n}^{(\Phi)} = 0$ for $\mu < n$ due to the orthogonality of $P_n^{(\Phi)}$ with respect to $\Phi \, d\omega$. If $\Phi = \Phi_m$ is a polynomial of degree m then we have the Christoffel-Fourier representation of $P_n^{(\Phi_m)}$ [8]:

$$\Phi_m P_n^{(\Phi_m)} = \sum_{\mu=n}^{m+n} \beta_{\mu,n}^{(\Phi_m)} P_\mu. \tag{6}$$

For the following, we need a result about the connection between the Fourier coefficients of $P_k^{(\Phi_m)}$ and $\Phi_m P_k^{(\Phi_m)}$ in terms of the P_μ.

Theorem 1. [7] Let $A_{m,n}^{(1)} := (\langle P_\mu, P_\nu \Phi_m \rangle_\omega)_{\mu,\nu=0,\ldots,n}$ have LU decomposition $A_{m,n}^{(1)} = L_n R_n$, and write

$$R_n^{-1} = \begin{pmatrix} \alpha_{0,0}^{(\Phi_m)} & \cdots & \alpha_{0,n}^{(\Phi_m)} \\ & \ddots & \vdots \\ 0 & & \alpha_{n,n}^{(\Phi_m)} \end{pmatrix}. \tag{7}$$

Then $P_k^{(\Phi_m)} = \sum_{\nu=0}^{k} \alpha_{\nu,k}^{(\Phi_m)} P_\nu$ for $k = 0, \ldots, n$.

Obviously this theorem also holds if we replace Φ_m by Φ. The choice of the LU decomposition influences the normalization of the orthogonal polynomials. If we use a Cholesky decomposition, the polynomials $P_k^{(\Phi_m)}$, $k = 0, \ldots, n$, are orthonormalized with respect to $\Phi_m \, d\omega$ [7]. The proof for general LU decompositions can be derived easily from there.

From (5) we obtain Φ if $\beta_{\mu,n}^{(\Phi)}$, $\mu = n, \ldots, \infty$, are known. For $P_k^{(\Phi)}$ given in (4) we get, using $a_{\nu,k}^{(\Phi)} = \langle P_\nu, P_k^{(\Phi)} \Phi \rangle_\omega / \langle P_\nu, P_\nu \rangle_\omega$,

$$\langle P_\nu, P_\nu \rangle_\omega a_{\nu,k}^{(\Phi)} = \sum_{j=0}^{\infty} \gamma_\nu \langle P_\nu, P_k^{(\Phi)} P_j \rangle_\omega$$
$$= \sum_{j=0}^{m} \gamma_\nu \langle P_\nu, P_k^{(\Phi)} P_j \rangle_\omega = \langle P_\nu, P_\nu \rangle_\omega a_{\nu,k}^{(\Phi_m)}$$

for all $m \geq 2n$ since $\langle P_\nu, P_k^{(\Phi)} P_j \rangle_\omega = 0$ for $\nu, k \leq n$, $j > 2n$. Using Theorem 1 we conclude that $P_k^{(\Phi)}$ is proportional to $P_k^{(\Phi_m)}$, $k = 0, \ldots, n$, $2n \leq m$, where Φ_m denotes the mth partial sum of Φ given in (3). This proves

Theorem 2. Let that Φ have a.e. in I an uniformly convergent Fourier expansion in terms of the P_ν, $\nu \in \mathbb{N}_0$. Let Φ_m denote the $m \geq 2n$th partial sum of this Fourier series. If $\Phi_m \, d\omega$ is a weight in I, then the orthogonal polynomials $P_n^{(\Phi_m)}$ resp. $P_n^{(\Phi)}$, which belong to $\Phi_m \, d\omega$ resp. $\Phi \, d\omega$, are proportional. Under a suitable normalization, $P_n^{(\Phi_m)} = P_n^{(\Phi)}$.

Theorem 2 says that $P_k^{(\Phi)}$, $k = 0, \ldots, n$, are orthogonal with respect to all partial sums $\Phi_m = \sum_{\nu=0}^{m} \gamma_\nu P_\nu$, $m = 2n, 2n+1, \ldots$, if $\Phi_m \, d\omega$ are weights, and that $P_k^{(\Phi_m)} = P_k^{(\Phi)}$ by a suitable normalization. From (5) and (6) it suffices to determine Φ_m, $m \geq 2n$, to obtain weights with respect to which $P_k^{(\Phi)}$, $k = 0, \ldots, n$, are orthogonal. In Theorem 3 we shall show that in general this statement remains valid if Φ_m has zeros or sign changes in I.

Since $\langle P_\mu, P_\mu \rangle_\omega \beta_{\mu,\nu}^{(\Phi_m)} = \langle P_\nu^{(\Phi_m)}, P_\mu \Phi_m \rangle_\omega = \sum_{j=0}^{\nu} \alpha_{j,\nu}^{(\Phi_m)} \langle P_j, P_\mu \Phi_m \rangle_\omega$, using (7), we have $\operatorname{diag}(\langle P_0, P_0 \rangle_\omega, \ldots, \langle P_n, P_n \rangle_\omega) \left(\beta_{\mu,\nu}^{(\Phi_m)} \right)_{\mu,\nu=0,\ldots,n} = A_{m,n}^{(1)} R_n^{-1}$
$= L_n$. In particular, for $n = m$ we have

$$\begin{pmatrix} \beta_{0,0}^{(\Phi_m)} & & 0 \\ \vdots & \ddots & \\ \beta_{m,0}^{(\Phi_m)} & \cdots & \beta_{m,m}^{(\Phi_m)} \end{pmatrix} = \operatorname{diag}\left(\frac{1}{\langle P_0, P_0 \rangle_\omega}, \ldots, \frac{1}{\langle P_m, P_m \rangle_\omega} \right) L_m. \tag{8}$$

§3. Calculating the Weights

Let $\{P_n\}_{n\in\mathbb{N}_0}$ and $\{P_\nu^{(\Phi)}\}_{\nu=0,\ldots,m}$ be given. We want to determine weights Φ with respect to which the $P_\nu^{(\Phi)}$, $\nu = 0,\ldots j$, are orthogonal for a $j \in \mathbb{N}_0$. Using Theorem 2 it suffices to determine the mth partial sum Φ_m of the Fourier expansion of Φ in terms of the P_μ with $m \geq 2j$. Φ_m can be obtained from (6), with $n = 0$. This means that we need only the elements $\beta_{0,0}^{(\Phi_m)},\ldots,\beta_{0,m}^{(\Phi_m)}$ of the first column in (8). If the $P_\nu^{(\Phi)}$ are given as in (4) and $\mathbf{P}_m := (P_0,\ldots,P_m)^T$, $\mathbf{P}_m^{(\Phi)} := (P_0^{(\Phi)},\ldots,P_m^{(\Phi)})^T$, then we have

$$\mathbf{P}_m = R_m^T \mathbf{P}_m^{(\Phi)} \tag{9}$$

where R_m^{-1} is given in (7).

Writing $\tilde{P}_\mu^{(\Phi)} := s_\mu P_\mu^{(\Phi)}$ where $\{\tilde{P}_\mu^{(\Phi)}\}_{\mu\in\mathbb{N}_0}$ are orthonormal with respect to the (unknown) weight $\Phi\,d\omega$, $D_m := \mathrm{diag}(s_0,\ldots,s_m)$, $\tilde{R}_m := D_m R_m$, we see that $A_m := \tilde{R}_m^T \tilde{R}_m$ is positive definite. Thus $\tilde{R}_m^T \tilde{R}_m$ is the Cholesky decomposition of A_m (uniquely determined modulo that one can change the sign for each element of $\tilde{R}_m$). Due to Mysovskih [7] it follows that A_m is a Gram matrix with coefficients $a_{i,\nu} = \langle P_i, P_\nu \Phi \rangle_\omega$, $i,\nu = 0,\ldots,m$. From (8) and $L_m = \tilde{R}_m^T = R_m^T D_m$ we have

$$\left(\tilde{\beta}_{0,0}^{(\Phi_m)},\ \tilde{\beta}_{1,0}^{(\Phi_m)},\ \ldots,\ \tilde{\beta}_{m,0}^{(\Phi_m)}\right)^T = s_0 \left(\frac{r_{0,0}}{\langle P_0,P_0\rangle_\omega},\ \frac{r_{0,1}}{\langle P_1,P_1\rangle_\omega},\ \ldots,\ \frac{r_{0,m}}{\langle P_m,P_m\rangle_\omega}\right)^T, \tag{10}$$

where $\tilde{\beta}_{\mu,0}^{(\Phi_m)}$ denote the Fourier coefficients of $\tilde{P}_0^{(\Phi_m)}\Phi_m$ in terms of the P_μ, $\mu = 0,\ldots,m$. Thus we get

$$\Phi_m = \frac{s_0}{\tilde{P}_0^{(\Phi_m)}} \sum_{\nu=0}^{m} \frac{r_{0,\nu}}{\langle P_\nu, P_\nu\rangle_\omega} P_\nu = \frac{1}{P_0^{(\Phi_m)}} \sum_{\nu=0}^{m} \frac{r_{0,\nu}}{\langle P_\nu, P_\nu\rangle_\omega} P_\nu$$

or $r_{0,\nu} = \gamma_\nu P_0^{(\Phi_m)} d_\nu$ by (3). Here, the factor $1/P_0^{(\Phi_m)}$ can be omitted since the weights are only determined modulo a constant factor as can be seen from the definition of orthogonality in (2). (Since $P_0^{(\Phi_m)}$ is a constant it is independent from m, Φ_m, respectively Φ). By using Theorem 2 we summarize

Theorem 3. *If Φ has a.e. on I an uniformly convergent Fourier series in terms of the $\{P_\mu\}_{\mu\in\mathbb{N}_0}$, then the polynomials $P_\nu^{(\Phi)}$, $\nu = 0,\ldots,j$, where $2j \leq m$, satisfy the orthogonality relation $\langle P_l^{(\Phi)},\ P_k^{(\Phi)}\hat{\Phi}_m\rangle_\omega = \hat{d}_l \delta_{k,l}$, $\hat{d}_l \neq 0$, $k,l = 0,\ldots,j$. Here,*

$$\hat{\Phi}_m = \sum_{\mu=0}^{m} \frac{r_{0,\mu}}{\langle P_\mu, P_\mu\rangle_\omega} P_\mu \tag{11}$$

converges for $m \to \infty$ on I a.e. uniformly to a function $\hat{\Phi}$ which is proportional to Φ, and $r_{0,\mu}$, $\mu = 0,\ldots,m$, are given by the first row of R_m in (9).

Since the functions with and without the hat are proportional, we omit the hat.

Generating Weight Functions 313

We remark that $\Phi_m\, d\omega$ may not be a weight in the classical sense since Φ_m may have, e.g., zeros or sign changes in I in particular if m is small. Furthermore, in general the algorithm cannot detect whether a system of polynomials form a system of orthogonal polynomials or whether a weight can be written in the form $\Phi\, d\omega$ or determine the spectrum.

For calculating the coefficients $r_{0,\mu}$, $\mu \in \mathbb{N}_0$, of Φ_m resp. Φ, it is not necessary to determine the whole matrix R_m. The subsequent algorithm to be derived bears a similarity to the modified Chebyshev algorithm for calculating recurrence coefficients of orthogonal polynomials (cf. [4, 10]) since we use the recurrence relations of two sequences of orthogonal polynomials in a similar manner. But a comparison of both algorithms shows immediately that they are different.

From (9) we have $r_{0,0} = P_0/P_0^{(\Phi)}$ and $P_1 = r_{0,1}P_0^{(\Phi)} + r_{0,2}P_1^{(\Phi)}$ or
$$(a_0 x + b_0)P_0(x) = r_{0,1}P_0^{(\Phi)}(x) + r_{0,2}(a_0^{(\Phi)} x + b_0^{(\Phi)})P_0^{(\Phi)}(x).$$

From this we obtain $r_{0,2} = \frac{a_0}{a_0^{(\Phi)}} r_{0,0}$ and $r_{0,1} = \left(b_0 - \frac{b_0^{(\Phi)}}{a_0^{(\Phi)}} a_0\right) r_{0,0}$. For $\nu \geq 1$ we use the recurrence relation (1), insert $P_k = \sum_{\mu=0}^{k} r_{\mu,k}P_\mu^{(\Phi)}$ from (9), use the recurrence relation for $xP_\nu^{(\Phi)}(x)$, and get

$$P_{\nu+1}(x) = a_\nu x \sum_{\mu=0}^{\nu} r_{\mu,\nu} P_\mu^{(\Phi)}(x) + b_\nu \sum_{\mu=0}^{\nu} r_{\mu,\nu} P_\mu^{(\Phi)}(x) - c_\nu \sum_{\mu=0}^{\nu-1} r_{\mu,\nu-1} P_\mu^{(\Phi)}(x)$$

$$= a_\nu \frac{r_{\nu,\nu}}{a_\nu^{(\Phi)}} P_{\nu+1}^{(\Phi)}(x) + \left[a_\nu \left(\frac{r_{\nu-1,\nu}}{a_{\nu-1}^{(\Phi)}} - \frac{r_{\nu,\nu}}{a_\nu^{(\Phi)}} b_\nu^{(\Phi)}\right) + b_\nu r_{\nu,\nu}\right] P_\nu^{(\Phi)}(x)$$

$$+ \sum_{\mu=1}^{\nu-1} \left[a_\nu \left(\frac{r_{\mu-1,\nu}}{a_{\mu-1}^{(\Phi)}} - \frac{r_{\mu,\nu}}{a_\mu^{(\Phi)}} b_\mu^{(\Phi)} + \frac{r_{\mu+1,\nu}}{a_{\mu+1}^{(\Phi)}} c_{\mu+1}^{(\Phi)}\right) + b_\nu r_{\mu,\nu} - c_\nu r_{\mu,\nu-1}\right] P_\mu^{(\Phi)}(x)$$

$$+ \left[a_\nu \left(-\frac{r_{0,\nu}}{a_0^{(\Phi)}} + \frac{r_{1,\nu}}{a_1^{(\Phi)}} c_1^{(\Phi)}\right) + b_\nu r_{0,\nu} - c_\nu r_{0,\nu-1}\right] P_0^{(\Phi)}(x).$$

Since $P_{\nu+1} = \sum_{\mu=0}^{\nu+1} r_{\mu,\nu+1} P_\mu^{(\Phi)}$, we can compare the Fourier coefficients on both sides. We see that we have to determine all elements $r_{\mu,\nu}$, $\mu \leq \nu$, $\mu+\nu \leq m$, to get $r_{0,\nu+1}$, $\nu = 0, \ldots, m-1$. Furthermore, we obtain

Theorem 4. *The coefficients $r_{\mu,\nu}$, $\mu = 0, 1, 2, \ldots, \nu$, $\nu = 0, 1, \ldots$, satisfy the recurrence relation*

$$r_{0,0} = P_0/P_0^{(\Phi)}, \qquad r_{0,1} = \left(b_0 - \frac{b_0^{(\Phi)}}{a_0^{(\Phi)}} a_0\right) r_{0,0}, \tag{12}$$

$$r_{\mu,\nu+1} = \begin{cases} a_\nu \frac{r_{\nu,\nu}}{a_\nu^{(\Phi)}}, & \mu = \nu+1, \\[4pt] a_\nu \left(\frac{r_{\nu-1,\nu}}{a_{\nu-1}^{(\Phi)}} - \frac{r_{\nu,\nu}}{a_\nu^{(\Phi)}} b_\nu^{(\Phi)}\right) + b_\nu r_{\nu,\nu}, & \mu = \nu, \\[4pt] a_\nu \left(\frac{r_{\mu-1,\nu}}{a_{\mu-1}^{(\Phi)}} - \frac{r_{\mu,\nu}}{a_\mu^{(\Phi)}} b_\mu^{(\Phi)} + \frac{r_{\mu+1,\nu}}{a_{\mu+1}^{(\Phi)}} c_{\mu+1}^{(\Phi)}\right) \\ \quad + b_\nu r_{\mu,\nu} - c_\nu r_{\mu,\nu-1}, & \mu = 1, \ldots, \nu-1, \\[4pt] a_\nu \left(-\frac{r_{0,\nu}}{a_0^{(\Phi)}} + \frac{r_{1,\nu}}{a_1^{(\Phi)}} c_1^{(\Phi)}\right) + b_\nu r_{0,\nu} - c_\nu r_{0,\nu-1}, & \mu = 0. \end{cases}$$

If we choose an arbitrarily $a_{-1}^{(\Phi)} \neq 0$ and set $r_{\mu,\nu} := 0$ for $\nu < 0$ or $\mu < 0$, using $r_{\mu,\nu} = 0$ for $\mu > \nu$, we can simplify (12) to

$$r_{0,0} = P_0/P_0^{(\Phi)},$$
$$r_{\mu,\nu+1} = a_\nu \left(\frac{r_{\mu-1,\nu}}{a_{\mu-1}^{(\Phi)}} - \frac{r_{\mu,\nu}}{a_\mu^{(\Phi)}} b_\mu^{(\Phi)} + \frac{r_{\mu+1,\nu}}{a_{\mu+1}^{(\Phi)}} c_{\mu+1}^{(\Phi)} \right) \qquad (13)$$
$$+ b_\nu r_{\mu,\nu} - c_\nu r_{\mu,\nu-1}, \quad \mu = 0, \ldots, \nu+1, \ \nu \in \mathbb{N}_0.$$

We remember that Φ_m resp. Φ are only determined modulo a constant factor. Thus we may assume that the $r_{\mu,\mu}$, $\mu = 0, 1, \ldots, \lfloor m/2 \rfloor$, are given initial values. Since R_m is regular we have $r_{\mu,\mu} \neq 0$. We use the recurrence (13) to determine $r_{\nu,\mu}$ for $\nu = 0, 1, \ldots, \mu - 1$, $\nu \leq m - \mu$, $\mu = 1, \ldots, m$.

Remark. By using $(R_m^T)^{-1}$ instead of R_m^T in (9) we can proceed in a similar manner to obtain $\varphi_l \in \Pi_l$ of a weight $d\tilde{\omega} = d\omega/\varphi_l$ where $\varphi = \lim_{l\to\infty} \varphi_l$ (uniformly convergence on I) and $\varphi_l = \sum_{\nu=0}^{l} \delta_\nu P_\nu \neq 0$ in I. Combining these algorithms we can approximate Φ/φ for a weight $\Phi\, d\omega/\varphi$ by a rational function Φ_m/φ_l where the coefficients of the nominator and denominator can be obtained recursively.

§4. Numerical Remarks

Since we only use a finite number of given orthogonal polynomials, they may be orthogonal with respect to multiple weights. In general the (unknown) weight $\Phi\, d\omega$ with respect to which all elements of $\{P_\nu^{(\Phi)}\}_{\nu \in \mathbb{N}_0}$ are orthogonal may differ from the weight $\Phi_m\, d\omega$ with respect to which the elements of $\{P_\nu^{(\Phi)}\}_{0 \leq \nu \leq \lfloor m/2 \rfloor}$ are orthogonal. Thus, it doesn't make sense to conclude anything about Φ after we have calculated Φ_m or to estimate $\|\Phi_m - \Phi\|_\infty$ numerically.

Instead of this it is more convenient to calculate $P_\nu^{(\Phi_m)}$ by Theorem 1 or (9) (replace Φ by Φ_m there) and compare the (given) polynomials $P_\nu^{(\Phi)}$ with the calculated $P_\nu^{(\Phi_m)}$ ones for $\nu = 0, \ldots, n$. Regarding a suitable normalization we have the following conclusion. If $2n \leq m$, then by Theorem 2 both polynomials are identical (if there were no rounding errors) else $P_\nu^{(\Phi_m)}$ approximates $P_\nu^{(\Phi)}$ [11]. Since they are expanded in terms of P_ν, $\nu = 0, \ldots, n$, it is sufficient to compare the corresponding Fourier coefficients.

We have implemented and tested our algorithm in VM-PASCAL on an IBM-4341-computer. For simplicity we choose $I = [-1, 1]$, $P_n(x) := T_n(x)$, $n \in \mathbb{N}_0$, (Chebyshev polynomials of the first kind) i.e., $d\omega(x) = \frac{dx}{\sqrt{1-x^2}}$ in the following examples. We approximate Φ by Φ_m as described in this article. Then we calculate the corresponding polynomials $P_\nu^{(\Phi_m)}$ which approximate $P_\nu^{(\Phi)}$. Afterwards we go in the converse direction. Starting with the polynomials $\widetilde{P}_\nu := P_\nu^{(\Phi_m)}$ and with the known $\widetilde{P}_\nu^{(\widetilde{\Phi})} := P_\nu$, $\nu = 0, \ldots, m$, we calculate a weight with respect to which $\widetilde{P}_\nu^{(\widetilde{\Phi})}$, $\nu = 0, \ldots, n \leq \lfloor m/2 \rfloor$, are

Generating Weight Functions 315

orthogonal. In other words: We multiply $d\omega$ by Φ_m and divide it by itself afterwards. Hence, if we would use exact calculations, we would get back the original polynomials T_k, $k = 0,\ldots,n$. Writing $\widetilde{P}_k^{(\widetilde{\Phi})} = \sum_{\nu=0}^{k} \tilde{\alpha}_{\nu,k}^{(\widetilde{\Phi})} P_\nu$, we have $\tilde{\alpha}_{\nu,k}^{(\widetilde{\Phi})} = \delta_{\nu,k}$, $\nu = 0,\ldots,k$, $k = 0,\ldots,n$, theoretically. Therefore, the quality of the approximately calculated Fourier-coefficients $\tilde{\alpha}_{\nu,k}^{(\widetilde{\Phi})}$ can be measured by

$$E_{0/n}(\Phi_m) := \max_{k=0}^{n} E_k(\Phi_m)\ ,\qquad E_k(\Phi_m) := \max\left(\left|1-\tilde{\alpha}_{k,k}^{(\widetilde{\Phi})}\right|,\ \max_{\nu=0}^{k-1}\left|\tilde{\alpha}_{\nu,k}^{(\widetilde{\Phi})}\right|\right).$$

i) In Table 1 we use the $m = (2k+1)$th partial sum

$$\Phi_{2k+1}(x) := \tfrac{\pi}{2} - \tfrac{4}{\pi}\sum_{\nu=0}^{k}\tfrac{1}{(2\nu+1)^2}T_{2\nu+1}(x)\ ,\quad x \in [-1,1],$$

of the Fourier-Chebyshev-series of $\arccos(x)$.

ii) For the $m = 2k$th partial sum

$$\Phi_{2k}(x) := \tfrac{2}{\pi} + \tfrac{4}{\pi}\sum_{\nu=1}^{k}\tfrac{1}{1-4\nu^2}T_{2\nu}(x)\ ,\quad x \in [-1,1],$$

of the Fourier-Chebyshev-expansion of $\sqrt{1-x^2}$ we obtain the results given in Table 2.

n	m	$E_{0/n}(\Phi_m)$
10	11	1.26565 (−14)
	21	9.38138 (−15)
	31	8.75688 (−15)
	41	1.45439 (−14)
20	11	3.27506 (−9)
	21	2.74503 (−14)
	31	2.22461 (−14)
	41	7.37327 (−14)
30	11	3.98983 (−4)
	21	1.02609 (−8)
	31	6.20615 (−14)
	41	8.44047 (−14)
40	11	8.20789 (−1)
	21	3.25646 (−7)
	31	3.14451 (−8)
	41	1.19585 (−13)

Table 1

n	m	$E_{0/n}(\Phi_m)$
10	10	7.71605 (−15)
	20	5.45397 (−15)
	30	5.34295 (−15)
	40	5.30131 (−15)
20	10	3.23207 (−9)
	20	2.38184 (−14)
	30	2.28151 (−14)
	40	1.99563 (−14)
30	10	3.52909 (−4)
	20	5.51372 (−9)
	30	2.36179 (−14)
	40	4.30905 (−14)
40	10	1.79999 (0)
	20	1.22859 (−6)
	30	2.75389 (−8)
	40	9.25787 (−14)

Table 2

References

1. Askey, R. and M. Ismail, Recurrence relations, continued fractions and orthogonal polynomials, Mem. Amer. Math. Soc. Vol. 49, 1984, iv+108pp.

2. Chihara, T. S., *An Introduction to Orthogonal Polynomials*, Gordon and Breach, New York, 1978.

3. Dombrowski, J. and P. Nevai, Orthogonal polynomials, measures and recurrence relations, SIAM J. Math. Anal. **17** (1986), 752–759.

4. Gautschi, W., On generating orthogonal polynomials, SIAM J. Sci. Statist. Comput. **3** (1982), 289–317.

5. Geronimo, J. S. and W. Van Assche, Approximating the weight function for orthogonal polynomials on several intervals, J. Approx. Theory **65** (1991), 341–371.

6. Maki, D., On constructing distribution functions with applications to Lommel and Bessel functions, Trans. Amer. Math. Soc. **130** (1968), 281–297.

7. Mysovskih, I. P., On the construction of cubature formulas with fewest nodes, [Russian], Dokl. Akad. Nauk S.S.S.R. **178** (1968), 1252–1254, English translation in: Soviet Math. Dokl. **9** (1968), 277–281.

8. Uvarov, V. B., The connection between systems of polynomials Orthogonal with respect to different distribution functions, [Russian], Ž. Vyčisl. Mat. i mat. Fis. **9** (1969), 1253–1262, English translation in: U.S.S.R. Computational Math. and Math. Phys. **9** (1969), 25–36.

9. Van Assche, W., Asymptotics for orthogonal polynomials and three-term recurrenes, in *Orthogonal Polynomials: Theory and Practice*, P. Nevai (ed.), NATO ASI Series C: Mathematical and Physical Sciences, vol. 294, Kluwer Academic Publishers, Boston, 1990, pp. 435–462.

10. Wheeler, J. C., Modified moments and Gaussian quadratures, Rocky Mountain J. Math. **4** (1974), 287–296.

11. Wilson, M. W., Convergence of discrete analogs of orthogonal polynomials, Computing **5** (1970), 1–5.

M.-R. Skrzipek
FernUniversität-GHS Hagen
Fachbereich Mathematik
Postfach 940
D-58084 Hagen
Germany
Michael.Skrzipek@FernUni-Hagen.de

On Products of Positive Definite Functions

Hans Strauss

Abstract. In this paper we consider the problem of interpolation defined by linear functionals. This includes the problem of scattered data interpolation for multivariate functions (e.g., Hermite-Birkhoff interpolation). In order to solve the problem products of positive definite and reproducing kernels are used. The interpolating functions are also solutions of variational problems.

§1. Introduction

Let S be a subset of $\mathbb{R}^d$, $d \geq 1$, and let $F(S)$ be the class of real-valued functions $f : S \to \mathbb{R}$. Assume that H is a Hilbert space in $F(S)$ and $L_i : H \to \mathbb{R}$, $i = 1, \ldots, n$, are bounded linear functionals. Then we consider the following problem:

Given real numbers $z_1, \ldots, z_n$. Determine a function $f \in H$ such that $L_i f = z_i$, $i = 1, \ldots, n$.

In the literature the problem has often been studied for the case that $L_i f = f(x_i)$, $i = 1, \ldots, n$ for some given set of points $x_1, \ldots, x_n$ in S, i.e., the Lagrange interpolation problem. In particular, the radial basis function approach has been used (see [1, 3, 5] and the references therein). These methods have also been extended to the case of sums of radial basis functions (see e.g., [2, 8]).

Here we shall give another approach. We consider positive definite kernels and reproducing kernels (see Section 2) and use products of these functions. Let $(x_i, y_i)_{i=1}^n$ in $\mathbb{R}^2$ be given. Then the functions have the form

$$f(s,t) = \sum_{i=1}^{n} a_i K_1(s, x_i) K_2(t, y_i).$$

Lagrange interpolation using these functions has been considered in [7]. Moreover, it has been shown that the solutions of these interpolation problems are solutions of some variational problems. In this paper the results are extended to general linear functionals.

In Section 2 positive definite and reproducing kernels are defined and basic properties of these functions are studied. Moreover, some examples of such functions are given and a Hermite-Birkhoff interpolation problem is considered.

In Section 3 products of reproducing kernels are investigated. A general theorem concerning interpolation defined by linear functionals is obtained. Lagrange interpolation of scattered data in the multivariate case using products of positive definite kernels can easily be derived.

In Section 4 an extension of the interpolation problem is given and it is shown that the interpolating functions are also solutions of some variational problems.

§2. Positive Definite and Reproducing Kernels

Let S in $\mathbb{R}^d$, $d \geq 1$ be given. A function $P : S \otimes S \to \mathbb{R}$ is called a positive definite kernel (or PD-kernel) if $M = (P(x_i, x_j))_{i,j=1}^n$ is a positive definite matrix (or PD-matrix), i.e.,

$$\sum_{i,j=1}^n c_i c_j P(x_i, x_j) \geq 0$$

for all $\mathbf{c} = (c_1, \ldots, c_n)$ in $\mathbb{R}^n$ and for all choices of finite subsets $X = \{x_1, \ldots, x_n\} \subseteq S$. The function P is called a strictly positive kernel or SPD-kernel if the strict inequality is true for all nontrivial $\mathbf{c} \in \mathbb{R}^n$ and all choices of X with pairwise distinct points. Then M is called a strictly positive definite matrix (or SPD-matrix).

Let H be a Hilbert space in $F(S)$ endowed with the inner-product $(\cdot, \cdot)$. The function $K : S \otimes S \to \mathbb{R}$ is called a reproducing kernel of H if
(i) $K(\cdot, y)$ is contained in H for any $y \in S$,
(ii) $f(y) = (f, K(\cdot, y))$ for any $y \in S$ and $f \in H$.

The following properties are well known (see [4, p. 34–36]):

(a) A Hilbert space $H \subseteq F(S)$ has a reproducing kernel if and only if $Lf = f(y)$ is a bounded linear functional on H for every $y \in S$. Moreover, the reproducing kernel is unique.

(b) $K(x, y) = K(y, x)$ for all $x, y \in S$.

(c) A positive definite kernel P defined on a set S is associated with a Hilbert space H such that P is a reproducing kernel of H. This means that a positive definite and a reproducing kernel are the same thing.

Positive Definite Functions

We now consider linear functionals defined on a Hilbert space H. Let $L : H \to \mathbb{R}$ be a linear functional and let K be a reproducing kernel of H. Then $L^x K(x, y)$ means that y is held fixed and L is applied to $K(x, y)$ as a function of x. Similarly $L^y K(x, y)$ is defined. We then have

$$L^x L^y K(x, y) = L(L^y K(x, y)).$$

For completeness we show the well known result.

Theorem 1. *Let $H \subseteq F(S)$ be a Hilbert space with reproducing kernel K. Let L_i, $i = 1, \ldots, n$, be bounded linear functionals on H. Then the following is true:*

(a) *The matrix*
$$K = (L_i^x L_j^y K(x, y))_{i,j=1}^n$$
is a PD-matrix.

(b) *K is an SPD-matrix if and only if $L_1^y K(\cdot, y), , \ldots, L_n^y K(\cdot, y)$ are linearly independent functions.*

Proof: (i) Since L_i is a bounded linear functional there exists a function $h_i \in H$ such that $L_i f = (h_i, f)$ for all $f \in H$, $i = 1, \ldots, n$. Hence $L_i^y K(x, y) = (h_i, K(x, \cdot)) = (h_i, K(\cdot, x)) = h_i(x)$ for $x \in S$ and, therefore,

$$L_i f = (h_i, f) = (L_i^y K(\cdot, y), f), \quad i = 1, \ldots, n.$$

(ii) Let $\mathbf{a} = (a_1, \ldots, a_n) \in \mathbb{R}^n$ be given. Then it follows that

$$0 \leq (\sum_{i=1}^n a_i L_i^y K(\cdot, y), \sum_{j=1}^n a_j L_j^y K(\cdot, y))$$
$$= \sum_{i,j=1}^n a_i a_j (L_i^y K(\cdot, y), L_j^y K(\cdot, y)) = \sum_{i,j=1}^n a_i a_j L_i^x L_j^y K(x, y)$$

where the last equation follows from (i). Hence K is a PD-matrix.

If $L_1^y K(\cdot, y), \ldots, L_n^y K(\cdot, y)$ are linearly independent then the strict inequality is true for all nontrivial $\mathbf{a}$ and K is an SPD-matrix. □

Remark. Let P be a PD-kernel on S. Let be given pairwise distinct points $\{x_1, \ldots, x_n\}$ in S and $f \in F(S)$. Define linear functionals $L_i f = f(x_i)$, $i = 1, \ldots, n$. Then it follows from the above mentioned properties that there exists a Hilbert space H such that P is a reproducing kernel of H. Moreover, $L_1, \ldots, L_n$ are bounded functionals on H. Then we can apply Theorem 1(b) and obtain a Lagrange interpolation problem associated with an SPD-matrix K (see also [7]).

We shall now give some examples and references for positive definite and reproducing kernels.

Example. (a) A real-valued function f on $\mathbb{R}^n$ is said to be positive definite (or a PD-function) if $f(x-y)$, $x,y \in \mathbb{R}^n$ is a positive definite kernel. These functions have been characterized in a famous theorem of Bochner. For special examples of PD-functions and the theorem of Bochner see [6].

(b) Recently compactly supported PD-functions have been considered (see [9]).

(c) Many examples of Hilbert spaces with a reproducing kernel are given in [4, pp. 53].

Finally we give an example for special linear functionals.

Example. Let $I = [a,b]$ be given and let

$$W_m(I) = \{f : I \to \mathbb{R}, f^{(m)} \in L_1(I) \cap L_2(I)\}, \quad m \geq 1$$

be endowed with the inner product

$$(f,g) = \sum_{i=0}^{m-1} f^{(i)}(a)g^{(i)}(a) + \int_a^b f^{(m)}(x)g^{(m)}(x)dx.$$

This is a Hilbert space with reproducing kernel K (see [4, p. 59 and 7]).

Let points $\{x_1,\ldots,x_n\}$ in I be given and nonnegative integers $j_1,\ldots,j_n$ satisfying $0 \leq j_i \leq m-1$, $i = 1,\ldots,n$. Then the linear functionals $L_i f = f^{(j_i)}(x_i)$, $i = 1,\ldots,n$, are bounded. Moreover, it can be seen that $L_1^y K(\cdot,y),\ldots,L_n^y K(\cdot,y)$ are linearly independent. Hence, Theorem 1(b) can be applied and we obtain a Hermite-Birkhoff interpolation problem which has a unique solution.

§3. Product Functions

In this section we want to consider products of reproducing and positive definite kernel functions. Using this theory we can synthesize multidimensional functions as the product of lower-dimensional functions (e.g., a one-dimensional and a two-dimensional function).

It is well known that the product of two positive definite kernels is again a positive definite kernel. But under some slight assumptions we obtain stronger results.

Let sets $S_i \subseteq \mathbb{R}^d$ and Hilbert spaces $H_i \subseteq F(S_i)$ with reproducing kernels K_i be given for $i = 1, 2$. Let $S^* = S_1 \otimes S_2$ and $H = H_1 \otimes H_2$ be the direct product of H_1 and H_2. Then

$$K(s,t;x,y) = K_1(s,x)K_2(t,y), \quad ((s,t),(x,y)) \in S^* \otimes S^*$$

is the reproducing kernel of H (see [4, p. 45]).

Positive Definite Functions

Theorem 2. *Let these assumptions be given. Suppose that $L_i : H_1 \to \mathbb{R}$ and $M_i : H_2 \to \mathbb{R}$, $i = 1, \ldots, n$ are bounded linear functionals such that $(L_i, M_i) \neq (L_j, M_j)$ for all $i \neq j$. Moreover, assume that for all subsets $I, J \subseteq \{1, \ldots, n\}$ satisfying $L_{i_1} \neq L_{i_2}$ for all $i_1, i_2 \in I$, $i_1 \neq i_2$ ($M_{j_1} \neq M_{j_2}$ for all $j_1, j_2 \in J$, $j_1 \neq j_2$) the set of functions $(L_i^x K_1(s, x))_{i \in I}$ ($((M_j^y K_2(t, y))_{j \in J})$ are linearly independent, respectively. Then the following is true:*

(a) *The functions $L_1^x M_1^y K(\cdot; x, y), \ldots, L_n^x M_n^y K(\cdot; x, y)$ are linearly independent.*

(b) *The matrix*
$$K = (L_i^s M_i^t L_j^x M_j^y K(s, t; x, y))_{i,j=1}^n$$
is an SPD-matrix.

Proof: (a) Let the functionals $(L_i, M_i)_{i=1}^n$ be denoted as follows:

$$(\tilde{L}_{1,1}, \tilde{M}_1), ,\ldots, (\tilde{L}_{1,m_1}, \tilde{M}_1),$$
$$\vdots$$
$$(\tilde{L}_{k,1}, \tilde{M}_k), ,\ldots, (\tilde{L}_{k,m_k}, \tilde{M}_k).$$

Let the linear system be given

$$\sum_{j=1}^k \sum_{i=1}^{m_j} a_{ij} \tilde{L}_{j,i}^x \tilde{M}_j^y K(s, t; x, y) = 0, \quad ((s,t), (x,y)) \in S^* \otimes S^*.$$

It follows from the definition of K that

$$\sum_{j=1}^k \sum_{i=1}^{m_j} a_{ij} \tilde{L}_{j,i}^x K_1(s, x) \tilde{M}_j^y K_2(t, y) = 0.$$

Since the functions $\tilde{M}_j^y K_2(\cdot, y))_{j=1}^k$ are linearly independent by assumption it follows that
$$\sum_{i=1}^{m_j} a_{ij} \tilde{L}_{j,i}^x K_1(\cdot, x) \equiv 0, \quad j = 1, \ldots, k.$$

Since the functions $(\tilde{L}_{j,i}^x K_1(\cdot, x))_{i=1}^{m_j}$ are also linearly independent for $j = 1, \ldots, k$, it follows that $a_{ij} = 0$ for $i = 1, \ldots, m_j$ and $j = 1, \ldots, k$. This proves the assertion (a).

(b) By assumption the linear functionals L_i and M_i defined on H_1 and H_2 are bounded, respectively. Then it follows from (a) and Theorem 1 that K is an SPD-matrix. □

If we consider Lagrange interpolation the following result can be derived from Theorem 2 (see also [7]).

Proposition 3. Let $S_i \subseteq \mathbb{R}^d$ be given and let K_i be positive definite kernels defined on S_i, $i = 1, 2$. Set $S^* = S_1 \otimes S_2$. Assume that the functions $K_i(\cdot, u_1), \ldots, K_i(\cdot, u_k)$ are linearly independent for all choices of distinct points $u_1, \ldots, u_k$ in S_i, $i = 1, 2$. Let $\{(x_i, y_i)\}_{i=1}^n \subseteq S^*$ be given pairwise distinct points. Then
$$((K_1(x_i, x_j)K_2(y, y_j))_{i,j=1}^n$$
is an SPD-matrix.

Proof: Assume that the positive definite kernels K_i are associated with the Hilbert spaces H_i in $F(S_i)$ such that K_i is a reproducing kernel of H_i, $i = 1, 2$. Then $K(s, t; x, y) = K_1(s, x)K_2(t, y)$ is a reproducing kernel of $H = H_1 \otimes H_2$. Assume that $L_i f = f(x_i)$ and $M_i f = f(y_i)$, $i = 1, \ldots, n$. Since H_1 and H_2 contain a reproducing kernel the point evaluation functionals are bounded. Then the assertion follows from Theorem 2. □

Remark. (a) The functions which have been given in Section 2 can be used in order to get functions of higher dimension. For the construction distinct kernels can be used. Therefore we obtain a great variety of product functions.

(b) The Hermite-Birkhoff interpolation problem of Section 2 can also be extended to higher dimensional problems.

(c) The matrix K of Theorem 2 can be considered as a Hadamard product of two matrices which are PD-matrices but no SPD-matrices in general. Under some slight assumptions we obtain that K is even an SPD-matrix.

(d) Recently conditionally positive definite functions and sums of these functions have been studied extensively in order to solve interpolation problems. The situation for products of conditionally positive definite functions seems to be more complicated (see, e.g., [7], Remark 2.9).

§4. An Extended Problem

In this section we want to consider a space G which is a direct sum of a Hilbert space H and a finite dimensional subspace U. We now define on G only a semi-inner product. We also solve interpolation problems for this situation. Moreover, we show that our interpolating functions are solutions of a variational problem.

Theorem 4. Let $H \subseteq F(S)$ be a Hilbert space which contains a reproducing kernel K and let $U = \text{span}\{u_1, \ldots, u_m\}$ be an m-dimensional subspace in $F(S)$ such that $U \cap H = \{0\}$. Let $G = \text{span}\{U \cup H\}$. Assume that $(\,,\,) : G \otimes G \to \mathbb{R}$ is a semi-inner product such that $(\cdot, \cdot)$ represents on H the inner product of H and $(u, g) = 0$ for all $u \in U$, $g \in G$. Let $L_i : G \to \mathbb{R}$, $i = 1, \ldots, n$, $n \geq m$, be linear functionals which are bounded. Assume that the rank of the matrix $C = (\mathbf{c}_1, \ldots, \mathbf{c}_m)$ where
$$\mathbf{c}_i = (L_1 u_i, \ldots, L_n u_i)^T, \quad i = 1, \ldots, m$$

is equal to m. Then the following is true.

(a) Let $L_1^y K(\cdot, y), \ldots, L_n^y K(\cdot, y)$ be linearly independent functions. Then the linear system

$$\sum_{j=1}^n a_j L_i^x L_j^y K(x,y) + \sum_{j=1}^m b_j L_i u_j = y_i, \quad i = 1, \ldots, n$$

$$\sum_{j=1}^n a_j L_j u_i = 0, \quad i = 1, \ldots, m$$

has a unique solution $(a_j^*)_{j=1}^n$ and $(b_j^*)_{j=1}^m$ for all choices of real numbers $y_1, \ldots, y_n$.

(b) Let a functional $L : G \to \mathbb{R}$ be defined by

$$L(g) = (g, g), \quad g \in G.$$

Assume that $g \in G$ is an arbitrary function satisfying $L_i g = y_i$ for $i = 1, \ldots, n$. Let $(a_j^*)_{j=1}^n$ and $(b_j^*)_{j=1}^m$ be a solution of (a) and $g^* = \sum_{j=1}^m b_j^* u_j + \sum_{j=1}^n a_j^* L_j^y K(\cdot, y)$. Then

$$L(g) \geq L(g^*).$$

The strict inequality is true iff $g - g^* \in U$.

This result is an extension of Theorem 1.4 in [7].

Remark. (a) If we set $m = 0$ in this theorem then we obtain the situation of the previous sections. This means that the interpolating functions of these problems are solutions of some variational problems.

(b) The theorem also gives an extension of well known results for spline interpolation in the univariate case to scattered data interpolation for multivariate splines. In particular, the minimal properties of interpolating natural splines and similar splines are extended to the case of multivariate splines.

References

1. Dyn, N., Interpolation and approximation by radial and related functions, in *Approximation Theory VI*, C. Chui, L. Schumaker, and J. Ward (eds.), Academic Press, New York, 1989, pp. 211–234.
2. Dyn, N. and C. A. Micchelli, Interpolation by sums of radial functions, Numer. Math. **58** (1990), 1–9.
3. Powell, M. J. D., The theory of radial basis function approximation in *Advances in Numerical Analysis, Vol. II*, W. A. Light (ed.), Oxford Univ. Press, Oxford, 1990, pp. 105–210.

4. Saitoh, S., *Integral Transforms, Reproducing Kernels and Their Applications*, Longman, Essex, 1997.
5. Schaback, R., Multivariate interpolation and approximation by translates of a basis function, in *Approximation Theory VIII, Vol. 1: Approximation and Interpolation*, Charles K. Chui and Larry L. Schumaker (eds.), World Scientific Publishing Co., Inc., Singapore, 1995, pp. 491–514.
6. Stewart, J., Positive definite functions and generalizations, an historical survey, Rocky Mountain J. Math. **6** (1976), 409–434.
7. Strauss, H., On interpolation with products of positive definite functions, Numer. Algorithms **15** (1997), 153–165.
8. Sun, X., Solvability of multivariate interpolation by radial or related functions, J. Approx. Theory **72** (1993), 252–267.
9. Wu, Z., Compactly supported positive definite radial functions, Advances in Comp. Math. **4** (1995), 283–292.

Hans Strauss
Institut für Angewandte Mathematik
Universität Erlangen–Nürnberg
Martensstr. 3
91058 Erlangen, Germany
strauss@am.uni-erlangen.de

On the Jackson Theorem for Approximation by Algebraic Polynomials

Gancho Tachev

Abstract. A direct theorem for approximation by algebraic polynomials on finite interval is considered, and the dependence of the constants on the degree of the derivative is studied.

§1. Main Result

The aim of this paper is to establish a Jackson type estimate for the L_p-approximation of functions from certain weighted Sobolev spaces by algebraic polynomials in one variable on a finite interval. Special attention is paid to the dependence of the constants on the degree of the polynomials and of the used Sobolev norm.

Probably, there are hundreds of papers and books concerning such direct theorems, but we mention here only two applications, closely related to the above problem, and motivating it. One of them is the so called h-p version of the finite element method, where the mesh and the degree of elements are simultaneously refined and increased. To achieve the exact exponential rate of convergence we need as sharp as possible local estimates for approximations on rectangular subdomains. Having at hand such estimates we can optimize in two directions, first by choosing an appropriate mesh (i.e., to determine the side-lengths h_1, h_2 of each rectangle), and secondly by choosing the proper distribution of the degrees of the polynomial in x_1 and in x_2. For more details and recent results in the two- and three-dimensional case of h-p theory, see [3, 6, 7, 8, 9]. The case of L_2-norm is considered in [4].

Another field connected with our problem are some more complicated nonlinear approximation processes, as described in Section 5.4.3 of [5]. There, it was proposed to study the approximation by piecewise polynomial functions of fixed degree on arbitrary partitions of a domain into $\leq N$ rectangles.

The main result in this note is the following direct theorem for approximation of functions f in the Sobolev spaces $W_p^r(\varphi)$, $x \in [-1, 1]$.

Theorem 1. If $f \in W_p^r(\varphi)$, $1 \leq p \leq \infty$, and $n \geq 4r$ then

$$E_n(f)_p \leq \left(\frac{c}{n}\right)^r \|\varphi^r f^{(r)}\|_{L_p(A)} \tag{1}$$

with an absolute constant c, independent of f, n, and r.

Proof: Inequality (1) is the Jackson inequality which is an essential step to prove the well-known direct theorem for algebraic approximation on a finite interval. To prove (1), we actually follow the proof of (7.1) in [1] and evaluate on each step the dependence of the constants on the corresponding parameters.

For every function f represented by

$$f(x) = T(x) + \frac{1}{r!} \int_A f^{(r)}(t)(x-t)_+^{r-1} dt \tag{2}$$

with $T(x)$ — the Taylor polynomial of degree $r-1$ of f at -1, we approximate by the following polynomial of degree $\leq 4rn$

$$Q(x) = T(x) + \frac{1}{r!} \int_A f^{(r)}(t) Q_t(x) dt. \tag{3}$$

Here

$$Q_t(x) := \begin{cases} (x-t)^{r-1}, & t \leq t_n, \\ 0, & t > t_1, \\ (x-t)^{r-1} \int_{-1}^x \Lambda_k(u) du, & t_{k+1} < t \leq t_k,\ k = 1, \ldots, n-1, \end{cases}$$

and $t_k = \cos \theta_k$, $\theta_k := (2k-1)\pi/(2n)$, $k = 1, \ldots, n$ are the zeros of the Chebyshev polynomials C_n of degree n. For future use, set

$$\Lambda_k(x) := c_k \left(\frac{C_n(x)}{x - t_k}\right)^{4r},\ k = 1, 2, \ldots, n, \tag{4}$$

with c_k chosen so that $\int_A \Lambda_k(x) dx = 1$.

To prove (1) it is enough to consider only functions for which (2) holds (for details see p. 260 in [1]). We define

$$E_r(x,t) := |(x-t)_+^{r-1} - Q_t(x)|, \quad \delta_n(x) := \varphi(x)/n.$$

The crucial estimation is contained in the following

Lemma 1. *With an absolute constant $c > 0$ we have*

$$\int_A E_r(x,t)\delta_n(t)^{-r}\,dx \leq c^r, \quad t \in A, \tag{5}$$

and

$$\int_A E_r(x,t)\delta_n(t)^{-r}\,dt \leq c^r, \quad x \in A. \tag{6}$$

Lemma 1 is actually Lemma 7.2 in [1], here we additionally show the exponential estimate for the constants in the right-hand sides of (5) resp. (6). We postpone the proof of Lemma 1, and proceed with the argument for Theorem 1. As a straightforward corollary from [1] and the above Lemma 1, we see that

$$E_{4nr}(f)_{L_p(A)} \leq \left(\frac{c}{n}\right)^r \frac{1}{r!} \|\varphi^r f^{(r)}\|_{L_p(A)} \tag{7}$$

holds for all $f \in W_p^r(\varphi)$, $1 \leq p \leq \infty$, $n \geq r$. To complete the proof of Theorem 1, we use (7) and obtain

$$E_n < E_{4r[n(4r)^{-1}]} \leq \frac{c^r}{[n(4r)^{-1}]^r r!}\|\varphi^r f^{(r)}\|_p \leq (cn^{-1})^r \|\varphi^r f^{(r)}\|_p.$$

Proof of Lemma 1. We begin with some preliminaries. Since $\int_A \Lambda_k(u)du = 1$ we have

$$E_r(x,t) \leq |x-t|^{r-1}, \quad x,t \in A. \tag{8}$$

For $t \in I_k$ this can be improved as follows

$$E_0(x,t) = \left|(x-t)_+^0 - \int_{-1}^x \Lambda_k(u)\,du\right| \leq \begin{cases} \int_{-1}^x \Lambda_k(u)du, & x < t, \\ \int_x^1 \Lambda_k(u)du, & x \geq t. \end{cases}$$

It is not so difficult to observe that in the case $t \in I_k$, $x \notin I_k$ each of the integrals above does not exceed

$$c_k \int_{|x-t_k|}^\infty u^{-4r}\,du \leq c_k \frac{|x-t_k|^{-4r+1}}{4r-1}. \tag{9}$$

Following the proof of Lemma 7.1 in [1], after simple calculations we may assert that

$$c_k \leq (6\pi)^{4r} 9\pi^{-1} \delta_n^{4r-1}(t_k). \tag{10}$$

From (9) and (10) we get for $t \in I_k$, $x \notin I_k$, $k = 1,\ldots, n-1$

$$E_r(x,t) = |x-t|^{r-1} E_0(x,t) \leq CM(x,t), \tag{11}$$

where

$$M(x,t) := |x-t|^{r-1}|x-t_k|^{-4r+1}\delta_n(t_k)^{4r-1}, \quad t \in I_k,$$
$$C := (6\pi)^{4r}9(\pi(4r-1))^{-1}. \tag{12}$$

Here and later, by c (or C) we denote absolute positive constants which may differ at each occurrence. Lastly we note that

$$E_r(x,t) = 0 \quad \text{if } x \leq t, t \geq t_1, \text{ or if } x \geq t, t \leq t_n. \tag{13}$$

With these preliminaries at hand, we are ready to prove Lemma 1. According to the symmetry of $E_r(x,t)$, it is enough to prove (5) only for $t \in [0,1]$, or (6) for $x \in [0,1]$, respectively. If $t \in [t_1,1]$ then by (8), and by using

$$1 - t \leq 1 - t_1 = 2\sin^2 \frac{\pi}{4n} \leq \frac{\pi^2}{8n^2},$$

we see that the integral (5) can be estimated from above by

$$\delta_n(t)^{-1}(1-t)^r \leq n^r(1-t)^{r/2} \leq \left(\frac{\pi}{2\sqrt{2}}\right)^r.$$

Let $t \in I_k$ for some fixed $k = 1, \ldots, n-1$, $t \geq 0$. We consider two cases: $|t-x| \leq \delta$ and $|t-x| > \delta$, where $\delta = \lambda \delta_n(t_k)$ and the constant λ is sufficiently large such that $[t_{k+2}, t_{k-1}] \subset [t-\delta, t+\delta]$ for every $t \in I_k = [t_{k+1}, t_k]$. Using (7.7) in chapter 8 in [1], some simple computations show that it is enough to take $\lambda = 2(3\pi)^2$.

In the case $|x - t| \leq \delta$ we use (8) here and (7.7) in [1], and obtain the upper bound $(3\lambda)^r$ in (5). For the second case $|x - t| > \delta$, we note that $x \notin I_k$. This implies that there exists an interval I_j between t and x, and, consequently,

$$|x - t| \leq C|x - t_k|, \tag{14}$$

for the $t \in I_k$ under consideration, where the constant C is independent of k.

Now we estimate $E_r(x,t)$ by (11), $|x-t|$ by (14), and $\delta_n(t)$ by $3\delta_n(t_k)$. We get an upper bound

$$C\delta_n(t_k)^{3r-1}\int_{|x-t|\geq \delta} |x-t|^{-3r}\,dx \leq C^r. \tag{15}$$

This completes the proof of (5). The estimation for (6) is analogous. We only need to follow step by step the proof of Lemma 7.2 in [1] (see p. 263), and to evaluate on each step the constants at most exponential growth. Finally, we may assert that the absolute constant c in Lemma 1 is bounded by $(6\pi)^4$. Lemma 1 is established.

Remark. The more difficult problem is to prove the two-dimensional analogon of Theorem 1. We will consider the functions from the Sobolev classes

$$W_p^{s_1,s_2} = \left\{ f : \frac{\partial^{s_i} f}{\partial x_i} \in L_p(\Omega), i = 1, 2 \right\},$$

where $1 \leq p \leq \infty$ and the smoothness parameters s_1, s_2 are non-negative integers. The following construction is similar to the one proposed in [2]. For $f \in W_p^s(\varphi)$, Theorem 1 yields an algebraic polynomial $P_n(f)$ of degree n such that

$$\|f - P_n(f)\|_{L_p(A)} \leq (cn^{-1})^s \|\varphi^s f^{(s)}\|_{L_p(A)}$$

with $0 \leq s \leq n/4$, and c is an absolute constant independent of s, n, and f.

For $f \in W_p^{s_1,s_2}$ we denote by P_{n_i} the polynomial operator P_n applied with respect to the variable x_i, $i = 1, 2$. Let $P_{n_1,n_2} f := P_{n_1}(P_{n_2} f)$. If we can prove that the operator P_{n_i}, or some other polynomial operator, yielding the estimation (1), is uniformly bounded, i.e.,

$$\|P_{n_i}(f)\|_{L_p(Q)} \leq C \|f\|_{L_p(Q)}, \tag{16}$$

where C is an absolute constant, independent of f, n_i, $i = 1, 2$, then we may estimate as follows:

$$\|f - P_{n_1,n_2}(f)\|_{L_p(Q)} \leq \|f - P_{n_1}(f)\|_{L_p(Q)} + \|P_{n_1}(f - P_{n_2}(f))\|_{L_p(Q)},$$

and, together with (16) for $i = 1$, the proof of two-dimensional analogon to Theorem 1 easily follows.

References

1. De Vore, R. A. and G. G. Lorentz, *Constructive Approximation*, Springer, New York, 1993.
2. Ditzian, Z. and V. Totik, *Moduli of Smoothness*, Springer, New York, 1987.
3. Guo, B. and I. Babuska, The h-p version of the finite element method, Part 1: The basic approximation results, J. Comput. Mech. **1** (1986), 21–41.
4. Tachev, G., Approximation by algebraic polynomials on rectangles, Proc. of IDoMat98, 1998, submitted.
5. Oswald, P., *Multilevel Finite Element Approximation, Theory and Applications*, Teubner, Stuttgart, 1994.
6. Elschner, J., On the exponential convergence of some boundary element methods for Laplace's equation in non-smooth domains, in *Proc. Conf. on Boundary Value Problems and Integral Equations on Non-Smooth Domains*, Luminy, 1993, pp. 69–80.

7. Babuska, I., Advances in the p and h-p versions of the finite element method. A survey, Int. Ser. Numer. Math. **86** (1988), 31–46.
8. Gui, W. and I. Babuska, The h,p and h-p versions of the finite element method in 1-dimension, Numer. Math. **49** (1986), Part 1: 577–612, Part 2: 613–657, Part 3: 659–683.
9. Guo, B., The h-p version of the finite element method for solving boundary value problems in polyhedral domains, in *Proc. Conf. on Boundary Value Problems and Integral Equations on Non-smooth Domains*, Luminy, 1993, pp. 101–120.

Gancho Tachev
Department of Mathematics
University of Architecture
1 Hristo Smirnenski blvd.
1421 Sofia
Bulgaria
gtt_fte@bgace5.uacg.acad.bg

On Approximation of Cauchy-Type Integrals by Sequences of Rational Functions with Preassigned Poles

Genrikh Ts. Tumarkin

Abstract. Necessary and sufficient conditions are established for the possibility of approximation to Cauchy-type integrals by sequences of rational functions with poles specified by a table, in general simply-connected domains with rectifiable boundaries. Also we investigate the case when the kth approximant r_k is a sum of special rational functions (defined similar to an expansion into series of Faber polynomials) with poles from the kth row of the given table. We describe the method of obtaining such results. Only the outlines of some proofs are given.

§1. Statements of the Problems and Results

Denote by $K_p(G)$, $p \geq 1$, the space of Cauchy-type integrals (CTI) $f(z)$ with norm defined by

$$f(z) = \int_\gamma \frac{\omega(\zeta)d\zeta}{\zeta - z}, \omega(\zeta) \in L_p(\gamma), \quad \|f\|^p_{K_p(G)} = \inf \int_\gamma |\omega(\zeta)|^p |d\zeta|. \qquad (1)$$

Here, infimum is taken over all functions $\omega(\zeta)$ from $L_p(\gamma)$ that appear in the representation (1) for $f(z)$. We consider the problem of approximation of $f(z) \in K_p(G)$ inside the domain G with rectifiable boundary γ by sequences of rational functions $\{r_k(z)\}$ with poles given by a table $\{\alpha_{k,j}\}$. We also investigate the problem when the kth rational approximant (RA) $\tilde{r}_k(z)$ is a sum of special rational functions, defined similar to the partial sums of an expansion as a Faber Polynomial (FP) series, with poles from the kth row of the given table. We obtain that if the corresponding sequence $\{\tilde{r}_k(z)\}$ of special rational approximants (SRA) converges to $f(z)$ inside G, then this sequence converges in the norm of $K_p(G)$. If the given table of poles does not provide the uniform approximation to any CTI inside G, the class of approximable functions is fully described. If the domain's boundary is regular in the Ahlfors sense, the sequence so constructed also converges in the mean on γ to the boundary values (BV) of $f(z)$. We assume that G is bounded, and denote by D the unit disk with boundary Υ. Also G^- denotes the complement of the closure of G; $w = \psi(z), (\psi(\infty) = \infty)$, the conformal mapping of G^- onto $|w| > 1$; $z = \varphi(w)$, the inverse conformal mapping. Analytic and meromorphic functions from all classes, considered in this paper, have BV – nontangential limits a.e. on the boundaries. We use the same notations for classes of the functions and their BV. Publications [3, 4, 5, 7, 8, 9] contain the general facts used in this study.

First, we consider the problem (1) of approximation CTI by sequences

$$r_k(z) = \frac{c_{k,n} z^{n-1} + \cdots + c_{k,0}}{(z - \alpha_{k,j_1}) \cdots (z - \alpha_{k,j_n})}, \qquad (2)$$

where poles are from the kth row $\{\alpha_{k,j}\}$, but the complex coefficients $c_{k,j}$ are a priori arbitrary. In (2) the poles are finite.

Then we consider the problem (2), where approximants $\tilde{r}_k$ with poles from the kth row of the table are partial sums of series of rational functions, analogous to FP.

Theorems 1 and 2 to follow relate to problem (1).

Theorem 1. *For the possibility to approximate any CTI (1) in $K_p(G), p \geq 1$, by sequences of rational functions r_k with poles $\{\alpha_{k,j}\}$, the following condition is necessary and sufficient:*

$$\lim S_k = \infty, S_k = \sum_j (1 - \frac{1}{|\beta_{k,j}|}), \quad \beta_{k,j} = \psi(\alpha_{k,j}). \qquad (3)$$

To formulate results in the non-closeness ($\liminf S_k < \infty$) case, for a given table we introduce a subclass N_α of the Nevanlinna class $N(G^-)$. Functions $\omega^-(z)$ of this subclass are characterized by the condition: The product $\omega^-(z) B_\alpha^-(z)$ belongs to the Smirnov subclass $E_1(G^-)$ of analytic functions in the domain G^-, and equal to 0 at ∞. Here $B_\alpha^-(z)$ is analytic in G^- (corresponding to the table $\{\alpha_{k,j}\}$) such that $|B_\alpha^-(z)| \leq 1, z \in G^-, \ln|B_\alpha^-(\varphi(w))| = \limsup \ln |b_k(w)|$, where $b_k(w)$ denotes the Blaschke product with zeroes at all of the points $\beta_{k,j} = \psi(\alpha_{k,j})$ from the kth row of the given table.

Theorem 2. *Let the table $\{\alpha_{k,j}\}$ satisfy the condition $\liminf S_k < \infty$. Then a function $f(z)$ from class $K_p(G), p \geq 1$, is approximable by sequences $\{r_k(z)\}$ with poles $\{\alpha_{k,j}\}$ in $K_p(G)$ if and only if the function $\omega(\zeta)$ in the representation (1) coincides a.e. on γ with the boundary values of the function $\omega^-(z)$, which is meromorphic in G^- and belongs to N_α.*

Now we can formulate some corollaries of Theorems 1 and 2.

Corollary 1. *If $f(z) \in K_p(G)$ can be approximated in $K_{p_1}(p > p_1 \geq 1)$ by a sequence of rational functions $\{r_k^{(1)}(z)\}$, then there exists a sequence with the same poles as $\{r_k^{(1)}(z)\}$, converging to $f(z)$ in $K_p(G)$.*

Corollary 2. *In the case of non-closeness, all functions $f(z)$, that can be approximated in $K_p(G)$, are pseudoanalytically continuable into the domain G^-.*

The notion of pseudoanalytical continuation was introduced by H. Shapiro in [10] and used by the author [11] earlier in an implicit form. See also results of M. Djrbashian, the author, and some others in Approximation Theory; and R. Douglas, H. Shapiro, A. Shields, et al. in the theory of backward shift operators.

Consider now problem (2). In [14] we investigated the approximation of CTI $f(z) \in K_2(G)$ inside domain G by SRA sequences with preassigned poles given by the table. An SRA with poles from the kth row of the table $\{\alpha_{k,j}\}$ is defined similarly to the definition of FP (see also [1, 2, 6, 11]). These papers study certain problems connected with the expansions of CTI $f(z) \in K_p(G)$ in the series of SRA in a particular case of poles, given by a sequence of points. The initial point of all these constructions is the orthogonal Takenaka-Malmquist-Walsh (TMW) system of rational fractions on $|w| = 1$ with fixed poles [16].

To construct an SRA system with given poles $\{\alpha_j\} \subset G^-$, we use $\{\Phi_k(w)\}$ – the TMW system with poles $\beta_j = \psi(\alpha_j)$. In [11] the author introduced the system $\{M_k^*(z)\}$ of rational functions with poles β_j, as principal parts of

$$\Phi_k(\psi(z))\sqrt{\psi'(z)}. \tag{4}$$

We used such systems in [15] for solving the problem of approximation to $f(z) \in K_2(G)$ by sequences of SRA $\{r_k^*\}$ with poles prescribed by the kth row of the table $\{\alpha_{k,j}\}$. The explicit formulas for the approximants r_k^* were obtained. They are similar to partial sums of Faber series:

$$r_k^*(z) = \sum_j c_{k,j} M_{k,j}^*(z). \tag{5}$$

Here the coefficients $c_{k,j}$ are expressed as integrals over γ, containing $\omega(\zeta)$ and simple functions, defined by the poles of $M_{k,j}^*(z)$. In [15] we found the necessary and sufficient conditions for convergence inside the domain G (for any given $f(z) \in K_2(G)$) of the constructed sequence $\{r_k^*\}$ to the function $f(z)$, that is the same as the condition of closeness in Theorem 1. In the non-closeness case, we got results, analogous to Theorem 2, that describe the class of functions uniformly approximable inside the domain G by $\{r_k^*(z)\}$; however, in the necessary condition we assumed additionally that $G^- \in S$.

Here we obtain results for any domain with rectifiable boundary γ. In this general case we modify the system $\{M_k^*(z)\}$, replacing it with the system $\{\tilde{M}_k(z)\}$, by introducing an additional factor $\{I[\psi(z)]\}^{-1/2}$ to functions in (4). Here $I(w)$ is the inner factor in the representation of $\varphi'(w) \in H_1, \varphi'(w) = I(w)Q(w)$. It is obvious that for $G^- \in S$, we have $\tilde{M}_k(z) = M_k^*(z)$. From now on, instead of $\{r_k^*(z)\}$, we use the sequences $\{\tilde{r}_k(z)\}$ analogous to (5).

Theorem 3. *(The closeness case).* For the possibility of approximation of any CTI $f(z) \in K_2(G)$ by sequences $\{\tilde{r}_k(z)\}$ uniformly inside domain G, the condition (3) is necessary and sufficient.

Theorem 4. *(The non-closeness case).* If $\liminf S_k < \infty$, then a function $f(z) \in K_2(G)$ is uniformly approximable inside domain G by sequences $\{\tilde{r}_k\}$ if and only if the function $\omega(\zeta)$ from (1) coincides a.e. on γ with BV of a function $\omega^-(z)$ in the subclass $N_\alpha \subset N(G^-)$.

We use Theorems 1–4 in proving the following result:

Theorem 5. *If the sequence $\{\tilde{r}_k\}$ constructed for $f(z)$ uniformly converges to $f(z) \in K_2(G)$ inside G, then it also converges to $f(z)$ in $K_2(G)$.*

Note. Let the boundary of G satisfy the Ahlfors regularity condition:

$$\text{measure}\{\gamma \cap \{|z - \zeta_0| < \varrho\}\} < C\varrho, \zeta_0 \in \gamma, \varrho > 0.$$

Then using well-known results of A. Calderon and G. David, in addition to convergence in $K_p(G), p > 1$, the sequences of rational functions considered here also converge to BV of $f(z)$ on γ in $L_p(\gamma)$.

§2. Isometric Correspondence of Cauchy-type Integral Spaces

Now we introduce an important tool to establish results from Section 1 – the isometric mapping of the Banach space $K_p(G)$ onto $K_p(D)$. Such isometric mappings are induced by the conformal mapping of the domain G^- onto $|w| > 1$. To any function $f(z) \in K_p(G)$ there corresponds $Tf = F(w) \in K_p(D)$:

$$f(z) = K_\gamma^\omega(z) \to T_p f = F(w) = K_\Upsilon^\Omega(w), \qquad (6)$$

where

$$\Omega(t) = \tau_p[\omega(\zeta)] = \omega[\varphi(t)]\{\varphi'(t)\}^{1/p}\{I(t)\}^{1/q}, 1/p + 1/q = 1. \qquad (7)$$

Here we denote by $I(t)$, the BV of the inner function in the factorization $\varphi'(w)$. For the inverse transform T_p^{-1} we have:

$$F(w) = K_\Upsilon^\Omega(z) \to T_p^{-1} F = f(z) = K_\gamma^\omega(z);$$

$$\omega(\zeta) = \tau_p^{-1}\Omega = \Omega[\psi(\zeta)]\{\psi'(\zeta)\}^{1/p}\{I[\psi(\zeta)]\}^{-1/q}. \qquad (8)$$

It is easy to see, as $|I(t)| = 1$ a.e. on Υ, that for the corresponding $\Omega(t)$ and $\omega(\zeta)$ in (7) and (8), we have $\|\omega\|_{L_p(\gamma)} = \|\Omega\|_{L_p(\Upsilon)}$.

We now verify the correctness of the definitions, by proving that using all admissible functions in the representation (1) of $f(z) \in K_p(G)$, we get, according to (6), the same element $F(w) \in K_p(D)$. A similar verification is necessary for the inverse transform T_p^{-1}.

First, we describe the sets of all $\{\Omega(t)\}$ and $\omega(\zeta)$ that may be used in the CTR $F(w) = K_\Upsilon^\Omega(w)$ and $f(z) = K_\gamma^\omega(z)$. Let $F(w) = K_\Upsilon^{\Omega_1}(w) = K_\Upsilon^{\Omega_2}(w)$. Then,

$$\int_\Upsilon \frac{\Omega_1(t) - \Omega_2(t)}{t - w} dt \equiv 0, \quad |w| < 1. \tag{9}$$

From (9) we deduce that $\Omega_1(t) - \Omega_2(t)$ is BV of a function $\Phi(w) \in H_1\{|w| > 1\}$. Here, $\Phi(w) \in H_p\{|w| > 1\}$ means that $\Phi(1/w) \in H_p(D)$. Because $\Omega_1(t) - \Omega_2(t) \in L_p(\Upsilon)$, we conclude by using the properties of H_p that $\Omega_1(t) - \Omega_2(t)$ is BV of $h_p^-(z) \in H_p\{|w| > 1\}$ with $h_p^-(\infty) = 0$. We denote by H_p^- the set of all such $h_p^-(z)$. Because H_p is closed, H_p^- is a closed subspace of $L_p(\Upsilon)$. Thus the set $\{\Omega(t)\}$, corresponding to $F(w)$, is a coset $\Omega_0 + H_p^-$ in the quotient Banach space $L_p(\Upsilon)/H_p^-$, and the norm in $K_p(D)$ is equal to the quotient norm

$$\|F\|_{K_p(D)} = \|\Omega_0 + H_p^-\|_{L_p(\Upsilon)} = \inf \|\Omega_0(t) + h_p^-(t)\|, h_p^- \in H_p^-.$$

For $f(z) \in K_p(G)$, the set of all $\omega(\zeta)$ in (1) can be expressed as

$$\{\omega(\zeta)\} = \omega_0(\zeta) + \{a_p^-\}, \text{ where } \int_\gamma \frac{a_p^-(\zeta)d\zeta}{\zeta - z} \equiv 0, \quad z \in G. \tag{10}$$

From Smirnov's results, it follows that $a_p^-(\zeta)$ in (10) is BV of an analytic function $a_p^-(z) \in E_1(G^-), a_p^-(\infty) = 0$, besides $a_p^-(\zeta) \in L_p(\gamma)$. It can be easily shown that the subclass A_p^- of all such functions is closed in $L_p(\gamma)$.

Thus, we have proved that the set of all $\{\omega\}$ in the CTR (1) may be expressed by the formula: $\{\omega(\zeta)\} = \omega_0(\zeta) + A_p^-$. Consequently, $K_p(G)$ may be considered as a quotient Banach space $L_p(\gamma)/A_p^-$ with the norm

$$\|f\|_{K_p(G)} = \|\omega_0 + A_p^-\| = \inf \|\omega_0 + a_p^-\|_{L_p(\gamma)}, a_p^- \in A_p^-.$$

Indeed, by the transforms (6) and using, (8) the zero elements of both quotient spaces $L_p(\gamma)/A_p^-$ and $L_p(\Upsilon)/H_p^-$ correspond to each other. For any $a_p^-(\zeta)$ we verify that $\tau_p(a_p^-) = h_p^-(t)$. Also,

$$\tau_p(a_p^-) = \{a_p^-[\varphi(t)]\varphi'(t)\}\{Q(t)\}^{1/q} \tag{11}$$

(because $\varphi'(w) = I(w)Q(w)$). The first factor in the product (11) is BV of a function from H_1^-. From $a_p^-(z) \in E_1(G^-)$ it follows that $a_p^-[\varphi(w)]\varphi'(w) \in H_1\{|w| > 1\} \subset N^+\{|w| > 1\}$. Thus each of the two factors in (11) belongs to the Smirnov class $N^+\{|w| > 1\}$. (We use the facts that $H_p \subset N_\alpha$ and $Q(w) \in N^+$, being an outer function.) Then the product (11) belongs to the class N^+, and, from the Smirnov's theorem, is the BV of a function from $H_p\{|w| > 1\}$; it is 0 at ∞ because $a_p^-(\infty) = 0$. Thus $\tau_p(a_p^-) = h_p^-(t)$. Analogously, $\tau_p^{-1} h_p^- = a_p^-$.

We formulate the properties of the mapping in Lemma 1, adding the information about the existence and uniqueness of extremal elements.

Lemma 1. *The conformal mapping $w = \psi(z)$ of the domain G^- onto $|w| > 1$ induces an isometric transform τ_p of the space $L_p(\gamma)$ onto the space $L_p(\Upsilon)$. This gives simultaneously: 1) The isometry A_p^- onto H_p^-. 2) The isometry T_p of the space $K_p(G)$ onto $K_p(D)$ (the spaces $K_p(G)$ and $K_p(D)$ are isometrically isomorphic respectively to the quotient spaces $L_p(\gamma)/A_p^-$ and $L_p(\Upsilon)/H_p^-$. 3) The following relations are valid for all $f(z) \in K_p(G); f(z) = K_\gamma^\omega(z)$ and $F(w) = Tf$):*

$$\|f\|_{K_p(G)}^p = \int_\gamma |\tilde{\omega}(\zeta)|^p |d\zeta| = \inf \int_\gamma |\omega(\zeta)|^p |d\zeta| = \inf \int_\gamma |\omega + a_p^-|^p |d\zeta| =$$

$$\inf \int_\Upsilon |\tilde{\Omega}(t) + h_p^-|^p |dt| = \int_\Upsilon |\tilde{\Omega}(t)|^p |dt| = \|F\|_{K_p(D)}^p. \quad (12)$$

The existence and uniqueness of the element $h_p^-(t)$ for which infimum in (12) is attained may be proved similar to the same fact for H_p in the disk [7]. The analogous fact for the extremal element among a_p^- is deduced from the isometry.

We use the established isometry T_p in the problem of approximation of CTI by rational functions. This isometry transforms the set of rational functions $\{r_k(z)\}$ with preassigned poles inside domain G^- onto the set of rational functions $\{R_k(w)\}$ with poles in $|w| > 1$. The following lemma holds

Lemma 2. *For any rational function $r(z)$ with poles $\{\alpha_j\}$ inside G^-, the T_p-image is a rational function $R(w)$ with poles $\{\beta_j\}, \beta_j = \psi(\alpha_j)$. A similar conclusion is valid for the inverse transform T_p^{-1}.*

To prove the first statement, it is sufficient to verify the possibility of the representation of the meromorphic inside $|w| > 1$ function as

$$T_p[r(z)] = r[\varphi(w)][\varphi'(w)]^{1/p}[I(w)]^{1/q} = R(w) + \rho(w), \rho(w) \in H_p^-. \quad (13)$$

Here $R(w)$ is a sum of all principal parts on the expansions of (13) near its poles (also including constant member at ∞). The function $\rho(w)$ belongs to H_p^- because $\varphi'(w) \in H_1\{|w| > 1\}, |I(w)| \leq 1$.

To prove the second statement, we establish a representation:

$$T_p^{-1}[R(w)] = R[\psi(z)]\{\psi'(z)\}^{1/p}\{I[\psi(z)]\}^{-1/q} = r(z) + \mu(z), \mu(z) \in A_p^-. \quad (14)$$

First, we verify that $\{\psi'(z)\}^{1/p}\{I(\psi(z))\}^{-1/q} \in E_1(G^-)$. Lemma 2 immediately follows from (13) and (14).

Now we can reduce the problem of approximation for $f(z) \in K_p(G)$ by rational functions with poles in G^- to a similar problem for $F(w) \in K_p(D)$. From Lemmas 1–2, we have

$$\|f - r\|_{K_p(G)} = \|F - R\|_{K_p(D)}, \quad F = T_p f, R = T_p r. \quad (15)$$

Rational Approximation of Cauchy-Type Integrals

Consider the case $p > 1$. According to Riesz's theorem for any $p > 1$, we have $\Omega(t) = \Omega^+(t) + \Omega^-(t)$, where $\Omega^+(t)$ and $\Omega^-(t)$ are BV of functions from classes H_p and H_p^- respectively. Besides,

$$\|\Omega^+(t)\|_{L_p} \leq C_p \|\Omega(t)\|_{L_p}.$$

Thus, we can use CTI for $F(w) \in K_p(D)$ with $\Omega^+(t)$ instead of $\Omega(t)$. Using (15) and the fact that $F(w) - R(w) = K_\Upsilon^{\Omega^+ - R}$, we get

$$\|F - R\|_{K_p(D)} \leq \|\Omega^+(t) - R(t)\|_{L_p(\Upsilon)}. \qquad (16)$$

Note, that

$$\|F - R\|_{K_p(D)} = \inf \|\Omega^+ - R + h_p^-\|_{L_p(\Upsilon)} = \|\Omega^+(t) - R(t) + \tilde{h}_p^-\|_{L_p(\Upsilon)}. \quad (17)$$

Here $\tilde{h}_p^-$ is the extremal element in relation (12). Riesz's theorem then gives, for H_p-components of functions $\Omega^+(t) - R(t) + \tilde{h}_p^-(t)$,

$$\|\Omega^+(t) - R(t)\|_{L_p(\Upsilon)} \leq C_p \|\Omega^+(T) - R(t) + \tilde{h}_p^-(t)\|_{L_p(\Upsilon)}. \qquad (18)$$

From (16), (17), and (18), we get

$$\|f - r\|_{K_p(G)} \leq \|\Omega^+(t) - R(t)\|_{L_p(\Upsilon)} \leq C_p \|f - r\|_{K_p(G)}. \qquad (19)$$

Using the relation (19), we reduce the problem of the approximation of $f(z)$ by rational functions in $K_p(G)$ to a similar problem for $\Omega^+(t)$ in $L_p(\Upsilon)$. We have established previously the correspondence between rational functions $r(z)$ and $R(w)$. After that it is possible to use our previous results [12, 13, 14] describing the properties of approximable functions on Υ in the spaces L_p by rational functions with preassigned poles.

The page limitation of the paper does not allow us to give detailed proofs of all results mentioned in Section 1 (especially, in the case $p = 1$, when we cannot apply Riesz's theorem). Besides using isometry T_p, for $p = 1$, we apply some other methods and results, including those used in [12, 13], along with some properties of the TMW series. So it is possible to establish the properties of SRA for $f(z) \in K_p(G)$ (not only for $p = 2$ as in the theorem 5). We also get results about the convergence in G^- of rational approximants RA for $f(z) \in K_p(G)$ in the non-closeness case. All the results can be extended to the approximation CTI in finitely-connected domains. Detailed proofs will be given in subsequent papers. Additional references can be found in the papers cited.

References

1. Djrbashian, M. M., On expansions of analytic functions in series of rational functions with given set of poles, Izv. Akad. Armenii, Ser. Fiz. Mat. **10** (1957), 21–29.

2. Djrbashian, M. M., Expansions of systems of rational functions with fixed poles, Izv. Akad. Armenii, Mat. **2** (1967), 3–31.

3. Duren, P. L., *Theory of H^p Spaces*, Academic Press, New York, 1970.

4. Garnett, J. B., *Bounded Analytic Functions*, Academic Press, New York, 1981.

5. Hoffman, K., *Banach Spaces of Analytic Functions*, Prentice-Hall, Englewood Cliffs, 1982.

6. Kazarian, K. S. and V. M. Martirosian, Basicity of systems of Faber-Djrbashian rational functions in Smirnov spaces and their subspaces, Izv. Akad. Nauk Armenii, Mat. **26** (1991), 26–52.

7. Koosis, P., *Introduction to H^p Spaces*, Cambridge University Press, Cambridge, 1980.

8. Nikolski, N. K. and V. P. Khavin, The results of V. I. Smirnov in complex analysis and their further development, in *Smirnov V. I., Selected Papers. Complex Analysis. Mathematical Theory of Diffraction*, Leningrad Univ. Press, Leningrad, 1988, pp. 111–145.

9. Privalov, I. I., *Boundary Properties of Analytic Functions*, Gostechizdat, Moscow, 1950.

10. Shapiro, H. S., Weighted polynomial approximation and boundary values of analytic functions, in *Contemporary Problems in Theory of Analytic Functions*, Nauka, Moscow, 1966, pp. 326–335.

11. Tumarkin, G. Ts., The decomposition of analytic functions in series of rational fractions with a given set of poles, Izv. Akad. Nauk Armenii, Ser. Fiz. Mat. **14** (1961), 9–31.

12. Tumarkin, G. Ts., Approximation with respect to various metrics of functions defined on the circumference by sequences of rational functions with fixed poles, Izv. Akad. Nauk SSSR, Ser. Mat. **30** (1966), 721–766. Engl. transl. in; AMS Transl. (2) **77** (1968), 183–233.

13. Tumarkin, G. Ts., Description of a class of functions admitting an approximation by fractions with preassigned poles, Izv. Akad. Nauk Armenii, Ser. Mat. **1** (1966), 85–109.

14. Tumarkin, G. Ts., Conditions for the convergence of boundary values of analytic functions and approximation on rectifiable curves, in *Contemporary Problems in Theory of Analytic Functions*, Nauka, Moscow, 1966, pp. 283–295.

15. Tumarkin, G. Ts., Approximation of functions analytic in a simply connected domain and representable by Cauchy-type integral by sequences of rational fractions with poles prescribed by a given matrix, in *Research on Linear Operators and Theory of Functions*, Zap. Nauch. Sem. LOMI, V. 170, Nauka, Leningrad, 1989, pp. 254–273.

16. Walsh, J. L., *Interpolation and approximation by rational functions in the complex domain*, Col. Publ., V. 20, AMS, Providence, 1969.

Sharp Error Estimates for Multivariate Positive Linear Operators

Shayne Waldron

Abstract. A *sharp* pointwise error estimate is given for multivariate positive linear operators which reproduce linear polynomials. It is applied to a number of operators including the multivariate Bernstein operators, and the recently introduced Bernstein-Schoenberg type operators of Dahmen, Micchelli, and Seidel.

§1. Introduction and the Main Result

If $L : C[a,b] \to C[a,b]$ is a positive linear operator which reproduces linear polynomials, then it is known (see, e.g., DeVore [3, Th.2.5, p. 39]) that there is the sharp error estimate

$$|f(x) - Lf(x)| \leq \frac{1}{2} L\big((\cdot - x)^2\big)(x) \, \|D^2 f\|_\infty, \qquad \forall f \in C^2[a,b]. \tag{1}$$

In this paper we give the multivariate generalization of (1). The emphasis is on the previously unobserved fact that this straightforward generalization gives *sharp* multivariate bounds, which are then applied to a number of operators of interest.

In the remainder of this section we present the main result. In Section 2, some examples are considered. These include multivariate Bernstein operators (like linear interpolation at the vertices of a simplex and bilinear interpolation at the vertices of a rectangle), and the recently introduced Bernstein-Schoenberg operators based on blossoming. In Section 3, we give a more general Korovkin-type theorem which also covers the case of multivariate positive linear interpolation operators which do not reproduce linear polynomials.

For simplicity, we let K be a compact subset of $\mathbb{R}^s$ and consider maps L on $C(K)$ and $C^2(K)$. In the case of locally defined maps (for unbounded regions K), or other function spaces (like Sobolev spaces) appropriate modifications of the results below can be made. As usual, L is positive means $Lf \geq 0$ whenever $f \geq 0$, while L reproduces the linear polynomials means that $Lp = p$ whenever p belongs to Π_1 (the linear polynomials).

To determine the sign of one of the constants occurring in Theorem 2, we need the following result from the 'folklore' of positive linear operator theory.

Proposition 1. *Suppose that Ω is a convex subset of $\mathbb{R}^s$ and $L : C(\Omega) \to C(\Omega)$ a positive linear operator which reproduces the linear polynomials. Then*

$$f \text{ is convex} \implies f \le Lf. \qquad (2)$$

Proof: Suppose that f is convex and $x \in \Omega$. Since f is convex, it is possible to choose a linear polynomial p with $p(x) = f(x)$ and $p \le f$. When f is C^1 at x this is simply the tangent plane to f at x. Since L is positive and reproduces p,

$$p \le f \implies p = Lp \le Lf,$$

and so

$$f(x) = p(x) \le Lf(x),$$

as supposed. □

This 'shape property' is given for the Bernstein-Schoenberg operator in Goodman [4, Th. 1, p. 442], and it is presumably known for other operators as well. It also holds in the infinite dimensional setting when Ω is a convex compact subset of a locally compact space (use inequality (1.5.4) of Altomare and Campiti [1]).

Next we present the main result. For $f \in C^2(K)$ we define the semi-norm

$$\|D^2 f\|_{\infty, K} := \sup_{x \in K} \sup_{\substack{u_1, u_2 \in \mathbb{R}^n \\ \|u_i\|=1}} |D_{u_1} D_{u_2} f(x)|,$$

where $D_y f$ is the derivative of f in the direction y. This measures the maximum size of the second derivative of f over K.

Theorem 2. *Suppose that K is a compact convex subset of $\mathbb{R}^s$ and $L : C(K) \to C(K)$ is a positive linear operator which reproduces the linear polynomials. Then, for every $x \in K$ there is the sharp pointwise error estimate*

$$E(f, x) := |f(x) - Lf(x)| \le \frac{1}{2} S(x) \|D^2 f\|_{\infty, K}, \qquad \forall f \in C^2(K), \qquad (3)$$

where the nonnegative function $S := S_L : K \to \mathbb{R}^+$ is defined by

$$S(x) := L(\|\cdot - x\|^2)(x) = E(\|\cdot - x\|^2, x) = E(\|\cdot\|^2, x) = L(\|\cdot\|^2)(x) - \|x\|^2. \quad (4)$$

There is equality in (3) for f from the space of quadratics

$$Q := \Pi_1 \oplus \text{span}\{\|\cdot\|^2\} = \text{span}\{\|\cdot - c\|^2 : c \in \mathbb{R}^n\}. \qquad (5)$$

Proof: Let $T_{1,x} f$ be the linear Taylor interpolant to f at x, i.e., the linear polynomial which matches the value and first order derivatives of f at x. From the (univariate) integral error formula for Taylor interpolation

$$R_{1,x} f(y) := f(y) - T_{1,x} f(y) = \int_0^1 (1-t) D_{y-x}^2 f(x + t(y-x)) \, dt, \qquad y \in K,$$

it follows that
$$|R_{1,x}f| \leq \frac{1}{2}\|\cdot-x\|^2\|D^2f\|_{\infty,K}. \tag{6}$$

Since L reproduces $p = T_{1,x}f$, applying it to $T_{1,x}f - f = -R_{1,x}f$ gives
$$T_{1,x}f - Lf = -L(R_{1,x}f). \tag{7}$$

Since L is positive, using (7) and (6) we obtain
$$|T_{1,x}f - Lf| \leq L(|R_{1,x}f|) \leq L(\frac{1}{2}\|\cdot-x\|^2\|D^2f\|_{\infty,K}) = \frac{1}{2}\|D^2f\|_{\infty,K}L(\|\cdot-x\|^2). \tag{8}$$

Evaluating (8) at x gives
$$|f(x) - Lf(x)| \leq \frac{1}{2}L(\|\cdot-x\|^2)(x)\|D^2f\|_{\infty,K} = \frac{1}{2}S(x)\|D^2f\|_{\infty,K},$$

which is (3). This is sharp for $f := \|\cdot-x\|^2$, and hence for any quadratic polynomial from Q (since L reproduces Π_1). Since $\|x-x\|^2 = 0$ and L reproduces Π_1, the first two equalities in (4) follow immediately. The third, that
$$E(\|\cdot\|^2, x) := \big|\|x\|^2 - L(\|\cdot\|^2)(x)\big| = L(\|\cdot\|^2)(x) - \|x\|^2$$
follows from Proposition 1 and the fact that $\|\cdot\|^2$ is convex. □

It is an immediate consequence of (3) that there is the *sharp* error estimate
$$\|f - Lf\|_{L_\infty(K)} \leq \frac{1}{2}C_L\|D^2f\|_{\infty,K}, \quad \forall f \in C^2(K), \tag{9}$$

where
$$C_L := \max_{x \in K} S(x). \tag{10}$$

After proving Theorem 2, it was pointed out to the author that the estimate (3) is given in Raşa [5, (10)] in the setting where $\mathbb{R}^s$ is replaced by a real inner product space, with $S(x)$ taken to be $L(\|\cdot\|^2)(x) - \|x\|^2$. It occurs as a special case of a more general estimate (involving c-convex functions), but it is not observed that this case is sharp.

§2. Examples of Sharp Estimates

In this section the function $S := S_L$ of Theorem 2 is computed for several operators L, and hence the sharp error estimate (3) is obtained.

Example 1. Let T be a (nondegenerate) simplex in $\mathbb{R}^s$, with vertices V, and corresponding barycentric coordinate functions $(\lambda_v)_{v \in V}$. The multivariate Bernstein operator of degree n on this simplex, $B_n := B_{n,T} : C(T) \to C(T)$, $n = 1, 2, \ldots$, is defined by

$$B_n f(x) := \sum_{v_1 \in V} \sum_{v_2 \in V} \cdots \sum_{v_n \in V} f\Big(\frac{v_1 + \cdots + v_n}{n}\Big) \lambda_{v_1}(x) \cdots \lambda_{v_n}(x). \tag{11}$$

This operator is positive and reproduces the linear polynomials, and so Theorem 2 can be applied. It can be shown (see comments below) that

$$S(x) := S_{B_n}(x) := B_n(\|\cdot - x\|^2)(x) = \frac{1}{n}(R^2 - \|x - c\|^2), \qquad (12)$$

where c is the center and R the radius of the (unique) sphere containing V. Hence B_n satisfies the sharp error estimate, for $x \in T$, that

$$|f(x) - B_n f(x)| \le \frac{1}{2n}(R^2 - \|x - c\|^2)\|D^2 f\|_{\infty,T}, \qquad \forall f \in C^2(T), \qquad (13)$$

and, in particular, the sharp error estimate, of the form (9), that

$$\|f - B_n f\|_{L_\infty(T)} \le \frac{1}{2n}(R^2 - d^2)\|D^2 f\|_{\infty,T}, \qquad \forall f \in C^2(T), \qquad (14)$$

where
$$d := \text{the distance of } c \text{ from } T = \min_{x \in T} \|x - c\|.$$

The operator $B_1 = B_{1,T}$ is the map of linear interpolation at the vertices V of T (interpolation by linear polynomials). Its estimates (13) and (14) were recently proved in Waldron [6] by using an integral representation of the error. At the time, these were the only known sharp pointwise error estimates for a multivariate interpolation operator, and the role of the positivity of B_1 in obtaining them was not fully appreciated.

By taking the result proved in [6] that $S_{B_1}(x) = R^2 - \|x - c\|^2$, and the formula for $S_{B_n}(x)$, when T is a standard simplex, given in Altomare and Campiti [1, p. 315], one can conclude (12). The formula for $S_{B_n}(x)$, denoted by $\sigma_{n,x}^2$, is of the form of $1/n$ multiplying a quadratic polynomial (namely $R^2 - \|x - c\|^2$ for the standard simplex), and is obtained through a probablistic interpretation of S_{B_n} (see [1] for further details). Alternatively, since $S_{B_n}(x) = B_n(\|\cdot\|^2)(x) - \|x\|^2$ and B_n is degree reducing, S_{B_n} is a quadratic polynomial vanishing at V, and hence is a scalar multiple of $R^2 - \|\cdot - c\|^2$. The scalar multiplier of $1/n$ can be determined by considering the restriction of (3) to any edge of T which is then just the error for the univariate Bernstein operator.

Example 2. Let B_n, B_m be the univariate Bernstein operators of degrees n, m defined on the intervals $[a, b]$, $[c, d]$, respectively (cf. (11)), i.e.,

$$B_n f(x) := \sum_{k=0}^{n} f(v_k)\, p_k(x), \quad x \in [a, b],$$

$$B_m f(y) := \sum_{j=0}^{m} f(w_j)\, q_j(y), \quad y \in [a, b],$$

where

$$v_k := \frac{ka + (n-k)b}{n}, \quad p_k := \lambda_a^k \lambda_b^{n-k}, \quad w_j := \frac{jc + (m-j)d}{m}, \quad q_j := \lambda_c^j \lambda_d^{m-j}.$$

Then the bivariate tensor product Bernstein operator of coordinate degree (n,m) on the rectangle $R := [a,b] \times [c,d]$, $B_{n,m} := B_{n,m,R} = B_n \otimes B_m : C(R) \to C(R)$, is defined by

$$B_{n,m}f(x,y) := \sum_{k=0}^{n}\sum_{j=0}^{m} f(v_k, w_j) p_k(x) q_j(y), \quad (x,y) \in R. \quad (15)$$

This operator is positive and it reproduces the bilinear polynomials (which contain the linear polynomials). Using the fact that the univariate Bernstein operator reproduces constants, we compute that

$$\begin{aligned}
S_{B_{n,m}}(x,y) &:= B_{n,m}(\|\cdot - (x,y)\|^2)(x,y) \\
&= \sum_{k=0}^{n}\sum_{j=0}^{m} \left\{(v_k - x)^2 + (w_j - y)^2\right\} p_k(x) q_j(y) \\
&= \sum_{k=0}^{n}(v_k - x)^2 p_k(x) + \sum_{j=0}^{m}(w_j - y)^2 q_j(y) \quad (16) \\
&= B_n(|\cdot - x|^2)(x) + B_m(|\cdot - y|^2)(y) \\
&= S_{B_n}(x) + S_{B_m}(y) \\
&= \frac{1}{n}(x-a)(b-x) + \frac{1}{m}(y-c)(d-y).
\end{aligned}$$

From (16) sharp pointwise error estimates for $B_{n,m}$ can be obtained. For example, if L is the map of bilinear interpolation at the vertices of the unit square, i.e., $B_{1,1}$ with $R := [0,1]^2$, then for $(x,y) \in [0,1]^2$ we have the sharp estimate

$$|f(x,y) - Lf(x,y)| \leq \frac{1}{2}\{x(1-x) + y(1-y)\} \|D^2 f\|_{\infty, [0,1]^2}, \quad (17)$$

for all $f \in C^2(R)$.

Example 3. It is possible to define the tensor product of positive linear operators (see [1, p. 32]). This tensor product is a positive operator, and it reproduces the linear polynomials if each of its factors does so. The construction, an abstract version of (15) which relies on associated families of Radon measures, is technical. Hence we provide only a brief outline of it and the corresponding general form of (16). The reader should consult [1, p. 32] for full details.

Let $L_i : C(K_i) \to C(K_i)$, $i = 1, \ldots, p$, where $K_i \subset \mathbb{R}^{s_i}$ is compact and convex, be a (finite) collection of positive linear operators which reproduce $\Pi_1(K_i)$ (the linear polynomials on K_i). The tensor product

$$L := \bigotimes_{i=1}^{p} L_i : C(K) \to C(K), \qquad K := \prod_{i=1}^{p} K_i \subset \mathbb{R}^{s_1} \times \cdots \times \mathbb{R}^{s_p}$$

is a positive linear operator (K is a compact convex region). It reproduces $\Pi_1(K_1) \otimes \cdots \otimes \Pi_1(K_p)$ which contains $\Pi_1(K)$. Using properties of the tensor product and the fact that each L_i reproduces constants one can argue, as in (16), that

$$\begin{aligned} S_L(x_1, \ldots, x_p) &:= L(\| \cdot -(x_1, \ldots, x_p) \|^2)(x_1, \ldots, x_p) \\ &= L_1(\| \cdot -x_1 \|^2)(x_1) + \cdots + L_p(\| \cdot -x_p \|^2)(x_p) \\ &= S_{L_1}(x_1) + \cdots + S_{L_p}(x_p), \end{aligned} \qquad (18)$$

which is the general form of (16).

Here is a specific example. Suppose that $L : C(K) \to C(K)$ is the map of interpolation from $\Pi_1(\mathbb{R}^2) \otimes \Pi_1(\mathbb{R})$ at

$$V := \{(0,0,0), (1,0,0), (0,1,0), (0,0,1), (1,0,1), (0,1,1)\}$$

the vertices of the triangular prism K in $\mathbb{R}^3$. With $K =: T \times I \subset \mathbb{R}^2 \times \mathbb{R}$, this map is of the form $L = B_{1,T} \otimes B_{1,I}$, and hence it is positive and reproduces the linear polynomials. Thus, by (18) and (12), we obtain the sharp estimate

$$|f(x,y,z) - Lf(x,y,z)| \leq \frac{1}{2} \{x(1-x) + y(1-y) + z(1-z)\} \|D^2 f\|_{\infty, K}, \qquad (19)$$

for all $f \in C^2(K)$.

Example 4. Here we briefly consider the Bernstein–Schoenberg operators recently introduced by Dahmen, Micchelli, and Seidel [2] (also see Goodman [4]). These multivariate operators, which are based on blossoming, are locally defined and positive. They reproduce the linear polynomials. Hence, with appropriate modifications, we can apply Theorem 2 to them. They generalize the Bernstein operators of Example 1 and certain variation diminishing spline operators of Schoenberg. Following the notation of [4, p. 442] we define the Bernstein-Schoenberg operator

$$\mathcal{S}_n f(x) := \sum_{I \in J} \sum_{\alpha \in \Gamma_n} f(\zeta_\alpha^I) B_\alpha^I(x), \qquad x \in \mathbb{R}^s.$$

The proof of Theorem 3 of [4] shows that

$$\begin{aligned} S(x) := S_{\mathcal{S}_n}(x) &:= \mathcal{S}_n(\| \cdot -x \|^2)(x) \\ &= \sum_{I \in J} \sum_{\alpha \in \Gamma_n} \|\zeta_\alpha^I - x\|^2 B_\alpha^I(x) \\ &= \sum_{I \in J} \sum_{\alpha \in \Gamma_n} \|\frac{1}{n} \sum_{j=0}^{s} \sum_{l=0}^{\alpha_j - 1} x^{i_j, l} - x\|^2 B_\alpha^I(x) \\ &= O(1/n), \qquad n \to \infty, \end{aligned} \qquad (20)$$

where the 'constant' in the big O depends on the geometry of the triangulation and clouds defining $\mathcal{S}_n$ near the point x, and it is possible to choose a constant which works for all x from a given compact subset of $\mathbb{R}^s$. From (20) and (3) the convergence results of [4] can be obtained. In light of the special case (12), finer estimates of (20) might be possible.

Example 5. The sharpness of (3) implies certain saturation results. For example, if B_n is the multivariate Bernstein operator of Example 1, then (14) implies that

$$\|f - B_n f\|_{L_\infty(T)} = O(1/n), \quad n \to \infty, \quad \forall f \in C^2(T),$$

while for $f \in Q$ (see (5)),

$$|f(x) - B_n f(x)| = \frac{1}{n} C_{x,f}, \quad x \in T,$$

where

$$C_{x,f} := \frac{1}{2}(R^2 - \|x - c\|^2)\|D^2 f\|_{\infty,T},$$

with $C_{x,f} > 0$ when $f \in Q \setminus \Pi_1$ and $x \in T \setminus V$. In other words, B_n has saturation order $1/n$ at every point $x \in T \setminus V$. Similarly, by (16), the bivariate tensor product Bernstein operator $B_{n,m}$ of Example 2 has saturation order $1/n + 1/m$.

The general result is that a family (L_k) of multivariate positive linear operators that reproduces the linear polynomials has saturation order $S_{L_k}(x)$ at the point x. This can also be deduced from [1, Th. 6.2.5].

§3. A Korovkin-Type Theorem

For positive linear operators which possibly do not reproduce the linear polynomials, the proof of Theorem 2 can be adapted to obtain the following quantitative Korovkin-type theorem for C^2-functions. Let e_i denote the linear polynomial $y \mapsto y_i$.

Theorem 3. Suppose that K is a compact convex subset of $\mathbb{R}^n$, and $L : C(K) \to C(K)$ is a positive linear operator. Then, for every $x \in K$, there is the pointwise error estimate

$$E(f, x) := |f(x) - Lf(x)| \leq |f(x)| E(1, x) + |L(D_{\cdot - x} f(x))(x)|$$

$$+ \frac{1}{2} E(\|\cdot - x\|^2, x) \|D^2 f\|_{\infty, K}, \quad \forall f \in C^2(K), \tag{21}$$

where the second term on the right can be estimated by either of

$$\left|L(D_{\cdot - x} f(x))(x)\right| \leq \|\nabla f(x)\| \sqrt{\sum_{i=1}^n E(e_i - x_i, x)^2} \tag{22}$$

or

$$\left|L(D_{\cdot - x} f(x))(x)\right| \leq \|\nabla f(x)\| E(\|\cdot - x\|, x). \tag{23}$$

Proof: Similar to that for Theorem 2. □

If, in addition, L reproduces the linear polynomials, then (21) reduces to the sharp estimate of Theorem 2. Similar estimates to (21) involving moduli of continuity can be found in [1, §5.1, p. 265].

Acknowledgments. Thanks to Professors Altomare and Raşa for their useful comments.

References

1. Altomare, F. and M. Campiti, *Korovkin-type Approximation Theory and its Applications*, Walter de Gruyter & Co, Berlin, 1994.
2. Dahmen, W., C. A. Micchelli, and H.-P. Seidel, Blossoming begets B-spline bases built better by B-patches, Math. Comp. **59** (1992), 97–115.
3. DeVore, R. A., *The Approximation of Continuous Functions by Positive Linear Operators*, Springer-Verlag, Berlin, 1972.
4. Goodman, T. N. T., Asymptotic formulas for multivariate Bernstein-Schoenberg operators, Constr. Approx. **11** (1995), 439–454.
5. Raşa, I., Approximation of twice differentiable functions by positive linear operators, Anal. Numér. Théor. Approx. **14** (1985), 131–135.
6. Waldron, S. The error in linear interpolation at vertices of a simplex, SIAM J. Numer. Anal. **35** (1998), 1191–1200.

Shayne F. D. Waldron
Department of Mathematics
University of Auckland
Private Bag 92019
Auckland
New Zealand
waldron@math.auckland.ac.nz
http://www.math.auckland.ac.nz/~waldron/

Realizable Approximation Bounds for a Solvable Neural Network

Sumio Watanabe

Abstract. This paper studies the relation between function approximation bounds for a learning system and the functional measure from which a target function is randomly taken. It is proved that the adaptive learning system has almost the same bounds as the fixed one in the Wiener type measure, but that the former has the smaller bounds than the latter in the competitive or quantum measure.

§1. Introduction

Neural networks and radial basis functions are now being used in many function approximation problems such as pattern recognition, robotic control, and time series prediction. Their essential form is given by

$$f_N(x) = \sum_{n=1}^{N} a_n \varphi(b_n, x),$$

where φ is some nonlinear function, and $\{a_n, b_n\}$ is a set of parameters. Although they play a central role in information processing systems, it is not yet clarified in what kinds of environments they work effectively. To answer this question, Barron [1] proved that, if the Fourier transform of the target function $g(x)$ satisfies an L^1 type condition, then f_N can approximate g by optimizing both $\{a_n\}$ and $\{b_n\}$ with the approximation bounds $(1/N)^{1/2}$, which is independent of the dimension of x. On the other hand, it is proved by Mhaskar [3] that, if $g(x)$ satisfies an L^2 type condition, or in other words, if it is contained in the Sobolev space, then the approximation bounds depend on the dimension of x, which is attained by optimizing only $\{a_n\}$.

In this paper, in order to clarify the difference between them, we compare two function approximation systems under the condition that the target function is randomly taken from a functional probability measure. One system is an adaptive approximation system, which optimizes both $\{a_n\}$ and $\{b_n\}$ for a given target function. The other is a fixed approximation system, which uses the optimally fixed $\{b_n\}$ and optimizes $\{a_n\}$ for a target function.

Based on the above framework, we show that the two systems have almost the same efficiency under the Wiener type measure, but that the adaptive system is more effective than the fixed one under the competitive or quantum measure.

§2. Approximation Bounds for An Orthonormal System

Let $H = L^2([0,1)^D)$ be the space of the complex-valued square integrable functions on the set $[0,1)^D$ in the D-dimensional Euclidian space. H is a Hilbert space with the inner product

$$(f,g) = \int_{[0,1)^D} \overline{f(x)} g(x) dx,$$

and the induced norm $\|\cdot\| = (\cdot,\cdot)^{1/2}$. Let $\{e_j(x); j \in \mathbb{Z}\}$ be a complete orthonormal system (CONS) in the one dimensional space, $L^2([0,1)^1)$. A CONS in H can be made by $e(n,x) = e_{n_1}(x_1)\, e_{n_2}(x_2) \cdots e_{n_D}(x_D)$ where $n = (n_1, n_2, \cdots, n_D) \in \mathbb{Z}^D$ ($\mathbb{Z}^D$ is the D-dimensional lattice) and $x = (x_1, x_2, ..., x_D) \in [0,1)^D$. A function $g \in H$ can be represented with coefficients $\{g_b\}$,

$$g(x) = \sum_{b \in \mathbb{Z}^D} g_b\, e(b,x), \qquad g_b = (e(b,\cdot), g).$$

We consider a function approximation system $f_N(a,b,x)$ which is defined by

$$f_N(a,b,x) = \sum_{j=1}^N a_j\, e(b_j, x) \quad (a = \{a_j\}, b = \{b_j\}),$$

where $\{a_j \in \mathbb{C}, b_j \in \mathbb{Z}^D; j = 1, 2, ..., N\}$ is the set of parameter. We have the following two theorems.

Theorem 1. *Assume that there exists an $\alpha > 0$ such that*

$$M_\alpha(g) \equiv \sum_{m \in \mathbb{Z}^D} |g_m| |m|^\alpha < \infty.$$

Then, for a sufficiently large N,

$$\inf_{a,b} \|g - f_N(a,b,\cdot)\| \leq \frac{2^{1/2+\alpha/D} M_\alpha(g)}{N^{1/2+\alpha/D}}.$$

Proof: This proof is based on the method in [1]. First, we consider the case $\alpha = 0$. Let us define a probability distribution on $\mathbb{Z}^D$ by $p(n) = |g_n|/M_0(g)$. Then $\sum_{n \in \mathbb{Z}^D} p(n) = 1$. The expected value of $f(n)$ by this distribution is referred to as $E_n\{f(n)\}$. By using a notation $k(n,x) = \{g_n M_0(g)/|g_n|\}e_n(x)$, we have two equations,

$$E_n\{k(n,x)\} = \sum_{n\in\mathbb{Z}^D} k(n,x)p(n) = \sum_{n\in\mathbb{Z}^D} g_n e_n(x) = g(x),$$

$$\int E_n\{|k(n,x)|^2\}dx = E_n\{M_0(g)^2\}\int |e_n(x)|^2 dx\} = M_0(g)^2.$$

The minimized square error $Er = \inf_{a,b} \|g - f(a,b,\cdot)\|^2$ is

$$Er = \inf_{a_1,b_1} \cdots \inf_{a_N,b_N} \int |\sum_{j=1}^N a_j\, e_{b_j}(x) - g(x)|^2 dx.$$

By replacing a_j by $g_{b_j}M_0(g)/(|g_{b_j}|N)$ and $\inf_{b_j}$ by E_{b_j},

$$Er \le E_{b_1}\cdots E_{b_N}\left\{\int |\frac{1}{N}\sum_{j=1}^N k(b_j,x) - g(x)|^2 dx\right\}.$$

By using the central limit theorem,

$$Er = \frac{1}{N}\int E_n\{|k(n,x)-g(x)|^2\}dx \le \frac{1}{N}\int E_n\{|k(n,x)|^2\}dx = \frac{M_0(g)^2}{N},$$

which shows the theorem for the case $\alpha = 0$. For the case $\alpha > 0$, by the definition $\hat{g}_{N/2} \equiv \sum_{|n|\ge (N/2)^{1/D}} g_n e_n(x)$, we have $Er^{1/2} \le \inf_{a,b}\|\hat{g}_{N/2} - f_{N/2}(a,b,\cdot)\|$.

$$Er^{1/2} \le \frac{M_0(\hat{g}_{N/2})}{(N/2)^{1/2}} = \frac{1}{(N/2)^{1/2}}\sum_{|n|\ge (N/2)^{1/D}} |g_n||n|^\alpha \cdot \frac{1}{|n|^\alpha} \le \frac{M_\alpha(g)}{(N/2)^{1/2+\alpha/D}}.$$

□

Theorem 2. *Assume that* $L_\alpha(g) \equiv \{\sum_{m\in\mathbb{Z}^D} |g_m|^2|m|^{2\alpha}\}^{1/2} < \infty$ *with* $\alpha > 0$. *If the set* $\{b_j \in \mathbb{Z}^D; j=1,2,...\}$ *is ordered as* $|b_j| \ge |b_{j'}|$ $(j > j')$, *then for sufficiently large* N

$$\inf_a \|g - f_N(a,b,\cdot)\| \le \frac{L_\alpha(g)}{N^{\alpha/D}}.$$

On the other hand, there exists a function $g(x)$ *which satisfies the condition that* $L_\alpha(g) < \infty$ *and that for an arbitrary* $\epsilon > 0$ *(even if b is optimized)*

$$\inf_{a,b} \|g - f_N(a,b,\cdot)\| \ge \frac{L_\alpha(g)}{N^{(\alpha/D)+\epsilon}}.$$

Proof: We define $Er(b) = \inf_a \|g - f_N(a,b,\cdot)\|^2$. Then

$$Er(b) = \sum_{n>N^{1/D}} |g_n|^2 \le \frac{1}{N^{2\alpha/D}}\sum_{n>N^{1/D}} |g_n|^2|n|^{2\alpha} \le \frac{1}{N^{2\alpha/D}} L_\alpha(g)^2.$$

For the latter half, there is a sample $|g_n|^2 = \dfrac{1}{1+|n|^{D+2\alpha+2D\epsilon}}$. □

Note that Theorems 1 and 2 correspond to the results by Barron [1] and Mhaskar [3], respectively.

§3. Functional Measures and Approximation Bounds

Let μ be a probability measure on $H = L^2([0,1]^D)$, and

$$E_g\{F(g)\} = \int_H F(g)d\mu(g)$$

be the expected value of the function F.
We define two approximation bounds.

$$Fix(N) = \inf_b E_g\{\inf_a \|g - f_N(a,b,\cdot)\|\}, \qquad (1)$$

$$Adap(N) = E_g\{\inf_b \inf_a \|g - f_N(a,b,\cdot)\|\}. \qquad (2)$$

It immediately follows that $Fix(N) \geq Adap(N)$. Note that bounds (1) show the efficiency of the system which uses the optimally fixed CONS and optimizes the coefficients for a target function. Bounds (2) are for the system which optimizes both the CONS and the coefficients for a target function.

Definition. *Let $a(n)$ and $b(n)$ be sequences on $\mathbb{Z}^D$. The sequences $a(n)$ and $b(n)$ are called* in the same order *(denoted by $a(n) \cong b(n)$), if and only if there exist $c_1, c_2 > 0$, $N > 0$ such that, for an arbitrary $|n| > N$, $c_1 b(n) \leq a(n) \leq c_2 b(n)$.*

First, we consider a case $E_g\{|g_n|\} \cong E_g\{|g_n|^2\}^{1/2}$. In this case, we can show $Adap(N) \approx Fix(N)$ based on some assumptions.

Theorem 3. *Assume that $E_g\{|g_n|\} \cong E_g\{|g_n|^2\}^{1/2} \cong (1/|n|)^{\alpha+(D/2)}$ with $\alpha > 0$. Then*

$$Fix(N) \cong \frac{1}{N^{\alpha/D}}.$$

Proof: For sufficiently large N, $Fix(N) \cong E_g\{(\sum_{|n| \geq N^{1/D}} |g_n|^2)^{1/2}\}$. By using Minkowski's and Cauchy-Schwarz's inequalities,

$$\sum_{|n| \geq N^{1/D}} E_g\{|g_n|\}^2 \leq E_g\{(\sum_{|n| \geq N^{1/D}} |g_n|^2)^{1/2}\}^2 \leq \sum_{|n| \geq N^{1/D}} E_g\{|g_n|^2\}$$

which implies $\dfrac{Const.}{N^{\alpha/D}} \leq Fix(N) \leq \dfrac{Const.}{N^{\alpha/D}}$. □

In order to analyze $Adap(N)$, we define an event $\mathcal{A}_N(\gamma)$ by

$$\mathcal{A}_N(\gamma) = \{g \in H \; ; \; \max_{|n| \leq (\gamma N)^{1/D}}^{N} |g_n| \leq \sup_{|n| \geq (\gamma N)^{1/D}} |g_n|\}$$

where $\max_{k \in S}^N f(k)$ denotes the N-th largest value of $f(k)$ in the set S. Note that, if $\gamma_1 < \gamma_2$, then $\mathcal{A}_N(\gamma_1) \supset \mathcal{A}_N(\gamma_2)$.

Realizable Approximation Bounds

Theorem 4. Let μ be a measure on H which satisfies the following three conditions. (1) Coefficients $\{g_n\}$ are independent. (2) $E_g\{|g_n|^2\}^{1/2} \cong E_g\{|g_n|\} \cong (1/|n|)^{\alpha+(D/2)}$ with $\alpha > 0$. (3) There exists a sequence $\gamma_N \to \infty$ such that $\lim_{N\to\infty} \mu\{\mathcal{A}_N(\gamma_N)\}(\gamma_N)^{2\alpha/D} = 0$. Then,

$$\frac{1}{(\gamma_N N)^{\alpha/D}} \leq Adap(N) \leq \frac{1}{N^{\alpha/D}}.$$

Proof: The latter inequality is derived from Theorem 3. For the former, we define two functions,

$$h_L(g) = \left\{\sum_{|n|>L^{1/D}} |g_n|^2\right\}^{1/2}, \quad k(g) = \inf_{|b|\leq(\gamma N)^{1/D}} \left\{\sum_{n\in\mathbb{Z}^D \ominus b} |g_n|^2\right\}^{1/2}$$

where $\mathbb{Z}^D \ominus b$ is the set $\{n \in \mathbb{Z}^D, n \neq b_j (j = 1, 2, ..., N)\}$. It follows that $h_{(\gamma_N N)}(g) \leq k(g) \leq h_N(g)$. Let $\chi(g)$ be the characteristic function for the event $\mathcal{A}_N(\gamma_N)$. Then

$$Adap(N) \geq E_g\{(1-\chi(g))\inf_{a,b} \|g - f_N(a,b,\cdot)\|\} \geq E_g\{k(g)\} - E_g\{\chi(g)k(g)\}$$

$$\geq E_g\{k(g)\} - E_g\{\chi(g)^2\}^{1/2} E_g\{k(g)^2\}^{1/2} = E_g\{h_{(\gamma_N N)}(g)\}$$

$$-\mu(\mathcal{A}_N(\gamma_N))^{1/2} E_g\{h_N(g)^2\}^{1/2} \geq \frac{1}{(\gamma_N N)^{\alpha/D}} - \frac{\mu(\mathcal{A}_N(\gamma_N))^{1/2}}{(N)^{\alpha/D}},$$

which completes the Theorem 4. □

Example 1. (Wiener Type Measure). The Wiener type measure is defined by $d\mu(g) \propto \exp(-(g,(1+(-\Delta/(2\pi)^2)^p)g))$, $(p > 0)$. Let us use the Fourier series for approximation,

$$f_N(a,b)(x) = \sum_{n=1}^{N} a_n \exp(2\pi i(b_n \cdot x)).$$

Each $g_n \in \mathbb{C}$ is subject to the probability distribution with a density function, $(1/(2\pi\sigma_n^2))\exp(-|g_n|^2/(2\sigma^2))$ $(\sigma_n^2 = 1/(1+|n|^{2p}))$. Therefore, $E_g\{|g_n|\} \cong \sigma_n$ and $E_g\{|g_n|^2\} \cong \sigma_n^2$. By using Theorem 3, $Fix(N) \cong (1/N)^{(2p-D)/(2D)}$. The cumulative distribution function of $|g_n|$ is $F_n(x) = 1 - \exp(-x^2/(2\sigma_n^2))$. We define two events $\mathcal{B}_N$ and $\mathcal{C}_N$. $\mathcal{B}_N$ is the event in which $\max_{|n|\leq(3N)^{1/D}} |g_n| \geq M_{(3N)^{1/D}}$ where $M_{|n|}$ is the median of $|g_n|$. Then by using the central limit theorem, there exists a constant $\eta > 0$ such that, for sufficiently large N, $\mu(\mathcal{B}_N) \geq 1 - \exp(-\eta N)$. $\mathcal{C}_N$ is the event in which $\sup_{|n|\geq(\gamma_N N)^{1/D}} |g_n| \leq M_{(3N)^{1/D}}$. Then there exists a constant $c_0 > 0$ such that

$$\mu(\mathcal{B}_N) = \prod_{|n|\geq(\gamma_N N)^{1/D}} F_n(M_{(3N)^{1/D}}) \geq \exp(-\exp(\log N - c_0 \gamma_N^{p/D})).$$

Because $\mathcal{B}_N$ and $\mathcal{C}_N$ are independent and $\mathcal{A}_N^c \supset \mathcal{B}_N \cap \mathcal{C}_N$, if $\gamma_N = (1/(2c_0)) \times (\log N))^{D/p}$, then

$$0 \leq \mu(\mathcal{A}_N)\gamma_N^{(2p-D)/(2D)} \leq (1 - \mu(\mathcal{B}_N)\mu(\mathcal{C}_N))\gamma_N^{(2p-D)/(2D)} \to 0.$$

By using Thereom 4, $Adap(N) \geq (1/N)^{(2p-D)/(2D)} \cdot (1/\log N)^{(2p-D)/(2p)}$.

§4. Competitive or Quantum Measure

Secondly, let us consider the case $E_g\{|g_n|\} \ll E_g\{|g_n|^2\}^{1/2}$. This case is realized by competitive or quantum measure, and we show $Adap(N) \ll Fix(N)$.

Theorem 5. Let $E_g\{|g_n|\} \leq 1/(1+|n|^{\alpha+D})$ with $\alpha > 0$. Then

$$Adap(N) \leq \frac{1}{N^{1/2+\alpha/D}}.$$

Proof: This theorem is immediately derived from Theorem 1. □

Theorem 6. Let $X_r(g) = \sum_{r \leq |n| < r+1} |g_n|^2$. If $E_g\{X_r(g)^{1/2}\} \cong 1/r^\beta$ with $\beta > 1$, then

$$\frac{1}{N^{(\beta-1/2)/D}} \leq Fix(N) \leq \frac{1}{N^{(\beta-1)/D}}.$$

Proof: For a sufficiently large N, $Fix(N) \cong E_g\{(\sum_{r=N^{1/D}}^{\infty} X_r(g))^{1/2}\}$, and

$$(\sum_{r=N^{1/D}}^{\infty} E_g\{X_r(g)^{1/2}\}^2)^{1/2} \leq E_g\{(\sum_{r=N^{1/D}}^{\infty} X_r(g))^{1/2}\} \leq \sum_{r=N^{1/D}}^{\infty} E_g\{X_r(g)^{1/2}\}$$

which shows the theorem. □

Example 2. (Competitive Measure). We divide $\{\mathbb{Z}^D\}$,

$$\mathbb{Z}^D = \bigcup_{r=0}^{\infty} S_r, \quad S_r = \{n \in \mathbb{Z}^D; r \leq |n| < r+1\},$$

and define the competitive measure μ by: (1) if $r \neq r'$, $\{g_n; n \in S_r\}$ and $\{g_n; n \in S_{r'}\}$ are independent, (2) the random variable $g^{(r)} = \{g_n; n \in S_r\}$ has the following density function with s, t ($t-s-1 > 0$).

$$p(g^{(r)}) = \frac{1}{Z} \delta(\sum_{n \in S_r} \Theta(|g_n|) - [r^s]) \prod_{n \in S_r} p_0(g_n),$$

where $\delta(x)$ is Dirac's delta function, $\Theta(x) = 0$ if $x = 0$ or 1 if $x > 0$, $[x]$ is the largest integer that is not larger than x, $p_0(z)$ is some density function on $\mathbb{C}$ with $E(|z|) \cong E(|z|^2)^{1/2} = 1/r^t$, and Z is the normalizing constant.

Realizable Approximation Bounds

By this measure,

$$E_g\{|g_n|\} = \frac{|n|^s}{|n|^{D-1}} \cdot \frac{1}{|n|^t} = \frac{1}{|n|^{D-1-s+t}}.$$

Then, by using Theorem 5, $Adap(N) \le (1/N)^{1/2+(t-s-1)/D}$.

$$E_g\{X_r(g)^{1/2}\} = \int (\sum_{j=1}^{[r^s]} |z_j|^2)^{1/2} \prod_{k=1}^{[r^s]} p_0(z_k) dz_k \cong |r|^{s/2} \cdot \frac{1}{|n|^t} \cong \frac{1}{|n|^{t-s/2}}.$$

By using Theorem 6, $(1/N)^{(t-s/2-1/2)/D} \le Fix(N) \le (1/N)^{(t-s/2-1)/D}$, which shows the effectiveness of adaptive learning system.

Theorem 7. *Assume that there exist constants $\alpha \ge \beta \ge 0$, $\epsilon > 0$ such that $E_g\{L_\alpha(g)^2\} < \infty$ and that*

$$\mu(\{g \in H; 0 < |g_n| < \frac{\epsilon}{1+|n|^\beta} \ (n \in \mathbb{Z}^D)\}) = 0.$$

Then, we have an inequality,

$$Adap(N) \le \frac{Const.}{N^{1/2+(\alpha-\beta)/D}}.$$

Proof: By the assumption, $|g_n| \le |g_n|^2(1+|n|^\beta)/\epsilon$ holds with probability one. Then, by combining

$$E_g\{M_{\alpha-\beta}(g)\} = E_g\left\{\sum_{n \in \mathbb{Z}^D} |g_n||n|^{\alpha-\beta}\right\} \le \frac{1}{\epsilon} E_g\left\{\sum_{n \in \mathbb{Z}^D} |(1+|n|^\alpha)|g_n|^2\right\}$$

$$\le \frac{2}{\epsilon} E_g\{L_\alpha(g)^2\} < \infty$$

with Theorem 1, we obtain Theorem 7. □

Example 3. (Quantum Measure). Let us consider the case that $|g_n|$ is digitized, or in other words, $|g_n|$ is a random variable on $\{k\epsilon; k \in \mathbb{N} \cup \{0\}\}$ with $\epsilon > 0$. Further, we assume that (1) $\{g_n; n \in \mathbb{Z}^D\}$ are independent, (2) the probability $Pr(\{|g_n| = k\epsilon\}) \propto 1/(1+|n|)^{k\delta}$ with $\delta > D$. In this measure, we define $N_r(g)$ by the number of elements of the set $N_r(g) = \#\{g_n; r \le |n| < r+1, \ |g_n| = \epsilon\}$. Then

$$E_g\{X_r^{1/2}\} = E_g\left\{\left(\sum_{r \le |n| < r+1} |g_n|^2\right)^{1/2}\right\} \ge E_g\left\{\left(\sum_{r \le |n| < r+1} \epsilon^2 \delta_{\epsilon,|g_n|}\right)^{1/2}\right\}$$

$$= \epsilon \ E_g\{N_r(g)^{1/2}\} \cong \frac{1}{r^{(\delta-D+1)/2}}.$$

Here, we used the fact that $N_r(g)$ is subject to $B(r^{D-1}, (1/r)^\delta)$. Then, by using the proof for the first half of Theorem 6, $Fix(N) \geq (1/N)^{(\delta-D)/(2D)}$. On the other hand, for $|n| \geq 1$

$$E_g\{|g_n|^2\} = \sum_{k=0}^{\infty} \frac{k^2 \epsilon^2}{(1+|n|)^{\delta k}} \cong \frac{\epsilon^2}{|n|^\delta}$$

which ensures $E_g\{L_0(g)^2\} < \infty$. We apply the Theorem 7 with $\alpha = \beta = 0$,

$$Adap(N) \leq \frac{Const.}{N^{1/2}}.$$

§5. Discussion

In the previous works by Barron and Mhaskar, it is shown that the approximation bounds by the adaptive or fixed system are related to the L^1 or L^2 type norm, respectively. In this paper, we tried to compare them based on the environmental probability measures. Roughly speaking, adaptive systems work effectively in the measures which have competitive coefficients and digitized values, while fixed systems do in those which have independent coefficients and continuous values.

§6. Conclusion

In this paper, we proved that layered neural networks are useful in the environments where the target function is randomly taken from the competitive or quantum measure. The future study is to establish the necessary and sufficient condition on the functional measure to ensure that the adaptive learning machine is effective.

Acknowledgments. This research was partially supported by the Ministry of Education, Science, Sports, and Culture in Japan, Grant-in-Aid for 09680362.

References

1. Barron, A. R., Universal approximation bounds for superpositions of sigmoidal function, IEEE Trans. on Information Theory **39** (1993), 930–945.
2. Barron, A. R., Approximation and estimation bounds for artificial neural networks, Machine Learning **14-1** (1994), 115–133.
3. Mhaskar, H. N., Neural networks for optimal approximation of smooth and analytic functions, Neural Computation **8** (1996), 164–177.
4. Vakhania, N. N., *Probability Distributions on Linear Spaces*, North Holland, New York, 1981.
5. Watanabe, S., On the essential difference between neural networks and regular statistical models, *Proc. of Int. Conf. on Information Sciences*, vol. 2, 1997, pp. 149–152.

TECHNOLOGY IN POSTWAR AMERICA

CARROLL PURSELL

TECHNOLOGY IN POSTWAR AMERICA

A History

COLUMBIA UNIVERSITY PRESS   NEW YORK

Columbia University Press
Publishers Since 1893
New York   Chichester, West Sussex

Copyright © 2007 Columbia University Press
All rights reserved

Library of Congress Cataloging-in-Publication Data
Pursell, Carroll W.
  Technology in postwar America : a history / Carroll Pursell.
    p. cm.
  Includes bibliographical references and index.
  ISBN 978-0-231-12304-4 (cloth : alk. paper)
  1. Technology—United States—History. 2. Technology—Social aspects   I. Title.

T21.P83 2007
609.73—dc22   2007001753

Every effort has been made to find and credit all holders of copyright for material included in this book. For further information please write the author c/o Columbia University Press.

Columbia University Press books are printed on permanent and durable acid-free paper.
Printed in the United States of America

c 10 9 8 7 6 5 4 3 2 1

For Matt & Becky

CONTENTS

Introduction ix

1. ARSENAL OF DEMOCRACY   1
2. THE GEOGRAPHY OF EVERYWHERE   20
3. FOREIGN AID AND ADVANTAGE   39
4. THE ATOM AND THE ROCKET   59
5. FACTORIES AND FARMS   78
6. "IT'S FUN TO LIVE IN AMERICA"   98
7. BRAIN DRAIN AND TECHNOLOGY GAP   118
8. FROM TECHNOLOGY DRUNK . . .   134
9. . . . TO TECHNOLOGY SOBER   155
10. A WIRED ENVIRONMENT   174
11. STANDING TALL AGAIN   193
12. GLOBALIZATION, MODERNITY, AND THE POSTMODERN   212

Notes 231
Bibliography 259
Index 271

INTRODUCTION

In 1941, media mogul Henry Luce, founder of both *Life* and *Time* magazines, famously proclaimed the American Century, calling on the American people "to accept wholeheartedly our duty and our opportunity as the most powerful and vital nation in the world and in consequence to exert upon the world the full impact of our influences."[1] Six decades later, Luce's "duty" and "opportunity" came to look very imperial indeed, as America came to be known as the "homeland," and its culture, including its technology, had been projected onto the rest of the world. As the world's only superpower, the United States devised and supported a regime of technology that both convinced and compelled its own citizens, as well as distant peoples, to adjust to a globalized economy, culture, and political order designed to be very much in the American nation's favor. The flowering of consumer technologies at home was intimately connected to those that allowed the country to pursue its interests around the world.

I have set out to tell a story about American technology since World War II: how it changed, why it took the form it did, and what it has meant to the country. When people say that we live in a technological age, they speak truly, but human beings have always lived in a technological age. Technology is, after all, one of the definitions of what it means to be human. In recent times, however, technology seems to have become an agent that has radically transformed our lives. This book, then, is about the machines and processes that have loomed so large in our lives over the past half a century.

But to say that this book is about technology in late twentieth–century America begs more questions than it answers, and the trouble begins with the word "technology" itself. As a number of observers have pointed out, it is essentially an empty term, imbued with different meanings to meet different needs. Leo Marx has noted that the word itself is a rather recent invention, introduced in the early twentieth century as machines and processes were more closely linked in complex networks, and science and technology, once agents of progress, became its measure instead.[2] The very emptiness of the concept made it a perfect vehicle to mystify any political economies that their beneficiaries wished to obscure. One important aim of this book, then, is to name the technologies with which we live—to show how the machines and processes came to be and why they exist at all. Technology, after all, is not something that merely happens to us; it is something that we have created for certain purposes, not always acknowledged.

A second problem with the term "technology" is that, in both popular and too often scholarly discourse, it is given independent agency. We hear that this or that technology had such and such an impact, that society reacted to it in certain ways. In other words, we posit a kind of technological determinism to explain why things change all around us. In this construction, technology develops out of its own internal logic, while society—all of us—scramble to accommodate these changes. Because progress, now defined as technological change, presumably cannot be stopped, it must be lived with. Another aim of this book is therefore to show that particular technologies flow from certain decisions made by people with specific goals that they want to accomplish. Other choices could have been made, things do not always turn out the way they were planned, and surprises abound; but human intention is always behind new technologies, and it is important to know who designed them and set them loose, and for what purposes.

Finally, while technologies have effects—indeed, they were designed to—these effects cannot be neatly divided into two piles labeled "good" and "bad." A certain effect can be and usually is good for some people and bad for others. A "labor saving" machine is good for its owners, but bad for the workers whose labor is "saved." That automobile bumpers are not designed to prevent damage in a collision as slow as five miles per hour is bad for car owners, but good for car manufacturers who keep costs down and for those who operate body-repair shops. In the end, to discover the why and so-what of any technology, the intentions of inventors and engineers, the aims of those for whom they work, and the results of their efforts have to be looked at carefully and specifically. At the same time, above and beyond the specifics lies a pattern of action, a set of institutions and operating assumptions that are relied upon to facilitate certain kinds of change. It is not a conspiracy, but a template for a political economy that makes certain assumptions about what constitutes progress, who deserves to succeed, and generally how the world works.

As Edward Said famously pointed out in his classic study of orientalism, "political society in Gramsci's sense reaches into such realms of civil society as the academy and saturates them with significance of direct concern to us."

Furthermore, he insists, "we can better understand the persistence and the durability of saturating hegemonic systems like culture when we realize that their internal constraints upon writers and thinkers were *productive*, not unilaterally inhibiting."[3]

I quote Said because I find a parallel in his formulation to the stories about technology told in this book. Some of the stories are about the political life of the nation: the space race, Star Wars, the Manhattan Project, foreign aid, and so on. Others are parts of our civil life: housing tracts, cars with tail fins, oral contraceptives, television, and farm tractors. One reading of Said suggests that our political and civil lives are intimately connected by our "saturating" technological culture. Political and civil concerns intermingle, and the technologies of both reinforce each other. Computer games and the Gulf and Iraq wars have mimicked each other to such an extent that American troops are trained with simulations, and Game Boys are an important recreational diversion at the front. Consumer technologies create a cultural understanding that helps to make political technologies acceptable and even seemingly inevitable, while those same political technologies provide both form and content for the technologies we find at home and at work.

Three final issues require clarification. First, not only do I believe that technology does not drive history, but I would argue that no other single abstraction does either. At various points in this book, I may implicate capitalism, militarism, or consumerism in the events being discussed, but I do not mean to imply that any of these alone drive the story any more than does technology. I believe that in the American context of the twentieth century, all of these are important shorthand labels for forces that intermingle and reinforce each other, in complex and often obscure ways.

Second, when I refer to gender issues, I do so from the understanding that gender—that is, the roles assigned by our society to each of us as female or male—is not only socially constructed, but always contingent and contested. In our American culture, technology is powerfully identified with masculinity ("toys for boys"), so my explicit references to gender often focus on the attempts to display or reinforce that masculinity. It seems unlikely that alone among the myriad aspects of our culture, technology is "neutral" in terms of gender, though it is often so portrayed.

Third, I realize that people outside of the United States, in reading this book, may be annoyed by my using the words "my" and "our" to refer to American objects and actions. I do this deliberately because I am not comfortable writing simply that "Americans" or "they" did this or that—especially when I am so often critical. In no way have I been somehow absent from my culture; I have been a part of it and share responsibility for it, and want to acknowledge that fact.

The twelve chapters of this book follow a roughly chronological path. Chapter 1 describes the events of World War II, not only through what were called the "instrumentalities of war," but also in how the government came to view the role of technology and science in American life, as both of which worked a profound change in the nation. The atomic bomb was surely the most dramatic technology employed, but rockets, jet engines, radar, and other electronic devices, as well as vast numbers of guns, tanks, and ships, were also critical. Behind all of it was an increased belief that technological progress, and the science that supported it, held the key to a stronger, richer, healthier, and happier America, and the federal government stepped forward to become the primary source and guarantor of this process.

Chapter 2 deals with the return of American servicemen and servicewomen from overseas, and American women from defense factories, mills, and shipyards, to what was widely understood to be a normal American life. The economic dislocations and gender disruptions of the war years were to be replaced as quickly as possible not with the status quo before the war, but with a prosperity earned by years of sacrifice. The fear that demobilization and canceling defense contracts would plunge the nation back into economic depression proved unwarranted, as pent-up demand for durable consumer technologies and continued government spending for such programs as GI benefits and interstate highways fuelled a postwar boom. Keynes met Ford, and the nation prospered. In 1949, Levittown, with its empty garages but houses with a full complement of appliances, was filled with young people ready to live the American dream.

Early in the postwar years, President Harry S Truman and President Dwight D. Eisenhower attempted to extend the reach of American technologies to all parts of the world outside of the Soviet bloc. Chapter 3 describes the Marshall Plan to integrate Western Europe into the structure of American economics, business, and technology, and the ambitious program to rebuild and democratize Japan. For the newly emerging nations of Africa and Asia, as well as those in Latin America, Truman proposed his Point Four program of technical aid for economic development. Crafted to combat communist designs in these areas, and based on development theory, these large-scale technological transfers led to the questioning both of American motives (dependency theory) and local consequences (a concern for appropriate technologies). The extension of American technological hegemony throughout the so-called free world was a great project in planetary engineering.

A social critic once claimed that whenever he heard the announcement of a new age—atomic, space, postindustrial, information—he held on to his wallet. Chapter 4 describes how the dramatic new technologies of World War II were taken over by new or reorganized government agencies, such as the Atomic Energy Commission, a reorganized Department of Defense, and in 1958, the National Aeronautics and Space Administration (NASA). In the context of the Cold War, the research on nuclear technologies and the development of both nuclear and conventional arms and missiles were all understood in terms of a race with the Soviet Union for development and deployment. Vast resources, in terms of both money and scientific and engineering talent, were diverted from civilian activities to the elaboration of often-baroque military technologies.

Chapter 5 examines how postwar America, as a country that still thought in terms of industrial mass production, increasingly applied that phrase to the farm as well as the factory. On the eve of World War II, rural electrification and the coming of the rubber-tired gasoline tractor provided the conditions for extending modern industrial technology to the work of the farm. With the additional commitment to chemical fertilizers and pesticides and the widespread use of irrigation even outside the arid areas of the West, farms became, to use Carey McWilliams' term, "factories in the field."[4] The cotton-picking machine had been perfected just before the war, and the tomato-picking machine just after. Together, they and other machines like them hastened the consolidation of corporate farms that were increasingly dependent on petroleum-based fuels and chemicals, large amounts of capital, and access to world markets.

It turned out, however, that the Industrial Revolution had not yet run its course either, even in the factories of the nation. In the 1950s, the term "automation" was coined to refer to significant increases in machine use and the "saving" of labor. Social commentators and labor unions both deplored the possibility of "workerless factories," a possibility that Kurt Vonnegut, Jr. explored in his first novel, *Player Piano* (1952). The increased mechanization of farms and the automation of factories not only increased productivity, but also changed the nature of the product and the calculus of human costs and benefits involved.

The massive and growing productivity of American farms and factories created what Lizabeth Cohen has called "a consumers' republic," in which shopping became the ultimate enactment of citizenship.[5] Chapter 6 investigates the claim of a 1947 pamphlet that "it's fun to live in America." Issued by Kiwanis International, the pamphlet showed Americans using cars, telephones, and radios, and nicely caught the easy conflation of consumer goods with the American way of life that characterized the postwar years. The mere filling of pent-up demand after World War II gave way in the 1950s and 1960s to a virtual cornucopia of consumer goods, from televisions to hula hoops, air conditioners and birth-control pills to Veg-O-Matics. Many of these, like television, FM radio, and domestic air conditioning, were actually in use on a very small scale before the war, but became ubiquitous afterward.

The new suburbs were virtual magnets for consumer durables, as each new homeowner needed a car, a full complement of household appliances, a garage with a lawnmower, and perhaps a home shop. Young men modified Detroit's cars into hot rods, and Chicano youth in particular chopped theirs into low riders. Partly through advertising, Americans adopted a popular version of what could be called technological determinism: new tools and toys could, it seemed, change one's life for the better. Richard M. Nixon expressed this aspect of American culture in his 1959 "kitchen debate" with Premier Nikita Khrushchev of the Soviet Union at the American exhibition in Moscow. In resolving their public disagreement over the politics of the production and social role of appliances, the two world leaders drank a toast to "the ladies," invoking a trope suggestive of the gendered hierarchy of technological production and consumption.

Chapter 7 shows that America, even at the height of its technological supremacy and confidence during the 1960s, needed to heavily tax the technological markets and talent of the rest of the world. American technological hegemony after World War II created problems and evoked concern abroad, even among allies of the United States. While the United States worked to restore and reform the economies and social infrastructures of Germany and Japan, its European allies only slowly recovered from their wartime devastation, and looked to America as both an example and a threat. Increasingly, American colleges and universities became destinations for students from overseas who sought advanced training in engineering, science, and medicine. Not surprisingly, many of them chose to stay in the United States, where professional opportunities and rewards were seen to be greater than those at home. This "brain drain" had the potential of crippling the reconstruction of war-torn societies and the building of new countries after the collapse of the European empires in Asia and Africa. Noting with alarm what they called the "technology gap" between themselves and the United States, European nations, most notably France, saw American hegemony as a threat to their own economies and cultures, and sought, in the words of former British Prime Minister Harold Wilson, to forge new futures for themselves in the "white heat" of science and technology.

A quarter century of fun and vanguardism in technology was not without its price. Chapter 8 looks at the accumulating problems associated with American technological supremacy, which, by the end of the 1960s, had led to something of a crisis of confidence. Rachel Carson's widely influential *Silent Spring* cast grave doubts on the miracle of DDT, one of the most ubiquitous and heralded of wartime advances. Ralph Nader's *Unsafe at Any Speed* questioned not only that most American of technologies, the automobile, but corporate irresponsibility as well. The threat of nuclear war was never far out of mind for the postwar generation, and the hot war in Vietnam, pitting as it did the full might of American arms against a small and hardly modern people, focused attention on the price to be paid for the kind of technological culture we adopted so enthusiastically. The first Earth Day, held in 1970, marked a convenient beginning to the environmental movement in this country, and an end to the period of unquestioned technological optimism.

The 1970s witnessed a remarkable variety of attempts to regain control of the country's technology, discussed in Chapter 9. Critics like Lewis Mumford, Paul Goodman, and Ivan Illich questioned the assumed progressive nature of the technological regime. The Nixon administration's efforts early in the decade to push through a program to build the supersonic transport (SST) led the then-Democratic Congress to establish an Office of Technology Assessment that could, at least in theory, advise lawmakers about the costs and benefits of new technological possibilities. Beginning on the private level, groups of social critics began to campaign for what they called "appropriate technologies" (AT). The OPEC-induced fuel shortages of the 1970s focused AT initiatives on energy sources, especially solar and wind-generated power.

Engineers and scientists created an electronic environment for World War II. The advances in this field generally, but especially the building of giant computers, sowed the seeds for an electronic environment for American society as well, which began in the 1970s, but became dominant in the 1980s and 1990s. Chapter 10 follows the development of electronic devices such as the transistor, which occurred in places like Bell Labs in New Jersey, but even more typically in the new research and development environment of Silicon Valley. The needs and visions of the Pentagon and large corporations such as IBM created an industry concentrating on large mainframes, of interest to relatively few researchers and strategists. When this work was joined with that of young outsiders like Steve Wozniak and Steve Jobs, however, the resulting personal computer truly revolutionized the way America works. By the year 2000, some sixty millionaires were being created in Silicon Valley alone each day.

Chapter 11 follows the resurgence of an assertive commitment to technology in the nation. In the 1970s, the broad national questioning of technology and its costs coincided with a U.S. defeat in Vietnam. If that defeat was associated with a perceived feminization of the country, as has been suggested, it paralleled the way in which appropriate technology and technology assessment had been characterized as unmanly. In the 1980s, the antidote for such gendered self-doubts proved to be the presidency of Ronald Reagan, who urged the nation to "stand tall again." The new president had his predecessor Jimmy Carter's solar panels removed from the White House roof and ordered a crash program called Star Wars to create a defense umbrella in space, beneath which America would be forever safe from nuclear attack. Sport utility vehicles and McMansions provided the individual and family-scale expressions of the renewal of masculinity and the search for a safe space.

Modernity is centrally about the collapsing of time and space, a project two centuries long that reached its apogee at the end of the twentieth century; it is analyzed in Chapter 12. The circulation of capital has become instantaneous; an automatic teller machine in London can not only access a bank account in America, but also give cash in pounds sterling and a receipt figured in dollars. The appropriate technology movement gave early voice to the objection that all places were not the same, all times not interchangeable, all solutions not universal, and that all values ought not to be commodified. At the end of the twentieth century, there was evidence aplenty that the core values of modernity were being undermined by the very technologies that had supported them. The periphery was challenging the center. Culturally, women, gays, and people of color were questioning the hegemony of white, straight, male culture. Geopolitically, developing and so-called "rogue" nations were gaining access to technologies of information and destruction that undercut the power and even threatened the survival of the United States and other "developed" countries. Modernity had always been contingent and contested, but its ability to absorb or deflect challenges appeared to be eroding dangerously. Its sturdy offspring, globalization, looked distressingly like its parent.

In 1968, writing a introduction to J.-J. Servan-Schreiber's popular book *The American Challenge*, Arthur Schlesinger, Jr. pointed out that in the author's opinion, "a strong and progressive Europe ... [was] vital to save the United States from the temptations and illusions of superpowership: Europe holds in her hands more than her own destiny."[6] In the early years of the new century, these words seem prophetic indeed. This book is intended to be ambitious in its coverage, but accessible in its presentation of this material. It is aimed primarily not at engineers and historians of technology, but to an audience of students and the general public. Because technology is a central fact of all of our lives, we are all responsible for trying to understand the many and complex ways in which we interact with it. Through the better understanding of our technology, we approach a better understanding of ourselves. It is my hope that this book contributes to that process.

TECHNOLOGY IN POSTWAR AMERICA

ONE

ARSENAL OF DEMOCRACY

The two decades between World War I and World War II were laced with discussions about technology and its place in American life. The word "technology" itself probably came into common use during this period to replace the older terms "machinery" or the "useful and mechanic arts." The new term was meant to stand for the aggregate of tools and machines that marked modern life, but its inclusiveness blurred the distinctions among the things to which it referred. Technology was in fact an empty category that could stand for anything or nothing; each American, hearing the word, could endow it with whatever facts or fantasies seemed important at the moment. The 1920s were marked by the proud claims of scientists, engineers, and their corporate or academic employers that technology had created a brave new world in which abundance would replace scarcity, wealth would replace poverty, and peace would reign throughout the world. The 1930s, on the other hand, with their mix of depression, unprecedented unemployment, and triumphant fascism, witnessed a widespread realization that labor saved represented jobs destroyed, production still had to be distributed and consumed, and vast technological power could be used for ill as well as for good.

The public debate that this realization triggered was cut short by the coming of World War II, waged this time in the Pacific as well as across the Atlantic. With men drafted into military service, and huge orders placed by both the American and Allied governments for weapons and other war materials, unemployment quickly gave way to labor shortages, and labor-saving seemed once again a socially desirable goal. The economy, which had hardly responded to years of New Deal policies, suddenly and positively reacted to massive government spending for defense. The machinery of mass production, first idealized and then demonized, now became the instrument of the Allied war effort.

Profound changes took place during World War II, not only in what were called the "instrumentalities of war," but also in how the government came to view the role of technology and science in American life. Technologically, the atomic bomb was surely the most dramatic new weapon, and it opened the way not only to nuclear power generation—the infamous "power too cheap to meter—but also fantasies of nuclear airplanes, the excavation of vast civil works projects, and a power source for artificial hearts. Jet propulsion promised a new era of commercial travel, rockets suggested a way to explore space, and the little-noticed new computers held possibilities then undreamed of. Besides these specific technologies, a general advance in electronics stood to benefit civilians through a host of applications.

New Research Regimes

Behind all of this was an increased belief that technological progress, and the science that supported it, held the key to a stronger, richer, healthier, and happier America. The federal government had always supported new technology and scientific research, but it had done so largely in-house, with its own resources and for its own specific purposes. However, after its wartime experience of commanding and controlling much of the nation's engineering and scientific talent, the government was willing to identify goals more proactively, providing resources to universities and corporations to do the work of national defense and economic growth.

In 1940, the year in which defense research was just beginning to be organized and funded, the nation as a whole spent an estimated $345 million on research and development. The several "estates of science" contributed quite unevenly to this sum. The largest amount, $234 million, or 68 percent of the total, came from private industry; 19 percent came from the federal government; 9 percent was contributed by colleges and universities; and 4 percent was distributed by state governments, private philanthropic foundations, and other institutions.[1]

Each of these estates expended its own money for its own purposes. During the Great Depression, some scientists had called on the federal government for both a larger measure of funding and more central planning for the research needs of the country to better coordinate the several estates, but the government adopted neither policy. The government was not convinced that scientific research had so strong a claim upon the public purse, industry feared government interference, and scientists in general were leery of bureaucratic control of their work.[2]

World War II changed the dynamic completely. Spurred by the new crisis, scientists and the federal government forged a new and far closer relationship.

During the war years, excluding money for nuclear programs, funds for research and development averaged $600 million, 83 percent of which was provided by Washington.³ The bulk of this money was funneled through the newly created Office of Scientific Research and Development (OSRD), headed by Vannevar Bush. Trained as an electrical engineer, Bush was highly placed in each of the estates of science. He was a vice president of the Massachusetts Institute of Technology, a coinventor of an important electromechanical analog computer, a founder of an industrial firm, chair of the government's National Advisory Committee on Aeronautics, a member of the National Academy of Sciences, and president of the philanthropic Carnegie Institution of Washington.

In consonance with the entire war effort, to avoid disrupting the prewar status quo, the OSRD carefully worked out grants and contracts to place war work in already existing centers of scientific excellence, getting work underway almost immediately. One consequence of this system was that the best scientists and best-equipped laboratories could make a maximum contribution to solving the technical problems of the war.

Another consequence was that the strong grew stronger and the weak grew weaker. Between 1941 and 1944, two hundred educational institutions received a total of $235 million in research contracts, but nineteen universities got three-quarters of that sum. Two thousand industrial firms received almost $1 billion in research contracts, but fewer than one hundred got over half of it. In 1939, Bell Labs—whose director, Frank B. Jewett, was on the OSRD—had only $200,000 worth of government contracts, accounting for just 1 percent of the laboratory's activities. By 1944, Bell's work for the government represented 81.5 percent of its activities, amounting to $56 million.⁴ It was the major reason that Bell was positioned to dominate the new and lucrative field of electronics after the war. The rise of the "contract state" got the government's work done and vastly increased the funding available for industrial and university scientists. It also greatly advantaged the corporate and educational giants of the nation.

The OSRD was arguably the most important new wartime technical agency, but it was certainly not the only one. The agency jealously protected its own responsibility for what Bush called the "instrumentalities of war"—that is weapons—but it could not prevent other new bodies from cropping up to take care of tasks that it had chosen to ignore. One such new body was the Office of Production Research, set up by the War Production Board (WPB).⁵

Bush's concentration on weapons had left open the whole problem of improving the machines and processes by which American industry was turning out staggering quantities of war material. Especially critical was the shortage of such strategic raw materials as the aluminum needed to make airplanes for the greatly enlarged military air arms. This was, of course, very close to the area of industrial research into which private industry had been expanding since the turn of the century, and which was rife with proprietary information, patent struggles, and the possibility of large shifts in market share. The mandarins of the OSRD were horrified that an agency under the direction of nonscientists, and enthusiastic New Dealers at that, might gain the power to destabilize the balance of competing industrial forces. But the pressure from within the Franklin D. Roosevelt administration finally led to action early in 1942.

Since before the Japanese attacked Pearl Harbor on December 7, 1941, Harold L. Ickes, secretary of the interior and a master at bureaucratic maneuvering, had been pushing the claim that a shockingly small proportion of the nation's scientists and engineers were being utilized for defense work through the elite clique associated with the National Academy of Sciences. At first, Bush was able to use this concern to expand his National Defense Research Committee in the spring of 1941, creating the OSRD and charging it with both technological development and medical research. But Ickes was not so easily put off. Setting up the WPB in 1942, coupled with Bush's continued refusal to let his agency become involved in matters of scarce materials such as rubber and aluminum, allowed the WPB to take on the problem.

The matter was triggered by the visit to Washington of two representatives from the Institutum Divi Thomae, a small Catholic research laboratory in Cincinnati, Ohio. They first visited Bush, but returned in March, this time accompanied by Dr. Sperti, and presented former congressional representative and ardent New Dealer Maury Maverick, now with the WPB, with a twenty-five page "Plan for the Organization of a Research Division (or bureau) of the War Production Board." Maverick excitedly told Donald Nelson, head of the WPB, that such an agency "*is an outstanding necessity*" (emphasis in original).[6] When Bush began to wonder if the OSRD should preempt this activity, Jewett, his colleague, warned him: "I'd shun it, like I would the seven-year itch." He continued with the advice to "keep out of it if you can and give the hard pressed boys all the help you can from the shore but don't go in swimming—the water is cold and there is a damnable undertow."[7]

As it turned out, Nelson turned to Maverick to look further into the matter, and Bush wrote to James B. Conant, chemist, member of OSRD, and president of Harvard University, that "the whole subject of industrial research in connection with strategic materials and the like seems to be in a terrible turmoil in WPB. I will tell you about it, but it is certainly getting thoroughly out of hand and onto dangerous ground."[8] At the same time, another form of the idea was put forward by Thomas C. Blaisdell, a member of the WPB planning committee. Writing to Nelson, he proposed a U.S. War Research Development Corporation that would be designed to accomplish "(1) the testing of new industrial processes; (2) the building of pilot plants; (3) the construction up to the stage of operation of processes which it has been decided shall be carried into full-bloom activity."[9]

What most alarmed Bush and his colleagues was that, to increase and improve wartime production, the New Dealers who were bringing these ideas forward were quite explicitly willing to tackle the matter of patent monopolies and other roadblocks to fully utilizing advances in industrial research. It looked like corporate privilege and private profits were to be put on the table for negotiation, and those like the men of the OSRD, who hoped to win the war without disturbing prewar business arrangements, were not happy. As it turned out, the 1942 Office of Production Research was a toothless tiger that, if it did little good, was unlikely to do much harm either.

The task of the OPR—to help decide how and from what things should be produced—was accomplished through five omnibus contracts that it had let with the National Academy of Sciences, the Manufacturing Engineering Committee of the American Society of Mechanical Engineers, the University of Illinois, the Armour Research Foundation, and the Research Corporation, a nonprofit organization in New York City. It had no laboratories of its own, and only seventy-seven employees by the end of 1944. Within these limits, it did what it could in such fields as the utilization of scrap metals, the production of light alloys, gasoline additives, and penicillin.[10]

At the same time, a national roster of scientific and specialized personnel was set up to mobilize more thoroughly the nation's scientists, engineers, and others with particular training and qualifications. The idea for such a list came, perhaps, from a similar registry that the Royal Society in England began in December 1939. An inventory of technical people who might be available for war work seemed like an obvious and good idea, satisfying at the same time an urge to be thorough and orderly and an assumption that in a democracy, all should be able to serve the common good. Using membership lists from the major technical societies and subscription lists from technical publications, questionnaires were sent out. When they were returned, the data was put onto punch cards.

The scale of the problem of enlisting trained personnel was significant. James B. Conant estimated that all told, only about 10,000 to 15,000 technical people worked for or through the OSRD. Meanwhile, the psychologist Leonard G. Carmichael, who headed the National Roster, guessed that there were "between 400,000 and 500,000 trained or qualified scientific men and women in the country" and that perhaps a quarter of them were not involved in war work. Not surprisingly, the definition of "war work" was a slippery one. The chemical industry claimed that all of their personnel were doing war work.

However, even the business-oriented journal *Fortune* worried that vitally needed technical people were falling through the net.[11]

The roster's Herculean effort was not much appreciated by the OSRD. James Phinney Baxter III, the official historian of that agency, remarked rather archly that the "punch cards were invaluable when one wished to know what American scientists spoke Italian, but as might be excepted the Roster was used less to obtain key men than the rank and file. Those charged with recruiting chemists and physicists for OSRD and its contractors knew the outstanding men in each field already and through them got in touch with many young men of brilliant promise."[12] The calm confidence, bordering on arrogance, that because the OSRD was made up of the "best people," they would automatically know who the other "best people" were, was no doubt based on more than a modicum of truth, but carried with it also a faint odor of aristocratic disdain for the scientific and technical masses.

The history of the National Inventors Council, another agency set up to deal with the technical problems and opportunities of preparedness, illustrates the way in which the presumptive sources of technological innovation shifted through time. In both the Civil War and World War I, some sort of government agency to encourage and vet ideas from the general public for new technologies had been established. In both cases, the results were mixed at best, but in 1940, it again appeared necessary to somehow provide for an anticipated outpouring of Yankee ingenuity, and the Commissioner of Patents suggested that the National Inventors Council be set up to give a preliminary screening to ideas sent to the Patent Office. In July 1940, Charles F. Kettering, then head of research and development at General Motors, was invited to chair the council.

The council set up twelve committees to deal with such areas as land transportation and armored vehicles, remote control devices, ordnance and arms. Each committee was chaired by a council member, but the real work was done by a small staff of fifty-five, of whom only nine were technically trained. Unsolicited ideas from the public took up most of their time, but they also tried to stimulate inventors to work in certain areas. However, the military did not want to signal to the Axis powers what its high-priority areas of interest were, and refused to tell the council anything about what they wanted. Equally handicapping, over a century of experience had demonstrated the difficulty of inventors making significant contributions in an area in which they had little real knowledge or experience.

As a result, the council's activities were modest in the extreme. With little funds and even less direction, it struggled to deal with the 208,975 ideas sent to it for evaluation. It found that only 6,615 (4.1 percent) of them had sufficient merit to be passed on for further work, and in the end, only 106 devices went into production, among which were a metal land mine detector and a mercury battery. According to a story circulated within the council, a "lone" inventor had brought them a workable proximity fuse, but that they had been ordered to turn it down as impractical. It turned out that it was in fact a tremendously important innovation, but again, worries about security, in addition to the fact that the OSRD was already working on such a device, robbed the council (as well as the inventor) of credit for its creation.[13]

Criticisms of Mobilization

In the run-up to World War II, the question was not simply what tools and machines would be produced; fundamentally, it was a matter of who would decide—and under what conditions—which old technologies would be produced, what new ones would be developed, and by whom. The power of the OSRD—its domination by the largest and most powerful scientific and technical providers in the country, and its enthusiastic support for President Roosevelt's strategy of dismissing the Dr. New Deal and relying on Dr. Win the War—led some New Deal stalwarts to question both the efficiency and the democracy of Bush's policies as the technical czar during the war. The *Nation*, a liberal journal, charged flatly in 1941 that "monopoly is the worst enemy of the technological progress so essential to national defense."[14] No doubt all would agree that winning the war was the supreme goal, but many individuals and institutions had one eye on the postwar period as well. Scientific and engineering research could be powerful solvents of the industrial status quo, and those who had the upper hand were not anxious to give any advantages to those who did not.

One example of what could happen was given by the political observer I. F. Stone in his 1941 book *Business As Usual: The First Year of Defense*. According to Stone, aluminum, which was absolutely critical to building the nation's air forces, was the one complete monopoly in the American economy, as the Aluminum Company of America (Alcoa) closely held the process for extracting the virgin metal from bauxite. In the late spring of 1941, the Reynolds Metal Company successfully ran tests of a new process at two new plants, built with loans from the Reconstruction Finance Corporation (RFC) and supplied with electrical power from the Tennessee Valley Authority (TVA) and the Bonneville project. The Office of Production Management (OPM) met the alleged breakthrough with skepticism. Stone quoted an OPM official as stating that "it wasn't the custom" in the agency to seek technical advice; rather, they sought guidance on "organizational" matters that they believed only Alcoa could supply. When agencies did seek outside technical advice, they quite reasonably turned to those scientists, engineers, and businesspeople with some background in whatever it was that needed attention. Not surprisingly, the "best" people often turned out to have close ties with the companies that dealt with those same issues—in this case, Alcoa.[15]

As early as the summer of 1940, however, attacks such as Stone's on the way that the country's technical capabilities were being mobilized led to a widely publicized report from sixteen "eminent engineers" who were identified as the National Defense Committee of the American Society of Mechanical Engineers. The committee reassured the public that "the program for national defense production is in good hands; its initiation on a large scale is now underway." Stone noted that most of these "engineers" were also businesspeople looking for defense contracts. Their number included such eminent persons as Charles E. Brindley, president of the Baldwin Locomotive works, T. A. Morgan, president of Sperry Gyroscope, and R. C. Muir, a vice president of General Electric.[16]

Bruce Catton, who after the war would become a best-selling historian of the Civil War, looked back in 1948 at how the war's industrial production had been run. He wrote, "It was one thing to agree that all scientific and technological resources must be used; it was quite another to determine exactly where those resources existed, how they were to be discovered, and in what way they could be harnessed." What Maury Maverick had been aiming for with his plan for the WPB was nothing less than "the removal of all the invisible impediments that stand between the age of scientific achievements and the fullest possible use of those achievements." What Maverick was "actually demanding," wrote Catton, "was production-for-use translated to the entire field of industrial research and technological advance."[17] It was a task far outside the prescriptions of Dr. Win the War.

The most sustained challenge to the OSRD came from Harley M. Kilgore, a Democratic senator from West Virginia who in 1942 became chair of the special Subcommittee on War Mobilization of the Senate Committee on Military Affairs, itself chaired by Senator Harry S Truman of Missouri. Before the end of the year, after extensive hearings, Kilgore had come up with Senate bill 2721, which called for "the full and immediate utilization of the most effective scientific techniques for the improvement of production facilities and the maximization of military output" by setting up an Office of Technological Mobilization, which would be given broad powers to draft facilities and intellectual property. Jewett warned Bush that the bill had been drafted by what he called "starry-eyed New Deal boys," a characterization that widened over the next few years to include a hint of Communist conspiracy.[18] S.B. 2721 failed to become law, but by the end of the war, Kilgore was already crafting plans for continuing governmental support of science along lines that continued to alarm the scientific elite.[19]

Postwar Planning

It was not clear to anyone whether World War II marked the end of the Great Depression of the 1930s, or was merely a brief respite from that extended economic disaster. As early as 1943, some scientists, military officers, and civilian administrators within the government began to plan for the extension of the emerging contract state, with a new governmental responsibility for supporting scientific research and development into the postwar period. Their motives were varied. Within the Navy Department, a small group of reserve officers known as the Bird Dogs laid plans to establish an Office of Naval Research to keep the fleet abreast of developing technologies.[20]

Within the Army Air Corps, such planners as General Lauris Norstad dreamed of increased contact with university-based intellectuals—not simply to benefit the military mind, but specifically to counter pacifism and antimilitarism in the academic community.[21] Scientists at work on wartime projects were anxious to return to their academic posts and their own research, though not so anxious to give up the federal funds to which they had become accustomed. Industries with strong research commitments were anxious to reap the market benefits of their wartime accumulation of research talent and information.

In the summer of 1945, Bush presented to the president a plan that he titled *Science the Endless Frontier,* which called for the demobilization of the OSRD, but the continued support of the nation's research and development with ample government funding. Roosevelt's had been the generation forcibly struck by the historian Frederick Jackson Turner's thesis that it had been the western frontier that had shaped the American character, institutions, and prosperity, and that the closing of that frontier, announced by the census of 1890, had presented the nation with a crisis. Science, Bush suggested, could function as the equivalent of the old geographical frontier, except that it would never close because the interrogation of nature would have no end. Congress, he urged, should establish a new National Research Foundation that would use tax monies to support, through grants and contracts, research in the military, medical, and industrial sciences.

Bush's argument was for greater peacetime support of science from the federal government, but he justified the plan by focusing on its presumed technological outcomes. Looking back at the war and the years leading up to it, he claimed that the nation's health, prosperity, and security all relied on technologies that had grown out of scientific research.[22] He cited penicillin and radar as outstanding examples of important scientific breakthroughs. On the home front, he claimed that in 1939, "millions of people were employed in industries which did not even exist at the close of the last war—radio, air conditioning, rayon and other synthetic fibers, and plastics are examples," and he held out the promise that after the war, "new manufacturing industries can be started and many older industries greatly strengthened and expanded if we continue to study nature's laws and apply new knowledge to practical purposes."[23] Toward this end, Bush urged that wartime research results should be declassified as soon as possible, that science education should be encouraged, perhaps even discharging scientifically talented military personnel early, and most of all, that federal funds for scientific research should be greatly expanded, made secure from destabilizing fluctuations, and distributed through his suggested foundation.

The wartime success of the OSRD had given evidence enough that targeted research, followed up by aggressive development, could produce startling technological successes in a remarkably short time, and there is no doubt that Bush firmly believed in what he was telling the president. At the same time, deploying technological promises as an argument for massive support of basic science had a history that was as much political as it was pragmatic. After World War I, some leading American scientists had plotted to win popular and political credit for the advances in the weaponry of that war, with the idea that their agenda and institutions, and not those of the country's engineers, should be given priority by a grateful nation and its government.[24] Even the phrase "research and development" gave the impression that research should, or perhaps even had to, happen ahead of development. Turning the phrase around—to say "development and research"—not only fell strangely on the ear, but felt foreign to the tongue. And yet if one wanted a particular new technology, one might give money to engineers who would begin to develop it, using research strategically to overcome any bottlenecks that might emerge. Having enjoyed the new largesse of taxpayers' money since 1940, however, the nation's scientists and the large elite corporations and universities for which they worked were not keen either to end or to share their wealth.

A controversy arose almost immediately over four main issues. Should the social sciences be included in the subsidy? Should funds be distributed in a traditional geographical pattern, or to centers and individuals of proven excellence? Should scientists have exclusive jurisdiction over the spending of this money—that is, should proposals be peer reviewed—or should they be held politically responsible for the spending of public money? Finally, should any patents that might result from the funded research be the property of the government that paid for it, or of the individual scientists—or more likely, their employers—who made the discoveries? The controversy over these points delayed the establishment of what came to be the National Science Foundation (NSF) until 1950.

A Postwar Research Regime

In the meantime, large segments of the nation's research efforts were orphaned, complicating the eventual task of coordination. Nuclear physics research, dramatically brought to the attention of the American people and the world in August 1945 over Hiroshima, had somehow to be continued and supported. Setting up the Atomic Energy Commission (AEC) in 1946 robbed the stillborn NSF of a very large segment of American science and future technological development. The National Institutes of Health (NIH), a small operation with vast paper authority in 1945, successfully took over unfinished medical research projects from the dissolved OSRD and made them the core of a new research empire. In 1946, the Bird Dogs managed to establish the Office of Naval Research (ONR) and began immediately to support basic research on the nation's campuses as part of the Navy's mission.

Finally, with medical, military, and nuclear research otherwise provided for, in 1950, Congress established the NSF, giving it responsibility for supporting basic research, science education, and science policy planning. The four points of dispute were thus settled. Supporting social sciences was to be permitted, but not specifically mandated. Funds would be distributed solely on the basis of scientific merit. Members of the governing National Science Board were to be nominated by the board itself, but appointed by the president. And, patents were to be retained by those doing the research.

The engine of this great research machine was designed between 1940 and 1945, but it was fueled by the developing Cold War. By 1955, the government had upped its research and development commitment to $3.308 billion. Of this total, $2.63 billion passed through the Department of Defense and another $385 million was distributed by the AEC. Health, Education and Welfare (HEW), the parent organization of the NIH, had parlayed a fear of disease into a runaway growth rate for itself, and was already spending $70 million a year. The NSF was budgeted at $9 million. Just ten years later, in 1965, the total research and development spending had risen to $14.875 billion—a 400 percent increase, of which the Department of Defense spent $6.728 billion. A newcomer to the Cold War, the National Aeronautics and Space Administration (NASA), formed out of the old National Advisory Committee on Aeronautics (NACA), accounted for $5.093 billion. The AEC spent $1.52 billion, much of which was for nuclear weapons. HEW had grown to $738 million, and the NSF to $192 million. Of this total of nearly $15 billion, only some 7 percent went to universities, another 5 percent to contract research centers, such as the California Institute of Technology's Jet Propulsion Laboratory and the University of California's Lawrence Radiation Laboratory, and the remaining 88 percent to other sources, mostly to private industry. It is little wonder that that in 1966, some 71 percent of the million and a half scientists and engineers in the nation worked for private industry. Looking more closely at the funds earmarked for university research in 1965, we discover that 42 percent came from HEW, 25 percent from the Department of Defense, 13 percent from the NSF, 8 percent from NASA, and 6 percent from the AEC and the Department of Agriculture.[25]

As it flourished in the mid-1960s, this rather large research establishment had several salient features. First, it depended heavily on federal funds; second, the bulk of these funds was directly tied to international competition with the Soviet Union; third, except for the Bureau of the Budget, the system was not coordinated or integrated by any science policy body; and fourth, the system was acceptable to both government agencies and to civilian, mainly industrial, scientists. The problems that arose over distribution of funds between scientific disciplines, and between science, engineering, and medicine, were muted by rapidly expanding budgets and the fact that most agencies relied heavily on peer-group evaluation by panels of established scientists. The iron fist of military domination was gloved in the velvet of ample funding and artistic freedom. Theodore von Karman, a Hungarian émigré scientist at the California Institute of Technology and the Jet Propulsion Laboratory, wrote characteristically that "some scientists worry a great deal about associations with the military.... I have never regarded my union with the military as anything but natural. For me as a scientist the military has been the most comfortable group to deal with, and at present I have found it to be the one organization in this imperfect world that has the funds and spirit to advance science rapidly and successfully."[26]

By 1970, however, all of these characteristics had proven to be liabilities as well as assets. Reliance on federal funds led to cutbacks in funding when the political needs of successive administrations required them. As the Cold War lost its grip, particularly on the nation's youth, both funds and justifications were threatened by reevaluation. With no clear science policy, it was possible for President Nixon to squeeze $300 million out of the research budget while adding $20 billion to the federal highway program.[27]

For all of the alarm and diversion, the perceived crisis of the 1970s was ample evidence of the revolution produced by the decisions made between 1940 and 1945. Pushing aside independent inventors and, to a significant degree, engineers, the nation's scientific elite had defined technology as the key to Allied victory, and scientific research as the key to technical innovation. National defense, public health, and the general welfare—increasingly defined in terms of economic production of consumer goods—were all seen as dependent on scientific research and development, the latter being a proper and worthy supplicant for public funds. In the years after the war, the federal government took on the responsibility of funding technological progress, a task that was to be done through institutions of scientific activity that were already established —in particular, large industries and universities. Through grants and contracts, the government outsourced the public's business to private interests. For a while, it looked like a win-win arrangement.

Producing the Material of War

While the search for new, science-based weapons was important, the immediate need in 1939 was most dramatically for old weapons, in staggering quantities. Historian Ruth Schwartz Cowan has suggested the dimensions of the problem. In 1939, the British Royal Air Force had 100,000 personnel, the German Luftwaffe had 500,000, and the United States Army Air Corps had 26,000, who in turn had 800 modern airplanes. The British had twice as many, and the Germans eight times as many—about 6,400 airplanes. Two years before the United States entered the war, President Roosevelt declared that the air force "must be immediately strengthened," and by the end of the war, the American aircraft industry had expanded to employ two million women and men and had an annual capacity of 100,000 planes.[28] It was a massive scaling up of America's technological capacity.

While this new understanding of the power of research and development, the role it could play in American life, and the mechanisms for its support and delivery were the most significant new technological legacy of the war years, they rested from 1940 to 1945 on a base of old-fashioned mass production. General Motors, the nation's largest corporation in 1970, did not list among the OSRD's twenty-five top industrial contractors in terms of dollars allocated, but its president, William S. Knudsen, was appointed chair of the National Advisory Defense Committee in 1940, and made co-director of the OPM a year later. By the time America entered the war, his company had already gained $1.2 billion in contracts for war material from the Allies. By the end of the war, it had produced $12.3 billion worth of arms. Perhaps surprisingly, only an estimated one-third of this massive outpouring was "comparable in form" to its former civilian products. Its first contract went to Chevrolet to make high-explosive artillery shells.[29]

Of course the United States and its allies also needed large numbers of motor vehicles, and in 1940, the army put out a call for a lightweight, four-wheel-drive vehicle. It was most satisfied by the design submitted by a small, nearly bankrupt company called American Bantam. Liking the design but not trusting Bantam's ability to deliver it in sufficient numbers, the army gave production contracts to both Ford and Willys-Overland. The latter copyrighted the popular name of the small truck—Jeep—and continued to make them after the war. The Jeep eventually became the first sport utility vehicle, a class that outsold even the family sedan by 2000.[30]

During World War II, Detroit was in its heyday as the Mecca of American mass production, but over the past generation, its lessons has been applied to a wide range of industrial activities, from the garment trade to food processing. The mass production system was deliberately and successfully designed to turn out vast quantities of goods, and with a guaranteed market and in many cases virtually guaranteed profits, the machine performed feats of productivity. Even before the war was over, a congressional committee attempted to estimate the scope of the entire enterprise. Calculating that 1939 was the last year largely unaffected by wartime needs, the committee found that the gross national product (GNP) rose by 75 percent between then and the end of 1943, or in terms of 1939 prices, from $89 billion to $155 billion. By 1944, war production accounted for 44 percent of the GNP.[31] The committee noted that "in the munitions industries the most important element [of increased production] has been the adoption of mass production methods which were not new advances but which were already well established in American industry." The moves were not dramatic, but they were effective: "designs were standardized, line production systems were developed, prefabrication and subassembly were increasingly employed, special-purpose machinery was introduced, and numerous other improvements were made in production methods."[32]

Much of the increase was accomplished by simply expanding facilities and adding workers and shifts. There were also significant improvements in the technologies of production, not all of which came from corporate or university laboratories. The Senate committee recognized that "most current developments can be traced to achievements made over a considerable period in the past. The war emergency, however, resulted in a great acceleration of the pace at which progress was made." Significantly, and perhaps contrary to what Bush and the rest of the OSRD scientists and engineers might have imagined, according to the committee "a surprisingly large proportion of the developments reported have been made by plant employees themselves, under employee suggestion systems worked out by labor-management committees, rather than by technicians."[33]

The report noted that one of the largest benefits of mass production—or as they called it, line production—was the opportunity for manufacturers to quickly train large numbers of workers who were not previously skilled in the tasks at hand. Nowhere was this more dramatically played out than in the wartime shipbuilding effort, in which tens of thousands of women were recruited to play the role of Rosie the Riveter, an icon appearing in posters encouraging women to join the war effort. These women, as well as many African Americans, were quickly taught the specific technological skills needed.[34]

As in the First World War, the federal government sought quickly to increase the pace of building both cargo and warships. There were nineteen private shipyards in the country before the war, and with government aid, twenty-one emergency shipyards were added. Among these were those of Henry J. Kaiser, a construction contractor whose firm was one of the so-called Six Companies to build Hoover Dam. Kaiser had teamed up with the construction firm Bechtel on that project, and in 1938, they began to think about taking advantage of what looked like a reviving shipbuilding industry. Most shipyards were on the East Coast, and neither Bechtel nor Kaiser had any knowledge of the industry, but their employees got the job of building a new yard for the recently organized Seattle-Tacoma Shipbuilding Company in Washington State.

The Six Companies joined the new Washington firm in 1939 to win a $9 million contract from the Maritime Commission to build five C-1 cargo ships. As historian Mark Foster has written, "Six Companies people constructed the shipways, watched the shipbuilding—and learned."[35] In December 1940, Kaiser became directly involved in shipbuilding with a shared contract to provide ships for Britain. The next month, in January 1941, a new American program to build Liberty ships was announced. Kaiser had already begun construction on a new shipyard in Richmond, California, and four months later, the first keel was laid down at that facility. In January, with the announcement of the Liberty ship program, Kaiser also began to build a new shipyard in Portland, Oregon. Construction on both yards began well before blueprints were ready, since capacity was seen as a magnet for further orders.

Over the next four years, the Kaiser shipyards employed nearly 200,000 workers. That most of the workers had no experience in shipbuilding, and many had never even worked in manufacturing before, suggested that production would be organized and carried out differently in the yards than in others. The ignorance of the workers was reflected at the highest levels, as Kaiser and his closest associates were also learning as they went. The few partners who did know the traditional way to build ships were horrified and soon dropped out.

The traditional way was to "lay a ship's keel, and then send hundreds of workers swarming into cramped quarters to perform many different functions" in bringing it to completion.[36] However, British and some East Coast yards had already experimented with subassemblies during the 1930s. When Kaiser seized upon the innovation, he was able to apply it to full advantage because, building his yards from scratch, he had plenty of room to lay out adequate areas for subassembly and ways to lay the keels to which the assembled pieces would be brought. Subassemblies were a critical part of Kaiser's ability to produce ships quickly—one was put together out of subassemblies in four days, fifteen hours, and twenty-six minutes—but other innovations more obviously their own helped as well.

Kaiser had sent one of his close associates to visit a Ford assembly plant in late 1940. While obvious features of mass production, such as the moving assembly line, were clearly inappropriate for shipbuilding, other parts were adapted. Kaiser realized that for most purposes, welding was as good as riveting, and easier to teach to the thousands of women who flocked to his yards. It was also less physically demanding, especially because the work to be done could be positioned so that the welders looked down rather than up at the joint to be sealed. In these new facilities too, workers could be taught new and quicker construction techniques in a shorter time than was taken in the older East Coast yards. A total of 747 vessels were launched during the war at that first Richmond shipyard alone. If Kaiser had not quite reproduced the mass production capacity of a Ford plant, he had helped revolutionized shipbuilding by borrowing heavily from Ford's model.

Kaiser's new construction techniques, pioneered in the shipyards themselves, were extended to the cities he had to construct adjacent to them for housing shipyard workers. Army cantonments had been built rapidly in World War I and again in World War II, but communities like Kaiser's Vanport City outside Portland, Oregon were privately built for civilian workers. With very low vacancy rates for housing in the Portland area, the Maritime Commission authorized Kaiser to build a new city of 6,000 units from scratch along the Columbia River. Land clearing began three days after the authorization, and within four months, the first seven hundred buildings were ready for occupancy. Nine months after construction began, Vanport contained 19,000 people and at least 40,000 were anticipated. The construction work proceeded rapidly, but the social innovations overshadowed the technological ones. To accommodate the needs of a workforce with such a large proportion of women, many of who were without partners, free bus service, child care, community kitchens where hot meals could be picked up and taken home, and similar amenities were provided for the residents.[37]

Although in the postwar imagination, Henry Kaiser's name was most widely associated with the mass production of ships, the congressional report *Wartime Technological Developments* listed numerous individual technical advances that, though relatively small, accumulated toward the dramatic speeding up of shipbuilding. The Moore Drydock Company was reported to be making subassemblies flat on the ground, and then using cranes to lift and swing them into place. The Sun Shipbuilding Company was launching its vessels at "an earlier stage of fabrication," then finishing them off in "wet basins." At the Federal Shipyard & Drydock Company, "Theodore Hanby designed a new type of die to form six, instead of the customary one, pipe hangers in a hydraulic press in one operation." In all, the committee listed forty-eight similar innovations, small and large, that helped add up to speedier ship production.[38]

Another area of technical advance in wartime production, less romanticized at the time, came in the area of quality control. Under the rubric of inspection, the committee report admitted that the twin factors of greatly increased production and the introduction of thousands of workers with no experience in the trade created the daunting problem of making sure that parts were up to standard dimensions. "Numerous inspection stations," it noted, "had to be integrated into production lines, and the proportion of inspectors to total employees has in many cases been considerably increased." Inspectors made use of both new and improved equipment, including X-ray machines, "electronic eyes," and gauges made from glass rather than steel.[39]

Although not specifically described among the fifty examples the report gave of improvement, newly developed "statistical methods for maintaining quality control have been widely applied," it noted, "and have contributed substantially to this objective."[40] This innovation, called statistical quality control (SQC), had been steadily developing in the United States and Great Britain during the interwar years, and was commonly understood to be the application of statistical methods to guarantee that production measured up to standards. Under the regime of Armory Practice jigs, gauges and fixtures were a critical component of the technological system that produced uniformity of parts, and if not rigorously applied, manufacturers ran the risk of producing vast numbers of defective units. The larger the number of items produced, the more time-consuming and onerous the constant checking became. A statistical method of checking for quality, through which only a few had to be actually tested to give confidence in the entire run, was obviously to be preferred, but the precise technique had to be carefully developed and taught to (often female) inspectors. Arguably the most famous of the SQC advocates was W. Edwards Deming, especially as he carried the technique to Japan beginning in 1947.

Reconversion

By the end of the war in the summer of 1945, the nation's industrial and technical elites had spun a dense web of mechanisms to further their work: grants and contracts, cost-plus techniques, dollar-a-year men, opportunities to work without compensation (WOC) while on leave from one's company, amortization schemes, relaxation of strict profit controls, programs to stockpile strategic materials, and research subsidies, among others. Prominent scientists and engineers, led by Vannevar Bush, tried to perpetuate their newly enjoyed federal largesse into peacetime, and the large corporations that had dominated wartime production were reluctant to give up the advantages they had enjoyed over the past five years. Their battlefront became the planning for reconversion to a peacetime economy.

The surrogate for this problem throughout the war had been the place of small businesses in industrial mobilization. By the middle of 1945, the Truman committee had identified the military's reluctance to deal with small firms in procurement matters, arguing that only large firms were "equipped with the plants and machinery, specially skilled workers, managerial know-how, financial stability, and established contracts with a wide variety of suppliers" necessary for successful war production. Furthermore, the committee explicitly rejected any responsibility to try to use wartime necessity to reform the structure and direction of the American economy.[41] Between June 1940 and the end of 1941, the one hundred largest companies received more than three-quarters of war contracts, a domination far exceeding that in the prewar years. In addition, large firms had been given $13 billion worth of new facilities. During the war, only one piece of legislation was passed that attempted to aid the small producer.[42]

If small industrial firms were disadvantaged in finding entry into war work, the possibility still remained that they could benefit from the gradual winding down of wartime production. Donald Nelson proposed that the small producers might be allowed to shift gradually from military to civilian production during the last months of the war, but both the military and big business strongly objected. The military entertained no suggestions that the war was won until all of the United States' enemies had surrendered, and the companies with the giant share of war work did not want to have to stand by and watch as their smaller competitors got an early start on filling pent-up civilian demand.[43] In 1944, the combined opposition of military and big business killed the idea: as the historian Barton J. Bernstein has put it, "through cunning maneuvering, these military and industrial leaders acted to protect the prewar oligopolistic structure of the American economy."[44]

The choice had never been simply to defeat the Axis or deliberately shape the economy. Those who led the scientific, technological, and industrial war effort purposefully chose to do both—that is, to channel innovation and production into paths that conserved and strengthened the economic status quo. For New Deal liberals such as Bruce Catton, the future would be shaped as much by how we chose to fight the war as by its eventual success. He angrily listed the failures of the war effort, as he saw them. Small business ("the *people's* businesses") was not protected and expanded; labor was blocked from full partnership; the dollar-a-year system was chosen over a system that would have put "the industrial effort closer to the hands of the people"; "the base for closely knit control of the American economy was not shifted by a new system to handle patent rights and to direct wartime scientific and technological research into new channels"; and a reconversion policy was chosen that locked all of the wartime advantages to big business into place for the postwar years. "We had won the war," he concluded pessimistically, "but we had done nothing more than that."[45] He was overlooking one important point, however: that the economic and social dislocations of the war prepared the way for a vast restructuring of the places where, and the ways in which, Americans lived in the early postwar years.

TWO

THE GEOGRAPHY OF EVERYWHERE

The return of 14 million American military personnel, including 350,000 women, from overseas—and the laying-off of many thousands of American women from defense factories, mills, and shipyards—was widely understood to define a return to normal life for the nation. The economic dislocations and gender disruptions of the war years were to be replaced as quickly as possible, not with the status quo before the war, but with a prosperity earned by years of sacrifice. The fear that demobilization and cancelled defense contracts would plunge the nation back into economic depression proved unwarranted, as pent-up demand for durable consumer technologies, and continued government spending for such programs as GI benefits and interstate highways, fuelled a postwar boom. Keynes met Ford, and the nation prospered. In 1949, Levittown, with its empty garages but houses with a full complement of appliances, was filled with young people ready to live the American dream.

Housing in 1940

If, as Le Corbusier said, "a house was a machine for living," by the eve of World War II, many such American machines had broken down. By 1990, Americans were spending $107 billion each year repairing, maintaining, and "improving" their homes, but fifty years before, 18 percent were simply "in need of major repair." Worse yet, a U.S. Census Bureau analyst writing in 1994 exclaimed, "It's really amazing how many young men went into . . . [World War II] from homes that had no plumbing."[1] The government undertook the first ever census of housing in 1940, and significant parts of the nation it described were still without what would later be thought of as normal amenities or necessities—though these were all, of course, domestic technologies that had been invented or introduced over the past century. By 1854, Boston water customers had 27,000 plumbing fixtures, including 2,500 "water closets" (toilets).[2] In 1940, 35 percent of American homes had no flush toilet in the house itself, 32 percent made do with an outside toilet or privy, and 3 percent had neither. Thirty-one percent had no running water in the house, 4 percent made do with an indoor hand pump, 21 percent had water within fifty feet of the house, and 5 percent had to go more than fifty feet for their supply. Thirty-nine percent had no indoor bathtub or shower. By 1950, after the beginnings of suburban housing development, some improvement was noted. By that year, the percentage of homes without indoor flush toilets had dropped from 35 to 25, those without bathtubs or showers from 39 to 27, and 70 percent now had hot and cold running water indoors.[3]

The problem of water and its uses was symptomatic of other technologies as well. Thanks in part to New Deal policies, 79 percent of all housing units had electrical lighting in 1940, and that number rose to 94 percent in 1950. At the same time, the percentage using kerosene or gasoline for light dropped from 20 to 6 percent. Units wired for lights could not necessarily support electrical appliances, which drew more current and needed heavier-gauge wire; in 1940, 44 percent of homes had mechanical refrigeration, 27 percent used ice boxes, and 29 percent made do with no refrigeration at all. By 1950, however, 80 percent of homes had mechanical refrigeration, only 11 percent relied on ice, and only 9 percent were without any means of refrigeration at all.[4]

One telling characteristic of the ten-year house census was that what was counted kept changing. In 1940, a relatively high 83 percent of homes had radios, doubling the figure from 1930; the figure then peaked at 96 percent in 1950 and declined to 92 percent in 1960. By that year, television, which had not been counted in 1940, and registered at 12 percent in 1950, had skyrocketed to 87 percent. In 1970, it stood at 96 percent. Television became so ubiquitous in the American home that the census stopped counting it in subsequent years. Clothes washers and dryers were not counted until 1960, and dishwashers only in 1970. Even telephones were not counted until 1960, when they were found in 78 percent of homes. Air conditioning was only counted beginning in 1960 as well.[5]

The proverbial kitchen sink was not enumerated until 1950, when 85 percent of homes boasted one. In defining a kitchen sink, the Enumerator's Reference Manual for that year made clear the hard technological reality of these amenities: "the kitchen sink," it explained, "must have a drain pipe which carries waste to the outside, and must be located inside this structure.... It need not have running water piped to it. Usually it is located in the kitchen; however, a sink (with a drain pipe to the outside) which is located in a hall, pantry, enclosed porch, or room adjacent to the kitchen, and used for washing dishes and kitchen utensils, is also a kitchen sink. A washbowl, basin, or lavatory located in a bathroom or bedroom is not a kitchen sink."[6]

Finally, in 1940, coal shovels and axes were essential household tools: 55 percent of homes were heated by coal and 23 percent with wood. Gas heating was widely used only in California and Oklahoma. By 1960, either oil or kerosene was used in 32 percent of homes. Utility-provided gas became popular in the 1950s and 1960s. Electricity then became an important source of heat in the 1970s and 1980s, used in 26 percent of homes by 1990.[7] Homes may have been havens from the world outside, but they were deeply imbedded in a broad technological network of utilities—and which utilities were chosen for use, as we shall see, had profound implications for the social and environmental fabric of the nation.

The inventory of American housing stock grew by 289 percent over the fifty years following 1940, though by only 157 percent in central cities. By far the largest growth—an increase of 519 percent—occurred in what the urban historian Dolores Hayden has called the nation's "sitcom suburbs."[8] In 1949 alone, there were 1.4 million housing starts, and in 1950, another 1.9 million, more than three times the prewar record. Part of this building boom was making up for the years of the Great Depression, during which activity had fallen off precipitously. Another part of the story was the great increase in marriages and babies during and immediately after the war. The numbers of women in the job force grew by 60 percent, and three-quarters of them were newly married. After a relatively low birth rate during the 1930s, a million more families were formed between 1940 and 1943 than would have been predicted for "normal" times. At the same time, the age at marriage dropped, and the birth rate grew from 19.4 to 24.5 per thousand. This flurry of marriage and births was accompanied by a tsunami of propaganda—from Hollywood, the government, popular songs, and magazines—insisting that the gender displacements of the war, especially the rush of women into the workforce, were only temporary, and that a return to traditional roles for women was an inevitable and desirable result of victory.[9]

Postwar Housing Shortage

Workers, especially women, were making more money during the war, but they were discouraged from spending it, both by official government policy and because consumer goods were simply not being made in their previous varieties and numbers. Although the 6.5 million women who entered the work force earned $8 billion, they were reminded that war production was directly linked to their inability to spend this hard-earned pay. One contemporary pamphlet claimed "that in 1942 alone . . . the cost of the war to consumers of the country is estimated at 3,500,000 automobiles, 2,800,000 refrigerators, 11,300,000 radios, 1,650,000 vacuum cleaners, and 1,170,000 washing machines. . . . The government is taking from the people what is needed to win the war. Consumers . . . must get along with what is left."[10] Because of this cornucopia of technology denied, postwar spending of dammed-up wealth was staggering. Between 1947 and 1950, Americans bought 21.4 million cars, 20 million refrigerators, 5.5 million electric stoves, and 11.6 million television sets. To house these tools, they also bought the 1 million homes that were being built each year.[11]

Returning veterans who, before the war, might have lived in bachelor accommodations or with their parents, now found themselves married and often with babies to provide space for. The historian Adam Rome has summarized the sometimes dramatic results of the housing shortage of the immediate postwar period. The government estimated that around 2.5 million young families still doubled up with relatives, but others who could not or would not exercise that option made do with a wide range of temporary expedients. In California, people were living in the unfinished fuselages of bombers, and a congressional report revealed that hundreds of thousands of veterans "could find shelter only in garages, trailers, barns, or chicken coops." One newspaper carried an ad for "Big Ice Box, 7 by 17 feet. Could be fixed up to live in," and on at least one occasion, "a crowd lined up at a funeral parlor to find the addresses of newly vacated apartments."[12] It was a huge market, which attracted some extraordinary builders.

In 1941, Congress explicitly stipulated that all of the wartime housing for workers built with government aid was to be torn down at the end of the war and not become a part of the nation's housing supply. Well aware of the impending housing crisis, Senator Robert Wagner introduced legislation to authorize the construction of public housing. Despite winning some support from conservatives concerned with preserving the family, however, the measure was blocked in the House Banking and Currency Committee when well-organized lobbies from the real estate, mortgage, and homebuilding industries opposed it. A similar measure was finally passed after President Harry S. Truman's victory in 1948, but by then, the era of large suburban developments was well underway.[13]

Sandwiched between the activities of public housing agencies with their multi-family buildings and the builders of the new suburban developments stood the more traditional middle-class relationship between one person or family that wanted a new home and an architect who could design it for them to their specifications. It was an arrangement in which the dreams of the client were made material through the technical skill of an expert, and for their part, architects could either express traditional aesthetics or innovate upon them. Some of the latter, like their clients, were innovative, but most were probably safely traditional. As with any other technological undertaking, designing and building a house was best left to experts. In the postwar years, however, this pattern—so logical for the middle and upper classes—was hardly suitable to building a million new houses a year for customers who only by buying a house could claim the status of being middle class. As a result, in the hands of the biggest developers, the best practices for home construction came to be defined in terms of fast and dirty, rather than design aesthetics or technological efficiency.

Mass Producing Homes

The postwar building of large number of homes, very quickly and using innovative "manufacturing" techniques, is usually said to have been introduced by William Levitt in the years after the war. Dolores Hayden, however, has noted that the less well-known developer Fritz B. Burns had anticipated many of the "mass production" techniques for building houses in his developments for Southern California aircraft workers in the late 1930s.[14] Reformers, including President Herbert Hoover, frequently expressed their dreams of turning the notoriously conservative building trades and businesses into a modern industry—dreams neatly expressed in the title of one Harper's magazine article title in 1934, "Houses Like Fords."[15] Burns built 788 identical houses that had two bedrooms and an area of 885 square feet. He accomplished this by first establishing a staging area where material was unloaded, and where his workers precut and joined subassemblies. The subassemblies were then transported to the site, where workers moved down the street to clear, pour cement, frame, and finish.

Burns also cut up-front costs by leaving some of the work for the new owners. Some finishing work was left undone, and even driveways were left unpaved. He sold trees and other plants, as well as material for fences, to the owners so that they could add these amenities as well. Later builders would hone the skill of leaving some of the work for others to do. Finally, as Hayden notes, the use of subassemblies to approach, if not actually gain, mass production using unskilled or semi-skilled labor mirrored exactly what the airplane manufacturers themselves were undertaking. Newly minted aircraft workers in the factories were also made suddenly middle-class by buying a home and acquiring a mortgage. As it would turn out, both social roles would prove difficult to pass down to the next generation.

Henry J. Kaiser undertook better known and more extensive worker housing at his innovative shipyards built at the beginning of the war. Vanport City, outside of Portland, Oregon, was a case in point. Within three days of receiving a go-ahead from Washington, Kaiser was clearing land for what would become a city of some 40,000 workers and their families. Just as every house is a bundle of networked technologies, so every city is a collection of components, including houses, stores, streets, churches, offices, and all of the other facilities needed for a society to function. Most, of course, have evolved over time with only a minimum of advance planning. At Vanport City, however, the great urban machine was designed all at once for maximum efficiency. The project architect was quoted as telling a reporter that "all my life I have wanted to build a new town, but—*not this fast*" (emphasis in original).[16] More than the construction techniques Kaiser used, his major innovations were in social engineering. He had to design for a workforce made up not of a father who went to work leaving his wife and kids behind, but for a richly diverse group of women and men, Black, Asian, and Hispanic as well as White, and including families often headed by single women. He also had to design for worker efficiency and the conservation of strategic materials. Workers relied on public transportation, and child-care centers were open twenty-four hours a day so that mothers could pick their children up after their shifts ended, bathe them, pick up a hot meal, and take home both child and meal on the bus.

In 1946, the magazine *American Builder* asserted, "the expansion of non-farm residential building in the period ahead will be facilitated by experience gained in the war housing program, in which both builders and a few general contractors built projects of several hundred units. With few exceptions, these projects were marked by more thorough planning of operations, more careful timing, and control of materials, and greater use of power-operated tools than were general in pre-war operative building."[17] Henry Kaiser was one of those builders who had learned from the war. In 1942, he talked with the industrial designer Norman Bel Geddes about building three-room, prefabricated steel-frame houses, and two years later, he told a Chicago meeting of the National Committee on Housing that after the war it was inevitable that "there [would] be a spreading out, with extensive requirements for separate dwellings."[18] Just two days after V-J Day, Kaiser joined in partnership with Fritz Burns to form Kaiser Community Homes. They decided to limit their activity to building two- and three-bedroom homes on the West Coast, where they had both done their major work.

Kaiser Community Homes built only a small percentage of the postwar homes that sprang up around the country, but the firm was one of a handful that pioneered new construction technologies and techniques. Kaiser had used subassemblies extensively to build ships during World War II, and now used them to construct paneling and partly finished porches and roofs in factories using stationary machinery. Dishwasher, garbage disposals, and plumbing fixtures were made in a converted aircraft plant in Bristol, Pennsylvania, then "pre-assembled" into standardized kitchens and bathrooms for installation into houses. Because Kaiser was active in a number of other areas, he was able to bypass other suppliers and use his own cement, gypsum, sand, gravel, and aluminum.

By 1950, Kaiser's other enterprises seemed more promising to him than housing, and he was already interested in building automobiles to challenge Detroit. In 1942, he commissioned R. Buckminster Fuller to design a futuristic "1945 model Dymaxion" car. Fuller, of course, also became famous for his geodesic dome structure, which he developed for the military but, in the 1960s, came to epitomize the house of choice for many hippie communes. All told, Kaiser and Burns probably built no more than 10,000 of their homes.[19]

Levittown

Arguably the most famous of postwar suburban developments, Levittown stood in stark contrast to Vanport City. Here, as Hayden explains it, "every family is expected to consist of male breadwinner, female housewife, and their children. Energy conservation is not a design issue, nor is low maintenance, nor is public transportation, nor is child care." In addition, as late as 1960, no African Americans lived among Levittown's 82,000 residents. Levitt claimed that it was a matter of business, not prejudice, and at any rate, the Federal Housing Administration (FHA) would not approve mortgages in integrated communities, nor to female-headed families.[20]

But in terms of construction techniques, William Levitt—called the "Suburb Maker" when he died in 1994—was as progressive as his social policies were reactionary.[21] In ecstatic prose, the July 3, 1950 issue of *Time* magazine described how 1,200 acres of potato fields on Long Island was turned into Levittown, a city of 40,000 souls occupying 10,600 homes, in a startlingly short period of time. *Time* was excited not only by the houses' rapid construction, but the way in which they were built. The magazine described trucks arriving with pallets of lumber, pipes, bricks, shingles, and copper tubing, dropped every one hundred feet. Ditching machines with toothed buckets on an endless chain dug a four-foot trench around a rectangle twenty-five feet by thirty-two feet every thirteen minutes. Then, building crews specializing in only one job completed a house somewhere in the development every fifteen minutes. Levitt and Sons had become, according to *Time*, "the General Motors" of housing.[22]

The analogy to mass production was obvious. The specialized crews of workers were each responsible for one of the twenty-six tasks into which the house-building work had been divided. Because the workers typically dealt with prefabricated subassemblies, many traditional carpenter skills were not needed, and because the workers were nonunion, they could be equipped with labor-saving tools like spray-paint guns. Levitt went further: like Ford before him, he vertically integrated, owning forest lands and operating his own lumber mill and nail-making plant. For other materials, he used his buying power to force suppliers to deal directly with his company, rather than conducting business through intermediaries.

The once popular Malvina Reynolds song "Little Boxes Made of Ticky Tacky" was written in response to the rows of identical houses climbing the hillsides of Daly City, California, but the image perhaps most often conjures up the homes of Levittown. Bill Griffin, the creator of the cartoon "Zippy the Pinhead," grew up in early Levittown, and has said of the houses, "since they all looked the same, I'd walk in and say 'Um, Mom?' And there would be a woman in the distance who would say, 'There's milk and cookies in the refrigerator,' and I'd just sort of sit down."[23] In actual fact, however, Levitt fully expected buyers to add on to and otherwise modify their homes, and they did. As with other kinds of technology, cars being an outstanding example, basic house designs are often altered as people try to live up to expectations of "normal" family life—or rebel against them.

In 1995, when the Levittown Historical Society was planning a fiftieth anniversary for the development, it could not find a single house in "original" shape that could appropriately be designated by the Town of Hempstead Landmark Commission. Owners had added Cape Cod styles, L-shaped wings, and second-floor porches and bay windows, covering over Levitt's cedar shingles and asbestos siding with vinyl. The ranch houses were "hardly recognizable as ranches. They have two floors, three floors, two-car garages. At least one," it was alleged, "has Greek columns and statues in front." Said representatives of the historical society, "It's as if, each of Levittown's 17,000 original houses was born a *tabula rasa*, and developed into its owner's version of the American dream."[24]

Technological Infrastructure

For Levitt and other land developers like him, the federal government made mortgages easily available through the FHA. Through the GI Bill, veterans could get a home for no money down, with payments often under a hundred dollars a month stretched over thirty years. At the urging of the electrical appliance industry, the FHA included the cost of household appliances as a part of mortgages. These same large companies then cut deals with developers to feature their appliances in the new houses. As early as 1929, the home economist Christine Frederick had argued that first-time home buyers were a certain market for appliances, and the all-electric home, which General Electric (GE) touted during the 1930s, received new impetus after the war.[25] The Levittown houses were not all-electric, but they did come equipped with GE electric stoves, Bendix washers, and Admiral televisions. In a sense though, the family car was the major appliance that most often went with the new house in the sitcom suburbs. As Levitt's houses were located far from jobs and most shopping, the car became as necessary as electric lights and a kitchen sink. Not surprisingly, Levitt did not bother to provide public transportation. If, as he declared, a man with a house was too busy to be a Communist, a man with a house and car was even less likely to have time for politics.

Nor was public transportation the only technology missing in Levittown. The engineering infrastructure required for cities of the scale at which Levitt built was almost entirely lacking. Forced to build roads, Levitt failed to articulate them with existing roads, and immediately dedicated them to the local governments to maintain. The company dug wells and installed pumps and distribution mains for water, but unloaded them on the adjoining towns as well. Worse yet, to save the expense of building a sewage system, Levitt gave each house a cesspool, a localized and privatized technology inferior even to septic tanks. Parks, schools, and even sidewalks were largely ignored as well. Levitt had created private spaces, not public ones.[26]

Other developers built ever more extensive suburban cities using production methods from both the auto industry and the wartime shipyards. In 1949, three of them—Ben Weingart, Mark Taper, and Louis Boyar—bought 3,500 acres of farmland between Los Angeles and Long Beach, California, and between 1950 and 1952, put up a hundred houses every workday, or five hundred in a week. Construction crews were made up of teams of thirty-five men, each team subdivided by task. In 1951, more than 4,000 workers were hired, mostly veterans unskilled in the building trades. Prewar subdivisions typically had five houses per acre; this one had eight. It took fifteen minutes to dig a foundation, and in just one 100-day period, crews poured concrete into 2,113 of them. So it went down the line: One crew finished its work to be replaced immediately—or as soon as the cement, plaster, or paint dried—by another. The workers may have moved from one unit to the next rather than standing in a line for the work to come to them, but as the *Los Angeles Daily News* correctly noted, it was a "huge assembly line" nonetheless. In all, 17,500 houses were built in the first three years. The sales office for the 1,100 square-foot houses opened on Palm Sunday in April 1950, and 25,000 potential customers lined up to buy them.[27]

While the California developers used many of the same mass-production techniques as Levitt had, they did some things differently, in part because Los Angeles County made them. The lots were of the minimum size that the county required; the county also insisted that the developers construct sewers. There were 105 acres of concrete sidewalks, 133 miles of paved streets, and 5,000 concrete poles for streetlights. One swimming pool was also dug grudgingly in a local park, and a company-owned golf course was leased to the county. Appliances were not automatically included, with the exception of a Waste King electric garbage disposal, a questionable amenity in a near-desert environment. Because first-time buyers, as we have seen, were unlikely to have other appliances, however, the company made O'Keefe and Merritt gas stoves, Norge refrigerators, and Bendix washers available for nine dollars per appliance per month. These extra costs could be added to the mortgage under FHA rules. The stainless-steel kitchen counter top and double sink came with the house.[28] As D. J. Waldie, a Lakewood native son, longtime city employee, and author of the fascinating memoir *Holy Land* noted, "Seventy-five percent of the buyers were purchasing their first house. They used credit to buy a car and furniture. By 1953, 98 percent of the households had a television set. No one had any money." They paid $10,290 for the house with options, and another $6,860 for the car and furniture.[29]

The type of infrastructural technologies, as well as their provision or absence, had large implications for the social, civil, and environmental health of the nation for generations to come. As Hayden has noted, "developers of the sitcom suburbs insisted that mass-producing two- and three-bedroom houses in distant, car-dependent tracts was not just one way, but the only way to house the nation." By the end of the century, roughly three-quarters of the country's housing had been built since 1940, and houses were getting larger as families got smaller. In 1980, two-thirds of our housing units were single-family detached houses, and by 1976, 84 percent had five or more rooms. At the same time, however, households were getting smaller; by 1980, a quarter were one person living alone, and another third consisted of only two people.[30]

The relative lack of public spaces, amenities, and services led to other civil dislocations. As we shall see, while instant cities of tens of thousands of newly arrived strangers cast a huge burden on local governments to provide services and infrastructure, they were a boon to the curators of private spaces in shopping strips and malls. Just as television stations sell their viewers to advertisers, so, as Waldie points out, "density is what developers sell to the builders of shopping centers."[31] The displacement of a civic by a consumer society was tied directly to the way our housing was designed. As Levitt reassured his buyers, "every house will have its own 'park' when all the trees are grown," and surely in Cold War America, a private park was to be preferred over a public one.[32]

Finally, there were environmental considerations. First, the 15 million homes put up during the 1950s alone were situated on land that was usually made ready for construction by large-scale machines. Earth-moving equipment was freshly adapted from wartime machines used by military construction units. One of the Levitt boys had served during the war with the Navy Construction Battalion—the Seabees—and was presumably well versed in the power of graders, backhoes, and especially bulldozers. It was common practice to simply strip and level the land to be built upon, filling in marshlands, digging into hillsides, and diverting creeks into underground conduits. In 1944, the magazine *Military Engineer* carried an advertisement for Rogers Hydraulic, Incorporated, titled "Mother Earth is Going to Have Her Face Lifted!" Beneath a drawing of the globe with unhappy looking female features capped by a bulldozer working from her cheek up to her scalp, the text asks: "Sounds like a rather ambitious undertaking, doesn't it, but that is more or less what is going to happen after this World Struggle is over. The Earth is in for a tremendous resurfacing operation. The construction, road building and grading jobs for Crawler type tractors in that not too distant period are colossal."[33]

Before the war, wetlands and steep slopes were barriers to most construction, but with the new heavy machinery that large-scale construction made profitable, even such inhospitable terrain could become home sites. Starting in the mid-1950s, over the next twenty years, nearly a million acres of "marshes, swamps, bogs and coastal estuaries" were converted to housing. There is no similar estimate for the number of hillsides carved into for housing, but the very popularity of the split-level house, designed in part for just such sites, is suggestive.[34] The environmental loss in biodiversity terms is staggering, but the direct social costs can also be counted as natural disasters: the widespread damage from floods, hurricanes, mudslides, and the like were utterly predictable once housing was built on large tracts of unsuitable land.[35]

Freshly denuded countryside was quickly covered with grass to provide each house with its obligatory lawn, or in Levitt's term, "park." These grassy areas were virtual technologies in themselves. The grass was not native, and often the result of careful breeding. Watering depended not only upon a water supply somewhere, but also on pipes, faucets, hoses, and sprinklers at the house site. Lawnmowers, increasingly powered by gasoline or electricity, were needed to keep the lawns trim, and during the second half of the 1940s, the number sold increased by 1,000 percent, from 139,000 in 1946 to 1.3 million in 1951.[36]

Edgers, trimmers, and other grooming tools added to the inventory. Then, in part thanks to the wartime development of DDT, pesticides, herbicides, fungicides, and other poisons were applied, and an entire industry of chemical lawn care arose to meet the need. Finally, to grow anything on land from which the topsoil had been scraped, a range of chemical fertilizers were developed. The 30 million lawns being cared for in 1960 covered a total of 3,600 square miles, and maintaining a modest home lawn took an estimated $200 and 150 hours of work each year.[37] One could hardly imagine a "natural" feature that was more massively artificial.

Chances were that the new lawn covered yet another necessary technology, the cesspool or septic tank. Because developers were reluctant to incur the expense of building a proper sewer system complete with treatment plants, private, on-site solutions were found. The cheapest was to simply dig a cesspool, an underground storage area where the family's waste merely accumulated. A more sophisticated and long-term technology was the septic tank, by which waste first entered a buried vault, and then ran out into shallow leach-lines where it would eventually become decontaminated. Which was used in Levittown has been disputed, but septic tanks were widespread in the new suburbs; one developer installed them in a tract of 8,000 homes. The government did not even count them until 1960, but it is estimated that there were about 4.5 million in 1945. Fifteen years later, 14 million were counted, and in 1958, metropolitan Phoenix had 70 percent of its homes on septic tanks.

With only two or three years of use, "a substantial number" of these septic tanks failed, creating a need for expensive repair work by the homeowner and widespread pollution of groundwater. Cesspools were even less reliable and effective. Many buyers apparently had no idea where their waste went, and were unprepared for the possible problems ahead. Some even made the situation worse by laying concrete patios over the vault leech lines, making periodic cleaning and repair virtually impossible. In one spectacular case, the FHA had to repossess an entire development of 1,000 homes because the systems failed and could not be repaired.[38] Eventually, local governments had to step in and impose new "sewer taxes" to create a proper system, and then force homeowners to hook up.

Appliances and Energy

Family homes not only produce sewage, but they consume energy. During the war, there had been well-publicized research on the use of solar power to heat home space and water, an on-site solution not unlike the use of septic tanks, but with more sustainable results. Unlike the latter, however, solar technologies, though well understood and successfully used in some areas of the country since the late nineteenth century, were quickly abandoned in peacetime.[39] The industries, heavily invested in the other, more traditional power sources, were eager to step forward and provide for the needs of the millions of new homes being built. The solid fuels of coal and wood were rapidly loosing ground to heating oil and natural gas.

The electric power industry, in cooperation with the manufacturers of electric appliances, were particularly aggressive, and their success was partly accounted for by the availability of new technologies, such as air conditioning.[40] Introduced without much success by Carrier in the 1930s, only a few homes used the technology by the end of the war. The government did not even bother to count installations until 1960. By that time, however, a million homes had central air conditioning, and a decade later, six million had room units. Because so many of the new housing developments were designed for young and working-class buyers, the extra cost of air conditioning meant either higher house prices or fewer square feet, neither of which was desirable. Nevertheless, by the mid-1950s, electrical utilities were noticing the impact of air conditioning. In 1956, Consolidated Edison in New York saw for the first time that peak loads began to occur during summer daytime hours.[41]

The increased use of energy presented both a problem and an opportunity for utilities. The problem was that if they increased capacity to meet peak demand—such as by building more power plants—there was the danger that it would sit idle during periods of low demand, as during long winter nights. "Balancing the load" had been an industry imperative since the late nineteenth century. The opportunity was to find ways to get customers to use more electricity during the low periods. The solution settled upon was electric heating. In technical terms, it was perhaps the worst way to use electricity, but its very inefficiency promised market growth. "We should view electric house heating as a principal means by which our rate of growth can be drastically increased," enthused the president of the Edison Electric Institute in 1958. Using electricity for heating would have another advantage as well: new housing would have to be built with heavier wiring to accommodate the loads, and old homes would have to be rewired. With this new, heavier wiring, additional appliances could be supported as well.[42]

Appliance manufacturers, particularly General Electric and Westinghouse, enthusiastically seconded the industry in its efforts. The campaign for "all-electric homes" was revived, advertising spending increased, and Medallion Homes were created to showcase the technology. Even the United Mine Workers came on board, as the dwindling use of coal to heat homes could, at least to some degree, be offset by coal used in new and enlarged power plants. It was a gold mine from the utilities' standpoint: in the mid-1950s the average home used about 3,000 kilowatt hours of electricity a year, while an all-electric home used between 20,000 and 30,000.[43] As with sewage disposal, the environmental consequences of increased power production was not part of accounting.

The problems of energy use were greatly exacerbated by the way houses in the new developments were designed. Small changes from traditional practice were made to accommodate the new technologies. In trying to show that air conditioning could pay for itself through construction shortcuts, Carrier listed the following, as described by Rome: "Eliminate the screened-in sleeping porch, and save $350. Eliminate the attic fan, and save $250 more. Since air conditioning eliminated the need for cross-ventilation, builders did not need to have windows in every wall, and windows cost $25 apiece. The windows also could be fixed in place, with no sashes or screens, at a savings of $20 a window. Eliminate ventilation louvers, and save another $125."[44] This strategy, of course, absolutely increased the need for air conditioning. Builders also were drawn to such cost-saving strategies as using concrete slab floors, drywall construction, low-pitched roofs, and exposed cathedral ceilings.[45] The effect of all of these modifications was to abandon years of design experience in how to deal with heat through passive solutions in favor of a technological fix, and to shift the cost of the asking price of the home from purchase price to utility bills over the life of the house—and onto the environment. These design decisions are evident in the premier home design of the postwar years, the California ranch house.

The federal Housing and Home Finance Agency announced in 1951, in the context of the Korean War, "when many of the materials normally used in dwelling construction are scarce, the conservation of materials, fuel and energy is essential. Therefore, it is doubly important now that climate assume its proper position as one of our natural resources." This effort to promote environmental design was anticipated in 1949, when *House Beautiful* began a campaign urging architects, builders, and their clients to take geography and climate seriously. Some parts of the country were cold and others hot; some were wet and others dry; some were sunny and others frequently cloudy. Locality mattered, it argued, and traditional regional variations in house design should not be ignored.[46]

The very essence of modernity, however, was to honor the universal and not the particular; its very hallmark was the collapsing of time and space. Thus, winter or summer made no difference, nor did north or south. Just as the detached, single-family dwelling was the universal type of house, so could it be situated anywhere in the nation without significant modification. If people wanted to pretend that they lived in California and adopt a California lifestyle, that could be accomplished by adopting the necessary technologies to keep it warm in the winter, cool in the summer, and generally to bring the outdoors inside—or perhaps more accurately, to muffle the distinction between inside and outside. The hegemonic belief that the proper domestic pattern was a breadwinning father, a housebound mother, and children dictated the need for a separate, private dwelling; and the way that a family was supposed to interact helped to make the ranch house attractive.

The home was to be the setting for the new family, dedicated to fun, recreation, and togetherness. The old separate gendered rooms were opened up as family life flowed seamlessly from kitchen to living room to family room to bedroom and patio. The home, with its flowing spaces, cathedral ceilings, sliding glass doors, and the ubiquitous picture window spoke of livability, comfort, and convenience, the perfect venue for the American dream of the perfect family. Patios and breezeways added a feeling of spaciousness even to small houses. Patios seemed to project living space into the yard. The avoidance of stairs allowed family life to flow effortlessly outside. The world outside, however, was pure artifice. Because the houses were so low, yard planting had to be low-growing or even miniature species. The perfect lawn, of course, was also the perfect view from the picture window. The proper design and attendant technologies could reproduce the California climate and the happy, family lifestyle that flourished there. To paraphrase the musician Frank Zappa, wherever you went, there you were.[47]

The houses of the sitcom suburbs, their design, and the technologies that filled them were not the way they were by accident or consumer preference, but through the hard work and cooperation of the real estate, banking, and construction industries, the electrical appliance manufacturers, and electric utilities, with the full support of federal, state, and local governments. It was a design that alarmed many social observers. The first fear was that it produced "a lonely crowd," to use the title of sociologist David Riesman's influential 1950 book. By the 1970s, its costs to the natural environment were being counted as well.

Both of these concerns could be echoed today in the very same suburbs. In 1996, Patricia Leigh Brown returned to her girlhood home in Highland Park, Illinois. It had been built as part of a development in 1953, and cost a very hefty $48,000. It was a 2,500-square-foot ranch house on a deep lot. By the time she returned, however, it was gone, a victim of "engulfment": it had been torn down to be replaced by a two-story home with twice the floor space, "*two two-car garages, a ceremonial courtyard, two walk-in closets bigger than our kitchen.*" It was being built not for a growing family as Levitt had envisioned, but for an empty-nest couple in their fifties.[48] The home, like other McMansions, as they came to be called, was not simply a space to be filled with the fruits of consumption; it was an object of desire in its own right.

Shopping Centers

The consumption that was so much a part of the lives of discharged veterans and former war workers, and which was so closely tied to their new homes and status as middle-class Americans, needed a commercial venue. Between 1947 and 1951, the Levitt family built seven small shopping strips that were largely unplanned, though at first profitable, adjuncts to their sprawling housing developments. Because the shopping strips were largely inadequate for the needs and desires of local residents, however, the resident traveled to other venues, nearby small towns, or back into the city to do their shopping.[49] A dedicated shopping precinct within or near the developments, however, had enormous potential to change the way America shopped, and could create large profits for their builders.

Some builders stepped forward almost immediately. Edward J. DeBartolo, who had graduated in civil engineering from Notre Dame and spent the war with the Army Corps of Engineers, built his first shopping center in his native Youngstown, Ohio in 1948. He was quoted as saying, "stay out in the country. That's the new downtown." Before he died in 1994, his firm had built more than 200 shopping malls in twenty states.[50] The real kickoff for such centers, however, proved to be the federal government's passage of legislation in 1954 that made possible "accelerated depreciation" for funds used to put up new business buildings. The new law allowed investors to rapidly write off new construction of such buildings, and they could claim any losses against unrelated income. Suddenly, shopping centers became lucrative tax shelters as well as places to purchase goods.[51]

Much of the consumer technology of the postwar years came together in the new shopping centers that were being constructed in the suburbs, far from the undesirable populations of the inner cities—the poor, people of color, prostitutes and other criminals—and close to the newly suburbanized middle class.[52] As D. J. Waldie has suggested, the developers of the large Los Angeles suburb of Lakewood, California made their real money not from selling the houses in the tract, but by using the new homeowners as potential customers to attract retail firms, such as the May Company department store, to occupy their central shopping center.[53] Easily accessible by car and surrounded by acres of asphalt parking lots, shopping centers gave suburbanites stores in which to buy their appliances and their televisions—or, in the early days, store windows in which to watch them. The centers became, in Lizabeth Cohen's phrase, the new community marketplaces of America. As Cohen points out, these centers had major effects on community life: "in commercializing public space they brought to community life the market segmentation that increasingly shaped commerce; in privatizing public space they privileged the right of private property owners over citizens' traditional rights of free speech in community forums; and in feminizing public space they enhanced women's claim on the suburban landscape but also empowered them more as consumers than as producers."[54]

The shopping center of the Lakewood, California development marked a large step forward from Levittown's shops. Its developers created sixteen small shopping precincts; each with an assortment of businesses—dry cleaners, barbers, beauty shops, and either drug stores or dime stores—and each located so that no house in the development was more than a half a mile away. As Lakewood had sidewalks, shoppers could presumably walk if necessary. The mall itself was in the center of the city, surrounded by a sea of asphalt containing 10,580 parking spaces and anchored by a May Company department store topped by a giant letter M facing in each direction. The shops met all of the needs of Lakewood's residents and became magnets for shoppers from other suburbs as well.[55] Older nearby suburbs such as Venice had been connected by interurban light-rail lines to the center of Los Angeles for movies and shopping trips. The people of Lakewood did not need such trips.

Northern New Jersey offered vivid evidence of the need for and influence of roads in suburban development. The urban areas in that part of the state were largely connected only by interurban rail lines before the war, but postwar road building, which preceded rather than followed suburbanization, changed that dramatically. The 118-mile New Jersey Turnpike and the 164-mile Garden State Parkway were both completed by the mid-1950s, and within a decade, Woodbridge Township, where the two intersected, had a population of 100,000. Just as the developing freeway system saw franchised motels, service stations, and family restaurants multiply at virtually every on- or off-ramp across the empty farmlands of the country, so the new shopping malls used highways to funnel customers into their stores.[56]

New regional shopping centers were epitomized by the Bergen Mall and Garden State Plaza in Paramus, New Jersey, the latter of which became the nation's largest shopping complex by 1957. It was built next to a planned exit of the Garden State Parkway, and its retail developers, R.H. Macy's and Allied Stores Corporation, "believed that they were participating in a rationalization of consumption and community no less significant than the way highways were improving transportation or the way tract developers were delivering mass housing." Rather like Walt Disney's theme parks, underground tunnels and delivery docks, hired security guards, air-conditioning, and piped-in music swaddled the shoppers away from the presumed unpleasantness of contemporary urban life, and made shopping trips, or just hanging out, a recreational as well as a commercial activity.[57]

The apotheoses of what Cohen calls "central sites of consumption" were the malls that followed. The West Edmonton Mall in Alberta, Canada covers 5.2 million square feet, and according to the *Guinness Book of Records*, was, when it was built, not only the world's largest shopping mall, but also had the largest indoor amusement park, the largest indoor water park, and the largest parking lot. Inside were "800 shops, 11 department stores, and 110 restaurants," along with a "full-size ice-skating rink, a 360-room hotel, a lake, a nondenominational chapel, 20 movie theaters, and 13 nightclubs." In the words of Margaret Crawford, "confusion proliferates at every level; past and future collapse meaninglessly into the present; barriers between real and fake, near and far, dissolve as history, nature, technology, are indifferently processed by the mall's fantasy machine."[58]

In 1956, the Southdale mall in suburban Minneapolis became the first enclosed shopping center. By the early 1990s, there were 28,500 others in North America, mostly in the United States. During those decades, the average time spent in trips to the mall went up from twenty minutes in 1960 to three hours in 1992. The last half of the twentieth century saw not only increased consumption of goods and services, in large part because of the prosperity of the period, but the increasing rationalization of consumption. American manufacturers had set the standard before the war with their articulation of mass production. After the war, just as farmers turned American agriculture into factory farming embodying much of the same logic, so retailers reshaped consumption and placed it in a context of leisure and spectacle. As Crawford concludes, "the world of the shopping mall—respecting no boundaries, no longer limited even by the imperative of consumption—has become the world."[59]

In an important and powerfully persuasive interpretation of American social, political, and cultural change over the last half of the twentieth century, historian Lizabeth Cohen has placed the American consumer firmly at the center of the revolution. "I am convinced," she has written, "that Americans after World War II saw their nation as the model for the world of a society committed to mass consumption and what was assumed to be its far-reaching benefits." However, consumption "did not only deliver wonderful things for purchase— the televisions, air conditioners, and computers that have transformed American life over the last half century. It also dictated the most central dimensions of postwar society, including the political economy . . . as well as the political culture." The old assumed distinction between the roles of citizen and consumer, so strongly reinforced during the war in the name of national survival, lost its power to define American society and behavior. "In the postwar Consumers' Republic," Cohen continues, "a new ideal emerged—the *purchaser citizen*—as an alluring compromise. Now the consumer satisfying material wants actually served the national interest."[60] It was this new purchaser citizen who President George W. Bush rallied after the terrorist attacks of September 11, 2001, not to delay gratification and pay more taxes to support a new war, but rather to help the country recover from its trauma by increasing consumer spending. In the years after World War II, the dream houses and the appliances to make them work began the profound change of turning citizens into consumers. By the end of the century, even voting (or not) seemed like just another market transaction, in which Americans, organized into focus groups, shopped around for what best suited them personally, whether it was lower gasoline prices or tax cuts. During these same years, the federal government undertook to create a similar culture worldwide.

THREE

FOREIGN AID AND ADVANTAGE

In 1941, the magazine mogul Henry Luce called on Americans "to accept wholeheartedly our duty and our opportunity as the most powerful and vital nation in the world and in consequence to exert upon the world the full impact of our influences."[1] That impact could be projected in many ways: through politics, economics, war, communications, and other vectors of power. As it turned out, many of these had in common an American desire and ability to impose its will through its myriad and marvelous technologies. Foreign aid to help rebuild Japan and Europe, especially Germany, and to support both the newly liberated nations born out of decaying Old World empires and the "underdeveloped" countries in our own hemisphere, commonly took the form of civil engineering, manufacturing, or agricultural projects and machines. Military hardware flowed freely to shore up a barrier to Soviet expansion, and American forces abroad were tied together with a vast network of electronic media. Demand for and access to raw materials worldwide was only part of the reason for what can best be called a dream of what historian Michael Smith has called planetary engineering.[2]

Aid to the Allies had been a critical feature of the American war effort, and between March 1941 and September 1946, the lend-lease program had spent $50.6 billion. After the war, a welter of programs were funded: $5.3 billion for foreign aid was appropriated in April 1948, and in October 1949, a Mutual Defense Assistance Act brought $1.3 in military aid to Atlantic Pact allies, joined by $5.8 billion in economic aid. In 1950, $3.1 billion was appropriated, including funds for the new Point Four program for technical aid to developing nations; in 1951, $7.3 billion was appropriated, including $4.8 billion for military aid; and in 1952, $6.4 billion was appropriated. Of all this, perhaps the most celebrated initiative was the European Recovery Program, better known as the Marshall Plan.

The Marshall Plan

The rise of Cold War hostilities with the Soviet Union in the late 1940s, breaking out into a hot war in Korea in 1950, has sometimes distracted our attention from the so-called "dollar gap" that threatened postwar America at the same time. Simply put, the rest of the world, including the defeated Axis powers, our exhausted allies, and the new nations emerging from European colonial rule, found it impossible immediately after the war to produce enough to earn the dollars necessary to increase their trade with the United States. In the face of this barrier to trade, the United States spent $22.5 billion in one or another sort of foreign aid schemes between 1945 and 1949.[3] This investment in recovery went far to accomplish its twin goals of preventing a communist takeover of additional territories and staving off economic depression at home. It was also a remarkable and successful policy shift for the United States, which had prided itself on remaining aloof from European problems and still had a strong isolationist cadre in the Republican Party.

On the largest scale, the worry over communist expansion—for which the obvious antidote was capitalist expansion—was matched by the simultaneous collapse of Great Britain's imperial role as an international player. Meanwhile, Italy and France were seen as "unstable," meaning that their working classes were resisting demands for austerity and harbored a strong communist presence. Finally, Germany was suffering from both moral and economic crises. All over Europe, the bitterly cold winter of 1946 to 1947 dashed any illusions that conditions were returning to the status quo before the war. Temporary aid funneled through the United Nations had not successfully restored economic conditions in either Germany or Austria. In view of these interlocking problems, Secretary of State George Marshall used the occasion of a speech at Harvard University on June 5, 1947 to announce what he formally called the European Recovery Program, but which became universally known as the Marshall Plan.

In the short term, Europeans, including both former friends and former enemies, needed food—particularly grain—and coal. In the long run, they also required such goods as fertilizers, steel, mining equipment, transport vehicles, and machine tools to rebuild their industrial economies.[4] The situation was exacerbated by Russian demands for reparations, among which they refused to count the many factories, plants, and sundry railroad equipment that they dismantled as they gained control of eastern German territory in the last days of the war. Some of it was shipped back to the Soviet Union, but some was left to rust across the countryside. An estimated 25 to 35 percent of German facilities were thus removed. The United States, of course, was busily removing as much rocket technology as it could find, along with the engineers who worked with it, but this was hardly a blow to Germany's ability to rebuild an export trade of value-added products.[5]

There was, of course, some opposition in the country to the whole idea of a reconstructed and resurgent Germany. That reluctance, however, seemed increasingly like a mere concession to the Russians, and congressional conservatives pushed consistently for an American policy that would lead to rebuilding a strong German economy. This opinion was ascendant by the 1950s, and when it was joined with a decision to rearm the defeated enemy, it assured a major flow of American technology into Europe as a whole, and Germany in particular. As always, American technology served American economic and political interests.

Reviving Japan

Like Germany, Japan had been devastated by the war; its economy, industrial base, and engineering infrastructure were either destroyed or severely disrupted. The American occupation of the postwar nation tried to balance punishing war criminals and instituting democratic reforms on the one hand, and maintaining the continuity of leaders and institutions to rapidly restore the economy on the other. A host of American experts, many of them engineers and other technologists, pitched in to help rebuild, reform, and modernize the productive technology of the shattered nation.

American technology and expertise were not new to the country. After the opening of Japan by the American Commodore Matthew Perry in 1853, the government of the so-called Meiji Restoration undertook a policy of modernization that included aggressively acquiring Western technologies. Unwilling to countenance the introduction of foreign-controlled firms, Japan bought machines and tools from overseas, licensed their use and manufacture, engaged foreign engineers as consultants, and sought in other ways to access the technology that it needed on terms that strengthened both the capabilities and the independence of the nation. One example of the success of the program was in textile production. Beginning in the 1870s, Japan imported spinning machines and expertise from the firm of Platt Brothers in Oldham, England. They then improved these and other machines, such as the Toyoda automatic loom, developed by Sakichi Toyoda, so that by the 1930s, Japanese looms were considered superior to those made in the West.[6]

Japan's economy, especially its industrial sector, was devastated during World War II. In the critical area of iron and steel production, one-seventh of the nation's steel-making capacity and a quarter of its iron-making capacity had been destroyed by military action. Much of what was left had been "overworked and poorly maintained" during the war, sources of raw materials were cut off, and two-thirds of the workers employed in the industry in 1944 had left by the end of 1945.[7] Because of old, worn-out, and antiquated technologies, Japanese steelworkers in 1947 produced only one-eighteenth as much steel per capita as did American workers. In addition, before the war, Japanese mills had depended heavily on imported scrap iron, some of which, even on the eve of hostilities, had been coming from the United States. Japan's old-fashioned open-hearth furnaces, invented in the West in the mid-nineteenth century, were major consumers of this scrap. After the war, visiting American engineers pointed out the deficiencies in what was left of the Japanese iron and steel industry, and two study groups from the Iron and Steel Institute of Japan visited the United States in 1950 to see for themselves what new technologies might be available.[8]

With the outbreak of the Korean War, the United States decided to buy steel and steel products from Japan in a deliberate effort to stimulate the rebuilding of the industry. Between 1951 and 1955, a steel industry "rationalization program" was directed by the famous Ministry of International Trade and Industry (MITI), which included foreign exchange loans to allow the industry to buy foreign technology abroad; MITI was also responsible for evaluating and licensing those technologies. By 1956, two-thirds of Japan's open-hearth capacity had been rebuilt, but the old technology did not solve the problems of small capacity, inefficiencies, and a continued need for scrap iron.

The solution, as it turned out, was a large-scale commitment to the relatively new technology of basic oxygen furnaces. Japan imported more complex machines, such as hot-strip mills, usually from either Germany or the United States. Japanese companies built less exotic machines themselves, though often from foreign designs and under the supervision of foreign engineers. Cutting prewar ties to German technology, occupation authorities supervised a shift to American technologies installed under the guidance of engineers from U.S. Steel, Armco, and other American firms.[9]

The careful supervision of MITI did not always guarantee happy results. Its most celebrated instance of shortsightedness involved its reluctance in 1953 to release $25,000 to a small firm called Tokyo Communications to enable it to license a new American technology. The device involved was the transistor, which Bell Laboratories had developed recently. It turned out that MITI had fairly judged the technology as being of major importance, but doubted that Tokyo Communications could exploit it properly. That company, however, went ahead with a contract and MITI eventually put up the money. The company grew not only the technology but itself as well, renaming itself Sony.[10]

A New Production Regime

An even more momentous transfer of American technology to Japan began in 1950 when the young engineer Eiji Toyoda began a three-month visit to Ford's River Rouge automobile plant in Detroit. His uncle had made the pilgrimage in 1929, when that shrine to technological vanguardism and modernity had drawn thousands from around the world to witness the miracle and sublime scale of mass production, or *Fordismo* as it was called abroad. Since its founding in 1936 by the Toyoda textile machinery family, Toyota Motor Company had produced by 1950 a total of only 2,685 cars, compared to the Rouge's daily total of 7,000. Young Toyoda reported back to Japan that he "thought there were some possibilities to improve the production system," a declaration of hubris that turned out to be true. After his return, he and his production chief Taiichi Ohno decided that what he had seen at the Rouge would have to be drastically modified to work in Japan. Those modifications amounted to nothing less than the celebrated Toyota Production System, better known by its generic name, lean production.[11]

It has been argued that even borrowing and modifying Detroit-style mass production was not enough to realize the production miracles that Japan witnessed in the last half of the twentieth century. By the time of Toyoda's visit to the Rouge, American authorities had begun to worry that the strong labor movement in Japan, with its demands for not just better wages, hours, and conditions, but some say in the setting of work rules and choices of technology as well, was posing a threat to the economic development of Japan that occupation officials fervently desired. In the United States, a bargain had long since been struck between most of organized labor and corporate management that growth in productivity was a technological fix that would not only avoid social strife, but also produce both profits and a high standard of living for workers. By the terms of this bargain, management reserved the right to make what technological and managerial changes it wished, so long as a portion of the benefits that flowed from increased efficiency were passed on to labor. If Japan was to become once again the "workshop of Asia," the workers' demands for justice and authority, and management's desire for "technological innovation, productivity, and profit in a competitive capitalist society" had to be reconciled.[12]

One American initiative took the form of helping the Japanese government to fund a Japan Productivity Center (JPC), to advance the idea that "growing the pie"—that is, expanding the economy —would obviate any need to quarrel over its division. Within the first two years of the JPC's existence, it sent fifty-three "small groups of managers and union leaders" to the United States to "learn the art of productivity." By the end of the 1950s, the JPC and other groups had brought back a host of techniques for quality control and production management that began to create a new labor-management environment, helping such firms as Toyota toward dramatic increases in efficiency.[13]

Ever since the 1920s, Japanese managers had attempted to implement some form of Frederick Winslow Taylor's scientific management system of worker efficiency, but as in the United States, the pure doctrine had proven difficult to install and enforce. In the postwar period, management took a number of smaller steps with the cumulative effect of, at long last, fulfilling the Taylorist dream of worker efficiency in Japan. The "dream of a technological fix for social problems," as Andrew Gordon has put it, "was shared by Americans in Japan and Americans at home," and led to importing modern systems of quality control (QC), which had been pioneered in the United States and Great Britain in the 1920s. As generally understood, QC was the application of statistical methods to guarantee that production measured up to standards.[14] Before the war, Japanese products were almost synonymous in the United States with shoddy work, and that widespread perception needed to be overcome as quickly and decisively as possible.

The first effort to introduce quality control techniques in Japan were made by the Civil Communications Section of the American occupation force, which worked to improve "precision and conformity to standards" in the reviving Japanese electronics industry. However, the Japan Union of Scientists and Engineers (JUSE) soon took a major role; it had worked informally during the war, was formally set up in 1946, and looked over new technologies that might best aid the growth of Japanese industry generally and its own organization and reputation for expertise. As one of its officers rhetorically asked in 1949, "We were deeply impressed by our recent finding that in Britain and the United States statistical quality control developed enormously during the war.... Why can we not have such refined techniques?" By 1950, JUSE saw QC as their way of "reviving science and perfecting technology." It was a movement, in the words of William M. Tsutsui, that was eventually "propelled in Japan by a dynamic alliance of top managers, engineering staff from the major industries, and prominent scientific scholars."[15]

Among the numerous management experts brought to Japan was W. Edwards Deming, who made his first trip there in 1947 and returned virtually every year from 1950 onward. At the invitation of JUSE in 1950, he began to spread the word about his ideas of QC systems.[16] Deming is often seen as the major vector for the introduction of SQC into postwar Japan, but it appears more likely that his enthusiasm, authority, and clear presentations mainly validated the movement and inspired its practitioners. Before too long, the worry developed that Deming's strong emphasis on the statistical part of the method had more appeal to engineers than to managers and workers, and that its very rigor and "scientific" precision worked against pragmatic gains on the shop floor.[17]

However, Deming was only one of a number of American experts who were brought to Japan to aid the cause of SQC. Lowell Mellen, a management consultant with only a high school education, had learned industrial training in Ohio and applied it to help increase American military production during World War II. After the war, he formed a company called Training Within Industry Incorporated (TWI) to continue his work. In 1951, he contracted with the U.S. Army to shift his activities to Japan where, according to one scholar, TWI became "the first salvo in management's battle to capture the hearts and minds of foremen." In Japan, Mellen was credited with introducing such techniques as written suggestion programs, methods of "continuous improvement," and generally more "enlightened" employee relations and training practices. By the mid-1980s, an estimated 60 percent of Japanese firms, including some of its largest and most successful, had instituted some form of small-group activities after Mellen's ideas.[18]

Yet another QC apostle was Joseph Juran, whom JUSE invited to Japan in 1954. Unlike Deming, Juran was not a statistician and deplored the fetishization of scientific precision at the expense of economic pragmatism. Emphasizing, like Mellen, the importance of enlisting the cooperation and enthusiasm of all elements of industry, Juran again touched on concerns already stirring within Japanese industrial circles, and helped focus a significant reorientation of QC in Japan. He was perhaps as important to the process as Deming himself.[19]

Since its opening up in the nineteenth century, Japan had been at pains to adopt and adapt Western technologies that would provide it with a modern industrial base. The U.S. policy after World War II to rebuild and modernize the devastated Japanese economy was done by drawing upon both Japanese traditional practices and the efforts of the American occupation forces, in large part by importing Western, and particularly American, technologies. Within two decades, basic oxygen steel furnaces, SQC, the transistor, and Toyota's improvements on Fordist mass production wrought what was universally hailed as an economic miracle. The dramatic rebuilding of Europe as an economic powerhouse, and the rise of Japan to industrial prominence far surpassing its prewar position, raised the important question of whether large infusions of American technology could work the same miracles in areas of the world that had yet to be modernized.

Aid to Developing Countries

In the years following World War II, the United States, like the Soviet Union, undertook a large-scale program of sending aid to newly developing nations. Most of the countries had until recently been colonies of European imperial powers. Others were so-called "underdeveloped" countries in Central and South America. U.S. foreign policy operated on the assumption that newly emerging nations that developed strong capitalist economies tied to world markets dominated by the United States could best provide the goods and services to their people that would inoculate them against the temptations of international communism. At the same time, U.S. domestic policy sought to maximize the opportunities for American corporations to invest in the newly emerging nations, and to use them as sources of natural resources as well as markets for surplus production.

These foreign and domestic policies found common support in what was called modernization theory.[20] Briefly put, this plan assumed that modernization was the engine that drove change and prosperity, and that it was best achieved—in all cases—by helping nations pass rapidly through what Walt W. Rostow called the "stages of economic growth." Pre-modern or traditional societies should be helped to acquire the "preconditions for take-off" that would lead to the stage of take-off itself. This would, in turn, lead to the "drive for maturity," which would result in the ultimate goal, an "age of high mass-consumption." Traditional societies, Rostow wrote, were "based on pre-Newtonian science and technology" and were characterized by a ceiling on productivity, the result of "the fact that the potentialities which flow from modern science and technology were either not available or not regularly and systematically applied."[21] During take-off, he believed, "new industries expand rapidly" and "new techniques spread in agriculture," which is thereby "commercialized."[22] No one put the matter more bluntly or optimistically than the British writer C.P. Snow in his famous essay *The Two Cultures and the Scientific Revolution:* "It is technically possible," he asserted in 1959, "to carry out the scientific revolution in India, Africa, South-east Asia, Latin America, the Middle East, within fifty years."[23]

Of course, industrial technologies have been imposed on many of these areas for some time, with little evidence of local prosperity. Beginning in the 1950s, and largely in Latin America, what came to be called dependency theory developed to oppose modernization theory. Dependency theory sought to explain why many developing countries had not become modern industrial nations a long century after Western Europe and the United States had industrialized, and despite the export of much of their new technology to other parts of the world. The theory made three critical claims about how trade and the transfer of technology worked in the real world. First, it pointed out that the world was divided into two kinds of nations: those that were central, metropolitan, and dominant, and those that were peripheral, dependent, and satellite. Second, forces centered outside the latter states were critical to their dependency. Third, the relationships between the dominant and dependent states were dynamic, with the dependency deliberately reproduced, reinforced, and intensified. The condition of dependency was thus not an accident, but the result of a deliberate effort on the part of dominant states. One specific expression of this view was made in 1970 by Luis Echeverria, then president of Mexico. Referring to Mexican businesses licensing new American technologies, he pointed out that "the purchase of patents and the payment of royalties is too expensive for underdeveloped countries. Scientific colonialism," he charged, "deepens the differences between countries and keeps systems of international domination in existence."[24]

The Point Four Program

The United States' foreign policy in this area began to take a new legislative shape in January 1949, when President Harry S Truman articulated his program for international aid. The president's broad call for foreign aid contained four points, only the last of which was directly concerned with technical aid. In June, Truman sent a plan to Congress to implement what came to be called his Point Four Program. The new nations in "parts of Africa, the Near East and Far East and certain regions of Central and South America," he claimed, must "create a firm economic base for the democratic aspirations of their citizens. Without such an economic base," he warned, "they will be unable to meet the expectations which the modern world has aroused in their peoples. If they are frustrated and disappointed, they may turn to false doctrines which hold that the way of progress lies through tyranny."[25] No one in Congress would have missed that Truman was invoking the threat of Soviet communism. Aid was to help prevent the new nations from being drawn into the Russian sphere of influence, and instead become a part of ours.

Truman then identified two types of aid that were needed. The first was "the technical, scientific, and managerial knowledge necessary to economic development." He was apparently thinking not only of "advice in such basic fields as sanitation, communications, road building, and government service," but also of surveys to discover and inventory natural resources. The second type of aid he called "production goods—machinery and equipment—and financial assistance." The "underdeveloped areas need capital," he pointed out, "for port and harbor development, roads and communications, irrigation and drainage projects, as well as for public utilities and the whole range of extractive, processing, and manufacturing industries."[26] Clearly Truman had in mind creating the kind of engineering infrastructure that the Western industrial nations had built up over centuries and, at least in the case of the United States, done so with significant amounts of foreign investment. To start the process, he recommended the appropriation of a modest $45 million.

As Congress took up Truman's plan, deep divisions were immediately obvious. The American business community was the most vociferous in its criticism, even though it had been amply consulted while the administration's bill was being drafted. Essentially, businesspeople had two demands: first, that the United States not work through the United Nations or other multilateral bodies in granting technical aid, and second, that the government limit its role to not much more than working out bilateral agreements with recipient nations and guaranteeing to protect business against any possible losses from war, revolution, nationalization, excessive taxation, or other such supposed dangers of new and unstable governments. Because business leaders, rather than economists or technologists, were the government's main source of information and political pressure, parochial American business interests rather than any real and informed concern for overseas development were paramount. The Act for International Development, signed into law on June 5, 1950, reflected the supposed interests of the target nations only as a shield against communist expansion. Before the end of the month, the outbreak of the Korean War focused attention largely on military aid, and the cooling down of the American economy in early 1950 meant that business was aggressively looking for new areas of profit.[27]

Secretary of State Dean Acheson put the administration's case most succinctly: "This legislation," he asserted, "is a security measure. And as a security measure, it is an essential arm of our foreign policy, for our military and economic security is vitally dependent on the economic security of other peoples.... Economic development will [also] bring us certain practical material benefits."[28] Despite the dramatic success of the Marshall Plan in Europe, some isolationist elements in the Congress still objected to such foreign involvement, but many in the business community accepted the goal and objected only to its means. The Chamber of Commerce of the United States insisted that "private industry has the industrial know-how. Government has not. The most effective assistance in industrial development abroad can be provided by skilled technicians of American corporations which are investing their funds."[29]

A compromise proposal at the time, to set up a study commission to look into the complexities of technical aid, was dismissed as merely a diversionary measure to delay any action, but for a time it was the focus of doubts about how best to use technology to promote development and modernity in general. Such questions as whether introducing modern technology might depress local food supplies or disrupt communal life were raised but rushed over. Senator Eugene Millikin asked: "Will the intrusion of these foreign-inspired programs and their operation on the ground accentuate the cleavages between races and classes? . . . Is it not a fact that the development of resource-poor areas may create more economic and social problems than sit solves?"[30]

In retrospect, according to a congressional report two decades later, the Point Four Program clearly had several distinct parts:

(1) Providing a recipient nation with U.S. technical experts to furnish advice and instruction in long- and short-range policy matters ranging from public administration to managerial organization and the development of improved rice strains;
(2) Executing demonstration projects;
(3) Providing equipment and materials for demonstration projects; and
(4) Bringing foreign nationals to the United States to receive technical training in American universities and Federal agencies.[31]

At the time the program was being first discussed, however, such distinctions were passed over quickly, and making the developing world safe for American capital appears to have been more thoroughly attended to. As the preamble to the Act for International Development had it, the scheme would work only if there was "confidence on the part of investors, through international agreements or otherwise, that they will not be deprived of their property without prompt, adequate, and effective compensation; that they will be given reasonable opportunity to remit their earnings and withdraw their capital; that they will have reasonable freedom to manage, operate, and control their enterprises; that they will enjoy security in the protection of their persons and property, including industrial and intellectual property, and nondiscriminatory treatment in taxation and in the conduct of their business affairs."[32] This elaborate concern for removing the risk for American private investment abroad left little room for any discussion of just what technologies were to be deployed for what technical purposes.

In November 1950, five months after the passage of the original Act for International Development, President Truman asked his advisory board for international development to consider "desirable plans to accomplish with maximum dispatch and effectiveness the broad objectives and policies of the Point Four program."[33] Significantly, the chair of that board was Nelson A.

Rockefeller, who styled himself simply as the president of the International Basic Economy Corporation of New York. The board broadly represented the influential centers of American power: agriculture, labor, the media, business, and professional engineering. In its report four months later, the group made two significant points to guide the program in the future. First, as Rockefeller put it in his forward, "we were confronted with two central problems, defense and development. It was clear that they were indivisible."[34] The second point was the belief that "the history of the United States is one of an expanding national economy with the productivity of labor increasing steadily at the rate of two percent a year throughout. The same must be true of the world community. It, too, requires an expanding economy that creates opportunity and increased earnings, better living conditions, hope and faith in their future. In such an atmosphere of expanding economic life," he concluded, "free men and free institutions can and will thrive and grow strong."[35]

It was a picture of a capitalist paradise that nicely justified and reinforced the U.S. desire to exploit new natural resources, markets, and labor sources, and it implied that the wholesale transfer of powerful Western technologies would produce the same results in the less-developed nations as it had in the more-developed nations. As one congressional report put it, "the technical assistance hypothesis was presented simplistically: technology delivered to the underdeveloped society ... [was supposed to be] easily grafted onto the society and economic progress follows automatically."[36]

This same report, however, first produced in 1969 and revised in 1971, also pointed out that "the experiences of U.S. technical assistance programs have ... shown that many forms of technology utilized in the industrialized nations cannot be easily assimilated and adopted by the less developed nations without considerable modification to their particular needs and capacities." Indeed, it continued, "it may be necessary to export (or even to invent) a technology which is appropriate to the industrializ[ing] countries in the early stages of their developmental process, such as wooden instead of steel farm implements, hand-powered washing machines, or progression to the farmer from the hoe to the animal-drawn plow instead of to the tractor. These differences," it concluded, "also lead to the conclusion that special technologies appropriate to the unique circumstances of the less developed country must be developed and diffused."[37] It was not clear, however, that many U.S. corporations were keen to export hand-powered washing machines, let alone teach poor people how to make their own.

In 1954, a study entitled *Fifty Years of Technical Assistance: Some Administrative Experiences of U.S. Voluntary Agencies* reported that most of the consultants interviewed believed that it was "more important to recognize constantly that innovations should be related to existing local needs rather than that they be determined largely a priori by the outside expert on the basis of his experience in his own home country."[38] Furthermore, the study warned that the "single-minded pursuit of substantive improvements without consideration of social conditions or of the social impact of the results ... or of the means used to obtain the results, may do more harm than good to indigenous people and may sharply limit the acceptance, retention, or humane use of improvements or innovations."[39]

The representatives of the voluntary agencies also noted that "people are by no means passive agents in absorbing new ideas from outsiders. They are apt to see in outsiders what they want to see and to absorb from the range of new practices and ideas presented by foreign project personnel, both off and on duty, those innovations which suit their own larger purposes."[40] However, although such indigenous agency was real, it was often no match for the power of development personnel. A popular account of Point Four, also published in 1954, acknowledged that the "rakes, forks and scythes [introduced] into Afghanistan, all of which the local blacksmith can make," had met with an "enthusiastic" response from Afghan farmers, and that these simple tools, together with the planting of cotton in rows rather than broadcast "meant greater production." At the same time, the author insisted that eventually tractors would have to be brought in, because as in the United States, farming would need to be mechanized so that the number of people engaged in agriculture could be drastically reduced.[41] The program was designed to force development through American business investment and the sale of American technology, so the choices of hoes or tractors were more likely to be made on the basis of what could be profitably sold, rather than what could be locally sustained.

Because local elites in the target countries often found their own advantage in the large-scale technologies that American public and private agencies offered, the development programs, justified in the United States by modernization theory, often found powerful local support as well. Large exports of raw materials, the mobilization of cheap labor, and the distribution of imported manufactured goods all worked best when facilitated by ruling cliques and politically connected local business people, who profited mightily from the new economy grafted onto the old. However, while the corporations that exported and fostered American technologies and their local collaborators enjoyed substantial gains from the aid programs, for the great majority of people in developing nations, the promise of an improved and modernized society remained merely a promise.[42] Attempts to think through what truly "appropriate" technologies would look like in a technical aid context became a major source of critique of American technologies during the 1970s.

Access to Raw Materials

One of the major goals of U.S. foreign aid, and economic development policies in general, was to give American industry access to natural resources worldwide, and American capital investment opportunities in them. In January 1951, Truman appointed a high-level President's Materials Policy Commission, which was instructed "to study the materials problem of the United States and its relation to the free and friendly nations of the world." He chose William S. Paley to chair the commission. Paley was the powerful chair of the board of the Columbia Broadcasting System (CBS), which he had run since 1928, and a founding director of Resources for the Future. In setting up the commission, Truman again emphasized the twin goals of so much of Cold War American policy: economic growth and prosperity at home, and fighting communism abroad. "By wise planning and determined action," Truman reminded the committee, "we can meet our essential needs for military security, civilian welfare, and the continued economic growth of the United States. We cannot allow shortages of materials to jeopardize our national security nor to become a bottleneck to out economic expansion."[43]

The problem was, as Alan M. Bateman puts it, "that of 32 chief critical minerals ... the United States is deficient in 18 and has a surplus of 9." This obviously meant that "the United States must look more and more to increasing quantities of minerals from foreign sources." Specifically, "the chief foreign sources that might meet the future needs of the United States are chiefly Latin America and the dominions and colonies of the western European nations."[44] One legislative response was the Stockpiling Act of 1946, which was based on the notion that the nation had to set aside large quantities of strategic materials in case supplies were to be cut off in times of hostilities or other crises. The act ostensibly addressed military concerns, but by 1952, the military wanted to get rid of the program because, perhaps not surprisingly, it was increasingly becoming a political football as commodity producers, processors, and traders attempted to corner markets, stabilize (or destabilize) prices, and generally place profit and market advantage ahead of national defense.[45] However, whether or not materials should be stockpiled, they certainly needed to be available, and that availability remained a factor in American aid abroad.

The Paley Commission acknowledged that "the areas to which the United States must principally look for expansion of its minerals imports are Canada, Latin America and Africa, the Near East, and South and Southeast Asia." Canada was a close ally, and Latin America had long been part of America's informal political and economic empire. But Africa and Asia were emerging from imperial rule, and were seen as places where American and Soviet patterns of development would contest for hearts and minds. The commission acknowledged that "the current United States programs of technical aid give assistance to some resource countries ... through geological and exploratory work by United States experts, through help in organizing local minerals and geologic bureaus, and through training local experts." The commission, however, recommended additional funding for these activities, and for giving "advice on mining technology." Additionally, in the commission's opinion, "wherever technical assistance is extended in these fields, the United States should seek assurances that the recipient country will promote conditions favorable to developing such resources as may be discovered."[46] Such conditions, of course, would have to include a secure environment for American business investment.

Export of Arms

The term "foreign aid" was perhaps most often thought of as offerings of food, civilian technology, and expertise, but it always contained a very large component of military technology. In 1949, the Mutual Defense Assistance Act provided $1.3 billion in military aid to eleven Atlantic Pact nations. In 1951, out of a foreign aid expenditure of $7.3 billion, $4.8 billion was for military aid. In 1952, out of a total bill of $6.4 billion, $3.1 billion was for military aid and another $2.1 billion for military bases at home and abroad, while only $1.2 billion was earmarked for "economic" aid.[47] In 1973, a report issued by the U.S. Arms Control and Disarmament Agency revealed that from 1961 to 1971, the arms trade worldwide amounted to $48.4 billion, of which the United States contributed $22.7 billion, and the Soviet Union $14.7 billion. India, Pakistan, Indonesia, Iran, Iraq, Lebanon, Cambodia, Finland, and Yugoslavia obtained munitions from both the United States and the Soviet Union. More closely held client states got special attention. South Vietnam got $5.2 billion worth of arms from the United States, triple the amount sent by the Soviet Union to North Vietnam. At the same time, Russia sent more arms to Cuba than the United States provided to all of the rest of Latin America.[48]

Arms sales by the U.S. government continued to rise. In 1971, total sales were only $1 billion, but by 1975, the figure stood at $11.6 billion, of which $9.1 billion was in government-to-government sales, and $2.5 billion, while still needing government approval, passed through commercial channels. Sometimes the pathway was less than direct. In 1967, a staff report from the Senate Committee on Foreign Relations noted that F-86 Sabrejets delivered to Venezuela had been purchased from West Germany, but had been manufactured in Italy under license from an American aerospace firm.[49] The rationale worked out in the 1960s was that such sales gave the United States influence over customer nations and put a damper on arms races and actual shooting wars. Many questioned whether such influence was more than theoretical, but there was no question that the sale of military technology added dollars to the American economy and improved the country's balance of payments.[50]

Grooming the Globe

The specific instances of exporting American technology during these postwar years each had its own cast of players and a different mix of motives. Not only was there no overarching plan or justification, but the contradictions inherent in the mix also made a rational approach probably impossible—or perhaps not even desirable. At the same time, the heady cocktail of technological vanguardism, capitalist expansion, messianic hubris, and generalized altruism created a regime that the historian Michael Smith, as mentioned above, has called planetary engineering. "By a kind of geophysical imperialism," Smith notes, "proponents of Big Science envisioned a planet as subservient to its masters' wishes as any conquered military opponent."[51]

When Winston Churchill, at the end of the war, noted that "America stands at this moment at the summit of the world," it seemed very much that Henry Luce's call for an American Century had been answered.[52] Going through the pages of the magazine *Popular Mechanics,* Smith found this dream of unlimited possibilities often described, as befit a newly emergent imperial power, in terms of colonization. Nuclear power would enable America to establish engineering footholds in all of the unlikely and inhospitable areas of the world: on the ocean's floor, at the poles, in deserts and jungles, deep underground and, of course, on the moon. The desire to explore was in part a search for natural resources, in part to break the bonds that tethered humankind to the thin membrane of air at the earth's surface, but also, one gets the impression, in part a grand and exhilarating adventure.

A half-century later, the sublime ignorance of inevitable environmental consequences is breathtaking. Project Plowshare was a government program in what the nuclear enthusiast Edward Teller called "geographic engineering." According to the program, "controlled" nuclear detonations would excavate harbors, dig canals, release deposits of natural gas deep within the earth, and blast highways through mountain ranges.[53] In the feverish rhetoric of the time, the conquest of Antarctica would "pull back the veil from a continent that was all mystery just a few years ago." A 1400-mile highway would "pierce the Amazon heartland," an area referred to as a "green hell."[54]

And planetary engineering was no mere dream of popular writers. The AEC issued a prize-winning film entitled *No Greater Challenge* in the 1960s. Maintaining that history was a record of challenges met and overcome, the film pointed out that much of the world's population lived in poverty and on the edge of starvation. At the same time, though some error in divine planning, vast areas of desert lay bordered by the world's oceans: the land there was fertile but dry, and the water was marred by salinity. To put this to rights, the AEC advocated building what it called "agro-industrial complexes" centered around nuclear power plants designed to desalinate the seawater for purposes of irrigation. Echoing the masculine rhetoric of *Popular Mechanics*, the land was said to be fertile but barren, almost waiting, even yearning, for man to water and plow it with his powerful tools and thus fulfill its destiny by making it fertile at last.

No real-life example of the kind of thinking engendered by planetary engineering better serves our purpose than the so-called Lower Mekong Basin Project, publicly proposed by President Lyndon B. Johnson on April 7, 1965. Presenting a study of this project, a congressional report emphasized that "the term [technology] means more than tools, manufacturing processes, and advanced engineering. It signifies the systematic, purposeful application of knowledge to modify an environment toward predetermined goals."[55] As part of a peace initiative aimed to quell the growing conflict in Vietnam, the United States would contribute $1 billion to an "ambitious, capital-intensive civil works program" that would develop not just Vietnam, but all four nations of Southeast Asia. According to the report, "the technological and engineering base of such a regional development scheme is . . . broad. It encompasses hydraulics, electric power, flood control, electronic communications, computer modeling, electrical industries, large demonstration farms, highway and bridge construction, fish and agricultural food processing, and many more fields of technological applications."[56]

Only days before making this appeal for "butter" in Southeast Asia, the president had already decided on guns as well. A part of the Johns Hopkins speech announced the continuation of the bombing of North Vietnam, but it was revealed only later that he had also just decided to commit American ground troops in South Vietnam. According to the report, the Mekong initiative "sought to present a constructive alternative to conflict. It sought to reassure the people of the United States that its leadership was seeking the peaceful alternative and stood ready to negotiate to this end." Furthermore, the very scale of the project appears to have been, in part, an effort to "diminish the resources and energy diverted to conflict in the region."[57]

Although a Mekong development scheme had been underway since the 1950s under American direction, Johnson's massive intervention was clearly inspired by his old New Deal support of the TVA. He bragged that "the vast Mekong River can provide food and water and power on a scale to dwarf even our own TVA," and as if to underline this parallel, at the end of 1966, he sent David E. Lillienthal, former general manager and chairman of the TVA and later a member of the AEC, to head "a regional planning mission" to Vietnam. Not surprisingly, his report was, as the RAND Corporation termed it, a recommendation for "a large-scale, capital-intensive water project." Even the romantic rhetoric of the 1930s engineering project was echoed: it was, in the words of Luce's *Life* magazine, "A Project to Harness the 'Sleeping Giant.'"[58]

The funding and contracts for planetary engineering projects came from a variety of sources. The American government and foreign countries contracted for them, often using aid funds or monies borrowed from the International Monetary Fund or World Bank—both of which grew out of the Breton Woods Conference of 1944 and began operations in 1946. The work itself, however, was often undertaken by large American engineering firms. Scholars have not spent much time untangling the histories of the firms, but what little we know suggests that they played a significant role in exporting American technology during the last half of the twentieth century. One such firm was Morrison Knudsen Corporation, which had been founded in 1912. By 2002, it was a part of the Washington Group International, which employed 35,000 people and had projects in forty-three states and thirty foreign countries.[59]

Brown and Root was a Texas company that had invested heavily in the political career of Lyndon B. Johnson long before he became president of the country. It had a rather successful run of government contracts, from the NSF's Project Mohole to drill through the earth's mantle, to building facilities in Vietnam to support American military activities during the war. Project Mohole, importantly, gave them valuable experience in deep-sea underwater drilling, which was a significant factor in their later prominence in offshore oil drilling and platform construction. By the end of the century, Brown and Root had been bought out by Halliburton Company, then headed by Richard B. Cheney, who would go on to become vice president under George W. Bush. By this time, Halliburton had constructed military bases in the Balkans and was building railways in both Britain and Australia.[60] Later, Halliburton was to become a prominent and controversial sole-bid contractor during the Iraq war and in the Bush administrations attempts to rebuild in Louisiana after hurricane Katrina.

An equally successful engineering firm operating on the global stage was Bechtel. An unofficial history of the firm is titled *Friends in High Places,* and such indeed seems to have been the case. President Johnson had favored Brown and Root, which constructed not only much of the infrastructure for the war in Vietnam, but other national projects such as the Space Center in Houston. When Nixon entered office, however, Bechtel, a California firm, was positioned to do better.[61] Between 1968 and 1974, Bechtel's gross revenues rose from $750,000 to $2 billion, partly from work on such high-profile projects as the Alaska pipeline and the Washington Metro subway system. Stephen (Steve) Bechtel, the patriarch of the company, received the valuable benefit of being appointed to the advisory committee of the U.S. Export-Import Bank, which financed a great deal of the nation's overseas activities. The elder Bechtel was a close associate of Henry Kearns, the president of the bank. It was not unusual for Bechtel first to advise the bank that a South American nation, for example, needed an oil pipeline, and then to advise the country to apply for a loan to build it, and then accept the contract for the work. One of Bechtel's successes was in convincing Kearns that the bank should aggressively loan money for nuclear power plants abroad, many of which Bechtel then built.[62] However, Bechtel's Washington power was best revealed under President Ronald W. Reagan. At one time, Bechtel's president, George Schultz, was secretary of state; Casper Weinberger, Bechtel's general counsel, was secretary of defense; and W. Kenneth Davis, who for twenty years had headed Bechtel's nuclear-development operations, was deputy secretary of energy, in charge of Reagan's program to greatly expand the industry. The planetary engineers were centrally placed to further their dreams of worldwide construction.

It was precisely this type of large-scale, capital-intensive engineering project that William Lederer and Eugene Burdick warned of in their popular 1958 novel, *The Ugly American*. In an unnamed Southeast Asian nation, American diplomatic and aid workers held themselves aloof from the common people and cooperated with the local elites to bring in American money for large engineering and military projects aimed at both "developing" and protecting the country. Set against them was the lone figure of Tom Knox, an expert in animal husbandry who roamed the jungle, learning the language, eating the local food, and finding simple and direct small solutions to the common technological problems of the local farmers. He built a simple pump for irrigation out of a broken bicycle. Homer Atkins, a self-described "heavy construction man" sent out from Washington, surprisingly supported small- over the large-scale technologies. To win the hearts and minds of the people and inoculate them against communist agents, he told the Vietnamese, "you don't need dams and roads.... Maybe later, but right now you need to concentrate on first things—largely things that your own people can manufacture and use."[63]

Technology that was immediately useful and adaptable for local people was designed to save souls one at a time. The overwhelming thrust of American technology policy abroad, however, was designed to convert the king. A host of small local projects no doubt made individual lives better by empowering people to choose and control their own technologies, but this program failed on a number of grounds important to powerful people. It did not offer opportunities for significant American capital investment, or export of American machines and tools. The empowerment of local peoples often worked against the political ambitions of local elites, and provided few opportunities for them to enrich themselves from aid funds. Local resources were used to develop local agriculture and often rural industry, rather than being exported to already industrialized nations. There was not much here to inspire American business or American politicians—nor, truth be told, the American voter. Such technologies lacked the glamour and potential profits of such vanguard postwar technologies as atomic bombs, nuclear power plants, and intercontinental ballistic missiles.

FOUR

THE ATOM AND THE ROCKET

After World War II, the war's dramatic new technologies were taken over by new or reorganized government agencies: the AEC, a reorganized Defense Department, and in 1958, NASA. During the Cold War, the research on nuclear technologies and the development of both nuclear and conventional arms and missiles were all understood in terms of a race with the Soviet Union for creation and deployment. Invention, rather than a negotiated compromise, became the preferred solution to the Soviet menace. Vast amounts of both money and scientific and engineering talent were diverted from civilian activities to the elaboration of often-baroque military technologies.

At the same time, the development of new technologies was said to be ushering in new ages, in relatively rapid succession. The Atomic Age was followed hard by the Space Age, which gave way in a few years to the Information Age. While these ages lasted, however, they created a fog of visionary rhetoric that tended to obscure the rather less noble goals of corporate profit, bureaucratic power, and national aggression. Those two great developments of World War II—the harnessing of atomic energy and the missile—became inexorably intertwined during the Cold War, and continued beyond the end of the century to fascinate and frighten policy makers who saw the world, first of the Soviet empire and then of nations emerging on the eastern periphery, as a dangerous and threatening place.

The Atomic Bomb

In 1956, Walt Disney enthusiastically proclaimed that "the atom is our future." "Atomic science," he continued, "began as a positive, creative thought. It has created modern science with its many benefits for mankind." More specifically, "the useful atom . . . will drive the machines of our atomic age."[1] Ironically, while such technologies as solar heating or electric automobiles were ridiculed as being far in the future even though they had many decades of use already behind them, the futuristic nature of nuclear technologies was celebrated as a hallmark of their desirability and even inevitability. Decades of science fiction had already familiarized readers with the endless and amazing possibilities of nuclear power when, on August 2, 1939, just days before the outbreak of World War II and six years before that war ended with the dramatic use of atomic bombs, the émigré physicist Albert Einstein wrote a letter to President Roosevelt warning that Nazi Germany might well be attempting to build nuclear weapons.

Roosevelt decided that the United States should explore the matter, and a uranium committee was set up to look into the possibilities. In 1941, responsibility was transferred to Vannevar Bush's OSRD, and the next year, the U.S. Army Corps of Engineers, arguably the government's premier manager of large-scale technology projects, was assigned the task of building an atomic bomb. The bomb had already been imagined, but inventing it was a complex group activity, involving massive infusions of money and trained personnel from dozens of different fields. Essentially, the cultures of academic science and industrial engineering not only had to be melded somehow, but both had to accommodate themselves to a military culture of discipline and chain of command.[2]

The common phrase for such activity is "research and development," but the exact opposite was closer to what went on in the famous Manhattan District facilities. Uranium was to be enriched at laboratories in Oak Ridge, Tennessee; plutonium was manufactured in Hanford, Washington; and the bomb itself was designed in Los Alamos, New Mexico. At all three facilities, scientists and engineers toiled to develop the new technology, and when new problems arose, as they always did, the scientists did whatever research they thought might solve the issue before them. Out of this stew of clashing personalities, deep uncertainty as to how to proceed, the thrill of invention, and strict secrecy came a successful test blast at Alamogordo, New Mexico, on July 16, 1945. Word of the success was sent to President Truman, who was attending the Potsdam Conference of Allied leaders: the code for a successful test was that "it's a boy." (Had it failed, word would have been sent that "it's a girl.") On August 6, Hiroshima was destroyed with a uranium bomb, and three days later, Nagasaki was destroyed with a plutonium bomb. Over 200,000 people were instantly killed in the two cities.

The Manhattan Project had fulfilled its mission and the OSRD was going out of business, but atomic power was not going away. The problem of managing atomic technology in the postwar years was met by setting up an independent governmental agency, the AEC, in 1946. It was deliberately and rather ostentatiously set up as a civilian agency, but its portfolio was almost completely dominated by military needs and wishes. It got to work immediately producing more atomic bombs: at the end of World War II, there was one left unused. By 1949, there were several dozen, and by the late 1980s, the nation's nuclear arsenal included some 80,000 nuclear warheads, some of them hundreds of times more powerful than those used on Japan. The doctrine of "massive retaliation" adopted by the Eisenhower administration, led to the concept of mutually assured destruction (MAD), and the building boom that the AEC oversaw as a part of the race to keep ahead of the Russians produced what was widely seen as nuclear overkill. There were, in fact, more weapons than targets.

Part of keeping ahead, especially after the Soviets tested their own atomic bomb in 1949, was to rush the development of a hydrogen bomb. Such a weapon would be three to ten times more powerful per pound than the existing fission bombs were, the materials would be much cheaper, and there would be no practical limit to the weapon's size or power. President Truman gave the go-ahead to develop the hydrogen bomb in January 1950, a device was tested on November 1, 1952, and the first real H-bomb tested was set off in the spring of 1954. By that time, the Soviets had already exploded a bomb of their own (in August 1953). The American nuclear arsenal added thousands of H-bombs by 1960.

Nuclear Power

President Eisenhower spelled out a new initiative in December 1953 when, in a speech before the United Nations, he presented his "Atoms for Peace" program. One result of the longing to find peaceful uses for atomic power, the program offered to share nuclear know-how and materials with other nations to use for medical and research purposes. The attempt to internationalize the atom was motivated by the desire to use the atom as one more thread to bind the "free world" together and to the United States, and the desire to make nuclear technology a measure of national power and international generosity.

Domestically, using the atom to generate electricity was another new direction for atomic technology. This too was to some extent motivated by the urgent desire, especially on the part of some of the prominent scientists involved in the Manhattan Project, to find positive uses for their work, which would hopefully outweigh the more obvious threat to human existence. Ironically, however, the first push to produce useable nuclear power came from the Navy, which wanted a submarine fleet to leapfrog over Soviet capabilities. Nuclear-powered submarines could operate underwater for very long distances, free of the frequent need to surface for air and to refuel. Besides, having watched the

Army grow in stature and influence through its involvement with the "nuclear future," the Navy wanted its own piece of the action.

In 1948, the Navy handed its program to Admiral Hyman Rickover. Like the Manhattan District's Leslie Groves, Rickover was a military officer driven to complete tasks that many of his colleagues thought were impossible. Rickover spent some time at Oak Ridge, picking up some knowledge of nuclear science and engineering. Four years later, he tested his first prototype reactor, and in 1954, he saw the launching of the Navy's first nuclear submarine, the *Nautilus*. It was a success of personal vision, excellent engineering, and rigid discipline. Nuclear propulsion was soon extended to aircraft carriers as well.[3]

Meanwhile, the AEC looked into nuclear-powered railroad trains, but gave up the idea because of the inherent radiation dangers. The same was true of the nuclear-powered airplane project, although it persisted throughout the 1950s. The amount of material needed to shield the crew was deemed too heavy, and radiation leakage into the exhaust was worrisome. The spilling of nuclear fuel in case of a crash was an additional problem, and at long last, hopes were dashed and careers cut off when the whole concept was abandoned.

The enthusiasm for using the intense heat of nuclear reactors to turn water into steam to drive turbines, which could then turn generators to create commercial amounts of electricity, was rampant in the 1950s. In 1954, Admiral Lewis Strauss, then chair of the AEC, rashly predicted that nuclear generation would make electricity "too cheap to meter," a bit of nuclear hyperbole that would haunt the power program into the present. The effort was a priority in the AEC, and in 1957, the first working commercial reactor was fired up outside of Pittsburgh, Pennsylvania—the very heart of coal country. Inaugurated by President Eisenhower, the reactor was a modified version of the one Rickover designed for submarines, and the contractor was Westinghouse, a major player in the naval program.

Two social problems had to be solved before public utilities would line up to build nuclear power plants. First, it was necessary to account for the masses of radioactive waste, both spent fuel and the heavily contaminated plant itself, which had to be dealt with somehow once its operating life was over. The AEC had no clear idea of what the costs would be or how the technical problems might be overcome, but it agreed to accept responsibility for the whole problem, thus relieving private corporations of unknown but obviously huge costs in the future. The other problem was liability.

Not surprisingly, given the great uncertainties and potential catastrophic risks of nuclear power plants, the insurance industry was reluctant to underwrite this new technology. An AEC study completed in 1957 estimated that a serious power plant accident near Detroit would cause $7 billion in property damage, kill thousands of people, and injure tens of thousands more. The potential for massive liability claims frightened off the insurance industry, but not Congress. That same year, it passed the Price-Anderson Act, which removed the hand of the market by legally limiting the liability of any one nuclear accident to $560 million.⁴

Some idea of the blinding enthusiasm for building nuclear power plants can be gleaned from the story of Consolidated Edison Company's proposal at the end of 1962 to build one in the heart of New York City.⁵ Everyone agreed that economies were realized when power was generated as close as possible to consumers, but the ConEd proposal tested the limits of AEC guidelines. By 1957, it was agreed that all reactors near population centers must be covered by a containment building, but the AEC had no useable formula for deciding the relationship between distance and adequate "engineered safety features." Over the years, a developing industry and changing technologies combined to keep the problem alive and unsettled. ConEd's power plant, along with Westinghouse's reactor, was to be located just a mile and a half from the United Nations complex, making the engineered safety features the heart of the approval question.

In the context of a growing national debate about the nuclear arms race and the public health effects of fallout from atmospheric testing, citizens in New York City began to question the cheerful safety assumptions of ConEd and the AEC. Without the safety factor of distance to rely upon, the utility's engineers proposed a containment structure of reinforced concrete 150 feet in diameter and seven feet thick. Barriers of carbon steel plate a quarter of an inch thick were added two feet apart, with the space filled with "low-density, pervious concrete." The pioneer nuclear scientist Glenn Seaborg, while not referring specifically to the Queens site, declared that he would not "fear having my family residence within the vicinity of a modern power plant built and operated under our regulations and controls."⁶

The total reliance on unprecedented engineering features to provide absolute guarantees of safety proved too much for the project. The AEC's regulatory staff expressed its strong reservations directly to ConEd, and the utility withdrew its application in 1964. It did so, however, without conceding the dangers of its plan, citing instead the recent signing of a contract for additional hydroelectric power from Canada. The case left the question of whether nuclear power plants could be placed in metropolitan areas unresolved, but it was important in mobilizing the first significant citizen opposition to the building of such facilities.

The ConEd plant in Queens was never built, but by the end of the century, 104 others were in operation, supplying about one-fifth of the nation's electricity. It was unclear, however, when or even if any more would ever be built. Citizen distrust of official safety assurances was a large part of the problem, but so were the delays and overruns in costs that made a mockery of construction and operating-cost estimates. Some argued that the cost overruns themselves were the result of needless safety and environmental restrictions, which in turn were a manifestation of public opposition.

Meanwhile, every few years, one or another plant came very close to suffering a catastrophic disaster. In 1966, the Enrico Fermi Fast Breeder Reactor in Detroit had a partial fuel meltdown. In 1975, the TVA's Brown's Ferry plant suffered a 17-hour fire in which one reactor had all its redundant safety systems impaired. Then, on March 28, 1979, Unit 2 of the Three Mile Island plant near Harrisburg, Pennsylvania suffered a partial core meltdown, and some radioactivity escaped the facility. Neighbors were evacuated, and the nation watched in suspense to see if the long-predicted disaster was finally unfolding. As it turned out, the situation was stabilized and proponents redoubled their efforts to sell atomic power as both inevitable and safe, this time through a newly set up U.S. Committee on Energy Awareness, financed by the nuclear industry with an advertising budget of $25 million to $30 million.

In the Soviet Union, near Kiev, the long-imagined accident really did occur, on April 28, 1986. Radioactivity from the Chernobyl power plant traveled over a thousand miles; 135,000 people were evacuated, and much of eastern and middle Europe received some contamination. American defenders of the technology denied that American plants could ever suffer the same fate because American plants were designed differently, but the worries continued. The Soviets had placed a statue of Prometheus outside the Chernobyl plant, reminding us of the words of the chorus in response to Prometheus's boast that he had given fire to the human race. "What?" the chorus cries, "Men, whose life is but a day, possess already the hot radiance of fire?" "They do;" Prometheus replies, "and with it they shall master many crafts."[7] The presence of this statue took on sardonic and tragic meanings. The Chernobyl plant will need to be kept sealed off for many tens of thousands of years before people can again go near it. Human beings, "whose life is but a day," can hardly grasp, much less plan for, this new contingency.

Other miracles of technological efficiency were imagined for peaceful uses of atomic power. It is said that with new technologies, one cannot know what is impossible until one has established what is possible. Project Plowshare was a case in point. Begun by the AEC in 1957, the project was based on the belief that nuclear blasts, beneath the surface or above ground, could be used to move large amounts of dirt safely and efficiently in civil engineering jobs. Harbors and canals could by excavated, a nuclear-blasted cut through the mountains east of Los Angeles was proposed to make way for a new freeway, and natural gas and oil could be "mined" from deep underground. Enthusiasm for Project Plowshare was centered at the Lawrence Livermore National Laboratory, and the lab's most prominent physicist, Edward Teller—the putative father of the H-bomb—was one of the project's most vocal backers. Teller suggested that a river that "flows uselessly to the sea" could be diverted by explosions to irrigate "desert wastes." The astonishing lack of concern for any environmental effects to the land was matched by a sublime disregard of the dangers of radioactive fallout. Teller called all of it "geographic engineering"; his university colleague, Glenn Seaborg, also a champion of Project Plowshare, called it "planetary engineering."[8] Only a few schemes reached the testing stage, and none were actually undertaken.

The Popular Atom

Technology in the United States has long been thought of as democratic, not only because it supposedly benefits the common people, but also because it is thought to be largely accessible to everyone, through purchase and use but also intellectually. For most of the nation's history, people could understand where technology came from, how it worked, and what it could be used for. This was no longer true of many important tools and processes by the end of the nineteenth century, but the post–World War II nuclear technology seemed particularly magical in the sense that though it appeared to work, most people could not explain how, and certainly most people would never own their own reactor or atomic bomb.

The struggle to impose what Michael Smith has called a kind of "male domesticity" on the atom, undertaken by such magazines as *Popular Mechanics,* proved to be a real challenge. The single activity in the long and secret production process of nuclear technology that was open to the average citizen was the discovery of uranium deposits and their mining for ore. *Popular Mechanics* wedded the familiar do-it-yourself urge with the frontier myth of striking it rich. The magazine urged readers to build their own Geiger counters; "it sometimes takes no more than enterprise, luck and a little do-it-yourself talent to make a full-fledged uranium king," it proclaimed. Adventure, recreation, and patriotism combined to draw Americans out to the hills to prospect for uranium.[9]

The government was the only legal market for uranium, and in March 1951, it set off the boom by doubling the price it was offering and adding a $10,000 bonus for each newly opened mine. It also put out guidebooks, built access roads, constructed mills that would purchase and process the ore, and made geological reports readily available. An estimated 2,000 miners descended upon the Colorado Plateau, some using airplanes fitted out with a "scintillation counter," but most had little more than a Geiger counter and a pickax. They staked out claims under the 1872 Mining Act, one of which, the Mi Vida mine in Utah, yielded a hundred million dollars' worth of ore. In the best tradition of the Old West, the successful prospector of the Mi Vida mine, Charlie Sheen, built a sumptuous mansion and bought out a bank that had once refused to lend him $250. He also built a mill for processing uranium ore, and employed 300 workers and 60 truckers to haul the material.[10]

The boom, of course, had its downside. Utah's loose regulation of stock schemes meant that large amounts of worthless shares of nearly nonexistent companies circulated widely. Much worse were the dangers to the health of the 6,000 or so uranium miners, most of whom were local Mormon workers or Navaho Indians. The AEC was aware of the dangers involved, but insisted that mine safety was a state problem and warned that any adverse publicity might curtail uranium production and thus weaken America's defenses. By 1966, an estimated 97 miners had died, and it was not until 1990 that President George H.W. Bush signed the Radiation Exposure Compensation Act. The uranium-mining industry died out in the 1980s, killed off by overproduction, the winding down of the Cold War, the moratorium on building nuclear power plants, and the discovery of large reserves of ore in Canada, Russia, and the former Belgium Congo. The government sold its United States Uranium Enrichment Corporation in 1998.[11]

Aside from prospecting for uranium, average Americans could participate in the nuclear age by taking part in civil defense activities. Building a bomb shelter in one's suburban backyard was one such project. In large cities, shelters tended to be public—the basements of tall buildings or the subway stations, similar to those made familiar by dozens of wartime films of the London blitz. In the suburbs, people were once again often required to substitute private for public infrastructure.

Irving Janis, a Yale University psychologist under contract with the RAND Corporation, prepared a study in 1949 titled "Psychological Aspects of Vulnerability to Atomic Bomb Attacks." While doubting that any ordinary construction materials were likely to deflect radiation, he found some advantages in home shelters. He thought that the "constructive activity" of building them would "contribute to the feeling that 'I am really able to do something about it,'" thus working against the "strong feelings of insecurity about impeding atomic bomb attacks." Even the private nature of the structures would be a positive force, because it would lessen the citizen's reliance on the government and thus undercut any tendency to blame the government for any perceived vulnerability. Finally, being a sterling example of Michael Smith's "male domesticity," the shelters would be invested with "considerable symbolic value as an anxiety reducing feature of the environment."[12] *Popular Mechanics* suggested a range of shelters, from a $30 project involving 100 sandbags to an elaborate $1,800 shelter dug under the patio.[13] How much food and water to stock, and whether or not one would be ethically obliged to share shelters with feckless neighbors who had neglected to build their own, were popular topics of discussion.

Civil defense tactics were perhaps most often aimed at children. The infamous "duck and cover" exercises taught to children in school were only part of a much larger effort to prepare for children's safety. Because conservatives attacked public education in the early 1950s, charging it with everything from watered-down curricula to leftist propaganda, educational professionals were eager to find ways to tie their enterprises more closely to national political agendas. Seeking more federal funding for education in an era when national defense was a touchstone to the federal budget, educators sought to emphasize the critical role of public schools in civil defense.[14]

In the main, this turned out to be less a matter of new curricula than it was of extracurricular activities and the architectural design of school buildings. While children were given virtually no facts about the bomb and its likely effects (in an effort not to scare them or their parents), such "target" cities as New York, Los Angeles, Chicago, and Philadelphia instituted atomic air-raid drills in late 1950 and early 1951. By early 1952, New York City had passed out metal identification tags—similar to military dog tags—to all children from kindergarten to the fourth grade. An article in the *Journal of the National Education Association* warned that tattooing children was of questionable use because of the possibility of the skin being burnt off. Especially after President John F. Kennedy declared mass fallout shelters a priority in 1962, new school buildings were designed for "double duty" as educational centers and as places to care for "the well, injured, contaminated and dead" in case of attack. Such design features as few or no exterior windows better equipped the schools to double as bomb shelters. As one historian has written, "the congeries of good and professional motives that affected school civil defense programs taught a generation to equate emotional maturity with an attitude of calm acceptance toward nuclear war."[15]

The federal government's role in developing nuclear technologies extended beyond merely putting up the money, passing enabling legislation, and writing supportive regulations. According to historian Michael Smith, "nuclear power is above all a product of federal promotional objectives in the 1950s and 1960s." In part, not surprisingly, the promotional campaign tried to counter the very sensible public realization that nuclear technology held unprecedented potential dangers. In part, however, the campaign sought to create a need that it wanted the nuclear industry to fill. "The postwar nuclear policy," according to Smith, "retained the Manhattan Project's penchant for secrecy, its equation of nuclear objectives with national identity, its sense of urgency, and its faith in the ability of government scientists to find last-minute solutions to any technical problems."[16]

With the reality of nuclear overkill available to both the United States and the Soviet Union, in Smith's words, "both sides increasingly sought symbolic deployments, employing nuclear technology as an emblem of military and ideological superiority." Thus, most advocates of nuclear power "viewed nuclear-generated electricity as a propaganda weapon in the ideological struggle between the United States and the Soviet Union."[17] A revised Atomic Energy Act was passed in 1954 that encouraged rather than forbade the civilian spread of nuclear technology and materials. As only public skepticism about all things nuclear stood in the way of the rapid expansion of power production, the government undertook an advertising campaign to sell the peaceful atom.

Part of the campaign was carried on by the Atomic Industrial Forum (AIF), established after a suggestion by the AEC administrator. Industry and government agreed that it would be easier to convince the public to accept accidents rather than to avoid them completely: a key was to encourage the public's faith in the ability of "experts" to solve problems. Using a combination of secrecy and advertising, the nuclear lobby moved forward with such products as the film *The Atom and Eve*, celebrating electrical appliances produced in 1965 by New England utilities, and Walt Disney's television production and children's book, *Our Friend the Atom* (1956). Glenn Seaborg declared that "in a homely sense, the atom is like one of those old many-sided jackknives that can do almost anything a kid would want to do—whittling, screwdriving, bottle-opening and so on."[18]

Seaborg's invoking of a "homely sense"—not just of jackknives, but "old" ones at that—and tying it to things "kids would want to do" is a prime example of the "male domesticity" that Smith identified as a central trope of pro-nuclear propaganda. The boy with his universal tool joined "our friend the atom" and the dancing housewife of *The Atom and Eve* in the Atomic Age world, in which Eisenhower's peaceful atom enhanced rather than threatened the American way of life. "For most Americans," Smith wrote, "this translation of 'unlimited power' from the international arena to the household removed the atom from a social environment in which they felt helpless, to the familiar trappings of a consumer culture where their choice of products conveyed the illusion of control."[19]

A strange story unfolded in suburban Detroit in 1995, in which an adolescent boy named David Hahn was inspired by campaigns to sell science in general, and atomic science in particular, to a public of all ages. His first text was the *Golden Book of Chemistry Experiments,* published in 1960, but soon he gained the notion of making a collection of samples of each of the elements in the periodic table. Many of them were easily gotten, but the radioactive elements created problems. By assiduous reading and correspondence with industry and government officials, however, he learned that he could extract thorium-232 from the mantles of commercial gas lanterns. Thorium was only two steps below uranium in the table, and had a radioactive half-life of 14 billion years. He extracted polonium from electrostatic brushes he bought through the mail.

At his father's urging, he decided to become an Eagle Scout, and undertook to earn a merit badge in atomic energy. The appropriate Scout pamphlet explained that "getting useful energy from atoms took several ... years. People could not build reactors in their homes." He accepted this as a challenge, and began to build what he hoped would be a breeder reactor in his parents' backyard potting shed. Word got out, and on June 26, 1995, personnel from the Environmental Protection Agency arrived in protective clothing and spent the next three days tearing down the shed and digging up the surrounding soil, packing it in large containers, and hauling it away. The budding scientist joined the Navy and was assigned to duty on a nuclear-powered aircraft carrier. There were limits to how successfully the nuclear age could be domesticated, but not to how it could be deployed for national aggrandizement.[20]

And the Rockets' Red Glare

Missile technology was another, and closely related, postwar arena in which the United States chose to challenge the Soviet Union, not only to achieve military superiority, but also, in the very grandest terms, to demonstrate to the world that the U.S. commitment to technological vanguardism made the United States a superior civilization. The Space Age, with its dramatic space race as a set piece, vied with the Atomic Age as a marker of the claim that the future had been given its thrilling contours by American know-how and character.

The putative father of American rocketry was Robert Goddard, one of those inventors who was famous for being neglected. A physicist, Goddard "designed, patented and tested almost every feature of the modern rocket," but also represented the kind of nineteenth-century individualism that refused to become embedded in the sort of bureaucratic team with the management skills and military financing that quickly became the norm in the field of missile technology.[21]

Goddard began thinking about rockets as a means of interplanetary space travel in 1909, took out patents for a "rocket apparatus" in 1914, and in 1917, got a small grant from the Smithsonian Institution to continue his work. After working with the U.S. Army Signal Corps during World War I, he returned to his experiments, and with other small grants from Clark University—where he was on the faculty—the Carnegie Institution of Washington, and the aviation philanthropist Daniel Guggenheim, he and his assistants attempted 38 launches between 1930 and 1941. Thirty-one of these had at least some measure of success, and he built ever-larger rockets, up to twenty-two feet in length and eighteen inches in diameter. He received no further subsidies, however, and died in August 1945, just as the atomic bombs were marking the end of World War II.[22] Despite Goddard's early work, the American rocket program actually grew out of the German advances of World War II.

Wernher von Braun, the head of the German missile program, was a member of the Nazi party and an SS officer, as well as a passionate advocate of rocket development. Research and development was carried on at Peenemunde on the Baltic coast, but rockets were assembled by slave labor from the Dora concentration camp. More than 20,000 workers died building the rockets—many more than perished from their use. First flown in October 1942, the German V-2 missile was forty-six feet long, carried a warhead of 2,200 pounds, and flew at 3,500 miles an hour. Before the end of the war, 1,155 V-2 missiles had been launched against Britain, and 1,675 more against Antwerp and other targets on the continent. Because they traveled so fast, they arrived without warning and could not be defended against. They were ideal weapons of terror against civilian targets.[23]

Not surprisingly, the United States was anxious to get the weapon itself, a desire that nicely paralleled von Braun's wish to escape prosecution as a war criminal and be paid to continue his rocket work. He calculated that the United States was likely to support him better than would the Russians, and surrendered to American forces, which also managed to take possession of a sizeable number of actual rockets. This coup was invoked years later after the successful Soviet orbiting of *Sputnik* by the somewhat cynical canard that the Russians must have gotten better Germans than did the Americans. Von Braun, who became a hero in America, was set up with facilities in Texas after the war, and between 1946 and 1951, sixty-seven of the confiscated German V-2s were test-fired at White Sands Proving Ground in New Mexico. In 1950, von Braun and his team were moved to a permanent facility in Huntsville, Alabama.

Five years later, President Eisenhower announced that the United States would participate in the International Geophysical Year (IGY) by launching a "small unmanned satellite" for scientific purposes. Von Braun and his team at Huntsville were passed over for the task, as their work for the Army was considered too important to interrupt with a merely scientific project. The task of launching *Vanguard* was given to the Navy, which, while giving it a low priority as befit an essentially civilian effort, was not averse to getting involved in missile development. *Vanguard* was ready for its first test in 1957. In December, it collapsed on the launch pad, destroying its four-pound payload.

By then, however, the Russians had astonished the world by orbiting a 186-pound scientific satellite called *Sputnik I* on October 4. Panic swept the government and the nation as a whole, both because the United States was not used to having its own technologies so publicly trumped, and because the success had sinister implications for the future Soviet capacity to deliver intercontinental ballistic missiles (ICBMs) with nuclear payloads. The government turned to the von Braun team, and on January 31, 1958, an Army missile orbited the country's first satellite; it weighed 14 kilos compared with the Soviets' 500. Some national pride could be salvaged, perhaps, by the fact that not long after Russia introduced *Sputnik,* the Ford Motor Company brought out its new Edsel.

The panic around *Sputnik* was focused on the supposedly second-rate nature of America's scientific capabilities and the slow development of missile research and development in particular. On the first score, Eisenhower established the famous President's Science Advisory Committee (PSAC) to advise him, staffing it with the best of the nation's scientific establishment. The National Defense Education Act (NDEA), passed in 1958, pumped money into scientific and engineering education in colleges and universities.

To bring a new sense of urgency to the problem of missile development, the old (1915) National Advisory Committee on Aeronautics (NACA) was revamped and given responsibilities for rocket research. Congress established the new NASA in July 1958, and its first administrator, T. Keith Glennan, began its operations on October 1, almost a year after the *Sputnik* shock. Glennan immediately authorized Project Mercury to orbit an astronaut around the Earth, and started working to capture von Braun and his team, which finally joined NASA in 1960.

A Race to Space

At first, NASA's work was somewhat dispiriting to those who were keen to catch up to the Russians. In 1959, seven out of seventeen launches failed, and the first satellite to work as planned was *Explorer 6*, not launched until August of that year. Astronaut Alan Shepard became the first American in space when he took a suborbital ride in a Mercury capsule on May 5, 1961, but again the Russians had gotten there quicker and better: Yuri Gagarin had been launched into Earth orbit on April 12. Just twenty days after Shepard's flight, President Kennedy announced that the nation would land an astronaut on the moon before the end of the decade.

The decision to go to the moon was complicated, more political and symbolic than strategic or scientific. During the 1960 election campaign, Kennedy had asserted that the incumbent Republicans had allowed a dangerous "missile gap" to develop, and although when he received secret briefings as president he discovered that this was not so, he was publicly committed to doing something dramatically different. The disastrous Bay of Pigs debacle of mid-April, in which an ill-conceived invasion of Cuba failed spectacularly, also led the new president to want to give the public something positive, even visionary, to think about. Finally, Kennedy believed that something had to be done to recapture America's image as the world's leading technological powerhouse. The Apollo program absorbed NASA's attention for eleven years and cost $25.4 billion.[24] The seven astronauts, already chosen in April 1959, became somewhat sanitized American heroes: boyishly clean-cut, ordinary, adventuresome, brave, and worthy role models for American youth. In April 1967, three of them were burned to death in a capsule sitting on a launching pad.

Finally, astronauts landed on the moon on July 20, 1969 using *Apollo 11*'s lunar module. The accomplishment was hailed abroad as well as at home as a transcendent moment in human history, but it was over almost before it began. President Nixon called the *Apollo 11* voyage the most significant week in the history of Earth since the creation. But although "President Kennedy's objective was duly accomplished," wrote the director of NASA's Ames Research Center in 1987, "the Apollo program had no logical legacy." It was a technological dead end. One reporter likened the whole race to the moon to a dog chasing a car. The reporter had once slowed down and stopped when a dog chased his car. The dog, somewhat uncertain what to do once it had caught the car, hesitated, marked it as dogs will, and then walked away.[25]

The Apollo program was essentially wrapped up by the end of 1972, and NASA moved on to its next big program, developing a space shuttle that had been sold to President Nixon earlier that year. It was conceived of as a comparatively inexpensive, reusable vehicle to travel back and forth from space in order to, for example, orbit and retrieve military reconnaissance satellites. The space shuttle *Columbia* made its first successful orbit in 1981, but the shuttle never met the cost and usability criteria initially set out for it. It was so complex that it took vast number of hours to keep it working safely and preparing it for the next flight. Like the Apollo program, the shuttle was a spectacular stunt that had little or no payoff. The loss of the *Challenger* on January 28, 1986, apparently because the weather was too cold for the O-rings used to seal joints, cast the whole program in doubt. The tragedy of the deaths of the regular astronauts was compounded by the death of a schoolteacher who had been placed on board strictly as a publicity stunt. The disaster led to a two-year halt in shuttle launches. On February 1, 2003, the *Columbia*, which had been the first to fly in 1981—making it twenty-two years old—broke up on reentry, killing all on board and stranding American astronauts on the space station.

The shuttle's major use came to be in line with its original purpose—supplying goods and personnel to the orbiting space station, which is still under construction. President Reagan sought to reinvigorate the space program, and gave the go-ahead to build what was conceived of as a giant orbiting laboratory and jumping-off place for lunar and planetary explorations. The space station *Freedom*, as it was called, was planned as an elaborate piece of machinery, and other nations were invited to take part in building and staffing it. Initially planned to cost $8 billion, within five years the projected cost had blown out to $24 billion. This coincided with an equally dramatic expansion of the national debt, and at that point, managers began to scale back on *Freedom*'s configuration, which in turn led to questions as to whether the whole thing was worthwhile.[26] It was hailed as inaugurating "a permanent human presence in space," and being a "base camp" "halfway to anywhere," but such romantic rhetorical flourishes hardly hid that the space station was another of those inventions of which necessity was not the mother.

However, NASA's ability to marshal such enormous resources largely on the basis of images of power and vanguardism was won, not given. As Michael Smith has noted, after World War II, there emerged "a dizzying bureaucratic network of government agencies, committees, and aerospace contractors, with NASA personnel serving as coordinators."[27] There were real vested interests at stake. Aerospace was perhaps the best-defined sector of the military-industrial complex, and as such, it linked corporate well-being with busy factories and well-paying jobs, and through them, to every congressional district in the country. All of the money that the NASA programs cost represented salaries and contracts with real people and firms. The agency spent $3.8 billion with businesses in 1967, and among their top ten contractors were five aerospace firms, along with IBM, General Motors, and General Electric.[28] Another industry, however, was also deeply implicated: the media in general, and advertising in particular. As Smith points out, the trick was to "portray American use of technology as benign, elegant, beyond the earthbound concerns of military and diplomacy. To succeed fully, the manned space program had to project an image directly contradicting its origins."[29]

Image making was, of course, a well-understood technology in itself. As Smith shows, the model that NASA adopted was that used to sell that other wildly successful American icon, the automobile. Smith invokes the term "commodity scientism" to suggest the media's power to instill in the public a "superstitious belief in the power of science and technology."[30] The growing gap between process and product, which had resulted from a half century of increasingly complex and opaque technologies, as well as the rise of the word "technology" itself as a free-floating signifier, meant among other things that large parts of the public could be literally "sold" a wide array of gadgets and machines based almost solely on the constructed meanings attached to them.

The *Apollo 11* press kit contained 250 pages of acronyms and charts to help reporters translate NASA's space jargon, but did not mention the mission's purpose or social significance.[31] Well-known techniques of automobile advertising—unveiling (the new model), transivity (the transfer of attributes from the object to the consumer), and helmsmanship (the mastery over the object and its environment)—combined to make the event itself more important than its origins or purpose. The public could not go along on the space voyages, but they could participate in the spectacle, thrill at the transfer of power and competence that went with the jargon, and share the technological mastery of the astronauts who were portrayed as just like the rest of us. Alan Shepard claimed that "a capsule is quite a bit like an automobile," and reportedly when he inspected the *Redstone* rocket that would carry him into space, he "sort of wanted to kick the tires."[32] The conflation of a space program with no real purpose and the complex, breakdown-prone design of the cars being produced by Detroit was not promising, but it was deeply appropriate.

While the public focused its attention on NASA's extravaganzas, the Pentagon was quietly going about its own space programs. Beginning immediately after World War II, each of the services, including the newly independent Air Force, began programs to guarantee their role in space warfare. In 1946, the Army Air Force contracted Consolidated Vultee Aircraft to develop an intercontinental missile that resulted in the workhorse Atlas ICBM of the 1950s. In November 1956, the Department of Defense stopped the Army from further work on ICBMs, giving land-based systems to the Air Force and water-based systems to the Navy. The latter then developed the Polaris, designed to be fired from submarines. The Air Force continued with the Atlas and quickly followed with the Titan and Thor intermediate-range missiles. The Thor was designed for deployment in Europe, and with an Agena second-stage rocket, launched the Air Force's first satellite into orbit on February 28, 1959. The Army, also arguing for an intermediate-range missile to be used in Europe, finally developed its Jupiter at Huntsville under von Braun's leadership.[33]

One famous missile, the Patriot, illustrates the ambiguity of declaring any one of them a "success." The Patriot was intended to bring down enemy aircraft, but when Iraq invaded Kuwait in August 1990, the missile was pressed into service, after hasty modifications, to intercept Iraqi Scud missiles. The trick was to "identify incoming missiles, track them, determine their trajectory in a few seconds, fire a rocket at an estimated point of arrival, guide the interceptor to the point allowing it to make last minute adjustments, and explode a warhead as it approaches the incoming missile, all at a closing speed of 8,000 miles per hour."[34]

By the beginning of 1991, the United States claimed 100 percent success for the Patriot, and President George H.W. Bush had stood in the plant that made them, backed by massed American flags, to celebrate their success. By April 1992, however, it was possible to claim only one clear success in Saudi Arabia, and possibly another in Israel. One source of confusion was that the early, optimistic estimates were based on "interceptions," which implied "kills" but really only meant that, according to one general, that they had "crossed paths" in the sky. The longing for scientific accuracy in counting the Patriot's successes was frustrated by conflicting definitions, faulty or inadequate information, and the political needs of the commentators to show that the program was worthwhile. Meanwhile, the Scuds were so inaccurate that they could not be counted on to fall within two or three kilometers of their intended targets. But one larger issue at stake in the disagreement over the Patriot program was whether or not it was evidence that a Star Wars missile-defense system would work if built.[35]

For rocket and missile programs, satellites were arguably as important payloads as nuclear warheads, especially because the latter could be used only under the most extreme circumstances, and the former came to be relied upon on a continual basis. Contracts for spy satellites had been let even before *Sputnik I*. The Central Intelligence Agency launched its first in 1960, and over the next decade, it put a total of 120 into orbit. In the late 1960s, an Air Force satellite was intercepting Soviet telephone conversations.[36] Such satellites grew larger and larger over time, and their use became so routine that civilian versions in geosynchronous orbits by the end of the century could direct the ordinary American driver to the nearest fast-food outlet or motel.

One of the most successful NASA programs has been the robotic exploration of space, and especially the solar system. The agency has long been criticized for sending astronauts into space where they accomplish little that instrumentation could not do better and much cheaper. *Viking I* and *Viking II* landed on the surface of Mars in 1976. Whether the hunt for the elusive "life on Mars" was worth the expense or not was debatable. Called by one critic a higher form of stamp collecting, the program, and those to Jupiter and Saturn (launched in 1977 to arrive in 1980) at least had the virtue of gaining real scientific information. The $2 billion Hubble Telescope was launched in 1990, and its spectacular photographs of deep space phenomenon immediately made it a national favorite.[37]

Walter McDougall has categorized the four claims for revolutionary discontinuity that have been made for the Space Age. Profound dislocations were said to have been made in "(1) international politics, (2) the political role of science and scientists, (3) the relationship of the state to technological change, and (4) political culture and values in nations of high technology."[38] On the first of these issues, McDougall found that "the international imperative stimulated the rapid development of space technology, but it was not in turn transformed by it." Early hopes that the United Nations might set rules for the use of space were undercut by the realization of both the United States and Soviet Union that neither really wanted to be constrained by international controls. Rather than political agreement on limits, both sides tried to invent technological solutions to the problems that technology created. One contributing complication, of course, was that the same technologies were used for space science and military purposes.[39]

McDougall's conclusion on the second issue is much like the first: "the Space Age introduced science and technology to the political arena," he wrote, "but it did not transform politics or usher the scientists to power."[40] Scientists and engineers had been active in Washington since at least World War I, and because knowledge is power, and politics is all about power, there has been a growing quid pro quo for many years. The nation's chemists came out of World War I in a position of popular acclaim, which carried over into campaigns to get corporations to fund more industrial research. After World War II, physicists seemed to hold the key to health, prosperity, and security. The Space Age both fed upon and reinforced this popular view.

On the third point, McDougall sees "a revolution in *governmental* expectations that occurred in the decade after 1957.... What seems to have happened in the wake of Sputnik was the triumph of a technocratic mentality in the United States that extended not only to military spending, science, and space, but also to foreign aid, education, welfare, medical care, urban renewal, and more."[41] State command and control of continuous technology change became the norm in America as it had been in the Soviet Union, and while its numerous manifestations were never coordinated as advocates of an explicit and transparent science policy would have preferred, it manifested itself in both the nuclear and space fields, and, as McDougall suggests, in many others as well.

Finally, on the fourth point, McDougall sees "the emergence of a 'revolutionary centrism' offering technological, not ideological or social, change to play midwife to a future as secure and bountiful, but less threatening, than that offered by either a socialist Left or a laissez faire Right." He concludes, "There *is* such a thing as a Space Age," and as part of its legacy, "the entire drift of industrial democracies toward a materialistic, manipulative approach to public policy under the post-Sputnik infatuation with technique." He quotes the former Democratic presidential candidate Adlai Stevenson as asserting in 1965 that "science and technology are making the problems of today irrelevant in the long run, because our economy can grow to meet each new charge placed upon it.... This is the basic miracle of modern technology.... It is a magic wand that gives us what we desire!"[42] By the early twenty-first century, with its resurgence of fundamentalism and terror both at home and abroad, those happy days of a "revolutionary centrism" seem long left behind.

In 1970, James Carey and John Quirk, warning of the "Mythos of the Electronic Revolution," wrote that "at the root of the misconception about technology is the benign assumption that the benefits of technology are inherent in the machinery itself so that political strategies and institutional arrangements can be considered minor." As they observed about the utopian dreams of peace and democracy surrounding the anticipated triumph of electricity, however, "technology finally serves the very military and industrial policies it was supposed to prevent."[43] The same, unfortunately, was true of the atomic and rocket technologies. Far from making war impossible ever again, and producing power too cheap to meter, the nuclear age is ever more dangerous, as both war and looming environmental consequences not only persist, but also manifest irreversible characteristics. Missiles remained linked to the delivery of nuclear weapons, and promises of space colonies to relieve Earth's problems are revealed as hopelessly naïve. Both technologies deliver useful though dangerous results, but hardly the promised utopia of peace and prosperity. Contrary to Stevenson's optimistic prediction, the problems of 1965 have not proven to be irrelevant.

FIVE

FACTORIES AND FARMS

World War II saw the apotheosis of modern industrial might. Mass production had allowed the United States and its allies to outproduce the Axis powers, an accomplishment that provided the allied powers an indispensable base from which to launch their military operations. However, the Industrial Revolution had not yet worked its course, even in the factories of the United States. In the 1950s, the term "automation" was coined to refer to significant increases in using automatic machines and "saving" labor. Social commentators and labor unions both deplored the possibility of "workerless factories," which Kurt Vonnegut, Jr. explored in his first novel, *Player Piano* (1952). But even automation, with its growing array of robots, was not the end of the process. Beginning first in Japan, a new industrial regime, christened "lean production," set new standards of efficiency and quality, in part by standing old techniques of mass production on their heads.

Postwar America was still a country that was famous for its industrial production, and for the first time, the term could apply to the farm as well as the factory. On the eve of World War II, rural electrification and the coming of the rubber-tired gasoline tractor provided the conditions for extending modern industrial technology to the work of the farm. With the additional commitment to chemical fertilizers and pesticides, and the widespread use of irrigation even outside the arid areas of the West, farms became, to use Carey McWilliams' term, "factories in the field." The cotton-picking machine had been perfected just before the war, and the tomato-picking machine just after. They and other machines like them hastened the consolidation of corporate farms, increasingly dependent on petroleum-based fuels and chemicals, high capitalization, and world markets. The opposite side of the coin, not surprisingly, was decline for family farms and the marginalization of a food regime that placed taste and nutrition, if not ahead of efficiency and profit, at least as values that could deliver the economic goods.

Mass Production

When Henry Ford built his first car at the turn of the last century, handcrafted methods applied by artisans was still the prevailing method of manufacture, despite a century of industrial revolution. By 1913, Ford had borrowed the technique of making and using interchangeable parts, adopted pre-hardened steel tools, and began to simplify the assembly of parts. In that year, he introduced his justly famous moving assembly line at the new Highland Park plant, and made Detroit—and the name Ford—synonymous with modern mass production. Then, with the opening of his River Rouge factory in 1927, Ford carried his innovations to their apparent limit: dedicated machine tools, interchangeable parts, and the moving assembly line were now joined by engineers nearly as specialized as the production workers, and a vertically integrated productive system that included Ford's own rubber plantations, iron foundries, transportation facilities, railroads, and a fleet of ships. Meanwhile, at General Motors, Alfred P. Sloan developed the selling of cars on credit, the annual model change, and the multidivisional structure of corporate governance. With the coming of industrial unionism with the United Auto Workers in the late 1930s, the industrial regime of mass production appeared to be complete.

During World War II, this massive and powerful engine of production performed miracles of scale. Detroit's auto industry produced tanks, jeeps, trucks, and myriad other items. In Southern California, a flourishing aircraft industry would soon be transformed into an aerospace industry. Steel mills increased production, and the aluminum-processing industry grew disproportionately. Between the world wars, American industry and its military had learned to cooperate and adjust to each other's forms of organization and procedural systems. In an important way, soldiers, sailors, and aviators were engaged in an industrial-scale act of mass consumption through destruction, which exactly mirrored the mass production of the nation's factories. Although Ford had established plants in England and Germany in 1931, mass production had not yet taken hold in Europe, and was seen the world over as not just an American innovation, but as a veritable symbol of America itself.

After the war, production returned to civilian needs, and in 1955 alone, three companies—Ford, General Motors, and Chrysler—accounted for 95 percent of the more than 7 millions cars sold in the United States. Over the years, however, Detroit—the avatar of American industry—had grown technologically conservative in product design. Meanwhile, during the 1960s and 1970s, a resurgent European auto industry finally began to adopt the technologies of mass production and created such new features as front-wheel drive, disc brakes, fuel injection, unitized bodies, five-speed transmissions, and more efficient engines. Detroit concentrated instead on consumer amenities such as air conditioning, power steering, sound systems, automatic transmissions, ever more powerful engines, and, above all, styling.[1]

Automation

In the important area of production technology, however, the American automobile industry was again a leader. In 1946, D.S. Harder, then vice president of manufacturing for Ford Motor Company, coined a new word to describe the work processes to take place in the firm's new Cleveland engine-manufacturing plant. As part of his review of the layout of the plant's equipment, he called for "more materials handling equipment between these transfer machines. Give us more of that automatic business ... that 'automation.'"[2] Ford established what appears to have been the first automation department the next year, staffing it with what the company called "automation engineers."

The following year, the word appeared in print in the pages of the influential trade journal, *American Machinist*. The author, Rupert Le Grand, offered what was the first definition of this new phenomenon. Automation was, he said, "the art of applying mechanical devices to manipulate work pieces into and out of equipment, turn parts between operators, remove scrap, and to perform these tasks in timed sequence with the production equipment so that the line can be put wholly or partially under pushbutton control at strategic stations."[3] When the Ford plant located just outside of Cleveland, Ohio opened in 1950, it was called the nation's first automated factory, though it employed over 4,500 workers. Certain activities, especially the shaping of cylinder blocks, marked a significant advance in automatic processes that captured the imagination. In 1952, John Diebold published his pioneering study *Advent of the Automatic Factory*, in which he wrote that the new word referred to both "automatic operations and the process of making things automatic," the latter involving "product and process redesign, the theory of communications and control, and the design of machinery." It was, he concluded, "a distinct area of industrial endeavor, the systematic analysis and study of which will yield fruitful results."[4]

From the beginning, the term "automation" has meant many things to many people. The most common use of the word seemed to be to refer to any process that had been made more automatic than it was before. As often as not, it was used loosely and interchangeably with "mechanization," or simply "technological change." It seems reasonable, however, to limit the word to describing the grouping of automatic production machines, transfer machines to move work pieces between them, and some kind of control system that regulates the entire process, including itself. The first two elements go back to the nineteenth century. Oliver Evans' "automatic" flour-mill machinery at the turn of the nineteenth century could work nearly unattended, as endless screws and conveyor belts carried the grain, then flour, from one processing device to the next. Feedback mechanisms, by which a machine could regulate itself, were as old as thermostats and governors, such as that on James Watt's steam engines. The availability of computers, however, successfully tied together the separate elements of the new automated systems. By 1966, some 15,000 to 20,000 computers were at work in industry, some of the first being used to monitor and log data in electrical power-generating plants and oil refineries.

Automation as a concept was joined in 1948 by another new word, "cybernetics," coined by Norbert Wiener, a professor at the Massachusetts Institute of Technology (MIT). In his book *Cybernetics,* Wiener defined cybernetics as "control and communication in the animal and the machine."[5] His 1950 book, *The Human Use of Human Beings,* however, most caught the public imagination. In it, he predicted that "it will take the new tools 10 to 20 years to come into their own. A war would change this overnight." The result, he believed, would be a social disaster: "Let us remember that the automatic machine ... is the precise economic equivalent of slave labor.... It is perfectly clear that this will produce an unemployment situation, in comparison with which the present recession and even the depression of the thirties will seem like a pleasant joke."[6]

The prospect could hardly be starker, and in 1952 Kurt Vonnegut, then a copywriter for General Electric, published his first darkly comic novel, *Player Piano.* Set not too far in the future, that war that Wiener had predicted would speed up automation had taken place. Society had been reengineered to the point where only two classes existed—a small number of engineers who ran the country through one large industrial corporation (the president was a vacuous figurehead who looked good on television) and an army of the unemployed, that the government kept occupied with New Deal-type busywork. Workerless factories turned out consumer items that were allocated on the basis of programs worked out by a giant computer. It was a fascist dystopia, and one that was a reasonable extrapolation of mid-nineteenth-century American industrial (and political) conditions.

As the popular press increasingly covered the promises and dangers of automation, a spirited defense of the practice was made in a study commissioned by the Machinery and Allied Products Institute and its affiliated organization, the Council for Technological Advancement. The study was published in 1966 under the title *The Automation Hysteria.*[7] The book confronted and refuted allegations that automation would inevitably lead to mass unemployment. After noting that the word "automation" was used in a wide variety of situations to describe an equally varied range of practices, the author sought to define it as simply the latest wrinkle in the long march of technological change that had been going on for over a century. Despite fifteen years of "alarmist doctrine," that study noted with satisfaction that most Americans, and especially the federal government, had rejected the false claims of massive unemployment, and "approved of technological progress in general (including automation), and even favored its acceleration, but recognizing the need for ameliorating the personal hardship it entails."[8] The argument was as old as the Industrial Revolution itself: through the benign intervention of some invisible hand, technological change always creates more jobs than it destroys, while hardships are individual and only temporary.[9]

The federal government was alarmed not so much that automation might cause massive unemployment, but that the public fear of that possibility might slow down technological change, and decided to study the matter itself. In December 1964, President Johnson set up a National Commission on Technology, Automation, and Economic Progress. Its report more or less came to the same conclusion as the study by the Machinery and Allied Products Institute: automation was merely one form of technological change. It was satisfied, as one reviewer put it, that while "during the last five years, 'automation' has indeed displaced workers, and quite probably at the 20,000-per-week clip so often cited in alarm ... most of those displaced have secured other jobs."[10] Two years before, Ben Seligman, the director of the Department of Education and Research for the Retail Clerks International Association, had already rejected this conclusion. "Meanwhile," he wrote, "the technology of our age—fostered by the state, which at the same time has responded with such pathetic inadequacy to the problems it has created—continues to spread with undiminished force, altering traditional work relationships and dispossessing hundreds of thousands of people in every passing year."[11]

Not content to rely upon the cheerful predictions of industry-sponsored studies, organized labor sought to protect itself—usually, as the phrase had it, by "buying a piece of the machine." Following a long union tradition of trading shop-floor control for better wages, hours, and conditions—especially job security—workers sought to share the benefits of, rather than oppose, the installation of various forms of automation. Starting in 1954, with an eye toward the next year's round of bargaining, the United Auto Workers' leader Walter Reuther laid the question of automation squarely on the negotiating table.[12]

On the West Coast, the notoriously belligerent longshoremen's union, headed by the tough and effective Harry Bridges, faced a similar problem.

Under a work regime that dated back hundreds of years, workers walked into and out of ships loading or unloading sacks of coffee beans or stalks of bananas carried on their shoulders. One longshoreman, Reg Theriault, left a moving narrative of their work rhythms. Assigned to a given pier by a dispatcher, upwards of 150 men would work a ship, operating in teams, covering for each other, and congregating in small workers' restaurants and coffee shops at the head of each pier. By the time he wrote about it in 1978, the union had reached an agreement with the maritime industry to accept containerization, which had been introduced in 1956. Henceforth, cargo would pass through West Coast ports sealed in large steel containers. These would arrive at and leave the docks stacked on trucks, from which they were loaded or unloaded by giant cranes. Theriault worked with just one other longshoreman, who tended the other side of the truck. With the depopulation of the docks, the restaurants and coffee shops closed down as well, and the closest of either was three to four miles away. The camaraderie of longshoring culture had left as well.[13] The new cargo-handling technology was hardly the kind of automation imagined by Diebold, much less Wiener's cybernetics, but the purpose and the effects were similar.

A quarter-century after Theriault wrote, another step in dockside automation was taken in Harry Bridges' native Australia. There, after a violent and unsuccessful attempt in 1998 by one shipper to replace union "wharfies" with scab labor (trained in Persian Gulf states), management completely replaced the dispatcher who had assigned jobs to the workers and was a powerful and relatively independent force in the union culture. Instead, by 2003, a computerized system of work allocation cut the process of assigning workers to ships from six hours to fifteen minutes, and work assignments were automatically sent to the workers' mobile telephones.[14]

Another example of automation was that of the automated butcher. In 1973, the American Federation of Labor and Congress of Industrial Organization (AFL-CIO) Meat Cutters Union publicly resisted the use by California's Lucky Stores of machines that cut up pork for sale in their supermarkets. The union had a list of complaints, including that the machines were dangerous, that the resulting cuts of meat contained bone dust and bits, and that because the pork had to be frozen before the machine could cut it up, Lucky Stores was deceiving its customers by claiming that the meat was fresh. For its part, the company sued the union for $10 million in an attempt to stop them from leafleting at the stores. The pork was only chilled down to 28 degrees Fahrenheit, Lucky claimed, and because pork did not actually freeze until it reached 26 degrees, the union's claim that the pork was frozen was false. In fighting the union's claim that the meat was soaked in a preservative, the company insisted that the liquid nitrogen bath used to lower the temperature of the meat, while it did kill surface germs, evaporated and left no residue. The union, according to Lucky, was simply trying to save jobs—one person at the machine would replace fifteen others, according to the union—by resisting technological progress. One industry official claimed that machines to cut other kinds of meat were on the way, and that "the retail butcher is on the way out."[15]

Elsewhere in the supermarket, companies had been thinking about automating checkout counters since the 1950s, but not until the Universal Product Code (the now-familiar bar code) was established in 1973 was such a process feasible. By 1975, an estimated 30 to 40 sites were using the scanning technology on an experimental basis, but the other 20,000 or so large supermarkets in the country represented a potentially lucrative market. Three years later, consumer and union resistance threatened to halt the spread of checkout scanner. Consumers were said to want each item in the stores to continue to be individually priced. Unions worried that "an end to price marking would threaten some 1.7 million employees currently holding clerk and stock jobs." In California, Vons supermarkets distributed leaflets to customers that declared, "We're always working hard to develop new ways to serve you better. That's why we're introducing our new computerized Fastcheck System. Once you check it out, we're sure you'll find checkout faster, easier, more accurate than ever before."[16]

In the end, consumers accepted checkout scanners, and the item pricing issue turned out to be smaller than imagined. Two other problems however, were hardly mentioned at the time, and have had less attention even now. First, the number and frequency of items scanned became another tool by which management could measure and discipline labor. Workers were required to set a certain number of scans per hour, a presumed indicator of worker efficiency. And from the customer who paid by swiping a credit card, or used one of the increasingly popular store-membership cards, stores collected and then sold a wealth of data about what, where, and when specific items were purchased, and by whom. When correlated with demographic information already in one's credit record, or filled out on a membership form, marketers gained important and specific information about consumer behavior.

Industrial Robots

Another aspect of technological change, closely related to automation, was the increasing use of robots in industry. The word entered the English language only in the 1920s through the play *R.U.R.* by Karel Capek, and came from the Czech term for labor. In the play, the robots were machines designed not only to do the work of human beings, but to look like them as well. As early as 1928, an article in *The Literary Digest* noted that "routine work is done more and more by machines. This is a familiar idea. But some of these machines are assuming an appearance that caricatures the human form, are speaking with the human voice, and performing tasks that have hitherto required the services of creatures that think." Linking these to early computers—the Great Brass Brain of the U.S. Coast and Geodetic Survey, and the Product Integraph of Vannevar Bush at MIT—the article closed with the reassuring words of "a high official" of the New York Edison Company: "the mechanical man and his ultimate universal practical application will rid humanity of much drudgery and thousands of uncongenial tasks."[17] The mechanical man still looked more like R2D2 and C3PO from *Star Wars* than the much more ambiguous "replicants" of the film *Blade Runner*.

When robots did actually come into common use in American industry after the war, they looked nothing like the machines of the imagination. By 1971, robots were being used largely in the auto industry, where they were deployed on the assembly line, especially for spot welding. Unions were said to be relatively unconcerned because, in the well-worn tradition, the robots were eliminating jobs that were "hot, dirty, exhausting, tedious, dangerous and otherwise undesirable." Nonetheless, the words of one robot fabricator gave pause. The robot, he said, "won't suffer from fatigue, will not need rest periods or coffeebreaks, and while it never will replace an artisan it will do a job exactly as taught." The reporter who interviewed him caught the spirit, and ended his description with the promise that "the robot never gets a cold, doesn't mind heat and dust, seldom experiences an inefficient Blue Monday and never lifts its arm to carry a picket sign. That remains a human trait, a prerogative of unions. And it could have a lot to do with the robot's development."[18]

By the mid-1970s, an estimated 6,000 robots were at work, doing "heavy lifting, welding, die-casting and paint-spraying," especially in the auto and electrical industries. The specter of robots taking over people's jobs was still present, but still marginalized by optimistic promises of new jobs created and the resigned belief that progress could not be stopped. Perhaps some of the anxiety was displaced into popular cultural attitudes, such as that embodied in the new card game, Robot. The game was, in fact, the familiar game Old Maid, with the Old Maid card replaced by the Robot. It was a win-win solution: the misogynist stereotyping of unmarried women was eliminated and replaced by the perfect substitute, because, as an advertisement put it, "no one wants to be a ROBOT." A more ambivalent reaction was elicited by a headline from Tokyo that read, "Factory Robot Stabs Worker To Death." The victim, Kenji Urada, had apparently stepped across a safety barrier and activated the robot, "whose arm stabbed him in the back."[19] It was a dramatic and particularly riveting example of labor displacement.

Lean Production

Not only was Japan somewhat ahead of America in adopting robots for factory work, but it was the original site of a radically new industrial regime: the next form of industrialization, which threatened to replace mass production just as mass production had replaced the reliance on handcrafted methods. Lean production, as it was called, grew out of the work of the Toyota Motor Company to avoid the disadvantages of mass production by reintroducing some of the aspects of handcrafted operations.[20]

The story began in 1950, when Eiji Toyoda, a young Japanese engineer, visited that shrine of American mass production, Henry Ford's River Rouge plant. At that time, the plant was producing 7,000 cars a day, whereas the visitor's Toyota Motor Company had produced only 2,685 in the thirteen years of its existence. With the important assistance of Toyota's chief production engineer, Taiichi Ohno, the company began to rethink the way cars were made. For one thing, the use of dedicated tools and dies, typical of American production, made sense only in terms of the very large volume of cars produced. The hundreds of massive stamping presses, dedicated to making only one part over and over, sometimes for years, was beyond both Toyota's means and needs. By developing a method of changing dies every few hours, and having production workers do the changing, Ohno was able by the late 1950s to reduce the time needed to change dies from a full day to three minutes, and could do so without die-changing specialists. The result was a need for fewer presses and a more flexible production process.

Another innovation was in quality control. The introduction of W. Edwards Deming's statistical methods of quality control and of quality circles was only part of it. A mutual loyalty between the company and the workers was a key element, but all of these were important because they made possible a higher degree of individual responsibility and competence than was found in American auto plants. In the United States, the compulsion to "move the metal" meant that only supervisors had the authority to stop the sacred assembly line, and defective cars were moved along to the end of the line, where they were reworked by a large number of people employed solely to make repairs on defective units. One consequence was that in an American plant, roughly one-fifth of the floor space and a quarter of the total hours required to produce a car were spent on reworking them. Under the regime of lean production, virtually none of this was necessary, because each worker on the assembly line had the power and the obligation to stop production until the cause of any problem was discovered and solved.

Another reform was in reorganizing parts and subassembly supply. Only 15 percent of the total manufacturing time elapsed at the final assembly line; the rest was used to make the 10,000 parts and put them together into the 100 major subassembly components that would eventually become a car. Toyota again charted a middle ground between the chaos of individual suppliers and the extreme vertical integration of American firms. Suppliers were brought together and made thoroughly familiar with each other's standards and methods. The final step was to innovate the famous just-in-time inventory system, by which parts and subassemblies were not stockpiled, but delivered only as needed. The very name "just-in-time" (translated from the Japanese word *kanban*) gives some hint of the seemingly precarious nature of the enterprise—the sense that everything had to go just right for everything to come out right. Again, quality was key. Defective units were not produced in large quantities, or warehoused for a long period of time before being discovered as defective.

Despite the potential to displace mass production in many industrial arenas in America, the Toyota method of lean production, which took the company twenty years to perfect, has not yet been widely adopted. An agreement between General Motors and Toyota went some distance in that direction, but even in that case, the experiment was soon attenuated. It was widely suspected that the particular nature of labor relations in Japanese industry in particular, but also the general workings of Japanese culture, made transferring lean production to America highly problematic. Still, some aspects of the system can be detected in odd corners of American production. By the end of the twentieth century, the corporatization of American universities had led to something like untenured, just-in-time lecturers being used to replace the fixed and relatively inflexible senior and tenured faculty members trained in one specialty. The latter professor is seen as the functional equivalent of the American automotive production engineer who someone from the Harvard Business School discovered had spent his entire career designing automobile door locks.

The Farm as Factory

If Toyota led the way to reintroducing diversity, complexity, and flexibility into the factory, American agriculture was busily erasing the same values in an attempt to reproduce the classic characteristics of industrial mass production on the farm. Since the mid-nineteenth century, agriculture had lagged behind manufacturing in the hallmarks of industrialization: large-scale enterprises, a proletariat workforce, mechanization, rationalization, specialization, and a coordinated assault by science and technology on traditional methods. Indeed, even while such infrastructure as the telegraph, the railroad, and ocean-going steamships were integrating American farmers into world markets, many Americans rejected the industrial model of farming, as the family farm and some measure of self-sufficiency were hailed as important rural virtues.

During the first half of the twentieth century, however, the industrial model made headway as ideology. International Harvester, the farm machinery manufacturer, used the advertising slogan "Every Farm a Factory," and Carey McWilliams, social critic and longtime editor of the *Nation*, caught the public's imagination with his exposé of farm labor condition in California, titled *Factories in the Field*. What tied together the discrete technological changes—the tractors, rural electrification, the large irrigation projects, the innumerable research and development projects at the land-grant state agricultural schools—was an industrial ideal, to use historian Deborah Fitzgerald's phrase. She argues that, "although individual technologies, particular pieces of legislation, new sorts of expertise, the availability or disappearance of credit opportunities are all key to understanding what happened in twentieth-century agriculture, it is essential to grasp the overarching logic of change that was taking place in bits and pieces and the industrial system that was being constructed across the country."[21]

The gross results of the triumph of industrial farming after World War II were graphically laid out in the *Statistical Abstract of the United States*. Between 1940 and 1969, the farm population of the country dropped from just over 30 million to just under 10 million. The number of the nation's farms fell from over 6 million to only about 3 million. During these same years, the average size of farms rose from just over 150 acres to well over 350.[22] Between 1987 and 1997, the total number of farms was still shrinking and in terms of size, the only category actually increasing in number—from 67,000 to 75,000—were farms of over 2,000 acres. In 1997, there were 84,002 corporate farms in the country, averaging 1,565 acres in size. "Family" farms were still evident however: the very largest farms in the country, with an average of 5,571 acres, were held by family-held corporations with eleven or more stockholders.[23]

Another set of statistics shows the ripple effect of industrial agriculture. Like the family farm, small towns have been idealized in America, but as the former disappear, so do the latter. Many of the towns sprang up to service the surrounding farms, providing churches, schools, government offices, and banks, as well as cafes, movie theaters, service stations, and retail shops. As the nearby farms disappear, so do their customers, and the towns shrink and often die. Between 1950 and 1988, the percentage of Americans living in rural areas shrank from 43.9 percent to 22.9 percent. As a demographer for the Department of Agriculture remarked in 1990, "a lot of these little places just aren't needed any more."[24]

The Brave New Chicken

The growth of large corporate farms was a direct result of the industrialization of agricultural production, which, as in the manufacturing sector, gave a distinct advantage to operators that could marshal the capital to expand operations by increasing acreage, accessing cheap labor, buying new and larger machinery, adopting the latest chemical fertilizers and pesticides, and even achieving tight vertical integration. All of this can be seen in the story of what the historian William Boyd has called "the industrial chicken."[25] Again, statistics highlight the scope of the story: between 1950 and 1999, American production of broiler chickens increased by 7 percent a year, until by the end of the century, over 40 billion pounds a year were sold, and the average American ate more than eighty pounds each per annum. The bird that supported this vast enterprise increased in market weight from 2.8 pounds in 1935 to 4.7 pounds in 1995. During that same period, the time it took for a chicken to go from newly hatched to that weight fell from 112 days to 47, and the pounds of feed it took to grow each broiler fell from 4.4 to 1.9. The broiler chicken was not only a market triumph, but also "one of the more thoroughly industrialized commodities in American agriculture." Indeed, "through technologies of confinement and continuous flow, nutrition and growth promotion, and breeding and genetic improvement, the barnyard chicken was made over into a highly efficient machine for converting feed grains into cheap animal-flesh protein."[26]

The small number of giant firms that produced these birds were known as integrators because they integrated vertically, using their own "hatcheries, feed mills, and processing operations," while at the same time, and just as importantly, "farming out" the actual raising of the chickens to contract growers who took most of the chances of failure. Because of the need for contract farmers and cheap labor, the industry concentrated in the South. The Delmarva Peninsula on the East Coast produced 600 million chickens annually, also producing, inevitably, an equally massive pollution problem: 1.5 billion pounds of manure, more than the waste from a city of 4 million people.

This industrial style of chicken production has been exported abroad, and it is from England that we have the following detailed look at how it works.[27] The factory farm is located in bucolic Oxfordshire, one of 100 owned by the same company. In a two-month cycle, the farm received 200,000 day-old chicks, and after six weeks, they send them off for slaughter. The chickens are broilers, specifically bred to have large breasts, which are the preferred part of the bird. The other kind of chicken, designed to lay eggs, is an entirely different type. The broilers on this farm are kept in eight sheds; each shed is 200 feet long and 80 feet wide, and houses 26,000 birds. They live their short but productive lives in cages, within a space of about nine by nine inches. The lights are kept on 23 hours a day, while computers monitor the feed, water, temperature, and ventilation. As one of the managers proudly points out, "the industry's really gone forward in the last five to 10 years. It's a lot more scientific now. We have more technical knowledge than we ever had."

From this farm, the chickens are trucked to the slaughterhouse, where 15,000 chickens are killed and processed every hour. An overhead conveyor belt acts as a kind of factory disassembly line, not unlike the one that is said to have inspired Henry Ford. Hung upside down, the birds are dipped into a bath of electrified water that is supposed to stun them. Then, they are passed by whirling blades that cut off their heads. Just as in an American automobile plant, a worker stands by to "rework" the defective carcasses that have been only partially beheaded. Artificial pluckers with rubber fingers massage the feathers off, a machine drills a hole in the backside, another machine scoops out the entrails, and yet another vacuums the insides while the feet and gizzards are cut off. A digital camera takes a picture of each bird, and a computer analyzes the image for signs of damage: "it doesn't get tired," a proud worker remarks. "It's the same on Friday afternoon as it is on Monday. It takes out human error."

In two hours, the chickens emerge, some cut especially for McDonald's or KFC, but most headed for supermarkets, many with marinade flavors like garlic and lemon or herbs injected into the skin, wrapped in plastic, and hardly touched by human hands. It is a River Rouge of poultry; the product is as thoroughly reworked by science and engineering as the processing machinery, and hardly more natural.[28]

Farm Machinery

Like the chicken, the tomato has been transformed since World War II into a predictable cog in the machine of industrial agriculture. In this case, the early attempt to invent a machine to pick the fruit was set aside, and instead, a fruit was invented to accommodate the machine. In 1949, two professors at the

University of California's agricultural campus at Davis set out to build a tomato-picking machine. The wartime shortage of labor, which affected farms as well as factories, had been solved by a temporary federal *bracero* program, designed to bring Mexican farm workers north to do contract work. G. C. Hanna, the professor of vegetable crops who had initiated the picker project, was quoted in the *Wall Street Journal* as being worried about what would happen when the program died: he had "seen nationality after nationality in the field, and I felt that someday we might run out of nationalities to do our hard work."[29] The *bracero* program did end in 1965, and the specter of union organization among the remaining American farm workers was unsettling to California growers. That same year, Cesar Chavez threw his National Farm Workers Association into an ongoing grape strike in Delano, California.

By 1959, Hanna and his partner, Coby Lorenzen, a professor of agricultural engineering, had produced both their new tomato and their new machine to pick it. Much more had to be done to make the machine available to farmers, however. Large agricultural machinery companies seemed uninterested in the machine, so the university licensed its manufacture to a small, local machine maker. These were to be used for processing tomatoes—to be turned into catsup and tomato sauce—but processors had to be talked into lowering their standards for the fruit they would accept at the canneries. Because one characteristic of the new tomato was that it had to be picked while it was green so that it was firm enough to stand up to the mechanical fingers that tore up the vines, an agricultural chemical firm stepped forward with ethereal, a gas that turned the green tomatoes red while in storage. According to its advertising slogan, "ethereal helps Nature do what Nature does naturally." It was one of the 4 million chemicals listed in Chemical Abstracts, maintained by the American Chemical Society in 1965. Of this number, 50,000 were thought to be a part of ordinary, everyday life, but the total did not include the 1,500 active ingredients found in pesticides, nor the 5,500 chemicals that were added to food after it left the farm to improve flavor, appearance, or shelf-life.[30]

Finally, growers had to be convinced to use the machines, and farms had to be consolidated to provide sufficient acreage to make the machine's use economically feasible. Traditional methods of growing tomatoes were worse than useless. Because a picker pulled up the vines by the roots, it could only pass through the field once. All the fruit, therefore, had to be ready at once, which meant that they had to set at the same time, which required the plants to blossom at the same time. This acute timing necessitated the precise application of fertilizers and irrigation. Contract farmers were given computer software that planned all this out.

Like the lettuce picker, which appeared in fields at about the same time, the tomato picker changed the nature and scale of this type of fieldwork. Rather than passing through the field at intervals, bent over, cutting off the ripe fruit, a crew of sixteen, including a driver, a supervisor, and fourteen sorters, rode on the machine as it passed down the rows. The sex best suited for the job was in dispute. Some claimed that men suffered from motion sickness, while others worried that women talked and argued too much, rather than paying attention to the sorting. Either way, labor costs were lowered, and migrant workers, predominantly Mexican-American, were put out of work.

As had always been true of agriculture, different crops in different parts of the country were brought into the industrial system at different times and to varying degrees. Rural electrification between the wars made milking machines feasible, and a complex system of price controls and subsidies was firmly in place as well. By the beginning of the twenty-first century, however, the latest technology—robot milking machines—was yet to make a significant appearance. Only 10 such machines were used on American farms, compared with 65 in Canada and nearly 1,600 in Europe. "Rising labor costs, problems with conventional milking methods and a desire for more flexibility" were cited as reasons why a farmer might pay $150,000 for one of the machines. In one description, it was said that "cows are admitted to the milking area by computer, and a robotic arm does the job of washing the teats and placing the milking cups, using a laser to guide it. No person," it concluded, "needs to be present." As no two cows are quite the same, each has a transponder on its collar that the machine reads as the animal enters the milker. The locations of the teats are stored in the computer database so that the cups can be properly fitted to each cow. A mechanical override is provided so that farmers can take over any part of the operation, but the freedom of being able to stay in bed beyond 5:30 a.m. was a part of the machine's attraction. This particular kind, manufactured by Lely in the Netherlands, is suggestively named the Astronaut.[31]

Focusing on another sector of the nation's agriculture, the *New York Times* editorialized in 2002 that "factory farms have become the dominant method of raising meat in America. Agribusiness loves the apparent efficiency that comes with raising thousands of animals in a single large building where they are permanently confined in stalls or pens." Called confined animal feeding operations (CAFOs), they are operated by a small number of very large companies that seek to "control the production from birth through butchering and packaging."[32] The *New York Times* was deploring particularly the vast accumulations of manure that inevitably accumulated but were poorly controlled, as well as the "precarious narrowing of genetic resources" that supported these operations, but the cost of factory farms was much greater than just these factors.

Genetically Modified Organisms

Most of the examples above involve some degree of modification of plant or animal species, usually through some updated version of the ancient art of selective breeding. A quantum leap beyond this, however, is to use biotechnology to modify species. Essentially, this involves splicing genes from one species into another to introduce specific characteristics. Corn and soybeans, for example, have been modified to resist pesticides (made by the same company that developed and patented the new plants) or produce their own pesticides. We now find "flounder genes in strawberries, mice genes in potatoes, cow genes in sugarcane and soy, chicken genes in corn."[33] These genetically modified organisms (GMOs), as they are called, are touted to increased food production and thus provide hope for the starving millions of the world, but it apparently does not work quite that way.

In the six years between 1996 and 2002, GMOs have come to represent a growing proportion of the crops raised in the United States: 34 percent of corn, 75 percent of soy, 70 percent of cotton, and 15 percent of canola. One company, Monsanto, provides 90 percent of the GMO seed grown around the world.[34] Growers agree not to replant the seed a second year until they have signed a new contract and bought new seed. One Iowa farmer described his place in the agro-industrial complex: "My job," he said, "is the production end of this assembly line. We're just a small little cog in the wheel. . . . What we're concerned with is production agriculture."[35]

At least one Department of Agriculture study suggests that yields from GMOs are no higher than those from the seeds they replace, and that it is even declining. In essence, the GMOs are marketed not because they are better, but because Monsanto and other firms have invested heavily in the research and development of the proprietary seeds. This radical and widespread tampering with evolution, however, is bound to have ripple effects that are largely unknown and even outside the immediate interests of either the companies or the federal government. Both players insist that the GMOs are safe, but the consuming public, it turns out, is not altogether convinced.

There is no way to tell for certain, but according to an estimate by the Grocery Manufacturers of America, perhaps 70 percent of the food found in groceries contains GMOs. Traditional corn and soy is mixed with the genetically modified product in the market, and few processors can be sure of what they are using. The story has its share of ironies: Gerber, the maker of baby foods, has banned GMOs from its products. Gerber, however, is owned by Novartis, a Swiss pharmaceutical firm that makes genetically modified corn and soy seeds. McDonald's will not use modified potatoes for its fries, but cooks them in GMO vegetable oils.[36]

Whether tracing the evolution of the technologies of broiler production, tomato picking, or beef and pork manufacture, the story is startlingly similar. Large-scale, integrated, machine- and therefore capital-intensive production regimes marshaled scientific research, engineering development, business rationalization, vertical integration, political clout, economic subsidies, cheap labor, marketing innovations, and smart advertising to reform American agriculture. Not all sectors of rural production were changed, of course, just as not all factories managed to fully institute mass production technologies. But the ideal was there, providing a convenient and even compelling rationale for choosing one path over another.

Organic Farms and Farmers' Markets

No system, however successfully hegemonic, is ever without pockets of resistance, and the factory farm's list of things to do could be as easily used as a list of practices to avoid. Over the last three decades of the twentieth century, two traditional agricultural practices have steadily gained headway by rejecting the central tenants of agribusiness: production on organic farms (at least initially), and marketing at farmers' markets. Both challenged the need to adopt an industrial model and the desirability of following the modernist plan of using increasingly sophisticated science and technology to bend nature to man's will. Indeed, both rejected an easy acceptance of what was perhaps modernity's central triumph—the obliteration of time and space. Organic farmers who sold their product through local farmers' markers, and the customers who were willing to pay a premium to support them, insisted on what seemed like an old-fashioned and even quixotic preference for the here and now—accepting seasonal and local restraints as well as bounties. The asparagus to be had out of season in the supermarket was available only by way of a massive technological commitment to transportation technology to bring it north from Mexico, where it was likely grown with equal parts chemical application, machine planting and harvesting, and exploited labor. Organic farming had once been the only kind practiced, and even after the coming of factory farms, was still preserved, often more out of principle than necessity by some agriculturalists. During the 1970s, linked to the emerging environmental movement, the drive to discover and use only appropriate technologies, and a renewed interest in natural foods and a "diet for a small planet," some small farmers and back-to-the-land activists decided to abandon at least chemical fertilizers and pesticides, and among the more radical, to go back to using draft animals instead of tractors. Chickens would be free-range rather than confined to cramped cages, cheeses would be handmade, and heirloom tomatoes would be preferred over the industrialized variety.

One Japanese-American California peach grower's father, fresh from of a federal internment camp during World War II, had been talked into chemical-based farming by both chemical salesmen and government extension workers. His son, however, was faced years later with having to pull out a peach variety that, while excellent in taste, did not market well because of its cosmetic appearance and a short shelf life. Instead, David Masamoto decided to go organic and await the results. Through direct marketing to upscale consumers, including the growing number of restaurants, such as Berkeley's Chez Panisse, that specialized in fresh, local, seasonal, and organic ingredients, Masamoto not only saved his trees, but also inspired others to follow suit.[37] By the end of the century, organic products had become the fastest expanding sector of American agriculture—growing 20 percent a year during the 1990s.[38] The program of "safeguarding the ecological integrity of local bioregions; creating social justice and equality for both growers and eaters; and cultivating whole, healthful food," proved wildly attractive and successful.[39] Critics, however, believed that the more recent taking-over of much of organic production by large corporate growers was a betrayal, not a triumph, for the movement.[40]

Farmers' markets have a similar history. Every sizeable city in the country had at least one and often several produce markets, many dating from colonial times. By the 1960s, the markets were disappearing from the urban landscape as land prices escalated and more people shopped for food in supermarkets. The large market that once stood on San Francisco's Market Street, with its stalls selling fruit and vegetables, but also cheeses, meat, flowers, and other agricultural offerings, was typical. Soon after the middle of the century, it was torn down for more intensive construction, and not replaced until the opening of a refurbished ferry building—itself the victim of changing transportation technologies—filled with food stalls in the summer of 2003. Between these two structures, however, the city continued to be served by open-air markets, operating one or more days a week, at which local farmers sold to local customers, brokers were banned, new foods were discovered, and personal relationships were forged. During the 1970s, farmers' markets proliferated across the country, sometimes though not always with the aid and encouragement of local and state governments, and often in the face of their active opposition.

The symbiotic relationship between organic farmers and farmers' markets was intensified at the end of the century when the rapidly growing sales of organic products increasingly available in supermarkets attracted the attention of industrial agriculture. To standardize the definition and marketing of "organic" products, in 2002, the federal government issued a national organic rule (NOR) which established the categories of "100 percent organic," "certified organic," and "made with organic ingredients."[41] Activists fought back industry attempts, at first accepted by the Department of Agriculture, to permit such practices as fertilizing organic crops with recycled industrial sludge containing heavy metals, but the whole effort in itself made it easier for corporate food giants to enter the field.

Organic junk food met federal standards, but not those of activists who refused to measure organic in terms of federal regulations instead of social reform. Animals that spent their entire lives clamped in stalls, but fed organic grain and producing milk shipped halfway across the continent, were not acceptable to the organic pioneers. As one critic put it, "large food makers are not about to cede such enormous profit potential to small farmers producing whole, healthful, locally grown foods. Instead, they've opted to channel consumer demand ... in a direction that poses no threat to the industrial foundations upon which modern food empires have been built."[42] Grimmway Farms in California began in 1968 as a roadside stand, but by 2002, it was producing 9 million carrots a day, which were sold in 30 countries around the world. Eighteen thousand of their acres devoted to carrots are organic, but they are "planted, harvested, washed, sliced and packaged by machine—keeping prices low."[43]

As industrial agriculture fills the shelves of supermarkets with "organic" foods that are such in only the narrowest of definitions, small, local, family farms have had to turn increasingly to direct sales to consumers, sometimes at roadside stands in the countryside, but now also at the growing number of farmers' markets. Organic farmers often work without technologies of a wide variety. But those they choose to employ are picked deliberately, justified practically, and used to advance a vision larger than progress and profit.

The Amish

Slowly, organic farming is being introduced to one group of farmers who have long been thoughtful and careful about technological choices for both farm and field. The Amish began as Swiss Anabaptists in 1693, but Mennonite congregations, from whom they split off, had come to Pennsylvania as early as

1683, and by 1900, there were an estimated 5,000 living in the country. In an era before most rural people had access to cars, tractors, electricity, and other industrial technologies, the Amish fitted in more easily with the prevailing technological habits. As these new "improvements" became available, however, the Amish carefully picked and chose between them, rejecting many of those that they believed would adversely affect their ways of living, and embracing those that seemed less of a threat.

One Amish craftsperson commented simply that "we try to keep the brakes on social change, you know, a little bit."[44] From outside the culture, not all of the choices made are obvious: In 1910, telephones were banned from homes, but in the 1980s, they were allowed in workshops—a choice, in fact, that even Alexander Graham Bell had made a century before. In 1923, tractors were banned for fieldwork, power lawn mowers in the 1950s, and computers in 1986. At the same time, washing machines with their own gasoline engines were allowed in the 1930s, hay balers in the 1950s, and modern kitchens in the 1980s.[45] The point is not that each of these choices was correct for all Americans—or even for all Amish—but that the community had a clear sense of who they were and what kind of lives they wanted to live, and made conscious collective choices about each technology presented.

By the end of the twentieth century, the combination of large families and the desire to have sons follow along in an agricultural tradition led to the dividing and then subdividing of farms until they became too small to support a farm family. In Ohio, which has several large Amish communities, advocates of organic farming were working to make the Amish aware that by changing from conventional practices to organic, higher prices and growing demand promised to raise the income that could be produced from each acre. Amish farmers, already using draft animals rather than tractors to cultivate their fields, and valuing farm families as much as family farms, appeared to be ready candidates for adopting organic regimes. It was a hopeful example of the power of oppositional cultures to adapt and prevail. On the other hand, the Amish lifestyle continued to contrast dramatically with the growing emphasis in the mainstream American culture on having "fun" based on enjoying the full menu of consumer technologies.

SIX

"IT'S FUN TO LIVE IN AMERICA"

In 1947, Kiwanis International issued the first of what it planned to be a series of leaflets "prepared in the interest of the American Way of Life."[1] The Cold War was just beginning, but it was clear that if the American way of life was worth saving from communism, Americans should be clear about what needed to be defended. The cover, in red, white and blue, carried the title "It's Fun to Live in America," and featured three drawings. At the top, given pride of place, was a family-sized convertible, with Dad driving, Mom at his side, and a backseat full of kids. Second, a family was gathered around a large console radio, listening to music. At the bottom, a woman in an apron was talking on the telephone. Inside, the reader was told that there was one car for every five Americans, but only one for every 1,130 Russians. Similarly, there was one radio for every two-and-a-half Americans, but only one for 48 Russians, and a telephone for every five Americans, but only one for every 100 Russians. The leaflet also compared Americans to the British and French, but it was the Russians who were the most badly off. The lesson was clear.

As Joanne Jacobson has written, "the fifties may have been the last great moment when Americans entrusted their dreams of transformation to the material world.... The material world," she continues, "became a theater of transformation. On the glowing, capacious stage of *things*—cars, hula hoops, rockets—our destiny of motion was revealed." As she also notes, "in the postwar years rationed hunger was let loose on a whole new world of goods."[2] The pent-up demand for goods—both to replace those worn out by wartime use and those never available before—was matched by unspent money saved out of the comparatively good wages of war production. One object of consumer desire was a home of one's own, usually in the new suburbs materializing across the land. But those houses needed to be filled, not just with families, but also with things.

Appliances

Between the end of the war and 1950, consumer spending rose 60 percent. When this is broken down, however, the importance of the new home as a site of new goods is remarkable. Spending on home furnishings and appliances rose 240 percent in those same years. Young people who had rented before and during the war, or lived with parents, were likely to have very little that was suitable for use in Levittown or Lakewood. New homes had equally new yards, and the sale of lawnmowers was up 1,000 percent in those years: 139,000 were sold in 1946, and 1.3 million in 1951.[3]

In describing Lakewood, California, when he was a boy, D.J. Waldie noted that in 1946, his house and the one across the street had no telephones, and both his mother and the mother in that other house had no access to cars during the day. The houses in the new development, however, had appliances available at extra cost. As Chapter 2 relates, the developers offered kitchen ranges, refrigerators, and washers, adding the cost to the price of the house, leading to an increase of only nine dollars a month in the mortgage. The newly developed garbage disposal was included in the original price. In Levittown, the developers began to build television sets into one wall of the living room, and although this was not done in Lakewood, by 1953, some 98 percent of the homes had a set.[4]

And increasingly, on the kitchen shelves and in the refrigerator, food was kept in Tupperware containers.[5] Earl Tupper worked for a New England plastics manufacturer that encouraged its employees to experiment and tinker using company resources. In 1939, he set up his own company, and during the 1940s, he was designing kitchenware using a new kind of plastic developed for wartime use. In 1947, he made his great breakthrough, patenting the "Tupper seal," which made a distinctive burp as the air was expelled. Well designed, brightly colored, and performing as promised, Tupperware was sold at highly ritualized parties that became important social gatherings for housebound wives and mothers recently arrived in housing tracts that were, as yet, hardly communities. Alison J. Clarke, a historian of Tupperware, terms the parties as "the ideal home-based networking opportunity for a newly displaced population."[6]

Other kitchen necessities, though perhaps with less staying power, were the Veg-O-Matic and its many siblings. Samuel J. Popeil came to Chicago from New York in 1945 and went into the business of selling ordinary kitchen tools. In the same spirit of exploration that motivated Tupper, however, he soon produced his own devices, first the Citrex juicer, then the Toastette sandwich-pie maker, and then a host of others. The first of his more famous line of machines was the Chop-O-Matic, which used rotating steel blades with chop vegetables. At first, he sold the machines at dime-store demonstration tables, but in 1958, his son Ronald made a five-minute ad for television, using the same fast patter he used to pitch the device in stores. The iconic Veg-O-Matic, which worked more or less on the same principle, arrived in 1963; eventually 11 million were sold, again through (often late-night) television. These early infomercials were classics of the genre, and the excited cry—"But wait! There's more!"—came to epitomize both the medium and the message of progress through even the most modest of technologies.[7]

Food

An alternative to the family meal made from an out-of-the-can recipe (the canned tuna in canned mushroom soup was a classic), or lovingly chopped with a Veg-O-Matic, was the frozen TV dinner. Frozen foods were available commercially in the 1930s, but not until Maxson Food Systems manufactured its Strato-Plates in 1945 did frozen meals come on trays. Maxson, which sold to airlines and the military for in-flight meals, pulled out of the business, but food engineers continued to try to improve the product. It was discovered that gravy kept meat moist, but thickeners had to be added to the gravy to keep it from curdling. Vegetables had to be under-cooked so they did not turn to mush while the meat was heated sufficiently.[8]

In 1955, C. A. Swanson & Sons, a subsidiary of Campbell Soup Company, brought out TV Dinners, and gave their trade name to the entire genre of frozen meals. Meant to be eaten by the whole family (mother did not need to spend hours in the kitchen preparing the meal) while watching television, the trays were even suggestively shaped like TV screens. As one observer noted, "the TV dinner allowed real-life families to gather around a television set and watch TV families gathered around a dinner table."[9] Swanson and their competitors sold 70 million dinners in 1955 and 214 million in 1964. By the mid-1990s, that figure had reached 2 billion.

Cake mixes were another cooking aid that promised to be labor-saving, but ran the risk of alienating housewives at the same time. By the 1930s, consumers were buying enough bread and pastries from bakeries and grocery stores that home sales of flour was falling. The flour industry began in that decade to pre-mix ingredients for sale as cake mixes, to which the cook needed to add only water. Sales increased threefold between 1947 and 1948, but by 1953, up to 80 percent of cakes baked at home were still made from scratch. Some resistance was eventually overcome when dried eggs were left out of the mix, and cooks could add real eggs and thus feel more personally involved in the resulting product. Even so, labor-saving was catching on. Factory workers had been told for years that machines only replaced hard, dirty, drudgery on the job, leaving the worker with more free time and energy for the creative aspects of their work. Similarly, a 1961 cookbook chirped that "no longer plagued by kitchen-maid chores that have been taken over by the food manufacturers, anyone can become an artist at the stove."[10] The cookbook's author, Poppy Cannon, who had also written the popular *Can-Opener Cookbook*, had a decade before presented her friend Alice B. Toklas, Gertrude Stein's partner in Paris, with a number of boxes of cake mix. Toklas declared them delicious, especially the yellow cake and the devil's food.

The explicit ideology of the TV Dinner was, and still is to some extent, that the members of the family should eat meals together, under the assumption that this would strengthen the American family—and that strong families were bulwarks against a dangerous world in general and the communist menace in particular. The same point was made in 1951 when the Czekalinski family assembled on the west side of Cleveland, Ohio for a photograph that was to appear in the DuPont company's magazine *Better Living*. The venue was the cold-storage locker of the Great Atlantic and Pacific Tea Company warehouse on West Ninth Street, and the room was filled with a cornucopia of foods: 578 pounds of meat, 131 dozen eggs, 1,190 pounds of vegetables, 31 chickens hanging from a rack, 698 bottles of milk, and much more—two-and-a-half tons altogether. The cost of the food in 1951 was $1,300, and was supposed to represent what the Czekalinski family consumed in one year. The photograph of this "typical" American family and their bounty of food was used to illustrate an article entitled "Why We Eat Better," and was manipulated graphically to show how much better things were in America than in England, Belgium, Germany, Poland, or China. The proud image was reprinted in *Life* that same year, again in 1975 in the new company magazine *DuPont*, and once again in 1998 in *U.S. News & World Report*. Harold Evans, the editor of *U.S. News*, declared it to be "the image of the century."[11]

Kitchen Debate

The photograph had one other claim to fame: that it had been turned into a mural for display at the American National Exhibition in Moscow in 1959. Most famous as the scene of the "kitchen debate" between then-Vice President Richard Nixon and Soviet Premier Nikita Khrushchev, the exhibition had been agreed upon as a showcase for American science, technology, and culture. The overwhelming image of the show was "stylist domesticity." Americans bought three-quarters of the world's production of appliances in the 1950s, and they all seemed to be on display in Moscow. Eight hundred American manufacturers had goods on display, from twenty-two cars of the latest model to kitchen tools—five thousand of them—including pans, dishes, and small appliances. *Izvestia* was appalled: "What is this?" it asked, "A national exhibit of a great country, or a branch department store?"[12]

Fresh from its triumphal space launch of *Sputnik* just two years before, the Soviet projection of its technological prowess was very different, even truculently so. The Soviet exhibition, which had opened in New York just the month before, had its full complement of large, unstylish appliances turned out by state-run factories, but also heavy machinery, a model of a nuclear ice-breaker, and space capsules. In Moscow, one Russian observer exclaimed of the American display, "I see no plan in all this. The whole exhibition appeals only to bourgeois interests.... Your whole emphasis in on color, shape, comfort. We are more interested in the spirit behind things."[13]

But the lavish display of bourgeois desire did have a spirit behind it—even an ideology. As Karal Ann Marling has written, "to Nixon, the latest in kitchen consumerism stood for the basic tenets of the American way of life. Freedom. Freedom from drudgery for the housewife. And democracy, the opportunity to choose the very best model from the limitless assortment of colors, features, and prices the free market had to offer."[14] There was a powerful gender message here as well, though ironically one on which Nixon and Khrushchev could find common ground. Appliances, even for the kitchen, had a strong masculine association. They could be seen as substitutes for the (absent) man who both did the heavy work outside the home and earned the money for the technology, and may even have presented the appliances to his wife as a Christmas or birthday gift. Also, as Marling has pointed out, by not just lining up the appliances in rows as had the Soviets in New York, but contextualizing them in a "typical" though fantasy domestic setting, the United States was making a strong statement that the private, domestic—and therefore female—sphere was more important than the public, male arena. In America, citizenship was acknowledged and increasingly expressed through consumption rather than political action. The very aggressively masculinist bragging contest between the vice president and the premier in the kitchen of the "typical" six-room, $14,000 American home transplanted to Moscow grew heated, but was eventually settled amicably. At the end of the day, with glasses of California wine in hand, the two had the following exchange:

> K: We stand for peace [but] if you are not willing to eliminate bases then I won't drink this toast.

N: He doesn't like American wine!
K: I like American wine, not its policy.
N: We'll talk about that later. Let's drink to talking, not fighting.
K: Let's drink to the ladies!
N: We can all drink to the ladies.[15]

Do It Yourself

The spectacularly modern appliances that stocked the American house in Moscow were the very latest domestic technologies, but the small tools of an earlier time were enjoying a popular upsurge in the postwar era as well. With rising of home ownership, maintaining, repairing, and adding on to one's house became widespread.[16] Although the houses of Levittown had neither basements nor garages—the traditional sites of home workshops—the attics were unfinished, and it was assumed that the new homeowners would divide them into rooms as the family grew. While the "do-it-yourself" movement had been evident in the two decades before the war, in the 1950s it caught on with a new fervor. A 1952 issue of *House & Garden* pictured a woman, in fact a CBS-TV actress, wearing a dress "especially designed by Vogue Patterns" and "intended for today's 'handyman,' i.e., the little woman in the home." With pearl earrings and choker, but wearing sensible flat heels, the woman held a hammer, while displayed on the pegboard ("an excellent way to store them, and keep them handy too") were a saw, pliers, brace-and-bit, clamp, paint brushes, and, significantly, an electric drill.[17] The manufacturer Black & Decker had first introduced a hand-held electric drill in 1914, but few had sold, partly because of their expense. In 1946 they tried again, putting one on the market for $16.95. Over the next eight years, something like 15 million were sold, and by 1958, three-quarters of all "handymen" owned one.[18]

Despite the existence of *House & Garden's* gender-bending female "handyman," doing it yourself remained what it had become in the interwar period, "a major source of domestic masculinity." As Steven Gelber has noted, after the war "the house itself was becoming a hobby, both the location and the object of leisure time activity." The home-improvement activities reinforced masculine gender definitions. As Gelber also found, "men and tools were beginning to create a definitional loop; to be a man one used the tools, and using tools made one a man."[19] In the alienating environment of postwar suburbia, home tool use was one way in which a father could not only pass on mechanical skills to his son, but also teach by example what it was to be manly.

Hot Rods

Another form of do-it-yourself thinking that helped sustain masculinity in the postwar years was the hot-rod culture. Like hand tools, machines had long been generally associated with masculinity. In 1948, the magazine *Hot Rod* first appeared on the stands, celebrating the modification of "Detroit iron," the production cars that flowed off the lines of Ford, General Motors, and Chrysler. The popular conception of hot-rodders as street hooligans and juvenile delinquents was far from the mark. The (mostly) young readers of *Hot Rod* were exposed to a steady diet of not only news about activities and instruction on automobile mechanics, but a consistent ideology of H. F. Moorhouse has identified as "urgent prescriptions to labour, to strive, to plan, to exercise skill, to compete, to succeed, to risk: themes like those supposedly typical of some traditional 'work ethic' but now directed to unpaid time."[20] The litany of virtues traditionally associated with technology were all there: "activity, involvement, enthusiasm, craftsmanship, learning by doing, experimental development, display, and creativity," all in the service of mechanical modification, but often directed at such ancillary concerns as fuel and driving skills.[21]

The very specific and democratic materiality of the process is a useful corrective to the notion that postwar technology was all rockets, assembly lines, and satellites. *Hot Rod* gave many examples like the following: "Stuart has been developing the same engine, a 1934 Ford for the last eight years. He has constantly improved it and hopes to improve it even more in the future. At one lakes meet a rod went through the block, shattering a four by eight inch hole in the side of the engine. He salvaged the pieces, welded them together and welded that piece into the hole. Performance was not altered."[22] The former academic psychologist, Dr, P.E. Siegle, now working for the Maremont Automotive Products Corporation, put all of this in a largest context. Working on hot rods, he claimed, was "creative, educative, constructive, and masculine all of which are desirable elements in furthering the best in the American way of life."[23] After going through the pages of *Hot Rod*, one close student of the subject, H. F. Moorhouse, has emphasized that "the automobile is a machine, it is a technology: and people confront it, handle, know, master, and enjoy it in a way which is often very satisfying."[24]

Along the continuum of postwar car culture, lowriders turned old cars into a different sort of statement. If the hot rodders enacted "activity, involvement, display, and creativity" through souped-up engines and street races, the lowriders expressed the same virtues in pursuit of "customized cars that, by crafty design, rode low and slow, cruising gaudily in candy-colored glory just a few inches from the pavement of Los Angeles' wide boulevards."²⁵ The young Mexican-American owners used these rolling art displays to move slowly down the streets, radiating cool and pride in their Mexican and Catholic cultures. Mainstream Anglo-American culture—and especially the police—saw not artistry, craftsmanship, and a thoroughly American masculine love of the automobile, but ethnic defiance. Like the hot rod, the lowrider showed strikingly how a technology could be appropriated to ends that the makers had not imagined.

Detroit's Car Culture

Hot rodders and lowriders aside, most Americans saw little need to modify their cars, except through the commercially approved purchase of optional equipment, either from the dealer at the time of sale or later from other automotive-parts retail outlets. The pent-up demand for cars after the war, and the increasing need for them from young couples recently moved to the suburbs, created a growing commitment both to the vehicle itself and to all of the other elements of a national transportation system increasingly dominated by automobiles. The 1950s proved to be a fertile time for the invention and proliferation of everything from interstate highways to franchised places to eat and sleep. In 1945, a mere 70,000 cars were built, but in 1950, along with 1,154,000 new houses, American produced 6,665,000 cars. During the first ten years following V-J Day, the number of cars registered in the nation doubled, with 8,000,000 coming off the Detroit assembly lines in 1955 alone.²⁶

Considered strictly as machines (though, of course, no one did consider them strictly as machines), the automobiles were extremely conservative, featuring few of the innovations available on European cars, and generally characterized by poor engineering. Like all technologies, however, cars are fraught with meaning, and that was shaped to a remarkable degree by what was called styling. As one observer has noted, "a business once ruled by engineering took on the trappings of the dressmaker's salon."²⁷

The couturier-in-change—the creator of what one design historian called "populuxe," or luxury for all—was Harley J. Earl.²⁸ The stylist's father had been a Los Angeles carriage maker who did work for the Hollywood studios building chariots, stagecoaches, or other vehicles that were needed. He then turned to creating custom bodies for the automobiles of the stars. Soon, Earl's shop was bought by the wealthy Cadillac dealer and later broadcast magnate, Don Lee. Young Harley was put in charge of designs, and was introduced to Detroit auto executives who visited Southern California. In 1925, Earl was given the chance to design the body for the new LaSalle, to be produced by General Motors. His design was a smash success, and Earl was hired by Alfred P. Sloan to head General Motors' Art & Color Section. Using clay models, Harley Earl gradually began to reshape American automobiles into the lower, longer, wider cars that characterized the postwar years. He liked long, massive hoods, for example, because they suggested a larger engine and therefore more power, and between 1930 and 1958, he lowered the average height of General Motors' cars from 72 to 56 inches.[29]

Tail fins, which Earl introduced on the 1948 Cadillac, were one of his signature devices. They were inspired, apparently, by a wartime visit to see the sleek Lockheed P-38 Lightning fighter in 1941, and like the rear spoiler of a later day, tail fins soon appeared on every American car and stayed there, increasingly in a vestigial form, for the next two decades. Chrome was another Earl signature and, perhaps, his eventual downfall. By the late 1950s the average Detroit car carried forty-four pounds of chrome that served no particular engineering purpose. Its real role was to literally dazzle the public: Earl had his designers bevel the chrome to a 45-degree angle off the horizon so that it would catch the sun and reflect it back into the eyes of the beholder. Also by the late 1950s, however, Earl was beginning to produce parodies of his once advanced designs. He demanded that a hundred pounds of chrome be added—somewhere, perhaps anywhere—to the 1958 Buick. It marked the sad end of the overstuffed and overdesigned postwar automobile.[30] The ultimate humiliation, however, came in the fall of 1957 when, after a media campaign rumored to have cost $10 million, Ford unveiled its new Edsel. The profound market failure of this somewhat pointless and characterless car stood in embarrassing contrast to the more serious and successful Soviet technology unveiled within a month of the Edsel—the earth's first artificial satellite, *Sputnik*.

By the time of the Edsel debacle, however, new vehicles had already appeared on the American road. In 1955, 80,000 foreign cars were sold in the country. Typically smaller, faster, and better engineered, these cars gained a cache and a following that threatened the hegemony of the Detroit philosophy. Several American cars had already bucked the trend towards gigantism as well. After his successful career building Liberty ships, Henry J. Kaiser started making small cars, especially the Henry J., and at American Motors, George Romney had brought out the Rambler. Eighty thousand of the latter were sold in 1955. But in 1959 alone, 150,000 of the new, profoundly un-American Volkswagen Beetles were sold in the United States, and over the years, it became and remained phenomenally successful. Nonetheless, in 1970, the ten largest American corporations included three automobile manufacturers (General Motors was number one) and three oil companies.

The small-car trend was reinforced with the fuel shortage in 1973, when members of the Organization of the Petroleum Exporting Countries (OPEC) shut off petroleum supplies in retaliation for American support of Israel in the latest Near-East war. Suddenly, fuel efficiency was a virtue, and one that most American cars could not claim. The coincidence of fuel shortages with the passage of the Clean Air Act in 1970 meant a new challenge to American automotive engineering to produce fuel-efficient, nonpolluting cars. While the Big Three auto makers, Ford, Chrysler, and General Motors, protested that the technology could not be developed in the near future, and perhaps never at a price that the American consumer would be willing to pay, Japanese manufacturers, especially Honda and Toyota, began a dramatic expansion into the American market.

Of a piece with the reform movement of the 1960s and 1970s, the attack on the automobile, and American automobiles in particular, was fanned by muckraking books that questioned their progressive mission. John Keats' *Insolent Chariots* (1958) was one early example, but Ralph Nader's *Unsafe at Any Speed* (1965) raised the most controversy. The latter, which opened with a description of a fatal crash of a badly engineered Chevrolet Corvair, revealed Detroit's imperial indifference to questions of safety and the lengths to which General Motors would go to protect both its image and profits. In keeping with their long tradition of engineering conservatism, American companies resisted such available devices as seat belts. However, with Japanese and European makers providing not only an example but an alternative, Detroit had to come into line, albeit slowly and reluctantly. Still, the internal combustion engine that powered personal cars—the standard set early in the twentieth century— was hardly threatened by the century's end. Electric cars, hybrid electric-gasoline cars, methane and natural gas-powered cars, not to mention those powered by fuel cells, continued to be resisted. The fuel efficiencies gained in the 1970s were lost again with the rush toward enormously profitable SUVs in the 1990s. These large vehicles embodied the same corporate indifference to matters of safety and convenience. The bloated dreamboat of the 1950s was back.

Highways

The national dependence on personal transportation was not much questioned, either. Like most technologies, cars came embedded in a complex network of other supporting and enabling technologies. Few were more critical or problematic than the roads upon which cars could move about. The earliest horseless carriages had helped fuel the Good Roads Movement at the turn of the century, and the federal government had accepted responsibility for aiding road building nationwide in 1916. Over the years, however, it seemed that traffic kept at least one jump ahead of road building, and despite the expenditure of vast amounts of money by all levels of government, adequate roads remained a problem.

During the interwar years, various jurisdictions groped their way toward some solution. U.S. 1—the main highway along the East Coast—had been increased to four lanes between Baltimore and Washington in 1930, but abutting merchants had blocked any effort to ban left turns across traffic. Furthermore, businesses crowded right up to the edge of the traffic, and a thousand driveways connected directly to the road itself.[31] One way to avoid this hazardous situation had already been built in Woodbridge, New Jersey, where a now-familiar design—the cloverleaf—kept traffic moving. The first parkway, New York's Bronx River Parkway, was built in 1922, and anyone who traveled to Europe might have encountered the German autobahn, which was constructed in 1929.

During the 1930s, some experts, including the head of the federal Bureau of Public Roads, had suggested creating a vast nationwide network of high-speed expressways. Congressional legislation in 1944 contemplated a national system of interstate highways, partly to employ at least some of the soon-to-be-returning veterans. Insufficient money was appropriated by the wartime Congress, but in 1947, Maine moved ahead and opened a forty-seven-mile turnpike between Portsmouth and Portland. Pennsylvania planned extensions of its turnpike, which had been started just before the war. In 1952, the New Jersey Turnpike opened, cutting the driving time between New York City and Wilmington, Delaware from five hours to two. That same year Ohio issued bonds to begin its toll road, and soon Indiana followed suit. An expressway stretching from New England to Chicago was becoming a real prospect. All of these were, of course, toll roads.

Then, in 1954, President Eisenhower gave two aides responsibility for planning the speed-up of the federal highway program, and two years later, Congress overwhelmingly passed the Federal-Aid Highway Act, which gave rise to our present system of interstate highways. From the start, these were to be superhighways, built on an ample scale in the anticipation that they would be both heavily used and later expanded. The pavements of old highways were four to five inches thick; the 41,000 miles of new interstates were nine to ten inches thick, laid over a fifty-inch roadbed. Three-hundred-foot rights of way left room for expansion. Fleets of giant earth-moving equipment—bulldozers, dump trucks, graders, and others, including a paver that could lay an entire lane as it moved forward—were put to work.

California, and particularly Southern California, were popularly thought to be the epicenter of American car culture, and Los Angeles' Arroyo Seco freeway, which opened on the eve of the war, proved to be the precursor of many others. After the war, the city built seven more freeways, linking them with four-high interchanges that were seen as engineering wonders of the modern world. By 1990, most of California's freeways were part of the 42,795-mile interstate system, connecting major cities and providing ring roads around them. In some places, especially in Los Angeles, freeways that were intended to speed up traffic seemed merely to attract more cars, as commuters moved farther and farther out into former farmlands and drove farther and farther to get to work.

If highway speed within cities, even on interstates, seemed like an oxymoron, on the open road, cars went much faster. Within a few years, most states had raised their old fifty-five mile-per-hour speed limits to sixty-five, or even seventy-five. And safety improved: the number of deaths per 100 million vehicle miles dropped from more than five in the mid-1960s to fewer than two at the end of the century, even though vehicle miles traveled almost tripled. The increased use of safety devices on cars, especially seat belts, was one important cause, but the excellence of the interstates was another.[32]

Land prices rose dramatically along the rights of way, especially adjacent to exits, which seemed suspiciously numerous in some areas. Motels multiplied—20,000 were built between 1951 and 1964—as did service stations and fast-food restaurants.[33] The first Holiday Inn was built in the early 1950s, helping to prove the utility and profitability of franchise operations with distinctive signage that could be seen from a speeding car before it reached the exit ramp. The restaurant of Colonel Harlan Sanders, on Route 25 in Corbin, Kentucky, was put out of business when Interstate 75 was opened seven miles to the west in 1955. Using money from his first social security check, he began to franchise his secret chicken recipe. By 1960, he had two hundred fast-food eateries, and by 1963, six hundred. In 1954, when he was selling the Multimixer, which could make five milkshakes at once, Ray Kroc was struck by the fact that one establishment, McDonald's Famous Hamburgers in San Bernadino, California, had ordered eight of the machines. A visit convinced him that the virtual mass production of burgers, fries, and shakes—potentially forty at a time—along with reliably clean toilets, was a formula for much larger success. He won the right to use the McDonald's name and expertise, built his own unit, and by 1957, had franchised thirty-seven more. Kroc put up the giant Golden Arches at each site, and like the Holiday Inn boomerang, it became the guarantee of a dependable product no matter which exit ramp along those 41,000 miles was used.[34]

Some buildings themselves became signs of the new era of sculptured cars and long vacations. Wayne McAllister, called the "architect for a car culture," was known for his destinations both close and far away. Among the first were drive-in restaurants, as well as Southern California's fast-food icon, Bob's Big Boy. Among the more distant destinations were the Sands and El Rancho Hotel in Las Vegas, the last being the first theme resort on the strip. His designs, while capturing perfectly the modern mobility and "populuxe" of the era, also shared its ephemeral nature. Few of his creations survive.[35]

Even some of the freeways were not to last. In 1959, San Francisco was the scene of one of the most influential "freeway revolts," as protest groups forced the city's board of supervisors to stop the state's construction of the Embarcadero Freeway, which would have connected with the Bay Bridge and cut the city off from some of its best views and historic waterfront. In 1966, two more planned freeways were turned down, and eventually the stunted Embarcadero span was torn down. Instead, the city cooperated with the building of the Bay Area Rapid Transit (BART) subway, which connected San Francisco under the bay to Oakland, Berkeley, points north, east, and south, and eventually, to its international airport.[36]

In 1993, Los Angeles opened what was predicted to be that city's last freeway, the Century. It was an overblown example of its type, a bit like one of Harley Earl's last Buicks, though better engineered. Running 17.3 miles, eight lanes wide, through nine cities, it cost an unprecedented $2.2 billion. It also displaced 25,500 people, and much of the cost of the road was for court-ordered programs to mitigate that despoliation. The interchange that connects it to the San Diego Freeway covers 100 acres and stands seven stories high, with "5 levels, 7 miles of ramps, 11 bridges and 2 miles of tunnels." Carpool lanes are monitored and controlled by electronic sensors in the pavement, and a light-rail line runs down the median. It was designed to carry 230,000 vehicles each day.[37]

One ubiquitous result of the postwar "family" car and the freeway network was the family vacation. Taking long car trips, including across the whole of the continent, was not uncommon even in the early years of the twentieth century. Chambers of commerce all along the Lincoln Highway (U.S. 40) seized upon World War I as an excuse to argue that that highway should be paved and double-laned from New York to San Francisco. But Jack Kerouac and the "average" family alike made the road trip an icon of American culture after the war. For the first time, unions began to negotiate the two-week paid vacation as a normal and expected part of full-time employment, and by 1953, 83 percent of vacation trips were made by car.[38] Typically, perhaps, these early vacations were to the beach or the mountains, especially to the national parks in the West, which President Roosevelt had done so much to advertise during the 1930s. But even as the family vacation blossomed, new commercial destinations beckoned.

Fun Factories

In 1993, the magazine *Sunset* gave directions to what it called "Southern California's fun factories": for Knott's Berry Farm, "from State Highway 91, exit south on Beach Boulevard"; for Magic Mountain, "from I-5 exit on Magic Mountain Parkway"; and for Disneyland, "from Interstate 5 in Anaheim, take either the Harbor Boulevard or Katella Avenue exit."[39] In 1954, alongside the San Diego Freeway, which was then under construction, Walt Disney had begun to turn a 160-acre orange grove into an amusement park that he named Disneyland. It cost $11 million, and opened on July 17 the following year. Disneyland was designed, according to one scholar, "for the values of long-distance travel, suburban lifestyle, family life, the major vacation excursion, and the new visual culture of telecommunications."[40]

Hugely successful from the beginning—a million people visited it during its first seven weeks—it inspired a second attraction, Disney World, which opened near Orlando, Florida in October 1971. Within a decade, the two together drew more visitors than Washington, D.C. The Florida park was 27,000 acres, twice the size of the island of Manhattan.[41] Disneyland, Disney World, and Disney's the later Florida attraction EPCOT (Experimental Prototype Community of Tomorrow) embody twentieth-century American technology in three major ways. First, they depend on transportation technologies to deliver visitors to the parks. Second, they make use of a host of sophisticated technologies to make the visitors' experience mirror Disney's precise plan for them. And third, they quite explicitly preach a gospel of happiness through technological progress.

It is important to emphasize that the parks are, above all, "destinations." Michael Sorkin has insisted that "whatever its other meanings, the theme park [including Disney's] rhapsodizes the relationship between transportation and geography." As he points out, it is unnecessary for the average American family to go to Norway or Japan because those places can be reduced to "Vikings and Samurai, gravlax and sushi. It isn't that one hasn't traveled," he suggests, because "movement feeds the system, after all. It's that all travel is equal."[42] The trip to Disneyland is a trip to many places at once, some real, most not, but all transformed by simulation. Once inside the parks, and out of the inevitable pedestrian traffic jams, one rides through the Matterhorn, through a haunted house, or on Mr. Toad's Wild Ride. And people come by the millions. Orlando has more hotel rooms than Chicago, Los Angeles, or New York, and is the "number-one motor tourism destination in the world."[43] Disneyland itself became a mecca for not only American families in their station wagons full of kids, but visitors from overseas as well. On October 26, 1958, a Pan American World Airways Boeing 707 jet took off from New York for Paris, inaugurating an escalating frenzy of international jet airline travel. "We shrunk the world," exalted the pilot, Samuel Miller.[44] Any number of visitors from outside America found that a trip to Los Angeles, focusing on Disneyland, Las Vegas (a sort of adult theme park), and the Grand Canyon was the ideal American experience.

The technologies of the parks themselves are carefully thought out. As Sorkin has suggested, "one of the main effects of Disneyfication is the substitution of recreation for work, the production of leisure according to the routines of industry."[45] The senior vice president for engineering and construction of the parks was Rear Admiral Joseph "Can-Do Joe" W. Fowler, who had graduated from Annapolis in 1917 and taken a master's degree in naval architecture from MIT in 1921. He was in charge of engineering at both Disneyland and Disney World, and completed both on schedule.[46] The irony was that in attempting to maintain the illusion of old-time virtues and simplicity, the "real" world of modern technology had to be hidden. Built on something like a platform, the park rides atop a labyrinth of supply rooms and tunnels that allow both workers and garbage to be moved without being seen. In Disney's words, "I don't want the public to see the real world they live in while they're in the park."[47]

For his special effects, Disney's "imagineers" modified existing forms to create unique rides and experiences. In the Hall of Presidents, lifelike replicas nod and seem to agree with the bromides expressed by one of them. Swivel cars not only carried visitors through exhibits at a uniform and predetermined pace, but also twisted them to face the correct direction for each presentation. The monorail became a model for others, particularly those used at airports. To quote Sorkin again, "at Disney, nature is appearance, machine is reality."[48]

And finally, all of the Disney parks, but especially EPCOT, preach a gospel of progressive technological determinism. As Tom Vanderbilt puts it, "the message everywhere is of technology ascendant, a new religion offering salvation, liberation and virtual shopping in a crisply pixellated sheen. Disney has an almost millenarian faith in technology."[49] The historian Michael Smith has analyzed the message of EPCOT, and finds that after the questioning of technology that peaked in the 1970s, Disney wanted to "revive the public's faith in progress, and in technology as the principal agent of that progress." He noted that for Disney, individual choice figures in technological change at the consumption end of the process, but seems absent from the beginnings of design and production. Further, any social or environmental disruption caused by technological change is accounted for by a "radical discontinuity between the past and future." The "humorless perfection of the future" is juxtaposed with the "whimsically flawed" though "steadily improving" past, with the "the present poised uncomfortably in between." He concludes that "wrenched out of context in this way, events float free of causes and effects and history splinters into nostalgia" and "if decontextualized history is nostalgia, decontextualized technology is magic."[50] The American past, like Tommorrowland, is the result of not our own choices, technological or other, but the work of some invisible and benevolent hand.

Television

No technology has been linked more powerfully to Disneyland than television. In the larger sense, "Disneyland itself is a kind of television set, for one flips from medieval castles to submarines and rockets" as one moves through the magic space.[51] Less metaphorically, Disneyland (the park) was born at the same time as television, and is largely a result of Disney's television programs. Unlike most Hollywood movie moguls, Walt Disney neither feared nor underestimated the power and future of television. The new medium, not surprisingly, very much wanted access to Disney films and cartoons, and their creator decided that he would trade that access for the money to build Disneyland. In 1953, Disney made a deal with ABC, at that time the least popular of the three major networks, through which the latter would provide a $500,000 investment and guaranteed loans of up to $4,500,000 in exchange for one-third of Disneyland and a weekly television series. The programs began in October 1954, just three months after the park opened, and shared not only a name but the four segments of Adventureland, Frontierland, Fantasyland, and Tomorrowland.[52] The programs, of course, provided not only badly needed construction funds, but a powerful public relations and advertising medium for the park. The technologies were a seamless web. As Sorkin notes, "television and Disneyland operate similarly, by means of extraction, reduction, and recombination, to create an entirely new, antigeographical space."[53]

Modernity had been centrally about the collapsing of time and space, and for most of the nineteenth century, the railroad had symbolized the technological core of that transformation. At the same time, however, the telegraph began the shift from transportation to communication as its most striking manifestation. Followed by the telephone, radio, and motion pictures, the media came to represent the cutting edge of technologically mediated cultural change. Although visionaries imagined something like television early in the twentieth century, research and development on television did not produce a commercially viable version until the 1930s, when a small number of sets in New York City and London could pick up locally produced signals. Embedded in a culture dominated by radio and the movies, television had little appeal to consumers. In 1939, only 13 percent of Americans polled said they would buy a set for their homes.

Wartime advances in electronics technology helped to improve the pictures that could be transmitted, however, and by the end of the 1940s, television use was poised for growth that would make it arguably the technological icon of postwar America. In 1950, some 3.1 million sets were sold, and just two years later, 44 million people watched Lucy go to the hospital to have her baby, twice the number of people who watched Eisenhower inaugurated as president the next day. In 1955, 32 million sets were sold; by that time, two-thirds of American homes had televisions, and they were selling at a rate of 10,000 a day. By the end of the decade, 90 percent of homes had sets that were turned on an average of five hours per day.

The reason for this astonishing success probably had something to do with what the British cultural critic Raymond Williams called "mobile privitism."[54] People were both mobile and private, moving to the new suburbs and losing public spaces with the decline of inner cities and the rise of shopping malls. At the same time, they were experiencing a shifting definition of citizenship, from political participation to commodity consumption. Television provided a virtually real-time common experience of signs, a kind of virtual experience of community.

Not surprisingly, the results exacerbated the causes. Mass middlebrow magazines such as *Colliers, Look, Life,* and the *Saturday Evening Post* were wiped out while in 1954 alone, *TV Guide* gained 98 percent in circulation. Motion picture attendance shrank by half during the 1950s; in the first two years of that decade, 134 movie theaters closed in Los Angeles alone. Heralding the advent of what would one day be an almost necessary accessory to television sets, in 1955, Ampex demonstrated the first successful video recorder and player. By the end of the century, new Hollywood films would be out on video or DVD almost as soon as they were released to theaters, and some films earned more on tape than they ever would on the large screen.[55]

After its stunning success, television technology largely stabilized; large screens and the transition to color were the major advances. Unless, of course, one counts the Laff Box, already invented in the 1950s by Charles Douglass to add laughter to programs in order to enhance studio audience responses to live broadcasts—or, as time went on, to simulate an audience when none was present. By the turn of the century, the laugh machine was reduced to the size of a laptop computer and carried "hundreds of human sounds, including 'giggles, guffaws, cries, moans, jeers, ohs and ahs,'" forty of which could be mixed at any one time.[56]

Reproductive Technologies

Few technologies of the postwar years affected the everyday and intimate lives of Americans more than the controversial birth control pill. It became so ubiquitous and familiar that, like some celebrity superstars known only by their first names, it was called simply The Pill. By the end of the nineteenth century, chemicals in the body, soon to be given the general name of hormones, had begun to interest medical researchers. Those that were involved in reproduction were called "sex hormones," and in 1928, two American doctors named one of them progesterone, from the Greek meaning "in favor of" and "bearing offspring."[57] Researchers soon realized that because progesterone prevented ovulation, it could be used as a birth control mechanism, but because any American could be sent to prison for talking publicly about birth control, it was difficult to address the issue directly and openly. A Harvard endocrinologist, Fuller Albright, actually referred to "birth control by hormone therapy" in a 1945 essay entitled "Disorders of the Female Gonads," but the extremely limited supply of progesterone, added to the legal danger of any discussions of the subject except in coded words or discrete places, meant that future progress was far from certain.

At that point, two American women who had already defied the authorities on the topic of birth control became interested in the "hormone therapy." Margaret Sanger and her sister had opened the country's first birth-control clinic in 1916 and had served time in jail. By this time, her organization, the Birth Control League, had become the Planned Parenthood Federation. Sanger lacked the funds to back further research on hormones, but in 1951, a longtime suffragist friend and associate, Katherine Dexter McCormick, asked her what the birth-control movement most needed at mid-century. Sanger responded that a simple, cheap method of population control through contraception was the goal to aim for, and McCormick pledged the funds to carry the search forward.

McCormick, who came from a wealthy and prominent family, spent three years as a "special" student at MIT, then became a regular student in 1899 and graduated with a degree in biology in 1904. That same year, she married

Stanley R. McCormick, the youngest son of Cyrus Hall McCormick, the inventor of the reaper. With their combined fortunes, she was able to donate MIT's first on-campus dormitory for women and made the school the sole beneficiary of her $25,000,000 fortune when she died in 1967. Meanwhile, she devoted her attention and resources to helping Sanger.[58]

Sanger introduced McCormick to a physiologist, Gregory Pincus, who had a small research institute and had done work on the effects of progesterone on ovulation, but had lost his support from a drug company to pursue the subject. McCormick provided that support. Pincus soon met a Boston gynecologist, John Rock, who was already experimenting with regulating the menstrual cycles of some patients with doses of estrogen and progesterone by injection. He then contacted several drug companies and discovered one, Syntex, that could provide relatively inexpensive progesterone made from Mexican desert plants, and another, Searle, that had worked out an effective way to administer the hormone orally. Pincus talked Rock into quietly using oral progesterone on fifty Massachusetts women, and in 1955, the former announced to Planned Parenthood that not one of them had ovulated during the experiment. Because the whole subject was illegal in much of the United States, Pincus went to Puerto Rico, Haiti, and Mexico City, where precisely those women whom Sanger wanted to help lived, for large-scale trials. The results showed that the oral contraceptive had a failure rate of only 1.7 percent, far superior to the other commonly used technologies: diaphragms failed at the rate of 33.6 percent, and condoms at 28.3 percent. In 1957, Pincus received permission from the Food and Drug Administration (FDA) to use the pill, called Enovid, to treat menstrual disorders and repeated miscarriages. Within two years, a suspiciously large group of half a million women was taking The Pill. Despite opposition, particularly from the Catholic Church, the FDA in 1960 declared it safe for use as a contraceptive, and within six years, at least six million women were taking it—and that number was growing. In 1972, the Supreme Court declared that even unmarried women could have access to The Pill for contraceptive purposes, a victory not only for those who wished to use it, but also for the many feminists who had fought long and hard for their right to do so. Although The Pill was not the "cause" of the so-called sexual revolution of the 1960s and 1970s it was an important contributor, and the very act of not only legitimizing but normalizing public discussion of sexuality and reproductive issues led to further developments.

One unlooked-for development was the raising of issues of intellectual property rights due to the development of sophisticated family-planning technologies. The reproductive technologies that allowed doctors to use a husband's sperm to create a zygote with an egg from his wife—who had had a hysterectomy and was therefore unable to bear a child—and then to implant that zygote (created in vitro) in the womb of a second woman (for pay and with the proper legal contracts) led to the law case of *Johnson v. Calvert* before the California Supreme Court in 1993. The case was unusual not because the surrogate mother changed her mind and wanted to keep "her" baby, but because it suggested a new level of the commodification of a human being. Finding for the married couple, the court said that the mother who produced the egg had "intended to bring about the birth of a child," and that intention made her the natural mother. In a dissenting opinion, one justice rejected the implied analogy that "just as a[n] ... invention is protected as the property of the 'originator of the concept,' so too a child should be regarded as belonging to the originator of the concept of the child."[59] Just as the oral contraceptive became an integral part of the rethinking of sexuality, so other reproductive technologies, from in vitro fertilization to cloning, contributed to shifting the definition of identity itself.

SEVEN

BRAIN DRAIN AND TECHNOLOGY GAP

At the end of the twentieth century, statistics showed that the United States was exporting electronics to the rest of the world at a rate amounting to $181 billion, making the country what David Lazarus of the *San Francisco Chronicle* called "the tech toy store to the world." The leading markets were our closest neighbors: first Canada ($29.3 billion), then Mexico ($21.9 billion). Japan was the third-largest market ($16.1 billion) with the United Kingdom taking $10.8 billion. Of the next six countries, two were European and four Asian. American electronics manufacturers not only sold abroad—accounting for 26 percent of U.S. exports—but they invested there. A total of $12 billion had been invested in electronics in the United Kingdom, $7.5 billion in Singapore, $5.6 billion in Japan, and lesser amounts in Germany, France, Italy, and the Netherlands.[1]

As one French observer wrote in 1988, "success in technology seemed to come easy in America."[2] Decades earlier, America was at the height of its technological supremacy and confidence; the perceived American technological hegemony after World War II created problems for and evoked concern among even her closest allies abroad. While the nation worked to restore and reform the devastated economies and social infrastructures of Germany and Japan, its European allies only slowly recovered from the destruction of the war and looked to America as both an example and a threat—even as America came to worry that its own leadership was eroding as Europe and Japan began to recover.

Meanwhile, students from overseas increasingly enrolled in American colleges and universities, seeking advanced training in engineering, science, and medicine. Not surprisingly, many of these students chose to stay in the United States, where professional opportunities and rewards were seen to be greater than those at home. This "brain drain" had the potential to cripple the reconstruction of war-torn societies and the building of new nations from the collapse of European empires in Asia and Africa. Noting with alarm what they called the "technology gap" between themselves and America, European nations—most notably, France—saw American technological hegemony as a threat to their own economies and cultures, and sought, in the words of British prime minister Harold Wilson, to forge for themselves new futures in the "white heat" of science and technology.

European Alarm

The success story of American electronics at the end of the century would not have surprised the Atlantic Institute, a think tank set up in 1961 by what was then called the Atlantic Community. Established to "promote cooperation among the Atlantic countries in economic, political and cultural affairs," it had warned as early as 1966 that a gap existed between Europe and North America in some important science-based industries, and claimed that the gap was growing ever larger. It was a wake-up call to Europe from a group that included not only Europeans, but also prominent American business and political leaders. Its director-general was Walter Dowling, the former American ambassador to West Germany.[3]

The warning blended in nicely with what was widely seen as an "anti-American campaign" by the French president, Charles de Gaulle. A proud man leading a proud country, de Gaulle cast a jaundiced eye on what he saw as an American economic, cultural, and political hegemony, and was as keen to protect French *grandeur* as the French Academy was to protect the French language from corruption. For many people, the battlefield seemed to be culture in general, but by the 1960s, there was a growing technological aspect to the confrontation. Other nations might not be so disturbed by the threat to French language and culture, but, the editor of *Science* warned, when the spotlight turned to the threat of a technology gap, de Gaulle picked up support from the rest of Europe.[4]

By the end of 1966, the maneuvering for technological dominance, or at least survival, seemed to threaten the realignment of world power. Soviet Premier Alexei N. Kosygin, on a visit to Paris, warned that the United States was using scientific cooperation as a tool of political domination, and proposed that Russia and Europe form a technological alliance against American hegemony. De Gaulle responded by calling for more scientific and technological contacts between Western Europe and the Soviet Union. Prime Minister Harold Wilson of the United Kingdom recommended forming a Western European "technological community" to rival the United States, and the Italian foreign minister, Amintore Fanfani, urged the United States to undertake a "technological Marshall Plan."[5]

The Technology Gap

Science ticked off the aspects of the problem that were to dominate the debate over the next decade. First, Europe recognized that the large size of the market in America made proportionally large investments in research and development attractive. RCA, it was alleged, had spent $130 million to develop color television—something it was hardly likely to have done unless it expected to sell a very large number of sets. Second, not only was the American gross national product large, but the country invested a larger proportion of it in research and development than did Western Europe.

Two other factors that contributed to the Technology Gap, according to the journal, were less talked about in Europe. One was that while 40 percent of American young people in 1967 attended universities, only 8 percent did so in Europe. A second factor that was allegedly overlooked was that of social habits. Drawing upon a long tradition of democratic self-congratulation, *Science* claimed that in Europe, scientists waited for an assistant to make measurements, while in the United States, the scientist herself was willing to do the actual dirty lab work. The same tone permeated another social difference: in America, "top scientists" were willing to eat in cafeterias, while in Europe, they insisted on a "leisurely lunch" served on "a white tablecloth" and, presumably, including wine[6]

Debates over a technology gap had been largely limited to high government circles and scientists who hoped to convince their governments to increase research and development to combat it, until the publication in France of *Le Defi Americain* at the end of 1967. The author, J.-J. Servan-Schreiber, was the editor of the weekly *l'Express*, and his well-turned phrases lifted the issue to one of wide public awareness. Within four months of publication, the book sold 400,000 copies.[7] "In France," wrote historian Arthur Schlesinger, Jr., in the foreword to the American edition, "no book since the war, fiction or non-fiction, sold so many copies" so quickly.[8] That Schlesinger was picked to write the foreword to the American edition clearly indicated that the debate was a major cultural event, not simply an arcane argument among scientists, economists, and cabinet ministers.

For the American audience, Schlesinger was reassuring: Servan-Schreiber's analysis of America's role in Europe was both "subtler and more profound" than that of de Gaulle. While Servan-Schreiber warned that "if present tendencies continue, the third industrial power in the world, after America and Russia, could be, not Europe, but American industry in Europe," he believed that Europe could "escape American domination" through "federalism and social justice."[9] Calling the author "a European of the Kennedy generation and style," Schlesinger concluded, "a strong and progressive Europe, he rightly contends (only barely mentioning Vietnam), is vital to save the United States from the temptations and illusions of superpowership: Europe holds in her hands more than her own destiny."[10] It was a diagnosis that was to prove tragically accurate.

Servan-Schreiber began his book with the claim that in Europe, "we are witnessing the prelude to our own historical bankruptcy." The continent that had so recently lost its overseas empires was now in a war that was being fought not with armies, but "with creative imagination and organizational talent." His best example was the computer, a device he said had been "conceived" in Europe, with Germany taking the lead and Britain a close second. "The Americans," he claimed, "started late, made mistakes in their forecasts, were initially outpaced, and still have not solved all their problems. But they are surging ahead, and we have little time to catch up."[11] Servan-Schreiber may have been reading a different history, but he fixed on integrated circuits as the decisive advance that gave American producers the advantage.

Integrated circuits were important because they could be used in all kinds of machines, especially computers and missiles, but they were also important because, according to an estimate by the Organization for Economic Cooperation and Development (OECD), a firm had to sell a million of them a year to pay back its research costs. All of Europe, he claimed, could only use a quarter of that amount, while in the United States, three firms—Fairchild, Texas Instruments, and Motorola—were all marketing them. "Every European industrialist" he said, was studying the way IBM introduced its 360 series of computers. That giant had invested $5 billion over four years, a sum equal to what the American federal government was spending on its space program. To become a player in a game with these stakes, Europe would have to integrate its various national industries and greatly expand its use of circuits. "In no other industry," he concluded, "is colonization from abroad so oppressive." In what would become a major trope of this narrative, Japan, with its willingness to borrow, was said to have fared much better, and would catch up with America by 1980.[12]

Dramatic as the numbers were, Servan-Schreiber believed that the real problem was not one of technologies, but of human resources. Ten years ago, he said, people believed that the Soviet Union would overtake the United States in terms of standards of living, but clearly that was not going to happen. In America, "the ability to adapt easily, flexibility of organizations, the creative power of teamwork" all created a "confidence of the society in its citizens"—and *"this wager on man is the origin of America's new dynamism"* (emphasis in original).[13] This was hardly the opinion of what Servan-Schreiber unkindly called "the intellectuals of the older generation," and to make his point, he quoted Simone de Beauvoir: "in every country in the world, socialist or capitalist, man is crushed by technology, alienated from his work, enslaved and brutalized."[14] America was clearly the worst example of this degradation, but Servan-Schreiber wanted Europe to become even more like that country, at least insofar as it needed to adopt those aspects of American society that accounted for its technological dynamism—its business management style, its educational opportunities, its investment in research and development, and its adaptability, flexibility, and the willingness of its scientists and engineers to work together toward innovation. In his conclusion, he drove the point home yet again: "we are now beginning to discover what was concealed by 20 years of colonial wars, wars that dominated our thoughts and our behavior: the confrontation of civilization will henceforth take place in the battlefield of technology, science, and management."[15]

The Israeli sociologist Joseph Ben-David had just completed a study of much the same question of technological development, and had similar advice for the OECD: Europe should become more like the United States when it came to encouraging science and technology. In his influential book *Fundamental Research and Universities: Some Comments on International Differences*, Ben-David accepted the conventional wisdom that economic growth came from technological innovation, which in turn grew out of applied fundamental scientific research. The institutional arrangements that allowed this process to work so well in America went back to the turn of the twentieth century, with the rise of large research universities that combined basic scientific research and applied science: the physical sciences and engineering, or the life sciences and medicine.[16] Dael Wolfle, the publisher of *Science*, described the study as "a provocative and sure-to-be controversial analysis," and guessed that "in suggesting to European governments a means to narrow the gap, Ben-David has also offered the U.S. Government a challenging prescription for getting greater returns from the monies it invests in higher education and university research."[17]

The American Challenge kicked off a more public debate of the issue that raged for nearly a decade, on both sides of the Atlantic, with each new combatant emphasizing whichever piece of the argument most appealed, compelled, or represented its own interests. The political scientist Robert Gilpin pointed out one salient fact that the book largely overlooked. He argued that while "Europeans are almost unanimous in their opinion that the technology gap is real, is threatening to their long-term well being, and is widening," it was also true that "for Washington, on the other hand, the technology gap is an official non-issue." Evidence of American "leadership" was welcomed, especially because the areas that most concerned Europeans—computers, electronics, atomic energy, and aerospace—were those that America most wanted to keep to itself. Gilpin further noted that Europeans found it easier to cooperate with the United States than with each other, and that any integration of science and technology would have to wait until political integration was accomplished.[18]

In 1970, the Atlantic Institute issued a collection of essays titled *The Technology Gap: U.S. and Europe*. One essay focused on the influx of American companies into Europe.[19] Another pointed out that there were no European equivalents to the *Wall Street Journal* and *Business Week*, or even, it claimed, the *New York Times*, which were "in some respects matchless in calling attention to changes in the environment." Finally, it noted that there was an "income gap" and claimed that if national incomes in Europe caught up with those in the United States, the technology gap would "fade away."[20] Drawing on statistics published by Edward Denison, another essay presented a table of differences between the United States and European countries in terms of national income per employed person in 1960. All of Northwest Europe's the income was 41 percent of that in the United States, with a narrow variation between 35 percent in the Netherlands and 42 percent in Denmark. Italy outdid all others, with 60 percent.[21]

Antonie Knoppers, another contributor who had been at the 1966 meeting organized by the Atlantic Council in Geneva, admitted that four years on, "I and others have come to understand that the subject is much more complex than originally thought." Rejecting any notion of a "technological Marshall Plan," he believed that it was "more the phenomenon of the international corporation establishing itself in European states than the 'technology gap' itself which causes apprehensions." He also focused on the American phenomenon of small research-driven firms, established by technically trained entrepreneurs who found risk capital to turn ideas into marketable products. These firms, he believed, were often located near research universities. "Europe is weak here," he concluded.[22] Silicon Valley had not yet risen in the shadow of Stanford University, but that is clearly the sort of thing he had in mind.

By 1970, new initiatives were evident in Europe, and Gilpin's prediction that scientific and technological coordination would have to wait for political integration was put to the test. Some of Europe's "leading figures in industry, government, science and education" had created an informal network that worked out of the public spotlight. Their coup was the planned establishment of the International Institute for the Management of Technology, to be set up in a former monastery in Milan. The North Atlantic Treaty Organization (NATO) had run a science program since 1958, which the United States tried to nudge in a direction to help close the technology gap a decade later. A series of meetings was held that included American members, and resulted in a call for European nations to cooperate in establishing an analog to MIT.

The original idea had been to establish a comprehensive university, as Ben-David had described. But picking up the idea that the real gap was managerial rather than technological, the proposed institute was designed to be a management school that trained scientists and engineers for corporate management positions in European firms. West Germany offered some funding, and the United States, typically, offered everything but. IBM's European branch stood ready to contribute $10,000, and other firms were reported to be ready to do the same.[23]

Already by 1970, participants in the debate were becoming aware that the ground was shifting under the basic assumptions of the technology gap. Great Britain had always had something of a problem because it was never clear whether it was, or even wanted to be, a part of Europe. In 1963, Harold Wilson had taken over the Labour Party and started to position it for the 1964 general election. Labour had long wandered in the political wilderness, but Wilson brought it to victory—in part at least by his famous speech at the Scarborough party conference, in which he articulated the vision of a country revitalized by "the white heat of technology."[24]

By one measure of technological standing—the import and export of "know-how"—the United Kingdom had fared well since World War II. In 1971, it was the only European nation that sold more technological knowledge than it bought, and stood second only to the United States in this respect worldwide. But that ratio had begun to turn around only two years before, and income from sales of technological know-how abroad, while still climbing, was growing at a slower rate than they were in France, Germany and Japan. The whole picture was muddy, and any right policy had to be based on disaggregating the gross figures. The German "economic miracle" was based on the wholesale importation of technology, in part through licensing patents, a strategy unavailable to developing countries that could not afford the licensing fees. In Britain, there was apparently some "mistrust of foreign technology," but the general lesson seemed to be that a nation that wanted to be competitive should "both a borrower and a lender be."[25]

Growing American Fears

Britain was not alone in worrying about its worldwide position. One essay in the 1970 Atlantic Institute volume stated "that in the United States there currently is concern that American industry may be falling behind Europe technologically, and that this may partly explain its shrinking trade surplus."[26] Computers were a case in point. The good news was that the European market for computers in 1971 was growing at a rate of 20 percent a year, and American exports of data-processing equipment to that area had reached $1 billion a year. Seventy percent of the market belonged to IBM, but that company was feeling increasing heat from competitors. In part this was from other American firms, notably Honeywell, which had a larger presence abroad than at home.

European firms were also making inroads, often with government support. As early as 1965, the British minister of technology told the House of Commons that "the Government considers it essential there should be a rapid increase in the use of computers and computer techniques, and there should be a British computer industry." In France, President de Gaulle had also called for a national computer industry, and after several mergers, the Compagnie Internationale pour l'Informatique gained about 10 percent of the home market and joined with the American firm Control Data Corporation in studying closer integration. Both British and French companies out-sold IBM in Eastern Europe.[27]

As the Atlantic Institute observed, the balance of trade was a popular way to measure technological success. Far from resting on their laurels as world-technology leaders, many politically active American scientists and engineers, from both government and industry, viewed the situation through darkly tinted glasses. By 1971, there was a crescendo of concern in the United States that the country was "in danger of losing its preeminence in advanced technologies, particularly those technologies that are important in world trade."[28] West Germany and Japan, which were discouraged from rearming after World War II and therefore almost had to put their brains and wealth into civilian technologies, were the nations thought most likely to "eventually overtake the United States and gobble up a major share of the world market in high-technology products." It was a worst-case scenario, and curiously, government agencies pushed it harder than did corporations. The National Academy of Engineering, PSAC, and the Department of Commerce were by and large in agreement that a crisis was at hand.

Secretary of Commerce Maurice H. Stans declared flatly that "our technological superiority is slipping."[29] The key analyst at Commerce appears to have been Michael T. Boretsky, who specialized in world trade matters, particularly between the United States and Soviet Union; his views seemed to "pop up everywhere," according to *Science* magazine. According to Boretsky, during the 1950s and 1960s, the United States had trade deficits in raw materials and non-technology–intensive manufactured goods. In his opinion, the balance of trade in the agriculture sector was not clear, but in technology-intensive manufactured goods, the U.S. position was "rapidly deteriorating." Western Europe, Canada, and Japan were increasingly strong in this sector, and Japan had run a trade surplus with the United States since 1965, based largely on electrical and electronic goods, scientific instruments, and automobiles. Many economists disputed Boretsky's analysis, but those who chose to believe included some not entirely disinterested parties.

Since falling behind in high-tech products could be read as a sign that the government should spend more on research and development, it is not surprising that scientists helped to sound the alarm over trade. The director of the National Science Foundation, William D. McElroy, expressed his concern that "this country's going to get behind in the technological developments." The National Academy of Engineering held a symposium on technology and international trade, and its chair, John R. Pierce —who was also a research executive at Bell Labs—intoned that "today we are facing a technological challenge far more important to us and far more difficult to meet than the challenge of Sputnik."[30] A third voice of concern came from PSAC, which was chaired by Patrick E. Haggerty—board chair also of Texas Instruments—and boasted a who's who of American scientist-politicians and corporate representatives. Citing the by-then-familiar trade data, William D. Carey, a panel member and former director of the bureau of the budget, opined that "there are grounds to be deeply concerned about the intensity and diversity of what this country is doing in the field of technology."[31]

Many of the alarmists had a long record of arguing for ever-increasing federal support for research and development, either through direct government subsidies or such incentives as tax credits to corporations. The specter of a diminishing lead in the world trade of technology stimulated them once again to call for more support. One economist pointed out a potential flaw in this argument: "more research expenditures," he bluntly maintained, "do not seem to lead to more technical progress."[32] This position fundamentally challenged the whole postwar argument put forward by Vannevar Bush, that technology flows out of science and that it was necessary to plant seeds before hoping to pick fruit. The quarrel could be explained by the fact that either technical people did not understand economics, or that economists paid too little attention to technology. Both positions seemed all too plausible.

The responses of the government to this concern varied. Senator William Proxmire, a Democrat from Wisconsin who was a frequent thorn in the side of Big Science, took umbrage in 1974 at the activities of the Air Force Foreign Technology Division, which he said was secretly funded and used to recruit American defense scientists and engineers to collect data on overseas developments in science and technology.[33] A report the next year, issued by the

Subcommittee on Economic Growth of the Joint Economic Committee of the Congress and authored by Robert Gilpin, a professor of public and international affairs at Princeton University, declared that "technological innovation in the civilian industrial sector of our economy is at a critical point," and that "America's once unchallenged scientific and technological superiority has deteriorated." In part, he argued, this was because American innovations quickly diffused to "lower-wage economies," and partly because "European and Japanese economies" had "diminished what once appeared to be an unbreachable 'technological gap.'"[34]

Looking at the way the federal government had been distributing its research and development funds, Gilpin pointed out that they had, over the years, been concentrated on "'big science and technology' related to defense and prestige." The government would have to adopt an "overall national strategy" for science and technology, and Gilpin was ready with suggestions. First a correct policy would set targets and directions, and remain flexible. Second, "the coupling of technology and user demand" should take place at every level, from basic to applied research. Third, the government should "support and advance national capabilities," which boiled down to pumping more money into research universities. Fourth, agencies should be encouraged to support science and technology in universities and schools of engineering, rather than running their own laboratories, as had the Department of Agriculture. Gilpin's final admonition was that the government should concentrate on creating and supporting new industries, rather than wasting time and money trying to prop up old ones.[35]

If importing technology had helped Europe and Japan "catch up" to the United States, would it help if America did the same? The categories of "know-how," which moved so easily across national boundaries, included patents, licensing agreements, manufacturing rights, foreign direct investments, and the sale of high-tech products.[36] Sherman Gee, the head of the Technology Transfer Office of the Naval Surface Weapons Center presumably knew whereof he spoke when he complained that the "United States, which has focused mainly on space and defense efforts and only recently has had to confront problems associated with coupling more closely the nation's technology to economic needs, could therefore benefit from the experiences of these other nations." The federal government, he wrote, should "take on a new role as an intermediary and potential user of foreign technology."[37] Once again, the suggested cure ran up against the reluctance of the government to intrude too far into what was seen as the proprietary concerns of American companies, especially very large firms that might be expected to best take care of themselves and take a dim view of proactive government initiatives to import technology from abroad.

Yet another attack on American technological leadership came from foreign investors purchasing small high-tech companies in the United States, some of them prime targets for companies abroad who wanted to gain access to American technology in part because the firms had trouble raising venture capital at home. In 1978, a West German firm paid 73 percent above market price for a controlling interest in Advanced Micro Devices, an electronic equipment maker from California's Silicon Valley. "Every such acquisition," warned Senator Gaylord Nelson, "costs the United States jobs, profits and know-how. Cumulatively, foreign takeovers of high-technology firms put a dent in American financial and military leadership around the world by allowing foreign interests access to our most advanced technology."[38]

The largest technology gap of all, of course, was that between developed countries of the northern hemisphere, led by the United States, and the less-developed countries (LDCs) of the southern hemisphere. When the United Nations began planning a conference on science and technology for development, to be held in 1979, the fault lines within the United States, and between it and the LDCs, were made abundantly clear. The conference's concentration on the transfer of established technologies, most of which were proprietary to American corporations, to the LDCs found those corporations digging in against moves that might undermine their worldwide competitive advantages. At the same time, the science establishment objected to isolating particular technologies from the larger question of how science and technology in general might aid development. Former member of Congress Emilio Q. Daddario, by then president of the American Association for the Advancement of Science (AAAS), called for a policy that emphasized "creating needed technology which does not now exist and in strengthening capacities for scientific research and the use of scientific method." Finally, advocates of appropriate technologies wanted the conference to remain open to small-scale, locally appropriate and producible technologies rather than limiting transfers to large-scale, expensive off-the-shelf items. The nations of the southern hemisphere, however, wanted neither homemade solar cookers nor lessons in the scientific method. They wanted up-to-date northern technologies, and American corporations were not willing to hand them over.[39]

Brain Drain

One of the factors that kept LDCs from closing the technology gap with the United States and other developed nations was the so-called brain drain: the movement of technically trained people from less to more developed areas. In one of her novels of the South Asian diaspora, Bharati Mukherjee introduces the protagonist's former husband, Bishwapriya, as a California-based graduate of the Indian Institute of Technology, who "with his friend Chester Yee... developed a process for allowing computers to create their own time, recognizing signals intended only for them, for instantaneously routing information to the least congested lines."[40] Bishwapriya was fictional, but many thousands of other South Asians who flocked to work in Silicon Valley late in the century were very real indeed. So great was this instance of brain drain that in 2003, *San Francisco* magazine noted, under the headline "A Heat Wave Hits the South Bay," that "the tech boom of the 90s brought a migration of South Indians to Silicon Valley and the East Bay." The subsequent bursting of the technology bubble on Wall Street brought a virtual end to that influx, but as the magazine noted, "though the jobs that brought them here may have gone, the restaurants that cater to their appetites continue to thrive."[41]

However, as one observer of the brain drain puts it, "myth as well as reality links brain drain with the technology gap."[42] The two went hand in hand at least as far back as the Industrial Revolution. In the eighteenth century, the British parliament passed laws against expatriating workers in the newly mechanized textiles industry. Along with a ban on exporting plans, parts, and machines, the prohibition was intended to keep other nations from closing the new "technology gap" that was a critical part of Britain's rise to industrial preeminence.

The modern term "brain drain" apparently was coined by the British to describe their own loss of doctors, scientists, and engineers through emigration.[43] From the American perspective, the influx from Britain and other countries was welcomed. As the Cold War wore on, and especially after the Soviet launch of *Sputnik* in 1957, there was alarm in official circles that the United States needed many more technically trained people to keep ahead of the Soviets in science and technology. One strategy to close the gap was to train and utilize more Americans. New Math was introduced into schools to improve numeracy. The National Defense Education Act provided funds for college students who chose to major in engineering or the sciences. It was even belatedly recognized that if young women could be tempted into engineering schools, that alone could theoretically double the number of engineers coming out of the pipeline.

Even with all of these initiatives, however, the numbers were not enough to feed the seemingly insatiable demand created by rapidly expanding defense and space programs. It was sometimes alleged that the increase in NASA's budget one year was sufficient to hire every new Ph.D. engineer that graduated. Whether or not that was true, it was clear that the "demand" for scientists and engineers was not some fixed amount, but rose and fell with shifts in federal policy and the exigencies of the competition with the Soviets.[44] The gap was filled by taking in technically trained people from overseas.

In 1956, 5,373 scientists, engineers, and physicians arrived in the United States. In 1966, that number had risen to 9,534, of whom the largest number, 4,921, were engineers; 2,761 were physicians, and the remaining 1,852 were scientists. They were an increasing percentage of the "professional, technical, and kindred workers" who immigrated to the United States, and in these eleven years totaled 43,767. In terms of places of origin, in 1956, 2,419 arrived from Europe, 1,940 from North America (mainly Canada), 535 from Asia, 512 from South America, and 69 from Africa. Ten years later, Europe still sent the largest number, 3,612. Asia was second with 2,736, followed by North America with 2,591, South America with 807, and Africa with 129.[45]

The large increase in migrants from Asia was a direct result of U.S. immigration policy changes. Quotas had been introduced into immigration policy in the 1920s, but were modified by the Walter-McCarran Act of 1952, which gave preference within quotas to people of high education or exceptional abilities. In 1962, the law was liberalized so that those with immigration applications already on file could enter without regard to national quotas, causing the first jump in the entrance of Asian professionals. Then in 1965, a new immigration act removed quotas, while leaving the preference for the educated and skilled. In June 1967, there were an estimated 8,400 professional workers waiting for visas, of whom a third were thought to be scientists, engineers, or physicians.[46]

The brain drain from both Canada and Britain was significant, but it was assumed that those countries could weather the loss. Their reserves were larger to start with, they could undertake programs to lure immigrants back home, and perhaps most importantly, they too had an influx of trained people from LDCs. It was estimated that about half of the people entering the United States from Canada were not Canadians at all, and that many of those entering from Japan, a developed country, were in fact from Southeast Asia merely queuing up in Japan until they could enter the United States. People also came in significant numbers to the United States directly from LDCs, and their percentage of the total immigration of scientists, engineers, and physicians increased throughout the 1960s.[47] In 1956, 1,769 people in this group entered the country, making up 32.9 percent of the total inflow. By 1966 that number had reached 4,390, making up 46 percent of the total. While it was sometimes maintained that "brain drain is a symptom of underdevelopment," the hemorrhage of this already scare resource in nations which, as colonies, had been systematically discouraged from mass education, was bound to have a negative effect on efforts to engage in the "nation-building through science and technology" that the same observer advocated. Sixty percent of those from LDCs in 1966 came from the eighteen countries that had received the bulk of U.S. aid, including, in order of numbers, India, Columbia, South Korea, Turkey, and the Dominican Republic.[48] The Dominican Republic offered a startling example of what could happen in a worst-case scenario. In 1962, only 78 Dominican physicians came to the United States, but that number was a third of the new doctors for that country for the entire year. Forty-four Dominican engineers arrived in 1963, fully two-thirds of the new graduates. The 18 engineers from Chile were 20 percent of that country's new engineers, and Israel saw an astonishing 41.7 percent of its new doctors leave for America.[49]

Immigration of already trained professionals was not the only problem. In 1963, American colleges and universities enrolled 22,510 students from developing countries majoring in the sciences, engineering, and medicine. In 1966, while 4,390 scientists, engineers, and physicians arrived from LDCs, 28,419 students from these countries were enrolled in curricula in these fields. Some of these students remained in the United States, but even those who returned to their own countries did not always contribute to development as much as might be expected. According to one report, "many of their students, who do go home after study and research in the United States, are ill equipped for work in their own countries because the education and training they have shared in U.S. colleges and universities are naturally directed to the needs of this country." Furthermore, "the sophisticated specialization in study, research, and application in U.S. science, engineering, and medicine, is not what developing countries usually need from their scientific professionals." As one physician from the University of Oklahoma School of Medicine found, "all too commonly, foreign trainees spend months or even years learning techniques or skills which require equipment not obtainable in their native country."[50]

Even if immigrants left after graduation, while still here, they contributed significantly to America's technological lead. In 1964, 53 percent of the overseas science students and 41 percent of the engineering students were engaged in research and development, compared with only 35 percent and 27 percent of the American students in the same majors. Immigrants were much more likely to be working on postgraduate degrees. Because scientists and engineers from overseas were less likely to wind up in management, production, sales, and other nonscientific branches of corporations, they were disproportionately doing what they had been trained to do.[51]

European critics writing at the same time as Servan-Schreiber tended to share his gloomy assessment of continental social arrangements. Writing of the brain drain rather than the technology gap, D. N. Chorafas in 1968 put it a bit more forcefully but came to the same conclusion: "The brain drain from Europe is merely a symptom of a basic disease in the European economic system."[52]

Unless something was done, he warned, "Europe will earn a living from its technological industries but the United States will own them." After studying the process in forty-three countries, including in Africa, Asia, and South America, he concluded that "few have yet recognized the value of ability. As a result, the brains elude them and go to America."[53]

The core of the attraction was what he called "the availability of professional opportunity." The huge research and development effort in the United States was an insatiable machine that consumed more trained people than the country could provide on its own. "But she is not worrying," he observed, because "while the need for brains exists, the brains will gather in the United States, for the financial rewards, for the research opportunity, for professional horizons, for creative conditions."[54]

In 1967, a congressional staff study found that there were, within the Unites States, essentially four different views of the brain drain problem. The first view denied that there was any problem at all, and maintained that scholars had been roaming the world for hundreds of years and that, on balance, this was a good thing. The second position maintained that the brain drain was real but a boon, both to the intake country and the country of origin, because both benefited from the advances of science and technology. A third position welcomed the drain from already developed countries, but worried about that from LDCs. The latter group, the view held, should try to prevent the outflow of talent, hopefully by "developing." Those who held the fourth view were alarmed at the flow of talent out of LDCs, and thought that the United States should do something about it. The government could frankly admit that it was taking advantage of the migration to compensate for shortfalls in American professional workers, and plug visa loopholes that were making the situation worse. American corporations abroad could hire more professionals from the countries in which they operated, rather than bring Americans over to fill top jobs.[55]

It is probably accurate, though not very dramatic, to say that brain drain is a constant and ubiquitous activity that can have serious consequences for specific areas, people, and fields of work. In the United States, there has been a persistent flow of talent from rural areas and small towns to large cities, and in recent years, from the Midwest to both coasts. Cities of the Rust Belt, like Cleveland, Ohio, lose talented young people to Chicago and Seattle. Worldwide, recent doctoral graduates in biology from Britain come to the United States for postdoctoral training and stay on in faculty positions. Young Irish mathematicians and engineers come to America for college and stay on, in part to avoid the realities of a "mixed" marriage at home. Those who leave the United Kingdom are likely to be replaced by doctors, scientists, and engineers from the former colonies and present commonwealth countries. It is difficult to escape the belief, however, that the whole process is anything less that a subsidy that the weak pay to the strong.

The flow of scientific and engineering talent to the United States in the mid-twentieth century was undoubtedly one of the factors that created and maintained a technology gap between this country and the rest of the developed world. The brain drain, however was not simply a relic of the Cold War era. In the summer of 2003, the European Commission, the executive arm of the European Union, issued a report entitled *Researchers in the European Research Area: One Profession, Multiple Careers*. A key finding of the study was that of the Europeans who take a doctorate in the United States, 73 percent pursued careers there. This figure was up from 49 percent in 1990.[56] Perhaps an even more startling fact that was while the European Union produced more scientists and engineers per capita than did the United States, they made up a smaller portion of the work force. In the United States, there were 8.66 scientists and engineers for every thousand workers. In Japan, that number was 9.72. In the European Union, it was 5.36.

With the admission of ten new countries to the European Union in the spring of 2004, the region had twenty-five separate systems of higher education, each with its own rules and traditions for training, recruitment, hiring, promotion, and governance. In too many cases, the report points out, ambitious young researchers had to wait for senior faculty members to die before they could act independently, and moving from academia to industry and back again was difficult. The Commission provided guidelines to reform the academy, noting that "our university systems are much less flexible than in the U.S." It also suggested that stipends be used to lure European academics back from overseas, and that provisions be made to allow researchers to move from country to country within the European Union without losing job seniority or pension rights.[57] If the United States at the beginning of the new century was still "the tech toy store to the world," its culture of scientific and technological hyperactivity meant that it was still the land of opportunity, at least for the world's scientific, engineering, and medical elite—although there was a growing unease within the country itself that scientific and technological hyperactivity were not perhaps entirely good things, and might not be able to continue forever.

1. The "mass production" of ships during World War II at the Kaiser & Oregon Shipyards, c. 1943. (Oregon Historical Society, Or Hi 68758)

2. Lula Barber, Meta Kres, and Brendall outside a welding shop at the Bethlehem-Fairfield Shipyards, Baltimore, Maryland, 1942. (Photo courtesy of Veterans History Project, American Folklife Center, Library of Congress)

3. Levittown came to symbolize the sterility and mass conformity that were thought to characterize postwar suburban life. (Courtesy of the National Park Service)

MOTHER EARTH IS GOING TO HAVE HER FACE LIFTED!

SOUNDS like a rather ambitious undertaking, doesn't it, but that is more or less what is going to happen after this World Struggle is over. The Earth is in for a tremendous resurfacing operation.

The construction, road building and grading jobs for Crawler type tractors in that not too distant period are colossal. While our Plant is now engaged in essential work for the Armed Forces, we are not forgetting for a moment our duty and obligation to those who depend on Rodgers presses for quick repair of crawler tracks and other heavy machinery. Right *now*, all engaged in essential work are eligible for Rodgers Hydraulic Track and Universal Presses. Wire or write for full information and prices. *If it's a Rodgers, it's the best in Hydraulics.* Rodgers Hydraulic Inc., St. Louis Park, Minneapolis 16, Minnesota.

Manufacturers of:
UNIVERSAL HYDRAULIC PRESSES
TRACK PRESS EQUIPMENT
HYDRAULIC KEEL BENDERS
HYDROSTATIC TEST UNITS
POWER TRACK WRENCHES
HYDRAULIC PLASTIC PRESSES
PORTABLE STRAIGHTENER
FOR PIPE AND KELLYS

Rodgers HYDRAULIC Inc.

4. During World War II, some were already looking ahead to a postwar effort at "planetary engineering." In 1944, the magazine *Military Engineer* carried this advertisement for a manufacturer of earth-moving equipment eager to turn its attention to civilian projects. (Courtesy of Granite Fluid Power)

5. Workers at a Marshall Plan reclamation project in Holland. Photograph by George Rodger. (Still Picture Branch, National Archives at College Park, Maryland)

6. Project Plowshare held out the promise of neat, clean, almost surgical excavations using nuclear explosions. Even this artist's conception suggests some of the environmental disruption that could be expected to occur. (Courtesy Lawrence Livermore National Laboratory)

"All I'm Asking You to Do Is Grab Hold of the Tail"

7. In 1969, Herblock suggested that plans for a "thin" ABM system represented the proverbial "tiger by the tail." (A 1969 Herblock Cartoon, copyright by The Herb Block Foundation)

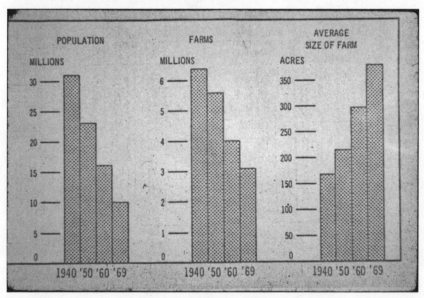

8. Between 1940 and 1969, the farm population dropped and the number of farms decreased, but their average size increased. (Redrawn from U.S. Department of Commerce, *Statistical Abstract of the United States, 1970* [Washington, 1970], 581)

9. After World War II, Kiwanis International launched a series of pamphlets in the "interest of the American Way of Life." The first, titled "It's Fun to Live in America," compared the number of cars, radios, and telephones per capita in the United States, Britain, France, and most pointedly, the Soviet Union. (Reprinted with permission of Kiwanis International)

10. A four-high Los Angeles freeway in 1954. (Copyright 1954, California Department of Transportation)

11. A hot rod on the streets of Oakland, California, 1973. Such alterations of common technologies were at the same time a hobby, an expression of personal affirmation, and a work of art. Photo by author.

12. In 2001, Henry J. Czekalinski recalled how his family was photographed in 1951 surrounded by what was supposed to be a year's supply of food. The image became an icon of the Cold War and was on display at the American pavilion in Moscow when the Nixon-Khrushchev Kitchen Debate was held in 1959. (Courtesy of Cleveland Magazine. Photograph by Alex Henderson)

TO HIM THAT HATH...

13. Beggar-your-neighbor had been the name of the game in technology since at least the beginning of the early modern period. In this case, as in the past, people with technological skills tended to go where they were most appreciated and best paid. (*Punch*, Feb. 8, 1967. Reproduced with the permission of Punch Ltd., www.Punch.co.uk)

14. In 1970, Ron Cobb, whose work appeared in many of the alternative newspapers of the day, gave his assessment of where development, growth, and progress were taking America. (From *Raw Sewage* [Los Angeles, 1970]. With permission of Ron Cobb)

16. The logo of the California Office of Appropriate Technology, in 1977, gave something of the flavor of the cultural times from which it sprang.

15. The early years of the American space program witnessed a large number of failed launches. This Vanguard rocket was intended to launch a satellite during the International Geophysical Year in 1957. It lost thrust after two seconds and had to be destroyed. (Courtesy National Aeronautics and Space Administration)

17. Solar water heater and windmill at the appropriate technology demonstration farm of the New Alchemy Institute, Falmouth, Massachusetts, 1975. Photo by author.

18. This front cover of the journal *Technology Assessment*, which appeared in 1974, captured the fear that technology assessment might, at least in some cases, lead to technology arrestment. (Reprinted by permission of the publisher, Taylor & Francis Ltd., www.tandf.co.uk/JOURNALS)

19. A Bill Mauldin cartoon from 1976 suggests solar power as an alternative to the nuclear option. The debate was renewed with the rise in oil prices in 2006. (Reprinted with permission of the Bill Mauldin Estate)

20. Australian parents pull the plug on the family TV to try to reverse the influence of a globalized America cultural hegemony on their son. (*The Weekend Australian*, May 25–26, 2002. Copyright © Andrew Weldon)

21. The Pentagon's Advanced Research Projects Agency (ARPA) brought together a group of universities, think tanks, and government agencies in the early 1970s to form the first stage of the now ubiquitous internet. (*Science* 175 [March 10, 1972], iv)

22. Governor Arnold Schwarzenegger and the notorious Hummer both symbolized the rage for masculinity that was seen as an antidote to America's defeat in Vietnam. (AP via AAP, photo by Joe Raymond © 2005 The Associated Press)

23. The graphic violence of many computer games, variously interpreted as either educational or desensitizing, held appeal for many young males. (Courtesy of John Shakespeare, *Sydney Morning Herald*)

EIGHT

FROM TECHNOLOGY DRUNK...

A quarter century of fun and vanguardism in technology was not without its price: A series of problems associated with American technological supremacy accumulated over the years, and by the end of the 1960s, it led to something of a crisis of confidence. Rachel Carson's widely influential *Silent Spring* cast grave doubt on the miracle of DDT, one of the most ubiquitous and heralded of wartime advances. The threat of nuclear war was never far from the minds of those in the postwar generation, and the hot war in Vietnam, pitting as it did the full might of American arms against a small and hardly modern people, all focused attention on the price to be paid for the technological culture we had adopted so enthusiastically. The first Earth Day, held in 1970, marks a convenient beginning to the environmental movement in this country, and an end to the period of unquestioned technological optimism.

At the end of the 1960s, the San Francisco longshoreman Eric Hoffer, reputed to be President Lyndon Johnson's favorite "philosopher," published his evaluation of what was going wrong in America, what was going right, and what should be the way forward. "It was the machine age," he insisted, "that really launched the man-made creation. The machine was man's way of breathing will and thought into inanimate matter. Unfortunately," he added, "the second creation did not quite come off." As a young migrant farm worker, he recalled, "I knew nature as ill-disposed and inhospitable. . . . To make life bearable I had to interpose a protective layer between myself and nature. On the paved road, even when miles from anywhere, I felt at home." The task ahead, then, was large but obvious: "One would like to see mankind spend the balance of the century in a total effort to clean up and groom the surface of the globe—wipe out the jungles, turn deserts and swamps into arable land, terrace barren mountains, regulate rivers, eradicate all pests, control the weather, and make the whole land mass a fit habitation for man."[1] Hoffer's vision nicely captured the virile and arrogant faith in technology as a sovereign remedy for society's ills that alarmed a growing number of Americans.

Cold War Fears

Though it was often camouflaged or just ignored, no issue of the postwar era was more significant than that of the threat of mass annihilation from a nuclear war. "Contemporary culture," editorialized the *New Boston Review* in 1981, "grows in a dark place, beneath the shadow of the nuclear threat."[2] The fear grew out of the surprise and power of the bombs of Hiroshima and Nagasaki, but was greatly enhanced by the August 1949 announcement that the Soviet Union had tested its own nuclear weapon. President John F. Kennedy's urging that all Americans build bomb shelters in their yards, the "duck-and-cover" campaign aimed at school children, and the periodic alarms and diversions such as the Berlin Blockade (1948–49) and the Cuban missile crisis (1962) all reminded Americans that they and their country could disappear at any moment. Popular culture bore much of the burden of interpreting the threat, and such excitements as the silent spring described by Rachael Carson, the sighting of unidentified flying objects (UFOs), and even the suburban invasion of crabgrass readily became surrogates for the Soviet nuclear threat.

The film historian Peter Biskind nicely captured the paranoia of the time with his description of the film *The Day the Earth Stood Still*: "It's a fine spring weekend in Washington, D.C. The year is 1951, and American boys are still fighting in Korea, six thousand miles away. The Rosenbergs are in jail, waiting to be executed, and before the year is out, a bootblack in the Pentagon will be questioned seventy times by the FBI for once giving ten dollars to the Scottsboro Boys defense fund." The iconic science-fiction film featured a flying saucer landing on the Capitol Mall where the alien Klaatu, protected by the robot Gort, comes to warn the world about the absolute necessity of peace. As Biskind puts it, it is not the alien culture that is dystopian: "their message is that Earth (including the U.S.A.), far from being the last word in civilization, as centrists thought it was, exists in a state of virtual barbarism." It is "only on Earth, the world of disharmony and intolerance ... that technology and humanity, the head and heart, are at odds."[3] The Cold War, with its ever-present nuclear threat, had given birth to weapons of mass destruction, and it was unclear whether mere mortals, even Americans, were smart and good enough to prevent the Armageddon that they had made possible.

The comprehensive but amorphous threat of nuclear technology was made particular and personal in 1959, when it was discovered that because of the atmospheric testing of nuclear weapons by the United States, the level of the radioactive substance strontium 90 absorbed into the bones of children under four years of age had doubled in 1957 alone. The St. Louis physiologist Barry Commoner helped to form the Committee for Nuclear Information, thus beginning his long career as an environmental activist. That same year, the National Committee for a Sane Nuclear Policy (SANE) was founded, and in 1962, it recruited the revered baby doctor, Benjamin Spock, to the cause of banning atmospheric testing.[4] Spock went on to play a prominent role in both the civil rights and anti-Vietnam War movements.

The political economy of America's commitment to a weapons culture was provided with a name by none other than the president himself. In his farewell address to the nation, on January 17, 1961, President Dwight Eisenhower warned that there existed a "conjunction of an immense military establishment and a large arms industry" which, he claimed, was "new in the American experience." The "total influence," he went on, "—economic, political, even spiritual—is felt in every city, every State house, every office of the Federal government." While conceding that both defense forces and an arms industry were needed, he warned that "in the councils of government, we must guard against the acquisition of unwarranted influence, whether sought or unsought, by the military-industrial complex. The potential for the disastrous rise of misplaced power exists and will persist."[5]

The phrase "military-industrial complex" proved to be an enduring one, subverting the liberal belief that American was kept on an even keel by countervailing forces. The old constitutional ideal of a Newtonian polity, in which farmers balanced merchants, small states balanced large, and the North balanced the South, was challenged by the possibility that forces could converge as well. Commentators pointed out that the complex included members of Congress, labor unions, hawkish clergy, most of the press, and the scientific and technological mainstreams. The idea of the military-industrial complex explained and even came to stand for the postwar arms race, which ate up so much of the nation's scientific and engineering talent, as well as its wealth.

Arms Race

The outlines of the developing Cold War and the arms race are depressingly familiar. The famous National Security Council Paper 68 (1950) made explicit that the U.S.-U.S.S.R. conflict was the decisive one, that it was a fight to the finish, and that the American policy was to "contain" the Soviet Union. In 1950, the bill for containment was $13 billion, but by the time Eisenhower became president in 1953, the annual defense budget was up to $50 billion. The new administration was split between those who thought excessive defense spending would destroy the economy and those who professed that no defense budget was excessive if it was necessary. Eisenhower chose to try to protect the economy, and the nation's defense policy thus came to depend heavily on nuclear weapons and the means of delivering them as a cheaper alternative to conventional weapons. Advocates called it "massive retaliation," but the former head of General Motors, now Secretary of Defense Charles ("Engine Charlie") Wilson, was closer to the mark when he characterized the policy as one of "more bang for the buck." The realization in 1955 that Washington, D.C. would receive only a fifteen-minute warning of a Soviet nuclear attack was a sobering reminder that the choice was between deterrence and mutually assured destruction (MAD). The president knew further that America would suffer 50 million casualties if an attack occurred.

In the opinion of one central player, however, the race to develop weapons of mass destruction, the means to deliver them, and a measure of defense against them, either through an early warning system or the ability to shoot down both bombers and missiles, was not a race against the Soviet Union at all. Herbert York, a nuclear physicist who had worked on the first hydrogen bomb and became Eisenhower's first director of defense research and engineering, came to understand the arms race as one that the United States ran only against itself. "Many of the technologists," he explained, "used their own narrow way of viewing things to figure out what the Russians ought to have done next. They decided then that since the Russians were rational (about these things anyway), what they ought to have done next was what they must now be doing, and they then determined to save us from the consequences of this next Russian technological threat."[6]

This leap-frogging over our own accomplishments led, York believed, to "the accumulation of unnecessarily large numbers of oversized weapons." Furthermore, "our unilateral decisions have set the rate and scale for most of the individual steps in the strategic-arms race. In many cases we started development before they did and we easily established a large and long-lasting lead in terms of deployed numbers and types. Examples include the A-bomb itself, intercontinental bombers, submarine-launched missiles, and MIRV [multiple independently targetable reentry vehicles]."[7] It was a massively funded technological project which, by its very success, undermined the national security it was supposedly designed to guarantee.

If the terrifying technology of massive retaliation presented unbelievably high risks, the alternative arms policy, pursued at the same time, presented very high costs. "Flexible response," or having a whole range of weapons available, was designed to make it possible for the United States to accumulate smaller and more varied weapons than those in the nuclear arsenal, and to have them for use against Soviet surrogate states or forces. The war in Vietnam gave the country a clear look at what that arsenal looked like. Helicopter gunships; high-flying bombers dropping conventional explosives, napalm, and Agent Orange; killing zones monitored by remote-sensing devices that could be used to call in carpets of "smart" bombs; computer software designed to rationally select high-payoff targets—all were unleashed upon a small and poor nation of which most Americans had probably never heard. Paul Dickson, a student of the "electronic battlefield" wrote that in the late 1960s he felt "awe for the technological virtuosity involved, the challenge of mounting it, and the combination of Dr. Strangelove and Buck Rogers—things that go zap in the night and all that." He also felt "anger focused on the use to which it was being put as a death-dealing instrument in Southeast Asia and, on a more general level, the slick, casual kind of war of the future in which killing would be programmed like a department store billing system."[8]

Coincident with and in some ways helping to shape the student movements of the 1960s and 1970s, the Vietnam War became the horrible example of the destructive and dehumanizing effect of a quarter century of American technological development. It was a war machine, made up of companies that made the weapons and ordinance, the members of Congress who represented districts in which those companies had plants, the Pentagon, its allies in the press and pulpit, and a scientific establishment that had grown fat from Pentagon grants and contracts, as well as others. When student activists descended upon arms plants asking workers to join them in protesting the escalating conflict, the workers put on badges saying "Don't Bite the War That Feeds You."

DDT and the Environmental Movement

The war on communists was mirrored by, and intertwined with, a war against nature. One of the technological marvels to come out of World War II was the pesticide DDT. The great wartime success of DDT had led not only to its celebrity after the war, but to a reinforcement of the entire military analogy of a war against "bugs" of all sorts—an all-out assault on those enemies that threatened not only our way of life, but our very lives.[9] However, a century of using pesticides, both organic and inorganic, had established clear indications that residues of arsenic, lead, fluorine, and other chemicals on foodstuffs were injurious to consumer health.[10] Most of this concern had been with human health, but in 1962, Rachel Carson, a government biologist, drew attention to the disastrous effects of the new DDT on the environment in general, and particularly on songbirds.

Rachael Carson was not so much a prophet of doom as a chronicler of disaster. "A grim specter has crept upon us almost unnoticed," she announced. Life on Earth had evolved over many millions of years, but "only within the moment of time represented by the present century has one species—man—acquired significant power to alter the nature of his world." During "the past quarter century this power has not only increased to one of disturbing magnitude but it has changed in character. The most alarming of all man's assaults upon the environment is the contamination of air, earth, rivers, and sea with dangerous and even lethal materials. This pollution is for the most part irrecoverable; the chain of evil it initiates not only in the world that must support life but in living tissues is for the most part irreversible. In this now universal contamination of the environment, chemicals are the sinister and little-recognized partners of radiation in changing the very nature of the world—the very nature of its life."[11] Her prose, a combination of scientific knowledge and humane understanding, caused an uproar. Even President Kennedy is said to have stayed up late one night reading her articles in the *New Yorker*, which preceded publication of the book; in the morning, he sent a memo to his staff demanding that Carson's work be refuted or confirmed.

Not surprisingly, agricultural experts and chemical-industry officials were not enthusiastic. Even before the book appeared, the General Council for Velsicol Chemical Corporation wrote to the publisher warning that not only was the book "inaccurate and disparaging," but linked to "sinister influences" that wished to reduce agricultural production in the United States and Europe to the level of that "behind the Iron Curtain." The director of the New Jersey Department of Agriculture deplored that fact that "in any large scale pest program we are immediately confronted with the objection of a vociferous, misinformed group of nature-balancing, organic-gardening, bird-loving, unreasonable citizenry that has not been convinced of the important place of agricultural chemicals in our economy." And in a meeting of the Federal Pest Control Review Board, one member asked of the author, "I thought she was a spinster. What's she so worried about genetics for?" The other board members, it was reported, "thought this was very funny."[12]

Silent Spring survived the attacks of those with a vested interest in DDT and other chemical pollutants, and became one of the founding texts of the new environmental movement that arose after World War II. That movement, as historian Adam Rome has shown, grew out of three important developments of the 1960s: "a revitalization of liberalism, the growing discontent of middle-class women, and the explosion of student radicalism and countercultural protest."[13]

The Crisis of Liberalism

Prominent American liberals, such as John Kenneth Galbraith, Arthur Schlesinger, Jr., and Adlai Stevenson, addressed the growing imbalance between the nation's privatized material affluence and its impoverished public sector. In 1956, Schlesinger noted that "our gross national product rises; our shops overflow with gadgets and gimmicks; consumer goods of ever-increasing ingenuity and luxuriance pour out of our ears" while the social and engineering infrastructure of society was in decay. Two years later, Galbraith, in his best-selling book *The Affluent Society*, wrote that "the family which takes its mauve and cerise, air-conditioned, power-steered, and power-braked automobile out for a tour passes through cities that are badly paved, made hideous by litter, blighted buildings, billboards, and posts for wires that should long since have been put underground." This family, he thought, "may reflect vaguely on the curious unevenness of their blessings. Is this, indeed, the American genius?" The world of the affluent society was thoroughly familiar with radioactive dairy products, detergent foaming out of kitchen faucets, and skies obscured by smog. The private sector was providing technologies with consequences that the public sector was not adequately addressing.[14]

Ever since the late nineteenth century, many women who looked forward to greater political activism and equality looked back to essentialist arguments to legitimize their efforts. In the 1960s, some women's organizations, such as the League of Women Voters, lent support to the environmental and peace movements, often as "concerned housewives" or mothers with a "love for children." Such organizations helped "make Rachael Carson's *Silent Spring* both a best seller and a political force." Women Strike for Peace organized demonstrations to "protest the hazards of the arms race." The historian Ruth Schwartz Cowan has suggested that this willingness to reject extreme technological claims and consequences may not be an entirely recent development. To the extent that comfort and competence with technology is the mark of the "real" American, she suggests, "we have been systematically training slightly more than half of our population to be un-American." It may be true, she suggested in 1975, "that the recent upsurge of 'antiscience' and 'antitechnology' attitudes may be correlated very strongly with the concurrent upsurge in women's political consciousness. This is not to say that all of the voices that have been raised against the SST [supersonic transport] and atomic power plants and experimentation on animals have been female, but only that a surprisingly large number of them have been."[15]

The young people in the movement and involved in the counterculture were, as Rome noted, those who had grown up in the shadow of the bomb, felt the fear of the Cuban missile crisis of 1962, and witnessed on nightly television the horrors of the Vietnam War. However, while up to half a million Americans spent some time on communes during the late 1960s and early 1970s, those activists associated with the so-called New Left initially had little to say about environmental issues. When in the fall of 1964, Mario Savio climbed atop a police car on the University of California campus at Berkeley, in a scene that captured the spirit and reality of the new Free Speech Movement, he called upon the assembled students to throw their bodies "on the machine."[16] By then, that machine was already becoming more than just the military-industrial complex, or indeed the "plastic" society created by a blindly metastasizing capitalism. In Vietnam, a full eighth of the land had been covered with chemical defoliants, giant bulldozers had cleared terrain to deny it to enemy forces, and the notorious napalm had been used to destroy ground cover as well as village peoples. It seemed a clear example, as the term had it, of ecocide. For "many intellectuals, therefore," according to Rome, "the movement to end the war and the movement to protect the environmental became aspects of one all-encompassing struggle."[17] For all of these, "the machine" was both a metaphor and a material reality.

Supersonic Transport

The fight over whether or not to develop an SST was one of the defining controversies in the efforts both to limit technological development and use environmental arguments to do so. Supersonic flight had been achieved on October 14, 1947, when Charles Elwood "Chuck" Yeager took the Bell *X-1* through the so-called sound barrier. For years, large commercial airplanes had developed out of military aircraft, but when the Air Force and Navy attempted to scale up the *X-1*, they encountered major problems. Seeking a replacement for the B-52, the workhorse of the bomber command, the Air Force issued specification for the B-70 supersonic bomber in 1954. Five years later, the program was drastically scaled back, in large part because the need for such a plane in an era of intercontinental ballistic missiles was widely questioned.

At the time, both the French and British governments were contemplating building supersonic passenger planes, and the head of the Federal Aviation Agency, realizing that technologically, the erstwhile B-70 would have been superior to both, urged the president to recommend an American commercial version. Though boasting some superior technologies, it was estimated that the civilian version would cost a billion dollars to develop and carry between fifteen and one hundred passengers.[18]

When Kennedy became president, he adopted the idea of an American SST as a part of his effort to "get American moving again." He announced his support at a speech at the Air Force Academy in 1963, adopting the rhetoric of technological vanguardism that by then was quite common. Commitment to the SST, he proclaimed, was "essential to a strong and forward-looking Nation, and indicates the future of the manned aircraft as we move into a missile age as well."[19] Unlike its military prototype, such a plane would have to sustain itself in a commercial environment and, perhaps to the surprise of many, it turned out to have to perform in a natural environment as well. Although the next seven years saw numerous studies that uncovered problems both with specific technologies and commercial viability, both President Lyndon B. Johnson and Richard M. Nixon continued to support the concept. Then in 1970, Nixon established an Office of the Supersonic Transport in the Department of Transportation.

The timing could hardly have been worse for the SST, as the opening of the new office coincided with the first Earth Day. The environmental problems with the aircraft had been known, and commented upon, for years. The most discussed was the issue of the sonic boom an SST would inevitably create. The common estimate seemed to be that the plane in flight would be followed by a rolling boom perhaps fifty miles wide along the surface of the earth. The government's argument, as late as 1967, was that the magnitude of the boom and people's reactions to it could only be assessed by collecting data after actually building and flying the aircraft. Critics pointed out that there was ample data already available. In 1964, test flights with military planes were made over Oklahoma City. Afterward, 27 percent of the people surveyed said that they could not "learn to live" with the noise.[20] Secretary of Transportation John A. Volpe remarked dismissively that "clearly the SST will not be flying over the United States supersonically," and with contracts already promised to General Electric for engines and to Boeing for airframes, the prospect of "tens of thousands of employees laid off and an aviation industry that would go to pot" was unattractive.[21] Russell E. Train, head of the President's Council on Environmental Quality, admitted that the SST would be three to four times louder during takeoff and landing than existing planes, but a member of PSAC told a congressional committee that the plane would produce as much noise as fifty Boeing 747s taking off at once. The sonic boom, it turned out, would be only part of the problem.[22]

Representative Sydney Yates (D-Ill.) predicted that "someday they will tell us that the sonic boom is the sound of progress." Because factory smoke had been traditionally seen as a similar marker of progress and prosperity, Yates may have been right, but there were other problems as well. There was the "side-line noise" of takeoffs, and also the fact that flying at 70,000 feet, where it was virtually windless, the SST would deposit large amounts of water vapor and waste gases, leading to a "sun-shielding cloud cover."[23] An umbrella organization, A Coalition Against the SST, was organized by the Sierra Club, the National Wildlife Federation, and other groups, including two labor unions and a conservative taxpayer's group. Another member, the Citizens League Against the Sonic Boom, was headed by a researcher at the Cambridge Electron Accelerator and specialized in educational work, including producing newspaper advertisements and the *SST and Sonic Boom Handbook*, which had a print run of 150,000 copies. On a more fundamental level, the Nobel physicist Murray Gell-Mann advocated that, though in the past whatever technologically could be done was done, it was now time for some acts of what he called "technological renunciation" for the sake of the environment, and suggested that the SST eminently qualified.[24]

Despite protests by the administration that the SST was still vital to American interests, the combination of high cost, commercial uncertainty, and environmental concern sank the project, first with a negative Senate vote in the fall of 1970, and then with a final refusal to appropriate funds a few months later. It was in the nature of such projects, however, that they could and often did rise again after the opposition had retired. Late in 1972, the White House spokesperson John Ehrlichman told a Seattle audience that "the SST isn't dead," but David Brower, the president of Friends of the Earth, a few days later spoke to a different Seattle audience and vowed "an all-out effort" against any such revival. Congress continued to appropriate "research" funds as "a hedge against the future," but this too was typical of a range of presumably canceled technological projects that could best be described as legislatively undead. As late as 1980, supporters were still arguing for extending SST research activities.[25]

Ironically, by that time, one of the major arguments for the SST had all but disappeared. There had long been warnings that the joint Anglo-French Concorde and the Russian Tupolev-144 would give the United Kingdom, France, and the Soviet Union an advantage over the United States. In 1967, it had been predicted that the Concorde would enter commercial service in 1971, and that its success would kill the American aircraft industry, because the world's airlines would rush to buy Concordes over slower American airplanes. In fact, the Concorde left London for its first flight to the United States in January 1976, lifting off with what was described as an "excruciating" and "absolutely murderous" noise in the takeoff area.[26] Four years later, the Concorde production lines were shut down, and all of the planes either sold off at a loss or given away to state-owned British and French companies. It was a cautionary tale, made explicit by the director of science news for the *New York Times*. "The supersonic transport," he wrote in 1976, "along with other awesome technologies, like nuclear power and genetic engineering, is perhaps a symbol of a re-emerging imperative: look very carefully before leaping. Because something can be done, it does not necessarily follow that it must be done; and it should not, in any case, be done without considerable thought."[27]

Agribusiness and Industrial Foods

Few aspects of the environment were more personal, intimate, and immediate than the food we eat, and the food Americans ate and the agricultural regime under which it was produced became an important battleground during the 1970s. From early in the century, the ideological imperative in American agriculture had been, as historian Deborah Fitzgerald has chronicled, that every farm should be a factory.[28] Scientifically formulated pesticides and fertilizers—usually based on petroleum products—new and more powerful machines, and the efficiencies of specialization and mass production were all pushed as the modern way to produce food and fabric. In 1972, Jim Hightower, a populist writer and commentator who was, surprisingly, the one-time secretary of agriculture for Texas, explained how the industrial ideal had been brought to agriculture, and roundly condemned its results.

Calling his critique "Hard Tomatoes, Hard Times," Hightower outlined the workings of what he called the "land-grant college complex." This interlocking group of agencies was made up of the land-grant colleges, first financed by the federal government in 1862; the federally-financed state agricultural experiment stations authorized in 1887; and the network of agricultural extension workers that operated under the Smith-Lever Act passed in 1914. Together, they acted as a great research and educational engine that worked, he charged, solely to benefit the nation's expanding agribusiness, and against the interests of farm workers, small family farms, and the populations of rural communities. The statistics he marshaled were disturbing. In 1972, the land-grant complex spent about three-quarters of a billion dollars, but the results were badly skewed. Forty-seven percent of the nation's farmers earned less than $3,000 a year; since 1940, more than three million farms had disappeared, and in 1972, were doing so at the rate of 2,000 a week. Farms operated by African-Americans had fallen from 272,541 in 1959 to 98,000 in 1970. Each year, 800,000 people left rural homes for cities.[29]

Hightower argued that the complex was fascinated by "technology, integrated food processes and the like," and that "the largest-scale growers, the farm machinery and chemical input companies and the processors are the primary beneficiaries." It turned out that creating "genetically redesigned, mechanically planted, thinned and weeded, chemically readied and mechanically harvested and sorted, food products," was accomplished with machines and methods available only to those producers who had the acreage to use and the capital to invest in the new technologies. As "food rolls off the agribusiness assembly line," industrialized agriculture also generated "rural refugees, boarded-up businesses, deserted churches, abandoned towns, broiling urban ghettoes and dozens of other tragic social and cultural costs."[30]

Oddly enough, farm workers won a small victory when the so-called short hoe, or *el cortito*, as the largely Mexicano and Mexican-American field workers called it, was outlawed in 1975. The hoe, with a handle only twelve inches long, was issued by growers in California who believed that because the tool required workers to bend over to use it, the workers could better see and control the hand-chopping needed for weeding and spacing crops. After hearing testimony from doctors that 14 percent of workers developed permanent back injuries using the short hoe, the California Division of Industrial Safety voted to ban the implement.[31]

Hightower's critique was based in large part on the social pollution of mechanized agribusiness and its government benefactors, but other critics, feeling the beginnings of the environmental movement, focused on pollution of the ecosystem. At the beginning of 1970, a conference sponsored by the Cornell University College of Agriculture brought together scientists and engineers from a variety of sources to evaluate and seek solutions to agricultural pollution. There was apparently general agreement that contaminants from animal waste and chemical pesticides and fertilizers were threatening the nation's health, but there was less agreement on what should be done about it. A contamination expert from the Federal Water Pollution Control Administration said that "agriculture has done an excellent job of raising productivity ... but, not such a good job of handling its waste." By contrast, a representative of the chemical industry charged that attacks on pesticides came from "scientists who have turned into emotionally unstable prophets of doom."[32]

One often-cited source of pollution was the growing use of feedlots to fatten cattle. As the argument went, in the past, herds had roamed the grasslands, scattering their droppings across a wide area and returning some of the nutrients in grass back to the land. With feedlots, great numbers of cattle were penned into a relatively small area and fed grain to prepare them for slaughter. This concentrated the manure as well, and great mounds of it leached into the groundwater, sometimes polluting lakes, rivers, and urban water supplies. Meanwhile, fields deprived of the manure had to be treated with chemical fertilizers to keep them productive. In 1972, around 80 percent of American beef cattle were finished in feedlots, unlike in Latin America and Europe, where grass-fed beef still predominated. One lot in Colorado contained 100,000 animals. Bought when they weighed only 700 pounds, they were fed a precise and computer-controlled diet of a cereal made up of steamed, rolled corn and green alfalfa, spiked with vitamins and other additives. After 140 days of this diet and no exercise, each animal weighed between 1,000 and 1,500 pounds, and was ready for slaughter. Another feedlot, also with 100,000 animals, operated just a few miles down the road.[33] By 1980, 135 million metric tons of grain was fed to livestock, ten times the amount eaten by the American population. One group of scientists estimated that a return to grass-feeding of cattle would save 60 percent of the energy inputs to the process, and 8 percent of the land area used. This would result in a reduction by half of the animal protein produced in the country, but with rising obesity rates, some considered that an advantage rather than the reverse.[34]

The effects of agricultural chemicals on the nation's land and water was a sufficient scandal in the 1970s, but the Nader-organized Agribusiness Accountability Project, headed by Jim Hightower, discovered that they were actually being tested on farm workers in California. It was well known that when workers took to the fields too soon after they were sprayed, they developed symptoms of nerve damage, vomiting, dizziness, and sweating. California had recently increased the time before workers could reenter a field from seven to thirty days, and two chemical companies, seeking to build a case that the new, longer wait was not necessary, hired field hands, including women and children, who agreed to allow themselves to be sprayed while working. They received $3.50 for each blood sample taken, but were given the money only if they lasted through the entire course of the experiment. The companies claimed that the workers understood what they were agreeing to, and that, "in fact, the conditions were not unlike normal worker conditions."[35]

Much of the criticism of agricultural technology focused on either the social or environmental damage it caused, but there was also a growing dissatisfaction with the food it produced. In 1972, Hightower wrote that "while this country enjoys an abundance of relatively cheap food, it is not more food, nor cheaper food and certainly not better food than that which can be produced by a system of family agriculture."[36] Two years later, to make the point that large corporations controlled not only the growing of food, but its processing as well, he produced a menu for a "Dinner a la Corp." For an appetizer, he suggested "sautéed mushrooms by Clorox, wrapped in bacon by ITT." The salad was made with "Dow Chemical lettuce and Gulf & Western tomatoes"; the entrée was "turkey by Greyhound, ham by Ling-Temco-Vought," and after vegetables and drinks, the meal was completed by a dessert of "chocolate cream pie by ITT, pudding by R.J. Reynolds, ice cream by Unilever, almonds by Tenneco."³⁷

Such large, multinational, diversified companies made their money by, as economists would say, "adding value." A potato that sold loose in the grocery for only a few cents could be washed, peeled, cooked, mashed, extruded, sliced, fried, salted, vacuum-sealed in a metal can, then advertised widely. After all of this technological manipulation, the potato could be expected to be very expensive indeed. "In the produce bins of neighborhood supermarkets glisten mounds of waxy cucumbers," warned one critic, "green tomatoes, plastic bags of radishes, ultra-orange oranges, perfectly formed apples and pears," and "strawberries are available in winter." The trade-offs, he complained, "have been freshness, taste and ripeness," and while strawberries are available in winter, there was "no season when they tasted like strawberries." The food industry, he concluded, had "bred for symmetry, sprayed away blemishes, wrapped up shelf life and assured itself high profits. Consumers and small farmers are left holding the plastic bag."³⁸

Hightower described the next logical step in an article entitled "Fake Food Is the Future." The "new farm," he wrote in 1975, "is the laboratory, and the American provider is a multinational, multiproduct food oligopoly." He quoted the head of the Chemurgic division of Central Soya Company as predicting that "we might all be able to exist and flourish on a diet of three adequately compounded pills a day," and cited the warning of an ITT executive that "eventually, we'll have to depend on artificial food to feed the world's population."³⁹ The scientifically formulated food capsule enjoyed a less-than-enthusiastic welcome from the public, but showed up in forms that were not as obvious, the result of processes far from the public view. The makers of Baby Ruth and Butterfinger candy bars cover their products not with chocolate, but with a "synthetic substitute derived from cotton." Mounds and Almond Joy candy bars, on the other hand, were coated with "an undisclosed brown substance"; the president of the company making the candy declared, "I'm not sure exactly what it is." Hightower's best example was a product of International Foods called Merlinex—or, as it was advertised, "instant anything." It was used as "a substitute for real ingredients, extruding, baking, shaping, flavoring, stretching, and doing almost anything else to it. It is, Hightower charged, "the silly putty of the food world." More seriously, the federal government itself undermined confidence in the food supply by having to eventually ban additives that had been accepted for years—"cyclamates, MSG in baby food, DES in cattle feed, and some synthetic colors"—and the list went on.⁴⁰

In 1974, Americans ate an estimated 21 to 25 percent fewer dairy products, fruit, and vegetables that they had as recently as 1954, and 70 to 80 percent more "sugary snacks and soft drinks." Also in 1974, a writer for *Science* magazine reported that while the federal government spent $20 million a year on nutrition education, food manufacturers spent $3 billion on advertising their highly processed "overpriced products, made from cheap basic materials, from which they derive their greatest profits." Despite "new regulations on food additives, and advertising and food labeling requirements," she charged, "people have less and less idea of what in fact they are eating as supermarket shelves are inundated with literally thousands of new, highly processed products of questionable nutritive value."[41] What was obvious was that there was no national policy about nutrition, and as a result, a diet of highly machined and artificially compounded foods posed a growing health threat to Americans despite the bragging of the Department of Agriculture and the food industry that the United States was the best-fed country in the world. By the beginning of the new century, the problem of obesity, worldwide but especially in the United States, was gaining increasing attention.

The Triple Revolution

In March 1964, a group including many leaders of the nation's intellectual left, and apparently under the umbrella of the Center for the Study of Democratic Institutions in Santa Barbara, California, constituted themselves as an ad hoc committee to issue a short manifesto entitled *The Triple Revolution*. Stepping back from the particular issues of agribusiness, SST, Vietnam, and DDT, they crafted a short document that they hoped would frame a larger discussion and stimulate national debate. It was sent to President Johnson and congressional leaders, published in the pages of the journal *Liberation*, and reissued as a pamphlet. The committee included such well-known names as Erich Fromm, a psychoanalyst; Michael Harrington, author of *The Other America*; Tom Hayden, then a member of the Students for a Democratic Society (SDS); Robert L. Heilbroner, an economist; H. Stuart Hughes, a historian at Harvard University; Linus Pauling, a Noble laureate in chemistry; Gerald Piel, an editor of *Scientific American*; and Norman Thomas, a socialist leader. The committee commended the president for his recently declared War on Poverty, but warned that it hardly went far enough, and that the social, military, and economic changes the committee identified "will compel, in the very near future and whether we like it or not, public measures that move radically beyond any steps now proposed or contemplated."[42] The three revolutions were in cybernetics, weaponry, and the demand for human rights—especially of African-Americans for full equality. Furthermore, and most importantly, the three revolutions were interlinked.

The first revolution the committee identified was that resulting from the computer and automated production, or what they called cybernetics. The committee considered it the most significant of the three revolutions because it was, they believed, reorganizing the economic order of things. By greatly increasing the potential of production, and doing so while requiring fewer and fewer producers, it was destroying the old link between production and consumption. Traditionally, workers produced goods and were paid wages for their labor, thus enabling them to consume their own production. Increasingly, production without people was making some mechanism other than wages and salaries necessary to distribute wealth. "Surplus capacity and unemployment have thus co-existed at excessive levels," the committee charged, and as a result "a permanently depressed class in developing in the U.S." Their urgent demand was for a massive increase in public spending through improvements in education, a large public-works program, the building of low-cost public housing, the development of rapid transit and other forms of public transport, construction of a public power system, rehabilitation of decommissioned military bases, and a revision of tax codes to provide the needed funding.[43]

All this was happening, of course, just as the civil-rights movement was putting the issue of racism and inequality toward African-Americans on the political agenda in dramatic terms. That community's demand for jobs, however, was bound to be frustrated, not simply by racist restrictions, but also by the increasing pace at which jobs were disappearing. Cybernation, the committee claimed, was falling hardest upon African-Americans, who were disproportionately unemployed. Until some new way of distributing the productive wealth of the country was devised, the goals of full equality were impossible to meet.[44]

What was called the "weaponry revolution" was given relatively short shrift. The committee's basic point was that revolutionary new weapons had made war impossible to win, and that the continued dedication of such a large part of the federal budget to weapons—10 percent of the gross national product when space programs were added in—could not be justified or maintained at its present level. The committee pointed out that President Johnson was already beginning to cut back, especially by closing military bases. Because such a large proportion of the country's employment was in the defense industry—between six and eight million people directly, and as many indirectly—cutbacks threatened to exacerbate already high levels of unemployment. The defense cuts were on top of the cutbacks attributed to cybernation, and both made employment opportunities for African Americans bleak indeed.

Calls for Reform

If Eric Hoffer expressed the technological optimism and vanguardism of the period, a corporal's guard of intellectuals, working and writing individually, gained an audience for more cautionary understandings of the role of technology in late twentieth-century America. Ivan Illich, for example, published a short book in 1973 titled *Tools for Conviviality,* which he followed the next year with *Energy and Equity*. Illich was born in Vienna in 1926, and during the 1950s was a parish priest serving a Puerto Rican community in New York City. During the 1960s, he founded centers for "cross-cultural studies," first in Puerto Rico and then in Cuernavaca, Mexico.

Tools for Conviviality began with the idea that although machines had been introduced originally to take the place of slaves, "the evidence shows that, used for this purpose, machines enslave men" instead. Needed were "procedures to ensure that controls over the tools of society are established and governed by political process rather than by decisions by experts." What he called "a convivial society" would "be the result of social arrangements that guarantee for each member the most ample and free access to the tools of the community." "Convivial tools," he continued, "are those which give each person who uses them the greatest opportunity to enrich the environment with the fruits of his or her vision.... Tools today cannot be used in a convivial fashion." They can only "foster conviviality to the extent to which they can be easily used, by anybody, as often or as seldom as desired, for the accomplishment of a purpose chosen by the user."[45]

On the other hand, according to Illich, "certain tools are destructive no matter who owns them" because "destructive tools must inevitably increase regimentation, dependence, exploitation, more impotence." Among these, he mentioned "networks of multilane highways, long-range, wide-band-width transmitters, strip mines, or [a particular *bete noire* of his] compulsory school systems." Illich made it clear that most of the large-scale technologies that characterized modern Western society were enslaving rather than liberating. "Most of the power tools now in use," he claimed "favor centralization of control," and therefore "the prevailing fundamental structure of our present tools menaces the survival of mankind."[46]

Energy and Equity (1974) drew upon the same argument to comment on the then-concurrent OPEC energy crisis. "High quanta of energy," he asserted, "degrade social relations just as inevitably as they destroy the physical milieu." Drawing upon the contemporary interest in appropriate technologies, Illich insisted that "for the primitive, the elimination of slavery and drudgery depends on the introduction of appropriate modern technology, and for the rich, the avoidance of an even more horrible degradation depends on the effective recognition of a threshold in energy consumption beyond which technical processes begin to dictate social relations." "A low energy policy allows for a wide choice of life styles and cultures," he wrote, and suggested that "participatory democracy postulates low energy technology. Only participatory democracy creates the conditions for rational technology."[47] His solution to the energy crisis was not to try to overwhelm it with more oil, but to rethink our need for oil in the first place.

The psychoanalyst Erich Fromm, who had been a member of the Ad Hoc Committee on the Triple Revolution, addressed "America's situation in 1968" with a book titled *The Revolution of Hope: Toward A Humanized Technology*. His argument was typically apocalyptic: "We are at the crossroads: one road leads to a completely mechanized society with man as a helpless cog in the machine—if not destruction by thermonuclear war; the other to a renaissance of humanism and hope—to a society that puts technique in the service of man's well-being." Except for the reference to thermonuclear war, it was a statement that could have been, and on more than one occasion, was uttered with despair in the early nineteenth century as the new Industrial Revolution, and what Thomas Carlyle called "mechanism," was ushering in modernity.[48]

A major difference from previous critics of modernity, however, was in Fromm's belief that "today, and increasingly so since the beginning of the First World War,... hope is disappearing fast in the Western World." This hopelessness was disguised behind masks of optimism or "revolutionary nihilism," and grew out of the fact that the entrenched "system" was driven by two imperatives: "if something could be done it must be done, and the principle of maximal efficiency and output." As a psychoanalyst, Fromm believed that "the tendency to install technical progress as the highest value is linked up not only with our overemphasis on intellect, but, most importantly, with a deep emotional attraction to the mechanical, to all that is not alive, to all that in manmade. This attraction to the non-alive, which is in its more extreme form an attraction to death and decay (necrophilia), leads even in its less drastic form to indifference toward life instead of reverence for life."[49]

His hopefulness grew out of his belief that while "the technetronic society may be the system of the future, . . . it is not yet here." Society must turn its back on those "unconscious values which . . . are generated in the social system of the bureaucratic, industrial society, those of property, consumption, social position, fun, excitement etc.," and live instead by the "official, conscious values" of the "religious and humanistic tradition: individuality, love, compassion, hope, etc." For Fromm, Human beings had to be reintroduced into planning, individuals had to become politically active at the grassroots level, consumption had to be brought under control, and "new forms of psychospiritual orientation and devotion" had to emerge. Finally, he advocated more "investments in the public sector," which he believed "have a two-fold merit: first, of fulfilling needs adapted to man's aliveness and growth; second, of developing a sense of solidarity rather than one of personal greed and envy."[50] It was the antithesis of the way of life built into and promoted by postwar suburbanization and the culture of global engineering.

In 1969, Paul Goodman, who styled himself simply an "author and educator," called for a "new reformation." He was speaking quite specifically of something analogous to Martin Luther's attempt to reform the Catholic Church in his day. The analogy was reasonable because of his belief that in the Western world, science had become the new religion. Goodman asked the rhetorical question in an essay titled "Can Technology be Humane?"; his answer was that it not only could, but was already humane by its very nature. In his powerful formulation, "whether or not it draws on new scientific research, technology is a branch of moral philosophy, not of science. It aims at prudent goods for the commonweal and to provide efficient means for these goals."[51]

According to Goodman, American "social priorities are scandalous: money is spent on overkill, supersonic planes, brand-name identical drugs, annual model changes of cars, new detergents, and color television, whereas water, air, space, food, health, and foreign aid are neglected. And much research" he added, "is morally so repugnant, e.g., chemical and biological weapons, that one dares not humanly continue it." Two hundred thousand engineers and scientists "spend all their time making weapons," and it was not sufficient to merely try to have them turn their talents to "solving problems of transportation, desalinization, urban renewal, garbage disposal, and cleaning up the air and water." The crisis was religious and historical, and demanded an entire rethinking of the way science was dealt with.[52]

"Every advanced country is overtechnologized," he charged, yet "no country is rightly technologized." This was true because "the chief moral criterion of a philosophic technology is modesty" and modern technology was anything but. "The complement to prudent technology," he added, "is the ecological approach to science," and because the ecological consequences of so many massive technological interventions were unknown, "the only possible conclusion is to be prudent; when there is serious doubt, to do nothing." To refrain from action, especially bold action, in late twentieth-century America, however, seemed disturbingly feminine, almost "Oriental" in its passivity. "But technological modesty, fittingness," according to Goodman, "is not negative. It is the ecological wisdom of cooperating with Nature rather than trying to master her."[53]

The problems were not inherent in science and technology, nor were they exclusively American or capitalist. But because they were not seen and judged as a part of moral philosophy, they were negative forces in the world. And finally, "the deepest flaw of affluent societies that has alienated the youth is not, finally, their imperialism, economic injustice, or racism, bad as they are, but their nauseating phoniness, triviality, and wastefulness, the cultural and moral scandal that Luther found when he went to Rome in 1510."[54]

The same year in which Paul Goodman called for a New Reformation, John McDermott, writing in the *New York Review of Books*, also saw the ways in which technology had become a religion in America; as the title of essay had it, technology was the "Opiate of the Intellectuals." The occasion for his essay was the publication of the fourth annual report of the Harvard University Program on Technology and Society, and especially its final essay by the program director Emmanuel G. Mesthene. It was the director, and a cohort of others who most commonly addressed the subject in public, who were not only technology's defenders, but advocates of the ideology that McDermott called *laissez innover*, defined as the faith that "technological innovation exhibits a distinct tendency to work for the general welfare in the long run."[55]

According to McDermott, Mesthene and his ilk had created and then knocked down straw-figures he claimed to be excessively utopian (Marx, Comte, and the Air Force) and as well as those with a dystopian vision ("many of our youth"), and defined *laissez innover* as a "middle course" that "consists of an extremely abstract and—politically speaking—sanitary view of technology." In fact, "as a summary statement of the relationship between social and technological change it obscures far more than it clarifies, but that," McDermott added, "is often the function and genius of ideologues." He charged that "*laissez innover* was now the premier ideology of the technological impulse in American society," one that he labeled "newly aggressive" and "right-wing."

The belief that the cure for technological ills was yet more technology, and that "only if technology or innovation (or some other synonym) is allowed the freest possible rein ... will the maximum social good be realized," played into the hands of the "scientific-technical decision-maker, or, more briefly and cynically," the "Altruistic Bureaucrat." Because "advanced technological systems are both 'technique intensive' and 'management intensive,'" and because "within integrated technical systems, higher levels of technology drive out lower, and the normal tendency is to integrate systems," such systems are nearly impossible to influence from the outside. In the end, "technology, in its concrete, empirical meaning, refers fundamentally to systems of rationalized control over large groups of men, events, and machines by small groups of technically skilled men operating through organizational hierarchy." As a result, the distance between the upper and lower social orders of the nation was growing: "Technology conquers nature," he concluded, "but to do so it must first control man."[56]

Illich, Goodman, and McDermott were hardly alone in their alarm. Lewis Mumford, Jacque Ellul, Herbert Marcuse, and other public intellectuals all wrote to alert the public to the growing power of our technologies and the urgent need to clarify just what we wanted them to do and how best to bring them under some sort of humane and democratic control. The MIT historian Elting Morison contributed to the debate with his own small book, titled *From Know-how to Nowhere: The Development of American Technology* (1974). His thesis was nicely encapsulated in the title itself. At the turn of the nineteenth century, the nation knew what large democratic goals it needed technological help to accomplish, but was hard-pressed to know how to go about it. By the late twentieth century, American technology could accomplish almost anything, but there was growing disagreement or even disinterest among the nation's citizens as to just what ends technology should be directed. His narrative began, as he said, "with those Americans who did not know how to run a level or build a lock but were determined to construct canals as the first support for what they called the 'stupendous' vision of 'a new American age.' Starting from that point in the wilderness, the line runs directly to the present necessity to reenact some even larger vision for the world."[57] Even as Morison wrote, however, efforts were being made on the edges of *laissez innover* to bring it under new controls.

NINE

... TO TECHNOLOGY SOBER

The years around 1970 witnessed a remarkable variety of attempts to regain control of the nation's technology, shifting it from one set of controls to another. The Nixon administration's efforts early in the decade to push through a program to build the SST led the Democratic Congress to establish an Office of Technology Assessment, which was designed to advise lawmakers on the costs and benefits of new technological possibilities. Beginning on the private level, groups of social critics began to campaign for what they called appropriate technologies (AT). The OPEC-induced fuel shortage focused AT initiatives on energy sources, especially solar and wind-generated power. It is hardly surprising that the crescendo of criticisms of technology led to demands that technology be brought under control somehow—or, alternatively, that its control be wrested from the corporate, military, and governmental agencies that had used it to move American society into directions not obviously in the public interest. How that might be done was debated, and attempted, at every level of American life.

Technology and the Economy

President Johnson weighed in early on the emerging concern over the nation's technology, especially the worry that the further spread of automation would lead to massive unemployment. In 1964, he appointed a National Commission on Technology, Automation, and Economic Progress, which issued its report in 1966. As might be expected, the commission relied heavily on a kind of economic rationalism to reassure its readers that things were not as bad as they seemed. The "most useful measure of [the pace of technological change] ... for policy purposes," it insisted, "is the annual growth of output per man-hour in the private economy"—that is, "efficiency." From 1909 to 1947, the growth of efficiency had been at 2 percent per annum, and from 1947 to 1965, it had been at 3.2 percent. In the commission's judgment, "this is a substantial increase, but there has not been and there is no evidence that there will be in the decade ahead an acceleration in technological change more rapid than the growth of demand can offset, given adequate public policies." It admitted that "technological change and productivity are primary sources of our unprecedented wealth, but many persons have not shared in that abundance. We recommend that economic security be guaranteed by a floor under family income." Specifically, it recommended "a program of public service employment, providing, in effect, that the Government be an employer of last resort, providing work for the 'hard-core unemployed' in useful community enterprises."[1]

The report was effectively buried, presumably because its message came dangerously close to "socialism," as one critic charged. Its call for a "floor under family income" seemed to echo the suggestion of the Ad Hoc Committee on the Triple Revolution for a guaranteed minimum income for the unemployed, unemployable, and underemployed. Ironically, President Nixon's proposal for a guaranteed national income was as close as the commission's ideas ever came to serious consideration. The economist Robert Lekachman, commenting on the report, expressed his hope that it would be widely read and debated, but conceded that "even in America, even in the 1960's, significant alteration in the way we conduct our lives seems mainly to come about through social and political conflict, extra-legal action, riots in the streets, and the death of martyrs."[2]

The Conversion from War to Peace

One way to keep the economy rolling along and address the perception that technology was being used in destructive and counterproductive ways was to redirect it, to make technology sober. But how could this be done? At the end of the century, Norman Augustine, chair and chief executive officer (CEO) of the aerospace giant Martin Marietta, admitted sarcastically that "our industry's record at defense conversion is unblemished by success." Noting that "rocket scientists can't sell toothpaste," he blamed it on the fact that "we don't know the market, or how to research, or how to market the product. Other than that," he concluded, "we're in good shape."[3]

"Rocket scientist" had entered the popular lexicon as a synonym for genius, and the idea that the nation's army of defense scientists and engineers, with their career managers, were a vast treasury of technical talent and innovation not to be squandered was firmly rooted in the nation's imagination. The phrase "If we can go to the moon, then why can't we ... ?" captured the conviction that all of the talent, assembled from the nation's universities and drawn from around the world, was a neutral tool, like technology itself; it could be turned to virtually any public or private purpose. If this tremendous force was going in the wrong direction, or had no direction, then it only had to be redirected toward new and more widely agreed-upon goals. If we no longer needed F/A-18s, then let the scientists and engineers make toothbrushes. The trouble was that, as the writer Joan Didion noted, "there was from the beginning a finite number of employers who needed what these people knew how to deliver, and what these people could deliver was only one kind of product."[4]

However, the conviction that the technical behemoth of defense and aerospace needed only to be redirected, not dismantled, was widely shared. In 1966, Steven Rivkin wrote a report that was published two years later as *Technology Unbound: Transferring Scientific and Engineering Resources from Defense to Civilian Purposes*.[5] Writing in January 1968, he proclaimed, perhaps somewhat naively, that "the continuance of the war in Vietnam is the single most prominent barrier to the reallocation of effort and resources needed to meet these crucial domestic needs," which the uprisings of the past year in urban America violently highlighted. President Johnson shared the belief that a reoriented military-industrial complex could meet the nation's needs, and not just its own. Johnson's 1967 economic report stated that "the resources now being claimed by the war can be diverted to peaceful uses both at home and abroad," and that through planning and foresight, "we need have no fear that the bridge from war to peace will exact a wasteful toll of idle resources, human or material."[6] The belief that at some point in the not-too-distant future, scientific and technological resources would be turned away from the military and freed up for civilian application was often asserted. It remained to be seen, however, whether the same scientists, engineers, managers, and corporations that made guns so well could also produce butter.

To help the process along, in 1965, Congress passed the State Technical Services Act declaring that "wider diffusion and more effective application of science and technology in business, commerce, and industry are essential to the growth of the economy," and particularly that "the benefits of federally financed research, as well as other research, must be placed more effectively in the hands" of business.[7] At the end of World War II, Vannevar Bush, in his report *Science the Endless Frontier*, had urged the rapid move of wartime research into civilian hands. During the Kennedy administration, J. Herbert Hollomon, the assistant secretary of commerce, also argued that such transfers were needed. Pointing out that the United States spent 2.8 percent of its gross national product on research and development, but that the bulk of that—2 percent—went to space, defense, and atomic energy, he charged that "government-sponsored activity is becoming increasingly dependent on a sophisticated science and technology peculiarly suited to very specialized military and space objectives, and thus more and more unlikely to be of important direct benefit to the economy." In 1963, he noted, the nation's supply of scientists and engineers engaged in research would increase by 30,000, "but the increase in support for research and development for space will require about the same number."[8] His proposal, according to *Science*, "was knifed by a House Appropriations Subcommittee and left for dead." It was revived two years later, however, and reintroduced. This time it passed, and President Johnson, signing it into law, called it the "sleeper" of the eighty-ninth Congress.[9]

The original Kennedy plan, called the Civilian Industrial Technology (CIT) program, had been defeated by the near unanimous opposition of large corporations that had their own ongoing research programs. These firms, almost by definition the largest in their fields, argued that "CIT would simply be taking taxes from successful industries to promote research aimed at building up faltering competitors." The program was seen as the government financing industrial research, an activity that had traditionally been left to firms that were big and wealthy enough to take it on.[10]

In 1969, the CIT program was abandoned when a House subcommittee, at the insistence of its chair, cut out all funding for its activities. These had never amounted to more than $5 million a year, which, of course, had to be divided among the fifty states. That both NASA and the AEC had by that time started their own programs to "spin-off" research to industry also worked against the program. In the end, neither the radical notion of converting from war to peace research and development, nor the more modest plan to simply spin off a bit of the technology for civilian use, proved acceptable to those with the power to stop those efforts.

Social Responsibility in Engineering

It might have seemed to outside observers that engineers were, at least in some sense, in control of the nation's technology, but it did not appear that way to some of them. Engineers maintained a professional ideal of technical competence operating in a self-regulating and independent manner, fundamentally responsible to the public interest. Since at least the turn of the century, they had been uncomfortable with the knowledge that most practitioners were in fact salaried employees of large bureaucratic corporations or government agencies. Some of their numbers were drawn towards the social movements of the 1960s, and the charge that technology was central to the problems of the period struck home with particular force.

In the spring of 1971, a group of engineers, many of whom were connected to universities and concentrated in the New York City area, formed themselves into The Committee for Social Responsibility in Engineering (CSRE). They were stimulated to do so, they claimed, by the "misuse of technical talent, for destructive or frivolous purposes, while the lights are literally going out in our cities, coupled with rising engineering unemployment." Their concerns were a pastiche of complaints, characterized by a belief that federal spending for technological research, and therefore their own efforts, were being both misdirected and unnecessarily limited. At a meeting in March 1971, held concurrently with but separate from the annual conference of the Institute for Electrical and Electronic Engineers (IEEE), the committee's agenda included a host of questions: "What can be done for the engineer who is forced to choose between giving up his job and carrying out an assignment he feels is reprehensible, such as doing weapons-related research? Should engineering societies attempt to influence governmental policies on technology? If so, in what direction? How can engineering jobs be created that will serve society better than SST's or ABM's? Why is research on such a vital problem as controlled thermonuclear fusion, which might solve our energy problems without damaging the environment, funded on such a miserly basis? Why aren't there crash programs to develop the technology needed for quality mass transit systems? Is there a role for engineering unions?"[11] The mixture of conscience (weapons research), technological utopianism (thermonuclear fusion), and status anxiety (jobs and unionization) made the group direct descendents of those with similar concerns in the progressive era, when the beginnings of a new machine age seemed to go hand in hand with a greatly expanded but increasingly constrained and corporatized engineering practice.[12]

The CSRE continued for a number of years, holding its own meetings, leafleting at meetings of the large engineering societies, sending contingents of engineers to anti-Vietnam rallies and peace marches, and publishing a journal, *Spark*. As Francis J. Lavoie, senior editor for the journal *Machine Design*, enthused, "the engineer of the 70's is marching, picketing, and striking. He's [*sic*] speaking out on issues that affect him, as an individual or as a member of society. If he isn't yet doing it as loudly as some would wish, he's at least serving notice that he's concerned—and that he intends to voice his concern." This activism was not always confined to critiquing technology. Noting that "not all 'activist' groups are long-haired, young, and radical," he cited "the newly formed International Association for Pollution Control . . . [which was] made up of interdisciplinary professionals, all with a basic degree in engineering." The Lavoie went on to mention other groups as well as CSRE: Scientists and Engineers for Social and Political Action (SESPA), and the Committee on Technology and Society, which was eventually made an official division of the American Society of Mechanical Engineers.[13]

In the progressive era, the conservation movement had posed a special challenge to engineers, who were seen as doing the dirty work of corporations but also had a special opportunity to sell their expertise and presumed impartiality as the only practical route to using natural resources more efficiently. In the early years of the environmental movement, this dual challenge came to the fore for professionals again.[14] Issuing the challenge "builders of the environment, unite!" the executive director of the American Society of Civil Engineers in 1970 declared that "it is time that we converted pointless, inhibitive emotion into constructive, cooperative promotion."[15] In February of the same year "Engineering... Environmental Design for the 1970s" was chosen as the theme for the twentieth annual National Engineers Week, and three months later, the American Institute of Chemical Engineers created a new environmental division to, among other things, "further the application of chemical engineering in the environmental field."[16] Clearly, for the engineering profession, the corrective to worries about the pace, scale, and direction of technology was more, not less engineering.

The Office of Technology Assessment

The fight over the SST had made it painfully obvious to Congress that in any such contest over technology, it was woefully outclassed by the administration in residence at the other end of Pennsylvania Avenue. The White House could call upon an army of bureaucratic "experts" to back up its claims, while Congress had to construct a reaction out of lobbyists, the perceived interests of their constituents, and the testimony of witnesses before the relevant committees. This imbalance was exacerbated after the election of President Nixon in 1968, when a Republican White House faced off against a Democratic Congress.

Already in the spring of 1967, Emilio Daddario (D-Conn.) had introduced a bill into the House of Representatives that would "provide a method for identifying, assessing, publicizing, and dealing with the implications and effects of applied research and technology." According to the bill, Congress "hereby finds that, although such applied research and technology offers vast potentials, it also frequently produces undesirable by-products and side effects." Were the bill to become law, it would establish a board to assess the likely effects of any proposed new technology that the government was planning to back, and then find ways to encourage the "good" while mitigating the "bad."[17] It took five years of work, but late in 1972, Public Law 92-484 finally established a congressional Office of Technology Assessment (OTA). In the interim, Daddario kept up the pressure by holding hearings on the matter and commissioning studies from the National Academy of Public Administration, the National Academy of Engineering, and the National Academy of Sciences. The report of the last, made in 1969, was the most significant, both because of its source and because its unanalyzed assumptions gave a clear insight into the thinking of the nation's leading scientific elite at the time.

First, the authors assumed that technological change, conceived of as an amorphous aggregate, would inevitably continue; it could only be managed, not prevented. Second, the authors confessed that "our panel starts from the conviction that the advances of technology have yielded and still yield benefits that, on the whole, vastly outweigh all the injuries they have caused and continue to cause." This ringing endorsement was based not on any rigorous data, but entirely on an almost religious faith in progress. Third, the authors maintained that "technology as such is not the subject of this report, much less the subject of this panel's indictment. Our subject, indeed, is human behavior and institutions, and our purpose is not to conceive ways to curb or restrain otherwise 'fix' technology but rather to conceive ways to discover and repair the deficiencies in the process and institutions by which our society puts the tools of science and technology to work." Fourth, the panel emphasized that "it is therefore crucial that any new mechanism we propose foster a climate that elicits the cooperation of business with its activities. Such a climate cannot by maintained if the relationship of the assessment entity to the business firm is that of policeman to suspect." On the subject of the military, which by the panel's estimates accounted for more than half of all government spending on research and development, it noted stiffly that it did "not propose to pass judgment on military technology or on the decision-making mechanisms regarding it."[18] The positions of the study were striking in that they were decided upon a priori, rather than growing out of the study itself. They marked the limits of the panel's thinking, beyond which it would not venture.

Part of the explanation was surely the academy's unwillingness to alienate the two most powerful drivers of new technology, business and the military. This was closely related to a fear that any strong commitment to real assessment would play into the hands of what was perceived as a dangerous and irrational public attack against science and technology, which the panel thought to be underway within the country's population at large. "It is entirely possible," the academy warned, "that, frightened by the untoward side-effects of technological change and frustrated by their inability to 'humanize' its direction, people with much power and little wisdom will lash out against scientific and technological activity in general, attempting to destroy what they find themselves unable to control."[19] Just who these frustrated and frightened people with so much power might be was not revealed.

Similar worries were expressed from other quarters. An editorial writer for the *Los Angeles Times* claimed that "some businessmen, especially, fear that the attempt to [assess technologies] ... will have a sharp and negative impact on technological development in this country."[20] The cover illustration of the first issue of *Technology Assessment*, the official journal of the new International Society for Technology Assessment, pictured a slide rule made inoperative by a large padlock fastened through it. The editors explained that "the slide rule with a padlock on it portrays the concept that Technology Assessment, should be just that, Technology Assessment, *not* Technology *Arrestment* or *Harassment*."[21] The Nixon administration expressed the same concern, even using the same phrase, on the eve of the 1972 presidential election. Asserting that "technology assessment is a vital component of technological progress," it warned that "it should not become an instrument of technological arrestment."[22] William J. Price, of the Air Force Office of Scientific Research, was obviously worried about student unrest around the country, but thought that technology assessment could be "a vital force in society in counteracting the public backlash against science and technology. Even more important," he added, "this activity could provide both the catalyst, and a forum, through which the concern and good will of a rapidly growing number of persons, especially in the younger generation, could be harnessed to improve the quality of life in the United States and our relationship to the rest of the world."[23]

The OTA, with Daddario as its first director and Senator Ted Kennedy as the powerful chair of its governing board, got off to a shaky start at the end of 1972. With only a small budget and staff, it was instructed to respond to requests from members of Congress to look into various high-profile technologies that were likely to come before Congress for support. The OTA staff was too small to start its own investigations, or even collect its own data. Instead, contracts were let with companies and nonprofit organizations, such as universities, to make the actual assessments, which the staff would then put into a form that Congress could use. One continuing weakness was that the information upon which the assessments were based, almost necessarily, had to come from the private and public interests that were most involved, and therefore had the most directly at stake in the outcomes.

Though the agency was formally nonpartisan, it was always viewed with suspicion by conservatives, especially those in the Republican Party.[24] In 1995, acting on their promise to "downsize" the federal government, House Republicans eliminated the OTA's comparatively tiny budget of $22 million. The agency was temporary saved after three close votes, ironically through the efforts of the head of its governing board, a Republican from New York and former CEO of Corning, Incorporated. But by September 1995, the OTA's offices were piled with packing cases, and the next month it was gone. The agency's demise was so rapid and unexpected that some ongoing studies had to be abandoned. Again ironically, one was "on the role of the United States in United Nations peacekeeping operations and how to protect against weapons of mass destruction that fall into the hands of third parties like terrorists or small nations that do not have the means to build such arms."[25]

Lurking behind the notion of technology assessment was a largely ignored assumption that once the assessment was completed, some hard choices had to be made about which technologies to encourage and which to discourage. Presumably, that choice would be based at least in part on what one wanted to accomplish. One weakness of much public thinking about technology assessment had been the belief that technologies could be labeled as either good or bad, while in reality those two categories bled into one another, so that most technologies were "good" and "bad" at the same time, depending upon who wanted to do what, and to whom.

The Appropriate Technology Movement

The AT movement, which flourished between the late 1960s and the early 1980s, was an attempt to better fit the means of technology with its ends. The movement had its first advocates in the foreign policy field of technical aid to newly emerging economies, but soon was applied to domestic concerns as well.[26] If some newly formed nations were underdeveloped, it was argued, the United States was demonstrably overdeveloped. The idea was not to abandon technology, but to carefully select and develop technologies that would have positive social, political, environmental, cultural, and economic effects.

California was the center of the AT movement in many ways. In 1969, Nancy and John Todd, along with friends, formed the New Alchemy Institute, which attempted to "create an alternative dynamic to the very destructive one that we were currently engaged in, as members of industrial society."[27] The next year, the institute moved from San Diego to Cape Cod in Massachusetts, near the Woods Hole Oceanographic Institution, where they had gotten new jobs. Working on a 12-acre farm, the institute concentrated on the areas of food, energy, and shelter, building an A-frame greenhouse, solar-heated tanks for fish farming, a windmill, and other structures. Until it closed in 1992, the farm served as an educational site, demonstrating classic AT technologies.

Also in 1969, the Community for Environmental Change was incorporated in California, and in 1974, the group launched the Farallones Institute, headed by Sim Van der Ryn, a professor of architecture at the University of California at Berkeley. Within a year, the institute was operating two sites: a farm in Sonoma County, which among other things trained Peace Corps workers, and an Integral Urban House in the industrial flatlands of Berkeley. At the second site, Van der Ryn and his associates attempted to see how far they could successfully go in cutting an urban house off from the engineering infrastructure, which, they believed, served both the individual and the environment poorly. "The emergence of adaptive, small-scale patterns of habitat," Van der Ryn wrote, "requires the development of new kinds of technology. We need methods and equipment that are: 1) cheap enough to be accessible to nearly everyone, 2) simple enough to be easily maintained and repaired, 3) suitable for small-scale application, 4) compatible with man's needs for creativity, and 5) self-educative in environmental awareness."[28]

When Jerry Brown was elected governor of California in 1974, he drew many of the people on his new administrative team from the Zen Center in San Francisco, including Van der Ryn, who he appointed state architect. Besides such projects as designing the green Gregory Bateson office building in Sacramento, Van der Ryn talked Brown into letting him establish an Office of Appropriate Technology (OAT) in the architect's office. Although operating on a very small budget, Van de Ryn saw OAT as a "counterweight to the tendency of present State law and procedures to subsidize and favor large-scale, expensive and wasteful forms of technology over more modest and frugal ones."[29] During his two terms in office, Brown pushed such measures as wind farms, tax credits for solar technologies, light-rail construction, methane car fleets for the state, and farmers' markets. During these years, California became an AT showcase, demonstrating how governments could begin to reshape their technological priorities.

In 1976, Representative George E. Brown of California read Governor Brown's executive order setting up OAT into the *Congressional Record*, and expressed "the hope that each member of Congress will become interested and informed on this new approach to technological growth," which he defined as "less complex, less capital intensive, more labor intensive, frequently smaller, more decentralized, and environmentally benign type of technology appropriate to the specific needs of a community or particular area of the country, or world."[30] Congress had already taken note of the movement, pushing its ideas on foreign aid agencies, the Energy Research and Development Administration, the NSF, and other agencies. Jimmy Carter, trained as a nuclear engineer, won the presidential election in that same year, and proceeded to back AT initiatives ranging from the establishment of a National Center for Appropriate Technology, which was funded in 1977, through the planning of a world's fair in Knoxville, Tennessee to embody the theme "Energy Turns the World," to installing solar panels on the White House.

The apparent triumph of AT proved to be short lived. Upon succeeding Jerry Brown as governor of California, the Republican George Dukmajian abolished the OAT and cut off funds for the AT program that had been established at the University of California. The demise of the AT movement, and the failure of the federal government to ever undertake significant research and development efforts in the field, was in part due to the enormously powerful interests that were deeply invested—in terms of capital, professional careers, and expertise—in the technologies against which the AT movement worked. Also, however, over the decade of its publicity, AT had been successfully feminized in the popular culture: something small, as E. F. Schumacher had suggested in his classic text *Small is Beautiful*, and soft, as in the physicist Amory Lovins' popular book, *Soft Energy Paths*. While it lasted, however, the AT advocates served, in effect, as the technological arm of the rising environmental movement.

Consumer Revolt

In attempting to regain control of the machine, ordinary citizens had at least four institutions that they could use. One was the marketplace itself, in which people could theoretically vote with their dollars to support or reject any particular technology. This worked less well in practice than in theory, in part because the built environment had been designed around certain technologies and marginalized others. A resident of the postwar housing developments could choose to buy a Ford or a Chevrolet, but could hardly refuse to buy a car altogether.

A second recourse could be had to insurance. That industry, presumably, made a cold calculation as to the danger of a technology, then charged a rate commensurate with that danger. In extreme cases, insurance companies could refuse to insure a technology at all. This happened with nuclear power, and until the federal government stepped in with the Price-Anderson Act to limit liability in case of a nuclear accident, utilities could not afford to build power plants. Most people, however, lacked the political clout of the electrical utilities. Faced with exorbitant insurance premiums, they would simply forgo the offending technology.

A third mechanism was the government regulatory agency, a device much used since the progressive era to evaluate technological activities, establish rules and guidelines for their production and use, and educate the public about the relevant issues that concerned them. Of course, these agencies faced persistent charges that they were too friendly with the firms that they were supposed to regulate, that they were often underfunded and understaffed, and that politics often trumped science in making regulations. Nevertheless, it was difficult to see how the country could operate without them.

Finally, citizens had recourse to the courts, either to seek redress from injury, or to force compliance with existing laws and regulations. Americans had a reputation around the world for being a particularly litigious people, and taking issues to court was a long and familiar tradition. One could collect payment for pain and suffering caused by a faulty machine, or force the closure of a dangerously polluting mill.

All of these were put into play during what came to be called the consumer revolt of the 1970s, and one major nexus of concern and activity was the American icon, the automobile. In 1966, Daniel P. Moynihan, later a U.S. senator, documented what he called "The War Against the Automobile."[31] "A series of converging events," he wrote in 1966, "make it likely that the United States is, at long last, going to come to terms with a gigantic domestic problem: traffic safety." Some 4 to 5 million persons were being injured each year, resulting in 100,000 people with permanent disabilities. Another 50,000 people died. "Life is expensive; justice, elusive; death, inevitable," Moynihan admitted, and if nothing could be done about automobile safety, we would just have to live (or die) with it. But injuries were "unnecessary—or at least meliorable." There was talk of improving driver competence and behavior, but virtually the entire population drove. Instead, Moynihan suggested that car interiors should be redesigned to better protect passengers; the "attraction of this approach," he pointed out, is that it could be put into effect by changing the behavior of a tiny population—the forty or fifty executives who run the automobile industry."

There were "at least three clusters of reasons" that explained the failure to do anything about the problem up until then. First, he pointed to the well-documented "venality" of the automobile industry. For decades, engineering had taken a backseat to maximizing design and profit. This explained why defects continued to show up in new cars, and why companies hushed it up and did little about it. The second reason was that "we have not much wanted" to do anything. The car, he explained, "is a central symbol of potency and power: the equivalent of the sword, or the horse, or the spear of earlier ages. It is both a symbol of aggression, and a vehicle thereof. It is a sanctioned form of violence ... and a prime agent of risk-taking in a society that still values risk-taking."

His third "cluster of reasons" was the "failure of government." He admitted to having been briefly a government representative, in the 1950s, to the President's Committee for Traffic Safety, which used the presidential seal "with abandon" and spent government money. Its chair, he discovered, was chosen and paid by the automobile industry. Government-wide, there was little research being done into traffic safety. "Government regulation of the automobile," he pointed out, "began as a form of tax collection, upon which a layer of law enforcement was superimposed." Against this dismal record, Moynihan found two sources of activity, both of which became mobilized in the 1950s. First were physicians working through the American Medical Association and the American College of Surgeons. Accident traumas were, after all, a public health problem. The second group to step forward were lawyers—not, he noted, the American Bar Association, nor the prestigious law schools, but the despised claimants' attorneys working through their American Trial Lawyers Association.

No doubt the most famous lawyer to become involved was Ralph Nader, the man *Time* magazine called "The U.S.'s Toughest Customer."[32] In the fall of 1965, while 106 damage suits were pending against General Motors for defects in its Chevrolet Corvair, Nader published his soon to become best-selling exposé, *Unsafe At Any Speed*. His first chapter dealt with the fatal crash of a Corvair in Santa Barbara, California. Auto leaders immediately brushed off the accusation that they sacrificed safety for speed, style, and profit, but Nader was called to testify before a congressional committee. He appeared in February 1966, and the next month, made headlines again by charging that General Motors had private investigators checking into his life and background, and had even tried to compromise him with beautiful women of easy virtue.[33] This last allegation was never proved, but General Motors had hired an investigator, a former Federal Bureau of Investigation (FBI) agent who instructed his colleagues in a memo to find out "'what makes him tick,' such as his real interest in safety, his supporters, if any, his politics, his marital status, his friends, his women, boys, etc., drinking, dope, jobs—in fact all facets of his life."[34] On March 22, appearing before a Senate subcommittee investigating auto safety, James T. Roach, the president of General Motors, admitted that his company had investigated Nader, but denied any harassment. By the end of that month, copies of Nader's book were selling at 6,000 per week, and by the end of 1969, it had sold 55,000 hardback copies and more than 200,000 paperbacks. Nader's indictment of General Motors' technology and the company's response were headline news.

Corporate America, however, was not prepared to admit that it manufactured unsafe technologies. William B. Murphy, chair of the Business Council, stormed that "this country in on a safety kick. It is a fad, on the order of the Hula Hoop. We are going through a cycle of over-emphasis on safety."[35] That same month, May 1966, the journal *Car Life* editorialized that "the nation's legislative leaders are whipping up a magnificent case of mass hysteria. The automobile as currently designed and manufactured is patently unsafe. 'Safety,' these legislators would lead us to believe, can only be achieved after the expenditure of untold millions of tax dollars. 'Safety,' they say, means re-designing the car to make it crash-proof at any speed. . . . Any automobile can be safe," it concluded, "as long as its driver knows and understands its limitations, and does not exceed them."[36] Despite such attacks, President Johnson in September signed a Traffic Safety Act designed to come up with new standards for Detroit's 1968 models.

By 1970, the attack on the automobile had broadened into a comprehensive movement to protect consumers from unsafe products of any kind. As Paul Jennings, president of the International Union of Electrical, Radio, and Machine Workers wrote in an introduction to Nader's 1968 pamphlet, *Who Speaks for the Consumer?*, the goal should be that "all products offered are always safe to the consumer." He singled out workers, operating through their unions, as having the "right and the responsibility" to see to consumer safety, but others were involved as well.[37] In 1969, a Center for Law and Social Research was organized to aid consumers by appearing before federal regulatory bodies. The right of the public to participate in agency hearings had only been conceded three years before, when a Protestant church had intervened to object to the renewal of a license for a southern television station that had discriminated against African-Americans. The Center was preparing to make such interventions routine, and had picked as its first action to file a petition before the U.S. Court of Appeals seeking to force the secretary of agriculture to immediately ban the sale of DDT in the United States.[38]

That same month, the Magazine Publishers Association ran a full-page advertisement in *Good Housekeeping*, headlined: "There are some people around who want to spoil a woman's greatest indoor sport." The sport was shopping, and according to the association, "it's hard to believe, but there are some fairly influential people around today who think the government ought to do your shopping for you. . . . Restrictions," it concluded, "don't stimulate competition. They tend to make all products the same. And that's not the way this country got prosperous."[39] Warnings came also from Kay Valory, the consumer advisor to California governor Ronald Reagan. Calling Nader and his followers a threat to free enterprise, she suggested that consumers who wanted help against companies should turn for aid to the National Association of Manufacturers.[40]

In early 1970, when the Consumer Assembly held its fourth annual meeting in Washington, it could look back with some satisfaction at the progress made. A "host of Federal laws regulating areas ranging from auto safety and soap packaging to money lending and the purity of meat and poultry" had been passed. At least thirty-four states had set up consumer-protection agencies, as had a number of cities. New York City's mayor John Lindsay was being advised by the high-profile former Miss America, Bess Myerson Grant. Even the Nixon administration had been pressured into appointing a presidential assistant for consumer affairs. In addition, at least fifty private consumer groups had sprung up across the country. Nader, however, in giving the main address to the assembly meeting, appeared to despair of any real progress being made.[41]

A report issued by the National Commission on Product Safety in June 1970 showed just how hard it was to get government agencies to regulate consumer safety. Cumbersome procedures were set in place that slowed progress. Agencies sought cooperation rather than confrontation with the industries they were to regulate, they relied on industry lobbyists for the data they needed to make findings, and they were routinely underfunded. The FDA's Bureau of Compliance, which was "responsible for enforcing the banning or labeling of thousands of hazardous products," had only one full-time employee. If strict rulings were made, there was an expectation that the affected industry would tie the agency up in the judicial process for years and might pressure friendly members of Congress to cut agency appropriations even further.[42]

One further attempt at reform was made in proposed legislation to allow consumer class-action suits against companies that produced and sold defective or dangerous technologies. Even President Nixon's consumer advisor testified in favor of the move—an action that the White House immediately regretted—but a panel from the American Bar Association went on record opposing the reform. That all nine members of the panel were from large law firms representing corporate clients cast immediate doubt on their opposition.[43]

The business community did not depend solely upon the ABA to denounce class actions again defective products. The head of the New York State Better Business Bureau charged that consumer activists used "harassment, hunch, prejudice and emotion," and were pushing business "into a deepening decline." The auto industry, he said, had been "savaged" by Congress, and warned that something he called "synthetic consumerism" only made things worse because it ignored "facts." The head of the American Advertising Federation expressed this same commitment to facts rather than emotion. The consumer movement, in his opinion, was dangerously substituting "emotion for reason and judgment on issues involving sales and marketing practices." Class action suits in this area, he warned, "could result in needless harassment of business rather than actually redressing legitimate consumer grievances."[44] This common appeal to facts was very much part of scientific and technological thinking, but seized upon and exploited by the business community. The argument also had distinct gender overtones: women, who were seen as the consumers, were essentially emotional, while manufacturing and business in general were the domain of men, who were essentially rational.

Neo-Luddites

Compared with the mostly ameliorating efforts undertaken by citizen and professional groups, as well as by governments at all levels, the efforts of those who were loosely grouped under the banner of neo-Luddism were strikingly radical. The term had long since acquired negative connotations of violent, irrational action aimed at destroying modern technology, and therefore undoing progress. It came from a group of artisans in the British textile trades who, during 1811 and 1812, had managed to smash and burn an estimated £100,000 worth of property, including the newly invented devices that had mechanized their trades and were putting them out of work. The harsh response of the British, both governments and vigilantes, was, in the words of one recent writer, "the greatest spasm of repression Britain ever in its history used against domestic dissent."[45]

Like other marginalized groups in late twentieth-century America, some who sought to expose the profoundly political nature of technology in a culture that persisted in seeing it as either neutral or gloriously progressive embraced the term Luddite, using it as a weapon to shock. The historian David F. Noble, during his tenure with the Smithsonian Institution, tried to mount an exhibit on the Luddites, publishing a manifesto titled *Progress Without People: In Defense of Luddism*. For a frontispiece, he selected a photograph of himself inspecting "the only Luddite sledgehammer still in existence."[46] Kirkpatrick Sale wrote in 1995 of the "technophobes and techno-resisters" who were "attempting to bear witness to the secret little truth that lies at the heart of the modern experience: whatever its presumed benefits, of speed or ease or power or wealth, industrial technology comes at a price, and in the contemporary world that price is rising and ever threatening."[47] He cited a *Newsweek* survey that seemed to show that up to a quarter of the population was, by a very broad definition of the Luddite stance (it was hardly a movement), part of a group ranging from radical environmentalists who were prepared, like the Luddites, to burn and smash, to those who had been automated out of a job—also like the Luddites. Just as the Luddites of the early nineteenth century had succeeded in placing the "machine question" on the political agenda is England, so the aim of their heirs in the late twentieth century was to move beyond the "quiet acts" of personal withdrawal that the social critic Lewis Mumford had advocated, and attempt to make the "culture of industrialization and its assumptions less invisible."[48]

The closest thing to a manifesto for the neo-Luddites was drafted in 1990 by Chellis Glendinning, a New Mexico psychologist who had authored *When Technology Wounds: The Human Consequences of Progress* (1990). "Neo-Luddites," she wrote, "are 20th century citizens—activists, workers, neighbors, social critics, and scholars—who question the predominant modern worldview, which preaches that unbridled technology represents progress. Neo-Luddites have the courage to gaze at the full catastrophe of our century: the technologies created and disseminated by modern Western societies are out of control and desecrating the fragile fabric of life on Earth."[49] She identified three "principles of Neo-Luddism;" first, "Neo-Luddites are not anti-technology;" second "all technologies are political;" and third "the personal view of technology is dangerously limited." Her proposed "Program for the Future" involved dismantling six technologies that she considered particularly destructive and a search for "new technological forms": she favored "technologies in which politics, morality, ecology, and technics are merged for the benefit of life on Earth."[50]

Unfortunately for neo-Luddites, the term was also attached to the most violent and notorious antitechnology fanatic of the day: Theodore Kaczynski, the so-called Unabomber. Between 1978 and 1995, Kaczynski mailed or delivered sixteen package bombs to scientists, academics, and businesspeople, injuring twenty-three of them and killing three. The motives behind the serial killer remained unknown until 1993, when he began to contact the press. Two years later, the *Washington Post*, in cooperation with the *New York Times*, published his 35,000-word manifesto, titled "Industrial Society and Its Future."[51] His brother recognized his handwriting from the document and identified him to the authorities. Kaczynski was arrested in his remote Montana cabin, and in 1998, he was convicted of murder and sentenced to life in prison.[52]

Once Kaczynski was arrested, it was discovered that he had been a brilliant young undergraduate at Harvard University, and had gone on to earn a doctorate in mathematics at the University of Michigan and teach at the University of California, Berkeley. Judgments abound over why he became a killer, ranging from a troubled childhood to the brutal psychology experiments for which he volunteered at Harvard to the arrogance drawn from his success in the abstract culture of mathematics. He was also, of course, thought to be insane—specifically, a paranoid schizophrenic. Of greater interest in the context of neo-Luddism, however, was not his uniqueness, but the very ordinary and perhaps even banal nature of his views on industrial society. Copies of his manifesto were circulated among academic historians of science in the hope of finding some thread that would connect the ideas expressed with a particular individual, but the common reaction was that it was a competent restatement of "anti-technology views [that] were common among students in the 1970s, when the Vietnam War turned so much technology to violent ends."[53] As the journalist Kirkpatrick Sale put it, "he is not a nut. He is a rational man and his principal beliefs are, if hardly mainstream, entirely reasonable." That said, he also guessed that the author was "measurably unbalanced," with a "disturbing obsession with power."[54]

Kaczynski's argument is not easily summarized, in part because it is not tightly constructed. A sampling of his views, however, gives the flavor of his beliefs: "The Industrial Revolution and its consequences have been a disaster for the human race.... We therefore advocate a revolution against the industrial system.... Industrial-Technological society cannot be reformed.... This implies revolution, not necessarily an armed uprising, but certainly a radical and fundamental change in the nature of society.... It would be better to dump the whole stinking system and take the consequences.... The technophiles are taking us all on an utterly reckless ride into the unknown."[55]

Of course, the majority of neo-Luddites were not homicidal autodidacts, or even fundamentally antitechnology. "There has never been a movement that simply and unthinkingly hated machines and set about destroying them," wrote Theodore Roszak in introducing a 1997 Sierra Club book entitled *Turning Away From Technology: A New Vision for the 21st Century*. "Neo-Luddites," he went on, "know that true progress—improvements in the quality of life, not the quantity of goods—never grows from machines, but from the judgment and conscience of our fellow human beings.... By way of an alternative [to technological enthusiasm], Neo-Luddites opt for prudence and the human scale. Theirs is a simple program: scale down, slow down, decentralize, democratize."[56] It was a formula to which many in late twentieth-century America could subscribe.

It was not, however, a formula that the defenders of the nation's scientific and technological culture could easily accept. Writing about what he called the "Science Backlash on Technoskeptics," Andrew Ross, a professor of American studies and himself a target of anger and derision, pointed out in 1995 that the so-called culture wars had a "science-wars" component. He quoted the Harvard biologist E.O. Wilson as telling an audience that "multiculturalism equals relativism equals no supercollider equals communism."[57] Ross's belief was that "we live and work today in conditions and environments that demand a critique of technoscience as a matter of course, an exercise of minimal citizenship that will address public anxiety about the safety of everything from the processed food we consume to the steps toward a biologically engineered future." He noted that not for the first time, "a hue and cry has been raised about the decline of science's authority, or its imperilment on all sides by the forces of irrationalism."[58]

The charge that critics of late-century technology, like Ross, usually lacked technical expertise was no doubt true, but also irrelevant. The issue being raised was not so much how to redesign the machine but how to design a society that was nurturing and just. That is a question in which all citizens of a democracy are expert.

TEN

A WIRED ENVIRONMENT

The electrical engineers and physicists who attempted to create "an electronic environment for war," to use historian Hunter Dupree's phrase, built better than they knew. Not only are smart bombs, fly-by-wire airplanes, battlefield laptops, and even the videophones of embedded reporters all descendents of that World War II–era effort, but the domestic environment has been transformed as well. Computers, transistors, and integrated circuits insinuated themselves into every crevice of daily life, doing much more than the complex mathematical calculations for which they were originally designed. By the end of the century, the newly wired society was already giving way to a wireless culture.

Early Computers

Before 1950, no machine would qualify as a modern computer, but like most devices, computers had a long prehistory. In 1802, the Jacquard loom was programmed with rigid cards, in which holes had been carefully arranged, to weave beautifully intricate patterns automatically. The loom seems to have influenced the British visionary Charles Babbage in the design of his Analytical Engine in the first half of the nineteenth century, though the machine was not actually built until the twentieth century. In the United States, Herman Hollerith invented a machine to handle census data on punched cards, which could replace twenty human "computers" (usually female mathematicians). Hollerith's devices were used to manipulate data from the 1890 census, and in 1896, Hollerith organized the Tabulating Machine Company. Through a merger in 1911, the company was renamed the Computer-Tabulator-Recording Company (C-T-R), which in 1924 was renamed as International Business Machines (IBM).[1] By this time, the United States was the leading manufacturer of business machines, with C-T-R (tabulators), National Cash Register (cash registers), Burroughs Adding Machine Company (adding machines), and Remington Rand (typewriters), all of which were to be major players in the post–World War II computer story.

Perhaps the first well-known "computer," Vannevar Bush's analog Differential Analyser was developed in the late 1920s and immediately dubbed the "Giant Brass Brain" by the press.[2] An electromechanical device, it and copies modeled on it worked during and after the war; one was used at the University of California, Los Angeles to solve problems in the aviation industry. An even better-known machine, the Mark I, was designed by Howard Aiken beginning in 1937, in cooperation with IBM. Aiken took his doctorate at Harvard in 1939, and stayed on as a faculty member, talking his colleagues into grudging support of his new machine. Unveiled in 1944, the Mark I relied on electromagnetic relays rather than vacuum tubes, because these relays were basic features of IBM business machines.[3] It was a striking piece of machinery: Fifty-one feet long, eight feet high, and three feet wide, it weighed five tons and had 760,000 parts, including 2,200 counter wheels and 3,300 relay components, all connected by 530 miles of wire. Data was entered by means of paper tape and retrieved on a standard IBM electric typewriter.

By this time, Aiken was working for the Navy, and the Mark I was officially a unit of the Bureau of Ships. It ran almost constantly during 1944 and 1945, working on a variety of military problems including one that John von Neumann had undertaken for the secret atomic bomb project at Los Alamos. The computer worked at Harvard for another fourteen years after the war, being abandoned finally in 1959. Because it relied on electromagnetic relays, its design had little direct influence on later computers, but it had dramatically proved that computers not only worked, but were very, very useful.[4]

A computer using vacuum tubes rather than electromagnetic relays began operations in 1945 at the University of Pennsylvania's Moore School of Electrical Engineering. It was the first electronic rather than mechanical machine of its kind. Named the Electronic Numerical Integrator and Computer (ENIAC), this computer had been funded by U.S. Army Ordinance to work on wartime problems for the Ballistic Research Laboratory, and was the result of design efforts by J. Presper Eckert and John W. Mauchly. ENIAC was another giant, measuring fifty by thirty feet, and using 18,000 vacuum tubes and so much electricity that when it was turned on, lights were said to dim all over Philadelphia. Because tubes burned out with distressing regularity and especially when the computer was first turned on, its handlers tended to just leave it on whether it was working or not.[5]

Meanwhile researchers at MIT began Project Whirlwind in 1944, which proved to be pivotal in computer development. Originally simply an analog control device for a flight simulator, it grew in both ambition and cost. After being almost terminated in 1949, it was saved by the decision to make it the central computer control for the Pentagon's Strategic Air Ground Environment (SAGE), a major attempt to provide an air defense system for the developing Cold War. SAGE was important for at least two reasons: it provided, as historian Jeffrey Yost has explained, "analog to digital conversion equipment, improving near-term computer reliability, and facilitating real-time computing, all of which would be utilized in scientific, medical, and other applications." It was, as he also points out, a prime example of a large, expensive, military-funded computer project that failed in its military mission but proved valuable to civilian applications.[6]

Computers for Business

In 1951, about half a dozen computers of one type or another were in operation, the most important probably being the Universal Automatic Computer (UNIVAC), an improved version of ENIAC that made headlines when CBS used it on election night in 1952 to predict the choice of Dwight Eisenhower as president based on early returns. The machine was dressed up with dramatic flashing Christmas lights for the occasion, and was the visual star of the program. Its prediction was so swift, however, that CBS staff did not trust it and waited until its prediction was confirmed before announcing the winner.

Described by Paul Ceruzzi as "the Model T of computers," the IBM 650 became available in 1953 and proved to be the machine that broke into the business market, with thousands being sold. A relatively small and inexpensive machine for common business purposes was a revelation after a decade of monster machines built at universities for solving mathematical problems for science. Before the success of the 650, assumptions were rife that the world would need only a handful of computers. Aiken had predicted that four or five would take care of American needs, and a physicist in Britain said in 1951 that "we have a computer here in Cambridge; there is one in Manchester and one at the [National Physical Laboratory]. I suppose there ought to be one in Scotland, but that's about all."[7]

Ceruzzi attributes this failure of vision—by some of the smartest and most knowledgeable people in the field—to three causes. First there was "a mistaken feeling that computers were fragile and unreliable." The frequent blowing of vacuum tubes by the ENIAC was apparent evidence for this worry. Second, those who designed and used computers in the early days were scientists and engineers in university settings who had needed to crunch very large numbers, and built machines to do just that. That other people might have different jobs for computers apparently did not occur to them. Third, the computer pioneers simply did not understand the machines themselves, how they worked, or how much and what kind of work they could do. Surely, they assumed, the need to hire a large staff of mathematicians to help program the machine sufficiently blocked its wide dissemination, especially if, after the initial problem had been solved, the owner had no other large problems to follow on. How many people needed to do 30 million operations in one day, the capacity of ENIAC?[8]

Programs and Software

During the 1960s, IBM's S/360 series of computers solidified the company's position as the country's leading computer manufacturer, to such an extent that its competitors—Remington Rand, Burroughs, Control Data, Digital Equipment Corp., NCR, RCA, and General Electric—came to be called the Seven Dwarfs. A major step in making the computer a business machine was the realization that programs did not have to be fed in on punched paper tape for every use, because general-purpose programs could be stored in the machine's memory alongside whatever data it contained. Eventually one could program a computer simply by choosing from a menu of commands. Developing programs that responded to simple commands made the computer accessible to nearly anybody. At this point, of course, the pressure was on to write more and better programs, and software became equal in importance to hardware.[9]

Software seemed to exist in a perpetual state of crisis. When IBM introduced its System 360 project, the software had not received the attention it deserved, and its operating system (OS/36) was delivered late, took up too much memory, was subject to cost overruns, and did not work very well.[10] Serious users had to learn programming languages such as FORTRAN, COBOL, and ALGOL. Sometimes the subordination of software to hardware proved costly. The Y2K problem grew out of an early concern to take shortcuts on software, in this case shortening code to a two- rather than four-digit date (e.g., 68 rather than 1968) to save computer time and storage, which seemed to be much larger issues than any distant millennium.

Slowly, programmers tried to upgrade their status by calling themselves software engineers. They developed a field in which women found employment more easily than they did in the field of hardware, no doubt partly because of the already lower status of the work, and perhaps also because women, hired to do calculations, had been the original "computers." Eventually, such giant firms as Microsoft, as well as many smaller ones, learned how to create an almost campus-like atmosphere that encouraged and rewarded software development.[11]

The ascendancy of IBM in the computer world was originally based on very large mainframes, which were still mostly found in government bureaus—particularly NASA and the Pentagon—large corporate headquarters, and on university campuses. At these last sites, they were usually grouped in computer centers, where the university administration kept its payroll and student records, and individual researchers scheduled time on the machines to run their cards punched with data. These mainframes gave material form to fantasies of "giant brains" and firmly associated computers, in the popular imagination, with large bureaucratic organizations that had Orwellian capabilities for social control. It was not until the rise of Silicon Valley, in California on the San Francisco Peninsula, that the computer became personal.

Terman at Stanford

Historian James Williams has traced the evolution of California's Santa Clara Valley from The Valley of Heart's Delight, so named when it was covered with orchards of apricots, prunes, and cherries, to Silicon Valley.[12] Arguably the most important engine of this transformation was Stanford University, which had been founded in Palo Alto in 1885. More specifically, it was the Stanford engineering school, which had cut its teeth on helping solve the problem of California's comparative lack of power to drive any kind of industrial base. As it happened, the hydroelectric potential of the Sierra Nevada mountain range forced the rapid development of the electrical engineering know-how necessary to transmit electricity over long distances. University professors forged strong bonds with engineers from the electrical utilities, students received practical as well as academic educations, and state-of-the-art facilities were built at Stanford. Similarly, shipping interests in San Francisco were eager to make use of the new technology of radio to improve communications with ships at sea. One local enterprise, the Federal Telegraph Company, founded in 1911 by a recent Stanford graduate and financed in part by faculty as well as the president of the university, established its headquarters near the campus in Palo Alto. Together these two commercial imperatives helped create not only a thriving electrical engineering program at Stanford, but also one that was comfortable with encouraging practical applications of their work.

In 1910, Stanford hired Lewis M. Terman, a psychologist most famous for his soon-to-be-developed Stanford-Binet intelligence test. His son, Frederick E. Terman, matriculated at Stanford, taking a degree in chemical engineering in 1920. After a stint at Federal Telegraph, the younger Terman returned to Stanford and took another degree, this time in electrical engineering. By 1924, he had received his doctorate in that field from MIT, where he studied under Vannevar Bush. After returning home, he accepted a part-time teaching job in the new radio laboratory in Stanford's electrical engineering building. Over the years, he moved up the administrative ladder ,and by the late 1920s had come to the conclusion that "Stanford is in an excellent strategic position to initiate a pioneer movement that will make this the national research center of electrical engineering."[13]

Terman became head of the electrical engineering department in 1937, and expanded his role as a facilitator of cooperation and technical development. One prewar story is particularly significant. As Williams tells it, a Stanford graduate, William W. Hansen, returned to the university's physics department in 1934. There, he developed a device he called a rhumbatron that could accelerate protons. At this same time, a young pilot named Sigurd Varian talked his brother Russell into helping him try to develop navigation and detection equipment for airplanes. The brother, as luck would have it, was a Stanford physics graduate who had been Hansen's roommate when they were students. He suggested that the rhumbatron might be useful in their navigation research. The brothers moved to Stanford and received research support of lab space and $100 a year, in exchange for half of any patents that resulted. Soon they had developed a new electron tube they called the klystron, which quickly received support from the Sperry Gyroscope Company.

Two of Terman's students, William Hewlett and David Packard, were brought back to Stanford, the latter on a fellowship funded out of the patent-pool money received from multigrid tubes invented by Charles Litton, and the same pool that held the klystron patents. In 1937, Terman encouraged the two to start a company to commercialize an audio-oscillator Hewlett had developed.[14] Along with Litton and Varian, Hewlett-Packard eventually became central to the development of Silicon Valley.

At this point, World War II broke out, and Terman was called by his former mentor Vannevar Bush, now head of the OSRD, to take over the Radio Research Laboratory at Harvard. During his wartime tenure, Terman speculated about the postwar shape of technology. "The war had made it obvious to me," he later said, "that science and technology are more important to national defense than masses of men. The war also showed how essential the electron was to our type of civilization." He also saw that after the war "there would be, for the first time, real money available to support engineering research and graduate students. This new ballgame would be called sponsored research." Terman's vision and entrepreneurial drive tied the new situation to his old vision of Stanford's preeminence. His first grant after returning to Palo Alto was $225,000 from the newly established ONR: out of it flowed a Nobel Prize and what Terman called "today's nationally recognized research program in engineering."[15]

Nor did Stanford have to do all the research work itself. In the early 1950s, the university decided to develop a forty-acre corner of the campus as a site for light industry. Varian Associates was the first to express an interest, and in part because Terman was a Varian director, the idea developed that the entire park should be settled by industries connected somehow to the university's research. Hewlett-Packard moved to the new Stanford Industrial Park in 1956; it was soon joined by Ampex and Lockheed's Space and Missile Division.[16]

By 1960, the park had grown to 650 acres and was home to more than forty firms. Nor was it the only local venue for research and development. On the eve of World War II, NACA had opened the Ames research center in nearby Sunnyvale. After NACA was folded into the new NASA in 1958, it became a center of aerospace research and development. By that time, there were also at least forty-three other federal military or research installations scattered about the Bay Area.[17] Some of these dated back to the mid-nineteenth century, evidence of just how long and successfully local politicians and business leaders had sought federal installations to seed economic growth.

Among other things, all of this military-inspired research and development led to a "blue sky dream" for thousands of postwar families. One was that of writer David Beers, whose memoir of that name tells the story of his father, an engineering graduate and Navy test pilot who took a job at Lockheed and moved his young family to the Valley of Heart's Delight. They did all of the postwar things, buying a lot in a subdivision, choosing the house plan, trying to grow the perfect lawn, and collecting all the necessary appliances, from a bicycle for David to a car for the whole family. The son later discovered that his father worked on "black budget" aerospace projects, having spent his whole life since his undergraduate days on the public payroll as a part of the Cold War military-industrial complex. For his family, it was mainly a sunny, happy suburban life, but for the father, the blue-sky dream had been clouded over with regrets and feelings of frustration around his secret weapons work.[18]

Transistor

Eventually the aerospace "tribe," as Beer describes it, was overshadowed by a new culture of computer technologies. The latter had received a massive boost with the arrival in 1956 of William Shockley, who shared a Nobel Prize for the invention of the transistor. Like Terman, Shockley had spent his boyhood in

Palo Alto before going east, eventually to the Bell Telephone Laboratories in Murray Hill, New Jersey. There, with wartime work winding down in 1945, Shockley was made leader of a team working in solid-state physics. Telephone technology was based on vacuum tubes, which were used to control electron flow, but the increasing scale and complexity of the Bell network was threatening to overwhelm the existing technologies. Shockley's team experimented with semiconductor crystals and successfully got them to act like vacuum-tube amplifiers. The success was announced in New York City with great fanfare on June 30, 1948, but the advance was so new and seemingly simple that the *New York Times* carried the news on page 46.[19]

Bell itself was a major beneficiary of the new device, for one thing because of the energy saved from not having to heat up vacuum tubes to set off their stream of electrons; also, transistors did not burn out like the old, hot tubes. After Bell came demands from the military, which was keen to improve battlefield communications using walkie-talkies and other devices. By 1953, the Pentagon was funding half of the research and development work that Bell was doing on transistors. One major improvement resulting from this work was the junction transistor, developed in 1951, which made use of the internal characteristics of the crystal rather than depending on its surface effects.[20]

However, aside from Bell products, the transistor did not appear in commercial devices until 1952. The earliest application was their use in hearing aids made by the Sonotone Corporation, announced in December of that year. They were so successful that within a year, three-quarters of all hearing aids used transistors, and in 1954, about 97 percent did. That was also the year that Texas Instruments' Regency all-transistor radio made its appearance for the Christmas trade. Perhaps the greatest commercial breakthrough came about in 1955, when IBM used transistors in its 7090 computer. As historian Robert Friedel has suggested, however, "the microelectronics revolution might have remained a quiet and limited affair but for two great events." One was the orbiting of *Sputnik* by the Soviet Union on October 4, 1957, which immediately focused the nation's attention on space. The other was Shockley's move to the emerging Silicon Valley.[21]

Silicon Valley

The connections and moving about of pioneer figures in the field were striking, and Shockley is the center of only one of the nodes of innovation. Cecil H. Green, who died in 2003 at the age of 102, had taken a master's degree in electrical engineering from MIT and for a time worked at Raytheon, one of the founders of which had been Vannevar Bush. There, Green was an assistant to Charles Litton, who later founded Litton Industries. In 1932, Green joined Geophysical Service, Incorporated, a Texas company that did seismic oil explorations. With three others, he bought out the firm in 1941, and after the war, helped steer it into electronics, especially submarine detection and radar. A decade later, the company's name was changed to Texas Instruments, and in 1952, it entered the semiconductor field, bringing out the pocket-sized transistor radio in 1954. By 1958, Texas Instruments was working on integrated circuits.[22]

Meanwhile, Shockley had returned to the West Coast fresh from his triumphs at Bell, and began to attract some of the brightest and most inventive minds in electronics to his new firm, which would concentrate on semiconductors. Within two years, however, Shockley's eccentric and dictatorial management style had alienated seven of his best people, who decided to leave to set up their own company. An investment firm put them in contact with Fairchild Camera and Instrument Company of New Jersey, which expressed interest in financing them, but insisted that they find someone with managerial leadership to help them. They then enticed the only star of Shockley's who had not already agreed to follow them out of the company—Robert Noyce—and the eight established Fairchild Semiconductor in Mountain View.[23]

Fairchild immediately became an incubator of engineering talent in the field. Williams has noted that of the 400 semiconductor engineers at one meeting in Sunnyvale in 1969, all but two dozen had worked at Fairchild. By the 1970s, former Fairchild employees had founded forty-one semiconductor companies, many of which did not stray outside Silicon Valley.[24] Such coming together and breaking apart was not unique to semiconductors, nor to the late twentieth century. In the mid-1800s, the Armoury System of using dedicated machine tools to manufacture interchangeable parts had been carried out of federal armories, and from firm to firm and industry to industry, by skilled mechanics moving on and setting up their own companies. But this new example of restless inventiveness, linked to an entrepreneurial drive, was striking.

Microprocessors

The next breakthrough came in 1958, when Jean Hoerni worked out the planar method of manufacturing transistors, which consisted of layering thin slices of silicone and silicon dioxide on a wafer of silicone crystal. The next and perhaps most profound development came from Noyce, who worked out the integrated circuit. Jack Kilby had already begun putting entire electronic circuits on a semiconductor, in this case a piece of germanium. Noyce's contribution was to show how Kilby's hand process could be replaced by the planar process, thus making possible the virtual mass production of incredibly small and complicated circuits. He described his breakthrough in 1959; it first came to market in 1962, and in 1971, the first microprocessor was marketed by Intel, a new firm that Noyce organized.[25] The engineer who developed the microprocessor for Intel was Marcian "Ted" Hoff, who had joined the company fresh from his doctorate in electrical engineering from Stanford. His chip, the Intel 4004, was replaced by 1975 by the company's third generation, the 8080, which was at the core of the Altair, the first computer priced within the reach of the average buyer.[26] Intel also developed the microprocessor for the Japanese firm Busicom, which made hand-held calculators.[27]

Having been the first to build an integrated circuit, Texas Instruments was anxious to find a commercial use for it. They had been successful at popularizing their transistor with the Regency radio, and wanted a device that would do the same for its new integrated circuit. At that time, the only calculator that used a transistor was that made by Sharp, and it weighed fifty-five pounds and had to be plugged into a wall socket for power. Texas Instruments' Pat Haggerty, in 1965, wanted a battery-powered calculator small enough to be hand-held, and he turned to Jack Kilby to make it. However, while the radio used only four transistors, a calculator would need hundreds, and seemed to present a problem well suited for solution by integrated circuits. It was yet another example of an invention looking for a necessity.[28]

Kilby thought of his proposed device as "a slide-rule computer" and realized that every function—getting information in, making calculations with the information, getting it out, and powering the whole process—was going to have to be invented virtually from scratch. Jerry Merryman, who Kilby put on the job, started with the basics: "I thought, I guess it's got a button on it somewhere. Maybe it's got a plus sign on it, and if you push that button, it was going to add some stuff." He also pointed out that he needed a "keyboard that was thin, simple, cheap, reliable—work a million times."[29]

The calculator was to carry four integrated circuits, but manufacturing them turned out to be no less a problem than designing them. Texas Instruments made more than it needed for the project and then tested them one at a time, even making individual repairs by hand in the many that needed them. They finally got half of them to work well enough. Light-emitting diodes (LEDs) were just becoming available, but Kilby's team stuck with the tried and true strip of tape with the solution printed on it. They applied for a patent on September 29, 1967, which was granted in 1975. By that time, they had licensed calculator production to Canon, which brought out its Pocketronic in 1970. The next year, Bowmar used Texas Instruments components to bring out the first calculator model that could in fact be fitted in a pocket. It also had a LED display and sold initially for $240, but by 1980, four-function pocket calculators were selling for less than ten dollars, and Texas Instruments was selling "scientific" models, with more functions, for $25 by 1976. It was the end of the slide rule.[30]

Games

If many people first encountered the computer in the form of a calculator, others happened across one in a corner bar. Pong, the first successful video game, had its start in 1958 through the efforts of William Higinbotham, an employee at the Brookhaven National Laboratory in rural New York. Higinbotham had worked on radar during the war and timing devices for early atomic bombs, but in the late 1950s, he was faced with entertaining visitors to Brookhaven. The lab's small analog computer could track missile trajectories and show them on the screen of an oscilloscope, and within a few days, Higinbotham had worked out the circuitry to allow visitors to not only track a "tennis ball" on the screen, but by twisting knobs, lob and volley as well. "I considered the whole idea so obvious," he later said, "that it never occurred to me to think about a patent."[31]

Pong was commercialized a few years later by Nolan Bushnell, founder of Atari Corporation. As an engineering graduate student at the University of Utah, he had come across a computer game called Spacewar, which had been created at MIT in 1962 to demonstrate the capabilities of a new Digital Equipment computer with a display. It was a "twitch" game, requiring immediate responses to the on-screen action. Not surprisingly, given that it was designed in the same month that John Glenn made the first orbital flight by an American, the game involved shooting down spacecraft as they zipped across the screen. The program for the game was preserved on paper tape, and copies quickly spread across the country. Bushnell saw it at Utah laboratory and tried to commercialize his own version of it (he failed), but it was everywhere. Stewart Brand, later editor of *The Whole Earth Catalog*, watched students playing it at the Stanford computer center: "They were absolutely out of their bodies," he reported, "like they were in another world. Once you experienced this, nothing else would do. This was beyond psychedelics."[32] It was high praise coming from one of Ken Kesey's Merry Pranksters, a traveling band of proto-hippies widely credited with popularizing LSD, and who helped launch the Grateful Dead.

Personal Computers

The burgeoning hacker culture among young students and engineers found its most important expression not in games, but in the development of the personal computer (PC). The coming of transistors and microprocessors helped to create a desktop machine that was marketed, especially by IBM, to businesses. The large corporate and government mainframes had already helped establish the popular reputation of the computer as an impersonal—even antipersonal—force in society. Even new business desktops fit easily into this category of management tools to deskill and control workers.[33] By 1977, however, an estimated 20,000 to 100,000 computers were found in homes, leading the *New York Times* to headline that "The Computer Moves From the Corporation To Your Living Room." The *Times* story warned however, that "not everyone who can afford one should run out to buy a home computer." For one thing, as it pointed out, they mostly came in kits and had to be assembled by the purchaser. For another, "you have to know something about programming a computer to use one."[34]

The progression from demonized mainframes to PCs happened very quickly. On July 25, 1972, Jack Frassanito received a patent for what has been claimed to be "the machine that is the direct lineal ancestor of the PC as we know it."[35] The Computer Terminal Corporation of San Antonio, Texas hired Frassanito in 1969 to design a "smart" machine that would have its own processor, so that it could stand alone, independent of any central unit. Intel and Texas Instruments were both induced to try to reduce circuitry to a microchip, but neither had accomplished this feat when CTC went public with its new terminal in 1970. The Datapoint 2200 had storage both internally and on cassette tapes, and a screen only twelve lines high, about the size of a standard IBM punch card. The idea was to "sell the new computer as a convenient electronic replacement for the mechanical card-punch machines then in use," but the people who bought them essentially reinvented them for other purposes that we now recognize as the domain of the PC.

The Intel 8008 chip came too late for the Datapoint 2200, but its improved version, the 8080 microprocessor, proved critical for PCs in general. It became a component of the Altair 8800, a kit costing $400 that was first advertised in *Popular Electronics* in January 1975.[36] That advertisement appears to have precipitated the formation of the famous Homebrew Computer Club, the Silicon Valley group that was a hotbed of PC development before its end in 1986.[37] Thirty-two computer enthusiasts came to its first meeting in the garage of a member in Menlo Park, but soon word had spread far enough that the group had to move to larger quarters. When the club held meetings in the auditorium of the Stanford Linear Accelerator, upward of 750 people attended.

The ethos of the group may be taken from the fact that in June, a paper-tape copy of Microsoft BASIC, the first commercial program from that small New Mexico firm, was stolen from a display of the Altair 8800 at a local hotel. MITS, the computer maker, had assumed that anyone buying its kit would do their own programming. Bill Gates, however, saw a market for someone willing to do the programming for them.[38] Copies were made and distributed free to members of the Homebrew Computer Club by Dan Sokol, a semiconductor engineer. Gates, then only twenty years old, was furious, and saw the event as obvious theft. Sokol had a different view: "Bill Gates owes his fortune to us. If we hadn't copied the tape, there would never have been an explosion of people using his software."[39] It was an early clash between the two powerful urges of Silicon Valley: on one hand, an entrepreneurial drive to become as rich as possible as fast as possible, and on the other, a self-proclaimed radically democratic desire to construct computer culture as liberating and freely available.

One of the first members of the Club was a Hewlett-Packer employee named Stephen Wozniak. While still an engineering undergraduate student at the University of California at Berkeley, he and his friend Steve Jobs had met John T. Draper, also known as Cap'n Crunch, a pioneer "phone phreak" who experimented with and exploited the telephone system. In 1969, Draper had pulled up to two isolated phone booths off of a freeway ramp in the Livermoore Valley of California, and with a cable from his Volkswagen microbus, sent a 2,600-cycle tone through one of the phones, gaining him access to international service. He placed a call to himself that traveled through Tokyo, India, Greece, South Africa, South America, New York, and finally to the booth next to his. He answered his own call and reportedly said: "Hello, test one, test two, test three." It was perhaps the ultimate example of medium over message. Soon, however, the phone phreaks produced their first product, the Blue Box, which gave its purchaser the same power to make free (and illegal) international calls that Cap'n Crunch had used. For a while, Wozniak and Jobs assembled Blue Boxes and sold them in the Berkeley dorms.[40]

It was only a year after the founding of the Homebrew Club that Wozniak bought a microchip and built his own prototype computer from parts borrowed from other computers. His friend Jobs suggested that they go back into business, this time making computers, and in 1976 Apple Computer, in the tradition of Hewlett-Packard, started in a garage in Cupertino. Within eight years, its sales reached over $1.5 billion.[41]

Adam Osborne was a self-publisher of books on computer topics who sold his company to McGraw-Hill in 1979. In 1981, he brought out his own computer. Although it weighed twenty-four pounds, it could be carried about, and that was enough to qualify it as a portable. Its real advantage, however, was that it eliminated the two problems that the *New York Times* had identified four years earlier: it did not have to be assembled from a kit, and it came with all of the important software already installed, including WordStar for word processing, the spreadsheet Super-Calc, and Bill Gates's BASIC. Osborne likened himself to Henry Ford in the early days of the automobile: "I give you 90 percent of what most people need." Many people apparently figured that they were buying the software and were getting the hardware free. Osborne's company was also a pioneer in the boom-and-bust mode of Silicon Valley enterprise. In 1981, he had 8,000 orders for his machines; in 1982, he had 110,000; and in 1983, he was bankrupt.[42]

Osborne's Henry Ford analogy was not far off the mark, but the welter of new devices and understandings of what was possible and what was desirable was so tangled and compressed into such a short period of time that clean time lines and accurate attributions are particularly difficult. A case in point is that of Doug Engelbart who, in December 1968, demonstrated his newly invented "mouse" at a conference in San Francisco. It has since been claimed that Engelbart and his colleagues at the Stanford Research Institute did more than just show off a mouse, for which there was, at the time, no very good use. They also "showed off the graphical user interface, integrated text and graphics and two-way video conferencing."[43]

Reminiscences thirty years later saw in all of this the largest of portents. "This demonstration set a brushfire burning that spread across the computing landscape," opined the "technology forecaster" Paul Saffo. Alan Kay, a Disney vice president who had been a graduate student at the University of Utah, at the time recalled that "for me, he was Moses opening up the Red Sea. It reset the idea of what it was to do personal computing. It wasn't just a place where you went to do a few important transactions. It was a place where you could spend all day." Kay went on to co-found Xerox PARC (Palo Alto Research Center) in 1970, where many of Engelbart's researchers migrated after their group lost funding. Xerox PARC has been credited with being the birthplaces of "the first personal computer, desktop publishing, laser printing and Ethernet," an early standard for local networks. Marc Andreessen, a co-founder of Netscape Communications in 1994, credited Engelbart with laying the groundwork for the Internet. Engelbart himself, looking back three decades, was still ambitious for his work: "I don't think the world has really bought it.... For me, it's been a totally serious goal—the idea of collective IQ."[44]

This almost messianic dream of a participatory democracy of computer users, a global village of liberated equals, is at the core of the early hacker vision that the anthropologist Bryan Pfaffenberger has claimed was "no revolution" at all. With their Blue Boxes and user-friendly icons, the hackers were among those "technological innovators [who] not only try to manipulate technology when they create a new artifact;... [but] also try to manipulate the *social* world for which the new artifact is intended." But, according to Pfaffenberger, "such strategies serve in the end to reproduce (rather than radically change) the social systems in which they occur."[45] By the end of the century, not only was the computer a place where you could spend all day, but a growing number of people actually did, either at work or at home, or working at home.

A good part of that time came to be spent either attending to email or surfing the web. If the hacker culture was one of "reconstitution," to use Pfaffenberger's phrase, then the Internet and the World Wide Web came out of a culture of "regularization." In the years after World War II, the military had an abiding concern to improve what they called 3C—command, control, and communication. The need was distributed throughout the military and for all operations, but took on particular urgency when one considered the aftermath of a nuclear exchange. If normal communications networks were destroyed— the telephone lines for instance—how would the commander in chief, or any other responsible person, send orders out into the radioactive waste?

The Internet

Concern in the United States about communications became acute after the Soviet Union orbited *Sputnik* in 1957. One of the Pentagon's responses was to set up the Advanced Research Projects Agency (ARPA), later called DARPA when "Defense" was added to the title. which in turn organized the Information Processing Techniques Office (IPTO) to worry particularly about 3C. The first head of IPTO, Joseph Carl Robnett Licklider, had, in Thomas P. Hughes's words, a "powerful vision of interactive, time-sharing, and networked computers."[46] Nor was his a narrow commitment to the 3C problems of the Pentagon. In a neat invocation of the military-industrial complex, Licklider declared that "what the military needs is what the businessman needs is what the scientist needs."[47]

In 1966, ARPA worked on putting together a network of the time-sharing computers scattered about the seventeen computer centers that it had funded around the country in academic, government, and industrial sites. Two years later, it solicited bids to connect four sites: the University of California, Los Angeles, the University of California, Santa Barbara, the Stanford Research Institute, and the University of Utah. Numerous components had to be designed, built, and connected, all within an environment of extreme uncertainty. The basic protocols for host-to-host communication—the network control program (NCP)—were not developed until 1972. One idea that had come out of the panic after *Sputnik* had been that of "packet switching," which would allow packets of information to seek out and even change the best routes around damaged communication lines.[48] Machines to route the message were called interface message processors (IMPs), the first of which were "debugged" Honeywell 516 computers.[49]

Finally, Lawrence Roberts, the project manager, decided to make a public demonstration of what was now called ARPANET. A first International Conference on Computer Communications was called in Washington, D.C. in October 1972. Over a thousand interested observers watched as computers of different manufacture communicated with each other from various sites on the network. Visitors were encouraged to log on themselves and exchange data and files. The demonstration was a great success, and as a result, according to Hughes, "computer engineers and scientists no longer considered it a research site for testing computer communications but saw it as a communications utility comparable to that of the telephone system."[50]

Just as the telephone began as a serious medium meant for business messages and emergencies rather than simple conviviality (Alexander Graham Bell refused to have one in his house), so ARPANET was intended strictly for exchanging scientific data. The network was already carrying email messages between researchers in 1973, though they were strictly limited to business. By the end of that year, however, three-quarters of the traffic on ARPANET was email. The telephone analogy was not the only one possible. Paul Baran of RAND who had early theorized on packet switching, apparently had the postal laws in mind when he allegedly warned that "you'll be in jail in no time." The first listserver was SF-LOVERS, a group of science-fiction enthusiasts. ARPA shut it down as inappropriate, but then relented when members argued that they were really just testing the mail capacity of the network.[51]

As is so often the case in the history of technologies, the project spilled over the borders envisioned and intended by both its makers and sponsors. In 1983, the Department of Defense recognized ARPANET's evolution and split it into two parts: MILNET continued to handle Pentagon business, and the remainder concentrated on serving computer-research people. When the NSF funded five supercomputers in 1986, they were organized into NSFNET. By this time, a new program, transmission control protocol/internet protocol (TCP/IP), released in 1973, allowed the interconnection of ARPANET with the others that had been set up. The resulting Internet was so powerful and successful that ARPANET was shut down in 1990 and NSFNET five years later.[52] With the organization of the World Wide Web, a body of software, and a set of protocols and conventions developed at CERN (the European Organization for Nuclear

Research) and the introduction of powerful search engines, the personal computer began to change the way information was shared and even understood.

Overlaying both the democratic hopes and reification of power were the hard facts of race and gender. It was soon apparent, and quite predictable, that there would be a "digital divide" among the races in America. Modems were of little use on Indian reservations, where telephones were scarce, and the unemployed, underemployed, and homeless in urban ghettoes and rural slums were hardly likely to be able to log on to enjoy the benefits of email and the Internet.[53] Similarly, while boys learned computer (if not social) skills through hours of gaming at the keyboard, their slash-and-kill adventures seemed to hold little appeal for girls who, even when they got "their own" software, found titles like Barbie Goes Shopping or other such "appropriate" programs.[54] Once again, new technologies tended to reflect rather than reform social and cultural reality. One striking exception is that of age: as far back as we can look, skill and knowledge have come with age, and living national treasures were most likely to be elderly artisans who almost literally thought with their hands. In the electronic age, at least to believe the popular lore, no one over thirteen can program a VCR (itself replaced by the DVD).[55] It gave a whole new dimension to the youth cult in America.

Cellular Phones

Computers did not, of course, completely define or exhaust the electronic environment of the late twentieth century. Al Gross was one of a host of inventors who found a useful role to play in World War II, and then went on to patent civilian devices. His walkie-talkie was demonstrated to the Office of Strategic Services during the war, and he developed a battery-operated, ground-to-air two-way radio for the services. His postwar patents were credited with "foretelling the advent" of cell phones, paging systems, and garage-door openers.[56] Martin Cooper made the first mobile-phone call in 1973. Working for Motorola, his team decided to forestall Bell Labs, which was interested in developing a car phone but was making sweeping claims about its allegedly unique ability with cellular technology. After several months' work, he was able to place a call with a hand-held, two-pound phone while standing on the sidewalk outside a Hilton hotel in midtown Manhattan. He placed the call to the head of Bell's car phone project.[57]

Descendents of Cooper's phone, used by 144 million Americans in 2003 and even larger percentages of the population in other countries around the world, turned out to be the savior of another device that had been around for many years. In 1931, AT&T had set up an experimental apparatus in New York that combined television and the telephone, allowing people to both see and talk to each other. After World War II, they returned to the concept, and by 1964, had a PicturePhone ready to show at the New York world's fair that year. Despite a favorable reaction from visitors, the product failed in the market and was withdrawn in 1972. AT&T tried again with the Video-Phone in the 1990s, but once again had to withdraw it from the market when sales disappointed. At the end of the century, it was widely ridiculed as the "Perpetual Next Big Thing," but by the early twenty-first century, the combination of camera and mobile phone was allowing people to take and transmit pictures, and even use it as a video camera, sending images to either computer storage or real-time viewing. Privacy concerns notwithstanding, it appeared that the PicturePhone had at last found a market.[58]

Smart Technology

The Next Big Thing might well prove that inventing a market is more difficult than inventing a new electronic device. At the end of the century, the lack of a market was clearly preventing the development of what MIT's Media Lab had termed "things that think." Engineering enthusiasts were at work on something called pervasive computing, which in many cases boiled down to smart appliances that would create a home environment in which, as one industrial designer put it, "the fridge talks to the coffee pot." The political philosopher Langdon Winner, echoing the irony of Ruth Schwartz Cowan's classic study *More Work for Mother*, warned that "as people add more and more time-saving, labor-saving equipment to their homes," their lives do not become simpler and easier. Instead "their days become more complicated, demanding and rushed." It was little more, he added, than catering to "the exotic needs of the world's wealthiest people."[59]

But one cannot stop progress, even if it isn't. Stanley Works was developing a line of smart tools, as opposed to what were called the "stupid tools" that built "the Great Pyramids and the Taj Mahal," and hung the pictures in the Metropolitan and the Louvre. Monticello, it was pointed out, was built with "stupid tools that did not contain a single microchip, laser, diode, liquid crystal display or ultra-sonic transducer."[60] The tone was ironic, but irony is the postmodern way of understanding. Perhaps not surprisingly, by 2002, it was in Japan where one could find the domestic-chip frontier. At a trade show in a Tokyo suburb, the Japan Electronic Industries Association, the Ministry of Economy, Trade and Industry (METI), and twenty firms showed off everything from a robot pet dog to a refrigerator that automatically ordered more beer when the last one was taken. The children's room was dominated by game consoles, and the elderly were monitored by an electronic thermos for tea that sent a message to a cell phone if it was not used within a certain period of time.[61]

When historians of technology address questions of technology and the environment, they usually follow the common practice of juxtaposing the natural with the artificial, however those terms are defined. The astonishing flourishing of electronics since World War II, however, makes it patently clear that technologies not only affect environments, but also help create them. The electronic environment in which we are all immersed may or may not be responsible for rising rates of brain tumors, for example, but it has profoundly influenced the ways in which we live. These changes are not always those intended or desired, and one can argue that the technologies are the result as much as the cause of these changes. But the new environment has been created, and continues to evolve, if often along deeply traditional lines.

ELEVEN

STANDING TALL AGAIN

In the 1970s, the broad national questioning of technology and its costs coincided with America's losing the war in Vietnam. If that defeat was associated with a perceived feminization of the country, as has been suggested, then it paralleled the way in which appropriate technology (AT) and technology assessment had been characterized as unmanly. In the 1980s, the antidote for such gendered self-doubts proved to be the presidency of Ronald Reagan, who urged the nation to "stand tall again."

In April 1979, Jimmy Carter was in the White House and solar panels were installed on the roof of the West Wing, just over the Cabinet Room. Visible to those passing along Pennsylvania Avenue, the panels were to provide hot water for the staff mess in the basement, saving an estimated $1,000 per year at 1979 prices. The installation was a vivid symbol of what AT advocates hoped many Americans would do to help wean the nation from its dependence on unsustainable energy sources, a goal that Carter shared.

Three years later, however, while repairs were being made to the roof of the West Wing, the panels were removed and placed in storage in a warehouse in Virginia. The thirty-two panels languished there until 1992, when they were acquired by Unity College in Maine, and installed to provide hot water for the school's cafeteria. It was an appropriate move, according to the campus development officer, because the college was "known for its programs in environmental sciences and natural-resources management." Again the symbolism reached to the very top of the federal government: although President Carter had supported AT initiatives, President Reagan created a political environment that was profoundly hostile to the movement.[1]

Vietnam

The removal of the solar panels was also a perfect example of what Susan Jeffords has called the "remasculinization of America," a process that took place during the last two decades of the twentieth century. The social movement of the 1960s and 1970s, she suggests, was experienced as a direct attack upon prevailing notions of masculinity. The resurgent civil rights movement, a new environmentalism, the consumer revolt, sexual freedom and The Pill, the spread of communes and other forms of hippie culture—and, of course, a revitalized feminism—all threatened entrenched forms of patriarchal privilege. The loss of the war in Vietnam, like all military defeats, was seen as a profound and humiliating feminization. On top of all this, two decades of an increasing criticism of what was seen as a technological society, culminating in the AT movement, which directly challenged the efficacy and even the rationality of technological vanguardism, put under siege what was arguably one of the most masculine of American values.

Jeffords traces the remasculinization of American culture through the films and books that represented an evolving understanding of the Vietnam War. Some of the "strategies" used to understand the war she identifies as "the shift from ends to means, the proliferation of techniques and technologies, the valorization of performance, the production and technologization of the male body as an aesthetic of spectacle, and the blurring of fact and fiction." She identifies "an intense fascination with technology that pervades Vietnam representation," and, more significantly, sees it as a technology "separated from its ostensible function."[2] It is technology as spectacle, partaking of the sublime.

In the years after the war, as Jeffords writes, after valorizing the men (though not the women) who fought in Vietnam, the next step in the remasculinization process was to bring that "separate world [of masculine virtue] 'home' so that the logic could be applied to an American society in which . . . 'masculinity had gone out of fashion.'" One of the ways in which masculinity was rescued was through the identification of the government, and in some sense law itself, as feminine.[3] The government was not only a "nanny state" but had, after all, tended to further or at least not stand up to the social movements of the 1960s and 1970s. It was widely claimed that the government itself had lost the war in Vietnam, which meant that the men who were sent there to fight could have won had politicians and bureaucrats not stymied them. The government also did not do enough to find and bring home those who were missing in action (MIA). As president, Reagan became both the model and the spokesperson for those who longed to see America "stand tall" again. The government, he famously proclaimed, was the problem, not the solution, and for the rest of the century, with the notable exception of the years of Bill Clinton's presidency, the political goal was to "starve the beast"—to cut taxes, run up deficits, and therefore make it impossible to afford the public services of the "nanny state."

A Rational World of Defense Intellectuals

The paradoxes and distractions inherent in the close identification of the military—both its personnel and weapons—with masculinity could be clearly seen through what Carol Cohn called the culture of "Sex and Death in the Rational World of Defense Intellectuals."[4] While attending a workshop on nuclear weapons, arms control, and strategic doctrine, Cohn discovered a specialized language that she called "technostrategic." Drawing on imagery from the larger American culture, the defense intellectuals used "sanitized abstraction" and "sexual and patriarchal" terms, paradoxically, as "domestic" images to talk about what ultimately can hardly even be thought about. The references to permissive action links (PAL), ballistic missile boost intercept (BAMBI), the "shopping list" for weapons production, various "menus of options," and "cookie cutter" models for nuclear attacks, gave a familiar and almost benign aura to matters of unimaginable horror.[5]

These homey invocations of the feminine worked side by side with images of male sexuality. Disarmament was dismissed as "get[ting] rid of your stuff," certain underground silos were chosen for the new MX missile because they were "the nicest holes," and lectures were filled with references to "vertical erector launchers, thrust-to-weight ratios, soft laydowns, deep penetration, and the comparative advantages of protracted versus spasm attacks—or what one military advisor . . . has called 'releasing 70 to 80 percent of our megatonnage in one orgasmic whump.'" The United States was said to need to "harden" its missiles because, "face it, the Russians are a little harder than we are."[6]

The point of the entire exercise, Cohn concluded, was "the drive to superior power as a means to exercise one's will." By learning and using the language, "you feel in control." The result is a kind of "cognitive mastery; the feeling of mastery of technology" that is available only to "articulate the perspective of the users of nuclear weapons, not that of the victim." The early days of the atomic bomb project were full of references to that fact that "man" had given "birth" to a new age, but in learning technostrategic language, men were reassured of their masculinity: "In learning the language," Cohn concludes, "one goes from being the passive, powerless victim [read feminine] to the competent, wily, powerful purveyor of nuclear threats and nuclear explosive power."[7] In the 1970s, the Smithsonian Institution opened its hugely popular National

Air and Space Museum, an institution that "goes beyond awe and amazement" to deliver a powerful lesson in "The Romance of Progress," as one critic wrote.[8] "The outline of a towering missile appears through the darkened glass of the window" as people enter, and "naturally" it is displayed standing on its guidance fins, pointed upwards. The emotions it invokes are fundamentally tied to the fact that it is our missile, pointed skyward to rain death and destruction on someone else, rather than hanging from the ceiling, "incoming" and pointing at us as it were. The visitor's perspective is that of the aggressor rather than the victim.

Missile Defense

In 1970, Herbert York, the nuclear physicist who had been intimately involved with the creation of the hydrogen bomb and was a scientific advisor in the Eisenhower administration, commented on the nature of weapons development. Advocates, he wrote, after describing the dire need for more "defense," "then promptly offered a thousand and one technical delights for remedying the situation. Most were expensive, most were complicated and baroque, and most were loaded more with engineering virtuosity than with good sense."[9] Little had changed a decade later when, on March 23, 1983, President Reagan proposed his Strategic Defense Initiative (SDI), a plan to build a system of space-based laser and particle-beam weapons to create a protective shield over the United States. It was to be expensive, complicated, and exhibiting an engineering virtuosity of perhaps impossible dimensions. It was not the first suggestion to build an anti–ballistic missile (ABM) system, but both its timing and its rich embodiment of technological fantasy made it a signal aspect of the nation's remasculinization.

The first ABM controversy began in 1969 when President Nixon publicly proposed his Safeguard system, but the idea had been around for as long as ICBMs were available. The basic idea was to use one missile, with a nuclear warhead, to intercept and deflect, incapacitate, or destroy incoming enemy missiles before they reached their target cities. By 1967, the U.S. Ballistic Missile Early Warning System, built around very large radar antennas, was thought to be able to provide a fifteen-minute warning of a Russian attack, and the North American Aerospace Defense Command (NORAD) in Colorado could then launch a response attack in ten seconds. In theory, using one missile to knock down another was obvious and simple enough, but the practical problems, including how to deal with decoys, were enormous.[10]

At the time, both the secretary of defense and President Johnson were on the record against deployment of the Nike-X as an ABM system, the latter saying that it would be merely "another costly and futile escalation of the arms race." Pressure was building, however, in a classic example of how the military-industrial complex could work. Some $2.4 billion had already been spent on research and development, and deployment was estimated to require another $30 billion, which would be spent primarily with twenty-eight companies. These firms, one journalist estimated, had about 1 million employees working in three hundred plants that were located in forty-two states and perhaps 172 congressional districts. This was a powerful network calling for continued employment, and to these could be added the populations of the twenty-five cities that the Pentagon was planning to protect with a "thin" system. If enough money was forthcoming, a "thick" system could protect an additional twenty-five cities. The convergence of interests among corporate executives, shareholders, union members, members of Congress, and citizens who would like to imagine that they were shielded from nuclear attack was a powerful one indeed.[11] That same year, President Johnson authorized the Sentinel system, which he claimed would protect the country against attacks from "rogue" missiles launched from China.

In the spring of 1969, the ABM debate became a public circus with President Nixon's announcement that he was ready to scrap Sentinel and deploy the new Safeguard system. Five years later, the system was ready for testing, but by then, a constellation of stars of the nation's scientific establishment had argued strenuously against it; it had survived in the Senate only by the tie-breaking vote of Vice President Spiro Agnew; it been redesigned primarily to protect American missile sites instead of cities; and the number of those sites was cut from seventeen to twelve, then to two, and finally to one. Its apparent success, after a twenty-year battle, was diminished only by the fact that, as one critic charged, it was "like a train that doesn't go anywhere." *Science* magazine concluded that "it is, if nothing else, a notable monument to Western technology and preoccupations, one which, like the funerary pyramids of ancient Egypt, will move future generations to marvel equally at the civilization's extraordinary technical skills and at its unswerving devotion to the mortuary arts."[12] The quarrel had been over the great cost, of course, but also the question of whether building it would destabilize the balance of power between the United States and the Soviet Union and add more fuel to the arms race. But, as York conceded, "the technological side of the arms race has a life of its own, almost independent of policy and politics."[13]

As it was finally built, Safeguard depended upon two large radar installations and two different missiles. The initial perimeter-acquisition radar would pick up Russian missiles coming over the North Pole, and Spartan missiles would be launched to bring them down. Since this would happen above the atmosphere, it would be difficult to sort out the actual warheads from various kinds of decoys, so the backup missile-site radar would track the objects after the atmosphere had begun to slow down the lighter decoys, and the Sprint missile, which traveled faster than a bullet, would then be targeted at the warheads. The remarkable Sprint was directed by Central Logic and Control, a program developed by Bell Laboratories and consisting of ten central processors. From 1970 to 1973, 10,000 professionals worked on the Safeguard design, including 2,000 computer programmers. The ABM was the only missile program that the Army was allowed to have—the Air Force had the ICBMs and the Navy its Polaris—so it too devoted considerable personnel to the program.[14] In 1972, before Safeguard was completed, the United States and Soviet Union entered into an ABM treaty that limited each country to having no more than one hundred ground-based ABMs, and these could only be used to protect the national capitals and missile sites.

Star Wars

Once moved into the White House in 1981, President Ronald Reagan began a massive arms buildup that was a key part of his plan to make the nation stand tall again. But when he made his February 23, 1983 speech suggesting what was essentially a new ABM system, even his advisors were caught off guard. During the course of a routine speech on defense matters, he announced a long-range research and development program to deal with the threat of Soviet ICBMs by making them "impotent and obsolete." He claimed that he had consulted with the Joint Chiefs of Staff, and White House staff explained that he was referring to space-borne laser and particle-beam weapons that could create an umbrella of safety over the entire country. Unlike all previous weapons initiatives, this one came from the top down, and was unrelated to any new perception of enemy technical advances.

In 1940, Reagan, then an actor, had played the part of an American secret agent in the film *Murder in the Air*. In that film, Russian spies try to steal plans for a secret weapon, an "inertia projector," which could shoot down aircraft by wrecking their electrical systems, but Reagan used the weapon to shoot the spies down instead. According to the script, the new weapon would "make America invincible in war and therefore the greatest force for peace ever invented." Nearly forty years later, while running for president, Reagan visited the NORAD command center in Cheyenne Mountain, Colorado, and was told that even if incoming Soviet missiles were picked up on radar, they could not be stopped. Later he told a reporter "I think the thing that struck me was the irony that here, with this great technology of ours, we can do all of this yet we cannot stop any of the weapons that are coming at us." After mentioning Soviet work on civil defense, he added that "I do think it is time to turn the expertise that we have in that field—I'm not one—but to turn it loose on what we do need in the line of defense against their weaponry and defend our population."[15]

The political background of the initiatives lay with a handful of the president's most conservative cold-warrior advisors. One was Lieutenant General Daniel O. Graham, who early in 1981 published an article in *Strategic Review* urging the development of a space-based defensive system using "off the shelf" technologies. A few months later, he formed a new organization, named High Frontier, to agitate for the project. Reagan appointees got the Pentagon's Defense Advanced Research Projects Agency (DARPA) to look at their plan for a three-stage defense system, but that agency reported that "we do not share their optimism in being able to develop and field such a capability within their timeframe and cost projections." After a meeting of High Frontier supporters at the conservative Heritage Foundation, they had a meeting with Reagan at which Edward Teller, the so-called father of the H-bomb, was present. All urged the president to "establish a strategic defense program modeled after the Manhattan Project." Teller's particular pet project was an X-ray laser that he claimed the Soviets were also developing.[16] It was this source, rather than the Pentagon, that accounted for the Star Wars speech.

The president's call fell largely on deaf ears. Cynics suspected that he had let his movie heroics bleed into his presidential policies. According to Herbert York and Sanford Lakoff, Richard Deluer, an under secretary of defense and the highest ranking technologist in the Pentagon, claimed that "a deployable space-based defense was at least two decades away and would require 'staggering' costs. To develop it, he added, eight technical problems would need to be solved, each of which was as challenging as the Manhattan Project or the Apollo Project." And even if it was possible to build, he concluded, it was likely not to work since "there's no way an enemy can't overwhelm your defense if he wants to badly enough."[17]

Something of the complexity of the technologies planned for the Strategic Defense Initiative (SDI) can be gleaned from the list of acronyms and abbreviations that prefaces the detailed study of the system by Lakoff and York. Ranging from AAA (anti-aircraft artillery) to VHSIC (very high speed integrated circuitry), and encompassing such items as FALCON (fission-activated light concept) and HF/DF Laser (hydrogen-fluoride/deuterium fluoride laser), the list contains 108 handy references.[18]

Over the course of twenty years, as one might expect, aspects of the system have been altered, some have been abandoned, and others have been added. In

1988, however, the OTA pictured a complex system that would attack incoming missiles in all four stages of their flight: boost, when it was lifting off and at its most vulnerable; post-boost, when it was maneuvering its separate warheads into their own trajectories; midcourse, when it was on inertial guidance far above the Earth's atmosphere; and terminal, when it was in reentry and rushing toward its target. To counter this attack, SDI would have sensor systems made up of a boost surveillance and tracking system, a space surveillance system, an airborne optical system, a terminal imaging radar, and a neutral particle beam to sort out the decoys. Infrared sensors were much relied upon, and the last had their own neutron detector satellite.

The weapons system itself was made up of six parts: space-based interceptors, or kinetic kill vehicles; space-based high-energy lasers (chemically pumped); ground-based free-electron lasers (with space-based relay mirrors); a neutral particle-beam weapon; a exoatmospheric reentry-vehicle interceptor system; and a high endoatmospheric defense interceptor.[19] It was a complex system made up of numerous components, none of which had been adequately tested, and most of which had not even been invented. One Army scientist working in the Star Wars program, Aldric Saucier, worried that SDI would shove aside less spectacular but more useful defense projects estimated in 1992 that Star Wars would wind up costing $1.37 trillion and still let 2 to 4 percent of launched enemy missiles get through.[20]

In March 1985, the Reagan administration finally began its campaign for the SDI, and with some technical specifics on the table, as Frances FitzGerald wrote, "discourse about strategic issues lifted off from reality altogether."[21] The Pentagon had already spent $2.3 billion on the project and was slated to spend another $31 billion over the next six years. Testifying before a congressional committee in December 1985, Alice Tepper Marlin, the executive director of the Council on Economic Priorities, claimed that while official estimates of the cost of a completed SDI (whatever it might look like) would be $100 billion and $500 billion, her organization believed that it would run anywhere from $400 billion to $800 billion. "The R&D stage alone," she pointed out, "is the largest military venture ever proposed."[22]

Indeed, whether SDI at this point was anything more than a research and development project was a matter of dispute. Some claimed that it was only an effort to find out whether the concept made sense, and whether the technology could be made to work. Others were already prepared to press on with plans for deployment. Some members of Congress were nervous about spending tens of billions of dollars if there was a possibility that the whole idea might prove infeasible. A prior commitment to deploy would force a continuation of work no matter what technical problems arose. There was also a belief that the administration's plan was to deliberately "create a constituency of contractors that would provide the program with enough momentum to make it difficult if not impossible for some future administration to scale back or curtail."[23]

Reagan himself never gave up on the Star Wars dream, but he did change his hostile and confrontational stance toward the Soviet Union, apparently became personally friendly with Soviet Premier Mikhail Gorbachev, and came to depend more on diplomacy than technology to deal with waning Cold War hostilities. The SDI program continued after he left office, but at a lower level than it might have, and some hoped that it would quietly fade away. By the year 2000, about $60 billion had already been spent on SDI research and development, and no workable system had yet been discovered. Nevertheless, Republican conservatives continued to press for more support and a commitment to deployment, and in 1998, about the time they attempted to impeach then-president Bill Clinton, he and Congress caved in. Former (and soon to be again) Secretary of Defense Donald Rumsfeld chaired a commission that found that "rogue states" might soon gain missiles. In January 1998, Clinton pledged to support financing for both more research and development and deployment of a missile defense system to protect the country from this new menace. Late in 1999, Clinton said that he would make a final decision of deployment the following summer (in the midst of the 2000 presidential campaign), and when he became a candidate for the presidency, George W. Bush issued a political call for pressing ahead with the full Reagan program.[24] In September of that year, just before the election and just after the latest missile tests had failed, Clinton said that he would leave the decision on deployment to his successor.

In December 2002, President Bush declared that the missile threat to the United States was such that he was planning to install the first anti-missile missiles in 2004; sixteen in Alaska, four in California, and another twenty at sea. This did not leave enough time to do the testing that was required by law of all new weapons systems, but the Bush administration argued that this was to be not a proper deployment but actually a kind of way to test the system in situ, as it were. Rumsfeld, again Secretary of Defense, told members of the Armed Services Committee that "I happen to think that thinking we cannot deploy something until everything is perfect, every 'i' is dotted and every 't' crossed, is probably not a good idea.... In the case of missile defense," he went on, "I think we need to get something out there, in the ground, at sea, and in a way that we can test it, we can look at it, and find out—learn from the experimentation with it."[25]

Polls since 1946 had shown that the American public firmly believed that scientists could, if they put their minds to it, develop some technological defense against nuclear missiles. Astonishingly, the polls also showed that most Americans thought the country already had such a defense.²⁶ What one scholar called "the warring images" that swirled around Star Wars suggest, in his opinion, "that arguments over missile defense have never been solely, or even primarily, about technology. Rather, they have always represented a clash between competing views of the fundamental nature of the nuclear age."²⁷

When Reagan's SDI was first derisively tagged "Star Wars," after the immensely popular film of that name, his administration tried to stop the identification, believing that it would invoke amusement and scorn from the American people. In fact, it seems likely that the opposite might have been the case. Like the film, the president's invocation of SDI tapped into deep wellsprings of American cultural history, evoking aspirations and emotions that floated free of actual Soviet intentions and capabilities, as well as any real or imagined technologies.²⁸

One of these cultural tropes was that of the frontier. At the end of World War II, Vannevar Bush had titled his report to the president *Science the Endless Frontier*, suggesting that just as the western frontier had provided abundant new resources and opportunities for enterprise, so could science, a half century after the original frontier had been declared closed, now make that same provision—and would do so forever into the future. Space too was framed as a frontier, providing not only new resources but, like the Old West, it was also a "space" in which new beginnings might be imagined.

At the same time, SDI was tied to what Edward Linenthal has termed "images of American vulnerability and deliverance."²⁹ The new United States, born out of a revolution against the mighty British Empire but also into a world of intensifying European rivalries, felt itself menaced both from across the Atlantic and from indigenous peoples and European colonists to the north, west, and south. Reagan put forward the SDI not only as a shield against the Soviet Union, but against the general threatening nature of the Cold War era. The invocation by its post-Reagan supporters as a defense against both "rogue" states (Iraq and North Korea were especially cited) and the rising military might of China tapped into what would come to be called the War on Terror.

The images of a beckoning frontier and hostile forces abroad both fed into the traditional American yearning for isolation from the troubles and threats of the world. The ideal action for many was not to reach out with negotiation, compromise, and cooperation, but to turn away with a mixture of fear and indifference. Reagan's epiphany at the NORAD base in Colorado led him to a nostalgic invocation of the country's old coastal defenses as some sort of continuing model for ensuring our security of the nuclear age.³⁰ Since the beginning of World War II, the old form of isolationism, so embraced by generations of conservative thinkers, had given way to a tradition of bipartisan support for an active American interventionism, but the core yearning was less successfully suppressed. Despite Reagan's suggestion that the technology might be freely shared, even with the Russians, SDI renewed the hope that America could go it alone.

The missile defense initiative also had an appeal for those who were looking for a "renewal" of American culture. The reliance on defense rather than the threat of retaliation appeared profoundly "pro-life" compared with the horror lurking behind a policy based upon mutually assured destruction (MAD). SDI's apparent moral stand on the sanctity of life recommended it to many in the clergy, with D. James Kennedy, a Presbyterian minister and the founder of the Center for Reclaiming America, invoking Nehemiah and the rebuilding of the wall around Jerusalem and adding that "we need to pray that the wall around America may be built again."[31] Part of the president's political appeal was his fatherly image, and his promise to make his family safe in their domestic tranquility was hugely appealing at a deep level.

In time, SDI became the centerpiece of Reagan's rearming of America. Part of any campaign for remasculinization required that the nation "stand tall again," and SDI certainly did that, with its strong commitment to the world of the defense intellectuals. But equally important was that the enemy be demasculinized, so that one could only be defined in terms of the other. By promising in his March 23 speech to make the enemy's nuclear weapons "impotent," he underlined the virility of America: its military capability, its technology, and its bedrock values.

A Manly Culture

President Reagan's military buildup, the largest ever in peacetime, was imposed on a nation already moving back toward a more manly and aggressive cultural stance. In 1974, the *New York Times* reported that "war toys, which have been keeping a low profile during the last five holiday seasons, mainly due to the Vietnam War and its aftermath, are making a comeback this season." For only $7.99, one could buy a Special Forces kit that contained four M48 tanks, two 105-millimeter howitzers, two thirty-caliber machine guns, two M47 tanks, two 155-millimeter self-propelled guns, and about seventy soldiers. There was still some consumer hostility toward military toys, however, enough that FAO Schwartz refused to stock them. At the same time, that chain's decision to carry a $12.95 Special Forces costume for boys, including a helmet, camouflage jumpsuit, and canteen, pointed to the way in which meanings could transcend categories. The costume, a store spokesperson said, "is not weaponed, and the

Army does exist, and we feel it is part of a child's education." One father interviewed, an investment manager from Baltimore, bought one for his 9-year old son. "I don't worry that it will hurt him in any way," he said. "Wearing the uniform may have a war connotation, but it also has a connotation of belonging to a group."[32]

Even the ten-year-old best-selling war toy, the G.I. Joe doll, was justified not so much as a killing machine but as a more generic masculine prototype. Its manufacturer, Hasbro Industries, purported not to consider the doll a war toy at all. "G.I. Joe," a spokesperson explained, "is an adventurer. He's up in the air these days, he's under the sea, he's in the desert, he's in the jungle. He's much more than just a soldier."[33] These imagined terrains of masculine adventure were precisely the same as those marked for technological exploitation by the postwar agenda for what Michael Smith called planetary engineering.[34]

If masculinity was spilling out of the military into more general cultural forms, its embodiment in the soldier was in some ways being undermined by weapons technology. The soldier and astronaut John Glenn had insisted that the Mercury space capsule have a window, even though its "pilots" were in fact not really in charge of flying it, and had no real need to see where they were going. The former bomber pilots who dominated the Air Force high command for years after World War II were notorious for their conviction that the manly art of actually flying an airplane was a necessary qualification for leadership. The growing dominance of ICBMs, however, threatened that linkage.

News reports at the turn of the new century indicated that the highly-touted revolution in military affairs (RMA) was based on a new arsenal of high-tech weapons, produced by new corporations, and forming a new segment of the military-industrial complex. Furthermore, the new weapons were said to be creating "an identity crisis" for American soldiers, especially the ground troops of the Army. Much of the Army's work was being taken over by "unmanned" planes and missiles, while in general, "speed and stealth take precedence over size and weight." This observation, made on the eve of the war with Iraq, was perhaps a bit premature, but the direction of events was clear enough.[35] Apparently, new weapons could sometimes undermine as well as reinforce military masculinity.

Attacking Appropriate Technology

The AT movement was an easy target of this campaign for remasculinization. For a decade, it had challenged the most powerful economic interests of the country—the petroleum companies, the automobile industry, public utilities, agribusiness, and others—all in the name of sustainability and good stewardship. Even while it was making small though potentially important gains in its advocacy of a different technology, which was more sustainable and gentle on the land, it had been consistently painted as feminine. The titles of two of the movement's most important books, *Soft Energy Paths* (1977) by Amory Lovins and *Small Is Beautiful* (1973) by E.F. Schumacher, seem to put them deliberately at odds with the ideals of virility, domination, and violence that characterized the fantasies of American masculinity. Schumacher himself made the obvious link between AT and domesticity, calling supporters of the dominant technological culture "the people of the forward stampede" and those who advocated appropriate technologies "the home-comers." This link was reinforced by oral interviews with farmers in Wisconsin who were trying to practice sustainable agriculture. The farmers believed that AT made viable family farms, and that family farms sustained farm families. When the Society of Women Engineers gave its first ever achievement award in 1952 to Dr. Maria Telkes, it noted that her lifelong work with solar energy had been in a field "which has not been developed nearly as fast as, for instance, nuclear energy. She has been known to remark wistfully, 'You see, sunshine isn't lethal.'"[36]

Opponents of the movement made the same connections, but gave them a sinister spin. The Canadian architect Witold Rybczynski published a book on AT in 1991 that he titled *Paper Heroes*. Invoking a familiar canard against the feminine, he claimed that AT advocates appealed to "emotion" rather than to "reason," a long-assumed masculine virtue. In language likely to cause dismay if not anxiety among Schumacher's "people of the forward stampede," he accused AT advocates of wanting to "withdraw" from the normal culture. Furthermore, these advocates used arguments that brought to mind "Oriental attitudes," an identification at once with the exotic, the erotic, and the feminine. Ironically, some women in the movement were at the same time charging that the dominant patterns of patriarchal privilege were as perfectly alive and well in AT circles as they were in other parts of society.[37]

Masculinity is always contested and contingent, and it is also ironic that some traditional aspects of American masculinity were widely admired among AT followers. One was that identified early in American history with the patrician gentleman, who ideally combined power with restraint, practicing discipline toward himself and generosity toward others. The other ideal was that of the preindustrial artisan, whose proud masculinity flowed from his productive work, his skill, his self-reliance, and his ownership and mastery of tools. By the end of the nineteenth century, both of these had been joined, and largely replaced, by the figure of the self-made and self-indulgent entrepreneur who bought, bullied, and blustered his way to the top. Although AT made many enemies purely on the basis of its threat to the dominant political economy of the country, its identification with the feminine also made it vulnerable to cultural attacks that had little to do with the technology itself.

The end of the AT movement did not mean an end to the technologies it sponsored, nor to the hostility directed against them. The attitude expressed by one former executive with the California utility Pacific Gas and Electric Company, that "real men build power plants with big smokestacks on top," continued to be linked to a parallel hostility to alternative power sources and disdain for conservation.[38] Federal support of energy efficiency and renewable energy was highest at the end of the Carter administration, before Reagan ended most of it. It rose again under Presidents George H.W. Bush and Bill Clinton, but by 2001, spending had dropped again to a third of what it had been twenty years before. Three months into George W. Bush's administration, his energy task force, chaired by Vice President Dick Cheney, was already set to cut support for energy-efficiency programs (setting standards for washing machines, refrigerators, and so forth) by another 15 percent.[39]

Conservation and efficiency were particularly hit hard. While the first George W. Bush budget slashed spending on these technologies, and Vice President Cheney was declaring that "conservation may be a sign of personal virtue, but it is not a sufficient basis for a sound, comprehensive energy policy," studies by scientists from the government's own national laboratories were reporting that new technologies could reduce the growth in demand for electricity by 20 to 47 percent. The Bush estimate, concentrating on supply rather than consumption, was that the nation needed to build a new large power plant every week for the next twenty years to keep up with growing demand.[40] In an editorial, the *New York Times* claimed that by using "existing technology," a federally mandated automobile fuel efficiency of 40 miles per gallon would save 2.5 billion barrels of oil each day by the year 2020—almost exactly as much as the nation then imported each day from the Persian Gulf.[41]

The attack on "alternative" technologies continued across the board. Late in 2002, the Bush administration went to court to support auto makers in their attempt to nullify a California clean air requirement that would, in effect, force them to produce electric ("zero-emission") vehicles. That same year, the United States used its influence at a United Nations conference on environmental problems, held in Johannesburg, to strip the conference report of all timetables and targets for the worldwide use of clean, renewable energy sources. Instead, it was reported, "U.S. officials said they prefer voluntary partnerships with business and other groups." For the year 2001, it turned out, the use of renewable energy in the United States fell 12 percent, to its lowest level in twelve years.[42]

Nor did the paths not taken always lead to the soft and small technologies that the AT movement admired. Late in 2002, a group of scientists, from a number of corporate and academic bodies, boldly called for the by-now-familiar "research effort as ambitious as the Apollo project to put a man on the moon." Such an effort, assumed to require tens of billions of dollars, would seek to improve existing technologies but also "develop others like fusion reactors or space-based solar power plants."[43] It was a bold and virile intervention in a field that had for years been dismissed as lacking hard thinking and practical solutions.

Superjumbo Jets

Passenger airliners are among the most polluting forms of transportation. The 16,000 planes in use generate an estimated 600 million tons of carbon dioxide every year. The Airbus A380 was designed to do something about that. At the end of the century, it looked as though the old competition to build the first supersonic jet was to be played out again with a race for the first superjumbo passenger plane. Boeing, which long dominated the market for wide-body jets, had delivered 1,353 of its workhorse 747s, and the European firm Airbus was determined to break that virtual monopoly with a new generation of planes. In June 1994, it began engineering-development work on what was described as a megajet. Boeing, at first inclined to accept the challenge, dropped out of the race quickly and began work on its own new 7E7 Dreamliner, a double-aisle jet carrying about 250 passengers and able to access more of the world's airports.[44]

Called by one travel writer "a 747 on Steroids," the new Airbus airplane had a wingspan of 261 feet, 10 inches, a length of 239 feet, 6 inches, and stood 79 feet high. As one description had it, it would be as long as a football field and the wings would hang "well beyond the sidelines." Although Airbus spoke of 555 passengers, with such amenities as bars, lounges, and even a casino, it could actually seat 880, and probably would. The plane's maximum takeoff weight of 1.2 million pounds would be carried on twenty wheels, its engines would develop 75,000 pounds of thrust, and it would have a range of over 10,000 miles.[45] Despite the oversized dimensions, a spokesperson for Airbus claimed that the A380 would not only make less noise than a 747, but it would be "more fuel efficient than a small car." The calculation was based on carrying a certain number of passengers for a certain number of miles, and if the casinos and cocktail lounges diminished that number, the calculated efficiency would sink. According to Simon Thomas, head of a London-based environmental research firm, "Better technology alone is not going to solve this problem."[46]

The dimensions of the A380, while spectacular enough in their own right, have consequences beyond the technology of the aircraft itself. Airports must deal with the logistics of unloading up to 880 passengers and shepherding them through passport control and customs, all the while encouraging them to stop at duty-free shops. And this only becomes a logistical problem if the airport can manage to land the plane at all. It was estimated that an airport would have to spend $100 million to upgrade taxiways, gates, baggage handling, and customs and immigration facilities. Los Angeles International Airport (LAX) was planning to build an entirely new terminal for the A380 that would allow two of them to park side by side, an impossibility even at their largest existing facility.[47] It was a startling example of the ways in which scaling up one component in a technological system warps the entire infrastructure.

Sport Utility Vehicles

Not all of the super-sizing of technologies and rejecting of sustainable-energy options was operating only at the government and corporate levels, though government and corporate policies were critical in even the personal and private sphere of consumerism. The American consumer's version of the Airbus A380 is surely the rapidly multiplying sport utility vehicle (SUV), and particularly the Hummer, the civilian analog to the military's Humvees.

The SUV had its roots in that World War II icon, the Jeep.[48] The Army had wanted a light truck to replace horses since World War I, but the coming of war to Europe again in 1939 gave the project a new urgency. The military's preferred design, which had been developed by American Bantam, a small company of uncertain production capabilities, was turned over to Ford and Willys-Overland for production. Willys registered the name Jeep as a trademark and continued to make them after the war, though after the war, other automobile companies refused to sell Willys the molded sheet metal for Jeep bodies that they has made during the war. Instead, Willys used the facilities of a metal-stamping factory that had previously shaped steel for washing machines. Although the rest of the industry was beginning to move into the organic, curved, almost voluptuous designs of postwar Detroit cars, the Jeep was frozen in a time warp of squarish, box-like fenders, hood, and other components. In 1946, Willys began producing its Jeep Station Wagon, which was offered with four-wheel drive beginning in 1949. The only competition came from Land Rover in Britain and Toyota in Japan, which had been turning out

Jeeps for the United States during the Korean War and now began making its own Land Cruiser.

Willys, which had been purchased by the World War II shipbuilder Henry J. Kaiser, was sold again in 1969 to American Motors, which began to reengineer the Jeep, especially to minimize its deadly tendency to roll over, making it one of the most dangerous cars on the road. The old CJ5, as it was called, was being sold as a work vehicle; the last Kaiser Jeep catalog in 1969 showed it pulling up a stump. Company executives realized that very few people really needed a vehicle with four-wheel drive, but nevertheless, American Motors marketers found "that there were many Americans living in cities who admired the Jeep's military heritage, liked its utilitarian image and wanted to ape the automotive fashions of the horses and hunting set of Nantucket and other wealthy enclaves." They especially liked the idea of four-wheel drive, which sounded like it might come in handy someday. As Jeep sales multiplied four times during the 1970s, *Time* called it "a macho-chic machine."[49]

To a large degree, the explosion of SUVs was a direct result of the corporate manipulation of government policies. Light trucks had been protected by a tariff imposed by President Johnson in 1964 as a retaliatory move against Europeans countries, which had restricted imports of frozen American chickens. In addition, light trucks, assumed to be used by tradespeople and small merchants, were largely exempted from the stricter safety, mileage, and pollution regulations that Detroit faced in marketing family cars. The result was that automobile manufacturers found it easier and more profitable to design technologies that would qualify for exemption rather than trying to meet the new standards. "We made damn sure they were classified as trucks, we lobbied like hell," one AMC executive later recalled, and in 1973, the Environmental Protection Agency classified Jeeps as such.

Other manufacturers soon drove through the same loophole. To get sufficient height off the ground to qualify as "an automobile capable of off-road operation," and to save money by using already standard components, they put new bodies on old pickup chassis. To bring the gross vehicle weight to the over 6,000 pounds required for trucks, they built up the suspension systems so that heavier loads could be carried. Although these new vehicles were gas guzzlers by any definition except the government's, as light trucks, they were exempt from the extra taxes imposed on other cars that did not meet the government standards for efficiency. Because they were technically off-road vehicles, they were allowed to have bumpers higher than those on cars. To cap it off, trucks are exempted from the luxury tax on cars costing over $30,000. "Small business owners," including doctors, dentists, lawyers, and other professionals, can deduct most of their purchase prices within the first year of use. Cheap to manufacture but loaded with luxury features, the SUV proved extremely profitable for the auto industry. The 1984 Jeep Cherokee, considered the first real SUV, was soon followed by others from all the major manufacturers.

The timing of the SUV was important not only in terms of federal safety and efficiency regulations, but also in terms of the culture of remasculinization. SUV drivers in general were often criticized: Stephanie Mencimer, a writer for the *Washington Monthly*, claimed that "unlike any other vehicle before it, the SUV is the car of choice for the nation's most self-centered people; and the bigger the SUV, the more of a jerk its driver is likely to be."[50] Not surprisingly, at least some drivers of Chrysler's Hummers see themselves not as jerks, but as patriots. One owner, the founder of the International Hummer Owner's Group, was quoted as saying that "in my humble opinion, the H2 is an American icon. Not the military version by any means, but it's a symbol of what we all hold so dearly above all else, the fact we have the freedom of choice, the freedom of happiness, the freedom of adventure and discovery, and the ultimate freedom of expression. Those who deface a Hummer in words or deeds," he claimed, "deface the American flag and what it stands for."[51]

Another owner made the military connection even more explicit. "When I turn on the TV," he said, "I see wall-to-wall Humvees, and I'm proud." The troops were "not out there in Audi A4s.... I'm proud of my country, and I'm proud to be driving a product that is making a significant contribution." One general sales manager at a Hummer dealership in Milwaukee was careful to explain that "I don't have people coming in here in camouflage and a beret," but admitted that the appeal of the Hummer was "testosterone."[52] His customers had apparently not seen the *New Yorker* cartoon showing a disgusted driver behind the wheel of his SUV, talking on his cell phone and telling someone that he would never have spent so much money on his vehicle if he had known that Viagra was about to hit the market for only a couple of dollars.

Robert Lang and Karen Danielson, two real-estate observers, noted that "the new 'supersized American dream' includes fully loaded SUVs and, yes, monster houses."[53] According to the National Association of Home Builders, the average American house size had grown from 1,900 square feet in 1987 to 2,300 in 2001—an increase of 20 percent. In 1988, only 11 percent of newly built homes were over 3,000 square feet, but by 2003, that figure was up to 20 percent. The lots to accommodate these home had shrunk 6.5 percent between 1987 and 2002, and the average size of households was shrinking as well; from 3.1 persons in 1970 to 2.59 in 2000. Often called McMansions, these houses, with their three- or four-car garages, called "Garage Mahals," were "not just a place to live,... but an assertion of the American sense of identity and statement of prosperity."[54] Certainly conspicuous consumption was nothing new, but the aggressive imposition of "lifestyle" technologies on space and other resources seemed, in the last decades of the century, a part of a seamless web of technological dominance that helped make Americans, per capita, the most prodigal with resources and the most polluting of any other nation. And in a globalized world, American hegemony spread its culture and technologies into every corner of the earth.

TWELVE

GLOBALIZATION, MODERNITY, AND
THE POSTMODERN

In the last decade of the twentieth century, a piece of graffiti appeared proclaiming that "Time and Space are so not everything happens at once, and not all to you." It was an entirely convincing and comforting formulation, but it was already becoming archaic. The annihilation of time and space, the project begun by modernity two centuries ago, had reached the point by the twenty-first century where it was almost literally true that everything happens at once, and it all happens to us.

The French sociologist Jean Baudrillard wrote in 1988 that "America is the original version of modernity."[1] It was, perhaps, a late reiteration of what the historian Thomas Hughes called the "Second Discovery of America," by which he meant the insight of that cohort of European intellectuals and avant-garde artists who found in the America of the early twentieth century not a virgin land of endless frontiers, but a technological artifact that brilliantly provided the machine vocabulary of a nascent modernism.[2] It was hardly in the American grain at the time to make bold cultural claims in the arts, but the nation could insist that it was tied closely to technology, and that technology was in turn tied closely to modernity.

Exactly what the connection is between technology and modernity has never been clear or precise. The meanings of both words are continually argued about and fought over, and because both were significantly redefined during the period to which they are now applied, the nature of their relationship cannot be considered outside of the contexts that give them meaning.

The Globalization of Technology

As early as 1988, the National Academy of Sciences issued a report entitled *Globalization of Technology*. "The effects of technological change on the global economic structure," it began, "are creating immense transformations in the way companies and nations organize production, trade goods, invest capital, and develop new products and processes." The goal, as the academy saw it, was to "sustain and improve world growth and improve growth per capita." The happy consequence of "arrangements such as transnational mergers and shared production agreements" was that technology could be harnessed "more efficiently, with the expectation of creating higher standards of living for all involved" in both developed and developing nations. The dynamism was assigned to companies, but governments became involved because development "sometimes conflicts with nationalistic concerns about maintaining comparative advantage and competitiveness."[3]

Linking technology with the hope for economic growth was nothing new, either at home or abroad. As Mary Lowe Good, under secretary of technology for the U.S. Department of Commerce, explained it in 1996, however, "much of the technology needed for national growth is developed and managed by multinational companies whose markets, operations, and sources of capital are distributed throughout the world." Speaking before the American Physical Society, she pointed out the same disjunction as had the National Academy of Sciences: "Technologies," she explained, "in the form of products, know-how, intellectual property, people and companies ... are being traded, transferred, hired, bought and sold on a global basis." As a result, competition took place on two levels: "First in the competition between companies. Second is the competition between nations to attract and retain the engines of wealth creation that increasingly skip around the globe looking for the best opportunities." Her example was South Korea, which was making massive investments in the hope of becoming a G7 nation by 2001.[4] A decade later, she could have referred to the large investment Singapore was making in stem-cell research facilities to attract American researchers held back by faith-based government policies at home.

The results of technology-driven global commerce were everywhere. Some were intended but illegal, such as the $13 billion lost to American software developers from pirated copies in 1993 alone. Some were legal but unintended, such as the Australian viewers of American crime programs who sometimes dialed 911 in emergencies rather than the Australian number, 000. Still others were both legal and intended, such as the "thousands of engineers in India" whom Under Secretary Good identified as "designing computer chips for America's leading firms, and beaming these designs overnight to California and Texas via satellite."[5]

The Case of India

The English-speaking and technically trained workforce of India seemed to have a particular attraction for American firms, in large part because of the cost of labor. American technical support personnel for a firm like Dell might make $12 an hour; their replacements in India make about a third of that. By 2003, the California health maintenance organization (HMO) Kaiser Permanente was outsourcing its computer operations to India, including the handling of its payroll information on 135,000 employees, members' personal data, and patients' records. Gary Hurlbut, the company's vice president for information technology was quoted as saying that "we're trying to move a good part of system maintenance offshore and free up employees for new opportunities." He did not say what those might be. So attractive is this particular part of the global workforce that an estimated half of the Fortune 500 companies were using Indian workers for computer services.[6]

The disappearance of manufacturing jobs that so changed the American economy in the late twentieth century were followed by service jobs early in the twenty-first century. In 2003, around 27,000 computer jobs and over 3,000 in architecture had been sent offshore. The most commonly exported jobs were software development, customer call services, "back-office" accounting, and even "product development." One study that year estimated that some 3.3 million jobs might be sent overseas by 2015, 70 percent of them to India, which was already graduating 2 million from college every year.[7]

The experience of India with globalization is out of scale, like everything else about that large and extraordinary country. The novelist and activist Arundhati Roy points out that while "India is poised to take its place at the forefront of the Information Revolution, three hundred million of its citizens are illiterate."[8] In a striking illustration of her point, she writes that "every night I walk past road gangs of emaciated labourers digging a trench to lay fiber-optic cables to speed up our digital revolution. In the bitter winter cold, they work by the light of a few candles." She calls "the modern version of globalization," a "process of barbaric dispossession on a scale that has few parallels in history." Conceding that it is marketed as being about the "eradication of world poverty," she asks if it is not instead "a mutant variety of colonialism, remote controlled and digitally operated."[9]

Roy cites the case of the power equipment manufacturing company Bharat Heavy Electricals (BHEL), which once exported its equipment worldwide. Forced into a joint venture with Siemens and General Electric, the firm now acts as little more than a provider of cheap labor, turning out German and American equipment and technology. She charges that "India's rural economy, which supports seven hundred million people, is being garroted" by the demands of American farmers, acting through the World Trade Organization (WTO), that India drop the agricultural subsidies that allow that sector to continue functioning. A world-dam business, which Roy estimates accounts for $32 billion to

$46 billion a year, has displaced over 30 million people in India during the last half century, half of them desperately poor Dalit and Adivasi peoples.[10]

With something like 3,600 big dams, India has more than all but two other countries. Additionally, 695 more dams are under construction, making up 40 percent of those being built worldwide. Jawaharlal Nehru famously called them "the temples of modern India," and while Roy admits that they "began as a dream," she claims that they have "ended as a grisly nightmare." By becoming export credit agencies—and therefore having their home countries insure their investments in "unstable" economic and political environments around the world—developers, according to Roy, "can go ahead and dig and quarry and mine and dam the hell out of people's lives without having to even address, never mind answer, embarrassing questions." In 1993, Enron signed the first contract for a private power project in India. In exchange for building the power plants, the Maharashtra State Electricity Board agreed to pay the energy company $30 billion.[11]

Disasters

Another aspect of technological globalization as empire is what Naomi Klein has called "disaster capitalism." The old empires had considered the New World, no matter how many indigenous people it contained, to be an empty space to be set upon with European technology and filled with European peoples living under European social structures and conditions. Disasters, both "natural" and those socially triggered, are, Klein suggests, "the new *terra nullius*."[12] Karl Marx's phrase "all that is solid melts into air" characterizes both capitalism and modernity; tearing down to rebuild is an imperative of both systems. Working hand in hand, the United States and such transnational organizations as the World Bank have made technological infrastructures the focus of reorganization around the world. In 2004, the United States set up an Office of the Coordinator for Reconstruction and Stabilization to draw up plans for the "post-conflict" reorganization of twenty-five countries not yet in "conflict." To create new "democratic and market-oriented" states out of the old ones, it might be necessary, according to the office's first head, to do some "tearing apart." This might well include, he added, helping to sell off "state-owned enterprises that created a nonviable economy."[13]

What Klein calls the "democracy building" industry is very large indeed. Halliburton, an engineering firm, has contracts amounting to $10 billion in Iraq and Afghanistan. Bearing Point, an American consulting firm that specializes in giving advice on selling off assets and running government services at the same time, made $342 million in profits in 2002. "Post-conflict" countries are now receiving 20 to 25 percent of the World Bank's total lending. In Afghanistan, years of fighting the Soviet invasion, civil war between warlords, and the American-led invasion left the built environment devastated. A trust fund for rebuilding the country is administered by the World Bank, which has refused to give funding to the Ministry of Health to build hospitals, instead directing the money to nongovernmental organizations (NGOs) that run their own private clinics on three-year contracts. In addition, according to Klein, the World Bank has mandated "'an increased role for the private sector' in the water systems, telecommunications, oil, gas and mining and has directed the government to 'withdraw' from the electricity sector and leave it to 'foreign private investors.'"[14]

"Natural" disasters also provide opportunities for globalization. Immediately after Hurricane Mitch, in October 1998, the Honduran government, at the insistence of the World Bank and International Monetary Fund (IMF), passed laws "allowing the privatization of airports, seaports and highways, and fast-tracked plans to privatize the state telephone company, the electric company and parts of the water sector," all to get the aid funds flowing. During the same period, Guatemala announced that it would sell its telephone system, and Nicaragua did the same, throwing in its electric company and petroleum sector. After the devastating 2004 tsunami in Indonesia and neighboring countries, World Bank loans sought not to rebuild the coastal fishing villages that had been wiped out, but rather tourist resorts and commercial fish farms. It also suggested that the governments might want to consider privatizing the destroyed roads systems. As the U.S. Secretary of State Condoleezza Rice was quoted as saying just weeks after the tsunami, the disaster provided "a wonderful opportunity" that "has paid great dividends for us."[15]

The process of globalization was driven primarily through direct political pressure from the G7 nations, particularly the United States. Additional pressure was funneled through the World Bank, WTO, and IMF. By the end of the 1990s, the IMF was operating and recommending policies in eighty countries around the world.

Sovereign states have always insisted on the right to go to war if their national interests are threatened. During the nuclear standoff of the Cold War, the United States and Soviet Union mostly resorted to wars within or between client states. The defeat of the United States in Vietnam, however, taught some lessons about the limits of this strategy. First, such wars were to be fought with professional, not conscripted, armies; second, they should be short, perhaps even instantaneous, to prevent the perception of things dragging on; and third, they should be, as summarized by Manuel Castells, "clean, surgical, with destruction, even of the enemy, kept within reasonable limits and as hidden as possible from public view, with the consequence of linking closely information-handling, image-making, and war-making."[16]

War in the Middle East

The first Gulf War was the very model of how post-Vietnam wars should work. During the agonizing, seven-year war between Iran and Iraq, Western countries helped keep the fighting going so that neither country could threaten the Middle East's oil supply to the West. The United States and Russia supplied Saddam Hussein in Iraq and Israel backed Iran, while Spain sold chemical weapons to both sides, according to Castells. When, after the war, Iraq reasserted claims to Kuwait, the American-led forces struck swiftly and decisively with what has been called the "100 hours' denouement."[17] The Air Force and Navy attacked Iraqi forces with such overwhelming and accurate firepower that the half a million Allied ground forces were largely unnecessary. Though smart bombs accounted for only 7 percent of the bombs used, they garnered most of the headlines, and television brought into American homes the same exciting visuals, images, and bloodless heroics of video games. In 1991, the editor of the journal *Science* noted that "valor and heroism are the focus of novels about war, but history has shown that, from bows and arrows to laser guided missiles, technology is decisive if it is very one-sided."[18]

That one-sidedness had not worked so well in Vietnam, but it did in Iraq in 1991, and was the basis for the shock-and-awe strategy of the United States in its second go at Iraq in 2003. On the eve of the second war, it was reported that "what concerns many officers [in the Pentagon] is how the debate over visions of empire overemphasize the role of technology and ignores certain variables that just might make 'empire building' a bit more complex—and consequential—than many imagine."[19] The widespread faith in military technology was accompanied by a desire to keep personnel levels as low as possible, but with two-thirds of American Special Forces already spread over eighty-five different countries, troops were thin on the ground anyway. One former Special Forces officer pointed out that while in Operation Desert Storm—the first Gulf War—a multiplying of favorable circumstances made super-weapons look good, "technology contributes virtually nothing to complex civil-military operations, like recent ones in Haiti, Somalia, Bosnia, and Kosovo." Even the vaunted command and control interconnectedness had its downside, as commanders far from the scene of action could micromanage operations that might be better left to the officers at the scene.[20]

The combination of America's new, more aggressive, and independent global reach, and a chosen dependence on technologies that allowed American forces to maintain surveillance over wide areas and strike at a distance, suggested a change in the placing and types of U.S. military bases around the world. Most overseas bases at the end of the century had been constructed and staffed to contain the Soviet Union during the Cold War. Decommissioning older bases in Germany, Japan, and South Korea, for example, would free up tens of thousands of troops for other assignments, though not necessarily at the new bases being planned in Eastern Europe, Central Asia and the Caucasus, the Persian Gulf, Africa, and the Pacific. The new facilities were announced in 2005 as being of two types: "forward operating sites," which would have logistical facilities such as airfields or ports and stockpiles of weapons, but only a small number of military technicians; and "cooperative security locations," which would be maintained by contractors of host-nation personnel and used only in times of crisis. Besides these, however, are the fourteen "enduring bases" being constructed in Iraq even before the "insurgency" was put down.[21]

Deploying American power in a more assertive way, countering the perceived passivity that flowed from the nation's Vietnam experience, implied new allies, and weapons as well as bases. According to a report from the Defense Department, America's "role in the world depends upon effectively projecting and sustaining our forces in distant environments where adversaries may seek to deny US access."[22] The increasing power and influence of China, the danger of terrorism, and the pursuit of oil all lay behind this realignment. The last, of course, was hardly a new goal. The United States, by the end of the century, had elaborated in great detail a technological lifestyle based on abundant cheap petroleum, and as early as 1980, President Jimmy Carter had sworn to keep Persian Gulf oil flowing "by any means necessary, including military force."[23]

Empire

One of the great utopian goals of modernism was to create just the sort of world-integrating process we now call globalization. Expanding transportation and communications routes served both commerce and the growing empires of the West, including the empire of the United States. Until the nineteenth century, European domination had taken root within a thin membrane of settlement stretched along the coastlines of Asia, Africa, Australia, and the Americas. With shallow-draft gun boats, Gatling guns, railroads, and the telegraph, Europeans drove indigenous peoples farther and farther into the interiors of their continents, and ancient civilizations, like that in China, was forced to give up "concession" areas, within which their sovereignty no longer prevailed.

America expanded its empire by buying or conquering contiguous territory: the Louisiana Territory, Florida, and then nearly half of Mexico. By early in the twentieth century, discontinuous lands were added: Alaska and the Virgin Islands by purchase, Hawaii by "annexation," and Puerto Rico and the Philippines by conquest from Spain. The great European powers expanded overseas early, and by the early twentieth century, the British Empire comprised nearly a quarter of the world. The end of World War I saw some significant reordering of these captive lands, however—and then, after World War II, all of these great empires, except the Russian and American, collapsed.

The end of empire, however, did not lead inevitably to the spread of democracy or even self-rule in all of these areas. New regimes of control were enacted that were justified, in part, by the old dream of globalization, harnessed as always to competing desires for universal uplift, economic advantage, and military dominance. One aspect of this globalization is the rise of what is widely called "the networked society," a label attached particularly to the use and influence of electronic media that have helped to create a new American empire.

Communications and Empire

Herbert Schiller focused on this networked society as early as 1971, in his book *Mass Communications and American Empire*. Noting that "if free trade is the mechanism by which a powerful economy penetrates and dominates a weaker one, the 'free flow of information' . . . is the channel through which life styles and value systems can be imposed on poor and vulnerable societies." Since the end of World War II, he maintained, "the American technological supremacy, and its leadership in communications in particular, has been receiving wider and wider appreciation inside the domestic business, military and governmental power structures."[24]

After the war, the American radio broadcasting industry, long organized as a private and commercial rather than a public and educational medium, was easily folded into new military communications initiatives. A substantial portion of the radio spectrum was reserved for government use, and of that, the military services used an estimated three-fifths. In 1946, an Armed Forces Communications and Electronics Association was formed, as it said, to further "the military-civilian partnership concept." In 1967, the industry sold over 60 percent of its output to the government, principally to the armed services. That year, eighteen electronics and communications forms were among the top fifty industrial defense contractors. Because immediate communications with "trouble spots" around the world was of critical concern to the government, a national communications system was established, and President Kennedy made the secretary of defense its executive agent.[25]

In 1945, Arthur C. Clarke, the British engineer and science fiction writer, had suggested that satellites launched into geosynchronous orbits could be used to transmit communications to widely separated points on the earth. By the 1960s, the military and two large corporations were seriously looking into communications satellites to expand America's global outreach. One company, the telecommunications giant AT&T, had decided that geosynchronous orbits were too complex for its needs, which were to merely supplement their land-based cables. Instead, AT&T was pursuing a system of random orbiting satellites, thus shifting the issue of engineering complexity from the satellites to the ground receivers that had to find and track them.

In 1958, the military also turned to communications satellites, and while initially looking into both synchronous and nonsynchronous types, after 1960, it concentrated on the former because of its interest in using smaller, less complicated ground receiving stations. The Pentagon soon realized that it lacked the resources to develop such a complicated and sophisticated technology, and abandoned the project, called Advent. By then, however, Hughes Aircraft Corporation had taken up the idea to try to reestablish its military capabilities after the Air Force cancelled its contract to develop a long-range interceptor. By 1960, the Hughes engineers were confident that they could solve the problems involved with geosynchronous satellites.[26]

T. Keith Glennan, the first director of NASA, firmly supported private enterprise and was anxious to see AT&T, as the largest telecommunications provider, develop a satellite system. Other members of President Eisenhower's administration, however, as well as some members of Congress, were less sanguine that the goals of profit maximization and national Cold War policy could be so easily melded. To counter the propaganda successes of the Soviet Union and its aggressive space program, many in Washington seized upon a global, satellite-based, and American-dominated communications system as an important way to provide the tangible benefits of American space technology to the peoples of the newly liberated, developing countries of the world, particularly in Africa and Asia. As things stood, the new nations were dependent on land-based and underwater cables laid by their former colonial masters, which of course ran directly from London or Paris.[27]

According to historian Hugh R. Slotten, a 1960 congressional report began the call for "a single global system, benefiting the entire world but also serving the cold war interests of the United States." The benefits could be substantial:

the military would have its access to communications picked up by relatively small, mobile receivers; cables tying former colonies to Europe would be replaced by satellites tying them to America; the United States would be seen as a technologically accomplished and generous benefactor of the world's underprivileged; and not least, as one member of Congress put it, "the nation that controls world-wide communications and television will ultimately have that nation's language become the universal tongue."[28]

In the summer of 1961, President Kennedy invited "all nations of the world to participate in a communications satellite system, in the interest of world peace and closer brotherhood among peoples of the world."[29] The following year, Congress passed the Communications Satellite Act of 1962, which enabled the formation of a private corporation to oversee the system. In 1963, the Communications Satellite Corporation (COMSAT) was set up, and in 1964, joined with agencies in other countries to form the International Telecommunications Satellite Organization (INTELSAT). *Intelsat I*, called "Early Bird," was launched in 1965, providing one path of telecommunications service between the United States and Europe. By 1977, the system had grown to include ninety-five nations with eight larger satellites and 150 ground stations providing 500 paths to eighty counties. One station in West Virginia used the Atlantic *Intelsat IV* to connect to thirty-one other stations located in such counties as Iran, Egypt, Rumania, Peru, Israel, the Soviet Union, Zaire, Portugal, South Africa, and Sweden.[30]

The flurry of institution building was partly an effort to keep AT&T from controlling the system. A geosynchronous system, with its relatively small, inexpensive, and mobile ground stations, was important to the Pentagon and the Department of State. But some legislators and government officials worried that the telecommunications giant, which had a history of resisting extending service to underdeveloped areas even within the United States, would now be reluctant to service developing nations, instead concentrating on the more lucrative European market. Additionally, AT&T was heavily invested in the older technology of undersea cables, the latest of which had been laid down only in 1956. President Kennedy considered speed of completion and the accompanying prestige as more important to the nation than profit, and called for a global system to cover the entire world, "including service where individual portions of the coverage are not profitable."[31] The federal government was the largest user of AT&T and other international facilities, and it wanted influence in directing the new system.

When COMSAT was finally set up as a private corporation, half of the stock was reserved for international carriers, and the rest was to be sold to the American public. No one carrier was allowed to have more than three representatives on the fifteen-person board of directors; six spots would be reserved for public shareholders, and three were to be appointed by the president. Thus, it was hoped, the influence of AT&T would be successfully watered down. When COMSAT undertook to entice foreign providers to join INTELSAT, it found European systems, many run by national post offices, reluctant to adopt the new satellite system, precisely because it created an American-controlled alternative communications route to their old colonies. Besides, like AT&T, the European systems already had heavily invested in cable systems. Once on board, they reinforced AT&T's technological conservatism and profit orientation, and resisted making decisions not based on "purely commercial" considerations. At home, the Federal Communications Commission refused to let COMSAT compete with AT&T, ruling that it could lease its services only to that company and three other international carriers.[32]

One scholar has judged that "the final system generally favored 'developed' countries, especially the United States and its nonclassified military communications." Finally, in 1966, President Johnson had to step in and order the State Department and other federal agencies to take "active steps" to see that earth stations were constructed in "selected less-developed countries." By 1969, when astronauts walked on the Moon for the first time, the event was watched on television via the INTELSAT system by five hundred million people in forty-nine countries.[33]

Agriculture

Although globalization is most dramatically seen through the success of communications technologies, it has also transformed the growing transport, marketing, and eating of foodstuffs and other agricultural crops. In stark contrast to the ideals of such groups as appropriate technology and slow foods advocates, a quick survey of an American supermarket at the turn of the century would show fresh fruits and vegetables shipped in from all over the world for the American table. In 1988, produce was transported 1,518 miles on average, mostly by truck, to stores in Chicago. Only 21 percent of the over ten quadrillion BTU of energy used in the food system is involved in growing it. The rest is divided between transportation (14 percent), processing (16 percent), packaging (7 percent), retailing (4 percent), and home refrigeration and preparation (32 percent). Refrigerated ships have carried food (initially, mostly meat) to the United States since the late nineteenth century, but by the end of the twentieth, refrigerated jumbo jets sixty times more energy intensive accounted for a growing proportion of the international food trade.[34]

More problematic is the spread of genetically modified organisms (GMOs) across the globe. During the six years between 1996 and 2002, GMO seeds had become so widespread that they accounted for 34 percent of American corn, 75 percent of soy, 70 percent of cotton, and 15 percent of canola crops. Five American companies control 75 percent of the patents for GMO crops, and one of them, Monsanto, supplies 90 percent of the GMO seed planted around the world. Because the seeds for GMO crops are patented, farmers cannot set some of their harvest aside for planting the next year; they are forced to buy more. For good measure, companies producing GMOs have bought up most of the seed companies in the United States. Local plant diversity and ancient farming knowledge are destroyed to favor a single, proprietary crop. The modified plants are designed to be part of large-scale, chemical, energy, and capital-intensive agricultural systems. As Gyorgy Scrinis, an Australian researcher, put it, "genetic-corporate agriculture is in fact a system for feeding on the world rather than for feeding the world."[35]

Monsanto's patented Bt corn has been genetically altered to produce its own pesticide, killing any corn borers that might attack it. Using Bt means that farmers can apply less pesticide on the crop. The same farmers can also plant Monsanto's Roundup Ready soybean seeds, which have been altered to be immune from the effects of Monsanto's herbicide Roundup. Both products together enable farmers to apply pesticide to their soybean fields to kill weeds without harming the crop itself. Other crops have also been altered: flounder genes have been placed in strawberries, mouse genes in potatoes, cow genes in sugarcane, cow genes and chicken genes in corn.[36]

One concern has been the migration of GMO crops, such as Bt corn, from where it is deliberately planted into areas where it is not wanted. In Iowa, it is turning up in fields of what is supposed to be organic corn, a designation that precludes GMO seeds. Farmers are losing their organic accreditization and, therefore, receiving less for their crops on the market. More startling, despite a Mexican law forbidding the planting of GMO corn, it has been discovered in fields around Oaxaca, far from the border with the United States. In that area, there are more than sixty varieties of local corn grown, a precious biodiversity that is now threatened. One industry scientist saw the bright side of this invasion: "If you're the government of Mexico," he said, "hopefully you've learned a lesson here and that is that it's very difficult to keep a new technology from entering your borders, particularly in a biological system."[37] The battle may already have been lost by the small Mexican corn farmer, however. European farmers get 35 percent of their income from government subsidies, and North American farmers, 20 percent. Even after investing in technologies such as satellite imaging to measure fertilizer application, farmers north of the border can afford to export their food crops at prices below cost. As the *New York Times* noted, for the small Mexican farmer, "these days modernity is less his goal than his enemy."[38]

The notion that the Mexican corn farmer or the peasant in Arundhati Roy's India should find modernization the enemy, not the goal, should surprise no one. Faust, the quintessential modern hero, sold his soul to Mephistopheles not for wealth, but for the power to do good. The large-scale reclamation project he undertook required killing the elderly couple whose cottage stood in his way. Those early victims of progress stand for the millions around the world today who are finding globalization destructive to their livelihoods, their cultures, and their very lives. But Faust's project, which required dredging wetlands and filling coastal waters, had environmental consequences as well.

Environment

Marshall Berman has called Johann Wolfgang von Goethe's *Faust* "the first, and still the best, tragedy of development." The emperor gives Faust and Mephisto, in Berman's word, "unlimited rights to develop the whole coastal regions, including carte blanche to exploit whatever workers they need and displace whatever indigenous people are in their way." The murder of the elderly couple is merely a part of what Berman calls "a collective, impersonal drive that seems to be endemic to modernization: the drive to create a homogenous environment, a totally modernized space, in which the look and feel of the old world have disappeared without a trace."[39]

Social space can be readily modernized, but natural space is not so easily erased. One of the most dramatic and dangerous consequences and evidence of globalization is the massive environmental damage done to the Earth in the form of rapidly diminishing biodiversity. but also in the phenomenon of global warming.

The idea that human activity could produce climate changes is very old. In the nineteenth century, the idea that cutting down forests could increase rainfall received wide currency in the United States. In 1896, a Swedish scientist, for the first time, suggested that the build-up of carbon dioxide gas in the Earth's atmosphere could raise the ambient temperature of the entire planet. This "greenhouse" effect was only one among many hypotheses, however, and not even the most plausible. In the 1930s, it was noted that the climate of North America was warming, and although most scientists thought it must be evidence of some long-term natural process, the amateur G.S. Callendar saw it as evidence that the greenhouse effect was taking hold.[40] Increased support of research on weather and the seas by the military during the Cold War allowed scientists to research the subject more thoroughly, and in 1961, it was determined that the amount of carbon dioxide in the atmosphere was indeed increasing each year.

Mathematical models of climate change were undertaken, but so many arbitrary assumptions had to be made that skeptics abounded. Nevertheless, as models improved, data accumulated, and a growing appreciation of the complexity of the processes took hold, by the end of the twentieth century, there was virtual agreement that global warming was happening, and at an accelerating rate. The greenhouse effect was identified as a major driver of the process, and there were warnings that if carbon dioxide emissions were not cut dramatically and immediately, disastrous climatic changes could be expected. Even a few degrees of warming would accelerate the melting of the polar ice caps, raising the level of the world's oceans and flooding many of the globe's largest and most populous cities.

The carbon dioxide problem was twofold. First, the rapid stripping of the world's forest cover, most recently in such areas as the Amazon and Southeast Asia, reduced the plants available to absorb the gas. More importantly, the continued burning of fossil fuels, both coal (largely for generating electrical energy) and petroleum (importantly in cars and other technologies of transportation) was pumping vast and increasing quantities of the gas into the atmosphere. Political and industrial leaders across the world tended to deny that any problem existed, citing scientific uncertainty as a reason for doing more research rather than taking steps to reduce emissions.

Among those who took the issue seriously, many advocated some sort of technological fix to ameliorate the problem. These ran the gamut from better insulation in homes, long-lasting light bulbs to save electricity, and using solar and wind power to produce it in the first place, to bringing back the "nuclear option," finding ways to "clean" coal before it was burned, or burying the carbon dioxide in the emissions. The range of technologies put forward by the AT movement of the 1970s, and mainly rejected at the time were still available, often in cheaper, more efficient forms than a generation earlier. Arguments that these could not be developed and brought on line soon enough to avert disaster only underscored the missed opportunities of the past. The continued enthusiasm of political leaders to provide massive subsidies for nuclear, coal, and petroleum energy sources, while refusing to subsidize alternative technologies, effectively keeps society locked into the same old systems that have created the problem in the first place.

In 1997, world leaders meeting in Kyoto, Japan agreed to cut the emissions of carbon-rich gases by 5.2 percent by 2010, compared with their 1990 totals.

By 2005, Russia had signed on, bringing the total number of committed countries to thirty-eight. At the time of the conference, developing countries, most notably China, India, and Brazil, were exempted from emission quotas on the grounds that they needed time to catch up economically with the already developed countries. It was perhaps a fatal omission, not least because it was seized upon by the new administration of President George W. Bush as the reason he was pulling the United States out of the agreement. The argument was simple and classic: If the United States adhered to Kyoto quotas while some large developing countries did not have to, it would be harmful to the American economy. The U.S. actions regarding the Kyoto agreement were only one indication of many that not only was the American government not going to encourage the use of alternative technologies, but it was also intent on backing away from steps already taken. In the name of the economy, efficiency standards for appliances were dropped, emission standards for cars were loosened, and subsidies for alternative technologies were abandoned.

The complete withdrawal from the Kyoto process tended to isolate the United States, with the president of the European Commission charging pointedly that "if one wants to be a world leader, one must know how to look after the entire Earth and not only American industry." European leaders who had called a meeting in Sweden to discuss issues of biodiversity and sustainable development found themselves instead deploring the American retreat, while pledging to persist in their plans to cut emissions.[41] Four years later, it appeared that a significant grassroots rebellion in the United States was making its own commitment to Kyoto. Halfway through 2005, some 150 American cities had officially decided to meet the Kyoto-style targets, and even some states were joining in.[42] Still, for many, the specter of underdeveloped nations continuing to race toward modernity—building large dams and electric power plants, buying more automobiles and cutting down their forests—while America was asked to turn away from such traditional markers of progress, sat ill. If America was the original version of modernity, were we now to give up that vanguard position?

Techno-Orientalism

Writing in 2004 from California's Silicon Valley, the *New York Times* columnist Thomas L. Friedman reported "something I hadn't detected before: a real undertow of concern that America is losing its competitive edge vis-à-vis China, India, Japan and other Asian tigers."[43] It was not simply the ubiquitous fact of cheap labor, but a menu of other problems: Asian nations were offering tax holidays and national health insurance to researchers; the new Department of Homeland Security was making it difficult for foreigners to gain U.S. visas, threatening the brain-drain from other countries upon which the United States had depended since World War II; fewer Americans were graduating in science and engineering, and fewer students were coming from overseas to study here; and finally, American corporations were simply moving their research and development to India and China. The possibility of a westward migration of the technological frontier was alarming both economically and culturally.

The concept of orientalism fits nicely at the intersection of postmodernism and globalization.[44] Building on the classic notion of orientalism as the Western construction of an exotic "other" in the "East," techno-orientalism refers to the cultural construction of an "oriental," and especially in this case Japanese, relationship with modern technology.

The West created the concept of the Orient when it invaded, conquered, studied, and exploited it. As the other, the Orient was quite specifically and necessarily not us. Like most binaries, East and West defined each other— Americans and Europeans knew that they were of the West because they were not of the East. So powerful was this identification that for nearly two centuries, Australians, living almost within sight of Indonesia but proudly following British understandings, referred to the Asian landmass as the Far East. Only recently has it admitted that it might more accurately be called the Near North.

Since so-called Orientals were the opposite of us, they were characterized by traits quite unlike our own. We were straightforward, they were shifty. We were rational, they were superstitious. We were trustworthy, they were unreliable. We were vigorously masculine, they were softly effeminate. And perhaps most importantly, we were good at technology and they were not: we invented, they copied. The West was modern and the Orient was premodern. Finally, the West was imagined as predominantly white, while the people of the East were "colored." Race therefore was a powerful marker of technological skill. As one student of the subject confidently claimed in 1896, "looking at this campaign of progress from an anthropological and geographic standpoint, it is interesting to note who are its agents and what its scenes of action. It will be found that almost entirely the field [of invention and technological progress] lies within a little belt of the civilized world between the 30th and 50th parallels of latitude of the western hemisphere and between the 40th and 60th parallels of the western part of the eastern hemisphere, and the work of a relatively small number of the Caucasian race under the benign influences of a Christian civilization."[45]

Premodern countries could be expected and even encouraged to become modern, but for Japan, after spending a century between being "opened up" by Commodore Perry and being rebuilt after World War II in an industrial apprenticeship, to then appear on its own terms to have outstripped the West in entire areas of technological endeavor, was unprecedented and unlooked for. As has been suggested, "what it has made clear are the racist foundations of western modernity. If it is possible for modernity to find a home in the Orient, then any essential, and essentializing, distinction between East and West is problematized." And to the extent that advanced technology is the marker of Western modernity, "the loss of its technological hegemony may be associated with its cultural 'emasculation.'"[46]

Worse yet, while in the years before World War II, "made in Japan" was almost synonymous in America with flimsy, shoddy, and cheap, by the end of the century, in areas such as automobiles, electronics, media, and robotics, Japan had come to seem the very center of high technology and cultural postmodernity. But old prejudices are as often transformed as shattered. "The association of technology and Japaneseness," write David Morley and Kevin Robins, "now serves to reinforce the image of a culture that is cold, impersonal and machine-like, an authoritarian culture lacking emotional connection to the rest of the world. The *otaku* generation—kids 'lost to everyday life' by their immersion in computer reality—provides a strong symbol of this."[47] In this context, the words of Jean Baudrillard seem very much to the point: "In the future, power will belong to those people with no origins and no authenticity, who know how to exploit that situation to the full."[48] The West has moved from thinking of the "Oriental other" as premodern to seeing them as postmodern, in both cases invoking the excitement and fear of the exotic and erotic. It is only the prospect of having to share the common ground of modernity that seems so destabilizing to our own self-understanding as to be unthinkable.

Vanguard or Rear Guard?

In his book *Networking the World*, Armand Mattelart writes of the utopian dream, coming out of both the Enlightenment and nineteenth-century liberalism, that the worldwide spread of communications technologies would create a "universal mercantile republic" which "had the vocation of bringing together the entire human species in an economic community composed of consumers." Citing Fernand Braudel's concept of world-economy, he reminds his readers that "networks, embedded as they are in the international division of labor, organize space hierarchically and lead to an ever-widening gap between power centers and peripheral loci." His conclusion is sobering. "As the ideal of the universalism of values promoted by the great social utopias drifted into the corporate techno-utopia of globalization, the emancipatory dream of a project of world integration, characterized by the desire to abolish inequalities and injustices in the name of the imperative of social solidarity, was swept away by the cult of a project-less modernity that has submitted to a technological determinism in the guise of refounding the social bond."[49]

For the half century following World War II, the United States gloried in its self-imagined vanguard position as the wealthiest, most powerful, and technologically innovative nation on Earth. But as scholars such as the geographer David Harvey suggest, sometime in the 1970s, modernity was overtaken with something else, called only postmodernity because its real nature has yet to reveal itself clearly.[50] It has to do, almost certainly, with a shift from the undisputed hegemony of the center towards an empowerment of the periphery, from a search for the universal to a new respect for the particular, from totalizing to partial projects, from cultural hierarchy to something more like anarchy, and a focus less on Us and more on Others. Corresponding technologies, such as alternative energy sources, have been developed and continue to evolve that answer to these new world conditions, but so far America has tended to shun them, clinging instead to the technologies that are thought to be responsible for our wealth, our power, and our cultural identity. If they no longer serve, however, it is time to harness our vaunted entrepreneurship, willingness to innovate, and technological imaginings in the service not of stagnation, but of the future.

In 1889, Mark Twain published his last great novel, *A Connecticut Yankee in King Arthur's Court*. It was a parable of progress, with a protagonist, Hank Morgan, who was transported back to the sixth-century England of King Arthur. Morgan had worked his way up to become the "head superintendent" at the great Colt arms factory in Hartford, Connecticut, and bragged that he had there "learned my real trade; learned all there was to it; learned to make everything; guns, revolvers, cannon, boilers, engines, all sorts of labor-saving machinery. Why, I could make anything a body wanted—anything in the world, it didn't make any difference what; and if there wasn't any quick new-fangled way to make a thing, I could invent one—and do it as easy as rolling off a log." With a mix of idealism and self-interest, Morgan styled himself "Sir Boss" and undertook to reform the culturally backward and economically underdeveloped kingdom by introducing those two great prides of American culture, democratic government and industrial technology.

In the end, Sir Boss discovered that it was easier to destroy the social and cultural infrastructure of another part of the world than to remake it in America's image. Ensconced in a cave protected by an electrified fence and Gatling guns, he and his young mechanics successfully fought off an army of knights defending the old order, but then he collapsed and the narrative was taken over by his head mechanic, Clarence. "We were in a trap, you see," said Clarence, "—a trap of our own making. If we stayed where we were, our dead would kill us; if we moved out of our defenses, we should no longer be invincible. We had conquered; in turn we were conquered. The Boss recognized this; we all recognized it.... Yes, but the boss would not go, and neither could I, for I was among the first to be made sick by the poisonous air bred by those dead thousands. Others were taken down, and still others. Tomorrow—."[51]

NOTES

Introduction

1. "The American Century," 1941, quoted in Herbert I. Schiller, *Mass Communications and American Empire* (Boston: Beacon Press, 1971), 1.

2. Leo Marx, "The Idea of 'Technology' and Postmodern Pessimism," in *Does Technology Drive History? The Dilemma of Technological Determinism,* ed. Merritt Roe Smith and Leo Marx (Cambridge, MA: MIT Press, 1994), 237–257.

3. Edward W. Said, *Orientalism* (New York, Vintage, 1979), 11, 14.

4. Carey McWilliams, *Factories in the Field* (Boston: Little, Brown, 1939).

5. Lizabeth Cohen, *A Consumer's Republic: The Politics of Mass Consumption in Postwar America* (New York: Alfred A. Knopf, 2003).

6. Arthur Schlesinger, Jr., "Foreword," J.-J. Servan-Schreiber, *The American Challenge* (New York: Atheneum, 1968), vii.

1. Arsenal of Democracy

1. C. E. Kenneth Mees and John A. Leermakers, *The Organization of Industrial Scientific Research,* 2d ed. (New York: McGraw-Hill, 1950), 16.

2. See Carroll Pursell, "The Anatomy of a Failure: The Science Advisory Board, 1933–1935," *Proceedings of the American Philosophical Society* 109 (December, 1965), 342–351; and "A Preface to Government Support of Research and Development: Research Legislation and the National Bureau of Standards, 1935–41," *Technology and Culture* 9 (April 1968), 145–164.

3. Mees and Leermakers, *The Organization of Industrial Scientific Research,* 16.

4. U.S. Senate Subcommittee on War Mobilization, *The Government's Wartime Research and Development, 1940–44.* Part II.—Findings and Recommendations. 97th Congress, 1st sess. (January 23, 1945), 20–22.

5. This story is told in Carroll Pursell, "Science Agencies in World War II: The OSRD and Its Challengers," in *The Sciences in the American Context: New Perspectives,* ed. Nathan Reingold (Washington, D.C.: Smithsonian Institution, 1979), 370–372.

6. Quoted in Pursell, "Science Agencies," 370.

7. Quoted in Pursell, "Science Agencies," 370.

8. Quoted in Pursell, "Science Agencies," 371.

9. Quoted in Pursell, "Science Agencies," 371.

10. U.S. Senate Subcommittee on War Mobilization, *The Government's Wartime Research and Development, 1940–44*. Part I—Survey of Government Agencies, 227–228, 268–271.

11. See Carroll Pursell, "Alternative American Science Policies during World War II," in *World War II: An Account of its Documents*, ed. James E. O'Neill and Robert W. Krauskopf (Washington, D.C.: Howard University Press, 1976), 155.

12. Quoted in Pursell, "Science Agencies," p. 368.

13. U.S. National Inventors Council, *Administrative History of the National Inventors Council* (Washington, D.C.: National Inventors Council, 1946).

14. "Mellon's Aluminum Company of America," *The Nation*, February 1941, 143.

15. I. F. Stone, *Business As Usual: The First Year of Defense* (New York: Modern Age Books, 1941), 59, 67, 83, 95.

16. Stone, *Business As Usual*, 157–158.

17. Bruce Catton, *The War Lords of Washington* (New York: Harcourt, Brace, 1948), 127–128.

18. Quoted in Pursell, "Science Agencies," 373.

19. See *The Politics of American Science, 1939 to the Present*, 2nd ed., ed. James C. Penick, Carroll W. Pursell Jr., Morgan B. Sherwood, and Donald C. Swain (Cambridge, MA: MIT Press, 1972), 102–106.

20. See The Bird Dogs, "The Evolution of the Office of Naval Research," *Physics Today* 14 (August 1961): 30–35.

21. Percy McCoy Smith, *The Air Force Plans for Peace, 1943–1945* (Baltimore: Johns Hopkins University Press, 1970), 110.

22. The following discussion is based on Vannevar Bush, *Science the Endless Frontier: A Report to the President on a Program for Postwar Scientific Research* (Washington, D.C.: Government Printing Office, 1945). Reprinted by the National Science Foundation, July 1960.

23. Vannevar Bush, *Science the Endless Frontier*, 10.

24. Carroll Pursell, "Engineering Organization and the Scientist in World War I: The Search for National Service and Recognition," *Prometheus* 24 (September 2006): 257–268.

25. John Walsh, "LBJ's Last Budget: R&D Follows a 'Status Quo' Pattern," *Science* 163, January 24, 1969, 368.

26. Theodore von Karman, *The Wind and Beyond* (Boston: Little, Brown, 1967), 351–352.

27. Jerome B. Weisner, "Rethinking Our Scientific Objectives," *Technology Review* 71 (January 1969): 16.

28. Ruth Schwartz Cowan, *A Social History of American Technology* (New York: Oxford University Press, 1997), 256–257.

29. James J. Flink, *The Automobile Age* (Cambridge: MIT Press, 1988), 275.

30. Keith Bradsher, *High and Mighty: SUVs—the World's Most Dangerous Vehicles and How They Got That Way* (New York: Public Affairs, 2002), 6–7.

31. U.S. Senate Subcommittee on War Mobilization, *Wartime Technological Developments*. 79th Congress, 1st sess. (May 1945), 1.

32. U.S. Senate Subcommittee on War Mobilization, *Wartime Technological Developments*, 4.

33. U.S. Senate Subcommittee on War Mobilization, *Wartime Technological Developments*, 2.

34. U.S. Senate Subcommittee on War Mobilization, *Wartime Technological Developments*, 4.

35. Mark S. Foster, *Henry J. Kaiser: Builder in the Modern American West* (Austin: University of Texas Press, 1989), 69.

36. Mark S. Foster, *Henry J. Kaiser: Builder in the Modern American West*, 83.

37. Mark S. Foster, *Henry J. Kaiser: Builder in the Modern American West*, 76–77.

38. U.S. Senate Subcommittee on War Mobilization, *Wartime Technological Developments*, 286–297.

39. U.S. Senate Subcommittee on War Mobilization, *Wartime Technological Developments*, 17–18.

40. U.S. Senate Subcommittee on War Mobilization, *Wartime Technological Developments*, 18, 153–164.

41. Quoted in Jim F. Heath, "American War Mobilization and the Use of Small Manufacturers, 1939–1943," *Business History Review* 46 (Autumn 1972): 298–299.

42. Jim F. Heath, "American War Mobilization," 307–308.

43. Bruce Catton, *The War Lords of Washington*, 245ff.

44. Barton J. Bernstein, "America in War and Peace: The Test of Liberalism," in *Towards A New Past*, ed. Barton J. Bernstein (New York: Pantheon, 1968), 295.

45. Bruce Catton, *The War Lords*, 307, 310.

2. The Geography of Everywhere

1. F. John Devaney, *Tracking the American Dream: 50 Years of Housing History from the Census Bureau: 1940 to 1990* (Washington, D.C.: U.S. Department of Commerce, 1994), 2, 35; quotation from "Comforts of U.S. Homes Up Sharply since WWII," *Plain Dealer* (Cleveland), September 13, 1994.

2. Maureen Ogle, *All the Modern Conveniences: American Household Plumbing, 1840–1890* (Baltimore: Johns Hopkins University Press, 1996), 4.

3. F. John Devaney, *Tracking the American Dream*, 35–36.

4. F. John Devaney, *Tracking the American Dream*, 37.

5. F. John Devaney, *Tracking the American Dream*, 37–38, 41.

6. Quoted in F. John Devaney, *Tracking the American Dream*, 39.

7. F. John Devaney, *Tracking the American Dream*, 40.

8. F. John Devaney, *Tracking the American Dream*, 11; Dolores Hayden, *Building Suburbia: Green Fields and Urban Growth, 1820–2000* (New York: Pantheon, 2003).

9. Elaine Tyler May, *Homeward Bound: American Families in the Cold War Era* (New York: Basic Books, 1988), 59 and *passim*.

10. Quoted in Elaine Tyler May, *Homeward Bound: American Families in the Cold War Era*, 74.

11. Susan M. Hartmann, *The Home Front and Beyond: American Women in the 1940s* (Boston: Twayne, 1982), 8.

12. Adam Rome, *The Bulldozer in the Countryside: Suburban Sprawl and the Rise of American Environmentalism* (Cambridge: Cambridge University Press, 2001), 18.

13. American Public Works Association, *History of Public Works in the United States, 1776–1976* (Chicago: American Public Works Association, 1976), 535.

14. Taken from Dolores Hayden, *Building Suburbia*, 268, footnote 13.

15. Cited in Adam Rome, *The Bulldozer in the Countryside*, 27, footnote 26.

16. Quoted in Dolores Hayden, *Redesigning the American Dream: The Future of Housing, Work, and Family Life* (New York: Norton, 1984), 3.

17. Quoted in Gail Cooper, *Air-Conditioning America: Engineers and the Controlled Environment, 1900–1960* (Baltimore: Johns Hopkins University Press, 1998), 151.

18. Quoted in Mark S. Foster, *Henry J. Kaiser*, 132.

19. Mark S. Foster, *Henry J. Kaiser*, 132–134, 142; for the geodesic dome, see Alex Soojung-Kim Pang, "Whose Dome Is It Anyway?" *American Heritage Invention and Technology* 11 (Spring 1996): 28, 30–31.

20. Dolores Hayden, *Redesigning the American Dream*, 6–7; Dolores Hayden, *Building Suburbia*, 135.

21. Richard Perez-Pena, "William J. Levitt, 86, Pioneer of Suburbs, Dies," *New York Times*, January 29, 1994.

22. *Time* quoted in Adam Rome, *The Bulldozer in the Countryside*, 16.

23. Quoted in Tom McIntyre, "The Geography of Nowhere," *San Francisco* 44 (November, 1997), 122.

24. Evelyn Nieves, "Wanted in Levittown: One Little Box, With Ticky Tacky Intact," *New York Times*, November 3, 1995.

25. Adam Rome, *The Bulldozer in the Countryside*, 38.

26. Dolores Hayden, *Building Suburbia*, 136–137.

27. Cited in D. J. Waldie, *Holy Land: A Suburban Memoir* (New York: St. Martin's Press, 1996), 11, 34.

28. Dolores Hayden, *Building Suburbia*, 139; D. J. Waldie, *Holy Land*, 37–91.

29. D. J. Waldie, *Holy Land*, 38, 91.

30. Dolores Hayden, *Building Suburbia*, 152; Dolores Hayden, *Redesigning the American Dream*, 12.

31. D. J. Waldie, *Holy Land*, 12.

32. Quoted in Adam Rome, *The Bulldozer in the Countryside*, 122.

33. *Military Engineer* 36 (May 1944).

34. Adam Rome, *The Bulldozer in the Countryside*, 121.

35. See Ted Steinberg, *Acts of God: The Unnatural History of Natural Disaster in America* (New York: Oxford University Press, 2003).

36. Adam Rome, *The Bulldozer in the Countryside*, 42.

37. Virginia Scott Jenkins, *The Lawn: A History of an American Obsession* (Washington, D.C.: Smithsonian Institution Press, 1994), 99; see also Ted Steinberg, *American Green: The Obsessive Quest for the Perfect Lawn* (New York: Norton, 2006).

38. Adam Rome, *The Bulldozer in the Countryside*, 88–89, 92.

39. See Ken Butti and John Perlin, *A Golden Thread: 2500 Years of Solar Architecture and Technology* (Palo Alto: Cheshire Books, 1980), *passim*.

40. See Gail Cooper, *Air-Conditioning America*, *passim*.

41. Adam Rome, *The Bulldozer in the Countryside*, 69, 73.

42. Quoted in Adam Rome, *The Bulldozer in the Countryside*, 73.

43. Adam Rome, *The Bulldozer in the Countryside*, 73–75.

44. Adam Rome, *The Bulldozer in the Countryside*, 71.

45. Gail Cooper, *Air-Conditioning America*, 152.

46. Adam Rome, *The Bulldozer in the Countryside*, 59, 60–61.

47. See Clifford E. Clark, Jr., "Ranch-House Suburbia: Ideals and Realities," in *Recasting America: Culture and Politics in the Age of the Cold War*, ed. Lary May (Chicago: University of Chicago Press, 1989), 171–191.

48. Patricia Leigh Brown, "In a Clash of Decades, A House Surrenders," *New York Times*, November 14, 1996.

49. Dolores Hayden, *Building Suburbia*, 135–136, 140.

50. Adam Bryant, "Edward J. DeBartolo, Developer, 85, Is Dead," *New York Times*, December 20, 1994.

51. Thomas W. Hanchett, "U.S. Tax Policy and the Shopping-Center Boom of the 1950s and 1960s," *American Historical Review* 101 (October 1996): 1083.

52. An excellent analysis of shopping centers is found in Lizabeth Cohen, "From Town Center to Shopping Center: The Reconfiguration of Community Marketplaces in Postwar America," in *His and Hers: Gender, Consumption, and Technology*, ed. Roger Horowitz and Arwen Mohun (Charlottesville: University Press of Virginia, 1998), 189–234.

53. D. J. Waldie, *Holy Land*, 12.

54. Lizabeth Cohen, "From Town Center to Shopping Center," 194.

55. Dolores Hayden, *Building Suburbia*, 140.

56. Lizabeth Cohen, *A Consumers' Republic: The Politics of Mass Consumption in Postwar America* (New York: Alfred A. Knopf, 2003), 198.

57. Lizabeth Cohen, "From Town Center to Shopping Center," 195, 197.

58. Lizabeth Cohen, "From Town Center to Shopping Center," 197; Margaret Crawford, "The World in a Shopping Mall," in *Variations on A Theme Park: The New American City and the End of Public Space*, ed. Michael Sorkin (New York: Noonday Press, 1992), 3–4.

59. Margaret Crawford, "The World in a Shopping Mall," 21, 7, 14, 30.

60. Lizabeth Cohen, *A Consumers' Republic*, 7–8.

3. Foreign Aid and Advantage

1. The American Century," 1941, quoted in Herbert I. Schiller, *Mass Communications and American Empire* (Boston: Beacon Press, 1971), 1.
2. Michael L. Smith, "'Planetary Engineering': The Strange Career of Progress in Nuclear America," in *Possible Dreams: Enthusiasm for Technology in America*, ed. John L. Wright (Dearborn, MI: Henry Ford Museum and Greenfield Village, 1992), 111-123.
3. Curt Cardwell, "Is Economics Behind the New Bush Doctrine?" History News Network, October 7, 2002, available at http://hnn.us/articles/1003.html (accessed September 4, 2006).
4. Charles S. Maier, "Introduction," in *The Marshall Plan and Germany: West German Development within the Framework of the European Recovery Program*, ed. Charles S. Maier (New York: Berg, 1991), 7, 12.
5. Charles S. Maier, "Introduction," 18.
6. Rudi Volti, *Technology Transfer and East Asian Economic Transformation* (Washington, D.C.: American Historical Association, 2002), 9–10.
7. Leonard H. Lynn, *How Japan Innovates: A Comparison with the U.S. in the Case of Oxygen Steelmaking* (Boulder: Westview Press, 1982), 43–44.
8. Leonard H. Lynn, *How Japan Innovates*, 45–47.
9. Leonard H. Lynn, *How Japan Innovates*, 48–50.
10. Rudi Volti, *Technology Transfer*, 15–16.
11. See James P. Womack, Daniel T. Jones, and Daniel Roos, *The Machine That Changed the World* (New York: Rawson Associates, 1990); quotation on 49.
12. Andrew Gordon, "Contests for the Workplace," in *Postwar Japan as History*, ed. Andrew Gordon (Berkeley: University of California Press, 1993), 373–374.
13. Andrew Gordon, "Contests for the Workplace." 376.
14. William M. Tsutsui, *Manufacturing Ideology: Scientific Management in Twentieth Century Japan* (Princeton: Princeton University Press, 1998), 191.
15. William M. Tsutsui, *Manufacturing Ideology*, 192, 194–195, 196.
16. Laura E. Hein, "Growth Versus Success: Japan's Economic Policy in Historical Perspective," in *Postwar Japan as History*, ed. Andrew Gordon (Berkeley: University of California Press, 1993), 109.
17. William M. Tsutsui, *Manufacturing Ideology*, 198, 201–202.
18. William M. Tsutsui, *Manufacturing Ideology*, 230; also see "Lowell Mellen, Helped Revive Postwar Japan," *Plain Dealer* (Cleveland); Andrew Gordon, "Contests for the Workplace," 387.
19. William M. Tsutsui, *Manufacturing Ideology*, 205.
20. This discussion is taken from Carroll Pursell, "Appropriate Technology, Modernity and U.S. Foreign Aid," *Science and Cultural Diversity: Proceedings of the XXIst International Congress of the History of Science*, I (Mexico City: Mexican Society for the History of Science and Technology and the National Autonomous University of Mexico, 2003), 175–187.
21. W.W. Rostow, *The Stages of Economic Growth: A Non-Communist Manifesto* (New York: Cambridge University Press, 1960), 4.

22. W.W. Rostow, *The Stages of Economic Growth*, 8.
23. C.P. Snow, *The Two Cultures and the Scientific Revolution* (New York: Cambridge University Press, 1961 [1959]), 48.
24. Quoted in Robert Gillette, "Latin America: Is Imported Technology Too Expensive?" *Science* 181, July 6, 1970, 41.
25. House Document no. 240, 81st Congress, 1st sess.1949 quoted in Henry Steele Commanger, ed., *Documents of American History*, 7th ed. (New York: Appleton-Century-Crofts, 1963), 558.
26. Henry Steele Commanger, ed., *Documents of American History*, 559.
27. *Technical Information for Congress*. Report to the Subcommittee on Science, Research and Development of the Committee on Science and Astronautics, House of Representatives. 92nd Congress, 1 sess. (April 15, 1971), 62–63.
28. Quoted in *Technical Information for Congress*, 64.
29. Quoted in *Technical Information for Congress*, 66.
30. Quoted in *Technical Information for Congress*, 68.
31. *Technical Information for Congress*, 69.
32. Copy of Act for International Development (Public Law 535), appended to *Partners in Progress: A Report to President Truman by the International Development Advisory Board* (New York: Simon and Schuster, 1951), 94.
33. *Partners in Progress*, 91–92.
34. *Partners in Progress*, ii.
35. *Partners in Progress*, ii.
36. *Technical Information for Congress*, 71.
37. *Technical Information for Congress*, p. 70.
38. Edwin A. Bock, *Fifty Years of Technical Assistance: Some Administrative Experiences of U.S. Voluntary Agencies* (Chicago: Public Administration Clearing House, 1954), 6.
39. Edwin A. Bock, *Fifty Years of Technical Assistance*, 3.
40. Edwin A. Bock, *Fifty Years of Technical Assistance*, 4.
41. Jonathan B. Bingham, *Shirt-Sleeve Diplomacy: Point 4 in Action* (New York: John Day, 1954), 77, 78.
42. See William Easterly, *The White Man's Burden: Why the West's Efforts to Aid the Rest Have Done So Much Ill and So Little Good* (New York: Penguin, 2006).
43. U.S. President's Materials Policy Commission, *Resources for Freedom, Vol. I, Foundations for Growth and Security* (Washington, D.C.: Government Printing Office, 1952), iii, iv.
44. Alan M. Bateman, "Our Future Dependence on Foreign Minerals," *The Annals of The American Academy of Political and Social Science* 281 (May 1952): 30, 25.
45. Glenn H. Snyder, *Stockpiling Strategic Materials: Politics and National Defense* (San Francisco: Chandler, 1966), 5, 267, 270.
46. U.S. President's Materials Policy Commission, *Resources for Freedom*, 60, 73–74.
47. Richard B. Morris, ed., *Encyclopedia of American History* (New York: Harper and Brothers, 1953), 390.

48. Reported in "International Traffic in Arms Hits $48 Billion," *Los Angeles Times*, October 23, 1973.

49. Reported in Tom Lambert, "U.S. Arms Sales Seen as Blow to Aims Abroad," *Los Angeles Times*, January 30, 1967.

50. Fred Kaplan, "Still the Merchants of Death," *The Progressive* 40 (March 1976): 22.

51. Quoted in Michael L. Smith, "'Planetary Engineering,'" 111.

52. Quoted in Michael L. Smith, "'Planetary Engineering,'" 112.

53. Michael L. Smith, "'Planetary Engineering,'" 119.

54. Quoted in Michael L. Smith, "'Planetary Engineering,'" 120.

55. House Committee on International Relations, *Science, Technology, and American Diplomacy: An Extended Study of the Interactions of Science and Technology with United States Foreign Policy* (Washington, D.C.: Government Printing Office, 1977), note on 365.

56. House Committee on International Relations, *Science, Technology, and American Diplomacy*, 365.

57. House Committee on International Relations, *Science, Technology, and American Diplomacy*, 366, 368.

58. House Committee on International Relations, *Science, Technology, and American Diplomacy*, 367, 410, 412, 370.

59. "Washington Group Names Regional Executive for Europe, Africa, Middle East," available at http://www.wgint.com (accessed October 16, 2006).

60. The best record of Brown and Root can be found in Dan Briody, *The Halliburton Agenda: The Politics of Oil and Money* (Hoboken: John Wiley, 2004).

61. Laton McCartney, *Friends in High Places: The Bechtel Story: The Most Secret Corporation and How It Engineered the World* (New York: Simon and Schuster, 1988), 152.

62. Laton McCartney, *Friends in High Places*, 152–164.

63. William J. Lederer and Eugene Burdick, *The Ugly American* (New York: Norton 1958), especially chapters 14 and 17.

4. The Atom and the Rocket

1. Walt Disney, "Foreword," in *The Walt Disney Story of Our Friend the Atom*, ed. Heinz Haber (New York: Simon and Schuster, 1956), 11, 10.

2. M. Joshua Silverman, "Nuclear Technology," in *A Companion to American Technology*, ed. Carroll Pursell (Malden: Blackwell Publishing, 2005), 305.

3. This discussion draws heavily upon M. Joshua Silverman, "Nuclear Technology," 298–320.

4. M. Joshua Silverman, "Nuclear Technology," 313.

5. Taken from George T. Mazuzan, "'Very Risky Business': A Power Reactor for New York City," *Technology and Culture* 27 (April 1986): 262–284.

6. Quoted in George T. Mazuzan, "'Very Risky Business,'" 276.

7. Michael L. Smith, "Advertising the Atom," in *Government and Environmental Politics: Essays on Historical Developments since World War Two*, ed. Michael J. Lacey

(Baltimore: The Woodrow Wilson Center Press and Johns Hopkins University Press, 1991), 258–259; Aeschylus, *Prometheus Bound and Other Plays* (New York: Penguin, 1961), 28.

8. Michael L. Smith, "'Planetary Engineering,'" 119.

9. Quoted in Michael L. Smith, "'Planetary Engineering,'" 114.

10. Tom Zoellner, "The Uranium Rush," *Scientific American Invention and Technology* 16 (Summer 2000): 56–63.

11. Tom Zoellner, "The Uranium Rush."

12. Quoted in Paul Boyer, *By the Bomb's Early Light: American Thought and Culture at the Dawn of the Atomic Age* (New York: Pantheon, 1985), 332.

13. Michael L. Smith, "'Planetary Engineering,'" 115.

14. JoAnne Brown, "'A Is for Atom, B Is for Bomb': Civil Defense in American Public Education, 1948–1963," *Journal of American History* 75 (June 1988): 69–71.

15. JoAnne Brown, "'A Is for Atom, B Is for Bomb,'" 90.

16. Michael L. Smith, "Advertising the Atom," pp. 234, 235.

17. Michael L. Smith, "Advertising the Atom," pp. 236, 237.

18. Quoted in Michael L. Smith, "Advertising the Atom," 247.

19. Michael L. Smith, "Advertising the Atom," 247.

20. This story is told in Ken Silverstein, *The Radioactive Boy Scout: The True Story of a Boy and His Backyard Nuclear Reactor* (New York: Random House, 2004), quote on 131.

21. Barton C. Hacker, "Robert H. Goddard and the Origins of Space Flight," in *Technology in America: A History of Individuals and Ideas*, 2d ed., ed. Carroll W. Pursell, Jr. (Cambridge: MIT Press 1990): 263.

22. Barton C. Hacker, "Robert H. Goddard and the Origins of Space Flight," 266.

23. Roger D. Launius, "Technology in Space," in *A Companion to American Technology*, ed. Carroll Pursell (Malden: Blackwell Publishing, 2005), 277.

24. Roger D. Launius, "Technology in Space," 284.

25. Quoted in Roger D. Launius, "Technology in Space," 285; William Hines, "The Wrong Stuff," *The Progressive* 58 (July 1994): 20.

26. Roger D. Launius, "Technology in Space," 292.

27. Michael L. Smith, "Selling the Moon: The U.S. Manned Space Program and the Triumph of Commodity Scientism," in *The Culture of Consumption: Critical Essays in American History, 1880–1980*, ed. Richard Wightman Fox and T.J. Jackson Lears (New York: Pantheon, 1983), 201.

28. Carroll W. Pursell, Jr., ed., *The Military-Industrial Complex* (New York: Harper & Row, 1972), 318.

29. Michael L. Smith, "Selling the Moon," 178.

30. Michael L. Smith, "Selling the Moon," 179.

31. Michael L. Smith, "Selling the Moon," 204.

32. Michael L. Smith, "Selling the Moon," 200.

33. Roger D. Launius, "Technology in Space," 278–279.

34. Harry Collins and Trevor Pinch, *The Golem at Large: What You Should Know About Technology* (Cambridge: University of Cambridge Press, 1998), 15.

35. Harry Collins and Trevor Pinch, *The Golem at Large*, 7–29.
36. Roger D. Launius, "Technology in Space," 280.
37. Roger D. Launius, "Technology in Space," 290–291.
38. Walter A. McDougall, "Technocracy and Statecraft in the Space Age—Toward the History of a Saltation," *American Historical Review* 87(October 1982): 1011. See also Walter A. McDougall, . . . *The Heavens and the Earth: A Political History of the Space Age* (New York: Basic Books, 1985).
39. Walter A. McDougall, "Technocracy and Statecraft in the Space Age," 1023.
40. Walter A. McDougall, "Technocracy and Statecraft in the Space Age," 1028.
41. Walter A. McDougall, "Technocracy and Statecraft in the Space Age," 1028, 1029.
42. Walter A. McDougall, "Technocracy and Statecraft in the Space Age," 1034, 1035, 1032.
43. James W. Carey and John J. Quirk, "The Mythos of the Electronic Revolution," *American Scholar* 6 (Summer 1970): 396.

5. Factories and Farms

1. James P. Womack et al., *The Machine That Changed the World*, 46.
2. Quoted in James Bright, "The Development of Automation," in *Technology in Western Civilization*, vol. 2, ed. Melvin Kranzberg and Carroll Pursell (New York: Oxford University Press, 1967), 635.
3. Quoted in James Bright, "The Development of Automation," II, 635.
4. Quoted in James Bright, "The Development of Automation," II, 637.
5. Quoted in James Bright, "The Development of Automation," II, 636.
6. Quoted in James Bright, "The Development of Automation," II, 636.
7. George Terborgh, *The Automation Hysteria* (New York: Norton Library, 1966), vii.
8. George Terborgh, *The Automation Hysteria*, 87–88.
9. For a detailed exploration of this debate during the Great Depression and after, see Amy Sue Bix, *Inventing Ourselves Out of Jobs? America's Debate over Technological Unemployment, 1929–1981* (Baltimore: Johns Hopkins University Press, 2000).
10. Robert Lekachman, "The Automation Report," *Commentary* 41 (May 1966): 67.
11. Ben B. Seligman, "Automation & the State," *Commentary* 37 (June 1964): 54.
12. James Bright, "The Development of Automation," II, 638.
13. Reg Theriault, *Longshoring on the San Francisco Waterfront* (San Pedro: Singlejack Books, 1978), *passim*.
14. Kelly Mills, "Hi-Tech Wins over Wharfie Culture," *The Australian*, May 27, 2003.
15. Harry Bernstein, "Court Battle Shapes Up over Automated Butcher Machines," *Los Angeles Times*, February 10, 1973.
16. William D. Smith, "Bidding to Automate the Check-Out," *New York Times*, December 14, 1975; "Computer Checkout May Be On Way Out," *Santa Barbara News-Press*, November 27, 1978; *Introducing Vons Fastcheck System*, undated leaflet in possession of author.
17. "Mechanical Men," *The Literary Digest* 99 (December 8, 1928): 20–21.

18. John Cunniff, "Robots Quietly Replacing Men," *Santa Barbara News-Press*, February 17, 1971.

19. "This Is a Robot Speaking: I May Put You Out of a Job," *Los Angeles Times*, November 3, 1976; advertisement in the calendar section of the *Los Angeles Times*, December 3, 1972; "Factory Robot Stabs Worker To Death," *San Francisco Chronicle*, December 9, 1981.

20. The following description is taken from James Womack et al., *The Machine that Changed the World*, especially 48–68.

21. Deborah Fitzgerald, *Every Farm a Factory: The Industrial Ideal in American Agriculture* (New Haven: Yale University Press, 2003), 4.

22. U.S. Department of Commerce, *Statistical Abstracts of the United States, 1970* (Washington, D.C.: Government Printing Office, 1970), 581.

23. U.S. Department of Commerce, *Statistical Abstract of the United States: 2001* (Washington, D.C.: Government Printing Office, 2001), tables 796 and 800.

24. Dirk Johnson, "Population Decline in Rural America: A Product of Advances in Technology," *New York Times*, September 11, 1990.

25. The following is based on William Boyd, "Making Meat: Science, Technology, and American Poultry Production," *Technology and Culture* 42 (October 2001): 631–664.

26. William Boyd, "Making Meat: Science, Technology, and American Poultry Production," 634, 637, 638.

27. This description is taken from Anthony Browne, "Ten Weeks to Live," *The Observer* (London), March 10, 2002.

28. For the possible relation between these farms and the threatened epidemic of bird flu, see "So Who's Really to Blame for Bird Flu," *Guardian*, June 7, 2006.

29. Quoted in James E. Bylin, "The Innovators," *Wall Street Journal*, June 17, 1968.

30. Carroll Pursell, *The Machine in America: A Social History of American Technology* (Baltimore: Johns Hopkins University Press, 1995), 291.

31. Ian Austen, "The Flexible Farmer Lets the Robot Do the Milking," *New York Times*, January 23, 2003.

32. Editorial, "The Curse of Factory Farms," *New York Times*, August 30, 2002.

33. Mark Shapiro, "Sowing Disaster? How Genetically Engineered American Corn Has Altered the Global Landscape," *The Nation* 275 (October 28, 2002), 15. The following account is mainly taken from this source.

34. Mark Shapiro, "Sowing Disaster?" 12.

35. Mark Shapiro, "Sowing Disaster?" 14.

36. David Barboza, "Modified Foods Put Companies in a Quandary," *New York Times*, June 4, 2000.

37. See David Mas Masumoto, *Epitaph for a Peach: Four Seasons on My Family Farm* (New York: Harper San Francisco, 1995).

38. Maria Alicia Gaura, "Withering Competition: Once-Thriving Small Organic Farms in Peril," *San Francisco Chronicle*, February 18, 2003.

39. Rich Ganis, "'Organic' Betrays its Roots," *Plain Dealer* (Cleveland), November 11, 2002.

40. See, e.g., Julie Guthman, *Agrarian Dreams: The Paradox of Organic Farming in California* (Berkeley: University of California Press, 2004).
41. Taken from Rich Ganis, "'Organic' Betrays its Roots."
42. Rich Ganis, "'Organic' Betrays its Roots."
43. Maria Alicia Gaura, "Withering Competition."
44. Quoted in Donald B. Kraybill, *The Riddle of Amish Culture* (Baltimore: Johns Hopkins University Press, 1989), 235.
45. Donald B. Kraybill, *The Riddle of Amish Culture*, 185–186.

6. "It's Fun to Live in America"

1. Kiwanis International, *It's Fun to Live in America* (Chicago: Kiwanis International, 1947).
2. Joanne Jacobson, "Exploding Plastic Inevitable," *The Nation* 269 (December 27, 1999): 30. Review of Alison J. Clarke, *Tupperware: The Promise of Plastic in 1950s America* (Washington, D.C.: Smithsonian Institution Press, 1999).
3. Adam Rome, *The Bulldozer in the Countryside*, 42.
4. D.J. Waldie, *Holy Land*, 38.
5. For Tupperware, see Alison J. Clarke, *Tupperware*.
6. Quoted in Joanne Jacobson, "Exploding Plastic Inevitable," 30.
7. David W. Dunlap, "But Wait! You Mean There's More?" *New York Times*, November 11, 1999.
8. Taken from Frederic D. Schwarz, "The Epic of the TV Dinner," *American Heritage: Invention & Technology* 9 (Spring 1994): 55.
9. Frederic D. Schwarz, "The Epic of the TV Dinner," 55.
10. Quoted in Laura Shapiro, "In the Mix," *Gourmet* 62 (August 2002): 115.
11. Henry J. Czekalinski, "How I Won the Cold War," *Cleveland* 30 (August 2001): 14, 16.
12. Quoted in Karal Ann Marling, *As Seen on TV: The Visual Culture of Everyday Life in the 1950s* (Cambridge: Harvard University Press, 1994), 243.
13. Quoted in Karal Ann Marling, *As Seen on TV*, 271.
14. Karal Ann Marling, *As Seen on TV*, 243.
15. Karal Ann Marling, *As Seen on TV*, 280–281.
16. Steven M. Gelber, "Do-It-Yourself: Constructing, Repairing, and Maintaining Domestic Masculinity," *American Quarterly* 49 (March 1997): 68. See also Steven M. Gelber, *Hobbies: Leisure and the Culture of Work in America* (New York: Columbia University Press, 1999).
17. "Today's Handyman Is the Little Woman-In-The-Home," *House & Garden* 101 (January 1952): 138.
18. Steven M. Gelber, "Do-It-Yourself," 97–98.
19. Steven M. Gelber, "Do-It-Yourself," 81, 83, 90.
20. H. F. Moorhouse, "The 'Work' Ethic and 'Leisure' Activity: The Hot Rod in Post-War America," in *The Historical Meanings of Work*, ed. Patrick Joyce (Cambridge: Cambridge University Press, 1987), 244.

21. H. F. Moorhouse, "The 'Work' Ethic and 'Leisure' Activity," 246. For strikingly similar characteristics of drag race culture, see Robert C. Post, *High Performance: The Culture and Technology of Drag Racing, 1950–1990* (Baltimore: Johns Hopkins University Press, 1994).
22. Quoted in H. F. Moorhouse, "The 'Work' Ethic and 'Leisure' Activity," 246.
23. Quoted in H. F. Moorhouse, "The 'Work' Ethic and 'Leisure' Activity," 251.
24. H. F. Moorhouse, "The 'Work' Ethic and 'Leisure' Activity," 257.
25. James Sterngold, "Making the Jalopy an Ethnic Banner; How the Lowrider Evolved from Chicano Revolt to Art Form," *New York Times*, February 19, 2000.
26. Karal Ann Marling, *As Seen on TV*, 134.
27. Karal Ann Marling, *As Seen on TV*, 136.
28. Cited in Karal Ann Marling, *As Seen on TV*, 141. Material on Earl is taken from Michael Lamm, "The Earl of Detroit," *American Heritage Invention and Technology* 14 (Fall 1998): 11–21.
29. Michael Lamm, "The Earl of Detroit," 16.
30. Karal Ann Marling, *As Seen on TV*, 141; Michael Lamm, "The Earl of Detroit," 16, 21.
31. Much of what follows is taken from T. A. Heppenheimer, "The Rise of the Interstates," *American Heritage Invention & Technology* 7 (Fall 1991): 8–18. For a more extended treatment, see Mark Rose, *Interstate: Express Highway Politics, 1941–1956* (Lawrence: The Regents Press of Kansas, 1979).
32. "Unsafe at Any Speed? Not Anymore" (graphs), *New York Times*, May 17, 2000.
33. For these roadside enterprises, see John A. Jakle and Keith A. Sculle, *Fast Food: Roadside Restaurants in the Automobile Age* (Baltimore: Johns Hopkins University Press, 1999); John A. Jakle and Keith A. Sculle, *The Gas Station in America* (Baltimore: Johns Hopkins University Press, 1994); and John A. Jakle, Keith A. Sculle, and Jefferson S. Rogers, *The Motel in America* (Baltimore: Johns Hopkins University Press, 1996).
34. Christopher Finch, *Highways to Heaven: The Auto Biography of America* (New York: HarperCollins, 1992), 237–241.
35. See William H. Honan, "Wayne McAllister, Architect for a Car Culture, Dies at 92," *New York Times*, April 3, 2000.
36. T. A. Heppenheimer, "The Rise of the Interstates," 14.
37. Robert Reinhold, "Final Freeway Opens, Ending California Era," *New York Times*, October 14, 1993.
38. Donna R. Braden and Judith E. Endelman, *Americans on Vacation* (Dearborn: Henry Ford Museum and Greenfield Village, 1990), 9, 44.
39. David Lansing, "What's New at the Parks?" *Sunset* 190 (May 1993): 28.
40. Margaret J. King, "Disneyland and Walt Disney World: Traditional Values in Futuristic Form," *Journal of Popular Culture* 15 (Summer 1981): 116.
41. Margaret J. King, "Disneyland and Walt Disney World," 121.
42. Michael Sorkin, "See You in Disneyland," in *Variations on a Theme Park: The New American City and the End of Public Space*, ed. Michael Sorkin (New York: Noonday Press, 1992), 210, 216.

43. Tom Vanderbilt, "Mickey Goes To Town(s)," *The Nation* 261 (August 28/September 4, 1995), 198.

44. Quoted in Miller's obituary, *San Francisco Chronicle*, September 12, 2001. Ironically, he died just two weeks before a pair of jumbo jets slammed into the towers of the World Trade Center.

45. Michael Sorkin, "See You in Disneyland," 228.

46. See Wolfgang Saxon, "Joseph Fowler, 99, Builder of Warships and Disney's Parks," *New York Times*, December 14, 1993.

47. Quoted in Margaret J. King, "Disneyland and Walt Disney World," 121.

48. Michael Sorkin, "See You in Disneyland," 223.

49. Tom Vanderbilt, "Mickey Goes To Town(s)," 200.

50. Michael L. Smith, "Back to the Future: EPCOT, Camelot, and the History of Technology," in *New Perspectives on Technology and American Culture*, ed. Bruce Sinclair (Philadelphia: American Philosophical Society, 1986), 70–72. See also Michael L. Smith, "Making Time: Representations of Technology at the 1964 World's Fair," in *The Power of Culture: Critical Essays in American History*, ed. Richard Wightman Fox and T.J. Jackson Lears (Chicago: University of Chicago Press, 1993), 223–244.

51. Quoted in Margaret J. King, "Disneyland and Walt Disney World," 128.

52. Raymond M. Weinstein, "Disneyland and Coney Island: Reflections on the Evolution of the Modern Amusement Park," *Journal of Popular Culture* 26 (Summer 1992): 147–149.

53. Michael Sorkin, "See You in Disneyland," 208.

54. Raymond Williams, *Television: Technology and Cultural Form* (New York: Schocken, 1975), 26.

55. Stewart Wolpin, "The Race for Video," *American Heritage Invention & Technology* 10 (Fall 1994): 62.

56. See "Charles Douglass, 93, Inventor of Laugh Track for TV, Dies," *New York Times*, April 26, 2003.

57. Much of the following is taken from Ruth Schwartz Cowan, *A Social History of American Technology*, 318–325.

58. Carla Lane, "A Transforming Influence: Katherine Dexter McCormick '04," *MIT Spectrum* (Spring 1994), available at http://web.mit.edu/mccormick/www/history/kdm.html (accessed September 5, 2006).

59. Quoted in Mark Rose, "Mothers and Authors: *Johnson v. Calvert* and the New Children of Our Imaginations," in *The Visible Woman: Imaging Technologies, Gender, and Science*, ed. Paula A. Treichler, et al. (New York: New York University Press, 1998), 221.

7. Brain Drain and Technology Gap

1. David Lazarus, "U.S. Leads World in Export of Tech Products," *San Francisco Chronicle*, March 13, 2000.

2. Jean-Claude Derian, *America's Struggle for Leadership in Technology* (Cambridge, MA: MIT Press, 1990), vii.

3. "'Brain Trust' Warns Europe Business Of Technology Gap," *Santa Barbara News-Press*, June 1, 1966.

4. Philip H. Abelson, "European Discontent with the 'Technology Gap,'" *Science* 155, February 17, 1967, 783.

5. Bryce Nelson, "Horning Committee: Beginning of a Technological Marshall Plan?" *Science* 154, December 9, 1966, 1307.

6. Philip H. Abelson, "European Discontent with the 'Technology Gap,'" 783.

7. Robert J. Samuelson, "Technology Gap: French Best Seller Urges Europe To Copy U.S. Methods," *Science* 159, March 8, 1968, 1086.

8. Arthur Schlesinger, Jr., "Foreword," in J.-J. Servan-Schreiber, *The American Challenge* (New York: Atheneum, 1968), vii.

9. Arthur Schlesinger, Jr., "Foreword," viii.

10. Arthur Schlesinger, Jr., "Foreword," xii, xi.

11. J.-J. Servan-Schreiber, *The American Challenge* (New York: Atheneum, 1968), xiii, 136.

12. J.-J. Servan-Schreiber, *The American Challenge*, 136–137, 138, 140, 142.

13. J.-J. Servan-Schreiber, *The American Challenge*, 251, 253. Emphasis in original.

14. J.-J. Servan-Schreiber, *The American Challenge*, 257.

15. J.-J. Servan-Schreiber, *The American Challenge*, 275.

16. Joseph Ben-David, *Fundamental Research and the Universities: Some Comments on International Differences* (Paris: OECD, 1968).

17. Dael Wolfle, "Universities and the Technology Gap," *Science* 160, April 26, 1968, 381.

18. Robert Gilpin, "European Disunion and the Technology Gap," *The Public Interest* 10 (Winter 1968): 43–44.

19. The Atlantic Institute, "The Technology Gap," in *The Technology Gap: U.S. and Europe*, ed. Richard H. Kaufman (New York: Praeger Publishers, 1970), 15. See note 22

20. Jean-Pierre Poullier, "The Myth and Challenge of the Technological Gap," in Richard H. Kaufman, ed., *The Technology Gap: U.S. and Europe* (New York: Praeger Publishers, 1970), 128, 126.

21. The Atlantic Institute, "The Technology Gap," 20. The figures came from Edward F. Denison, assisted by Jean-Pierre Poullier, *Why Growth Rates Differ: Postwar Experience in Nine Western Countries* (Washington: The Brookings Institution, 1967).

22. Antonie T. Knoppers, "The Causes of Atlantic Technological Disparities," in *The Technology Gap: U.S. and Europe*, ed. Richard H. Kaufman (New York: Praeger Publishers, 1970), 133, 135, 140–141.

23. D. S. Greenberg, "Son of Technology Gap: European Group Setting Up an Institute," *Science* 167, February 6, 1970, 850–852.

24. Geoffrey Goodman, "Harold Wilson: Leading Labour Beyond Pipe Dreams," *The Guardian*, May 25, 1995; David Edgerton, "The 'White Heat' Revisited: The British Government and Technology in the 1960s," *Twentieth Century British History* 7 (1996): 53–82.

25. Basil Caplan, "Knowhow: An Adverse Balance for Britain," *New Scientist and Science Journal* 50 (May 6, 1971): 326–327.

26. The Atlantic Institute, "The Technology Gap," 17.

27. Quoted in Alfred D. Cook, "Market Growth Put at 20% a Year," *New York Times*, March 14, 1971.

28. Philip M. Boffey, "Technology and World Trade: Is There Cause for Alarm?" *Science* 172, April 2, 1971, 37.

29. Quoted in Philip M. Boffey, "Technology and World Trade," 37.

30. Quoted in Philip M. Boffey, "Technology and World Trade," 38.

31. Quoted in Philip M. Boffey, "Technology and World Trade," 38.

32. Quoted in Philip M. Boffey, "Technology and World Trade," 40.

33. "Proxmire Hits AF Technical Spying Setup," *Los Angeles Times*, March 15, 1974.

34. Robert Gilpin, *Technology, Economic Growth, and International Competitiveness* (Washington: Government Printing Office, 1975), 1, 3. Report prepared for the use of the Subcommittee on Economic Growth of the Joint Economic Committee, U.S. Congress, July 9, 1975.

35. Robert Gilpin, *Technology, Economic Growth, and International Competitiveness*, 72–74.

36. The list is that of Sherman Gee, "Foreign Technology and the United States Economy," *Science* 187, February 21, 1975, 622.

37. Sherman Gee, "Foreign Technology and the United States Economy," 623, 626.

38. Gaylord Nelson, "Takeovers by Foreign Firms Drain Off U.S. Technology," *Los Angeles Times*, March 1, 1978.

39. Daddario quoted in John Walsh, "Issue of Technology Transfer Is Snag for 1979 U.N. Meeting," *Science* 198, October 7, 1977, 37.

40. Bharati Mukherjee, *Desirable Daughters* (New York: Hyperion, 2002), 23–24.

41. Scott Hocker, "A Heat Wave Hits the South Bay," *San Francisco* 50 (July 2003), 106.

42. Daniel Lloyd Spencer, *Technology Gap in Perspective: Strategy of International Technology Transfer* (New York: Spartan Books, 1970), 13.

43. Committee on Government Operations, *Brain Drain into the United States of Scientists, Engineers, and Physicians* (Washington: Government Printing Office, 1967), 1. Staff Study for the Research and Technical Programs Subcommittee, 90th Congress, 1st sess. (July 1967).

44. See, for example, Committee on Government Operations, *The Brain Drain into the United States*, 11.

45. Committee on Government Operations, *The Brain Drain into the United States*, 2, 4.

46. Committee on Government Operations, *The Brain Drain into the United States*, 1, 3, 4.

47. Committee on Government Operations, *The Brain Drain into the United States*, 5.

48. Committee on Government Operations, *The Brain Drain into the United States*, 5, 7; House of Representatives Committee on International Relations, *Science, Technology, and American Diplomacy* (Washington: Government Printing Office, 1977), vol. 2, 1316.

49. Committee on Government Operations, *The Brain Drain into the United States*, 7.

50. Committee on Government Operations, *The Brain Drain into the United States*, 8–9, 10.

51. Committee on Government Operations, *The Brain Drain into the United States*, 11–12.

52. D. N. Chorafas, *The Knowledge Revolution: An Analysis of the International Brain Market* (New York: McGraw-Hill, 1970), 13.

53. D. N. Chorafas, *The Knowledge Revolution*, 14, 15.

54. D. N. Chorafas, *The Knowledge Revolution*, 20.

55. Committee on Government Operations, *The Brain Drain into the United States*, 13–15.

56. Burton Bollag, "Europeans Propose Measures to Stem 'Brain Drain' to U.S.," *The Chronicle of Higher Education*, July 25, 2003.

57. Burton Bollag, "Europeans Propose Measures to Stem 'Brain Drain' to U.S."

8. From Technology Drunk...

1. Eric Hoffer, *The Temper of Our Times* (New York: Perennial Library, 1969), 42–43, 93–94, 114.

2. "As Our End Drifts Nearer... Editor's Afterward," *New Boston Review* (December 1981), xviii. Quoted in Paul Boyer, *By the Bomb's Early Light: American Thought and Culture at the Dawn of the Atomic Age*.

3. Peter Biskind, *Seeing Is Believing: How Hollywood Taught Us to Stop Worrying and Love the Fifties* (New York: Pantheon, 1983), 145, 157,158.

4. Paul Boyer, *Fallout: A Historian Reflects on America's Half-Century Encounter with Nuclear Weapons* (Columbus: Ohio State University Press, 1998), 82–83, 84.

5. Quoted in Carroll W. Pursell, Jr., ed., *The Military-Industrial Complex*, 206.

6. Herbert F. York, *Race to Oblivion: A Participant's View of the Arms Race* (New York: Simon & Schuster, 1970), 12.

7. Herbert F. York, *Race to Oblivion*, 230–231.

8. See Paul Dickson, *The Electronic Battlefield* (Bloomington: Indiana University Press, 1976), 195–196.

9. See James Whorton, *Before Silent Spring: Pesticides & Public Health in Pre-DDT America* (Princeton: Princeton University Press, 1974).

10. See Edmund P. Russell, III, "'Speaking of Annihilation': Mobilizing for War against Human and Insect Enemies." *The Journal of American History* 82 (March 1996): 1505–1529.

11. Rachel Carson, *Silent Spring* (Boston: Houghton Mifflin, 1962), 3, 6.

12. Quoted in Frank Graham, Jr., *Since Silent Spring* (Boston: Houghton Mifflin, 1970), 49, 55–56, 50.

13. Adam Rome, "'Give Earth a Chance': The Environmental Movement and the Sixties," *The Journal of American History* 90 (September 2003): 527.

14. Quoted in Adam Rome, "'Give Earth a Chance,'" 528, 529.

15. Ruth Schwartz Cowan, "From Virginia Dare to Virginia Slims: Women and Technology in American Life," *Technology and Culture* 20 (January 1979): 61, 62.

16. A chronology of these events is presented in Michael V. Miller and Susan Gilmore, eds., *Revolution at Berkeley: The Crisis in American Education* (New York: The Dial

Press, 1965), xxiv–xxix. A brief explanation by Savio, entitled "An End to History," appears on pp. 239–243.

17. Adam Rome, "'Give Earth a Chance,'" 542, 544, 546, 547.

18. *Technical Information for Congress*, Report to the Subcommittee on Science, Research, and Development of the Committee on Science and Astronautics, House of Representative, 92nd Congress, 1st sess. (Rev. ed. April 15, 1971), 690–692.

19. Quoted in *Technical Information for Congress*, 696.

20. Robert J. Samuelson, "The SST and the Government: Critics Shout into a Vacuum," *Science* 157, September 8, 1967, 1146–1151.

21. "Volpe Calls SST Essential to U.S.," *Washington Post*, April 1, 1970.

22. George Lardner, Jr., "Airports May Ban Noisy SST, Nixon Environment Aide Says," *Washington Post*, May 13, 1970.

23. Christopher Lydon, "SST: Arguments Louder Than Sonic Boom," *New York Times*, May 31, 1970.

24. Luther J. Carter, "SST: Commercial Race or Technology Experiment?" *Science* 169, July 24, 1970, 352–355.

25. "New Battle Expected by Opponents of SST," *Santa Barbara News-Press*, December 3, 1972; Marvin Miles, "Hopes for SST Are Dim But R&D Continues—Just in Case," *Los Angeles Times*, November 25, 1973; R. Jeffrey Smith, "SST Supporters Fly Above the Economic Fray," *Science* 208, April 11, 1980, 157.

26. Robert J. Samuelson, "The SST and the Government: Critics Shout into a Vacuum," 1147; "Concorde Compromise," *The Morning Call* (Allentown), January 27, 1976.

27. R. Jeffrey Smith, "SST Supporters Fly Above the Economic Fray," 157. John Noble Wilford, "Is Infinite Speed Really an Imperative?" *New York Times*, February 8, 1976.

28. Deborah Fitzgerald, *Every Farm a Factory*.

29. Jim Hightower, "Hard Tomatoes, Hard Times: Failure of the Land Grant College Complex," *Society* 10 (November-December 1972): 11, 14.

30. Jim Hightower, "Hard Tomatoes," 14, 18, 22.

31. "Farm Workers in California May Stand Up," *New York Times*, April 27, 1975.

32. David Bird, "Experts Debate Farm Pollution," *New York Times*, January 25, 1970.

33. John A. Jones, "The Beef Factory: Computers, Not Cowboys, Care for Steers," *Los Angeles Times*, May 24, 1972.

34. David Pimental et al., "The Potential for Grass-Fed Livestock: Resources Constraints," *Science*, February 22, 1980, 843.

35. "Farm Workers Used as Pesticide Guinea Pigs," *Los Angeles Times*, February 11, 1971.

36. Jim Hightower, "Hard Tomatoes," 22.

37. "Intelligence Report" *Parade*, March 10, 1974.

38. Alice Shabecoff, "The Agribusiness Market Gardens," *The Nation*, January 31, 1976, 114–115.

39. Jim Hightower, "Fake Food Is the Future," *Progressive* 39 (September 1975): 26.

40. Jim Hightower, "Fake Food is the Future," 26; Daniel Zwerdling, "Death for Dinner," *New York Review of Books*, February 21, 1974, 23.

41. Daniel Zwerdling, "Death for Dinner," 22; Constance Holden, "Food and Nutrition: Is America Due for a National Policy?" *Science*, May 3, 1974, 549, 548.

42. *The Triple Revolution* (Santa Barbara: The Ad Hoc Committee on the Triple Revolution, n.d.), 3. The pamphlet contains a complete list of the members and gives their affiliations.

43. *The Triple Revolution*, 7, 9, 11.

44. *The Triple Revolution*, 13.

45. Ivan Illich, *Tools for Conviviality* (New York: Harper & Row, 1973), 10, 12, 21, 22.

46. Ivan Illich, *Tools for Conviviality*, 26, 42, 45.

47. Ivan D. Illich, *Energy and Equity* (New York: Perennial Library, 1974), 3, 4, 5.

48. Erich Fromm, *The Revolution of Hope: Toward A Humanized Technology* (New York, 1968), vii.

49. Erich Fromm, *Revolution of Hope*, 23, 33–34, 44.

50. Erich Fromm, *Revolution of Hope*, 33, 90, 129.

51. Paul Goodman, "Can Technology be Humane?" Reprinted from *New Reformation: Notes of a Neolithic Conservative*, in *Western Man and Environmental Ethics*, ed. Ian G. Barbour (Reading, 1973), 229.

52. Paul Goodman, "Can Technology be Humane?" 227, 228–229.

53. Paul Goodman, "Can Technology be Humane?" 231, 233, 234.

54. Paul Goodman, "Can Technology be Humane?" 240–241.

55. John McDermott, "Technology: The Opiate of the Intellectuals." Reprinted in *Technology and the Future*, 4th ed., ed. Albert H. Teich (New York: St. Martin's Press, 1986), from *New York Review of Books*, July 1969, 99.

56. John McDermott, "Technology," 99, 100, 103, 104, 117.

57. Elting E. Morison, *From Know-How to Nowhere: The Development of American Technology* (New York: New American Library, 1974), 14.

9. . . . To Technology Sober

1. *Technology and the American Economy*, Report of the National Commission on Technology, Automation, and Economic Progress, I (February, 1966) (Washington, D.C.: Government Printing Office, 1966), 109, 110.

2. Robert Lekachman, "The Automation Report," 71.

3. Quoted in Joan Didion, *Where I Was From* (New York: Alfred A. Knopf, 2003), 137.

4. Joan Didion, *Where I Was From*, 137.

5. Steven R. Rivkin, *Technology Unbound: Transferring Scientific and Engineering Resources from Defense to Civilian Purposes* (New York: Pergamon Press, 1968).

6. Steven R. Rivkin, *Technology Unbound*, xi; Johnson quoted on p. xii.

7. Public Law 89–182, 89th Congress, September 14, 1965.

8. J. Herbert Hollomon, "Science, Technology, and Economic Growth," *Physics Today* 16 (March 1963): 38, 42.

9. D. S. Greenberg, "LBJ Directive: He Says Spread the Research Money," *Science* 149, September 24, 1965, 1485.

10. D. S. Greenberg, "Civilian Technology: Opposition in Congress and Industry Leads to Major Realignment of Program," *Science* 143, February 14, 1964, 660–661.

11. Press release from CSRE, March 22, 1971. In possession of author.

12. See Edwin T. Layton, *Revolt of the Engineers: Social Responsibility and the American Engineering Profession* (Cleveland: The Press of Case Western Reserve University, 1971).

13. Francis J. Lavoie, "The Activist Engineer: Look Who's Getting Involved," *Machine Design*, November 2, 1972, 82–88.

14. See Carroll Pursell, "Conservationism, Environmentalism, and the Engineers: The Progressive Era and Recent Past," in *Environmental History: Critical Issues in Comparative Perspective*, ed. Kendall E. Bailes (Lanham: University Press of America, 1985), 176–192.

15. "A Word With the Executive Director About . . . The Civil Engineer and the Conservationist," *Civil Engineering* 40 (February 1970): 37.

16. Carroll Pursell, "Conservationism, Environmentalism, and the Engineers: The Progressive Era and Recent Past," 186.

17. Quoted in Carroll Pursell, "Belling the Cat: A Critique of Technology Assessment," *Lex et Sciencia* 10 (October-December 1974): 133.

18. National Academy of Sciences, *Technology: Processes of Assessment and Choice* (Washington, D.C., July 1969), 3, 11, 15, 78-79.

19. Quoted in Carroll Pursell, "Belling the Cat," 139.

20. "Keeping Tabs on Technology," *Los Angeles Times*, April 21, 1972.

21. *Technology Assessment*, I (1974),1.

22. Daniel S. Greenberg, "Kennedy Moving Ahead with Technology Office Plans," *Science & Government Report*, November 15, 1972, 7.

23. *Technology Assessment*, Hearings before the Subcommittee on Science, Research, and Development of the Committee on Science and Astronautics, 91st Congress, 1st sess. (1969), 488.

24. See, for example, Constance Holden, "Conservatives Troubled About Course of OTA," *Science* 203, February 23, 1979, 729.

25. Colleen Cordes, "Congress Kills Technology Office After GOP Questions Its Effectiveness," *The Chronicle of Higher Education*, October 13, 1995, A33; quote from Warren E. Leary, "Congress's Science Agency Prepares to Close Its Doors," *New York Times*, September 24, 1995.

26. The following description is largely based on Carroll Pursell, "The Rise and Fall of the Appropriate Technology Movement in the United States, 1965–1985," *Technology and Culture* 34 (July 1993): 629–637.

27. Robert Gilman, "The Restoration of Waters," interview with John and Nancy Todd, *In Context* (Spring 1990): 42ff.

28. *The Farallones Institute* (Pt. Reyes: The Farallones Institute, n.d.), 4–7.

29. Sim Van der Ryn, "Appropriate Technology and State Government," memo dated August 1975, later issued as a pamphlet by OAT.

30. *Congressional Record*, May 27, 1976, 15876–15877.

31. The following is taken from Daniel P. Moynihan, "The War Against the Automobile," *The Public Interest* 3 (Spring 1966): 10–26.

32. See the special issue titled "The Consumer Revolt," *Time*, December 12, 1969.

33. A chronology of these events is given in "One Book That Shook the Business World," *New York Times Book Review*, January 18, 1970, 10.

34. Quoted in Elinor Langer, "Auto Safety: Nader vs. General Motors," *Science* 152, April 1, 1966, 48.

35. Quoted in "One Book That Shook the Business World," 10.

36. Editors, "Safety Hysteria," *Car Life*, (May 1966), 84.

37. Ralph Nader, *Who Speaks for the Consumer?* (n.p.: League for Industrial Democracy, 1968), 5.

38. David A. Jewell, "New Law Center to Aid Consumers," *Washington Post*, December 30, 1969.

39. Advertisement sponsored by the Magazine Publishers Association, "There Are Some People Around Who Want to Spoil Woman's Greatest Indoor Sport," *Good Housekeeping*, December, 1969.

40. "The Irate Consumer," *Newsweek*, January 26, 1970, 63.

41. "The Irate Consumer," 63.

42. Morton Mintz, "Wide Public Deception Laid To Agencies on Product Safety," *Washington Post*, June 3, 1970.

43. John D. Morris, "2 in House Deplore Bar Group's Consumer Stand," *New York Times*, April 19, 1970; Ronal Kessler, "'Class Action' Bill: Its Rise and Fall," *Washington Post*, April 27, 1970.

44. Peter Millones, "Harassment Laid to Consumerism," *New York Times*, February 18, 1970; William G. Cushing, "Class Action Suits Seen 'Harassment,'" *Washington Post*, March 25, 1970.

45. Kirkpatrick Sale, *Rebels Against the Future: The Luddites and Their War on the Industrial Revolution—Lessons for the Computer Age* (Reading: Addison-Wesley, 1995), 4.

46. David F. Noble, *Progress Without People: In Defense of Luddism* (Chicago: Charles H. Kerr, 1993).

47. Kirkpatrick Sale, "Lessons from the Luddites: Setting Limits On Technology," *The Nation*, June 5, 1995, 785.

48. Kirkpatrick Sale, "Lessons from the Luddites," 788.

49. Chellis Glendinning, "Notes toward a Neo-Luddite Manifesto," *Utne Reader*, March-April, 1990, 50.

50. Chellis Glendinning, "Notes toward a Neo-Luddite Manifesto," 51–53.

51. The piece was reprinted as *The Unabomber Manifesto: Industrial Society & Its Future* (Berkeley: Jolly Roger Press, 1995).

52. An excellent account of Kaczynski in given in Alston Chase, "Harvard and the Making of the Unabomber," *Atlantic Monthly*, June, 2000, 41–65.

53. John Schwartz, "Scholars Find Unabomber Ideas 'Unoriginal,'" *Plain Dealer* (Cleveland), August 16, 1995, reprinted from the *Washington Post*.

54. Kirkpatrick Sale, "Toward a Portrait Of the Unabomber," *New York Times*, August 6, 1995.

55. *The Unabomber Manifesto*, 3, 35, 47, 62.

56. Stephanie Mills, ed., *Turning Away From Technology: A New Vision for the 21st Century* (San Francisco: Sierra Club Books, 1997), vii, x.

57. Andrew Ross, "'Culture Wars' Spill Over: Science Backlash On Technosceptics," *The Nation*, October 2, 1995, 346.

58. Andrew Ross, "'Culture Wars' Spill Over," 350, 348.

10. A Wired Environment

1. This draws heavily upon Jeffrey R. Yost, "Computers and the Internet: Braiding Irony, Paradox, and Possibility," in *A Companion to American Technology*, ed. Carroll Pursell (Malden: Blackwell, 2005), 340–360.

2. "Mechanical Men," *Literary Digest* 99 (December 8, 1928): 20.

3. I. Bernard Cohen, "The Father of the Computer Age," *American Heritage Invention & Technology* 14 (Spring 1999): 59.

4. I. Bernard Cohen, "The Father of the Computer Age," 62–63.

5. Paul Ceruzzi, "An Unforseen Revolution: Computers and Expectations, 1935–1985," in *Imagining Tomorrow: History, Technology, and the American Future*, ed. Joseph J. Corn (Cambridge, MA: MIT Press, 1986), 191.

6. Jeffrey R. Yost, "Computers and the Internet," 344.

7. Quoted in Paul Ceruzzi, "An Unforseen Revolution," 190.

8. Paul Ceruzzi, "An Unforseen Revolution," 191, 195.

9. Paul Ceruzzi, "An Unforseen Revolution," 199.

10. Jeffrey R. Yost, "Computers and the Internet," 348.

11. Jeffrey R. Yost, "Computers and the Internet," 349–350. For an excellent memoir by a woman software engineer, see Ellen Ullman, *Close to the Machine* (San Francisco: City Lights Books, 1997).

12. This section draws heavily from James C. Williams, "Frederick E. Terman and the Rise of Silicon Valley," in *Technology in America: A History of Individuals and Ideas*, 2d ed., ed. Carroll W. Pursell, Jr. (Cambridge, MA: MIT Press, 1990), 276–291.

13. Quoted in James C. Williams, "Frederick E. Terman and the Rise of Silicon Valley," 281.

14. James C. Williams, "Frederick E. Terman and the Rise of Silicon Valley," 282–283.

15. Quoted in James C. Williams, "Frederick E. Terman and the Rise of Silicon Valley," 284–285.

16. James C. Williams, "Frederick E. Terman and the Rise of Silicon Valley," 287.

17. Roger W. Lotchin, *Fortress California, 1910-1961: From Warfare to Welfare* (New York: Oxford University Press, 1992), 347.

18. David Beers, *Blue Sky Dream: A Memoir of America's Fall from Grace* (New York: Doubleday, 1996), *passim*.

19. Robert Friedel, "Sic Transit Transistor," *American Heritage Invention & Technology* 2 (Summer 1986): 37. One member of the team, John Robinson Pierce, went on to involve Bell in a NASA project that in 1960 resulted in the launch of *Echo I*, the first

communication satellite. See Wolfgang Saxon, "John Robinson Pierce, 92, a Father of the Transistor," *New York Times*, April 5, 2002.

20. Robert Friedel, "Sic Transit Transistor," 38.

21. Robert Friedel, "Sic Transit Transistor," 39–40.

22. Paul Lewis, "Cecil H. Green, 102, Dies; Founder of Texas Instruments" *New York Times*, April 15, 2003.

23. James C. Williams, "Frederick E. Terman and the Rise of Silicon Valley," 288–289.

24. James C. Williams, "Frederick E. Terman and the Rise of Silicon Valley," 290.

25. Robert Friedel, "Sic Transit Transistor," 40.

26. James C. Williams, "Frederick E. Terman and the Rise of Silicon Valley," 290.

27. Jeffrey Yost, "Computers and the Internet," 23.

28. Mike May, "How the Computer Got Into Your Pocket," *American Heritage Invention & Technology* 15 (Spring 2000): 47.

29. Quoted in Mike May, "How the Computer Got Into Your Pocket," 48, 49.

30. Mike May, "How the Computer Got Into Your Pocket," 50, 53–54.

31. Frederic D. Schwarz, "The Patriarch of Pong," *American Heritage Invention & Technology* 6 (Fall 1990): 64.

32. Quoted in John Markoff, "A Long Time Ago, in a Lab Far Away...," *New York Times*, February 28, 2002.

33. See, e.g., Barbara Garson, *The Electronic Sweatshop: How Computers Are Transforming the Office of the Future into the Factory of the Past* (New York: Penguin, 1989). See also Evelyn Nakano Glenn and Charles M. Tolbert II, "Technology and Emerging Patterns of Stratification for Women of Color: Race and Gender Segregation in Computer Occupations," in *Women, Work, and Technology: Transformations*, ed. Barbara Drygulski Wright et al.(Ann Arbor: University of Michigan Press, 1987), 318–331.

34. David Gumpert, "The Computer Moves from the Corporation to Your Living Room," *New York Times*, February 4, 1977.

35. Lamont Wood, "The Man Who Invented the PC," *American Heritage Invention & Technology* 10 (Fall 1994): 64.

36. Jeffrey Yost, "Computers and the Internet," 24.

37. James C. Williams, "Frederick E. Terman and the Rise of Silicon Valley," 290.

38. John Markoff, "A Tale of the Tape From the Days When It Was Still Micro Soft," *New York Times*, September 18, 2000.

39. Quoted in John Markoff, "A Strange Brew's Buzz Lingers in Silicon Valley," *New York Times*, March 26, 2000.

40. Quoted in Bryan Pfaffenberger, "The Social Meaning of the Personal Computer: Or, Why the Personal Computer Revolution Was No Revolution," *Anthropological Quarterly* 61 (1988): 39.

41. James C. Williams, "Frederick E. Terman and the Rise of Silicon Valley," 290.

42. Quoted in John Markoff, "Adam Osborne, 64, Dies; Was Pioneer of Portable PC," *New York Times*, March 26, 2003.

43. Jamie Beckett, "The Inventor's Mouse Revolution Computer Gizmo Changed the World 30 Years Ago," *San Francisco Chronicle*, December 10, 1998.

44. Jamie Beckett, "The Inventor's Mouse." See also John Markoff, *What the Dormouse Said: How the 60s Counterculture Shaped the Personal Computer Industry* (New York: Viking, 2005).

45. Bryan Pfaffenberger, "The Social Meaning of the Personal Computer," 40, 45.

46. Thomas P. Hughes, *Rescuing Prometheus* (New York: Pantheon, 1998), 259.

47. Quoted in Thomas P. Hughes, *Rescuing Prometheus*, 265.

48. Edwin Diamond and Stephen Bates, "The Ancient History of the Internet," *American Heritage* 46 (October 1995): 40.

49. Thomas P. Hughes, *Rescuing Prometheus*, 288, 282.

50. Thomas P. Hughes, *Rescuing Prometheus*, 291–292.

51. Thomas P. Hughes, *Rescuing Prometheus*, 292; Edwin Diamond and Stephen Bates, "The Ancient History of the Internet," 42, 45.

52. Thomas P. Hughes, *Rescuing Prometheus*, 294; Edwin Diamond and Stephen Bates, "The Ancient History of the Internet," 42, 45.

53. On the "digital divide," see Donna L. Hoffman and Thomas P. Novak, "Bridging the Racial Divide on the Internet," *Science*, April 17, 1998, 390–391.

54. Don Oldenburg, "The Electronic Gender Gap," *Washington Post*, November 29, 1994; *New York Times*, February 27, 1997.

55. For an example, see V. David Sartin, "Young Teaching Old," *Plain Dealer* (Cleveland), November 17, 1995.

56. Obituary in *New York Times*, January 2, 2001.

57. Interview by Todd Wallack with Cooper in *San Francisco Chronicle*, April 3, 2003.

58. Lisa Guernsey, "The Perpetual Next Big Thing," *New York Times*, April 13, 2000.

59. Quoted in Lisa Guernsey, "Putting a Chip in Every Pot," *New York Times*, April 13, 2000.

60. Peter H. Lewis, "Smart Tools for the Home With Chips, Lasers, L.C.D.'s," *New York Times*, April 20, 2000.

61. "Japanese Smart Home Does All the Chores," *Sydney Morning Herald*, April 4, 2002.

11. Standing Tall Again

1. White House press release, June 20, 1979; Rex W. Scouten, White House curator, to author, July 20, 1992; "College Uses Panels Discarded by White House," *Chronicle of Higher Education*, July 1, 1992, A5.

2. Susan Jeffords, *The Remasculinization of America: Gender and the Vietnam War* (Bloomington: Indiana University Press, 1989), 1, 9.

3. Susan Jeffords, *The Remasculinization of America*, 168–169.

4. Carol Cohn, "Sex and Death in the Rational World of Defense Intellectuals," *Signs* 12, no. 4 (1987): 687–718.

5. Carol Cohn, "Sex and Death," 690, 697, 699.

6. Carol Cohn, "Sex and Death," 693.

7. Carol Cohn, "Sex and Death," 697, 704, 707.

8. Michal McMahon, "The Romance of Technological Progress: A Critical Review of the National Air and Space Museum," *Technology and Culture* 22 (April 1981): 285–286.

9. Herbert York, *Race to Oblivion*, 11.

10. Daniele Zele, "Missiles and Anti-Missiles," *The New Republic*, February 25, 1967, 16–18.

11. Frederic W. Collins, "$30 Billion for Whom? Politics, Profits and the Anti-Missile Missile," *The New Republic*, March 11, 1967, 13–15.

12. Nicholas Wade, "Safeguard: Disputed Weapon Nears Readiness on Plains of North Dakota," *Science*, September 27, 1974, 1140.

13. Herbert York, *Race to Oblivion*, 180.

14. Nicholas Wade, "Safeguard."

15. Quoted in Sanford Lakoff and Herbert F. York, *A Shield in Space? Technology, Politics, and the Strategic Defense Initiative: How the Reagan Administration Set Out to Make Nuclear Weapons "Impotent and Obsolete" and Succumbed to the Fallacy of the Last Move* (Berkeley: University of California Press, 1989), 7, 8.

16. Sanford Lakoff and Herbert F. York, *A Shield in Space?* 12.

17. Sanford Lakoff and Herbert F. York, *A Shield in Space?* 15.

18. Sanford Lakoff and Herbert F. York, *A Shield in Space?* xi-xv

19. Office of Technology Assessment, *SDI: Technology, Survivability, and Software* (Washington, D.C.: Government Printing Office, 1988), 74. Reproduced in Sanford Lakoff and Herbert F. York, *Shield in Space?* 93.

20. Aldric Saucier, "Lost in Space," *New York Times*, March 9, 1992.

21. Frances FitzGerald, *Way Out There in the Blue: Reagan, Star Wars and the End of the Cold War* (New York: Simon and Schuster, 2000), 17.

22. In *Impact of Strategic Defense Initiative [SDI] on the U.S. Industrial Base*. Hearing before the Subcommittee on Economic Stabilization of the Committee on Banking, Finance and Urban Affairs, House of Representatives, 99th Congress, 1st sess. (December 10, 1985), 71.

23. Sanford Lakoff and Herbert F. York, *A Shield in Space?* 34.

24. Frances FitzGerald, "The Poseurs of Missile Defense," *New York Times*, June 4, 2000.

25. Quoted in David Firestone, "Pentagon Seeking to Deploy Missiles Before Full Testing," *New York Times*, February 27, 2003. See also William D. Hartung and Michelle Ciarrocca, "Star Wars II: Here We Go Again," *The Nation*, June 19, 2000, 11–13, 15–16, 18, 20.

26. Frances FitzGerald, "The Poseurs of Missile Defense."

27. Edward T. Linenthal, "Warring Images Help to Fuel the Cultural Clash Over Missile Defense," *Chronicle of Higher Education*, August 4, 2000, B6.

28. This discussion draws upon Edward T. Linenthal, "Warring Images."

29. Edward T. Linenthal, "Warring Images."

30. Sanford Lakoff and Herbert F. York, *A Shield in Space?* 8.

31. Quoted in Edward T. Linenthal, "Warring Images."

32. Judy Klemesrud, "And the War Toys Are Rolling Along," *New York Times*, December 22, 1974.

33. Quoted in Judy Klemesrud, "And the War Toys Are Rolling."

34. Michael L. Smith, "'Planetary Engineering,'" 110–123.

35. Ian Mount, David H. Freedman, and Matthew Maier, "The New Military Industrial Complex," *Business 2.0*, March, 2003, available at www.business2.com/articles/mag/print/0,1643,47023.FF.html (accessed September 5, 2006).

36. Carroll Pursell, "The Rise and Fall of the Appropriate Technology Movement," 635, 636.

37. Witold Rybczynski, *Paper Heroes. Appropriate Technology: Panacea or Pipe Dream?* (New York: Penguin, 1991), 13

38. Susan Sward, "A Lost Opportunity That Worsened Crisis," *San Francisco Chronicle*, February 21, 2001.

39. Joseph Kahn, "Energy Efficiency Programs Are Set for Bush Budget Cut," *New York Times*, April 5, 2001.

40. Joseph Kahn, "U.S. Scientists See Big Power Savings from Conservation," *New York Times*, May 6, 2001.

41. Editorial, "Enlightenment on Energy," *New York Times*, October 22, 2001.

42. Katharine Q. Seelye, "White House Joins Fight Against Electric Cars," *New York Times*, October 10, 2002; Joseph B. Verrengia, "U.S. Opposes Renewable Energy Goals," *Plain Dealer* (Cleveland), August 28, 2002; Matthew L. Wald, "U.S. Use of Renewable Energy Took Big Fall in 2001," *New York Times*, December 8, 2002.

43. Andrew C. Revkin, "Scientists Say a Quest for Clean Energy Must Begin Now," *New York Times*, November 1, 2001.

44. Aircraft-Info.net, "Airbus A380," available at http://portal.aircraft-info.net/article8.html (accessed September 5, 2006); Michael Hennigan, "Launch of Airbus A380 May Doom the Boeing 747, An Icon of American Technology for a Generation," available at http://www.finfacts.com/cgi-bin/irelandbusinessnews/exec/view.cgi?archive=2&num=163 (accessed September 5, 2006).

45. Joe Sharkey, "Where Do You Park a 747 on Steroids?" *New York Times*, May 4, 2004; Michael Hennigan, "Launch of Airbus."

46. Stephan Lovgren, "Airbus Unveils A380 'Superjumbo' Jet," *National Geographic News*, January 18, 2005.

47. Joe Sharkey, "Where Do You Park?"; Stephan Lovgren, "Airbus Unveils A380 'Superjumbo' Jet."

48. The following description relies heavily on Keith Bradsher, *High and Mighty*.

49. Bradsher, *High and Mighty*, 20, quote on 22.

50. Stephanie Mencimer, "Bumper Mentality," *Washington Monthly*, December 20, 2002.

51. Quoted in Danny Hakim, "In Their Hummers, Right Besides Uncle Sam," *New York Times*, April 5, 2003.

52. Quoted in Danny Hakim, "In Their Hummers."

53. Quoted in Jennifer Evans-Cowley, "McMansions: Supersized Houses, Supersized Regulations," *Tierra Grande*, January, 2005, available at http://recenter.tamu.edu/tgrande/vol12-1/1713.html.

54. Jennifer Evans-Cowley, "McMansions."

12. Globalization, Modernity, and the Postmodern

1. Jean Baudrillard, *America* (London: Verso, 1988), 76.

2. Thomas P. Hughes, *American Genesis: A Century of Invention and Technological Enthusiasm, 1870–1970* (New York: Viking, 1989), 295.

3. H. Guy Stever and Janet H. Mutoyama, eds., *Globalization of Technology: International Perspectives* (Washington, D.C.: National Academy Press, 1988), 1, 4.

4. Mary Lowe Good, "Globalization of Technology Poses Challenges for Policymakers," *APS News,* July 1996 edition, available at http://www.aps.org/apsnews/0796/11568.cfm (accessed September 5, 2006).

5. Mary Lowe Good, "Globalization of Technology"; Angela Cuming, "Move for 000 Strip during US TV shows," *Sun-Herald* (Sydney), May 15, 2005.

6. David Lazarus, "Kaiser Exporting Privacy," *San Francisco Chronicle*, May 14, 2003.

7. Mary Vanac and Nicole Harris, "Sending Work to India," *Plain Dealer* (Cleveland), May 6, 2003; Amy Waldman, "More 'Can I Help You?' Jobs Migrate from U.S. to India," *New York Times*, May 11, 2003.

8. Arundhati Roy, *Power Politics* (Cambridge: South End Press, 2001), 25.

9. Arundhati Roy, *Power Politics*, 2, 13, 14.

10. Arundhati Roy, *Power Politics*, 47, 16, 26, 20.

11. Arundhati Roy, *Power Politics*, 63, 66, 62, 53.

12. Naomi Klein, "The Rise of Disaster Capitalism," *The Nation*, May 2, 2005, 11.

13. Quoted in Naomi Klein, "Disaster Capitalism," 9.

14. Naomi Klein, "Disaster Capitalism," 10.

15. Naomi Klein, "Disaster Capitalism," 10–11.

16. Manuel Castells, *The Rise of the Network Society,* 2d ed., (Oxford: Blackwell Publishers, 2000), 486.

17. Manuel Castells, *Network Society*, 490, 486.

18. Quoted in Carroll Pursell, *White Heat: People and Technology* (Berkeley: University of California Press, 1994), 167.

19. Jason Vest, "The Army's Empire Skeptics," *The Nation*, March 3, 2003, 27.

20. Quoted in Jason Vest, "Army's Empire Skeptics," 29.

21. Michael T. Klare, "Imperial Reach: The Pentagon's New Basing Strategy," *The Nation*, April 25, 2005, 13, 16.

22. Quoted in Michael T. Klare, "Imperial Reach," 17.

23. Quoted in Michael T. Klare, "Imperial Reach," 14.

24. Herbert I. Schiller, *Mass Communications and American Empire* (Boston: Beacon Paperback, 1971), 8–9, 12.

25. Herbert I. Schiller, *Mass Communications*, 57, 59, 51–52, 34–35, 44.

26. Hugh R. Slotten, "Satellite Communications, Globalization, and the Cold War," *Technology and Culture* 43 (April 2002): 317, 322, 323–324.

27. Hugh R. Slotten, "Satellite Communications," 328.

28. Quoted in Hugh R. Slotten, "Satellite Communications," 336, 329.

29. Quoted in Burton I. Edelson and Louis Pollack, "Satellite Communications," *Science*, March 18, 1977, 1125.

30. Burton I. Edelson and Louis Pollack, "Satellite Communications," 1125, 1127, 1128.

31. Quoted in Hugh R. Slotten, "Satellite Communications," 341.

32. Hugh R. Slotten, "Satellite Communications," 347.

33. Hugh R. Slotten, "Satellite Communications," 348, 349, 350.

34. "The Cost of Food Transportation," December 13, 2002, available at www.iptv.org; Danielle Murray, "Oil and Food: A Rising Security Challenge," May 9, 2005, www.earth-policy.org.

35. Mark Schapiro, "Sowing Disaster: How Genetically Engineered American Corn Has Altered the Global Landscape," *The Nation*, October 28, 2002, 12; Gyorgy Scrinis, "GM crops Will Not Help Feed the World," *Age* (Melbourne), July 8, 2003.

36. Mark Schapiro, "Sowing Disaster," 14, 15.

37. Quoted in Mark Schapiro, "Sowing Disaster," 18.

38. Tina Rosenberg, "Why Mexico's Small Corn Farmers Go Hungry," *New York Times*, March 3, 2003.

39. Marshall Berman, *All That Is Solid Melts Into Air: The Experience of Modernity* (New York: Penguin, 1988), 40, 68.

40. This discussion is based on Spencer Weart, *The Discovery of Global Warming* (Cambridge: Harvard University Press, 2004).

41. Quoted in "Europe Backs Kyoto Accord," BBC News, March 31, 2001.

42. Editorial, "It Pays to Be Green," *New Scientist*, June 4, 2005.

43. Thomas L. Friedman, "Losing Our Edge?" *New York Times*. April 22, 2004.

44. David Morley and Kevin Robins, "Techno-Orientalism: Futures, Foreigners and Phobias," *New Formations* 16 (Spring 1992): 136.

45. Robert Friedel, "Perspiration in Perspective: Changing Perceptions of Genius and Expertise in American Invention," in Robert J. Weber and David N. Perkins, eds., *Inventive Minds: Creativity in Technology* (New York: Oxford University Press, 1992), 9.

46. David Morley and Kevin Robins, "Techno-Orientalism," 147, 152.

47. David Morley and Kevin Robins, "Techno-Orientalism," 154.

48. Quoted in David Morley and Kevin Robins, "Techno-Orientalism," 141.

49. Armand Mattelart, *Networking the World, 1794–2000* (Minneapolis: University of Minnesota Press, 2000), 5, 98, 120.

50. David Harvey, *The Condition of Postmodernity: An Enquiry into the Origins of Cultural Change* (Oxford: Blackwell Publishers, 1991).

51. Mark Twain, *A Connecticut Yankee in King Arthur's Court* (New York: Signet, 1963 [1889]), 14–15, 318.

BIBLIOGRAPHY

Aeschylus. *Prometheus Bound and Other Plays*. New York: Penguin, 1961.
American Public Works Association. *History of Public Works in the United States, 1776–1976*. Chicago: American Public Works Association, 1976.
Atlantic Institute. *The Technology Gap: U.S. and Europe*. New York: Praeger Publishers, 1970.
Bateman, Alan M., "Our Future Dependence on Foreign Minerals," *The Annals of The American Academy of Political and Social Science* 281 (May 1952): 25–32.
Baudrillard, Jean. *America*. London: Verso, 1988.
Beers, David. *Blue Sky Dream: A Memoir of America's Fall from Grace*. New York: Doubleday, 1996.
Ben-David, Joseph. *Fundamental Research and the Universities: Some Comments on International Differences*. Paris: OECD, 1968.
Berman, Marshall. *All That Is Solid Melts Into Air: The Experience of Modernity*. New York: Penguin, 1988.
Bernstein, Barton J. "America in War and Peace: The Test of Liberalism." In *Towards A New Past*, ed. Barton J. Bernstein, 289–321. New York: Pantheon, 1968.
Bingham, Jonathan B. *Shirt-Sleeve Diplomacy: Point 4 in Action*. New York: J. Day, 1954.
Bird Dogs, "The Evolution of the Office of Naval Research," *Physics Today* 14 (August 1961): 30–35.
Biskind, Peter. *Seeing Is Believing: How Hollywood Taught Us to Stop Worrying and Love the Fifties*. New York: Pantheon, 1983.
Bix, Amy Sue. *Inventing Ourselves Out of Jobs? America's Debate over Technological Unemployment, 1929–1981*. Baltimore: Johns Hopkins University Press, 2000.
Bock, Edwin A. *Fifty Years of Technical Assistance: Some Administrative Experiences of U.S. Voluntary Agencies*. Chicago: Public Administration Clearing House, 1954.
Boyd, William, "Making Meat: Science, Technology, and American Poultry Production," *Technology and Culture* 42 (October 2001): 631–664.
Boyer, Paul. *By the Bomb's Early Light: American Thought and Culture at the Dawn of the Atomic Age*. New York: Pantheon, 1985.
Boyer, Paul. *Fallout: A Historian Reflects on America's Half-Century Encounter with Nuclear Weapons*. Columbus: Ohio State University Press, 1998.

Braden, Donna R. and Judith E. Endelman. *Americans on Vacation*. Dearborn: Henry Ford Museum and Greenfield Village, 1990.
Bradsher, Keith. *High and Mighty: SUVs—the World's Most Dangerous Vehicles and How They Got That Way*. New York: Public Affairs, 2002.
Bright, James. "The Development of Automation." In *Technology in Western Civilization*, vol. 2, ed. Melvin Kranzberg and Carroll Pursell, 635–655. New York: Oxford University Press, 1967.
Briody, Dan. *The Halliburton Agenda: The Politics of Oil and Money*. Hoboken: John Wiley, 2004.
Brown, JoAnne, "'A Is for Atom, B Is for Bomb': Civil Defense in American Public Education," *Journal of American History* 75 (June 1988): 68–90.
Bush, Vannevar. *Science the Endless Frontier*. Washington, D.C.: Government Printing Office, 1945.
Butti, Ken and John Perlin. *A Golden Thread: 2500 Years of Solar Architecture and Technology*. Palo Alto: Cheshire Books, 1980.
Carey, James W. and John J. Quirk. "The Mythos of the Electronic Revolution," *American Scholar* 39 (Spring 1970): 219–241 and (Summer 1970): 395–424.
Carson, Rachel. *Silent Spring*. Boston: Houghton Mifflin, 1962.
Castells, Manuel. *The Rise of the Network Society*, 2d ed. Oxford: Blackwell, 2000.
Catton, Bruce. *The War Lords of Washington*. New York: Harcourt, Brace, 1948.
Ceruzzi, Paul. "An Unforseen Revolution: Computers and Expectations, 1935–1985." In *Imagining Tomorrow: History, Technology, and the American Future*, ed. Joseph J. Corn, 188–201. Cambridge, MA: MIT Press, 1986.
Chorafas, D.N. *The Knowledge Revolution: An Analysis of the International Brain Market*. New York: McGraw-Hill, 1970.
Clark, Clifford E., Jr. "Ranch-House Suburbia: Ideals and Realities." In *Recasting America: Culture and Politics in the Age of the Cold War*, ed. Lary May, 171–191. Chicago: University of Chicago Press, 1989.
Clarke, Alison J. *Tupperware: The Promise of Plastic in 1950s America*. Washington, D.C.: Smithsonian Institution Press, 1999.
Cohen, I. Bernard, "The Father of the Computer Age," *American Heritage Invention & Technology* 14 (Spring 1999): 56–63.
Cohen, Lizabeth. "From Town Center to Shopping Center: The Reconfiguration of Community Marketplaces in Postwar America." In *His and Hers: Gender, Consumption and Technology*, ed. Roger Horowitz and Arwen Mohun, 189–234. Charlottesville: University Press of Virginia, 1998.
Cohen, Lizabeth. *A Consumers' Republic: The Politics of Mass Consumption in Postwar America*. New York: Knopf, 2003.
Cohn, Carol, "Sex and Death in the Rational World of Defense Intellectuals," *Signs* 12, no. 4 (1987): 687–718.
Collins, Harry, and Trevor Pinch. *The Golem at Large: What You Should Know About Technology*. Cambridge: Cambridge University Press, 1998.

Commanger, Henry Steele, ed. *Documents of American History*, 7th ed. New York: Appleton-Crofts, 1963.

Cooper, Gail. *Air-Conditioning America: Engineers and the Controlled Environment, 1900–1960*. Baltimore: Johns Hopkins University Press, 1998.

Cowan, Ruth Schwartz. "From Virginia Dare to Virginia Slims: Women and Technology in American Life," *Technology and Culture* 20 (January 1979): 51–63.

Cowan, Ruth Schwartz. *A Social History of American Technology*. New York: Oxford University Press, 1997.

Crawford, Margaret. "The World in a Shopping Mall." In *Variations on A Theme Park: The New American City and the End of Public Space*, ed. Michael Sorkin, 3–30. New York: Noonday Press, 1992.

Czekalinski, Henry J. "How I Won the Cold War," *Cleveland* 30 (August 2001): 14, 16.

Denison, Edward F., assisted by Jean-Pierre Poullier. *Why Growth Rates Differ: Postwar Experience in Nine Western Countries*. Washington: The Brookings Institution, 1967.

Derian, Jean-Claude. *America's Struggle for Leadership in Technology*. Cambridge, MA: MIT Press, 1990.

Devaney, F. John. *Tracking the American Dream: 50 Years of Housing History from the Census Bureau: 1940 to 1990*. Washington, D.C.: U.S. Bureau of the Census, 1994.

Diamond, Edwin and Stephen Bates. "The Ancient History of the Internet," *American Heritage* 46 (October 1995): 34–36, 38, 40, 42–45.

Dickson, Paul. *The Electronic Battlefield*. Bloomington: Indiana University Press, 1976.

Didion, Joan. *Where I Was From*. New York: Alfred A. Knopf, 2003.

Disney, Walt. "Foreword." In Heinz Haber. *The Walt Disney Story of Our Friend the Atom*. New York: Simon and Schuster, 1956.

Easterly, William. *The White Man's Burden: Why the West's Efforts to Aid the Rest Have Done So Much Ill and So Little Good*. New York: Penguin, 2006.

Edgerton, David, "The 'White Heat' Revisited: The British Government and Technology in the 1960s," *Twentieth Century British History* 7 (January 1996): 53–82.

The Farallones Institute. Pt. Reyes: The Farallones Institute, n.d.

Finch, Christopher. *Highways to Heaven: The Auto Biography of America*. New York: HarperCollins, 1992.

Fitzgerald, Deborah. *Every Farm A Factory: The Industrial Ideal in American Agriculture*. New Haven: Yale University Press, 2003.

FitzGerald, Frances. *Way Out There In the Blue: Reagan, Star Wars and the End of the Cold War*. New York: Simon and Schuster, 2000.

Flink, James J. *The Automobile Age*. Cambridge, MA: MIT Press, 1988.

Foster, Mark S. *Henry J. Kaiser: Builder in the Modern American West*. Austin: University of Texas Press, 1989.

Friedel, Robert. "Sic Transit Transistor," *American Heritage Invention & Technology* 2 (Summer 1986): 34–38, 40.

Friedel, Robert. "Perspiration in Perspective: Changing Perceptions of Genius and Expertise in American Invention." In *Inventive Minds: Creativity in Technology*, ed. Robert J. Weber and David N. Perkins. New York: Oxford University Press, 1992.

Erich Fromm, *The Revolution of Hope: Toward A Humanized Technology*. New York: Harper and Row, 1968.

Garson, Barbara. *The Electronic Sweatshop: How Computers Are Transforming the Office of the Future into the Factory of the Past*. New York: Penguin, 1989.

Gelber, Steven M. "Do-It-Yourself: Constructing, Repairing, and Maintaining Domestic Masculinity," *American Quarterly* 49 (March 1997): 66–112.

Gelber, Steven M. *Hobbies: Leisure and the Culture of Work in America*. New York: Columbia University Press, 1999.

Gilman, Robert. "The Restoration of Waters," interview with John and Nancy Todd, *In Context*, no. 25 (Spring 1990): 42–47.

Gilpin, Robert. "European Disunion and the Technology Gap," *The Public Interest* 10 (Winter 1968): 43–54.

Gilpin, Robert. *Technology, Economic Growth, and International Competitiveness*. Report prepared for the use of the Subcommittee on Economic Growth of the Joint Economic Committee, Congress of the United States. Washington: Government Printing Office, 1975.

Goodman, Paul. "Can Technology Be Humane?" In *Western Man and Environmental Ethics*, ed. Ian G. Barbour, 225–242. Reading, PA: Addison-Wesley, 1973.

Gordon, Andrew. "Contests for the Workplace." In *Postwar Japan as History*, ed. Andrew Gordon, 373–394. Berkeley: University of California Press, 1993.

Graham, Frank Jr. *Since Silent Spring*. Boston: Houghton Mifflin, 1970.

Guthman, Julie. *Agrarian Dreams: The Paradox of Organic Farming in California*. Berkeley: University of California Press, 2004.

Hacker, Barton C. "Robert H. Goddard and the Origins of Space Flight." In *Technology in America: A History of Individuals and Ideas*, 2d ed., ed. Carroll W. Pursell, Jr., 263–275. Cambridge, MA: MIT Press, 1990.

Hanchett, Thomas W. "U.S. Tax Policy and the Shopping-Center Boom of the 1950s and 1960s," *American Historical Review* 101 (October 1996): 1082–1110.

Hartmann, Susan M. *The Home Front and Beyond: American Women in the 1940s*. Boston: Twayne, 1982.

Hayden, Dolores. *Building Suburbia: Green Fields and Urban Growth, 1820–2000*. New York: Pantheon, 2003.

Hayden, Dolores. *Redesigning the American Dream: The Future of Housing, Work, and Family Life*. New York: Norton, 1984.

Heath, Jim F. "American War Mobilization and the Use of Small Manufacturers, 1939–1943," *Business History Review* 46 (Autumn 1972): 295–319.

Hein, Laura E. "Growth Versus Success: Japan's Economic Policy in Historical Perspective." In *Postwar Japan as History*, ed. Andrew Gordon, 99–122. Berkeley: University of California Press, 1993.

Heppenheimer, T.A. "The Rise of the Interstates," *American Heritage Invention & Technology* 7 (Fall 1991): 8–18.
Hightower, Jim. "Fake Food Is the Future," *The Progressive* 39 (September 1975): 26–29.
Hightower, Jim. "Hard Tomatoes, Hard Times: Failure of the Land Grant College Complex," *Society* 10 (November–December 1972): 11–14.
Hines, William. "The Wrong Stuff," *The Progressive* 58 (July 1994): 18–20.
Hoffer, Eric. *The Temper of Our Times*. New York: Perennial Library, 1969.
Hollomon, J. Herbert. "Science, Technology, and Economic Growth," *Physics Today* 16 (March 1963): 38–40, 42, 44, 46.
Hughes, Thomas P. *American Genesis: A Century of Invention and Technological Enthusiasm, 1870–1970*. New York: Viking, 1989.
Hughes, Thomas P. *Rescuing Prometheus*. New York: Pantheon, 1998.
Illich, Ivan. *Tools for Conviviality*. New York: Harper & Row, 1973.
Illich, Ivan D. *Energy and Equity*. New York: Perennial Library, 1974.
Jakle, John A. and Keith A. Sculle. *The Gas Station in America*. Baltimore: Johns Hopkins University Press, 1994.
Jakle, John A. and Keith A. Sculle. *Fast Food: Roadside Restaurants in the Automobile Age*. Baltimore: Johns Hopkins University Press, 1999.
Jakle, John A., Keith A. Sculle, and Jefferson S. Rogers. *The Motel in America*. Baltimore: Johns Hopkins University Press, 1996.
Jeffords, Susan. *The Remasculinization of America: Gender and the Vietnam War*. Bloomington: Indiana University Press, 1989.
Jenkins, Virginia Scott. *The Lawn: A History of an American Obsession*. Washington, D.C.: Smithsonian Institution Press, 1994.
Kaczynski, Ted. *The Unabomber Manifesto: Industrial Society & Its Future*. Berkeley: Jolly Roger Press, 1995.
Kaplan, Fred. "Still the Merchants of Death," *The Progressive* 40 (March 1976): 22–25.
King, Margaret J. "Disneyland and Walt Disney World: Traditional Values in Futuristic Form," *Journal of Popular Culture* 15 (Summer 1981): 116–139.
Kiwanis International. *It's Fun to Live in America*. n.p.: Kiwanis International, 1947.
Knoppers, Antonie T. "The Causes of Atlantic Technological Disparities," In *The Technology Gap: U.S. and Europe*, 133–147. New York: Praeger, 1970.
Kraybill, Donald B. *The Riddle of Amish Culture*. Baltimore: Johns Hopkins University Press, 1989.
Lakoff, Sanford and Herbert F. York. *A Shield in Space? Technology, Politics, and the Strategic Defense Initiative: How the Reagan Administration Set Out to Make Nuclear Weapons "Impotent and Obsolete" and Succumbed to the Fallacy of the Last Move*. Berkeley: University of California Press, 1989.
Lamm, Michael. "The Earl of Detroit," *American Heritage Invention & Technology* 14 (Fall 1998): 11–21.
Launius, Roger D. "Technology in Space." In *A Companion to American Technology*, ed. Carroll Pursell, 275–297. Malden: Blackwell, 2005.

Layton, Edwin T. *Revolt of the Engineers: Social Responsibility and the American Engineering Profession.* Cleveland: The Press of Case Western Reserve University, 1971.

Lederer, William J. and Eugene Burdick. *The Ugly American.* New York: Fawcett Crest, 1958.

Lekachman, Robert. "The Automation Report," *Commentary* 41 (May 1966): 65–71.

Lotchin, Roger W. *Fortress California , 1910-1961: From Warfare to Welfare.* New York: Oxford University Press, 1992.

Lynn, Leonard H. *How Japan Innovates: A Comparison with the U.S. in the Case of Oxygen Steelmaking.* Boulder: Westview Press, 1982.

Maier, Charles S., ed. *The Marshall Plan and Germany: West German Development within the Framework of the European Recovery Program.* New York: Berg, 1991.

Markoff, John. *What the Dormouse Said: How the 60s Counterculture Shaped the Personal Computer Industry.* New York: Viking, 2005.

Marling, Karal Ann. *As Seen on TV: The Visual Culture of Everyday Life In the 1950s.* Cambridge, MA: Harvard University Press, 1994.

Masumoto, David Mas. *Epitaph for a Peach: Four Seasons on My Family Farm.* San Francisco: Harper San Francisco, 1995.

Mattelart, Armand. *Networking the World, 1794–2000.* Minneapolis: University of Minnesota Press, 2000.

May, Elaine Tyler. *Homeward Bound: American Families in the Cold War Era.* New York: Basic Books, 1988.

May, Mike. "How the Computer Got Into Your Pocket," *American Heritage Invention & Tehcnology* 15 (Spring 2000): 47–51, 53–54.

Mazuzan, George T. "'Very Risky Business': A Power Reactor for New York City," *Technology and Culture* 27 (April 1986): 262–284.

McCartney, Layton. *Friends in High Places: The Bechtel Story: The Most Secret Corporation and How It Engineered the World.* New York: Simon and Schuster, 1988.

McDermott, John. "Technology: The Opiate of the Intellectuals." In Albert H. Teich, ed., *Technology and the Future,* 95–121. New York: St. Martin's Press, 1986.

McDougall, Walter A. "Technocracy and Statecraft in the Space Age—Toward the History of a Saltation," *American Historical Review* 87 (October 1982): 1010–1040.

McDougall, Walter A. . . . *The Heavens and the Earth: A Political History of the Space Age.* New York: Basic Books, 1985.

McMahon, Michal, "The Romance of Technological Progress: A Critical Review of the National Air and Space Museum," *Technology and Culture* 22 (April 1981): 281–296.

Mees, C.E., and John A. Leermakers. *The Organization of Industrial Scientific Research,* 2d ed. New York: McGraw-Hill, 1950.

Mills, Stephanie, ed. *Turning Away From Technology: A New Vision for the 21st Century.* San Francisco: Sierra Club Books, 1997.

Moorhouse, H.F. "The 'Work' Ethic and 'Leisure' Activity: The Hot Rod in Post-War America." In Patrick Joyce, ed., *The Historical Meanings of Work,* 237–309. Cambridge, MA: Cambridge University Press, 1987.

Morison, Elting E. *From Know-How to Nowhere: The Development of American Technology.* New York: New American Library, 1974.
Morley, David, and Kevin Robins, "Techno-Orientalism: Futures, Foreigners and Phobias," *New Formations* 16 (Spring 1992): 136–156.
Morris, Richard B., ed. *Encyclopedia of American History.* New York: Harper, 1953.
Moynihan, Daniel P., "The War Against the Automobile," *The Public Interest* 3 (Spring 1966): 10–26.
Mukherjee, Bharati. *Desirable Daughters.* New York: Hyperion, 2002.
Nader, Ralph. *Who Speaks for the Consumer?* N.p.: League for Industrial Democracy, 1968.
Glenn, Evelyn Nakano and Charles M. Tolbert II. "Technology and Emerging Patterns of Stratification for Women of Color: Race and Gender Segregation in Computer Occupations." In *Women, Work, and Technology: Transformations,* ed. Barbara Drygulski Wright, et al., 318–331. Ann Arbor: University of Michigan Press, 1987.
Noble, David F. *Progress Without People: In Defense of Luddism.* Chicago: Charles H. Kerr, 1993.
Ogle, Maureen. *All the Modern Conveniences: American Household Plumbing, 1840–1890.* Baltimore: Johns Hopkins University Press, 1996.
Pang, Alex Soojung-Kim, "Whose Dome is It Anyway?" *American Heritage Invention & Technology* 11 (Spring 1996): 28, 30–31.
Partners in Progress: A Report to President Truman by the International Development Advisory Board. New York: Simon and Schuster, 1951.
Penick, James L., Jr., Carroll W. Pursell, Jr., Morgan B. Sherwood, and Donald C. Swain, ed. *The Politics of American Science, 1939 to the Present,* 2d ed. Cambridge, MA: MIT Press, 1972.
Pfaffenberger, Bryan. "The Social Meaning of the Personal Computer: Or, Why the Personal Computer Revolution Was No Revolution," *Anthropological Quarterly* 61 (January 1988): 39–47.
Post, Robert C. *High Performance: The Culture and Technology of Drag Racing, 1950–1990.* Baltimore: Johns Hopkins University Press, 1994.
Poullier, Jean-Pierre. "The Myth and Challenge of the Technological Gap." In *The Technology Gap: U.S. and Europe,* ed. Atlantic Institute, 105-131. New York: Praeger Publishers, 1970.
Pursell, Carroll. "The Anatomy of a Failure: The Science Advisory Board, 1933–1935," *Proceedings of the American Philosophical Society* 109 (December 1965): 342–351.
Pursell, Carroll. "A Preface to Government Support of Research and Development: Research Legislation and the National Bureau of Standards," *Technology and Culture* 9 (April 1968): 145–164.
Pursell, Carroll, ed. *The Military-Industrial Complex.* New York: Harper and Row, 1972.
Pursell, Carroll. "Belling the Cat: A Critique of Technology Assessment," *Lex et Sciencia* 10 (October-December, 1974): 130–142.
Pursell, Carroll. "Alternative American Science Policies during World War II." In *World War II: An Account of its Documents,* ed. James E. O'Neill and Robert W. Krauskopf, 151–162. Washington, D.C.: Howard University Press, 1976.

Pursell, Carroll. "Conservationism, Environmentalism, and the Engineers: The Progressive Era and Recent Past." In *Environmental History: Critical Issues in Comparative Perspective*, ed. Kendall E. Bailes, 176–192. Lanham: University Press of America, 1985.

Pursell, Carroll. "Science Agencies in World War II: The OSRD and Its Challenges." In *The Sciences in the American Context: New Perspectives*, ed. Nathan Reingold, 359–378. Washington, D.C.: Smithsonian Institution, 1979.

Pursell, Carroll. "The Rise and Fall of the Appropriate Technology Movement in the United States, 1965–1985," *Technology and Culture* 34 (July 1993): 629–637.

Pursell, Carroll. *White Heat: People and Technology*. Berkeley: University of California Press, 1994.

Pursell, Carroll. *The Machine in America: A Social History of American Technology*. Baltimore: Johns Hopkins University Press, 1995.

Pursell, Carroll. "Appropriate Technology, Modernity and U.S. Foreign Aid," In *Science and Cultural Diversity: Proceedings of the XXIst International Congress of History of Science*, I, 175–187. Mexico City: Mexican Society for the History of Science and Technology and the National Autonomous University of Mexico, 2003.

Rivkin, Steven R. *Technology Unbound: Transferring Scientific and Engineering Resources from Defense to Civilian Purposes*. New York: Pergamon Press, 1968.

Rome, Adam. *The Bulldozer in the Countryside: Suburban Sprawl and the Rise of American Environmentalism*. Cambridge: Cambridge University Press, 2001.

Rome, Adam. "'Give Earth a Chance': The Environmental Movement and the Sixties," *The Journal of American History* 90 (September 2003): 525–554.

Rose, Mark. "Mothers and Authors: *Johnson v. Calvert* and the New Children of Our Imaginations." In *The Visible Woman: Imaging Technologies, Gender, and Science*, ed. Paula A. Treichler, Lisa Cartwright, and Constance Penley, 217–239. New York: New York University Press, 1998.

Rose, Mark H. *Interstate: Express Highway Politics, 1941–1956*. Lawrence: The Regents Press of Kansas, 1979.

Rostow, W.W. *The Stages of Economic Growth: A Non-Communist Manifesto*. New York: Cambridge University Press, 1960.

Roy, Arundhati. *Power Politics*. Cambridge: South End Press, 2001.

Russell, Edmund P., III. "'Speaking of Annihilation': Mobilizing for War against Human and Insect Enemies," *The Journal of American History* 82 (March 1996): 1505–1529.

Sale, Kirkpatrick. *Rebels Against the Future: The Luddites and Their War on the Industrial Revolution—Lessons for the Computer Age*. Reading: Addison-Wesley, 1995.

Schiller, Herbert I. *Mass Communications and American Empire*. Boston: Beacon Press, 1971.

Schwarz, Frederick D. "The Patriarch of Pong," *American Heritage Invention & Technology* 6 (Fall 1990): 64.

Schwarz, Frederick D. "The Epic of the TV Dinner," *American Heritage Invention & Technology* 9 (Spring 1994), 55.

Seligman, Ben B. "Automation and the State," *Commentary* 37 (June 1964): 49–54.

Servan-Schreiber, J. *The American Challenge.* New York: Atheneum, 1968.
Silverman, M. Joshua. "Nuclear Technology." In *A Companion to American Technology,* ed. Carroll Pursell, 298–320. Malden: Blackwell, 2005.
Silverstein, Ken. *The Radioactive Boy Scout: The True Story of a Boy and His Backyard Nuclear Reactor.* New York: Random House, 2004.
Slotten, Hugh R. "Satellite Communications, Globalization, and the Cold War," *Technology and Culture* 43 (April 2002): 315–350.
Smith, Michael L. "Advertising the Atom." In *Government and Environmental Politics: Essays on Historical Developments since World War Two,* ed. Michael J. Lacey, 233–262. Baltimore: The Woodrow Wilson Center Press and Johns Hopkins University Press, 1991.
Smith, Michael L. "Back to the Future: EPCOT, Camelot, and the History of Technology." In *New Perspectives on Technology and American Culture,* ed. Bruce Sinclair, 69–79. Philadelphia: American Philosophical Society, 1986.
Smith, Michael L. "Making Time: Representations of Technology at the 1964 World's Fair." In *The Power of Culture: Critical Essays in American History,* ed. Richard Wightman Fox and T.J. Jackson Lears, 223–244. Chicago: University of Chicago Press, 1993.
Smith, Michael L. "'Planetary Engineering': The Strange Career of Progress in Nuclear America." In *Possible Dreams: Enthusiasm for Technology in America,* ed. John L. Wright, 110–123. Dearborn: Henry Ford Museum and Greenfield Village, 1992.
Smith, Michael L. "Selling the Moon: The U.S. Manned Space Program and the Triumph of Commodity Scientism." In *The Culture of Consumption: Critical Essays in American History, 1880–1980,* ed. Richard Wightman Fox and T.J. Jackson Lears, 179–209. New York: Pantheon, 1983.
Smith, Percy McCoy. *The Air Force Plans for Peace, 1943–1945.* Baltimore: Johns Hopkins University Press, 1970.
Snow, C.P. *The Two Cultures and the Scientific Revolution.* New York: Cambridge University Press, 1961 [1959].
Snyder, Glenn H. *Stockpiling Strategic Materials: Politics and National Defense.* San Francisco: Chandler, 1966.
Sorkin, Michael. "See You in Disneyland." In *Variations on a Theme Park: The New American City and the End of Public Space,* ed. Michael Sorkin, 205–232. New York: Noonday Press, 1992.
Steinberg, Ted. *Acts of God: The Unnatural History of Natural Disaster in America.* New York: Oxford University Press, 2003.
Steinberg, Ted. *American Green: The Obsessive Quest for the Perfect Lawn.* New York: Norton, 2006.
Stever, H. Guy and Janet H. Mutoyama, eds. *Globalization of Technology: International Perspectives.* Washington, D.C.: National Academy Press, 1988.
Stone, I.F. *Business As Usual: The First Year of Defense.* New York: Modern Age Books, 1941.
Terborgh, George. *The Automation Hysteria.* New York: Norton Library, 1966.

Theriault, Reg. *Longshoring on the San Francisco Waterfront.* San Pedro: Singlejack Books, 1978.

Tsutsui, William M. *Manufacturing Ideology: Scientific Management in Twentieth Century Japan.* Princeton: Princeton University Press, 1998.

Twain, Mark. *A Connecticut Yankee in King Arthur's Court.* New York: Signet, 1963 [1889].

Ullman, Ellen. *Close to the Machine.* San Francisco: City Lights Books, 1997.

U.S. Congress, House of Representatives, Committee on International Relations. *Science, Technology, and American Diplomacy: An Extended Study of the Interactions of Science and Technology with United States Foreign Policy.* Washington, D.C.: Government Printing Office, 1977.

U.S. Congress, House of Representatives, Subcommittee of the Committee on Government Operations. *Brain Drain into the United States of Scientists, Engineers, and Physicians.* Staff Study for the Research and Technical Programs Subcommittee of the Committee on Government Operations. Washington, D.C.: Government Printing Office, 1967.

U.S. Congress, House of Representatives. *Technical Information for Congress.* Report to the Subcommittee on Science, Research and Development of the Committee on Science and Astronautics. 92nd Congress, 1st sess. (April 15, 1971).

U.S. Congress, Senate, Subcommittee on War Mobilization to the Committee on Military Affairs. *The Government's Wartime Research and Development, 1940–1944: Part I.—Survey of Government Agencies.* 79th Congress, 1st sess. (1945).

U.S. Congress, Senate, Subcommittee on War Mobilization to the Committee on Military Affairs. *The Government's Wartime Research and Development, 1940–1944: Part II.—Findings and Recommendations.* 79th Congress, 1st sess. (1945).

U.S. Congress, Senate, Subcommittee on War Mobilization of the Committee on Military Affairs. *Wartime Technological Developments.* 79th Congress, 1st sess. (May 1945).

U.S., Department of Commerce. *Statistical Abstracts of the United States, 1970.* Washington, D.C.: Government Printing Office, 1970.

U.S., Department of Commerce. *Statistical Abstracts of the United States, 2001.* Washington, D.C.: Government Printing Office, 2001.

U.S. National Commission on Technology, Automation, and Economic Progress. *Technology and the American Economy.* Report of the National Commission on Technology, Automation, and Economic Progress, I. Washington, D.C.: Government Printing Office, 1966.

U.S. National Inventors Council. *Administrative History of the National Inventors Council.* Washington, D.C.: National Inventors Council, 1946.

U.S. President's Materials Policy Commission. *Resources for Freedom, Vol. 1. Foundations for Growth and Security.* Washington: Government Printing Office, 1952.

Volti, Rudy. *Technology Transfer and East Asian Economic Transformation.* Washington, D.C.: American Historical Association, 2002.

Von Karman, Theodore. *The Wind and Beyond.* Boston: Little, Brown, 1967.

Waldie, D.J. *Holy Land: A Suburban Memoir.* New York: St. Martin's Press, 1996.

Weart, Spencer. *The Discovery of Global Warming.* Cambridge: Harvard University Press, 2003.

Weinstein, Raymond M., "Disneyland and Coney Island: Reflections on the Evolution of the Modern Amusement Park," *Journal of Popular Culture* 26 (Summer 1992): 131–165.

Weisner, Jerome B., "Rethinking Our Scientific Objectives," *Technology Review* 71 (January 1969): 15–17.

Whorton, James. *Before Silent Spring: Pesticides & Public Health in Pre-DDT America.* Princeton: Princeton University Press, 1974.

Williams, James C. "Frederick E. Terman and the Rise of Silicone Valley." In *Technology in America: A History of Individuals and Ideas,* 2d ed., ed. Carroll W. Pursell, Jr., 276–291. Cambridge, MA: MIT Press, 1990.

Wolpin, Stewart. "The Race for Video," *American Heritage Invention & Technology* 10 (Fall 1994): 52–62.

Womack, James P., Daniel T. Jones, and Daniel R. Roos. *The Machine That Changed the World.* New York: Rawson Associates, 1990.

Wood, Lamont. "The Man Who Invented the PC," *American Heritage Invention & Technology* 10 (Fall 1994): 64.

York, Herbert. *Race to Oblivion: A Participant's View of the Arms Race.* New York: Simon and Schuster, 1971.

Yost, Jeffrey R. "Computers and the Internet: Braiding Irony, Paradox, and Possibility," In *A Companion to American Technology,* ed. Carroll Pursell, 340–360. Malden: Blackwell, 2005.

Zoellner, Tom, "The Uranium Rush," *American Heritage Invention and Technology* 16 (Summer 2000): 56–63.

INDEX

Acheson, Dean, 48
Advanced Research Projects Agency (ARPA), 188, 189
Afghanistan, 215, 216
Agnew, Spiro, 197
agribusiness, 144–48
Agribusiness Accountability Project, 145
agriculture, and globalization, 222–24
agro-industrial complex, 55
aid to developing countries, 46–51
Aiken, Howard, 175, 176
Airbus 380, 207
Air Force Foreign Technology Division, 126
Alaska pipeline, 57
Albright, Fuller, 115
all-electric homes, 28, 32
Aluminum Company of America (Alcoa), 7
American Association for the Advancement of Science (AAAS), 128
American Century, ix
American Institute of Chemical Engineers, 160
American National Exhibition (Moscow), 101
American Society of Civil Engineers, 160
American Society of Mechanical Engineers: Committee on Technology and Society, 160; Manufacturing Engineering Committee of, 5; National Defense Committee of, 8

Ames Research Center, 180
Amish, 96–97
Ampex, 180
amusement parks, 111–13
Andreessen, Marc, 187
anti-ballistic missile (ABM), 159, 196, 198
appliances: and energy, 31–35; household, 99–100; smart, 191–92
Apollo Program, 72, 199, 206
Apple Computer, 186
appropriate technology, 51, 95, 128, 150, 163–65, 204–207, 225, 226
Armed Forces Communications and Electronics Association, 219
armory practice, 17, 182
Armour Research Foundation, 5
arms export, 53–54
arms race, 136–38, 197
Army Corps of Engineers, 60
Army Signal Corps, 70
ARPANET, 189
Arroyo Seco Freeway, 109
astronauts, 72
AT&T, 220, 221, 222
Atari Corp., 184
Atlantic Institute, 119, 123, 124, 125
atmospheric nuclear testing, 136
atom, the popular, 65–69
Atom and Eve, 68
atomic bomb, 59–64; test blast, 60

272 INDEX

Atomic Energy Commission (AEC), 11, 55, 60, 158
Atomic Industrial Forum (AIF), 68
Atoms for Peace, 61
automation, 78, 80–84
automobile: culture, 105–107; Edsel, 106; Volksvagen Beetles, 106; war against, 166

balance of trade, 125
Babbage, Charles, 174
Bateman, Alan M., 52
Baudrillard, Jean, 212, 228
Baxter, James Phinney, III, 6
Bay Area Rapid Transit (BART), 110
Bay of Pigs, 72
Bearing Point, 215
Bechtel, 15, 56, 57
Bechtel, Stephen (Steve), 57
Beers, David, 180
Bel Geddes, Norman, 25
Bell Laboratories, 3, 42, 126, 181, 190, 198
Ben-David, Joseph, 122, 124
Bergen Mall, 36
Berlin blockade, 135
Berman, Marshall, 224
Bernstein, Barton J., 18
Bharat Heacy Electricals (BHEL), 214
"Bird Dogs," 9, 11
birth control pill, 115–16, 194
Biskind, Peter, 135
Blade Runner, 85
Blaisdell, Thomas C., 4
Bob's Big Boy restaurant, 110
Boretsky, Michael T., 125, 126
Boyar, Louis, 28
Boyd, William, 89
brain drain, 128–33
Brand, Stewart, 184
Braudel, Fernand, 228
Bridges, Harry, 82, 83
Brindley, Charles F., 8
Bronx River Parkway, 108
Brower, David, 143

Brown and Root, 56
Brown, Rep. George E., 164
Brown, Gov. Jerry, 164, 165
Brown's Ferry, 64
Burdick, Eugene, 57
Bureau of the Budget, 12
Burns, Fritz B., 24, 25
Bush, Pres. George H. W., 66, 75, 206
Bush, Pres. George W., 38, 56, 201, 206, 226
Bush, Vannevar, 3, 4, 18, 85, 126, 157, 175, 179, 182, 202
Bushnell, Nolan, 184
"buying a piece of the machine," 82

calculator, 183–84
Calendar, G. S., 224
Cannon, Poppy, 101
Carey, William D., 126
Carlyle, Thomas, 151
Carmichael, Leonard G., 5
Carnegie Institution of Washington, 3, 70
Carson, Rachel, 134, 135, 138–39, 140
Carter, Pres. Jimmy, 165, 206, 218
Cary, James, 77
Castells, Manuel, 216
Catton, Bruce, 8, 19
cellular (mobile) phones, 190–91
Center for Law and Social Research, 168
Central Intelligence Agency (CIA), 75
Century Freeway, 110
Ceruzzi, Paul, 176
cesspools, 31
Challenger (space shuttle), 73
Chavez, Cesar, 91
Chemical Abstracts, 91
Cheney, Richard B., 56, 206
Chernobyl, 64
chickens, 89–90
Chorafas, D. N., 131
Citizens League Against the Sonic Boom, 143
civil defense, 66
Civil Rights Movement, 149, 194

INDEX   273

Civilian Industrial Technology (CIT) program, 158
Clarke, Alison J., 99
Clarke, Arthur C., 220
class-action suits, 169–70
Clean Air Act, 107
Clinton, Pres. Bill, 195, 201, 206
Coalition Against the SST, 143
Cohen, Lizabeth, 36, 37
Cohn, Carol, 195
Cold War fears, 135–36
Columbia (space shuttle), 72, 73
Command, Control, and Communication (3C), 188
Committee for Nuclear Information, 136
Committee for Social Responsibility in Engineering (CSRE), 159, 160
Committee on Technology and Society, 160
Commoner, Barry, 136
Communications Satellite Act, 221
Communications Satellite Corp. (COMSAT), 221, 222
computers, 81, 121, 125, 174–77; Altair, 183; Altair 8800, 185; Differential Analyser, 175; ENIAC, 175, 177; IBM 650, 176; Mark I, 175; UNIVAC, 176; personal, 185–88
Conant, James B., 4, 5
Concorde, 143
Confined Animal Feeding Operations (CAFO), 92–93
consumer revolt, 165–70, 194
containerization, 83
Control Data Corporation, 125
conversion to peace, 156–58
convivial tools, 150
Cooper, Martin, 190
Cowan, Ruth Schwartz, 13, 140, 191
Crawford, Margaret, 37
Cuban missile crisis, 135, 141
cybernetics, 81, 148, 149
Czekalinski family, 101

Daddario, Rep. Emilio Q., 128, 160, 162
dam business, worldwide, 214
Davis, W. Kenneth, 57
The Day the Earth Stood Still, 135
DDT, 138–39, 168
DeBartolo, Edward J., 35
De Beauvoir, Simone, 122
Defense Advances Research Projects Agency (DARPA), 199
defense intellectuals, 195–96
De Gaulle, Charles 119, 120, 125
Dell, 214
Deluer, Richard, 199
Deming, W. Edwards, 17, 44, 45, 86
Denison, Edward, 123
Department of Homeland Security, 227
dependency theory, 46
Dickson, Paul, 138
Dideon, Joan, 157
Diebold, John, 80
digital divide, 190
disaster capitalism, 215
Disney, Walt, 37, 59, 68, 111, 113
Disneyland, 111–12, 113
Disney World, 111
Do-It-Yourself, 103
Douglass, Charles, 114
Dowling, Walter, 119
Draper, John T. (Cap'n Crunch), 186
Dukmajian, Gov. George, 165
Dupree, Hunter, 174

Earl, Harley J., 105–106
Earth Day, 142
Echeverria, Luis, 47
Eckert, J. Presper, 175
Ehrlichman, John, 143
Einstein, Albert, 60
Eisenhower, Pres. Dwight D., 61, 71, 108, 136, 137, 176, 196, 220
electric drill, 103
electronic battlefield, 138
Embarcadero Freeway, 110

email, 189
empires, 218–19
energy crisis, 150
Energy Research and Development Administration (ERDA), 164
Engelbart, Doug, 187
Enrico Fermi Breeder Reactor, 64
Enron, 215
environmental movement, 138, 194
Environmental Protection Agency (EPA), 69, 209
European Union, 133
Evans, Oliver, 81
Experimental Prototype Community of Tomorrow (OPCOT), 111, 112

Fairchild Semiconductor, 121, 182
Fanfani, Amintore, 119
family-planning technologies, 116
Farallones Institute, 164
farm machinery, 90–93
farmers' markets, 95–96
farms: as factories, 87–89; organic, 94–96, 223
Faust, 224
Federal-Aid Highway Act, 108
Federal Communications Commission, 222
Federal Housing Administration (FHA), 26
Federal Telegraph Co., 178
Federal Water Pollution Control Administration, 145
feedlots, 145
Fitzgerald, Deborah, 88, 144
FitzGerald, Frances, 200
food, 100–101; cake mixes, 100; industrial, 144; TV dinners, 100
Food and Drug Administration (FDA), 116
Ford, Henry, 187
Foster, Mark, 15
Fowler, Rear Adm. Joseph W., 112
Frassanito, Jack, 185
Frederick, Christine, 28
Free Speech Movement, 141

Freedom (space station), 73
Friedel, Robert, 181
Friedman, Thomas L., 226
Fromm, Erich, 151
Fuller, R. Buckminster, 26

Gagarin, Yuri, 72
Galbraith, John Kenneth, 140
games (electronic), 184; Pong, 184; Spacewar, 184
Garden State Parkway, 36
Garden State Plaza, 36
Gates, Bill, 186
Gee, Sherman, 127
Gelber, Steven, 103
Gell-Mann, Murray, 143
Genetically Modified Organisms (GMO), 93–94, 223
geosynchronous satellites, 220–21
G.I. Joe doll, 204
Gilpin, Robert, 122, 123, 127
Glendinning, Chellis, 171
Glenn, John, 204
Glennan, T. Keith, 71, 220
globalization, 212–13
global warming, 224, 225
Goddard, Robert, 69–70
Good, Mary Lowe, 213
Goodman, Paul, 152
Gorbachev, Mikhail, 201
Gordon, Andrew, 44
Graham, Lt. Gen. Daniel O., 199
Grant, Bess Myerson, 169
Green, Cecil H., 181
Griffin, Bill, 27
Gross, Al, 190
Groves, Leslie 62
Guggenheim, Daniel, 70
Gulf War, 75, 217

Haggerty, Patrick E., 126, 183
Hahn, David, 69
Halliburton, 215

Halliburton Co., 56, 215
Hanna, G. C., 91
Hansen, William W., 179
Harder, D. S., 80
Harvey, David, 229
Hayden, Dolores, 22, 24, 26, 29
Hewlett, William, 179
Hewlett-Packard, 179, 180
High Frontiers, 199
Highland Park, 79
Hightower, Jim, 144, 145, 146, 147
highways, 107–111
Higinbotham, William, 184
Hiroshima, 11, 60, 135
Hoerni, Jean, 182
Hoff, Marcian "Ted," 183
Hoffer, Eric, 134, 150
Holiday Inn, 109
Hollerith, Herman, 174
Holloman, J. Herbert, 157
Homebrew Computer Club, 185, 186
Honeywell, 125
Hoover, Pres. Herbert, 24
hot rods, 104–105
housing, in 1940, 20–22; mass-produced, 24–26; postwar shortage, 22–24; technological infrastructure, 27–31
Hubble telescope, 76
Hughes Aircraft Corp., 220
Hughes, Thomas P., 188, 212
Hummer, 208, 210
Hurricane Katrina, 56
Hurricane Mitch, 216
Hussein, Saddam, 217
hydrogen bomb, 61

Ickes, Harold L., 4
Illich, Ivan, 150
India, and globalization, 213–15
industrial research, 4, 5, 76, 158
Institute for Electrical and Electronic Engineers (IEEE), 159
Institutum Divi Thomae, 4

Integral Urban House, 164
integrated circuits, 121, 183
Intel, 183, 185
Intelsat I, 221
International Business Machines (IBM), 121, 125, 174, 178, 185
International Telecommunications Satellite Organization (INTELSAT), 221, 222
International Geophysical Year (IGY), 71
International Monetary Fund, 56, 216
International Society for Technology Assessment, 162
Internet, 187, 188–90
Iran-Iraq War, 217
Iraq War, 56, 204, 215, 217

Jacobson, Joanne, 98
Jacquard loom, 174
Janis, Irving, 66
Japan, recovery, 41–42
Japan Productivity Center (JPC), 43
Japan Union of Scientists and Engineers, 44
Jeep, 14, 208, 209
Jeffords, Susan, 194
Jennings, Paul 168
jet airline travel, 114
Jet Propulsion Laboratory, 12
Jewett, Frank B., 3, 8
Jobs, Steve, 186
Johnson, Pres. Lyndon B., 55, 56, 82, 134, 142, 148, 149, 155, 157, 158, 168, 196, 197, 209, 222
Johnson v. Calvert, 117
Juran, Joseph, 45

Kaczynski, Theodore, 171–72
Kaiser, Henry J., 15, 25, 106, 209
Kaiser Permanente (HMO), 214
Kay, Alan, 187
Kearns, Henry, 57
Keats, John, 107
Kennedy, Pres. John F., 67, 72, 135, 139, 142, 220, 221
Kennedy, Sen. Ted, 162

Kettering, Charles F., 6
Khrushchev, Nikita, 101–102
Kilby, Jack, 183
Kilgore, Harley M., 8
Kitchen Debate, 101–103
Klein, Naomi, 215, 216
Knoppers, Arthur, 123
Knudsen, William S., 13
Kosygin, Alexei N., 119
Korean War, 42, 48, 209
Krok, Ray, 109
Kyoto accords, 225–26

Laissez innover, 153
Lakewood, Calif., 29, 35, 36, 99
Land-grant College Complex, 144
Lakoff, Sanford, 199
lawns, 30–31
Lawrence Livermore National Laboratory, 12, 65
lean production, 43, 78, 86–87
Lederer, William, 57
Lekachman, Robert, 156
Levitt, William, 24, 26
Levittown, 26, 36, 99, 103
liberalism, crisis of, 140–41
Liberty ships, 15
Licklider, Joseph Carl Robnett, 188
Lillienthal, David E., 56
Lindsay, John, 169
Linenthal, Edward, 202
Litton, Charles, 179, 182
Litton Industries, 182
Lockheed, Space and Missile Division, 180
Lorenzen, Coby, 91
Lovins, Amory, 165, 205
Lower Mekong Basin Project, 55
lowridres, 104
Luce, Henry, 39, 54

Maharashtra State Electricity Board, 215
Manhattan Project, 60, 61, 199
Marx, Karl, 215

Marlin, Alice Tepper, 200
Marling, Karal Ann, 102
Marshall, Gen. George, 40
Marshall Plan, 39, 40–41
Marx, Leo, ix
Masamoto, David, 95
masculinity, 195, 203–204
mass production, 13, 14, 16, 78, 79–80
Massachusetts Institute of Technology (MIT), 3, 81, 115, 116, 124, 175, 170, 181; Media Lab, 191
massive retaliation, 61, 137
Mattelart, Armand, 228
Mauchly, John W., 175
Maverick, Maury, 4, 8
McAllister, Wayne, 110
McCormick, Cyrus Hall, 116
McCormick, Katherine Dexter, 115–16
McCormick, Stanley R., 116
McDermott, John, 153
McDonald's restaurants, 109
McDougall, Walter, 76, 77
McElroy, William D., 126
McMansions (housing), 210
McWilliams, Carey, 78, 88
Mellen, Lowell, 45
Merryman, Jerry, 183
Mesthene, Emmanuel G., 153
microprocessors, 182–84
Microsoft, 177; BASIC, 186
military bases, 218
military-industrial complex, 73, 136, 141, 157, 180, 188, 197, 204
milking machines, 92
Millikin, Sen. Eugene, 49
MILNET, 189
Ministry of International Trade and Industry (MITI), 42
missile defense, 196–98
missiles, 69–71; *Agena*, 74; *Atlas*, 74; gap, 72; ICBMs, 71; *Jupiter*, 75; multiple independently targetable reentry vehicles (MIRV), 137; Nike-X, 196; *Patriot*, 75; Po-

laris, 74; *Redstone*, 74; *Spartan*, 197; *Sprint*, 198; *Titan*, 74; *Thor*, 74; V-2, 70
mobile privitism, 114
Modernization Theory, 46
Monsanto, 93, 223
Moon landing, 72, 222
Moorhouse, H. F., 104
Morgan, T. A., 8
Morison, Elting, 154
Morison Knudsen Corp., 56
Morley, David, 228
Motorola, 121, 190
Moynihan, Daniel P., 166
Muir, R. C., 8
Mukherjee, Bharati, 129
Mumford, Lewis, 171
Mutual Defense Assistance Act, 53
mutually assured destruction (MAD), 61, 137, 203
"Mythos of the Electronic Revolution," 77

Nader, Ralph, 107, 146, 167, 169
Nagasaki, 60, 135
National Academy of Science, 3, 4, 5, 161, 212
National Advisory Committee on Aeronautics (NACA), 3, 71
National Advisory Defense Committee, 13
National Aeronautics and Space Administration (NASA), 71, 72, 73, 74, 158, 180
National Air and Space Museum, 195–96
National Center for Appropriate Technology, 165
National Commission on Product Safety, 169
National Commission on Technology, Automation, and Economic Progress, 82, 155
National Committee for a Sane Nuclear Policy (SANE), 136
National Defense Education Act (NDEA), 71, 129
National Defense Research Committee (NDRC), 4
National Institutes of Health (NIH), 11
National Inventors Council, 6

National Roster of Scientific and Specialized Personnel, 5
National Science Foundation (NSF), 10, 11
National Security Council Paper 68, 136
Navy Construction Battalion (Seabees), 30
Nehru, Jawaharlal, 215
Nelson, Donald, 4, 18
Nelson, Sen. Gaylord, 128
Neo-Luddites, 170–73
Netscape Communications, 187
Network Control Program (NCP), 188–89
networked society, 219
New Alchemy Institute, 163–64
New Jersey Turnpike, 36, 108
New Left, 141
Nixon, President Richard M., 12, 72, 101–102, 142, 156, 160, 162, 169, 196
Noble, David F., 170
Norstad, Gen. Lauris, 9
North American Aerospace Defense Command (NORAD), 196, 198, 202
North Atlantic Treaty Organization (NATO), 123
Noyce, Robert, 182, 183
NSFNET, 189
nuclear power, 61–65

Office of Appropriate Technology (OAT), 164
Office of Naval Research (ONR), 9, 11, 180
Office of Production Management (OPM), 7, 13
Office of Production Research (OPR), 3, 5
Office of Scientific Research and Development (OSRD), 3, 60, 179
Office of the Coordinator for Reconstruction and Stabilization, 215
Office of the Supersonic Transport, 142
Office of Technology Assessment (OTA), 160–63, 200
Ohno, Taiichi, 43, 86
Organization for Economic Cooperation and Development (OECD), 121, 122

278 INDEX

Organization of the Petroleum Exporting Countries (OPEC), 107
Orientalism, 227
Osborne, Adam, 186, 187

Packard, David, 179
Paley, William S., 52
Perry, Com. Matthew, 41, 228
Pfaffenberger, Bryan, 187, 188
Pierce, John R., 126
Pill, the. *See* birth control pill)
Pincus, Gregory, 116
planetary engineering, 39, 54–58, 204
Planned Parenthood Federation, 115, 116
Point 4 Program, 39, 47–51
Popeil, Samuel J., 99
postwar, planning, 9–11; research regimes, 11–13
President's Council on Environmental Quality, 142
President's Materials Policy Commission, 52
President's Science Advisory Board (PSAC), 71, 142
Price, William J., 162
Price-Anderson Act, 63, 165
production regime, new, 43–45
Project Mercury, 71
Project Mohole, 56
Project Plowshare, 54, 64
Project Whirlwind, 175–76
Prometheus, 64
Proxmire, Sen. William, 126

quality control (QC), 44, 86. *See also* statistical quality control
Quirk, John, 77

Radiation Exposure Compensation Act, 66
raw materials, access to, 52–53
Reagan, Pres. Ronald, 57, 73, 168, 194, 196, 198, 199, 200, 201, 202, 203, 206
reconversion, 18–19
reform, calls for, 150–54

Regency all-transistor radio, 181, 182, 183
reproductive technologies, 115–17
Republican Party, 162
remasulinization of America, 194, 196
research and development, 10, 13, 120, 122, 126, 127, 131, 132, 157, 158, 161, 165, 180, 181, 197, 198, 200
Research Corporation, 5
Reuther, Walter, 82
Revolution in Military Affairs (RMA), 204
Reynolds, Malvina, 27
Reynolds Metal Company, 7
Rice, Condoleeza, 216
Rickover, Hyman, 62
Riesman, David, 34
River Rouge, 79, 86
Rivkiv, Steven, 157
Roach, James T., 167
Roberts, Lawrence, 189
Robins, Kevin, 228
robots, 78, 84–86
Rock, John, 116
Rockefeller, Nelson A., 49–50
Rome, Adam, 23, 33, 139, 141
Romney, George, 106
Rosie the Riveter, 15
Ross, Andrew, 172
Rostow, Walt W., 46
Roszak, Theodore, 172
Roy, Arundhati, 214, 215, 224
Rumsfeld, Donald, 201
Rybczynski, Witold, 205

Safeguard system, 196, 197, 198
Said, Edward, x
Sale, Kirkpatrick, 170, 172
Sanders, Col. Harlan, 109
Sanger, Margaret, 115–16
Saucier, Aldric, 200
Savio, Mario, 141
Schlesinger, Arthur, Jr., 120, 140
Schiller, Herbert, 219
Schultz, George, 57

Schumacher, E. F., 165, 205
Science the Endless Frontier, 9
scientific colonialism, 47
scientific management, 44
Scientists and Engineers for Social and Political Action (SESPA), 160
Scrinis, Gyorgy, 223
Seaborg, Glenn, 63, 65, 68
Seligman, Ben, 82
semiconductor, 181
Sentinel system, 197
septic tanks, 31
Servan-Schreiber, J.-J., 120, 121, 122, 131
sexual freedom, 194
Shockley, William, 180–81, 182
Shepard, Alan, 72, 74
shipbuilding, 15–17
shopping centers, 30, 35–37
short hoe (*el cortito*), 144
Silicon Valley, 123, 128, 178, 181–82, 185, 187, 228
Six Companies, 15
Sloan, Alfred P., 79, 106
Slotte, Hugh R., 220
Smith, Michael, 39, 54, 65, 67, 68, 73, 112, 204
Smithsonian Institution, 70
Snow, C. P., 46
Society of Women Engineers, 205
Social responsibility in engineering, 158–60
software, 177–78
Sokol, Dan, 186
sonic booms, 142
Sony, 42
Sorkin, Michael, 111, 112, 113
Southdale Mall, 37
Space Age, 76, 77
space exploration, 75
Space Race, 71–77
space shuttle, 72
Sperry Gyroscope Co., 179
Spock, Dr. Benjamin, 136
Sport Utility Vehicles (SUV), 107, 208–10
Sputnik, 70, 71, 76, 102, 129, 181, 188

spy satellites, 75
Stanford Research Institute, 187
Stanford University, 178, 170
Stans, Maurice, 125
Star Wars (film), 85
Star Wars, 75, 196, 198–203
State Technical Services Act, 157
Statistical Quality Control (SQC), 17
Stevenson, Adlai, 77, 140
Stockpiling Act of 1946, 52
Stone, I. F., 7
Strategic Defense Initiative (SDI). *See* Star Wars
Strategic Ground Environment (SAGE), 176
Strauss, Adm. Lewis, 62
Subcommittee on War Mobilization of the Senate Committee on Military Affairs, 8
Submarines (*Nautilus*), nuclear-powered, 61, 62
superjumbo jets, 207–208
supersonic transport (SST), 140, 141–44, 159, 160
Swanson, C. A. & Sons, 100

Taper, Mark, 28
Taylor, Frederick Winslow, 44
technological determinism, 229
technological Marshall Plan, 119
technology gap, 120–24
Technology Transfer Office of the Naval Surface Weapons Center, 127
techno-orientalism, 226–28
television, 113–15
Telkes, Maria, 205
Teller, Edward, 54, 65, 199
Tennessee Valley Authority (TVA), 56
Terman, Frederick E., 178–79
terrorist, attack (9/11), 38
terrorists, 163; war on, 202
Texas Instruments, 121, 126, 181, 182, 183, 184, 185
Theriault, Reg, 83
Three Mile Island, 64

Todd, Nancy and John, 163
tomato-picking machine, 90–92
Toyoda, Eiji, 43
Toyoda, Sakichi, 41
Toyota Motor Co., 43, 86
Traffic Safety Act, 168
Train, Russell E., 142
Training Within Industry Inc. (TWI), 45
transistor, 42, 180–81
Triple Revolution, 148–49, 156
Truman, Harry S, 8, 23, 47, 52, 61
tsunami, in Indonesia, 216
Tsutsui, William M., 44
Tupper, Earl, 99
Turner Thesis, 9
Twain, Mark, 229

Unabomber, 171
Unidentified Flying Objects (UFOs), 135
United Auto Workers, 79, 82
United Nations, 48, 61, 63, 76, 128, 163, 206
United States Export-Import Bank, 57
Universal Product Code, 84
University of Illinois, 5
uranium mining, 65–66
U.S. Ballistic Missile Early Warning System, 196

vacation, family, 110–11
Vanderbilt, Tom, 112
Van der Ryn, Sim, 164
Vanport City, 16, 25
Varian, Sigurd, 179
Varian Associates, 180
Veg-O-Matic, 99–100
video recorder, 114
Vietnam War, 55, 134, 138, 141, 157, 172, 194–95, 203, 216, 217, 218
Volpe, John A., 142
Von Braun, Wernher, 70, 71, 75
Von Karman, Theodore, 12

Vonnegut, Kurt, Jr., 78, 81
von Neumann, John, 175

Wagner, Sen. Robert, 23
Waldie, D.J., 29, 35, 99
Walter-McCarran Act, 130
War on Poverty, 148
War Production Board (WPB), 3
Washington Metro, 57
Weinberger, Casper, 57
Weingart, Ben, 28
West Edmonton Mall, 37
western frontier, 202
White Sands Proving Ground, 70
Wiener, Norbert, 81
Williams, James, 178, 182
Williams, Raymond, 114
Wilson, Charles (Engine Charlie), 137
Wilson, E. O., 172
Wilson, Harold, 119, 124
Winner, Langdon, 191
Wolfle, Dael, 122
Women Strike for Peace, 140
Works, Stanley, 191
World Bank, 56, 215, 216
World Trade Organization (WTO), 214, 216
World War II: criticism of mobilization, 7–9; research regimes, 2–7; wartime production, 13–17
World Wide Web (WWW), 188, 189
Wozniak, Stephen, 186

Xerox PARC (Palo Alto Research Center), 187

Yates, Rep. Sydney, 142
Yeager, Charles Elwood (Chuck), 141
York, Herbert, 137, 196, 197, 199
Yost, Jeffrey, 176

Zappa, Frank, 34